Probability and Mathematical Statistics (Continued)

WILLIAMS • Diffusions, Markov Processes, and Martingales, Volume I: Foundations

ZACKS • Theory of Statistical Inference

Applied Probability and Statistics

ANDERSON, AUQUIER, HAUCK, OAKES, VANDAELE, and WEISBERG • Statistical Methods for Comparative Studies

ARTHANARI and DODGE • Mathematical Programming in Statistics

BAILEY • The Elements of Stochastic Processes with Applications to the Natural Sciences

BAILEY • Mathematics, Statistics and Systems for Health

BARNETT and LEWIS • Outliers in Statistical Data

BARTHOLOMEW • Stochastic Models for Social Processes, *Second Edition*

BARTHOLOMEW and FORBES • Statistical Techniques for Manpower Planning

BECK and ARNOLD • Parameter Estimation in Engineering and Science

BELSLEY, KUH, and WELSCH • Regression Diagnostics: Identifying Influential Data and Sources of Collinearity

BENNETT and FRANKLIN • Statistical Analysis in Chemistry and the Chemical Industry

BHAT • Elements of Applied Stochastic Processes

BLOOMFIELD • Fourier Analysis of Time Series: An Introduction

BOX • R. A. Fisher, The Life of a Scientist

BOX and DRAPER • Evolutionary Operation: A Statistical Method for Process Improvement

BOX, HUNTER, and HUNTER • Statistics for Experimenters: An Introduction to Design, Data Analysis, and Model Building

BROWN and HOLLANDER • Statistics: A Biomedical Introduction

BROWNLEE • Statistical Theory and Methodology in Science and Engineering, *Second Edition*

BURY • Statistical Models in Applied Science

CHAMBERS • Computational Methods for Data Analysis

CHATTERJEE and PRICE • Regression Analysis by Example

CHERNOFF and MOSES • Elementary Decision Theory

CHOW • Analysis and Control of Dynamic Economic Systems

CLELLAND, BROWN, and deCANI • Basic Statistics with Business Applications, *Second Edition*

COCHRAN • Sampling Techniques, *Third Edition*

COCHRAN and COX • Experimental Designs, *Second Edition*

CONOVER • Practical Nonparametric Statistics, *Second Edition*

CORNELL • Experiments with Mixtures: Designs, Models and The Analysis of Mixture Data

COX • Planning of Experiments

DANIEL • Biostatistics: A Foundation for Analysis in the Health Sciences, *Second Edition*

DANIEL • Applications of Statistics to Industrial Experimentation

DANIEL and WOOD • Fitting Equations to Data: Computer Analysis of Multifactor Data, *Second Edition*

DAVID • Order Statistics, *Second Edition*

DEMING • Sample Design in Business Research

DODGE and ROMIG • Sampling Inspection Tables, *Second Edition*

DRAPER and SMITH • Applied Regression Analysis, *Second Edition*

DUNN • Basic Statistics: A Primer for the Biomedical Sciences, *Second Edition*

DUNN and CLARK • Applied Statistics: Analysis of Variance and Regression

ELANDT-JOHNSON • Probability Models and Statistical Methods in Genetics

ELANDT-JOI l Models and Data Analysis

continued on back

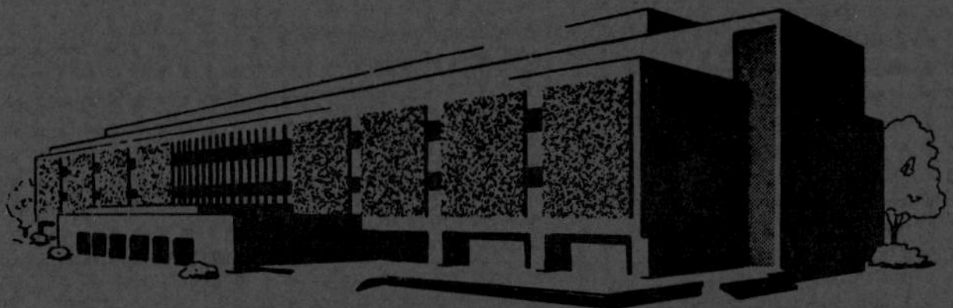

The Theory of Linear Models and Multivariate Analysis

The Theory of Linear Models and Multivariate Analysis

STEVEN F. ARNOLD

The Pennsylvania State University

John Wiley & Sons

New York · Chichester · Brisbane · Toronto

***Library of Congress Cataloging in Publication Data*:**

Arnold, Steven F 1944–
The theory of linear models and multivariate analysis.

(Wiley series in probability and mathematical statistics)
Bibliography: p.
Includes index.
1. Multivariate analysis. 2. Linear models (Statistics) I. Title.

QA278.A76 519.5′35 80-23017
ISBN 0-471-05065-2

Printed in the United States of America

10 9 8 7 6 5 4 3 2 1

To Rana, Timothy, and Christopher

Preface

Many models considered in applied statistics assume an underlying normal distribution. This book presents a detailed theoretical treatment of such models. It is intended to be used both as a reference book and as a textbook for a course in linear models and multivariate analysis. Among the models studied are the univariate linear model, the generalized linear model, the repeated measures model, random effects and mixed models, the correlation model, the multivariate one- and two-sample models, the multivariate linear model, and the discrimination model.

For each of these models there are procedures that are "optimal" in some sense. As would be expected sufficiency, estimation, hypothesis testing, and confidence intervals are discussed. In addition, the fairly general Huber condition is analyzed and for many of the models this condition is shown to guarantee that the sizes of the tests and confidence coefficients of the simultaneous confidence intervals are asymptotically unaffected by non-normality of the distributions. I am very sympathetic with Scheffé's (1959) well-known footnote on page 360, which says, "When I became aware that the nominal probability of type I error for the standard test of the equality of variances of two populations is invalidated by non-normality to the same order of magnitude as found in Table 10.2.1, I found little consolation in the optimum properties someone once established for that test (Scheffé, 1942)." The asymptotic validity of the "optimal" procedures derived in this book makes them not only "optimal" procedures, but also sensible procedures to use in practice. It is therefore quite important to know which procedures are asymptotically valid and which are not.

In considering many of the models in this book, I use a coordinate-free notation in which most of the statistics are expressed as projections on a particular subspace or as lengths of such projections. This notation leads to simpler and more easily interpretable expressions for the statistics, to simpler proofs of their properties, and to simpler derivations of particular formulas for special cases than the more traditional matrix notation.

The first chapter of the book contains a review of basic mathematical statistics together with some new definitions. There is also a more detailed introduction to invariance in estimation and testing problems. It is not my hope that the reader would understand invariance completely from this elementary treatment, but rather would develop an intuitive feel for the concepts from the many examples later in the book. In Chapters 2 and 3 the basic properties of projections and the multivariate normal distribution are derived. Chapters 4–12 treat the univariate linear model. These chapters are the core of the text. Later chapters show how to refine and extend the results for the univariate linear model to other more complicated models. Chapters 13–16 contain discussions of several models that are very similar to the univariate model (the generalized linear model, the repeated measures model, random effects and mixed models, and the correlation model). Chapters 17–22 discuss multivariate analysis. The use of a matrix normal distribution permits derivations that are nearly identical to those for univariate models. This matrix normal distribution plays the same role for multivariate problems that the vector normal distribution does for univariate problems. Finally, the Appendix contains the necessary matrix algebra for the book.

I wrote this book as a textbook for a course meeting for 75 minutes three times a week for 30 weeks. At Penn State this course is taken by all of our Ph.D. students and many of our better Master's degree students. The only prerequisites are a solid course in mathematical statistics and familiarity with the matrix algebra summarized in Section 1 of the Appendix.

I have included over 300 exercises, which are divided into three types. Type A exercises are numerical ones requiring substitution of small data sets into some of the procedures derived in the text. The same data sets are used for several different procedures to indicate the different calculations required for different models. Type B exercises ask for fairly straightforward derivations, whereas type C exercises are more difficult theoretical problems.

To keep the bibliography within reasonable bounds, I have listed only those papers specifically mentioned in the text. These references are of two different types. I have tried to find some of the more important historical papers and mention them in the appropriate chapters. In addition, at the end of many chapters there is a section labeled "Further Discussion" in which additional results are stated, together with sources for their proofs.

I have tried to write the book so that it can be used flexibly. For example, Chapters 12, 15, 16, 20, and 21 are not needed for later chapters and can therefore be skipped or reordered. In addition, Chapter 11 is only used in Sections 18.3 and 19.9, Chapter 13 is only used in Sections 15.6, 19.10, and 19.11, and Chapter 14 is only used in Sections 15.5, 18.7, and 19.7. Finally, the material in Sections 6.5, 7.7–7.9, 10.6, 17.6, 18.7, 19.10, and 19.11 can be skipped without any loss in continuity.

When I was younger, I had the very good fortune to learn statistics from two unusual teachers, Ingram Olkin and Charles Stein. I gratefully acknowledge their influence on my knowledge of and perspective on statistics. I hope their influence shows in this book.

I have also had many useful discussions with my colleagues at Penn State. These discussions have been quite helpful in correcting my thinking on many points, and I want to mention James Lynch, Thomas Hettmansperger, and particularly James Rosenberger who has labored mightily to help me understand how a "practical" person looks at these topics. I should also mention the constant encouragement and support given me by my chairman William Harkness.

In addition, I have been blessed with three very precise typists, without whose work this book would not have been possible. I thank Mary Lou Harkness, Bonnie Henninger, and Jeannine Volkert.

Finally, I would like to acknowledge the contributions of my students who have helped me refine my work. Particular thanks go to Jay Aubuchon, who contributed the graph comparing the risk functions of the James-Stein and ordinary least square estimators in Chapter 11.

STEVEN F. ARNOLD

State College, Pennsylvania
November 1980

Contents

The Theory of Linear Models and Multivariate Analysis

CHAPTER 1

Basic Statistical Definitions and Theorems

In this chapter we define the statistical concepts used in this book. The first three sections contain concepts and definitions that would be studied in a first-year course in mathematical statistics, together with some new definitions. Many of the results in these sections are stated without proof, since these results are usually covered in such a first-year course.

In Sections 1.5 and 1.6 we discuss the basic theory of invariance in hypothesis testing and invariance in estimation. The concept of invariance is used quite heavily to reduce the problems in this book to more manageable problems for which optimal procedures can be found. It is especially important that the reader understand the methods for finding uniformly most powerful (UMP) invariant tests and best invariant estimators. Therefore, particular attention should be paid to the examples is those sections. We often transform a particular problem to one that is easier to work with. In Section 1.7, we define such transformations and discuss their properties.

We assume throughout this book that the reader is familiar with the definitions and elementary properties of random variables, density functions, means, variances, and moment generating functions. These topics would be covered in most courses in mathematical statistics. A *random vector* is a vector whose components are random variables. Its *density function* is the joint density function of its components.

We define a *model* to be a $k \times 1$ random vector $\mathbf{X}$ taking values in a set $\chi \subset R^k$ and having density function $f(\mathbf{x}:\boldsymbol{\theta})$, where $\boldsymbol{\theta}$ is a p-dimensional vector of parameters taking values in $\Omega \subset R^p$. We call χ the *sample space* and Ω the *parameter space*. We assume that the sample space, the parameter space, and the function $f(\mathbf{x};\boldsymbol{\theta})$ are all known before we observe the random vector $\mathbf{X}$. It is our goal to try and make some inference about $\boldsymbol{\theta}$ after observing $\mathbf{X}$.

We often use the following elementary fact that is discussed in most textbooks in mathematical statistics. Let $\mathbf{X} \in \chi$ be a k-dimensional random vector having density function $f(\mathbf{x})$. Let $h(\mathbf{x})$ be an invertible function from χ to $\Xi \subset R^k$ such that h^{-1} has continuous partial derivatives in Ξ. Then $\mathbf{Y} = h(\mathbf{X})$ has density function $f(h^{-1}(\mathbf{y}))|J(\mathbf{y})|$, where $J(\mathbf{y})$ is the determinant of the $k \times k$ matrix whose ijth component is the partial derivative of the ith component of h^{-1} with respect to the jth component of $\mathbf{y}$. J is called the *Jacobian* of the transformation h^{-1}.

1.1. SUFFICIENCY AND COMPLETENESS.

Suppose we observe an $n \times 1$ random vector $\mathbf{X} \in \chi$ whose distribution depends on the $p \times 1$ parameter $\boldsymbol{\theta} \in \Omega$. A *statistic* $\mathbf{T}(\mathbf{X})$ is a $k \times 1$ function of $\mathbf{X}$ that does not depend on $\boldsymbol{\theta}$. A *sufficient statistic* $\mathbf{T}$ is a statistic such that the conditional distribution of $\mathbf{X}$ given $\mathbf{T}$ does not depend on $\boldsymbol{\theta}$. The idea is that a sufficient statistic contains all the information about $\boldsymbol{\theta}$ in the following sense. An observer who only observes $\mathbf{T}$ (and not the whole sample $\mathbf{X}$) can have a random mechanism turn out a random vector $\mathbf{Y}$ such that the conditional distribution of $\mathbf{Y}$ given $\mathbf{T}$ is the same as the conditional distribution of $\mathbf{X}$ given $\mathbf{T}$, since that conditional distribution does not depend on any unknown parameters and is therefore completely specified. The marginal distribution of $\mathbf{Y}$ is the same as the marginal distribution of $\mathbf{X}$. Therefore, the person who only observes $\mathbf{T}$ can generate a sample having the same distribution as one who observed $\mathbf{X}$. Therefore, all the information must be contained in $\mathbf{T}$. In the following theorem, and for the remainder of this section, it is assumed that all random vectors are either discrete or absolutely continuous and hence have density functions.

THEOREM 1.1 (Factorization). Let $\mathbf{X}$ have joint density function $f(\mathbf{x}; \boldsymbol{\theta})$. Then $\mathbf{T}(\mathbf{X})$ is a sufficient statistic if and only if $f(\mathbf{x}; \boldsymbol{\theta}) = h(\mathbf{x})$ $g(\mathbf{T}(\mathbf{x}); \boldsymbol{\theta})$ for some functions $h(\mathbf{x})$ and $g(\mathbf{t}; \boldsymbol{\theta})$.

A statistic $\mathbf{T}$ has a *complete family of distributions* if $E_{\boldsymbol{\theta}} h(\mathbf{T}) = 0$ for every $\boldsymbol{\theta} \in \Omega$ implies that $h(\mathbf{t}) = 0$ with probability 1. A *complete sufficient statistic* is a sufficient statistic whose family of distributions is complete. A complete sufficient statistic is the smallest sufficient statistic in that there is at most one unbiased estimator of any function $g(\boldsymbol{\theta})$. Therefore, in this sense, there is no redundancy. Let $V \subset R^k$ be a set. We say that $\mathbf{v} \in V$ is an *interior point* of V if there exists $\delta > 0$ such that the ball of radius δ about $\mathbf{v}$ is contained in V.

THEOREM 1.2 (Exponential criterion). Let $\mathbf{X}$ have joint density function $f(\mathbf{x}; \boldsymbol{\theta})$. Suppose $f(\mathbf{x}; \boldsymbol{\theta}) = c(\boldsymbol{\theta}) h(\mathbf{x}) \exp(\mathbf{Q}'(\boldsymbol{\theta}) \mathbf{T}(\mathbf{x}))$, where $\mathbf{Q}(\boldsymbol{\theta})$ is a function from Ω to R^k. If $\mathbf{Q}(\Omega)$ contains an interior point, then $\mathbf{T}(\mathbf{X})$ is a complete sufficient statistic ($\mathbf{Q}(\Omega) = \{\mathbf{Q}(\boldsymbol{\theta}) : \boldsymbol{\theta} \in \Omega\}$).

We now give an example of an exponential family in which $\mathbf{T}$ is not a complete sufficient statistic. This will demonstrate the need for some sort of topological condition like the assumption that $\mathbf{Q}(\Omega)$ has an interior point. Let $X_1, \ldots, X_n$ be independent with X_i normally distributed with mean θ and variance θ^2. Then this family is an exponential family and $\mathbf{T}'(\mathbf{X}) = (\sum_i X_i^2, \sum_i X_i)$, which is not complete, since $E[(n+1)\sum_i X_i^2 - (\sum_i X_i)^2] = 0$.

LEMMA 1.3. Let $\mathbf{T}(\mathbf{X})$ be a complete sufficient statistic, and let $\mathbf{S}(\mathbf{T})$ be an invertible function of $\mathbf{T}$. Then $\mathbf{S}$ is a complete sufficient statistic.

PROOF. The sufficiency of $\mathbf{S}$ follows from a straightforward application of the factorization criterion. The completeness follows from the definition of a complete family. Any function of a statistic with a complete family has a complete family. □

1.2. ESTIMATION AND CONFIDENCE INTERVALS.

An *estimator* of the $m \times 1$ vector $\boldsymbol{\tau}(\boldsymbol{\theta})$ is an $m \times 1$ statistic. An *unbiased estimator of* $\boldsymbol{\tau}(\boldsymbol{\theta})$ is an estimator $\mathbf{T}(\mathbf{X})$ such that $E_{\boldsymbol{\theta}}\mathbf{T}(\mathbf{X}) = \boldsymbol{\tau}(\boldsymbol{\theta})$. A *minimum variance unbiased estimator of* $\boldsymbol{\tau}(\boldsymbol{\theta})$ is an unbiased estimator $\mathbf{T}$ of $\boldsymbol{\tau}(\boldsymbol{\theta})$, such that if $\mathbf{S}$ is any other unbiased estimator of $\boldsymbol{\tau}(\boldsymbol{\theta})$ then $\operatorname{var}(S_i) \geqslant \operatorname{var}(T_i)$, $i = 1, \ldots, m$ where S_i and T_i are the ith components of $\mathbf{S}$ and $\mathbf{T}$.

THEOREM 1.4. (Lehmann-Scheffé). Let $\mathbf{S}(\mathbf{X})$ be a complete sufficient statistic. Let $\mathbf{T}(\mathbf{S})$ be an unbiased estimator of $\boldsymbol{\tau}(\boldsymbol{\theta})$. Then $\mathbf{T}(\mathbf{S})$ is the minimum variance unbiased estimator of $\boldsymbol{\tau}(\boldsymbol{\theta})$.

$\mathbf{T}(\mathbf{X})$ is a *maximum likelihood estimator* (*MLE*) for $\boldsymbol{\theta}$ if $f(\mathbf{x}; \mathbf{T}(\mathbf{x})) \geqslant f(\mathbf{x}; \boldsymbol{\theta})$ for all $\boldsymbol{\theta} \in \Omega$. If $\hat{\boldsymbol{\theta}}$ is a maximum likelihood estimator of $\boldsymbol{\theta}$, then $\boldsymbol{\tau}(\hat{\boldsymbol{\theta}})$ is defined to be the *maximum likelihood estimator of* $\boldsymbol{\tau}(\boldsymbol{\theta})$. It takes some work to show that this definition is consistent. See Mood, Graybill, and Boes (1974, pp. 284–286) for a discussion of this problem.

Let $\tau(\boldsymbol{\theta})$ be a real valued function of $\boldsymbol{\theta}$. A $1 - \alpha$ *confidence interval* for $\tau(\boldsymbol{\theta})$ is an interval $a(\mathbf{X}) \leqslant \tau(\boldsymbol{\theta}) \leqslant x(\mathbf{X})$ such that

$$P_{\boldsymbol{\theta}}(a(\mathbf{X}) \leqslant \tau(\boldsymbol{\theta}) \leqslant b(\mathbf{X})) = 1 - \alpha.$$

We note that this probability is computed over the distribution of $\mathbf{X}$. The parameter $\boldsymbol{\theta}$ is not random.

To define the remaining concepts used in estimation, we introduce loss functions. We return to the general problem of estimating the $m \times 1$ vector

$\boldsymbol{\tau}(\boldsymbol{\theta})$. Let $\mathbf{d} \in R^m$. A *loss function* is a real-valued function $L(\mathbf{d}; \boldsymbol{\theta})$. We think of $L(\mathbf{d}; \boldsymbol{\theta})$ as representing the loss when we estimate $\boldsymbol{\tau}(\boldsymbol{\theta})$ by $\mathbf{d}$ and $\boldsymbol{\theta}$ is the value of the parameter. Now, let $\mathbf{T}(\mathbf{X})$ be an estimator of $\boldsymbol{\tau}(\boldsymbol{\theta})$. The *risk function* of $\mathbf{T}$ is defined by

$$R(\mathbf{T}; \boldsymbol{\theta}) = E_{\boldsymbol{\theta}} L(\mathbf{T}(\mathbf{X}); \boldsymbol{\theta}).$$

That is, the risk function is the expected value of $L(\mathbf{T}(\mathbf{X}); \boldsymbol{\theta})$ where that expectation is taken over the distribution of $\mathbf{X}$. Let $\mathbf{T}_1(\mathbf{X})$ and $\mathbf{T}_2(\mathbf{X})$ be two estimators of $\boldsymbol{\tau}(\boldsymbol{\theta})$. Typically their risk functions would cross and $\mathbf{T}_1$ would have lower risk for some values of $\boldsymbol{\theta}$: whereas $\mathbf{T}_2$ would have lower risk for other values of $\boldsymbol{\theta}$. However, if $R(\mathbf{T}_1; \boldsymbol{\theta}) \leqslant R(\mathbf{T}_2; \boldsymbol{\theta})$ for all $\boldsymbol{\theta} \in \Omega$, then we say that $\mathbf{T}_1$ is *as good as* $\mathbf{T}_2$. If in addition, $R(\mathbf{T}_1; \boldsymbol{\theta}) < R(\mathbf{T}_2; \boldsymbol{\theta})$ for some $\boldsymbol{\theta} \in \Omega$, we say that $\mathbf{T}_1$ *is better than* $\mathbf{T}_2$. We say that $\mathbf{T}_1$ is an *inadmissible estimator* of $\boldsymbol{\tau}(\boldsymbol{\theta})$ if there exists an estimator $\mathbf{T}_2$ that is better than $\mathbf{T}_1$. If there is no estimator that is better than $\mathbf{T}_1$, we say that $\mathbf{T}_1$ is an *admissible estimator* of $\boldsymbol{\tau}(\boldsymbol{\theta})$. Let $\mathbf{T}(\mathbf{X})$ be an estimator of $\boldsymbol{\tau}(\boldsymbol{\theta})$, and let $m(\mathbf{T}) = \sup_{\boldsymbol{\theta} \in \Omega} R(\mathbf{T}; \boldsymbol{\theta})$. We say that $\mathbf{T}$ is a *minimax estimator* of $\boldsymbol{\tau}(\boldsymbol{\theta})$ if $m(\mathbf{T}) \leqslant m(\mathbf{T}^*)$ for all other estimators $\mathbf{T}^*(\mathbf{X})$. That is, $\mathbf{T}(\mathbf{X})$ is a minimax estimator if it minimizes the maximum risk. One aspect of the definitions in this paragraph is that they are tied to the particular loss function. It is quite possible that an estimator is admissible for some loss functions but not for others.

The loss functions considered in this book are quadratic loss functions. We now discuss such loss functions in the case where the function $\tau(\boldsymbol{\theta})$ being estimated is real-valued (i.e., in the case $m = 1$). We first look at the squared error loss function

$$L(d; \boldsymbol{\theta}) = (d - \tau(\boldsymbol{\theta}))^2.$$

The risk function for this loss function is the mean squared error,

$$R(T; \boldsymbol{\theta}) = E(T(\mathbf{X}) - \tau(\boldsymbol{\theta}))^2 = \mathrm{Var}(T(\mathbf{X})) + (E(T(\mathbf{X})) - \tau(\boldsymbol{\theta}))^2.$$

That is, the risk of an estimator $T(\mathbf{X})$ is just the variance of $T(\mathbf{X})$ plus the square of the bias of $T(\mathbf{X})$. Therefore, an estimator T that does well for squared error loss is one that has small variance and small bias. If $T(\mathbf{X})$ is unbiased the risk is just the variance of T, and therefore a unique minimum variance unbiased estimator is better than any other unbiased estimator for this loss function. However, the minimum variance unbiased estimator may be inadmissible if there is another estimator that is slightly biased but has a variance substantially smaller than the unbiased estimator. In this situation we may prefer the biased estimator; that is, we may be willing to have an estimator that is slightly biased if its variance is small. The mean squared error is often a useful measure of both the accuracy (lack of bias) and precision (small variance) of an estimator.

For most of the models considered in this book, we do not actually work with squared error loss, but multiply that loss function by a positive function $c(\boldsymbol{\theta})$; that is, we work with loss functions of the form

$$L^*(d;\boldsymbol{\theta}) = c(\boldsymbol{\theta})(d - \tau(\boldsymbol{\theta}))^2.$$

The risk function for L^* is given by

$$R^*(T;\boldsymbol{\theta}) = c(\boldsymbol{\theta})E(T(\mathbf{X}) - \tau(\boldsymbol{\theta}))^2 = c(\boldsymbol{\theta})R(T;\boldsymbol{\theta}).$$

We note first that $T_1(\mathbf{X})$ is better that $T_2(\mathbf{X})$ for the loss function L^* if and only if it is better for the loss function L, and therefore an estimator $T(\mathbf{X})$ is admissible for L^* if and only if it is admissible for L. Hence, so far as admissibility and ordering of rules is concerned, it does not matter whether we use the squared error loss L or the modified loss L^*. However, we often prefer a loss function of the form of L^* for the following reasons. First, the loss function L is dependent on the units of $\tau(\boldsymbol{\theta})$. By choosing $c(\boldsymbol{\theta})$ carefully, we can often eliminate this dependence on units. Second, it often happens that all procedures have maximum risk ∞ for the squared error loss function L, and therefore all procedures are minimax for that loss function. This difficulty is basically due to the dependence of L on the units of $\tau(\boldsymbol{\theta})$ and is eliminated when we scale the loss function to be unit free.

For an example we look at the one sample model in which we observe $X_1, \ldots, X_n$, independent where the X_i are normally distributed with unknown mean μ and unknown variance σ^2. Consider first estimating μ with the loss function

$$L^*\big(d;(\mu,\sigma^2)\big) = \frac{(d-\mu)^2}{\sigma^2}.$$

The minimum variance unbiased estimator and MLE of μ is $\bar{X} = \sum X_i/n$. The risk function for this estimator is easily seen to be l/n. In addition, it can be shown that $\bar{X}$ is an admissible estimator for μ and the only minimax estimator of μ for this loss function. (See Ferguson, 1967, pp. 141–142 and 176–178 for proofs of these facts when σ^2 is assumed known.) Note that if we had used the squared error loss function $L(d;(\mu,\sigma^2)) = (d-\mu)^2$, then $\bar{X}$ would have risk σ^2/n, and the maximum risk would be ∞. For this loss function all estimators have infinite maximum risk and hence all estimators are minimax.

Now, consider estimating σ^2 with the loss function

$$L^*\big(d;(\mu,\sigma^2)\big) = \frac{(d-\sigma^2)^2}{\sigma^4} = \left(\frac{d}{\sigma^2} - 1\right)^2$$

(Note that L^* has been normalized to be unit-free.) The minimum variance unbiased estimator of σ^2 is $S^2 = \sum(X_i - \bar{X})^2/(n-1)$ and the MLE is $(n-1)S^2/n$. A natural problem is to decide which of these estimators is better (at least for this loss function). For any $c > 0$, define $T_c(X) = c(n-1)S^2$. Using the fact that $(n-1)S^2/\sigma^2 \sim \chi^2_{n-1}$, we see that

$$R^*(T_c;(\mu,\sigma^2)) = E\left(\frac{c(n-1)S^2}{\sigma^2} - 1\right)$$

$$= c^2(n-1)(n+1) - 2c(n-1) + 1 \qquad (1.1)$$

(see Exercise B2). For the minimum variance unbiased estimator S^2, $c = 1/(n-1)$ and

$$R^*(S^2;(\mu,\sigma^2)) = \frac{n+1}{n-1} - 1 = \frac{2}{n-1}.$$

For the MLE, $(n-1)S^2/n$, $c = 1/n$ and

$$R^*\left(\frac{n-1}{n}S^2;(\mu,\sigma^2)\right) = \frac{(n-1)(n+1)}{n^2} - \frac{2(n-1)}{n} + 1$$

$$= \frac{2n-1}{n^2} \leqslant \frac{2}{n-1}.$$

Therefore, we see that for this model (and this loss function) the MLE is better than the minimum variance unbiased estimator. However, it is easily seen that the risk of T_c is minimized when $c = 1/(n+1)$ (see Exercise B2) and therefore, the estimator $(n-1)S^2/(n+1)$ is better than either the minimum variance unbiased estimator or the MLE and hence both the minimum variance unbiased estimator and MLE are inadmissible for this problem (with the loss function). It can also be shown that the estimator $(n-1)S^2/(n+1)$ is a minimax estimator for this problem, but is also inadmissible for it (see Stein, 1964).

We finish this section with a generalization of the Rao-Blackwell and Lehmann-Scheffé theorems which is proved in Ferguson (1967, pp. 121–122, 134–135). Let $h(\mathbf{U})$ be a function from R^m to R^1. We say that h is *convex* if for any $\mathbf{U}$ and $\mathbf{V}$ and any a such that $0 \leqslant a \leqslant 1$,

$$h(a\mathbf{U} + (1-a)\mathbf{V}) \leqslant ah(\mathbf{U}) + (1-a)h(\mathbf{V}).$$

THEOREM 1.5. Consider the problem of estimating $\boldsymbol{\tau}(\boldsymbol{\theta})$ with the loss function $L(\mathbf{d},\boldsymbol{\theta})$.

a. Let $\mathbf{S}(\mathbf{X})$ be a sufficient statistic and let $\mathbf{T}(\mathbf{X})$ be an estimator of $\boldsymbol{\tau}(\boldsymbol{\theta})$. Let $\mathbf{T}^*(\mathbf{S}) = E\mathbf{T}(\mathbf{X})\,|\,\mathbf{S}$. If $L(\mathbf{d};\boldsymbol{\theta})$ is a convex function of $\mathbf{d}$ for all $\boldsymbol{\theta}$, then $\mathbf{T}^*(\mathbf{S})$ is as good as $\mathbf{T}(\mathbf{X})$. If $\mathbf{T}(\mathbf{X})$ is unbiased, then so is $\mathbf{T}^*(\mathbf{S})$.

b. Let $\mathbf{S}(\mathbf{X})$ be a complete sufficient statistic and let $\mathbf{T}(\mathbf{S})$ be the minimum variance unbiased estimator of $\tau(\boldsymbol{\theta})$. If $L(\mathbf{d};\boldsymbol{\theta})$ is a convex function of $\mathbf{d}$ for all $\boldsymbol{\theta}$, then $\mathbf{T}(\mathbf{S})$ is as good as any unbiased estimator of $\tau(\boldsymbol{\theta})$.

All the loss functions considered in this book are convex functions of d. Part a of this theorem implies that for any estimator, there is an estimator based on the sufficient statistic that is as good. This reinforces the idea that nothing is lost if we reduce the data to a sufficient statistic. Part b implies that the minimum variance unbiased estimator is the best unbiased estimator for any convex loss function, since it is as good as any other unbiased estimator.

1.3. TESTING HYPOTHESES.

Let $\omega \subset \Omega$. In this section we consider the problem of testing that $\boldsymbol{\theta} \in \omega$ versus the general alternative $\boldsymbol{\theta} \in \Omega$. A *nonrandomized test* is a decision rule that for each $\mathbf{x} \in \chi$ decides whether to accept or reject the null hypothesis $\boldsymbol{\theta} \in \omega$. The set $C \subset \chi$ of points where the test rejects is called the *critical region* of the test. For many theoretical purposes, it is helpful to allow the possibility of randomized tests, although it is hard to imagine when one would actually be used. A *randomized test* is a test in which, after observing $\mathbf{X}$, a Bernoulli experiment with known probability of success $\phi(\mathbf{X})$ is performed. If the Bernoulli experiment is successful, the hypothesis is rejected. Otherwise it is accepted. The function $\phi(\mathbf{X})$, called the *critical function* of the test, is assumed to be a function from χ to $[0, 1]$, which does not depend on $\boldsymbol{\theta}$. We shall often write "the test ϕ" to mean "the test with critical function ϕ". If the test is a nonrandomized test with critical region C, then the critical function is the indicator function of the set C.

The *power function* of a test ϕ is the function $K_\phi(\boldsymbol{\theta}) = E_{\boldsymbol{\theta}}\phi(\mathbf{X})$. The power function represents the probability of rejecting, averaged over both $\mathbf{X}$ and the randomization. One advantage of introducing randomized tests is the following lemma.

THEOREM 1.6. Let $\mathbf{S}(\mathbf{X})$ be a sufficient statistic, and let $\phi(\mathbf{X})$ be a critical function. Let $\phi^*(\mathbf{S}) = E\phi(\mathbf{X}) \mid \mathbf{S}$. Then ϕ^* is a critical function that depends only on $\mathbf{S}$ that has the same power function as $\phi(\mathbf{X})$.

PROOF. Since $\mathbf{S}$ is sufficient, $\phi^*(\mathbf{S})$ does not depend on $\boldsymbol{\theta}$. Also

$$K_{\phi^*}(\boldsymbol{\theta}) = E_{\boldsymbol{\theta}}\phi^*(S) = E_{\boldsymbol{\theta}}(E\phi(X)|S) = E_{\boldsymbol{\theta}}\phi(X) = K_\phi(\boldsymbol{\theta}) \quad \square$$

This reinforces the notion that nothing is lost by looking at sufficient statistics. However the original test $\phi(\mathbf{X})$ may have been a nonrandomized test and the induced test $\phi^*(\mathbf{S})$ may be a randomized one.

The *size of a test* ϕ is the $\sup_{\boldsymbol{\theta} \in \omega} K_\phi(\boldsymbol{\theta})$. A test ϕ is *uniformly most power*

(UMP) size α if (a) ϕ has size α; and (b) if ϕ^* is any other test with size $\alpha^* \leqslant \alpha$, then $K_{\phi^*}(\boldsymbol{\theta}) \leqslant K_\phi(\boldsymbol{\theta})$, for all $\boldsymbol{\theta} \in \Omega - \omega$.

THEOREM 1.7. Let $T(\mathbf{X})$ be a real-valued statistic and let θ be a real-valued parameter. If $f(\mathbf{x};\theta)/f(\mathbf{x};\theta_o)$ is an increasing function of $T(\mathbf{x})$ for all $\theta > \theta_0$ (θ_0 fixed), then

$$\phi(\mathbf{X}) = \begin{cases} 1 & \text{if } T(\mathbf{X}) > k \\ 0 & \text{if } T(\mathbf{X}) \leqslant k \end{cases}$$

is UMP of its size for testing $\theta = \theta_0$ versus $\theta \geqslant \theta_0$.

There are two aspects of this theorem that should be emphasized. First, it is only applicable when the parameter θ is univariate and, secondly, it is only applicable for testing one-sided hypotheses. For the models considered in this book, UMP tests exist only when both these conditions are satisfied.

A test ϕ is said to be *unbiased* if $K_\phi(\boldsymbol{\theta}_0) \leqslant K_\phi(\boldsymbol{\theta}_1)$ for all $\boldsymbol{\theta}_0 \in \omega$, $\boldsymbol{\theta}_1 \in \Omega - \omega$. That is, ϕ is unbiased if the probability of rejecting when we should reject is always as large as the probability of rejecting when we should accept. A size α test ϕ is *inadmissible* if there exists another size α test ϕ^* such that $K_{\phi^*}(\boldsymbol{\theta}) \geqslant K_\phi(\boldsymbol{\theta})$ for all $\boldsymbol{\theta} \in \Omega - \omega$ with strong inequality for some $\boldsymbol{\theta} \in \Omega - \omega$. That is, ϕ is inadmissible if there is some other size α test that is more powerful than ϕ. If ϕ is not inadmissible, we say that ϕ is *admissible*.

We now define a minimaxity property for tests. Let D_α be the set of all size α tests. For all $\boldsymbol{\theta} \in \Omega - \omega$, define

$$\beta(\boldsymbol{\theta}) = \sup_{\phi \in D_\alpha} K_\phi(\boldsymbol{\theta})$$

$\beta(\boldsymbol{\theta})$ is called the upper envelope. For each $\phi \in D_\alpha$, define

$$A(\phi) = \sup_{\boldsymbol{\theta} \in \Omega - \omega} (\beta(\boldsymbol{\theta}) - K_\phi(\boldsymbol{\theta}))$$

$A(\phi)$ is called the maximum shortcoming of ϕ. A size α test ϕ_o is said to be *most stringent* if $A(\phi_0) \leqslant A(\phi)$ for all $\phi \in D_\alpha$. That is: ϕ_0 is most stringent if it minimizes the maximum shortcoming.

Define $\lambda(\mathbf{x}) = \sup_{\boldsymbol{\theta} \in \omega} f(\mathbf{x};\boldsymbol{\theta})/\sup_{\boldsymbol{\theta} \in \Omega} f(\mathbf{x};\boldsymbol{\theta})$. A test ϕ is a *likelihood ratio test* if

$$\phi(\mathbf{X}) = \begin{cases} 1 & \text{if } \lambda(\mathbf{X}) < k \\ 0 & \text{if } \lambda(\mathbf{X}) \geqslant k \end{cases}$$

$\lambda(\mathbf{X})$ is called the *likelihood ratio test statistic*.

1.4. NONCENTRAL DISTRIBUTIONS.

We now define the univariate distributions that are used in this book. The results in this section are stated without proofs, since these proofs are rather confusing and do not give much insight. Outlines of the proofs are given in the exercises.

We say that X has a *univariate normal distribution* with mean μ and variance σ^2 if X is a continous random variable with density

$$f(x;(\mu,\sigma^2)) = \frac{1}{\sqrt{2\pi}\,\sigma}\exp\left[-\frac{(x-\mu)^2}{2\sigma^2}\right], \qquad -\infty < x < \infty.$$

We write that

$$X \sim N_1(\mu,\sigma^2).$$

Let $X_1, \ldots, X_n$ be independent random variables such that $X_i \sim N_1(\mu_i, 1)$. Define

$$Y = \sum_{i=1}^{n} X_i^2 \qquad \text{and} \qquad \delta = \sum_{i=1}^{n} \mu_i^2.$$

We say that Y has a *noncentral* χ^2 *distribution* with n degrees of freedom and noncentrality parameter δ. We write that

$$Y \sim \chi_n^2(\delta).$$

(Note that this definition implies that the distribution of Y depends on the μ_i only through δ.) The density function of Y is given by

$$f(y;\delta) = \sum_{k=o}^{\infty} \frac{\exp\left(-\frac{\delta}{2}\right)\left(\frac{\delta}{2}\right)^k}{k!} \frac{y^{(m/2)+k-1}\exp(-y/2)}{\Gamma\left(\frac{m}{2}+k\right)2^{(m/2)+k}}, \qquad 0 < y < \infty. \tag{1.2}$$

We often write $U \sim a\chi_m^2(\delta)$ to mean that $u/a \sim \chi_m^2(\delta)$. If $\delta = 0$, we say that Y has a *central* χ^2*-distribution*. We say that K has a *Poisson distribution* with parameter θ if K has the discrete density function

$$f(k;\theta) = \frac{e^{-\theta}\theta^k}{k!}, \quad k = 0, 1, \ldots. \tag{1.3}$$

Lemma 1.8 follows directly from (1.2) and (1.3).

LEMMA 1.8. If $Y \mid K \sim \chi_{n+2K}^2(0)$, and K has a Poisson distribution with parameter $\delta/2$, then $Y \sim \chi_n^2(\delta)$.

In other words, a noncentral χ^2-distribution can be thought of as a central χ^2-distribution on a random number of degrees of freedom. This lemma can be used to find the mean and variance of the noncentral χ^2-distribution.

Now let X and Y be independent, $X \sim N_1(\mu, 1)$, $Y \sim \chi_n^2(0)$. Let

$$t = \frac{X}{\sqrt{Y/n}}.$$

We say that t has a *noncentral t-distribution* with n degrees of freedom and noncentrality parameter μ, and we write

$$t \sim t_n(\mu)$$

The density function of t is

$$g(t;\mu) = \frac{n^{n/2}(t^2+n)^{-(n+1)/2}}{\sqrt{\pi}\,\Gamma(n/2)2^{(n-1)/2}} \exp\left[-\frac{n\mu^2}{2(t^2+n)}\right]$$

$$\times \int_0^\infty \exp\left[-\tfrac{1}{2}\left(x - \frac{\mu t}{\sqrt{t^2+n}}\right)^2\right] x^n\,dx, \qquad -\infty < t < \infty. \quad (1.4)$$

If $\mu = 0$, we say that t has a *central t-distribution*.

Let X and Y be independent with $X \sim \chi_m^2(\delta)$, $Y \sim \chi_n^2(0)$. Let

$$F = \frac{X}{Y}\frac{n}{m}$$

We say that F has *noncentral F-distribution* with m and n degrees of freedom and noncentrality parameter δ. We write that

$$F \sim F_{m,n}(\delta).$$

The density function of F is given by

$$h()f;\delta) = \sum_{k=0}^{\infty} e^{-\delta/2} \frac{\left(\frac{\delta}{2}\right)^k}{k!} \frac{\Gamma\left(\frac{m+n}{2}+k\right)}{\Gamma\left(\frac{m}{2}+k\right)\Gamma\left(\frac{n}{2}\right)} \frac{\left(\frac{m}{n}\right)^{m/2+k} F^{m/2+k-1}}{\left(1+\frac{m}{n}F\right)^{(m+n/2+k)}}$$

$$0 < F < \infty. \qquad (1.5)$$

If $\delta = 0$, we say that F has a *central F-distribution*. The following lemma follows from (1.2), (1.4), and (1.5). It is really the only time that we use the noncentral t and F density functions.

LEMMA 1.9. Let $f(y;\delta)$, $g(t;\mu)$ and $h(F;\delta)$ be defined by (1.2), (1.4), and (1.5). Then $f(y;\delta)/f(y;0)$ is an increasing function of y for all $\delta > 0$, $g(t;\mu)/g(t;0)$ is an increasing function of t for all $\mu > 0$, and $h(F;\delta)/h(F;0)$ is an increasing function of F for all $\delta > 0$.

PROOF. See Exercise B4. □

The following notation is used throughout the text. We write t_n^α to be the upper α point of a central t-distribution on n degrees of freedom, $\chi_n^{2\alpha}$ to be the upper α point of a central χ^2-distribution on n degrees of freedom, and $F_{n,m}^\alpha$ to be the upper α point of a central F-distribution on n and m degrees of freedom. In other words, if $t \sim t_n(0)$, $\chi^2 \sim \chi_n^2(0)$, $f \sim F_{n,m}(0)$, then

$$P(t \leqslant t_n^\alpha) = P(\chi^2 \leqslant \chi_n^{2\alpha}) = P(F \leqslant F_{n,m}^\alpha) = 1 - \alpha$$

Now consider the testing problem in which we observe $F \sim F_{m,n}(\delta)$ and want to test that $\delta = 0$ against $\delta \geqslant 0$. By Theorem 1.7 and Lemma 1.9 the UMP size α test is to reject if $F > F_{m,n}^\alpha$. Similarly if we observe $t \sim t_n(\delta)$ or $Y \sim \chi_n^2(\delta)$ and want to test that $\delta = 0$ against $\delta \geqslant 0$, the UMP size α test is to reject if $t > t_n^\alpha$ or $Y > \chi_n^{2\alpha}$.

1.5. INVARIANCE IN HYPOTHESIS TESTING.

For most of the testing problems considered in this book, there is no UMP size α test. However, we can often find procedures that are UMP among all size α procedures that satisfy a certain property first suggested in a particular case by Hotelling (1931) and stated in a general way by Hunt and Stein (1946). In this section, we discuss that property and how to find such procedures.

Throughout this section we consider the testing problem in which we observe the random vector $\mathbf{X} \in \chi$, having density function $f(\mathbf{x};\boldsymbol{\theta}), \boldsymbol{\theta} \in \Omega$. We want to test that $\boldsymbol{\theta} \in \omega$ against $\boldsymbol{\theta} \in \Omega$. Let g be an invertible function from χ to χ. We say that this testing problem is invariant under g if there exists an invertible function $\bar{g}$ from Ω to Ω such that

(1) $\mathbf{Y} = g(\mathbf{X})$ has density $f(\mathbf{y};\, g(\boldsymbol{\theta}))$,
(2) $\bar{g}(\omega) = \omega$ (and hence $\bar{g}(\Omega - \omega) = \Omega - \omega$).

(If S is a set, we define $\bar{g}(\mathbf{S}) = \{\bar{g}(\mathbf{s}) : \mathbf{s} \in \mathbf{S}\}$.) We say that a procedure $\phi(\mathbf{X})$ is invariant under g if $\phi(g(\mathbf{x})) = \phi(\mathbf{x})$ for all $\mathbf{x} \in \chi$. In this book, we will be primarily concerned with invariant procedures for the following reason. Let $\mathbf{Y} = g(\mathbf{X})$, $\boldsymbol{\tau} = \bar{g}(\boldsymbol{\theta})$. Then $\mathbf{Y}$ has density $f(\mathbf{y};\boldsymbol{\tau})$. Furthermore, $\boldsymbol{\theta} \in \omega$ if and only if $\boldsymbol{\tau} \in \omega$, and $\boldsymbol{\theta} \in \Omega$ if and only if $\boldsymbol{\tau} \in \Omega$. Therefore, the testing problem in which we observe $\mathbf{X}$ is the same as the testing problem in which we observe **Y**. To be consistent, we should use the same procedure for the problem in which we observe **Y** as for the one in which we observe **X**. This implies that $\phi(g(\mathbf{X})) = \phi(\mathbf{Y})$ should equal $\phi(\mathbf{X})$ (i.e., that ϕ should be invariant).

The testing problems considered in this book are invariant under large collections of invertible functions. We now establish some elementary facts about those collections. Let h and k be two invertible functions from a space C to itself. The composition, $h \circ k$, of h and k is the invertible transformation given by $(h \circ k)(c) = h(k(c))$. We denote the inverse of $h(c)$ by $h^{-1}(c)$.

THEOREM 1.10. If a testing problem is invariant under invertible functions g_1 and g_2, it is invariant under $g_1 \circ g_2$, and $\overline{g_1 \circ g_2} = \bar{g}_1 \circ \bar{g}_2$. If a testing problem is invariant under the invertible function g, then it is invariant under g^{-1} and $\overline{g^{-1}} = \bar{g}^{-1}$.

PROOF. Let $\mathbf{Z} = g_2(\mathbf{X})$. Then $\mathbf{Z}$ has density $f(\mathbf{z}; \bar{g}_2(\boldsymbol{\theta}))$. Now, let $\mathbf{Y} = g_1(\mathbf{Z})$. Then $\mathbf{Y} = (g_1 \circ g_2)(\mathbf{X})$ has density $f(\mathbf{y}; (\bar{g}_1 \circ \bar{g}_2)(\boldsymbol{\theta}))$. Also $(\bar{g}_1 \circ \bar{g}_2)(\omega) = \bar{g}_1(\omega) = \omega$. Therefore, the testing problem is invariant under $g_1 \circ g_2$ and $\overline{g_1 \circ g_2} = \bar{g}_1 \circ \bar{g}_2$.

Now let **Y** have density $f(\mathbf{y}; \bar{g}^{-1}(\boldsymbol{\theta}))$ and let $\mathbf{Z} = g(\mathbf{Y})$. Then $\mathbf{Z}$ has density $f(\mathbf{z}; \bar{g}(\bar{g}^{-1}(\boldsymbol{\theta}))) = f(\mathbf{z}; \boldsymbol{\theta})$. Hence if $\mathbf{Z}$ has density $f(\mathbf{z}; \boldsymbol{\theta})$, then $\mathbf{Y} = g^{-1}(\mathbf{Z})$ has density $f(\mathbf{y}; \bar{g}^{-1}(\boldsymbol{\theta}))$. Finally, $\bar{g}(\omega) = \omega$, and hence $\omega = \bar{g}^{-1}(\bar{g}(\omega)) = \bar{g}^{-1}(\omega)$. Therefore, the problem is invariant under g^{-1} and $\overline{g^{-1}} = \bar{g}^{-1}$. $\square$

A collection H of invertible functions from a space to itself is called a *group* if it satisfies the following two conditions:.

(1) If $g_1 \in H$, $g_2 \in H$, then $g_1 \circ g_2 \in H$.
(2) If $g \in H$, then $g^{-1} \in H$.

Theorem 1.10 implies that the set of invertible functions that leave a testing problem invariant is a group as is the induced set of transformations on the parameter space..

Now let G be a group of invertible functions. We say that the testing problem is invariant under G if it is invariant under g for all $g \in G$. A critical function is invariant under G if it is invariant under G for all $g \in G$. Let $\bar{G}$ be the set of invertible transformations on the parameter space that is induced by G. From Theorem 1.10, again, we see that $\bar{G}$ is also a group. For many of the testing problems considered in this book, there is a group G and a critical function ϕ such that ϕ is the UMP size α test among all tests that are invariant under G. For the remainder of this section, we discuss the method used to find such critical functions.

Let G be a group of invertible functions from a set C to itself. A function $T(c)$ is called a *maximal invariant* if it satisfies the following two conditions:

(1) $T(g(c)) = T(c)$ for all $g \in G$, $c \in C$.
(2) If $T(c_1) = T(c_2)$, then there exists $g \in G$ such that $c_1 = g(c_2)$.

For any set C and any group G of invertible functions C to itself, there exists a maximal invariant. (See Exercise C1. This is the only place in the development of invariance where we use the fact that G is a group.)

LEMMA 1.11. Let C be a space, G a group of invertible functions from C to itself and $T(c)$ a maximal invariant under G. Then $h(g(c)) = h(c)$ for all $g \in G$, $c \in C$ if and only if there exists $k(t)$ such that $h(c) = k(T(c))$.

PROOF. Suppose $h(c) = k(T(c))$. Then $h(g(c)) = k(T(g(c))) = k(T(c)) = h(c)$. Conversely, suppose that $h(c) = h(g(c))$ for all g and c. Let $T(c_1) = T(c_2)$. We will be finished when we show that $h(c_1) = h(c_2)$. Since $T(c_1) = T\ (c_2)$, we see that $c_1 = g(c_2)$ for some $g \in G$. Therefore, $h(c_1) = h(g(c_2)) = h(c_2)$. □

Another way of stating this lemma is that any invariant function depends on c only through the maximal invariant $T(c)$, and is therefore really only a function of T.

Now consider a testing problem that is invariant under the group G, and let $\bar{G}$ be the induced group of invertible functions on the parameter space. Let $\mathbf{T}(\mathbf{X})$ be a maximal invariant under G and $\boldsymbol{\delta}(\boldsymbol{\theta})$ be a maximal invariant under $\bar{G}$. $\mathbf{T}(\mathbf{X})$ is called a *maximal invariant* for the testing problem and $\boldsymbol{\delta}(\boldsymbol{\theta})$ is called a *parameter maximal invariant*. (Note that any invertible function of a maximal invariant or parameter maximal invariant is also a maximal invariant or parameter maximal invariant.) The following theorem is the basic result of this section.

THEOREM 1.12. a. $\phi(\mathbf{X})$ is an invariant critical function if and only if there exists $\phi^*(\mathbf{T})$ such that $\phi(\mathbf{X}) = \phi^*(\mathbf{T}(\mathbf{X}))$ for all $\mathbf{X} \in \chi$.

b. The distribution of $\mathbf{T}$ depends only on $\boldsymbol{\delta}$.

PROOF. a. This follows directly from Lemma 1.11.

b. Let $h(\mathbf{t}; \boldsymbol{\theta})$ be the density of $\mathbf{T}$. Let $\mathbf{Y} = g(\mathbf{X})$. Then $\mathbf{Y}$ has density $f(\mathbf{y}; \bar{g}(\boldsymbol{\theta}))$. Now $\mathbf{T}(\mathbf{Y}) = \mathbf{T}(g(\mathbf{X})) = \mathbf{T}(\mathbf{X})$. Therefore, $h(\mathbf{t}; \bar{g}(\boldsymbol{\theta})) = h(\mathbf{t}; \boldsymbol{\theta})$. Hence, for each $\mathbf{t}$, $h(\mathbf{t}; \boldsymbol{\theta})$ is invariant under $\bar{G}$, and, therefore, by Lemma 1.11, $h(\mathbf{t}; \boldsymbol{\theta})$ depends on $\boldsymbol{\theta}$ only through $\boldsymbol{\delta}(\boldsymbol{\theta})$. □

As long as we are interested only in invariant rules, we may as well look at the problem, which is often simpler, in which we observe $\mathbf{T} = \mathbf{T}(\mathbf{X})$, the maximal invariant, having the induced distribution $k(\mathbf{t}; \boldsymbol{\delta})$, where $\boldsymbol{\delta}(\boldsymbol{\theta})$ is the parameter maximal invariant. We are now testing the null hypothesis that $\boldsymbol{\delta} \in \delta(\omega)$ versus the alternative that $\boldsymbol{\delta} \in \delta(\Omega)$. In many problems considered in this book there is a UMP size α test for this simpler problem. This test will therefore be the best size α test among all those that are invariant under G. Such a test is called a *UMP invariant size α test*.

In the following examples as well as in later sections of the book, we first reduce the model by sufficiency before looking at invariance. A justification for this approach is given in Exercise C9 in which it is shown that, for any invariant test based on the original sample, there is an invariant test based on the sufficient statistic that has the same power function and that a UMP invariant size α test for the model reduced by sufficiency is also UMP invariant size α for the original model. For a more detailed discussion of the relationship between invariance and sufficiency, see Hall et al. (1965).

EXAMPLE 1. Let $X_1, \ldots, X_n$ be independent random variables, with $X_i \sim N_1(\mu, \sigma^2)$. Then

$$\bar{X} = \frac{1}{n}\sum_i X_i, \qquad S^2 = \frac{1}{n-1}\sum_i (X_i - \bar{X})^2$$

is a complete sufficient statistic, $\bar{X}$ and S^2 are independent, and

$$\bar{X} \sim N_1(\mu, \sigma^2/n), \qquad (n-1)S^2 \sim \sigma^2 \chi^2_{n-1}(0)$$

a. We first consider testing that

$$H_0 : \mu = 0, \qquad \sigma^2 > 0$$
$$H_1 : -\infty < \mu < \infty, \quad \sigma^2 > 0.$$

This problem is invariant under the group G_1 of transformations of the form

$$g_a(\bar{X}, S^2) = (a\bar{X}, a^2 S^2), \qquad \bar{g}_a(\mu, \sigma^2) = (a\mu, a^2\sigma^2), \qquad a \neq 0.$$

Let

$$T_1(\bar{X}, S^2) = \frac{n\bar{X}^2}{S^2}, \qquad \delta_1(\mu, \sigma^2) = \frac{n\mu^2}{\sigma^2}.$$

Then T_1 is a maximal invariant, as the following argument shows:

(1) $T_1(g_a(\bar{X}, S^2)) = T_1(a\bar{X}, a^2S^2) = na^2\bar{X}^2/(a^2S^2) = n\bar{X}^2/S^2 = T_1(\bar{X}, S^2)$.

(2) Suppose that $T_1(\bar{X}, S^2) = T_1(\bar{X}^*, S^{*2})$. We need to show that there exists a such that $(\bar{X}, S^2) = g_a(X^*, S^{*2})$. If $T_1(\bar{X}, S^2) = T_1(\bar{X}^*, S^{*2})$,

then

$$\frac{\bar{X}^2}{S^2}=\frac{\bar{X}^{*2}}{S^{*2}}, \qquad \frac{|\bar{X}|}{|\bar{X}^*|}=\frac{S}{S^*}.$$

Let $a=\bar{X}/\bar{X}^*$. Then $(\bar{X},S^2)=(a\bar{X},a^2S^{*2})$.

Similarly, δ_1 is a parameter maximal invariant. In addition,

$$T_1 \sim F_{1,n-1}(\delta_1).$$

Since $\mu=0$ if and only if $\delta_1=0$, the reduced problem is the testing problem in which we observe $T_1 \sim F_{1,n-1}(\delta_1)$ and we are testing

$$H_0:\delta_1=0,$$
$$H_1:\delta_1 \geqslant 0.$$

By Theorem 1.7 and Lemma 1.9, a UMP size α test for this reduced problem is given by

$$\phi(T_1)=\begin{cases}1 & \text{if } T_1 > F^{\alpha}_{1,n-1}\\ 0 & \text{if } T_1 \leqslant F^{\alpha}_{1,n-1}\end{cases}$$

This test is therefore UMP invariant size α for the original problem. Note that this test is just the usual two-sided, one-sample t-test.

b. Now consider testing

$$H_0: \mu=0, \qquad \sigma^2>0$$
$$H_1: \mu \geqslant 0, \qquad \sigma^2>0.$$

This problem is not invariant under the whole group G_1 defined above, since if $a<0$, then $\bar{g}_a$ of the set where μ is nonnegative would be the set where μ is nonpositive. Hence, if $a<0$ then $\bar{g}_a$ would not map the parameter space $\Omega=\{(\mu,\sigma^2): \mu \geqslant 0, \sigma^2>0\}$ onto itself. However, the problem is invariant under the subgroup G_2 in which $a>0$. A maximal invariant and a parameter maximal invariant under G_2 are given by

$$T_2(\bar{X},S^2)=\sqrt{n}\,\frac{\bar{X}}{S}, \qquad \delta_2=\sqrt{n}\,\frac{\mu}{\sigma}.$$

(see Exercise B5). In addition:

$$T_2 \sim t_{n-1}(\delta_2).$$

We are testing

$$H_0:\delta_2=0,$$
$$H_1:\delta_2 \geqslant 0.$$

Therefore, by Theorem 1.7 and Lemma 1.9 a UMP invariant size α test is given by

$$\phi(T_2) = \begin{cases} 1 & \text{if } T_2 > t_{n-1}^{\alpha} \\ 0 & \text{if } T_2 \leqslant t_{n-1}^{\alpha} \end{cases}$$

Note that this test is just the usual one-sided, one-sample t-test.

We now consider the two-sample problems.

EXAMPLE 2. Let $X_1, \ldots, X_m$, $Y_1, \ldots, Y_n$ be independent, with $X_i \sim N_1(\mu, \sigma^2)$, $Y_i \sim N_1(\nu, \sigma^2)$. Then a complete sufficient statistic is given by

$$\overline{X} = \frac{1}{m}\sum_i X_i, \qquad \overline{Y} = \frac{1}{n}\sum_i Y_i,$$

$$S^2 = \frac{1}{m+n-2}\left(\sum_i (X_i - \overline{X})^2 + \sum_i (Y_i - \overline{Y})^2\right),$$

$\overline{X}$, $\overline{Y}$, and S^2 are independent, and

$$\overline{X} \sim N_1\left(\mu, \frac{\sigma^2}{m}\right), \qquad \overline{Y} \sim N_1\left(\nu, \frac{\sigma^2}{n}\right), \qquad (m+n-2)S^2 \sim \sigma^2 \chi^2_{m+n-2}(0)$$

a. We first consider testing that

$$H_0 : \mu = \nu, \qquad \sigma^2 > 0,$$

$$H_1 ; \infty < \mu < \infty, \qquad -\infty < \nu < \infty, \qquad \sigma^2 > 0.$$

This problem is invariant under the group G_3 of transformations of the form

$$g_{a,b}(\overline{X}, \overline{Y}, S^2) = (a\overline{X} + b, a\overline{Y} + b, a^2 S^2),$$

$$\bar{g}_{a,b}(\mu, \nu, \sigma^2) = (a\mu + b, a\nu + b, a^2\sigma^2) \qquad a \neq 0$$

A maximal invariant and a parameter maximal invariant for this problem are

$$T_3(\overline{X}, \overline{Y}, S^2) = \frac{(\overline{X} - \overline{Y})^2}{S^2\left(\frac{1}{m} + \frac{1}{n}\right)}, \qquad \delta_3(\mu, \nu, \sigma^2) = \frac{(\mu - \nu)^2}{\sigma^2\left(\frac{1}{m} + \frac{1}{n}\right)}$$

(see Exercise B5). Also,

$$T_3 \sim F_{1, m+n-2}(\delta_3)$$

and we are testing

$$H_0 : \delta_3 = 0,$$

$$H_1 : \delta_3 \geqslant 0.$$

Therefore, as before, the UMP invariant size α test rejects if T_3 is too large. Note that this test is just the two-sided, two-sample t-test.

b. Consider testing

$$H_0 : \mu = \nu, \qquad \sigma^2 > 0,$$
$$H_1 : \mu \geqslant \nu, \qquad \sigma^2 > 0.$$

This problem is not invariant under all of G_3, but only under the subgroup G_4 in which $a > 0$. Let

$$T_4(\overline{X}, \overline{Y}, S^2) = \frac{\overline{X} - \overline{Y}}{S\sqrt{\dfrac{1}{m} + \dfrac{1}{n}}}, \qquad \delta_4(\mu, \nu, \sigma^2) = \frac{\mu - \nu}{\sigma\sqrt{\dfrac{1}{m} + \dfrac{1}{n}}}.$$

Then T_4 is a maximal invariant as the following argument shows:

(1)

$$T_4\big(g_{a,b}(\overline{X}, \overline{Y}, S^2)\big) = T_4(a\overline{X} + b, a\overline{Y} + b, a^2S^2) = \frac{\left[a\overline{X} + b - (a\overline{Y} + b)\right]}{\left[aS\left(\dfrac{1}{m} + \dfrac{1}{n}\right)^{\frac{1}{2}}\right]}$$

$$= \frac{(\overline{X} - \overline{Y})}{S\left[\dfrac{1}{m} + \dfrac{1}{n}\right]^{\frac{1}{2}}} = T_4(\overline{X}, \overline{Y}, S^2).$$

(2) Suppose that $T(\overline{X}, \overline{Y}, S^2) = T(\overline{X}^*, \overline{Y}^*, S^{*2})$. Then

$$\frac{\overline{X} - \overline{Y}}{S} = \frac{(\overline{X}^* - \overline{Y}^*)}{S}, \qquad \frac{\overline{X} - \overline{Y}}{\overline{X}^* - \overline{Y}^*} = \frac{S}{S^*} = a > 0$$

and $\overline{X} - \overline{Y} = a(\overline{X}^* - \overline{Y}^*)$, $S^2 = a^2S^{*2}$. Let $b = \overline{X} - a\overline{X}^*$. Then $\overline{X} = a\overline{X}^* + b$, $\overline{Y} = \overline{X} - a(\overline{X}^* - \overline{Y}^*) = a\overline{Y}^* + b$. Therefore, $(\overline{X}, \overline{Y}, S^2) = g_{a,b}(\overline{X}^*, \overline{Y}^*, S^{*2})$.

Similarly δ_4 is a parameter maximal invariant. Also

$$T_4 \sim T_{m+n-2}(\delta_4)$$

and we are testing

$$H_0 : \delta_4 = 0,$$
$$H_1 : \delta_4 \geqslant 0.$$

Therefore, the UMP invariant size α test rejects for large T_4. Note that this test is just the usual one-sided, two-sample t-test.

Note that in these examples, even the two-sided problems are reduced to one-sided problems for which there are UMP tests. The following example shows that invariance does not always reduce the problem to one for which there is a UMP test.

EXAMPLE 3. Let $X_1, \ldots, X_m, Y_1, \ldots, Y_n$ be independent, with $X_i \sim N_1(\mu, \sigma^2)$, $Y_i \sim N_1(\nu, \tau^2)$. Then

$$\bar{X} = \frac{1}{m}\sum_i X_i, \qquad \bar{Y} = \frac{1}{n}\sum_i Y_i, \qquad S_X^2 = \frac{1}{m-1}\sum_i (X_i - \bar{X})^2,$$

$$S_Y^2 = \frac{1}{n-1}\sum_i (Y_i - \bar{Y})^2$$

is a complete sufficient statistic.

a. Consider first the one-sided problem in which we are testing

$$H_0 : -\infty < \mu < \infty, \qquad -\infty < \nu < \infty, \qquad \sigma^2 = \tau^2 > 0,$$

$$H_1 : -\infty < \mu < \infty, \qquad -\infty < \nu < \infty, \qquad \sigma^2 \geqslant \tau^2 > 0.$$

This problem is invariant under the group G_5 of transformations of the following form

$$g_{a,b,c}(\bar{X}, \bar{Y}, S_X^2, S_Y^2) = (a\bar{X} + b, a\bar{Y} + c, a^2 S_X^2, a^2 S_y^2), \qquad a \neq 0,$$

$$\bar{g}_{a,b,c}(\mu, \nu, \sigma^2, \tau^2) = (a\mu + b, a\nu + c, a^2\sigma^2, a^2\tau^2).$$

Let

$$T_5(\bar{X}, \bar{Y}, S_X^2, S_Y^2) = \frac{S_X^2}{S_Y^2}, \qquad \delta_5(\mu, \nu, \sigma^2, \tau^2) = \frac{\sigma^2}{\tau^2}$$

Then T_5 is a maximal invariant, δ_5 is a parameter maximal invariant, and

$$\frac{T_5}{\delta_5} \sim F_{m-1, n-1}(0).$$

(see Exercise B6). We are testing

$$H_0 : \delta_5 = 1,$$
$$H_1 : \delta_5 \geqslant 1.$$

Let $h(t_5; \delta_5)$ be the density of T_5. It is easily verified that $h(t_5; \delta_5)/h(t_5; 1)$ is an increasing function of t_5 for all $\delta_5 > 1$. Therefore, the UMP invariant size α test rejects if T_5 is too large.

b. Now consider the two-sided problem of testing

$$H_0 : -\infty < \mu < \infty, \qquad -\infty < \nu < \infty, \qquad \sigma^2 = \tau^2 > 0,$$

$$H_1 : -\infty < \mu < \infty, \qquad -\infty < \nu < \infty, \qquad \sigma^2 > 0, \qquad \tau^2 > 0.$$

This problem is invariant under G_5 but not any larger group. The maximal invariant is therefore T_5, and the parameter maximal invariant is δ_5. In the reduced problem, we are testing

$$H_0 : \delta_5 = 1,$$
$$H_1 : \delta_5 > 0.$$

The reduced problem is a two-sided problem, and therefore has no UMP test. Hence there is no UMP invariant size α test for the original problem.

For many problems considered in this book, it is possible to break up the group G leaving the problem invariant into several subgroups $G_1, \ldots, G_k$. That is, G is the smallest group containing $G_1, \ldots, G_k$. If this happens, we reduce the problem first by G_1, and find a maximal invariant T_1 under G_1. We then see how G_2 operates on T_1. We find a maximal invariant under G_2 for the problem that has already been reduced by G_1. We then apply G_3, and continue until we have applied all the groups. As an example of this type of reduction, consider Example 2a. This problem is invariant under the group G_3 of transformations of the form

$$g_{a,b}(\overline{X}, \overline{Y}, S^2) = (a\overline{X} + b, a\overline{Y} + b, a^2S^2), \qquad a \neq 0.$$

We can break this group into two subgroups, G_{31} and G_{32} where G_{31} consists of transformations of the form

$$g_b(\overline{X}, \overline{Y}, S^2) = (\overline{X} + b, \overline{Y} + b, S^2)$$

and G_{32} consists of transformations

$$g_a(\overline{X}, \overline{Y}, S^2) = (a\overline{X}, a\overline{Y}, a^2S^2).$$

We reduce the problem first by G_{31}. It is easily seen that a maximal invariant under G_{31} is

$$T_{31} = (\overline{X} - \overline{Y}, S^2).$$

We then see that G_{32} operates on T_{31} by

$$g_a(T_{31}) = g_a(\overline{X} - \overline{Y}, S^2) = \left[a(\overline{X} - \overline{Y}), a^2S^2\right], \qquad a \neq 0.$$

Then a maximal invariant under G_{32} (for the problem already reduced by G_{31}) is

$$T_3 = \frac{(\overline{X} - \overline{Y})^2}{S^2\left(\frac{1}{n} + \frac{1}{m}\right)}.$$

We note that the maximal invariant found by reducing first by G_{31}, then by G_{32} is the same as the maximal invariant found by reducing by all of G_3. Fortunately, for all the problems considered in this book, this sequential method is successful. In fact, as long as the problem reduced by the earlier groups is invariant under the later groups, the maximal invariant derived in the sequential method suggested above is also a maximal invariant under the group G containing $G_1, \ldots, G_k$ [See Lehmann (1959), p. 218, for details]. Therefore, in future sections of this book we list several groups leaving a problem invariant. We first reduce to a sufficient statistic, then find a maximal invariant under the first group. We then look at the problem reduced by the first group and find a maximal invariant under the second group and continue in this fashion. In the models considered later there are as many as six groups that leave a problem invariant, and it would be quite difficult to find a maximal invariant under the whole group G, but it is often easy to find the maximal invariant sequentially.

We now give two properties that relate invariant tests to unbiasedness and likelihood ratio tests.

THEOREM 1.13. a. A UMP invariant size α test is unbiased.

b. Suppose that $\mathbf{X}$ has a continuous family of densities and that g is a differentiable function for all $g \in G$. Then the likelihood ratio test is an invariant test.

PROOF. a. Let ϕ by a UMP invariant size α test and let ϕ^* be the critical function that is identically α. Then ϕ^* is a size α invariant test. Since ϕ is UMP invariant size α, it must be more powerful than ϕ^*. Hence

$$K_{\theta_1}(\phi) \geqslant K_{\theta_1}(\phi^*) = \alpha \geqslant K_{\theta_0}(\phi)$$

for all $\theta_1 \in \Omega - \omega$, $\theta_0 \in \omega$.

b. See Exercise C6. □

From part b of this theorem, we see that the UMP invariant size α test must be at least as powerful as the size α likelihood ratio test. Therefore, in future chapters we reduce the testing problems by invariance. If we find a UMP invariant test we stop. If there is no UMP invariant size α test, we consider several invariant size α tests including the likelihood ratio test.

As a final comment on invariant tests, we discuss the relationship between most stringent tests and invariance. Hunt and Stein (1946) showed under fairly general conditions on G that a test that is most stringent among invariant rules is most stringent among all rules. Since a UMP test is most stringent, a UMP invariant test is most stringent among invariant rules. Under the conditions of the Hunt-Stein theorem, therefore, a UMP invariant rule is most stringent. Unfortunately, the conditions of the Hunt-Stein theorem are not general enough to cover all the testing problems considered in this book, and for some of those problems it is not known whether the UMP invariant test is most stringent. See Lehmann (1959, pp. 326–347) for a more careful discussion of the relationship between invariance and most stringent tests.

1.6. INVARIANCE IN ESTIMATION.

We now discuss best invariant estimators. These estimators were first suggested by Pitman (1939) and are often called Pitman estimators.

We return to the problem of estimating the m-dimensional vector $\boldsymbol{\tau}(\boldsymbol{\theta})$

with the loss function $L(\mathbf{d};\boldsymbol{\theta})$. Let A be the set of possible values for $\boldsymbol{\tau}(\boldsymbol{\theta})$. ($A$ is often called the action space for this problem.) Let G be a group of transformations from χ to χ. We say that this estimation problem is invariant under G if for all $g \in G$, there exist invertible transformations $\bar{g}$ from Ω to Ω and $\tilde{g}$ from A to A such that:

(1) $\mathbf{Y} = g(\mathbf{X})$ has density $f(\mathbf{y};\ \bar{g}(\boldsymbol{\theta}))$,
(2) $\boldsymbol{\tau}(\bar{g}(\boldsymbol{\theta})) = \tilde{g}(\boldsymbol{\tau}(\boldsymbol{\theta}))$,
(3) $L(\tilde{g}(\mathbf{d});\ \bar{g}(\boldsymbol{\theta})) = L(\mathbf{d};\boldsymbol{\theta})$.

Let $\bar{G}$ and $\tilde{G}$ be the set of all transformations $\bar{g}$ and $\tilde{g}$. As in the last section $\bar{G}$ and $\tilde{G}$ are groups. We say that an estimator $\mathbf{T}(\mathbf{X})$ is invariant under G if

$$T(g(X)) = \tilde{g}(T(\mathbf{X}))$$

for all $g \in G$, $\mathbf{X} \in \chi$. Now, let $g \in G$ and Let $\mathbf{Y} = g(\mathbf{X})$, $\boldsymbol{\delta} = \bar{g}(\boldsymbol{\theta})$, $\mathbf{d}^* = \tilde{g}(\mathbf{d})$. Then $\mathbf{Y}$ has density $f(\mathbf{y};\boldsymbol{\delta})$. The loss $L(\mathbf{d}^*;\ \boldsymbol{\delta})$ for estimating $\boldsymbol{\tau}(\boldsymbol{\delta})$ with $\mathbf{d}^*$ is the same as the loss $L(\mathbf{d};\boldsymbol{\theta})$ for estimating $\boldsymbol{\tau}(\boldsymbol{\theta})$ with $\mathbf{d}$. Therefore, if we estimate $\boldsymbol{\tau}(\boldsymbol{\theta})$ by $\mathbf{T}(\mathbf{X})$ when we observe $\mathbf{X}$ we should estimate $\boldsymbol{\tau}(\boldsymbol{\delta}) = \tilde{g}(\boldsymbol{\tau}(\boldsymbol{\theta}))$ by $\mathbf{T}(\mathbf{Y})$ when we observe $\mathbf{Y}$. Therefore $\mathbf{T}(g(\mathbf{X})) = \mathbf{T}(\mathbf{Y})$ should equal $\tilde{g}(\mathbf{T}(\mathbf{X}))$. Hence, this argument suggests that to be consistent we should use invariant estimators.

We say that $\mathbf{T}(\mathbf{X})$ is the *best invariant estimator* of $\boldsymbol{\tau}(\boldsymbol{\theta})$ if $\mathbf{T}(\mathbf{X})$ is an invariant estimator and $\mathbf{T}(\mathbf{X})$ is better than any other invariant estimator. Note that the best invariant estimator depends on the loss function chosen for the problem

In the following example and in other sections of the book, we first reduce to a sufficient statistic before looking at invariance. A justification for this approach is given in Exercise C10 in which it is shown under conditions general enough for models in this book that the best invariant estimator for the problem reduced by sufficiency is also the best invariant estimator for the original problem.

We now illustrate the method for finding best invariant estimators with the one sample model discussed in Example 1 of Section 1.5. In that model we observe $X_1, \ldots, X_n$ independent, $X_i \sim N_1(\mu, \sigma^2)$. A sufficient statistic for this model is $(\bar{X}, S^2)$ defined in Example 1 of the last section. We first consider estimating μ with the loss function

$$L_1\big(d_1;(\mu,\sigma^2)\big) = \frac{(d_1 - \mu)^2}{\sigma^2}.$$

This estimation problem is invariant under the group G_1 of transformations

$$g_1(\bar{X}, S^2) = (a\bar{X} + b, a^2S^2), \qquad \bar{g}_1(\mu, \sigma^2) = (a\mu + b, a^2\sigma^2),$$
$$\tilde{g}_1(d_1) = ad_1 + b,$$

where $a > 0$. Note that

$$L_1(\tilde{g}_1(d_1); \bar{g}_1(\mu,\sigma^2)) = \frac{(ad_1 + b - (a\mu + b))^2}{a^2\sigma^2}$$

$$= \frac{(d_1 - \mu)^2}{\sigma^2} = L_1(d_1; (\mu,\sigma^2)).$$

An invariant estimator $T_1(\overline{X}, S^2)$ must therefore satisfy

$$T_1(a\overline{X} + b, a^2S^2) = aT_1(\overline{X}, S^2) + b,$$

for all $a \neq 0$, and all b, $\overline{X}$, and S^2. Now let $a = 1/S$, $b = -\overline{X}/S$. Then T_1 must satisfy

$$T_1(0,1) = \frac{1}{S}T_1(\overline{X}, S^2) - \frac{\overline{X}}{S}$$

or equivalently

$$T_1(\overline{X}, S^2) = \overline{X} + ST_1(0,1).$$

In addition, any estimator satisfying the above equation is invariant. We now find the invariant estimator with minimum risk. Note that

$$R(T_1; (\mu,\sigma^2)) = \frac{E(\overline{X} + ST_1(0,1) - \mu)^2}{\sigma^2}$$

$$= \frac{E(\overline{X} - \mu)^2}{\sigma^2} + \frac{2T_1(0,1)ESE(\overline{X} - \mu)}{\sigma^2} + \frac{[T_1(0,1)]^2 ES^2}{\sigma^2}$$

$$= R(\overline{X}; (\mu,\sigma^2)) + [T_1(0,1)]^2.$$

Therefore, if T_1 is an invariant estimator of μ, then its risk is minimized when $T_1(0,1) = 0$ and the best invariant estimator of μ for the loss function L_1 is $\overline{X}$.

Now consider estimating σ^2 with loss function

$$L_2(d_2; (\mu,\sigma^2)) = \frac{(d_2 - \sigma^2)^2}{\sigma^4}.$$

This problem is invariant under the group G_2 of transformations:

$$g_2(\overline{X}, S^2) = (a\overline{X} + b, a^2S^2), \qquad \bar{g}_2(\mu,\sigma^2) = (a\mu + b, a^2\sigma^2), \qquad \tilde{g}_2(d_2) = a^2 d_2$$

for all $a \neq 0$ and b. Note that

$$L_2(\tilde{g}_2(d_2); \bar{g}_2(\mu,\sigma^2)) = \frac{(a^2d_2 - a^2\sigma^2)^2}{a^4\sigma^4} = \frac{(d_2 - \sigma^2)^2}{\sigma^4} = L_2(d_2; (\mu,\sigma^2)).$$

In Exercise B7 you are asked to show that an estimator $T_2(\bar{X}, S^2)$ is invariant if and only if

$$T_2(\bar{X}, S^2) = T_2(0, 1)S^2$$

and that the best invariant estimator of σ^2 for the loss function L_2 is $(n-1)S^2/(n+1)$.

In both of the examples given above, an invariant estimator is completely determined by its value at one point, (0, 1), and all invariant estimators have constant risk function. In this situation, it is easy to find a best invariant estimator. We merely find the best value of the estimator at the one point. Since the risk is constant, that best value will not depend on the unknown parameters. In other more complicated situations (in which the estimator is not determined by its value at a particular point or the risk function of invariant estimators is not constant), best invariant estimators do not exist or they are difficult to find. In this book, we find best invariant estimators only when the invariant estimators are determined by their value at a particular point and the risk function of every invariant estimator is constant. (In Exercise C5, a sufficient condition is given to guarantee that all invariant rules are determined by the values at a particular point and that all invariant rules have constant risk.)

We have now considered three different methods for finding estimators: best invariant estimators, MLE's and minimum variance unbiased estimators. The next theorem relates these three methods.

THEOREM 1.14. a. Suppose that there is a complete sufficient statistic $\mathbf{S}$ and a minimum variance unbiased estimator $\mathbf{T}(\mathbf{S})$ of $\boldsymbol{\tau}(\boldsymbol{\theta})$. Suppose also that for all $g \in G$, $\tilde{g}$ has the form $\tilde{g}(\mathbf{d}) = \mathbf{A}\mathbf{d} + \mathbf{b}$ for some $m \times m$ invertible matrix $\mathbf{A}$ and $m \times 1$ vector $\mathbf{b}$. Then $\mathbf{T}(g(\mathbf{S})) = \tilde{g}(\mathbf{T}(\mathbf{S}))$ with probability 1.

b. Suppose that $\mathbf{X}$ has a continuous family of density functions and that g is differentiable for all $g \in G$. If the MLE of $\boldsymbol{\tau}(\boldsymbol{\theta})$ is unique, then it is invariant.

PROOF. a. Using the definition of invariance, we see that

$$E_{\boldsymbol{\theta}}\boldsymbol{T}(g(\boldsymbol{X})) = E_{\bar{\mathbf{g}}(\boldsymbol{\theta})}T(\boldsymbol{X}) = \boldsymbol{\tau}(\bar{g}(\boldsymbol{\theta})) = \tilde{g}(\boldsymbol{\tau}(\boldsymbol{\theta})).$$

Using the condition on $\tilde{g}$, we see that

$$E_{\boldsymbol{\theta}}\tilde{g}(\boldsymbol{T}(\boldsymbol{S})) = \tilde{g}(E_{\boldsymbol{\theta}}\boldsymbol{T}(\boldsymbol{S})) = \tilde{g}(\boldsymbol{\tau}(\boldsymbol{\theta})).$$

Therefore, $E_{\boldsymbol{\theta}}\tilde{g}(\mathbf{T}(\mathbf{S})) = E_{\boldsymbol{\theta}}\mathbf{T}(g(\mathbf{S}))$. Since $\mathbf{S}$ has a complete family of densities, $\tilde{g}(\mathbf{T}(\mathbf{S})) = \mathbf{T}(g(\mathbf{S}))$ with probability 1.

b. See Exercise C7. □

Part b of this theorem implies that the best invariant estimator is as good as the MLE and that if they are different, then the best invariant estimator is better, since it is better than any other invariant estimator. Using part a, we can show, by an argument similar to that in the proof of Theorem 4 of Chapter 6 of Lehmann (1959) that, under conditions general enough for models in this book, a minimum variance unbiased estimator must be equal to an invariant estimator with probability 1, or equivalently that there is an invariant minimum variance unbaised estimator. Therefore, unless the minimum variance unbiased estimator is equal to the best invariant estimator (with probability 1), the best invariant estimator is better. These facts are illustrated in the example above when estimating σ^2.

We now discuss the relationship between invariant estimators and minimaxity. Kiefer (1957) shows under fairly general conditions that the best invariant estimator is a minimax estimator. For a more elementary treatment of the relationship between invariance and minimaxity, see Ferguson (1967, pp. 166–176).

When considering testing problems in the remainder of the book, we look first for a UMP invariant test. If we can find one, we do not look any further (except in Chapter 12). If there is no UMP invariant test we find a maximal invariant and then look at several invariant tests, including the likelihood ratio test. However, in estimation problems we look at many different types of estimators, including some noninvariant ones. There are at least three important reasons for putting less emphasis on invariance in estimation than in testing. The first is that unless all invariant estimators are determined by their value at a particular point and have constant risk function, best invariant estimators do not exist or they are quite difficult to find. In these situations there seems to be little benefit to limiting discussion to invariant rules. (There is no nice algorithm for reducing an estimation problem by invariance comparable to the algorithm developed for testing problems in the last section.) The second reason is that the best invariant estimator is only defined for a particular loss function. The best invariant estimator is as good as the minimum variance unbiased estimator and MLE for that loss function, but may be considerably worse for other loss functions. (Note that Theorem 1.5 implies that the minimum variance unbiased estimator is the best unbiased estimator for any convex loss function.) The third reason, and probably the most important reason is that we shall see that for many estimation problems involving several parameters the best invariant estimator is inadmissible (even for the loss function it was derived for). In this situation, of course, MLE's and unbiased estimators are also inadmissible.

As a final comment in this section, we emphasize that the methods suggested in this section for finding best invariant estimators are quite different from those suggested in the last section for finding UMP invariant tests. In particular, in finding best invariant estimators there is no advantage in looking at maximal invariants.

1.7. TRANSFORMING A MODEL.

In this book, we often take a model that we want to study and change it to a model that is easier to analyze. Let the original model be one in which we observe the random vector $\mathbf{X} \in \chi$ having density $f(\mathbf{x}; \boldsymbol{\theta})$, $\boldsymbol{\theta} \in \Omega$. Let h and k be invertible functions from χ to Ξ and from Ω to Δ. Let $\mathbf{Y} = h(\mathbf{X})$, $\boldsymbol{\delta} = k(\boldsymbol{\theta})$, and let $f^*(\mathbf{y}; \boldsymbol{\delta})$ be the density of $\mathbf{Y}$. We call the model in which we observe $\mathbf{Y} \in \Xi$ having density $f^*(\mathbf{y}; \boldsymbol{\delta})$, $\boldsymbol{\delta} \in \Delta$ a *transformation* of the original model. We note that since h is invertible, observing $\mathbf{X}$ is equivalent to observing $\mathbf{Y}$. Similarly, since k is invertible, $\boldsymbol{\delta}$ is just a reparametrization of the model. So the transformed model should be essentially the same as the original model. In particular, if $\mathbf{T}(\mathbf{Y})$ is a sufficient statistic for the transformed model, then $\mathbf{T}(h(\mathbf{X}))$ should be a sufficient statistic for the original model. Similarly, if $\hat{\boldsymbol{\delta}}(\mathbf{Y})$ is the MLE of $\boldsymbol{\delta}$ in the transformed model, then $k^{-1}(\hat{\boldsymbol{\delta}}(h(\mathbf{X})))$ should be the MLE of θ for the original model. Finally, if $\phi(\mathbf{Y})$ is a "good" procedure for testing the null hypothesis that $\boldsymbol{\delta} \in k(\omega)$, then $\phi(h(\mathbf{X}))$ should be a "good" procedure for testing that $\boldsymbol{\theta} \in \omega$. In Exercise C4, these statements are made more precise, and suggestions are made for deriving them. In light of that exercise, for the remainder of the book we assume that the original model and the transformed model are the same. Once we have an optimal procedure for the transformed model, we can modify it in the obvious way to get an optimal procedure for the original model.

As an example of the ideas in the last paragraph, consider the one-sample model discussed in Example 1, in which we observe $X_1, \ldots, X_n$, independent with $X_i \sim N_1(\mu, \sigma^2)$. Suppose we want to test the null hypothesis that $\mu = 3$ against $-\infty < \mu < \infty$. We now transform this problem to one for which we know a UMP invariant size α test. Let $Y_i = X_i - 3$, $\nu = \mu - 3$. Then the Y_i are an invertible function of the X_i, and (ν, σ^2) is an invertible function of (μ, σ^2). In addition, $Y_i \sim N_1(\nu, \sigma^2)$. Therefore, the model in which we observe Y_i independent, $Y_i \sim N_1(\nu, \sigma^2)$ is a transformation of the original model. Furthermore, $\mu = 3$ if and only if $\nu = 0$. Therefore, testing that $\mu = 3$ against $-\infty < \mu < \infty$ in the original model is equivalent to testing that $\nu = 0$ against $-\infty < \nu < \infty$ in the transformed model. Let

$$S^2 = \frac{1}{n-1}\Sigma(Y_i - \bar{Y})^2 = \frac{1}{n-1}\Sigma(X_1 - \bar{X})^2, \qquad F = n\frac{\bar{Y}^2}{S^2} = n\frac{(\bar{X} - 3)^2}{S^2}.$$

Using Example 1a, we see that the UMP invariant size α test that $\nu = 0$ against $-\infty < \nu < \infty$ in the transformed model is given by

$$\phi(F) = \begin{cases} 1 & \text{if } F > F^{\alpha}_{1,n-1} \\ 0 & \text{if } F \leqslant F^{\alpha}_{1,n-1} \end{cases}.$$

In the light of Exercise C4, this procedure is also UMP invariant size α for the original problem. (Note that the group under which the original problem is invariant consists of transformations of the form $g(\overline{X}, S^2) = (a(\overline{X} - 3) + 3, a^2S^2)$, $a \neq 0$. The induced transformation is $\overline{g}\ (\mu, \sigma^2) = (a(\mu - 3) + 3, a^2\sigma^2)$.

EXERCISES

Type B

1. Let $X_1, \ldots, X_n$ be independent where X_i is normally distributed with unknown mean μ and unknown variance $\sigma^2 > 0$.

(a) Show that $(\Sigma X_i, \Sigma X_i^2)$ is a complete sufficient statistic for this model.

(b) Let $\overline{X} = \Sigma X_i / n$, $S^2 = \Sigma(X_i - \overline{X})^2/(n-1)$. Show that $(\overline{X}, S^2)$ is a complete sufficient statistic for this model.

(c) Find the minimum variance unbiased estimator of (μ, σ^2).

(d) Find the MLE of (μ, σ^2).

(e) Find the minimum variance unbiased estimator and MLE for μ^2.

(f) Find the likelihood ratio test for testing that $\mu = 0$ against $-\infty < \mu < \infty$.

(g) Find the likelihood ratio test for testing that $\mu = 0$ against $\mu \geqslant 0$.

(h) Find the UMP test for testing $\mu = 0$ against $\mu \geqslant 0$ when it is assumed that σ^2 is known, $\sigma^2 = 1$.

2. Verify (1.1). Show that $R(T_2; (\mu, \sigma^2))$ is minimized when $c = (n-1)/(n+1)$.

3. Let $Y \sim \chi_m^2(\delta)$. Find EY and $\operatorname{var} Y$.

4. Prove Lemma 1.9.

5. (a) Show that $T_2(\overline{X}, S^2)$ is a maximal invariant for Example 1b and show that it has the indicated distribution.

(b) Show that $T_3(\overline{X}, \overline{Y}, S^2)$ is the maximal invariant for Example 2a and that it has the indicated distribution.

6. (a) Show that T_5 is the maximal invariant for Example 3a and that it has the indicated distribution.

(b) Let $h(t_5; \delta_5)$ be the density function of T_5. Show that $h(t_5; \delta_5)/h(t_5; 1)$ is an increasing function of T_5 for all $\delta_5 > 1$.

7. Consider the problem of estimating σ^2 discussed in Section 1.6.

(a) Show that this problem is invariant under the group G_2 given there.

(b) Show that an estimator $T_2(\bar{X}, S^2)$ is invariant if and only if, $T_2(\bar{X}, S^2) = T_2(0, 1)S^2$.

(c) Show that the best invariant estimator has $T_2(0, 1) = (n-1)/(n+1)$. (See Exercise B2.)

Type C

1. Existence of maximal invariants. Let G be a group of invertible functions from a space C to itself. Define a relation $\equiv$ on C by $c_1 \equiv c_2$ if there exists $g \in G$ such that $c_1 = g(c_2)$.

(a) Show that $\equiv$ is an equivalence ration, that is, show that $\equiv$ satisfies the following:

(i) $c \equiv c$ for all $c \in C$,
(ii) $c_1 \equiv c_2$ implies that $c_2 \equiv c_1$,
(iii) $c_1 \equiv c_2$ and $c_2 \equiv c_3$ implies that $c_1 \equiv c_2$.

This equivalence relation $\equiv$ divides the set C into equivalent classes, that is, disjoint sets such that each point in C is in one class and such that c_1 and c_2 are in the same set if and only if $c_1 \equiv c_2$. Let D be the set of all the equivalence classes. Let $h(c)$ be the function from C to D that assigns to each point in C its equivalence class.

(b) Show that $h(c)$ is a maximal invariant.

2. Noncentral χ^2 and F density functions. Let $X_1, \ldots, X_n$ be independently distributed with $X_i \sim N_1(\mu_i, 1)$. Let $Y = \sum X_i^2$, $\delta = \sum \mu_i^2$.

(a) Without using (1.2), show that Y has moment generating function

$$M_y(t) = \frac{\exp(\delta t/(1-2t))}{(1-2t)^{n/2}}$$

(b) Show that Y has the density given in (1.2). (Hint: use Lemma 1.8 to find the moment generating function of a random variable having that density function.)

(c) Let

$$\frac{n}{n+2k} F|K \sim F_{m+2K, m}(0),$$

and let K have a Poisson distribution with parameter $\delta/2$. Without using (1.5), show that $F \sim F_{m,n}(\delta)$. (Hint: use Lemma 1.8 again.)

(d) Show that F has the density given in (1.5).

3. Noncentral t density. Let X and Y be independent with $X \sim N_1(\mu, 1)$,

$Y \sim \chi_n^2(0)$. Let

$$T = \frac{X}{\sqrt{Y/n}}, \qquad R = \sqrt{nX^2 + Y}.$$

(a) Show that

$$X = \frac{R}{\sqrt{T^2+1}}\,\frac{T}{\sqrt{n}}, \qquad Y = \frac{R^2}{T^2+1}.$$

(b) Find the joint density of T and R.

(c) Show that the marginal density of T is given in (1.4).

4. Transforming a model. Let $\mathbf{X} \in \chi$ be a random vector having density $f(\mathbf{x};\boldsymbol{\theta})$, $\boldsymbol{\theta} \in \Omega$, and let $\mathbf{Y} = h(\mathbf{X})$, $\boldsymbol{\delta} = k(\boldsymbol{\theta})$, where h and k are invertible functions and h has continuous partial derivatives. Let $\Xi = h(\chi)$, $\Delta = k(\Omega)$ and let $f^*(\mathbf{y};\boldsymbol{\delta})$ be the density function of $\mathbf{Y}$. The model in which we observe $\mathbf{Y} \in \Xi$ having density function $f^*(\mathbf{y};\boldsymbol{\delta})$, $\boldsymbol{\delta} \in \Delta$ we call the *transformed model*, and the model in which we observe $\mathbf{X} \in \chi$ having density $f(\mathbf{x};\boldsymbol{\theta})$, $\boldsymbol{\theta} \in \Omega$ we call the *original model*. The purpose of this problem is to show that optimal procedures for the transformed model are also optimal for the original model.

(a) Show that

$$f(\boldsymbol{x};\boldsymbol{\theta}) = |J(\boldsymbol{x})| f^*(h(\boldsymbol{x}); k(\boldsymbol{\theta})),$$

where $J(\mathbf{x})$ is the Jacobian of the transformation $h(\mathbf{x})$.

(b) Show that if $\mathbf{T}(\mathbf{Y})$ is a sufficient statistic for the transformed model, then $\mathbf{T}(h(\mathbf{X}))$ is a sufficient statistic for the original model.

(c) Show that if $\mathbf{T}(\mathbf{Y})$ is a complete sufficient statistic for the transformed model then $\mathbf{T}(h(\mathbf{X}))$ is a complete sufficient statistic for the original model.

(d) Show that if $\hat{\boldsymbol{\delta}}(\mathbf{Y})$ is the MLE for $\boldsymbol{\delta}$ in the transformed model, then $\hat{\boldsymbol{\theta}}(\mathbf{X}) = k^{-1}(\boldsymbol{\delta}(h(\mathbf{X})))$ is the MLE for $\boldsymbol{\theta}$ in the original model.

Now, let $\omega \subset \Omega$ and let $\omega^* = k(\omega) \subset \Delta$.

(e) Show that if $\phi(\mathbf{Y})$ is a size α procedure for testing that $\boldsymbol{\delta} \in \omega^*$ against $\boldsymbol{\delta} \in \Delta$, then $\phi(h(X))$ is a size α procedure for testing that $\boldsymbol{\theta} \in \omega$ against $\boldsymbol{\theta} \in \Omega$ for the original model.

(f) Suppose that the problem of testing that $\boldsymbol{\delta} \in \omega^*$ against $\boldsymbol{\delta} \in \Delta$ is invariant under the group G^* of transformations g^*. Show that the problem of testing that $\boldsymbol{\theta} \in \omega$ against $\boldsymbol{\theta} \in \Omega$ is invariant under the group G of transformation $g = h^{-1} \circ g^* \circ h$. (Hint: Let $\mathbf{U} = g^*(\mathbf{Y})$, $\mathbf{Z} = g(\mathbf{X})$. Then $\mathbf{U} = h(\mathbf{Y})$ and part a can be used. Also $\bar{g} = k^{-1} \circ \bar{g}^* \circ k$.)

(g) Show that if $T(\mathbf{Y})$ is a maximal invariant for the transformed model under G^*, then $T(h(\mathbf{X}))$ is a maximal invariant for the original model under G.

(h) Show that if $\phi(\mathbf{Y})$ is invariant for testing that $\boldsymbol{\delta} \in \omega^*$ against $\boldsymbol{\delta} \in \Delta$, then $\phi(h(\mathbf{X}))$ is invariant for testing that $\boldsymbol{\theta} \in \omega$ against $\boldsymbol{\theta} \in \Omega$, and that if $\phi(\mathbf{Y})$ is UMP invariant size α for the transformed model, then $\phi(h(\mathbf{X}))$ is UMP invariant size α for the original model.

(i) Show that if $\lambda(\mathbf{Y})$ is the likelihood ratio test statistic for testing that $\boldsymbol{\delta} \in \omega^*$ in the transformed model, then $\lambda(h(\mathbf{X}))$ is the likelihood ratio test statistic for testing that $\boldsymbol{\theta} \in \omega$ is the original model.

5. Let C be a set and H a group of transformation on C. We say that H *acts transitively* on C if for all $c_0 \in C$, $c \in C$, there exists $h \in H$ such that $c = h(c_0)$. Now consider an estimation problem that is invariant under the group G with induced group $\overline{G}$.

(a) Show that if G acts transitively on the sample space (reduced by sufficiency) then any invariant rule is determined by its value at a particular point. (Let $\mathbf{S}_0$ be a particular point and let $\mathbf{S}$ be an arbitrary point. Then there exists $g_{\mathbf{S}} \in G$ such that $S = g_{\mathbf{S}}(\mathbf{S}_0)$. Show that if $\mathbf{T}$ is invariant then $\mathbf{T}(\mathbf{S}) = \tilde{g}_{\mathbf{S}}(\mathbf{T}(S_0))$ so that $\mathbf{T}$ is completely determined by its value at $\mathbf{S}_0$.)

(b) Show that $R(\mathbf{T};\, \bar{g}(\boldsymbol{\theta})) = R(\mathbf{T};\boldsymbol{\theta})$ if $\mathbf{T}$ is invariant. [Verify that $R(\mathbf{T};\, \bar{g}(\boldsymbol{\theta})) = E_{\bar{g}(\boldsymbol{\theta})}L(\mathbf{T}(\mathbf{X});\ \bar{g}(\boldsymbol{\theta})) = E_{\bar{g}(\boldsymbol{\theta})}L(\tilde{g}^{-1}(\mathbf{T}(\mathbf{X}));\boldsymbol{\theta}) = E_{\boldsymbol{\theta}}L(\mathbf{T}(\mathbf{X});\boldsymbol{\theta}) = R(\mathbf{T};\boldsymbol{\theta})$. Note that if $\mathbf{X}$ has density $f(\mathbf{x};\, \bar{g}(\boldsymbol{\theta}))$, then $\mathbf{Y} = g^{-1}(\mathbf{X})$ has density $f(\mathbf{y};\boldsymbol{\theta})$ and that $L(\mathbf{d};\, \bar{g}(\boldsymbol{\theta}) = L(\tilde{g}^{-1}(\mathbf{d});\boldsymbol{\theta})$. Why?]

(c) Show that if $\overline{G}$ acts transitively on the parameter space then all invariant rules have constant risk.

6. Invariance of likelihood ratio tests. Assume the conditions of Theorem 1.13b. Let $g \in G$ and let $\mathbf{Y} = g^{-1}(\mathbf{X})$.

(a) Using the definition of invariance, show that $\mathbf{Y}$ has density $f(\mathbf{y};\, \bar{g}^{-1}(\boldsymbol{\theta})$. (Note that $\overline{g^{-1}} = \bar{g}^{-1}$.)

(b) Using the usual techniques for finding densities of transformed variables, show that $\mathbf{Y}$ has density $f(g(\mathbf{y});\boldsymbol{\theta})J_g(\mathbf{Y})$ where $J_g(\mathbf{y})$ is the absolute value of the Jacobian of g and does not depend on $\boldsymbol{\theta}$.

(c) Show that $f(g(\mathbf{x});\boldsymbol{\theta}) = f(\mathbf{x};\, \bar{g}^{-1}(\boldsymbol{\theta}))/J_g(\mathbf{x})$ for all $\mathbf{x} \in \chi$, $\boldsymbol{\theta} \in \Omega$.

(d) Show that

$$\sup_{\boldsymbol{\theta} \in \omega} f(g(\boldsymbol{x});\boldsymbol{\theta}) = \sup_{\boldsymbol{\theta} \in \omega} f(\boldsymbol{x};\boldsymbol{\theta})/J_g(\boldsymbol{x})$$

$$\sup_{\boldsymbol{\theta} \in \Omega} f(g(\boldsymbol{x});\boldsymbol{\theta}) = \sup_{\boldsymbol{\theta} \in \Omega} f(\boldsymbol{x};\boldsymbol{\theta})/J_g(\boldsymbol{x}).$$

[Hint: $\sup_{\boldsymbol{\theta} \in \omega} f(\mathbf{x};\boldsymbol{\theta}) = \sup_{\boldsymbol{\theta} \in \omega} f(\mathbf{x};\, \bar{g}^{-1}(\boldsymbol{\theta}))$. Why?]

(e) Show that the likelihood ratio test is invariant under g and hence is invariant under G.

7. Invariance of MLE's. Assume the conditions of Theorem 1.14b. Let $g \in G$.

(a) Show that $f(\mathbf{x};\ \bar{g}^{-1}(\boldsymbol{\theta})) = f(g(\mathbf{x}); \boldsymbol{\theta})J_g(\mathbf{x})$. (See problem C6.)

(b) Now, let $\hat{\boldsymbol{\theta}}(\mathbf{X})$ be a MLE of $\boldsymbol{\theta}$, and let $\hat{\hat{\boldsymbol{\theta}}}(\mathbf{X}) = \bar{g}^{-1}(\hat{\boldsymbol{\theta}}(g(\mathbf{X})))$. Verify the following equalities and inequalities: $f(\mathbf{x}; \hat{\hat{\boldsymbol{\theta}}}(\mathbf{x})) = f(g(\mathbf{x}); \hat{\boldsymbol{\theta}}(g(\mathbf{x})))J_g(\mathbf{x}) \geqslant f(g(\mathbf{x}); \boldsymbol{\theta})J_g(\mathbf{x}) = f(\mathbf{x};\ \bar{g}^{-1}(\boldsymbol{\theta}))$ for all $\mathbf{x} \in \chi$, $\boldsymbol{\theta} \in \Omega$.

(c) Show that $f(\mathbf{x}; \hat{\hat{\boldsymbol{\theta}}}(\mathbf{x})) \geqslant f(\mathbf{x}; \boldsymbol{\theta})$ for all $\mathbf{x} \in \chi$, $\boldsymbol{\theta} \in \Omega$, and therefore that $\hat{\hat{\boldsymbol{\theta}}}(\mathbf{X})$ is also a MLE of $\boldsymbol{\theta}$. [Note that as $\boldsymbol{\theta}$ ranges over Ω, $\bar{g}^{-1}(\boldsymbol{\theta})$ also ranges over Ω. . Why?]

(d) Therefore: $\boldsymbol{\tau}(\hat{\boldsymbol{\theta}}(\mathbf{X}))$ and $\boldsymbol{\tau}(\hat{\hat{\boldsymbol{\theta}}}(\mathbf{X}))$ are both MLE's of $\boldsymbol{\tau}(\boldsymbol{\theta})$. Since the MLE of $\boldsymbol{\tau}$ is unique $\boldsymbol{\tau}(\hat{\boldsymbol{\theta}}(\mathbf{X})) = \boldsymbol{\tau}(\hat{\hat{\boldsymbol{\theta}}}(\mathbf{X}))$. Show that $\boldsymbol{\tau}(\hat{\hat{\boldsymbol{\theta}}}(\mathbf{X})) = \tilde{g}^{-1}(\boldsymbol{\tau}(\hat{\boldsymbol{\theta}}(g(\mathbf{X}))))$.

(e) Show that $\boldsymbol{\tau}(\hat{\boldsymbol{\theta}}(g(\mathbf{X}))) = \tilde{g}(\boldsymbol{\tau}(\hat{\boldsymbol{\theta}}(\mathbf{X})))$ and hence that the MLE is invariant under G.

8. Sufficiency and invariance. In this problem and the next two, we justify reducing by sufficiency before looking at invariance. Let an estimation or testing problem be invariant under a group G of transformations on the sample space χ. Let $\mathbf{S}(\mathbf{X})$ be a sufficient statistic, and let Ξ be the set of possible values for $\mathbf{S}$. We say that $\mathbf{S}$ is *compatible* with G if for all $g \in G$, there exists g^* from Ξ to Ξ such that $\mathbf{S}(g(\mathbf{X})) = g^*(\mathbf{S}(\mathbf{X}))$ and the conditional distribution of $g(\mathbf{X})$ given $\mathbf{S} = S_0$ is the same as the conditional distribution of $\mathbf{X}$ given $\mathbf{S} = g^*(\mathbf{S}_0)$. (For example, in the one sample model if $g(X_1, \ldots, X_n) = (aX_1 + \mathrm{b}, \ldots, aX_n + \mathrm{b})$, then $g^*(\bar{X}, S^2) = a\bar{X} + \mathrm{b}, \mathrm{a}^2S^2$).) Let G^* be the set of g^*.

(a) Show that G^* is a group. [Hint: $g_1{}^* \circ g^*{}_2 = (g_1 \circ g_2)^*$.]

(b) Let $k(\mathbf{s}; \boldsymbol{\theta})$ be the joint density of $\mathbf{S}$. Show that the joint density of $g^*(\mathbf{S})$ is $k(\mathbf{s};\ \bar{g}(\boldsymbol{\theta}))$.

9. Sufficiency and invariance in testing. Now, consider the problem of testing that $\boldsymbol{\theta} \in \omega$ against $\boldsymbol{\theta} \in \Omega$. Suppose that this problem is invariant under G. Let $\mathbf{S}$ be a sufficient statistic that is compatible with G (see Exercise C8). By the reduced testing problem, we mean the problem in which we observe $\mathbf{S}$ having density $k(\mathbf{s}; \boldsymbol{\theta})$ and are testing that $\boldsymbol{\theta} \in \omega$ and $\boldsymbol{\theta} \in \Omega$.

(a) Show that the reduced problem is invariant under G^* (with induced group $\bar{G}$ on the parameter space).

(b) Show that if $\phi^*(\mathbf{S})$ is invariant under G^* then $\phi(\mathbf{X}) = \phi^*(\mathbf{S}(\mathbf{X}))$ is invariant under G.

(c) Now, let $\phi(\mathbf{X})$ be invariant under G. Show that $\phi^*(\mathbf{S}) = E\phi(\mathbf{X}) \mid \mathbf{S}$ is invariant under G^*.

$$\begin{aligned}\big[\phi^*(g^*(\mathbf{S}_0)) &= E(\phi(\mathbf{X})|\mathbf{S} = g^*(\mathbf{S}_0)) = E(\phi(g(\mathbf{X}))|\mathbf{S} = \mathbf{S}_0)\\ &= E(\phi(\mathbf{X})|\mathbf{S} = \mathbf{S}_0) = \phi^*(\mathbf{S}_0).\big]\end{aligned}$$

(d) Show that for any rule $\phi(\mathbf{X})$ that is invariant under G there is a rule $\phi^*(\mathbf{S})$ which is invariant under G^* that has the same power function. (Hint: See Theorem 1.6.)

(e) Show that if $\phi^*(\mathbf{S})$ is UMP invariant size α under G^*, then $\phi(\mathbf{X}) = \phi^*(\mathbf{S}(\mathbf{X}))$ is UMP invariant size α under G.

10. Sufficiency and invariance in estimation. Now consider the problem of estimating $\boldsymbol{\tau}(\boldsymbol{\theta})$ with the loss function L. Suppose that this problem is invariant under the group G of transformation on χ. Let $\mathbf{S}$ be a sufficient statistic that is compatible with G. Define the reduced problem in a way analagous to that in the last problem. Assume that the loss function $L(\mathbf{d}; \boldsymbol{\theta})$ is a convex function of $\mathbf{d}$ and that $\tilde{g}(\mathbf{d}) = \mathbf{Ad} + \mathbf{b}$ for all $\tilde{g} \in \tilde{G}$.

(a) Show that the reduced problem is invariant under the group G^* (with the induced groups $\bar{G}$ and $\tilde{G}$ on the parameter and action spaces).

(b) Show that if $\mathbf{T}^*(\mathbf{S})$ is an invariant estimator of $\boldsymbol{\tau}(\boldsymbol{\theta})$ under G^* then $\mathbf{T}(\mathbf{X}) = \mathbf{T}^*(\mathbf{S}(\mathbf{X}))$ is an invariant estimator of $\boldsymbol{\tau}(\boldsymbol{\theta})$ under G.

(c) Let $\mathbf{T}(\mathbf{X})$ be an invariant estimator of $\boldsymbol{\tau}(\boldsymbol{\theta})$ under G and let $\mathbf{T}^*(\mathbf{S}) = E\mathbf{T}(\mathbf{X}) \mid \mathbf{S}$. Show that $\mathbf{T}^*(\mathbf{S})$ is invariant under G^* $[\mathbf{T}^*(g^*(\mathbf{S}_0)) = E(\mathbf{T}(\mathbf{X}) \mid \mathbf{S} = g^*(\mathbf{S}_0)) = E(\mathbf{T}(g(\mathbf{X}) \mid \mathbf{S} = \mathbf{S}_0) = E(\tilde{g}(\mathbf{T}(\mathbf{X})) \mid \mathbf{S} = S_0) = \tilde{g}(E(\mathbf{T}(\mathbf{X}) \mid \mathbf{S} = S_0))$. Why?]

(d) Show that for any estimator $\mathbf{T}(\mathbf{X})$ that is invariant under G there is an estimator $\mathbf{T}^*(\mathbf{S})$ that is invariant under G^* that is as good as $\mathbf{T}(\mathbf{X})$. (See Theorem 1.5.)

(e) Show that if $\mathbf{T}^*(\mathbf{S})$ is the best invariant estimator for $\boldsymbol{\tau}(\boldsymbol{\theta})$ under G^* for the problem reduced by sufficiency then $\mathbf{T}(\mathbf{X}) = \mathbf{T}^*(\mathbf{S}(\mathbf{X}))$ is the best invariant estimator of $\boldsymbol{\tau}(\boldsymbol{\theta})$ under G for the original problem.

CHAPTER 2

Subspaces and Projections

For most of this book, the important statistics are expressed as projections on subspaces. This chapter therefore gives a short introduction to the theory of projections.

It is assumed that the reader is familiar with matrix and vector algebra. The following notation is used in this book. $\mathbf{I}$ is used for an identity matrix of any dimension, $\mathbf{0}$ is used for a $\mathbf{0}$ matrix of any size. $\mathbf{A}'$ is the transpose of $\mathbf{A}$. R^n is the set of all n-dimensional vectors. If $\mathbf{u}, \mathbf{v} \in R^n$, then $\langle \mathbf{u}, \mathbf{v} \rangle$ is used for the inner (dot) product of $\mathbf{u}$ and $\mathbf{v}$, and $\|\mathbf{u}\|$ is used for the length of $\mathbf{u}$. That is

$$\langle \mathbf{u}, \mathbf{v} \rangle = \mathbf{u}'\mathbf{v} = \mathbf{v}'\mathbf{u}, \qquad \|\mathbf{u}\| = \langle \mathbf{u}, \mathbf{u} \rangle^{\frac{1}{2}}.$$

2.1. SUBSPACES AND BASES.

We now give an elementary introduction to subspaces, dimension and bases. Let $\mathbf{v}_1, \ldots, \mathbf{v}_p \in R^n$. Then $\mathbf{u}$ is a *linear combination* of the $\mathbf{v}_i$ if there exists $a_1, \ldots, a_p \in R^1$ such that $\mathbf{u} = a_1\mathbf{v}_1 + \cdots + a_p\mathbf{v}_p$.

DEFINITION. Let V be a set, $V \subset R^n$. Then V is a *subspace* if V is closed under the operation of taking linear combination, that is, for all $\mathbf{v}_1, \ldots, \mathbf{v}_p \in V$, and for all $a_1, \ldots, a_p \in R^1$, $\mathbf{u} = a_1\mathbf{v}_1 + \cdots + a_p\mathbf{v}_p \in V$.

EXAMPLES

(1) The set consisting of the $\mathbf{0}$ vector is a subspace.
(2) Let $\mathbf{X}$ be an $n \times p$ matrix. Then $V = \{\mathbf{v} \in R^n : \mathbf{v} = \mathbf{X}\mathbf{a}$ for some $\mathbf{a} \in R^p\}$ is a subspace.
(3) Let $\mathbf{X}$ be an $n \times p$ matrix. Then $W = \{\mathbf{v} \in R^n : \mathbf{X}'\mathbf{v} = \mathbf{0}\}$ is a subspace.
(4) R^n is a subspace.

Let $\mathbf{v}_1, \ldots, \mathbf{v}_p \in V$, where V is a subspace. The $\mathbf{v}_i$ *span* V if every $\mathbf{v} \in V$ can be written as a linear combination of the $\mathbf{v}_i$. The $\mathbf{v}_i$ are *linearly independent* if $a_1\mathbf{v}_1 + \cdots + a_p\mathbf{v}_p = \mathbf{0}$ implies that $a_1 = \cdots = a_p = 0$.

DEFINITION. Let V be a subspace, $\mathbf{v}_1, \ldots, \mathbf{v}_p \in V$. The $\mathbf{v}_i$ are a *basis* for V if they are linearly independent and they span V. If, in addition, $\|\mathbf{v}_i\| = 1$ and $\langle \mathbf{v}_i, \mathbf{v}_j \rangle = 0$ for $i \neq j$, then the $\mathbf{v}_i$ are an *orthonormal basis* for V.

The following lemma summarizes some facts usually derived in a linear algebra course, and its proof is omitted.

LEMMA 2.1. a. Let V be a subspace. Then V has an orthonormal basis.

b. Any two bases for V have the same number of elements. This number is called the *dimension* of V.

c. If $\mathbf{v}_1, \ldots, \mathbf{v}_p$ form a basis of V, then every vector can be written in exactly one way as a linear combination of the $\mathbf{v}_i$.

d. If V has dimension p, $\mathbf{v}_1, \ldots, \mathbf{v}_p \in V$, and the $\mathbf{v}_i$ are linearly independent, then the $\mathbf{v}_i$ form a basis for V.

e. Let $\mathbf{X}$ be a $n \times p$ matrix of rank r, and let

$$V = \{\mathbf{v} \in R^n : \mathbf{v} = \mathbf{X}\mathbf{a} \text{ for some } a \in R^p\}$$

$$W = \{\mathbf{v} \in R^n : \mathbf{X}'\mathbf{v} = 0\}.$$

Then the dimension of V is r and the dimension of W is $n - r$.

f. If $W \subset V$ and the dimension of W equals the dimension of V, then $V = W$.

We write dim V for the dimension of V.

DEFINITION. The $n \times p$ matrix $\mathbf{X}$ is a *basis matrix* for V if the columns of $\mathbf{X}$ form a basis for V. $\mathbf{X}$ is an *orthonormal basis matrix* if the columns of $\mathbf{X}$ form an orthonormal basis for V.

LEMMA 2.2. Let $\mathbf{X}$ be a basis matrix for the p-dimensional subspace $V \in R^n$.

a. $\mathbf{X}$ is an $n \times p$ matrix of rank p and $\mathbf{X}'\mathbf{X}$ is invertible.

b. The vector $\mathbf{v} \in V$ if and only if $\mathbf{v} = \mathbf{X}\mathbf{b}$ for some vector $\mathbf{b} \in R^p$. The vector $\mathbf{b}$ is unique.

PROOF. a. This follows directly from the definition of a basis and from Theorem 3 of the appendix.

b. Let $\mathbf{X} = (\mathbf{X}_1, \ldots, \mathbf{X}_p)$. Let $\mathbf{v} \in V$. Then $\mathbf{v} = \sum b_i\mathbf{X}_i$ for some $b_i \in R^1$. Let $\mathbf{b} = (b_1, \ldots, b_p)'$. Then $\mathbf{v} = \mathbf{X}\mathbf{b}$. Conversely, suppose that $\mathbf{v} = \mathbf{X}\mathbf{b}$ for some $\mathbf{b} \in R^p$. Then $\mathbf{v}$ is a linear combination of the $\mathbf{X}_i \in V$. Since V is a subspace, $\mathbf{v} \in V$. Finally, suppose that $\mathbf{v} = \mathbf{Xb}$. Then $\mathbf{b} = (\mathbf{X}'\mathbf{X})^{-1}\mathbf{X}'\mathbf{v}$, so that $\mathbf{b}$ is unique. □

Examples in R^3

1. The only subspace of dimension 3 is R^3.
2. Subspaces of dimension 2 are planes through the origin.
3. Subspaces of dimension 1 are lines through the origin.
4. The only subspace of dimension $\mathbf{0}$ is the origin.

2.2. ORTHOGONALITY AND PROJECTIONS.

Let $\mathbf{u}, \mathbf{v} \in R^n$. Then $\mathbf{u}$ is orthogonal to $\mathbf{v}$, written $\mathbf{u} \perp \mathbf{v}$, if $\langle \mathbf{u}, \mathbf{v} \rangle = 0$. Let $\mathbf{u} \in R^n$, and let $V \subset R^n$ be a subspace. Then $\mathbf{u}$ is orthogonal to V, written $\mathbf{u} \perp V$, if $\langle \mathbf{u}, \mathbf{v} \rangle = 0$ for all $\mathbf{v} \in V$. Let U and V be subspaces of R^n. Then U is orthogonal to V, written $U \perp V$, if $\langle \mathbf{u}, \mathbf{v} \rangle = 0$ for all $\mathbf{u} \in U$, $\mathbf{v} \in V$.

DEFINITION. Let V be a subspace of R^n. The orthogonal complement of V, written $V^\perp$, is the set of all vectors orthogonal to V. Let $W \subset V$ be a subspace. Then V mod W, written $V \mid W$, is $V \mid W = V \cap W^\perp$. $V \mid W$ is the set of all vectors in V that are orthogonal to W.

LEMMA 2.3. $V^\perp$ and $V \mid W$ are subspaces.

PROOF. See Exercise B1 □

We are now ready for the last definition in this section.

DEFINITION. Let $\mathbf{y} \in R^n$, and V be a subspace of R^n. A projection of $\mathbf{y}$ onto V is a vector $\mathbf{v}$ such that (1) $\mathbf{v} \in V$; and (2) $\mathbf{y} - \mathbf{v} \in V^\perp$.

Equivalently, $\mathbf{v}$ is a projection of $\mathbf{y}$ onto V if $\mathbf{v}$ is in V and $\mathbf{y} - \mathbf{v}$ is orthogonal to V. In Theorem 2.5.b it is shown that $\mathbf{v}$ is the point in V that is closest to $\mathbf{y}$.

There are two immediate questions that come to mind with a definition of this type: (1) Given a vector $\mathbf{y}$ and a subspace V, is there a projection of $\mathbf{y}$ onto V? (2) Is the projection unique? The answer to both questions is "yes" as the following theorem shows.

THEOREM 2.4. Let V be a subspace of R^n and let $\mathbf{y} \in R^n$.

a. There exists a projection of $\mathbf{y}$ onto V.

b. The projection is unique.

c. If $\mathbf{X}$ is a basis matrix for V, then the projection of $\mathbf{y}$ on V is given by $\mathbf{X}(\mathbf{X}'\mathbf{X})^{-1}\mathbf{X}'\mathbf{y}$.

PROOF. a. Since every subspace has a basis, and hence a basis matrix, a follows from c.

b. Suppose $\mathbf{u}$ and $\mathbf{v}$ are both projections of $\mathbf{y}$ onto V. Then, by the definition of projection, $\mathbf{u}, \mathbf{v} \in V$. Since V is a subspace, $\mathbf{u} - \mathbf{v} \in V$. On the other hand, by the definition of projection $\mathbf{y} - \mathbf{u}, \mathbf{y} - \mathbf{v} \in V^\perp$. Since $V^\perp$ is also a subspace, $\mathbf{u} - \mathbf{v} = \mathbf{y} - \mathbf{v} - (\mathbf{y} - \mathbf{u}) \in V^\perp$. Therefore $(\mathbf{u} - \mathbf{v}) \perp (\mathbf{u} - \mathbf{v})$, and $\|\mathbf{u} - \mathbf{v}\|^2 = 0$. Hence $\mathbf{u} - \mathbf{v} = 0$.

c. Let $\mathbf{v} = \mathbf{X}(\mathbf{X}'\mathbf{X})^{-1}\mathbf{X}'\mathbf{y}$. By Lemma 2.2, $\mathbf{v} \in V$. We need to show that $\mathbf{y} - \mathbf{v} \in V^\perp$. Therefore, let $\mathbf{u} \in V$. By Lemma 2.2, $\mathbf{u} = \mathbf{Xc}$ for some vector $\mathbf{c}$. Hence $\langle \mathbf{y} - \mathbf{v}, \mathbf{u} \rangle = \langle \mathbf{y} - \mathbf{X}(\mathbf{X}'\mathbf{X})^{-1}\mathbf{X}'\mathbf{y}, \mathbf{Xc} \rangle = \mathbf{y}'\mathbf{Xc} - \mathbf{y}'\mathbf{X}(\mathbf{X}'\mathbf{X})^{-1}\mathbf{X}'\mathbf{Xc}$ $= \mathbf{y}'\mathbf{Xc} - \mathbf{y}'\mathbf{Xc} = 0$. Therefore $\mathbf{y} - \mathbf{v} \in V^\perp$. □

Hence, we have shown that projections exist, are unique, and are linear functions.

NOTATION. $\mathbf{P}_V\mathbf{y}$ is defined to be the projection of $\mathbf{y}$ on the subspace V. $\mathbf{P}_V$ is the linear function that assigns to each $\mathbf{y}$ its projection onto V. We also use $\mathbf{P}_V$ for the matrix of the function $\mathbf{P}_V$.

The following is a list of some elementary properties of $\mathbf{P}_V$.

THEOREM 2.5. a. Pythagorean theorem. If $\mathbf{v} \in V$, then $\|\mathbf{y} - \mathbf{v}\|^2 = \|\mathbf{y} - \mathbf{P}_V\mathbf{y}\|^2 + \|\mathbf{P}_V\mathbf{y} - \mathbf{v}\|^2$.

b. $\|\mathbf{y} - \mathbf{P}_V\mathbf{y}\|^2 \leqslant \|\mathbf{y} - \mathbf{v}\|^2$ for all $\mathbf{v} \in V$ with equality if and only if $\mathbf{v} = \mathbf{P}_V\mathbf{y}$.

c. $\mathbf{P}_V\mathbf{y} = \mathbf{y}$ if and only if $\mathbf{y} \in V$. $\mathbf{P}_V\mathbf{y} = \mathbf{0}$ if and only if $\mathbf{y} \in V^\perp$.

d. $\mathbf{P}_{V^\perp}\mathbf{y} = \mathbf{y} - \mathbf{P}_V\mathbf{y}$. $\|\mathbf{P}_{V^\perp}\mathbf{y}\|^2 = \|\mathbf{y}\|^2 - \|\mathbf{P}_V\mathbf{y}\|^2$.

e. Let $W \subset V$ be a subspace. Then $\mathbf{P}_W\mathbf{y} = \mathbf{P}_W(\mathbf{P}_V\mathbf{y}) = \mathbf{P}_V(\mathbf{P}_W\mathbf{y})$.

f. Let $W \subset V$ be a subspace. Then $\mathbf{P}_{V|W}\mathbf{y} = \mathbf{P}_V\mathbf{y} - \mathbf{P}_W\mathbf{y}$. $\|\mathbf{P}_{V|W}\mathbf{y}\|^2 = \|\mathbf{P}_V\mathbf{y}\|^2 - \|\mathbf{P}_W\mathbf{y}\|^2$.

g. If $V \perp W$, then $\mathbf{P}_V(\mathbf{P}_W\mathbf{y}) = \mathbf{P}_V\mathbf{P}_W\mathbf{y} = \mathbf{0}$. $\mathbf{P}_V\mathbf{P}_W = \mathbf{0}$.

h. $\mathbf{P}_V^2 = \mathbf{P}_V$, $\mathbf{P}_V' = \mathbf{P}_V$.

i. If $\dim V = p$ then $\operatorname{tr} \mathbf{P}_V = p$ (where $\operatorname{tr} \mathbf{P}_V$ is the trace of $\mathbf{P}_V$).

PROOF. We prove parts a, b, c, e, f, and i. Parts d, g, and h are left for homework.

a. $\mathbf{y} - \mathbf{P}_V\mathbf{y} \in V^\perp$ and $\mathbf{P}_V\mathbf{y} - \mathbf{v} \in V$. Therefore, $(\mathbf{y} - \mathbf{P}_V\mathbf{y}) \perp (\mathbf{P}_V\mathbf{y} - \mathbf{v})$, and $\|\mathbf{y} - \mathbf{v}\|^2 = \|\mathbf{y} - \mathbf{P}_V\mathbf{y} + \mathbf{P}_V\mathbf{y} - \mathbf{v}\|^2 = \|\mathbf{y} - \mathbf{P}_V\mathbf{y}\|^2 + \|\mathbf{P}_V\mathbf{y} - \mathbf{v}\|^2$.

b. This follows directly from part a.

c. Suppose $\mathbf{P}_V\mathbf{y} = \mathbf{y}$. Then $\mathbf{y} \in V$ by the definition of projection. Conversely suppose that $\mathbf{y} \in V$. Then $\mathbf{y} \in V$, and $\mathbf{y} - \mathbf{y} = \mathbf{0} \in V^\perp$. Therefore $\mathbf{y} = \mathbf{P}_V\mathbf{y}$. Now suppose that $\mathbf{P}_V\mathbf{y} = \mathbf{0}$. Then by the definition of $\mathbf{P}_V$, $\mathbf{y} = \mathbf{y} - \mathbf{P}_V\mathbf{y} \in V^\perp$. Conversely, suppose that $\mathbf{y} \in V^\perp$. Then $\mathbf{0} \in V$, $\mathbf{y} - \mathbf{0} \in V^\perp$. Therefore $\mathbf{P}_V\mathbf{y} = \mathbf{0}$.

e. Since $\mathbf{P}_W\mathbf{y} \in W \subset V$, by part c, $\mathbf{P}_V(\mathbf{P}_W\mathbf{y}) = \mathbf{P}_W\mathbf{y}$. Now, let $\mathbf{u} = \mathbf{P}_W(\mathbf{P}_V\mathbf{y})$. Clearly $\mathbf{u} \in W$, by definition of $\mathbf{P}_W$. To see that $\mathbf{y} - \mathbf{u} \in W^\perp$, let $\mathbf{w} \in W$. Then $\langle \mathbf{y} - \mathbf{u}, \mathbf{w} \rangle = \langle \mathbf{y} - \mathbf{P}_V\mathbf{y}, \mathbf{w} \rangle + \langle \mathbf{P}_V\mathbf{y} - \mathbf{P}_W(\mathbf{P}_V\mathbf{y}), \mathbf{w} \rangle$. Since

$\mathbf{w} \in W \subset V$, and $\mathbf{y} - \mathbf{P}_V\mathbf{y} \in V^{\perp}, \langle \mathbf{y} - \mathbf{P}_V\mathbf{y}, \mathbf{w} \rangle = 0$. Since $\mathbf{w} \in W$, and $\mathbf{P}_V\mathbf{y} - \mathbf{P}_W(\mathbf{P}_V\mathbf{y}) \in W^{\perp}$, $\langle \mathbf{P}_V\mathbf{y} - \mathbf{P}_W(\mathbf{P}_V\mathbf{y}), \mathbf{w} \rangle = 0$.

f. Let $\mathbf{u} = \mathbf{P}_V\mathbf{y} - \mathbf{P}_W\mathbf{y}$. Since $W \subset V$, $\mathbf{P}_W\mathbf{y} \in W \subset V$. Therefore $\mathbf{u} \in V$. We now show that $\mathbf{u} \in W^{\perp}$, and hence $\mathbf{u} \in V \mid W$. Let $\mathbf{w} \in W$. Since $\mathbf{P}_V = \mathbf{P}'_V$, $\mathbf{P}_W = \mathbf{P}'_W$,

$$\langle \mathbf{u}, \mathbf{w} \rangle = \langle \mathbf{P}_V\mathbf{y}, \mathbf{w} \rangle - \langle \mathbf{P}_W\mathbf{y}, \mathbf{w} \rangle = \langle \mathbf{y}, \mathbf{P}_V\mathbf{w} \rangle - \langle \mathbf{y}, \mathbf{P}_W\mathbf{w} \rangle$$
$$= \langle \mathbf{y}, \mathbf{w} \rangle - \langle \mathbf{y}, \mathbf{w} \rangle = 0,$$

since $\mathbf{w} \in W \subset V$. Thus $\mathbf{u} \in W^{\perp}$.

We now show that $\mathbf{y} - \mathbf{u} \in (V \mid W)^{\perp}$. Let $\mathbf{v} \in V \mid W$. Then

$$\langle \mathbf{y} - \mathbf{u}, \mathbf{v} \rangle = \langle \mathbf{y} - \mathbf{P}_V\mathbf{y}, \mathbf{v} \rangle + \langle \mathbf{P}_W\mathbf{y}, \mathbf{v} \rangle = 0 + 0 = 0,$$

since $\mathbf{y} - \mathbf{P}_V\mathbf{y} \in V^{\perp}$, $\mathbf{v} \in V$, $\mathbf{P}_W\mathbf{y} \in W$, $\mathbf{v} \in W^{\perp}$. Therefore, $\mathbf{u} \in V \mid W$ and $\mathbf{y} - \mathbf{u} \in (V \mid W)^{\perp}$, and $\mathbf{u} = \mathbf{P}_{V \mid W}\mathbf{y}$. Finally, $\|\mathbf{P}_V\mathbf{y}\|^2 = \|\mathbf{P}_W\mathbf{y} + \mathbf{P}_{V \mid W}\mathbf{y}\|^2 = \|\mathbf{P}_W\mathbf{y}\|^2 + \|\mathbf{P}_{V \mid W}\mathbf{y}\|^2$, since $\mathbf{P}_W\mathbf{y} \perp \mathbf{P}_{V \mid W}\mathbf{y}$.

i. $\operatorname{tr} P_V = \operatorname{tr} X(X'X)^{-1}X' = \operatorname{tr}(X'X)^{-1}X'X = \operatorname{tr} I = p$ [see (A9) of the appendix]. Note that the identity is $p \times p$. □

One of the most important aspects of the matrix $\mathbf{P}_V$ is that it does not depend on the particular basis matrix chosen. This follows from Theorem 2.4b. (A direct proof of this fact is suggested in the exercises.) Therefore in proving results about projections, we can often assume that the basis matrix is orthonormal. The following theorem summarizes some important properties of orthonormal basis matrices.

THEOREM 2.6. Let $\mathbf{X}$ be an orthonormal basis matrix for the subspace V. Then

a. $\mathbf{X}'\mathbf{X} = \mathbf{I}$.
b. $\mathbf{P}_V\mathbf{y} = \mathbf{X}\mathbf{X}'\mathbf{y}$.
c. $\|\mathbf{P}_V\mathbf{y}\|^2 = \|\mathbf{X}'\mathbf{y}\|^2$.

PROOF. Part a follows directly from the definition of orthonormal basis. Using part a, we see that $\mathbf{P}_V\mathbf{y} = \mathbf{X}(\mathbf{X}'\mathbf{X})^{-1}\mathbf{X}'\mathbf{y} = \mathbf{X}\mathbf{X}'\mathbf{y}$, and also that

$$\|\mathbf{P}_V\mathbf{y}\|^2 = \mathbf{y}'\mathbf{X}\mathbf{X}'\mathbf{X}\mathbf{X}'\mathbf{y} = \mathbf{y}'\mathbf{X}\mathbf{X}'\mathbf{y} = \|\mathbf{X}'\mathbf{y}\|^2. \quad \square$$

We now derive some results about $V^{\perp}$ and $V \mid W$. First we need the following lemma.

LEMMA 2.7. Let $\mathbf{x}_1, \ldots, \mathbf{x}_n$ be mutually orthogonal nonzero vectors. Then the $\mathbf{x}_i$ are linearly independent.

PROOF. Suppose that $\sum a_i\mathbf{x}_i = 0$. Then $0 = \|\sum a_i\mathbf{x}_i\|^2 = \sum a_i^2\|\mathbf{x}_i\|^2$, since the $\mathbf{x}_i$ are orthogonal. This implies that $a_i^2\|\mathbf{x}_i\|^2 = 0$. Since $\mathbf{x}_i \neq 0$, we see that $a_i = 0$, and hence that the $\mathbf{x}_i$ are linearly independent. □

LEMMA 2.8. Let $W \subset V \subset R^n$. Then
a. $\dim(V^\perp) = n - \dim(V)$;
b. $\dim(V \mid W) = \dim(V) - \dim(W)$.

PROOF. a. Let $\dim(V) = p$, $\dim(V^\perp) = k$. Let $\mathbf{x}_1, \ldots, \mathbf{x}_p$ be an orthonormal basis for V, and let $\mathbf{x}_{p+1}, \ldots, \mathbf{x}_{p+k}$ be an orthonormal basis for $V^\perp$. Then $\langle \mathbf{x}_i, \mathbf{x}_j \rangle = 0$, $1 \leqslant i \leqslant p$, $p+1 \leqslant j \leqslant p+k$, since $V \perp V^\perp$. Therefore the $\mathbf{x}_i$ are orthogonal and hence linearly independent. We now show that they span R^n. Let $\mathbf{u} \in R^n$. By Theorem 2.5.d, $\mathbf{P}_V \boldsymbol{\mu} = \mathbf{P}_V \mathbf{u} + \mathbf{P}_{V^\perp} \mathbf{u}$. However, $\mathbf{P}_V \mathbf{u} \in V$. Therefore, there exist a_i such that $\mathbf{u} = \sum_{i=1}^{p} a_i \mathbf{x}_i$, since $\mathbf{x}_1, \ldots, \mathbf{x}_p$ are a basis for V. Similarly $\mathbf{P}_{V^\perp}\mathbf{u} = \sum_{i=p+1}^{p+k} a_i \mathbf{x}_i$. Hence $\mathbf{u} = \sum_{i=1}^{p+k} a_i \mathbf{x}_i$, and the $\mathbf{x}_i$ span R^n. They are therefore a basis for R^n, and $p + k = n$.

b. See Exercise B5. □

LEMMA 2.9. a. $(V^\perp)^\perp = V$.
b. Let $W \subset V$. Then $V \mid (V \mid W) = W$.

PROOF. a. Let $\mathbf{v} \in V$, and let $\mathbf{u} \in V^\perp$. Then $\mathbf{u} \perp \mathbf{v}$. Therefore $\mathbf{v} \in (V^\perp)^\perp$. Hence $V \subset (V^\perp)^\perp$. Now let $\dim(V) = p$. Then $\dim(V^\perp) = n - p$, and $\dim((V^\perp)^\perp = n - (n - p) = p$. Therefore, $\dim(V) = \dim((V^\perp)^\perp)$. Hence $(V^\perp)^\perp = V$.

b. See Exercise B5. □

One often wants to check whether a given matrix $\mathbf{A}$ is a projection matrix. The following theorem gives easily checked conditions for this condition. Let $\mathbf{A}$ be an $n \times n$ matrix, and let V be the image of $\mathbf{A}$, that is $V = \{\mathbf{v} \in R^n : \mathbf{v} = \mathbf{A}\mathbf{c} \text{ for some } \mathbf{c} \in R^n\}$. Then V is a subspace.

THEOREM 2.10. $\mathbf{A} = \mathbf{P}_V$ if and only if $\mathbf{A}' = \mathbf{A}$ and $\mathbf{A}^2 = \mathbf{A}$. If $\mathbf{A} = \mathbf{P}_V$, then $\dim(V) = \operatorname{rank}(\mathbf{A}) = \operatorname{tr} \mathbf{A}$.

PROOF. Suppose that $\mathbf{A} = \mathbf{P}_V$. Then by Theorem 2.5h, $\mathbf{A}' = \mathbf{A}$, $\mathbf{A}^2 = \mathbf{A}$. Conversely, suppose that $\mathbf{A}^2 = \mathbf{A}$ and $\mathbf{A}' = \mathbf{A}$. Let $\mathbf{v} = \mathbf{A}\mathbf{y}$. We want to show that $\mathbf{v} = \mathbf{P}_V \mathbf{y}$. By the definition of V, $\mathbf{v} \in V$. Now let $\mathbf{w} \in V$. By the definition of V, again, $\mathbf{w} = \mathbf{A}\mathbf{c}$ for some c. Therefore,

$$\langle \mathbf{y} - \mathbf{v}, \mathbf{w} \rangle = \langle \mathbf{y} - \mathbf{A}\mathbf{y}, \mathbf{A}\mathbf{c} \rangle = \mathbf{y}'\mathbf{A}\mathbf{c} - \mathbf{y}'\mathbf{A}'\mathbf{A}\mathbf{c} = \mathbf{y}'\mathbf{A}\mathbf{c} - \mathbf{y}'\mathbf{A}\mathbf{c} = 0.$$

Hence, $\mathbf{y} - \mathbf{v} \in V^\perp$, and $\mathbf{v} = \mathbf{P}_V \mathbf{y}$. By Lemma 2.1e, $\dim(V) = \operatorname{rank} \mathbf{A} = \operatorname{tr} \mathbf{A}$ (see Theorem 2.5i.). □

We define an $n \times n$ matrix to be *idempotent* if $\mathbf{A}' = \mathbf{A}$ and $\mathbf{A}^2 = \mathbf{A}$. The theorem, therefore, says that $\mathbf{A}$ is a projection matrix if and only if it is idempotent.

EXERCISES.

Type A

1. Let V be the two-dimensional subspace of R^3 consisting of vectors $\mathbf{v} = (v_1, v_2, v_3)'$ such that $v_1 + 2v_2 + 3v_3 = 0$. Let W be the one-dimensional subspace of V in which $v_1 + v_2 + v_3 = 0$.

(a) Find basis matrices for V and $V^\perp$. Find $\mathbf{P}_V$ and $\mathbf{P}_{V^\perp}$ and show that $\mathbf{P}_V = \mathbf{I} - \mathbf{P}_{V^\perp}$.

(b) Find basis matrices for W and $V \mid W$. Find $\mathbf{P}_W$ and $\mathbf{P}_{V|W}$ and show that $\mathbf{P}_{V|W} = \mathbf{P}_V - \mathbf{P}_W$.

(c) Show that $\mathbf{P}_V^2 = \mathbf{P}_V$ and $\mathbf{P}_V\mathbf{P}_W = \mathbf{P}_W\mathbf{P}_V = \mathbf{P}_W$.

(Do not use Theorem 2.5 in this problem.)

Type B

1. Prove Lemma 2.3.

2. Show that $W \subset V$ if and only if $W^\perp \supset V^\perp$.

3. Let $\mathbf{X}_1$ and $\mathbf{X}_2$ be basis matrices for V.

(a) Show that there exists $\mathbf{A}$ invertible such that $\mathbf{X}_2 = \mathbf{X}_1\mathbf{A}$.

(b) Show by direct computation that

$$\mathbf{X}_2(\mathbf{X}_2'\mathbf{X}_2)^{-1}\mathbf{X}_2' = \mathbf{X}_1(\mathbf{X}_1'\mathbf{X}_1)^{-1}\mathbf{X}_1'$$

(i.e., that the projection matrix does not depend on the basis matrix chosen).

4. Prove parts d, g, and h of Theorem 2.5.

5. Prove part b of Lemmas 2.8 and 2.9.

6. (a) Let V and W be subspaces. Show that $V = W$ if and only if $\mathbf{P}_V = \mathbf{P}_W$.

(b) Use part a to give an alternative proof that $(V^\perp)^\perp = V$.

7. Show that $\|\mathbf{P}_V\mathbf{y}\|^2 \geqslant \|\mathbf{P}_W\mathbf{y}\|^2$ for all $\mathbf{y}$ if and only if $W \subset V$.

Type C

1. Let $U_1, \ldots, U_s$ be subspaces of R^n. Define the *sum* of the U_i written $\sum U_i$ to be the set of vectors of the form

$$\mathbf{v} = \sum \mathbf{u}_i, \qquad \mathbf{u}_i \in U_i.$$

(a) Show that $\sum U_i$ is a subspace.

(b) Show that $\sum U_i \subset V$ if and only if $U_i \subset V$ for all i.

(c) Show that $\sum U_i \perp V$ if and only if $U_i \perp V$ for all i.

(d) Show that $\|\mathbf{P}_{\sum U_i}\mathbf{y}\|^2 = \sum \|\mathbf{P}_{U_i}\mathbf{y}\|^2$ for all y if and only if $U_i \perp U_j$ for all $i \neq j$.

(e) Show that the dimension of $\sum U_i$ is less than or equal to the sum of the dimension of the U_i.

2. Let $U_i, \ldots, U_s$ be subspaces of R^n.

(a) Show that $\bigcap U_i$ is a subspace.

(b) Show that $\bigcap(U_i^{\perp}) = (\sum U_i)^{\perp}$.

(c) Show that $\sum(U_i^{\perp}) = (\bigcap U_i)^{\perp}$.

3. Generalized inverses. Let $\mathbf{U}$ be an $n \times r$ matrix. A *generalized inverse* of $\mathbf{U}$ is a matrix $\mathbf{T}$ such that $\mathbf{UTU} = \mathbf{U}$. A g-inverse of $\mathbf{U}$ is a matrix $\mathbf{T}$ such that $\mathbf{UTU} = \mathbf{U}$, $\mathbf{TUT} = \mathbf{T}$ and such that $\mathbf{TU}$ and $\mathbf{UT}$ are symmetric. In this problem we establish the existence and uniqueness of g-inverses. This, of course, establishes the existence of generalized inverses.

(a) Any matrix $\mathbf{U}$ has a g-inverse.
(Hint: Let $\mathbf{U}$ be an $n \times r$ matrix of rank p and let $\mathbf{U} = \mathbf{AB}$ where $\mathbf{A}$ is $n \times p$ of rank p and $\mathbf{B}$ is $p \times r$ of rank p. (See the proof of Lemma A9.) Then $\mathbf{A'A}$ and $\mathbf{BB'}$ are invertible. Let $\mathbf{T} = \mathbf{B'}(\mathbf{BB'})^{-1}(\mathbf{A'A})^{-1}\mathbf{A'}$. Show that $\mathbf{T}$ is a g-inverse.)

(b) Show that the g-inverse is unique.
(Hint: Let $\mathbf{T}_1$ and $\mathbf{T}_2$ be g-inverses of $\mathbf{U}$. Show first that $\mathbf{UT}_1 = \mathbf{UT}_2$ and $\mathbf{T}_1\mathbf{U} = \mathbf{T}_2\mathbf{U}$. Then show that $\mathbf{T}_1 = \mathbf{T}_1\mathbf{UT}_2 = \mathbf{T}_2$.)

4. Formula for projection when $\mathbf{X}$ does not have full rank. Let V be a p-dimensional subspace of R^n and let $\mathbf{X}$ be an $n \times r$ matrix whose columns span V (but are not linearly independent unless $p = r$). We call $\mathbf{X}$ a spanning matrix for V. Clearly $\mathbf{X'X}$ is not invertible.

(a) Show that $\mathbf{v} \in V$ if and only if there exists $\mathbf{b} \in R^r$ such that $\mathbf{v} = \mathbf{Xb}$.

(b) Show that $\mathbf{X'XC} = 0$ if and only if $\mathbf{XC} = 0$. (Hint: Let $\mathbf{X} = \mathbf{AB}$ where $\mathbf{A}$ is $n \times p$ of rank p and $\mathbf{B}$ is $p \times r$ of rank p. Then $\mathbf{X'XC} = \mathbf{0}$ implies that $(\mathbf{BB'})(\mathbf{A'A})\mathbf{BC} = \mathbf{0}$ which implies that $\mathbf{BC} = \mathbf{0}$ which implies that $\mathbf{XC} = \mathbf{ABC} = \mathbf{0}$.)

(c) Let $\mathbf{y} \in R^n$, $\mathbf{v} = \mathbf{XTX'y}$, where $\mathbf{T}$ is a generalized inverse of $\mathbf{X'X}$. Show that $\mathbf{v} = \mathbf{P}_V\mathbf{y}$ and hence that $\mathbf{P}_V = \mathbf{XTX'}$. (Hint: Show first that $\mathbf{X'X}(\mathbf{T'X'X} - \mathbf{I}) = \mathbf{0}$, then that $\mathbf{XT'X'X} = \mathbf{X}$ and then imitate the proof of Theorem 2.4c.)

CHAPTER 3

Properties of the Multivariate and Spherical Normal Distributions

In this chapter, we introduce the multivariate normal distribution and the spherical normal distribution, and we derive some elementary results about them. In Section 3.1, basic results about multivariate distributions are given. Properties of nonnegative definite matrices are discussed in Section 3.2 and the Mahalanobis distance is defined in Section 3.3. In Sections 3.4 and 3.5 the multivariate normal and spherical normal distributions are defined and their elementary properties derived.

3.1. MULTIVARIATE DISTRIBUTIONS.

Let $\mathbf{Y} = (Y_1, \ldots, Y_n)$ be an $n \times 1$ vector of random variables. We say that $\mathbf{Y}$ is an *n-dimensional random vector*. The *mean vector*, $E(\mathbf{Y})$, and *covariance matrix*, $\operatorname{cov}(\mathbf{Y})$, are defined by

$$E(\mathbf{Y}) = \begin{pmatrix} EY_1 \\ \vdots \\ EY_n \end{pmatrix}$$

$$\operatorname{cov}(\mathbf{Y}) = \begin{pmatrix} \operatorname{var}(Y_1) & \operatorname{cov}(Y_1, Y_2) & \cdots & \operatorname{cov}(Y_1, Y_n) \\ \operatorname{cov}(Y_2, Y_1) & \operatorname{var}(Y_2) & \cdots & \operatorname{cov}(Y_2, Y_n) \\ \vdots & \vdots & & \vdots \\ \operatorname{cov}(Y_n, Y_1) & \operatorname{cov}(Y_n, Y_2) & \cdots & \operatorname{var}(Y_n) \end{pmatrix}$$

That is, E(**Y**) is the $n \times 1$ vector whose ith component is the EY_i. cov(**Y**) is the symmetric $n \times n$ matrix whose ith diagonal element is var(Y_1) and whose (i, j)th off-diagonal element is cov(Y_i, Y_j). Note that var(Y_i) can be thought of as cov(Y_i, Y_i).

LEMMA 3.1. Let **X** be a random vector with mean vector $\boldsymbol{\mu}$ and covariance matrix $\boldsymbol{\Sigma}$.

a. Let $\mathbf{Y} = \mathbf{AX} + \mathbf{b}$. Then $E\mathbf{Y} = \mathbf{A}\boldsymbol{\mu} + \mathbf{b}$ and $\text{cov}(\mathbf{Y}) = \mathbf{A}\boldsymbol{\Sigma}\mathbf{A}'$.
b. $E\|\mathbf{X}\|^2 = \|\boldsymbol{\mu}\|^2 + \text{tr}\,\boldsymbol{\Sigma}$ (where $\text{tr}\,\boldsymbol{\Sigma}$ is the trace of $\boldsymbol{\Sigma}$).

PROOF. a. See Exercise B1.
b. Let

$$\mathbf{X} = \begin{pmatrix} X_1 \\ \vdots \\ X_n \end{pmatrix}, \qquad \boldsymbol{\mu} = \begin{pmatrix} \mu_1 \\ \vdots \\ \mu_n \end{pmatrix}, \qquad \boldsymbol{\Sigma} = \begin{pmatrix} \Sigma_{11} & \cdots & \Sigma_{1n} \\ \vdots & & \vdots \\ \Sigma_{n1} & \cdots & \Sigma_{nn} \end{pmatrix}$$

then $E\|\mathbf{X}\|^2 = E\sum_i X_i^2 = \sum_i EX_i^2 = \sum_i(\mu_i^2 + \Sigma_{ii}) = \|\boldsymbol{\mu}\|^2 + \text{tr}\,\boldsymbol{\Sigma}$. □

COROLLARY Let $\mathbf{a} \in R^n$. Then $\mathbf{a}'\,\text{cov}(\mathbf{Y})\mathbf{a} \geqslant 0$.

PROOF. Let $X = \mathbf{a}'\mathbf{Y}$. Then X is a one-dimensional random vector. Its covariance matrix is just its variance. Therefore,

$$0 \leqslant \text{var}(\mathbf{a}'\mathbf{Y}) = \mathbf{a}'\,\text{cov}(\mathbf{Y})\mathbf{a}. \quad \square$$

We say that a symmetric matrix **U** is *nonnegative definite* and write $\mathbf{U} \geqslant 0$ if $\mathbf{a}'\mathbf{U}\mathbf{a} \geqslant 0$ for all $\mathbf{a} \in R^n$. If $\mathbf{a}'\mathbf{U}\mathbf{a} > 0$ for all $\mathbf{a} \neq 0$, we say that **U** is *positive definite* and we write $\mathbf{U} > 0$. The corollary, therefore, implies that covariance matrices are nonnegative definite. In Section 3.4 we show that for any $n \times 1$ vector $\boldsymbol{\mu}$ and $n \times n$ nonnegative definite matrix $\boldsymbol{\Sigma}$, there exists **Y** normally distributed with mean variance $\boldsymbol{\mu}$ and covariance matrix $\boldsymbol{\Sigma}$. Therefore a matrix is a covariance matrix if and only if it is nonnegative definite.

LEMMA 3.2. $\text{cov}(\mathbf{Y}) \geqslant 0$. If **Y** has a continuous density function, then $\text{cov}(\mathbf{Y}) > 0$.

PROOF. $\text{cov}(\mathbf{Y}) \geqslant 0$ by the corollary above. Now suppose that cov(**Y**) is not positive definite. Then there exists $\mathbf{a} \neq 0$ such that $\text{var}(\mathbf{a}'\mathbf{Y}) = \mathbf{a}'\,\text{cov}(\mathbf{Y})\mathbf{a} = 0$. Then there would exist c such that $P(\mathbf{a}'\mathbf{Y} = c) = 1$. If **Y** had a continuous density function, then $P(\mathbf{a}'\mathbf{Y} = c) = 0$ for all **a** and c. □

Let $\mathbf{t} = (t_1, \ldots, t_n)'$ be a vector and let $\mathbf{Y} = (Y_1, \ldots, Y_n)'$ be a random vector. Then the *joint moment generating function* of **Y** is defined to be

$$M_{\mathbf{Y}}(\mathbf{t}) = E\left(\exp \sum_{i=1}^{n} Y_i t_i\right) = E(\exp(\mathbf{Y}'\mathbf{t})).$$

The following lemma is true for multivariate moment generating functions, but its proof is omitted.

LEMMA 3.3. If $\mathbf{X}$ and $\mathbf{Y}$ have the same joint moment generating function in some open set containing the origin, then they have the same distribution.

LEMMA 3.4. a. Let $\mathbf{X} = \mathbf{A}\mathbf{Y} + \mathbf{b}$. Then $M_{\mathbf{X}}(\mathbf{t}) = \exp(\mathbf{b}'\mathbf{t})M_{\mathbf{Y}}(\mathbf{A}'\mathbf{t})$.
b. Let $c \in R^1$. Let $\mathbf{Z} = c\mathbf{Y}$. Then $M_{\mathbf{Z}}(\mathbf{t}) = M_{\mathbf{Y}}(c\mathbf{t})$.
c. Let

$$\mathbf{Y} = \begin{pmatrix} \mathbf{Y}_1 \\ \mathbf{Y}_2 \end{pmatrix}, \qquad \mathbf{t} = \begin{pmatrix} \mathbf{t}_1 \\ \mathbf{t}_2 \end{pmatrix}$$

Then $M_{\mathbf{Y}_1}(\mathbf{t}_1) = M_{\mathbf{Y}}\binom{\mathbf{t}_1}{\mathbf{0}}$.
d. $\mathbf{Y}_1$ and $\mathbf{Y}_2$ are independent if and only if

$$M_{\mathbf{Y}}\begin{pmatrix} \mathbf{t}_1 \\ \mathbf{t}_2 \end{pmatrix} = M_{\mathbf{Y}}\begin{pmatrix} \mathbf{t}_1 \\ \mathbf{0} \end{pmatrix} M_{\mathbf{Y}}\begin{pmatrix} \mathbf{0} \\ \mathbf{t}_2 \end{pmatrix}.$$

PROOF. a.

$$\begin{aligned} M_{\mathbf{X}}(\mathbf{t}) &= E(\exp \mathbf{X}'\mathbf{t}) = E(\exp(\mathbf{b}'\mathbf{t} + \mathbf{Y}'\mathbf{A}'\mathbf{t})) \\ &= \exp(\mathbf{b}'\mathbf{t})E(\exp \mathbf{Y}'\mathbf{A}'\mathbf{t}) = \exp(\mathbf{b}'\mathbf{t})M_{\mathbf{Y}}(\mathbf{A}'\mathbf{t}). \end{aligned}$$

b. See Exercise B4.
c. $M_{\mathbf{Y}_1}(\mathbf{t}_1) = E\exp(\mathbf{Y}_1'\mathbf{t}_1) = E\exp(\mathbf{Y}_1'\mathbf{t}_1 + \mathbf{Y}_2'\mathbf{0}) = M_{\mathbf{Y}}\binom{\mathbf{t}_1}{\mathbf{0}}$.
d. If $\mathbf{Y}_1$ and $\mathbf{Y}_2$ are independent, then

$$\begin{aligned} M_{\mathbf{Y}}(\mathbf{t}) &= E(\exp(\mathbf{Y}_1'\mathbf{t}_1 + \mathbf{Y}_2'\mathbf{t}_2)) \\ &= E(\exp(\mathbf{Y}_1'\mathbf{t}_1 + \mathbf{Y}_2'\mathbf{0}))E(\exp(\mathbf{Y}_1'\mathbf{0} + \mathbf{Y}_2'\mathbf{t}_2)) = M_{\mathbf{Y}}\begin{pmatrix} \mathbf{t}_1 \\ \mathbf{0} \end{pmatrix} M_{\mathbf{Y}}\begin{pmatrix} \mathbf{0} \\ \mathbf{t}_2 \end{pmatrix} \end{aligned}$$

Conversely, suppose that the moment generating function of $\mathbf{Y}_1$ and $\mathbf{Y}_2$ factors in the above fashion. Let $\mathbf{X}_1$ and $\mathbf{X}_2$ be independent random vectors such that $\mathbf{X}_i$ has the same marginal density as $\mathbf{Y}_i$. Then, by part b,

$$M_{\mathbf{X}_1}(\mathbf{t}_1) = M_{\mathbf{Y}}\begin{pmatrix} \mathbf{t}_1 \\ \mathbf{0} \end{pmatrix}, \qquad M_{\mathbf{X}_2}(\mathbf{t}_2) = M_{\mathbf{Y}}\begin{pmatrix} \mathbf{0} \\ \mathbf{t}_2 \end{pmatrix}.$$

Therefore, since $\mathbf{X}_1$ and $\mathbf{X}_2$ are independent, the joint moment generating function of $\mathbf{X}_1$ and $\mathbf{X}_2$ is

$$M_{\mathbf{X}}(\mathbf{t}) = M_{\mathbf{Y}}\begin{pmatrix} \mathbf{t}_1 \\ \mathbf{0} \end{pmatrix} M_{\mathbf{Y}}\begin{pmatrix} \mathbf{0} \\ \mathbf{t}_2 \end{pmatrix} = M_{\mathbf{Y}}(\mathbf{t}).$$

By Lemma 3.3, $\mathbf{X}$ and $\mathbf{Y}$ have the same joint distribution, and hence $\mathbf{Y}_1$ and $\mathbf{Y}_2$ are independent. (In other words, this argument shows that $\mathbf{Y}_1$ and $\mathbf{Y}_2$ are independent because they have the same moment generating function as if they were independent.) □

3.2. NONNEGATIVE DEFINITE MATRICES AND THEIR SQUARE ROOTS.

From Lemma 3.2, we see that a covariance matrix must be nonnegative definite. If the random vector has a continuous density function, then the covariance matrix must be positive definite. We now state some elementary properties of nonnegative definite and positive definite matrices.

LEMMA 3.5. a. If $\mathbf{U} \geqslant 0$ then there exists a unique $\mathbf{V} \geqslant 0$ such that $\mathbf{U} = \mathbf{V}^2$. If $\mathbf{U} > 0$ then $\mathbf{V} > 0$.

b. If $\mathbf{U} > 0$ then $\mathbf{U}$ is invertible and $\mathbf{U}^{-1} > 0$.

c. If $\mathbf{U} \geqslant 0$ then $\mathbf{U}$ is invertible if and only if $\mathbf{U} > 0$.

d. If $\mathbf{U} \geqslant 0$ is a $p \times p$ matrix of rank r then there exists a $p \times r$ matrix $\mathbf{C}$ of rank r such that $\mathbf{U} = \mathbf{CC}'$.

e. Let $\mathbf{X}$ be an $n \times p$ matrix of rank r. Then $\mathbf{X}'\mathbf{X} \geqslant 0$. If $r = p$, then $\mathbf{X}'\mathbf{X} > 0$ and hence $\mathbf{X}'\mathbf{X}$ is invertible.

PROOF. See Theorems 3 and 5 of the appendix. □

Now, let $\mathbf{A} \geqslant 0$. We define the *square root* of $\mathbf{A}$ to be the unique matrix $\mathbf{B} \geqslant 0$ such that $\mathbf{A} = \mathbf{B}^2$. We write $A^{1/2}$ for the square root of $\mathbf{A}$. From Lemma 3.3a, we see that if $\mathbf{A} > 0$ then $\mathbf{A}^{1/2} > 0$. (This corresponds to the univariate case in which nonnegative numbers have nonnegative square roots and positive numbers have positive square roots.)

If $\mathbf{A} > 0$, then $\mathbf{A}^{-1} > 0$ and $\mathbf{A}^{1/2} > 0$. In addition

$$(\mathbf{A}^{-1})^{1/2} = (\mathbf{A}^{1/2})^{-1}$$

(see Exercise B3). We define

$$\mathbf{A}^{-1/2} = (\mathbf{A}^{1/2})^{-1}.$$

The following facts are used quite often in this book. If $\mathbf{A} > 0$, then

$$\mathbf{A}^{-1/2}\mathbf{A}\mathbf{A}^{-1/2} = \mathbf{I}, \qquad (\mathbf{A}^{-1/2})' = \mathbf{A}^{-1/2}, \qquad (\mathbf{A}^{1/2})' = \mathbf{A}^{1/2}. \tag{3.1}$$

(See Exercise B3.)

3.3. THE MAHALANOBIS DISTANCE.

In many of the models in this book, we observe an n-dimensional random vector $\mathbf{Y}$ such that $E\mathbf{Y} = \boldsymbol{\mu}$ and $\text{cov}(\mathbf{Y}) = \boldsymbol{\Sigma} > 0$, where $\boldsymbol{\mu}$ and $\boldsymbol{\Sigma}$ are restricted to lie in certain sets or have certain forms. We want to estimate $\boldsymbol{\mu}$. Let $\mathbf{d} \in R^n$. The loss function that we often use for such problems is

$$L(\mathbf{d}; (\boldsymbol{\mu}, \boldsymbol{\Sigma})) = (\mathbf{d} - \boldsymbol{\mu})'\boldsymbol{\Sigma}^{-1}(\mathbf{d} - \boldsymbol{\mu})$$

The square root of L was first suggested as a measure of the distance between $\mathbf{d}$ and $\boldsymbol{\mu}$ by Mahalanobis (1936) and is called the *Mahalanobis distance* between $\mathbf{d}$ and $\boldsymbol{\mu}$. We call the loss function the *Mahalanobis distance loss function.*

There are at least two reasons for choosing this loss function. The first reason is that we can find answers more easily for this loss function than for others we might try. (For example, the loss function $L(\mathbf{d},(\boldsymbol{\mu},\boldsymbol{\Sigma})) = \|\mathbf{d} - \boldsymbol{\mu}\|^2$ is more difficult to use.) The second reason has to do with invariance. Let $\mathbf{Y}^* = \mathbf{AY} + \mathbf{b}$ where $\mathbf{A}$ is an $n \times n$ invertible matrix and $\mathbf{b}$ is $n \times 1$. Suppose that $\boldsymbol{\mu}^* = \mathbf{A}\boldsymbol{\mu} + \mathbf{b}$, $\boldsymbol{\Sigma}^* = \mathbf{A}\boldsymbol{\Sigma}\mathbf{A}'$ are still in the space of permissible parameter values. Then $E\mathbf{Y} = \boldsymbol{\mu}^*$, $\text{cov}(\mathbf{Y}) = \boldsymbol{\Sigma}^*$. Let $\mathbf{d}^* = \mathbf{Ad} + \mathbf{b}$. It seems reasonable to require that the loss for estimating $\boldsymbol{\mu}^*$ by $\mathbf{d}^*$ when we observe $\mathbf{Y}^*$ should be the same as the loss for estimating $\boldsymbol{\mu}$ by $\mathbf{d}$ when we observe $\mathbf{Y}$. (In particular, $\mathbf{Y}^*$ may just be a change of units from $\mathbf{Y}$.) That is we want a loss function that satisfies

$$L(\mathbf{Ad} + \mathbf{b}; (\mathbf{A}\boldsymbol{\mu} + \mathbf{b}, \mathbf{A}\boldsymbol{\Sigma}\mathbf{A}')) = L(\mathbf{d}, (\boldsymbol{\mu}, \boldsymbol{\Sigma}))$$

for all $\mathbf{A}$ and $\mathbf{b}$ satisfying the above conditions. The loss function L satisfies this condition for all invertible $\mathbf{A}$ and all $\mathbf{b}$. In particular, the Mahalanobis distance is unit free. The loss function $L(\mathbf{d}; (\boldsymbol{\mu}, \boldsymbol{\Sigma})) = \|\mathbf{d} - \boldsymbol{\mu}\|^2$ does not satisfy this condition and is not unit free.

For an example, we return to the one sample model discussed in Sections 1.2 and 1.6 in which we observe X_i independent, $X_i \sim N_1(\mu, \sigma^2)$. We first consider the estimation of μ. Since $EX_i = \mu$ and the covariance matrix of X_i (which is just the variance of X_i) is σ^2, we see that the Mahalanobis distance loss function for this problem is just

$$L_1(d_1; (\mu, \sigma^2)) = (d_1 - \mu)\left(\frac{1}{\sigma^2}\right)(d_1 - \mu) = \frac{(d_1 - \mu)^2}{\sigma^2},$$

which is the loss function used in those sections for that problem. We now consider the estimation of σ^2. This problem does not quite fit into the above framework. However, suppose that we look only at $S^2 = \Sigma_i(x_i - \bar{x})^2/(n-1)$. Then $ES^2 = \sigma^2$ and the covariance matrix of S^2, which is just its variance, is $2\sigma^4/(n-1)$. Therefore, a Mahalanobis distance loss function for this problem is

$$L_2(d_2; (\mu, \sigma^2)) = (n-1)\frac{(d_2 - \sigma^2)^2}{2\sigma^4}.$$

The loss function that we used in Sections 1.2 and 1.6 is

$$L_2^*(d_2; (\mu, \sigma^2)) = \frac{(d_2 - \sigma^2)^2}{\sigma^4} = \frac{2}{n-1} L_2.$$

However, L_2^* is equivalent to L_2 in the sense that T_1 is better than T_2 for L_2 if and only if it is better than T_2 for L_2^*, T_1 is admissible for L_2 if and only if it is admissible for L_2^*, T_1 is a best invariant estimator for L_2 if and only if it is a best invariant estimator or for L_2^* and T_1 is minimax for L_2 if and only if it is minimax for L_2^*. Therefore, using L_2 is the same as using L_2^*.

3.4. THE MULTIVARIATE NORMAL DISTRIBUTION.

We now define the multivariate normal distribution and derive its basic properties. We want to allow the possibility of multivariate normal distributions whose covariance matrix is not necessarily positive definite. Therefore, we cannot define the distribution by its density function (see Lemma 3.2). Instead we define the distribution by its moment generating function. (The reader may wonder how a random vector can have a moment generating function if it has no density function. However, the moment generating function can be defined using more general types of integration. In this book, we assume that such a definition is possible but find the moment generating function by elementary means.) We find the density function for the case of positive definite covariance matrix in Theorem 3.9.

For motivation, we start with $Z_1, \ldots, Z_n$ independent random variables such that $Z_i \sim N_1(0, 1)$. Let $\mathbf{Z} = (Z_1, \ldots, Z_n)'$. Then

$$E(\mathbf{Z}) = 0, \qquad \operatorname{cov}(\mathbf{Z}) = \mathbf{I}, \qquad M_{\mathbf{Z}}(\mathbf{t}) = \prod \exp \frac{t_i^2}{2} = \exp \frac{\mathbf{t}'\mathbf{t}}{2}. \tag{3.2}$$

Let $\boldsymbol{\mu}$ be an $n \times 1$ vector and $\mathbf{A}$ an $n \times n$ matrix. Let $\mathbf{Y} = \mathbf{AZ} + \boldsymbol{\mu}$. Then

$$E(\mathbf{Y}) = \boldsymbol{\mu} \qquad \operatorname{cov}(\mathbf{Y}) = \mathbf{AA}'. \tag{3.3}$$

Let $\boldsymbol{\Sigma} = \mathbf{AA}'$. We now show that the distribution of $\mathbf{Y}$ depends only on $\boldsymbol{\mu}$ and $\boldsymbol{\Sigma}$. The moment generating function $M_{\mathbf{Y}}(\mathbf{t})$ is given by

$$M_{\mathbf{Y}}(\mathbf{t}) = \exp(\boldsymbol{\mu}'\mathbf{t}) M_{\mathbf{Z}}(\mathbf{A}'\mathbf{t}) = \exp\left[\boldsymbol{\mu}'\mathbf{t} + \frac{\mathbf{t}'(\mathbf{A}'\mathbf{A})\mathbf{t}}{2}\right] = \exp\left(\boldsymbol{\mu}'\mathbf{t} + \frac{\mathbf{t}'\boldsymbol{\Sigma}\mathbf{t}}{2}\right).$$

With this motivation in mind, let $\boldsymbol{\mu}$ be an $n \times 1$ vector, and let $\boldsymbol{\Sigma}$ be a nonnegative definite $n \times n$ matrix. Then we say that the n-dimensional random vector $\mathbf{Y}$ has an *n-dimensional normal distribution* with mean vector $\boldsymbol{\mu}$, and covariance matrix $\boldsymbol{\Sigma}$, if $\mathbf{Y}$ has moment generating function

$$M_{\mathbf{Y}}(\mathbf{t}) = \exp\left(\boldsymbol{\mu}'\mathbf{t} + \frac{\mathbf{t}'\boldsymbol{\Sigma}\mathbf{t}}{2}\right). \tag{3.4}$$

We write $\mathbf{Y} \sim N_n(\boldsymbol{\mu}, \boldsymbol{\Sigma})$. The following theorem summarizes some elementary facts about multivariate normal distributions.

THEOREM 3.6. a. If $\mathbf{Y} \sim N_n(\boldsymbol{\mu}, \boldsymbol{\Sigma})$, then $E(\mathbf{Y}) = \boldsymbol{\mu}$, $\operatorname{cov}(\mathbf{Y}) = \boldsymbol{\Sigma}$.

b. If $\mathbf{Y} \sim N_n(\boldsymbol{\mu}, \boldsymbol{\Sigma})$, c is a scalar, then $c\mathbf{Y} \sim N_n(c\boldsymbol{\mu}, c^2\boldsymbol{\Sigma})$.

c. Let $\mathbf{Y} \sim N_n(\boldsymbol{\mu}, \boldsymbol{\Sigma})$. If $\mathbf{A}$ is $p \times n$, $\mathbf{b}$ is $p \times 1$, then $A\mathbf{Y} + \mathbf{b} \sim N_p(\mathbf{A}\boldsymbol{\mu} + \mathbf{b}, \mathbf{A}\boldsymbol{\Sigma}\mathbf{A}')$.

d. Let $\boldsymbol{\mu}$ be any $n \times 1$ vector, and let $\boldsymbol{\Sigma}$ be any $n \times n$ nonnegative definite matrix. Then there exists $\mathbf{Y}$ such that $\mathbf{Y} \sim N_n(\boldsymbol{\mu}, \boldsymbol{\Sigma})$.

PROOF. a. This follows directly from 3.3 above.

b and c. See Exercise B5.

d. Let $Z_1, \ldots, Z_p$ be independent, $Z_i \sim N(0,1)$. Let $\mathbf{Z} = (Z_1, \ldots, Z_p)'$. It is easily verified that $\mathbf{Z} \sim N_p(\mathbf{0}, \mathbf{I})$. Let $\mathbf{Y} = \boldsymbol{\Sigma}^{1/2}\mathbf{Z} + \boldsymbol{\mu}$. By part b, above,

$$\mathbf{Y} \sim N_n(\boldsymbol{\Sigma}^{1/2}\mathbf{0} + \boldsymbol{\mu}, \boldsymbol{\Sigma}) . \quad \square$$

We have now shown that the family of normal distributions is preserved under linear operations on the random vectors. We now show that it is preserved under taking marginal and conditional distributions.

THEOREM 3.7. Suppose that $\mathbf{Y} \sim N_n(\boldsymbol{\mu}, \boldsymbol{\Sigma})$. Let

$$\mathbf{Y} = \begin{pmatrix} \mathbf{Y}_1 \\ \mathbf{Y}_2 \end{pmatrix}, \qquad \boldsymbol{\mu} = \begin{pmatrix} \boldsymbol{\mu}_1 \\ \boldsymbol{\mu}_2 \end{pmatrix}, \qquad \boldsymbol{\Sigma} = \begin{pmatrix} \boldsymbol{\Sigma}_{11} & \boldsymbol{\Sigma}_{12} \\ \boldsymbol{\Sigma}_{21} & \boldsymbol{\Sigma}_{22} \end{pmatrix},$$

where $\mathbf{Y}_1$ and $\boldsymbol{\mu}_1$ are $p \times 1$, and $\boldsymbol{\Sigma}_{11}$ is $p \times p$.

a. $\mathbf{Y}_1 \sim N_p(\boldsymbol{\mu}_1, \boldsymbol{\Sigma}_{11})$, $\mathbf{Y}_2 \sim N_{n-p}(\boldsymbol{\mu}_2, \boldsymbol{\Sigma}_{22})$.

b. $\mathbf{Y}_1$ and $\mathbf{Y}_2$ are independent if and only if $\boldsymbol{\Sigma}_{12} = \mathbf{0}$.

c. If $\boldsymbol{\Sigma}_{22} > 0$, then the condition distribution of $\mathbf{Y}_1$ given $\mathbf{Y}_2$ is

$$\mathbf{Y}_1 \mid \mathbf{Y}_2 \sim N_p\left(\boldsymbol{\mu}_1 + \boldsymbol{\Sigma}_{12}\boldsymbol{\Sigma}_{22}^{-1}(\mathbf{Y}_2 - \boldsymbol{\mu}_2), \boldsymbol{\Sigma}_{11} - \boldsymbol{\Sigma}_{12}\boldsymbol{\Sigma}_{22}^{-1}\boldsymbol{\Sigma}_{21}\right).$$

PROOF. a. Let $\mathbf{t}' = (\mathbf{t}_1', \mathbf{t}_2')$ where $\mathbf{t}_1$ is $p \times 1$. The joint moment generating function of $\mathbf{Y}_1$ and $\mathbf{Y}_2$ is

$$M_{\mathbf{Y}}(\mathbf{t}) = \exp\left(\boldsymbol{\mu}_1'\mathbf{t}_1 + \boldsymbol{\mu}_2'\mathbf{t}_2 + \tfrac{1}{2}(\mathbf{t}_1'\boldsymbol{\Sigma}_{11}\mathbf{t}_1 + \mathbf{t}_1'\boldsymbol{\Sigma}_{12}\mathbf{t}_2 + \mathbf{t}_2'\boldsymbol{\Sigma}_{21}\mathbf{t}_1 + \mathbf{t}_2'\boldsymbol{\Sigma}_{22}\mathbf{t}_2)\right).$$

Therefore,

$$M_{\mathbf{Y}}\begin{pmatrix} \mathbf{t}_1 \\ \mathbf{0} \end{pmatrix} = \exp(\boldsymbol{\mu}_1'\mathbf{t}_1 + \tfrac{1}{2}\mathbf{t}_1'\boldsymbol{\Sigma}_{11}\mathbf{t}_1), \qquad M_{\mathbf{Y}}\begin{pmatrix} \mathbf{0} \\ \mathbf{t}_2 \end{pmatrix} = \exp(\boldsymbol{\mu}_2'\mathbf{t}_2 + \tfrac{1}{2}\mathbf{t}_2'\boldsymbol{\Sigma}_{22}\mathbf{t}_2).$$

By Lemma 3.4c, we see that $\mathbf{Y}_1 \sim N_p(\boldsymbol{\mu}_1, \boldsymbol{\Sigma}_{11})$, $\mathbf{Y}_2 \sim N_{n-p}(\boldsymbol{\mu}_2, \boldsymbol{\Sigma}_{22})$.

b. We note that

$$M_{\mathbf{Y}}(\mathbf{t}) = M_{\mathbf{Y}}\begin{pmatrix} \mathbf{t}_1 \\ \mathbf{0} \end{pmatrix} M_{\mathbf{Y}}\begin{pmatrix} \mathbf{0} \\ \mathbf{t}_2 \end{pmatrix}$$

if and only if

$$\mathbf{t}_1'\boldsymbol{\Sigma}_{12}\mathbf{t}_2 + \mathbf{t}_2'\boldsymbol{\Sigma}_{21}\mathbf{t}_1 = 0.$$

Since $\boldsymbol{\Sigma}$ is symmetric and $\mathbf{t}_2'\boldsymbol{\Sigma}_{21}\mathbf{t}_1$ is a scalar, we see that $\mathbf{t}_2'\boldsymbol{\Sigma}_{21}\mathbf{t}_1 = \mathbf{t}_1'\boldsymbol{\Sigma}_{12}\mathbf{t}_2$.

Finally, $\mathbf{t}_1'\boldsymbol{\Sigma}_{12}\mathbf{t}_2 = 0$ for all $\mathbf{t}_1 \in R^p$, $\mathbf{t}_2 \in R^{n-p}$ if and only if $\boldsymbol{\Sigma}_{12} = 0$, and the result follows from Lemma 3.4d.

c. We first find the joint distribution of

$$\mathbf{X} = \mathbf{Y}_1 - \boldsymbol{\Sigma}_{12}\boldsymbol{\Sigma}_{22}^{-1}\mathbf{Y}_2 \quad \text{and} \quad \mathbf{Y}_2.$$

$$\begin{pmatrix} \mathbf{X} \\ \mathbf{Y}_2 \end{pmatrix} = \begin{pmatrix} \mathbf{I} & -\boldsymbol{\Sigma}_{12}\boldsymbol{\Sigma}_{22}^{-1} \\ \mathbf{0} & \mathbf{I} \end{pmatrix} \begin{pmatrix} \mathbf{Y}_1 \\ \mathbf{Y}_2 \end{pmatrix}$$

Therefore, by Theorem 3.6c, the joint distribution of $\mathbf{X}$ and $\mathbf{Y}_2$ is

$$\begin{pmatrix} \mathbf{X} \\ \mathbf{Y}_2 \end{pmatrix} \sim N_n\left(\begin{pmatrix} \boldsymbol{\mu}_1 - \boldsymbol{\Sigma}_{12}\boldsymbol{\Sigma}_{22}^{-1}\boldsymbol{\mu}_2 \\ \boldsymbol{\mu}_2 \end{pmatrix}, \begin{pmatrix} \boldsymbol{\Sigma}_{11} - \boldsymbol{\Sigma}_{12}\boldsymbol{\Sigma}_{22}^{-1}\boldsymbol{\Sigma}_{21} & \mathbf{0} \\ \mathbf{0} & \boldsymbol{\Sigma}_{22} \end{pmatrix} \right)$$

and hence $\mathbf{X}$ and $\mathbf{Y}_2$ are independent. Therefore, the conditional distribution of $\mathbf{X}$ given $\mathbf{Y}_2$ is the same as the marginal distribution of $\mathbf{X}$,

$$\mathbf{X} \mid \mathbf{Y}_2 \sim N_p(\boldsymbol{\mu}_1 - \boldsymbol{\Sigma}_{12}\boldsymbol{\Sigma}_{22}^{-1}\boldsymbol{\mu}_2, \boldsymbol{\Sigma}_{11} - \boldsymbol{\Sigma}_{12}\boldsymbol{\Sigma}_{22}^{-1}\boldsymbol{\Sigma}_{21}).$$

Since $\mathbf{Y}_2$ is just a constant in the conditional distribution of $\mathbf{X}$ given $\mathbf{Y}_2$ we have, by Theorem 3.6c, that the conditional distribution of $\mathbf{Y}_1 = \mathbf{X} + \boldsymbol{\Sigma}_{12}\boldsymbol{\Sigma}_{22}^{-1}\mathbf{Y}_2$ given $\mathbf{Y}_2$ is

$$\mathbf{Y}_1 \mid \mathbf{Y}_2 \sim N_p(\boldsymbol{\mu}_1 - \boldsymbol{\Sigma}_{12}\boldsymbol{\Sigma}_{22}^{-1}\boldsymbol{\mu}_2 + \boldsymbol{\Sigma}_{12}\boldsymbol{\Sigma}_{22}^{-1}\mathbf{Y}_2, \boldsymbol{\Sigma}_{11} - \boldsymbol{\Sigma}_{12}\boldsymbol{\Sigma}_{22}^{-1}\boldsymbol{\Sigma}_{21}). \quad \square$$

Note that we need $\boldsymbol{\Sigma}_{22} > 0$ in part c so that $\boldsymbol{\Sigma}_{22}^{-1}$ exists.

Until now in the development of properties of the multivariate normal distribution, we have worked with the moment generating function There are two reasons for this approach. The first reason is that if $\boldsymbol{\Sigma}$ is not positive definite, then there is no density function (see Lemma 3.2). The second reason is that the moment generating function is the natural tool to use to prove Theorems 3.6 and 3.7. If $\mathbf{Y} \sim N_n(\boldsymbol{\mu}, \boldsymbol{\Sigma})$ and $\boldsymbol{\Sigma}$ is positive definite, we say that $\mathbf{Y}$ has a *nonsingular multivariate normal distribution*. If $\boldsymbol{\Sigma}$ is not positive definite, we say that $\mathbf{Y}$ has a *singular multivariate normal distribution*. To allow the possibility of singular normal distributions, we cannot work with the density function.

We now give some examples of singular multivariate normal distributions. We then find the density function for nonsingular multivariate normal distributions.

EXAMPLE 1. Let $Y \sim N_1(\boldsymbol{\mu}, \boldsymbol{\Sigma})$. Then $\boldsymbol{\Sigma}$ is just a real number. Therefore, the only singular univariate normal distributions occur when $\boldsymbol{\Sigma} = 0$. By the definition of the multivariate normal distribution, we see that if $\boldsymbol{\Sigma} = 0$, then the moment generating function of Y is $M_Y(t) = \exp(\mu t)$, which is the moment generating function of a random variable such that $P(Y = \mu) = 1$. That is, Y is degenerate at μ. Therefore, the only univariate singular normal distribution is a discrete distribution that is degenerate at a point.

EXAMPLE 2. Let $\mathbf{Y} \sim N_2(\boldsymbol{\mu}, \boldsymbol{\Sigma})$. Let $\boldsymbol{\Sigma} = \begin{pmatrix} a & b \\ b & c \end{pmatrix}$. Then $\boldsymbol{\Sigma} \geqslant 0$ if and only if $a \geqslant 0$, $c \geqslant 0$, $ac - b^2 \geqslant 0$. $\boldsymbol{\Sigma} > 0$ if and only if $a > 0$, $c > 0$, $ac - b^2 > 0$. Therefore, all bivariate singular normal distributions are of the following four types; (1) Both $a = 0$ and $c = 0$ (and therefore $b = 0$); (2) $a = 0$, $c > 0$ (and therefore $b = 0$); (3) $a > 0$, $c = 0$ (and $b = 0$); or (4) $a > 0$, $c > 0$, $b^2 = ac$.

(1) Suppose first that $a = 0$, $b = 0$, $c = 0$. Then $\mathbf{Y} \sim N_2(\boldsymbol{\mu}, 0)$, and as in Example 1, $P(\mathbf{Y} = \boldsymbol{\mu}) = 1$. Hence $\mathbf{Y}$ is degenerate at $\boldsymbol{\mu}$.

(2, 3) If $a = 0$, $b = 0$, $c > 0$, then $Y_1 \sim N_1(\mu_1, 0)$, and $Y_2 \sim N_1(\mu_2, c)$, so that Y_1 is degenerate, but Y_2 is not. ALso, Y_1 and Y_2 are independent. (In fact a degenerate random variable is independent of any other random variable.) The third type is similar to this one.

(4) We now consider the fourth type. Suppose that $a > 0$, $c > 0$, but $b^2 = ac$. Then neither Y_1 nor Y_2 has a degenerate distribution. However, $\text{var}(cY_1 - bY_2) = 0$, so $P(cY_1 - bY_2 = c\mu_1 - b\mu_2) = 1$, and therefore all the probability lies on the line $cy_1 - by_2 = c\mu_1 - b\mu_2$.

EXAMPLE 3. Suppose that $\mathbf{Y} \sim N_3(\boldsymbol{\mu}, \boldsymbol{\Sigma})$. Then it can be shown that if $\boldsymbol{\Sigma} \not> 0$, then all the probability lies either in a point, a line, or a plane. Note that all these sets are subspaces that have been shifted (so that they no longer necessarily go through the origin). Such sets are sometimes called affine subspaces.

Before deriving the density function for the nonsingular multivariate normal distribution, we give a simple lemma.

LEMMA 3.8. Let $\mathbf{Y} \sim N_n(\boldsymbol{\mu}, \sigma^2 \mathbf{I})$, where $\mathbf{Y}' = (Y_1, \ldots, Y_n)$, $\boldsymbol{\mu}' = (\mu_1, \ldots, \mu_n)$ and $\sigma^2 > 0$ is a scalar. Then the Y_i are independent, $Y_i \sim N_1(\mu_i, \sigma^2)$ and

$$\frac{\|\mathbf{Y}\|^2}{\sigma^2} = \frac{\mathbf{Y}'\mathbf{Y}}{\sigma^2} \sim \chi_n^2\left(\frac{\boldsymbol{\mu}'\boldsymbol{\mu}}{\sigma^2}\right).$$

Proof. Let Y_i be independent, $Y_i \sim N_1(\mu_i, \sigma^2)$. The joint moment generating function of the Y_i is

$$M_{\mathbf{Y}}(\mathbf{t}) = \prod_{i=1}^{n} \left(\exp(\mu_i t_i + \tfrac{1}{2}\sigma^2 t_i^2)\right) = \exp(\boldsymbol{\mu}'\mathbf{t} + \tfrac{1}{2}\sigma^2 \mathbf{t}'\mathbf{t})$$

which is the moment generating function of a random vector that is normally distributed with mean vector $\boldsymbol{\mu}$ and covariance matrix $\sigma^2 \mathbf{I}$. Finally, $\mathbf{Y}'\mathbf{Y} = \Sigma Y_i^2$, $\boldsymbol{\mu}'\boldsymbol{\mu} = \Sigma \mu_i^2$ and $Y_i/\sigma \sim N_1(\mu_i/\sigma, 1)$. Therefore $\mathbf{Y}'\mathbf{Y}/\sigma^2 \sim \chi_n^2(\boldsymbol{\mu}'\boldsymbol{\mu}/\sigma^2)$ by the definition of the noncentral χ^2 distribution (see Section 1.4). □

We are now ready to derive the nonsingular normal density function.

THEOREM 3.9. Let $\mathbf{Y} \sim N_n(\boldsymbol{\mu}, \boldsymbol{\Sigma})$, with $\boldsymbol{\Sigma} > 0$. Then $\mathbf{Y}$ has density function

$$f_{\mathbf{Y}}(\mathbf{y}) = \frac{1}{(2\pi)^{n/2}|\boldsymbol{\Sigma}|^{1/2}} \exp\left(-\tfrac{1}{2}(\mathbf{y}-\boldsymbol{\mu})'\boldsymbol{\Sigma}^{-1}(\mathbf{y}-\boldsymbol{\mu})\right).$$

PROOF. We could derive this by finding the moment generating function of this density and showing that it satisfied (3.4). We would also have to show that this function is a density function. We can avoid all that by starting with a random vector whose distribution we know. Let

$$\mathbf{Z} \sim N_n(\mathbf{0}, \mathbf{I}), \qquad \mathbf{Z} = (Z_1, \ldots, Z_n)'.$$

Then the Z_i are independent and $Z_i \sim N_1(0, 1)$, by Lemma 3.8. Therefore, the joint density of the Z_i is

$$f_{\mathbf{Z}}(\mathbf{z}) = \prod_{i=1}^{n} \frac{1}{(2\pi)^{1/2}} \exp\left(-\tfrac{1}{2}z_i^2\right) = \frac{1}{(2\pi)^{n/2}} \exp\left(-\tfrac{1}{2}\mathbf{z}'\mathbf{z}\right).$$

Let $\mathbf{Y} = \boldsymbol{\Sigma}^{1/2}\mathbf{Z} + \boldsymbol{\mu}$. By Theorem 3.6c, $\mathbf{Y} \sim N_n(\boldsymbol{\mu}, \boldsymbol{\Sigma})$. Also $\mathbf{Z} = \boldsymbol{\Sigma}^{-1/2}(\mathbf{Y} - \boldsymbol{\mu})$, and the transformation from $\mathbf{Z}$ to $\mathbf{Y}$ is therefore invertible. Futhermore, the Jacobian of this inverse transformation is just $|\boldsymbol{\Sigma}^{-1/2}| = |\boldsymbol{\Sigma}|^{-1/2}$ (see Exercise B3). Hence the density of $\mathbf{Y}$ is

$$\begin{aligned} f_{\mathbf{Y}}(\mathbf{y}) &= f_{\mathbf{Z}}\left(\boldsymbol{\Sigma}^{-1/2}(\mathbf{y}-\boldsymbol{\mu})\right)\frac{1}{|\boldsymbol{\Sigma}|^{1/2}} \\ &= \frac{1}{|\boldsymbol{\Sigma}|^{1/2}(2\pi)^{n/2}} \exp\left(-\tfrac{1}{2}(\mathbf{y}-\boldsymbol{\mu})'\boldsymbol{\Sigma}^{-1}(\mathbf{y}-\boldsymbol{\mu})\right). \quad \square \end{aligned}$$

We now prove a result that is useful later in the book and is also the basis for Pearson's χ^2 tests.

THEOREM 3.10. Let $\mathbf{Y} \sim N_n(\boldsymbol{\mu}, \boldsymbol{\Sigma})$, $\boldsymbol{\Sigma} > 0$. Then

a. $\mathbf{Y}'\boldsymbol{\Sigma}^{-1}\mathbf{Y} \sim \chi^2(\boldsymbol{\mu}'\boldsymbol{\Sigma}^{-1}\boldsymbol{\mu})$.

b. $(\mathbf{Y} - \boldsymbol{\mu})'\boldsymbol{\Sigma}^{-1}(\mathbf{Y} - \boldsymbol{\mu}) \sim \chi_n^2(0)$.

PROOF. a. Let $\mathbf{Z} = \boldsymbol{\Sigma}^{-1/2}\mathbf{Y} \sim N_n(\boldsymbol{\Sigma}^{-1/2}\boldsymbol{\mu}, \mathbf{I})$. By Lemma 3.8, we see that

$$\mathbf{Z}'\mathbf{Z} = \mathbf{Y}'\boldsymbol{\Sigma}^{-1}\mathbf{Y} \sim \chi_n^2(\boldsymbol{\mu}'\boldsymbol{\Sigma}^{-1}\boldsymbol{\mu}).$$

b. See Exercise B7. $\square$

3.5. THE SPHERICAL NORMAL DISTRIBUTION.

For the first part of this book, the most important class of multivariate normal distribution is the class in which

$$\mathbf{Y} \sim N_n(\boldsymbol{\mu}, \sigma^2\mathbf{I}).$$

We now show that this distribution is spherically symmetric about $\boldsymbol{\mu}$. A rotation about $\boldsymbol{\mu}$ is given by $\mathbf{X} = \boldsymbol{\Gamma}(\mathbf{Y} - \boldsymbol{\mu}) + \boldsymbol{\mu}$, where $\boldsymbol{\Gamma}$ is an orthogonal matrix (i.e., $\boldsymbol{\Gamma}\boldsymbol{\Gamma}' = \mathbf{I}$). By Theorem 3.6, $\mathbf{X} \sim N_n(\boldsymbol{\mu}, \sigma^2\mathbf{I})$, so that the distribution is unchanged under rotations about $\boldsymbol{\mu}$. We therefore call this normal distribution [with $\mathrm{cov}(\mathrm{Y}) = \sigma^2 I$] the *spherical normal distribution*. If $\sigma^2 = 0$, then $P(\mathbf{Y} = \boldsymbol{\mu}) = 1$. Otherwise its density function (by Theorem 3.9) is

$$f_{\mathbf{Y}}(\mathbf{y}) = \frac{1}{(2\pi)^{n/2}\sigma^n} \exp\left(-\frac{1}{2\sigma^2}\|\mathbf{y} - \boldsymbol{\mu}\|^2\right).$$

By Lemma 3.8, we note that the components of Y are independently normally distributed with common variance σ^2. In fact, the spherical normal distribution is the only multivariate distribution with independent components that is spherically symmetric. (See Exercise C4.)

We now give some simple theorems about projections and spherical normal distributions.

THEOREM 3.11. Let $\mathbf{Y} \sim N_n(\boldsymbol{\mu}, \sigma^2\mathbf{I})$, $\sigma^2 > 0$. Then

a. $\mathbf{P}_V\mathbf{Y} \sim N_n(\mathbf{P}_V\boldsymbol{\mu}, \sigma^2\mathbf{P}_V)$.

b. If $V \perp W$, then $\mathbf{P}_V\mathbf{Y}$ and $\mathbf{P}_W\mathbf{Y}$ are independent.

PROOF. See Exercise B8. □

THEOREM 3.12. Let $\mathbf{Y} \sim N_n(\boldsymbol{\mu}, \sigma^2\mathbf{I}), \sigma^2 > 0$. Let V be a p-dimensional subspace of R^n. Then

$$\|\mathbf{P}_V\mathbf{Y}\|^2 \sim \sigma^2\chi_p^2\left(\frac{\|\mathbf{P}_V\boldsymbol{\mu}\|^2}{\sigma^2}\right).$$

PROOF. Let $\mathbf{X}$ be an orthonormal basis matrix for V. Let

$$\mathbf{Z} = \frac{1}{\sigma}\mathbf{X}'\mathbf{Y}, \qquad \boldsymbol{\nu} = \frac{1}{\sigma}\mathbf{X}'\boldsymbol{\mu}.$$

Since $\mathbf{X}'\mathbf{X} = \mathbf{I}$, $\mathbf{Z} \sim N_p(\boldsymbol{\nu}, \mathbf{I})$. By Lemma 3.8, we see that $\mathbf{Z}'\mathbf{Z} \sim \chi_p^2(\boldsymbol{\nu}'\boldsymbol{\nu})$. However,

$$\mathbf{Z}'\mathbf{Z} = \|\mathbf{Z}\|^2 = \frac{1}{\sigma^2}\|\mathbf{X}'\mathbf{Y}\|^2 = \frac{1}{\sigma^2}\|\mathbf{P}_V\mathbf{Y}\|^2,$$

by Theorem 2.6c. Similarly, $\boldsymbol{\nu}'\boldsymbol{\nu} = \|\mathbf{P}_V\boldsymbol{\mu}\|^2/\sigma^2$. □

COROLLARY. Let $\mathbf{Y} \sim N_n(\boldsymbol{\mu}, \sigma^2\mathbf{I})$, $\sigma^2 > 0$. If $\mathbf{A}$ is idempotent and rank $(\mathbf{A}) = p$, then

$$\mathbf{Y}'\mathbf{A}\mathbf{Y} \sim \sigma^2\chi_p^2\left(\frac{\boldsymbol{\mu}'\mathbf{A}\boldsymbol{\mu}}{\sigma^2}\right).$$

PROOF. This follows directly from Theorem 2.10. □

Although we do not use it in this book, the converse of this corollary is also true. That is, if $\mathbf{Y} \sim N_n(\boldsymbol{\mu}, \sigma^2\mathbf{I})$, and if $\mathbf{A}$ is a symmetric matrix such that $\mathbf{Y}'\mathbf{A}\mathbf{Y}/\sigma^2 \sim \chi_p^2(\boldsymbol{\mu}'\mathbf{A}\boldsymbol{\mu}/\sigma^2)$, then $\mathbf{A}$ is an idempotent matrix of rank p. See Hogg and Craig (1970, pp. 384–387) for a proof. By Theorem 2.10, therefore, the only quadratic forms that have χ^2 distributions are squared lengths of projections.

The final result in this section is often used to derive distribution theory for linear models.

THEOREM 3.13. Let $\mathbf{Y} \sim N_n(\boldsymbol{\mu}, \sigma^2\mathbf{I})$, $\sigma^2 > 0$, let $\mathbf{A}$ and $\mathbf{B}$ be $p \times n$ and $s \times n$ matrices and let $\mathbf{C}$ and $\mathbf{D}$ be $n \times n$ nonnegative definite matrices. Then

a. $\mathbf{AY}$ and $\mathbf{BY}$ are independent if and only if $\mathbf{AB}' = \mathbf{0}$
b. If $\mathbf{AC} = \mathbf{0}$, then $\mathbf{AY}$ and $\mathbf{Y}'\mathbf{CY}$ are independent.
c. If $\mathbf{CD} = \mathbf{0}$, then $\mathbf{Y}'\mathbf{CY}$ and $\mathbf{Y}'\mathbf{DY}$ are independent.

PROOF. a. Lemma 3.6c, we see that

$$\begin{pmatrix} \mathbf{AY} \\ \mathbf{BY} \end{pmatrix} = \begin{pmatrix} \mathbf{A} \\ \mathbf{B} \end{pmatrix} \mathbf{Y} \sim N_{t+s}\left(\begin{pmatrix} \mathbf{A}\boldsymbol{\mu} \\ \mathbf{B}\boldsymbol{\mu} \end{pmatrix}, \sigma^2 \begin{pmatrix} \mathbf{AA}' & \mathbf{AB}' \\ \mathbf{B}'\mathbf{A} & \mathbf{BB}' \end{pmatrix} \right).$$

Therefore, by Lemma 3.7b, we see that $\mathbf{AY}$ and $\mathbf{BY}$ are independent if and only if $\mathbf{AB}' = 0$.

b. Let $\mathbf{C}$ have rank r and let $\mathbf{E}$ be an $r \times n$ matrix of rank r such that $\mathbf{C} = \mathbf{E}'\mathbf{E}$. Then $\mathbf{AC} = \mathbf{0}$, and hence $\mathbf{ACE}' = \mathbf{AE}'\mathbf{EE}' = \mathbf{0}$. However, $\mathbf{EE}'$ is an $r \times r$ matrix of rank r and is invertible. Therefore, $\mathbf{AE}' = 0$. Hence $\mathbf{AY}$ and $\mathbf{EY}$ are independent by part a. Therefore, $\mathbf{AY}$ and $(\mathbf{EY})'(\mathbf{EY}) = \mathbf{Y}'\mathbf{E}'\mathbf{EY}$ $= \mathbf{Y}'\mathbf{CY}$ are independent.

c. See Exercise B9. □

It should be mentioned that idempotent matrices are nonnegative definite (see the exercises) so parts b and c apply to idempotent matrices. Parts b and c are actually true for arbitrary symmetric matrices $\mathbf{C}$ and $\mathbf{D}$. The converses are also true. See Hogg and Craig (1970, pp. 388–393) for a proof of part c and its converse in this more general setting.

EXERCISES.

Type A

1. Suppose that

$$\mathbf{Y} = \begin{pmatrix} Y_1 \\ Y_2 \\ Y_3 \\ Y_4 \end{pmatrix} \sim N_3\left(\begin{pmatrix} 1 \\ 2 \\ 3 \\ 4 \end{pmatrix}, \begin{pmatrix} 4 & 3 & 2 & 1 \\ 3 & 3 & 2 & 1 \\ 2 & 2 & 2 & 1 \\ 1 & 1 & 1 & 1 \end{pmatrix} \right)$$

(a) Find the joint distribution of $X_1 = Y_1 + Y_2 + Y_3 + Y_4$ and $X_2 = Y_1 - Y_2 - Y_3 + Y_4$.

(b) Are they independent? How is X_1 distributed?

(c) Find the marginal distribution of $(Y_1, Y_2)'$ and the joint distribution of $(Y_1, Y_3)'$.

(d). Find the conditional distribution of $(Y_1, Y_2)'$ given $(Y_3, Y_4)'$.

(e). Find the conditional distribution of $(Y_1, Y_2)'$ given Y_3.

Type B

1. Prove Lemma 3.1.

2. Let $\mathbf{A}$ be an idempotent matrix. Show that $\mathbf{A} \geqslant 0$ and that $\mathbf{A} > 0$ if and only if $\mathbf{A} = \mathbf{I}$. (Hint: If $\mathbf{A} > 0$ then $\mathbf{A}$ is invertible.)

3. Let $\mathbf{A} > 0$. Show the following

$$(\mathbf{A}^{-1})^{1/2} = (\mathbf{A}^{1/2})^{-1}, \qquad \mathbf{A}^{-1/2}\mathbf{A}\mathbf{A}^{-1/2} = \mathbf{I}, \qquad (\mathbf{A}^{-1/2})' = \mathbf{A}^{-1/2},$$

$$(\mathbf{A}^{1/2})' = \mathbf{A}^{1/2}, \qquad |\mathbf{A}^{-1/2}| = |\mathbf{A}|^{-1/2}$$

(Hint: $|\mathbf{A}| = |(\mathbf{A}^{1/2})^2| = |\mathbf{A}^{1/2}|^2$.)

4. Prove part b of Lemma 3.4.

5. Prove parts b and c of Theorem 3.6.

6. Let

$$\begin{pmatrix} \mathbf{Y}_1 \\ \mathbf{Y}_2 \end{pmatrix} \sim N_n\left(\begin{pmatrix} \boldsymbol{\mu}_1 \\ \boldsymbol{\mu}_2 \end{pmatrix}, \begin{pmatrix} \boldsymbol{\Sigma}_{11} & \boldsymbol{\Sigma}_{12} \\ \boldsymbol{\Sigma}_{21} & \boldsymbol{\Sigma}_{22} \end{pmatrix}\right)$$

where $\mathbf{Y}_1$ and $\boldsymbol{\mu}_1$ are $p \times 1$ and $\boldsymbol{\Sigma}_{11}$ is $p \times p$. Suppose that $\boldsymbol{\Sigma}_{22}$ is not positive definite. Let $\mathbf{A}$ be a matrix such that $\mathbf{A}\boldsymbol{\Sigma}_{22} = \boldsymbol{\Sigma}_{12}$. Show that

$$\mathbf{Y}_1 | \mathbf{Y}_2 \sim N_p(\boldsymbol{\mu}_1 + \mathbf{A}(\mathbf{Y}_2 - \boldsymbol{\mu}_2), \boldsymbol{\Sigma}_{11} - \mathbf{A}\boldsymbol{\Sigma}_{22}\mathbf{A}').$$

(Hint: How are $\mathbf{Y}_1 - \mathbf{A}\mathbf{Y}_2$ and $\mathbf{Y}_2$ distributed?)

7. Prove part b of Theorem 3.10.

8. Prove Theorem 3.11. (Hint: How is $\binom{\mathbf{P}_V \mathbf{Y}}{\mathbf{P}_W \mathbf{Y}}$ distributed?)

9. Prove part c of Theorem 3.13.

10. Let $\mathbf{Y} \sim N_n(\boldsymbol{\mu}, \sigma^2\mathbf{I})$, where $\boldsymbol{\mu}' = (\theta, \ldots, \theta)$. Let $\mathbf{Y}' = (Y_1, \ldots, Y_n)$. Then the Y_i are independent, $Y_i \sim N_1(\theta, \sigma^2)$. Let

$$\overline{Y} = \frac{1}{n}\sum_{i=1}^{n} Y_i, \qquad S^2 = \frac{1}{n-1}\sum_{i=1}^{n}\left(Y_i - \overline{Y}\right)^2$$

(a) Let e be an $n \times 1$ vector of 1's. Show that $\overline{Y} = \mathbf{e}'\mathbf{Y}/n$ and that

$$(n-1)S^2 = \mathbf{Y}'\left(\mathbf{I} - \frac{1}{n}\mathbf{ee}'\right)\mathbf{Y}.$$

(b) Using Theorem 3.13, show that $\overline{Y}$ and S^2 are independent.

(c) Using the corollary to Theorem 3.12, show that $(n-1)S^2 \sim \sigma^2\chi^2_{n-1}(0)$.

(d) Using Theorem 3.6b, show that $\overline{Y} \sim N_1(\theta, \sigma^2/n)$.

(e) Now let V be the one-dimensional subspace spanned by **e**. Show that

$$\mathbf{P}_V\mathbf{Y} = \begin{bmatrix} \overline{Y} \\ \vdots \\ \overline{Y} \end{bmatrix}, \qquad \|\mathbf{P}_{V^\perp}\mathbf{Y}\|^2 = (n-1)S^2.$$

Reinterpret the results of parts b–d in the light of part e.

Type C

1. Let $\mathbf{Y}' = (\mathbf{Y}_1', \ldots, \mathbf{Y}_k')$ where Y_i is $p_i \times 1$. Suppose that **Y** has a multivariate normal distribution. Show that the $\mathbf{Y}_i$ are mutually independent if and only if they are pairwise independent.

2. Let $\mathbf{Y} \sim N_n(\boldsymbol{\mu}, \boldsymbol{\Sigma})$, $\boldsymbol{\Sigma} > 0$.

(a) Let **A** be a symmetric matrix of rank p such that $\mathbf{A\Sigma A} = \mathbf{A}$. Show that

$$(\mathbf{Y} - \mathbf{c})'\mathbf{A}(\mathbf{Y} - \mathbf{c}) \sim \chi^2_p((\boldsymbol{\mu} - \mathbf{c})'\mathbf{A}(\boldsymbol{\mu} - \mathbf{c})).$$

(Hint Let $\mathbf{X} = \boldsymbol{\Sigma}^{-1/2}(\mathbf{Y} - \mathbf{c})$. Then $(\mathbf{Y} - \mathbf{c})'\mathbf{A}(\mathbf{Y} - \mathbf{c}) = \mathbf{X}'\boldsymbol{\Sigma}^{1/2}\mathbf{A}\boldsymbol{\Sigma}^{1/2}\mathbf{X}$ and $\boldsymbol{\Sigma}^{1/2}\mathbf{A}\boldsymbol{\Sigma}^{1/2}$ is idempotent if and only if $\mathbf{A\Sigma A} = \mathbf{A}$.)

(b) Show that Theorems 3.10 and 3.12 follow directly from the above results.

(c) Show that **CY** and **DY** are independent if and only if $\mathbf{C\Sigma D}' = 0$.

(d) Let $\mathbf{D} \geqslant 0$. Show that **CY** and $\mathbf{Y}'\mathbf{DY}$ are independent if $\mathbf{C\Sigma D} = 0$.

(e) Show that if $\mathbf{A\Sigma A} = \mathbf{A}$, then $\mathbf{A} \geqslant 0$.

3. Let **Y** be an n-dimensional random variable such that $\mathbf{t}'\mathbf{Y}$ has a normal distribution for all $\mathbf{t} \neq 0$.

(a) Show that **Y** has a finite mean vector and finite covariance matrix. (Hint: Let $\mathbf{t}_i' = (0, \ldots, 0, 1, 0, \ldots, 0)$ to get finite variances and means and use Cauchy-Schwartz to get finite covariances.)

(b) Let $\boldsymbol{\mu} = E\mathbf{Y}$, $\boldsymbol{\Sigma} = \text{cov}(\mathbf{Y})$. Show that $\mathbf{t}'\mathbf{Y} \sim N_1(\mathbf{t}'\boldsymbol{\mu}, \mathbf{t}'\boldsymbol{\Sigma}\mathbf{t})$. (See Lemma 3.1.)

(c) Show that $\mathbf{Y} \sim N_n(\boldsymbol{\mu}, \boldsymbol{\Sigma})$. (Hint: $M_{\mathbf{Y}}(\mathbf{t}) = M_{\mathbf{t}'\mathbf{Y}}(1)$. Why?)

4. Let $\mathbf{Y} \in R^n$ be a random vector with moment generating function

$M(\mathbf{t})$, $\mathbf{t} \in R^n$. Suppose that $\mathbf{\Gamma Y}$ has the same distribution as $\mathbf{Y}$ for all orthogonal matrices $\mathbf{\Gamma}$.

(a) Show that $M(\mathbf{\Gamma t}) = M(\mathbf{t})$ for all orthogonal matrices $\mathbf{\Gamma}$.

(b) Show that $\|\mathbf{t}\|^2$ is a maximal invariant under the group G of transformations $g(\mathbf{t}) = \mathbf{\Gamma t}$, where $\mathbf{\Gamma}$ is orthogonal. (See Theorem 11 of the Appendix.)

(c) Show that $M(\mathbf{t}) = h(\|\mathbf{t}\|^2)$ for some function h. (See Lemma 1.11.)

(d) Suppose that, in addition, the components of $\mathbf{Y}$ are independent. Show that $h(u + v) = h(u)h(v)$ for $u \geqslant 0$, $v \geqslant 0$. [Let $\mathbf{Y} = (Y_1, \ldots, Y_n)'$. Then $M_{Y_1}(t_1) = h(t_1^2)$, $M_{Y_2}(t_2) = h(t_2^2)$ and $M_{Y_1, Y_2}(t_1, t_2) = h(t_1^2 + t_2^2)$.]

(e) Show that $h(u) = \exp c\mu/2$ for some c. (See Feller, 1957 p. 413.)

(f) Show that $c \geqslant 0$. (Note that $c = EY_1^2$. Why?)

(g) Show that $\mathbf{Y} \sim N_n(\mathbf{0}, c\mathbf{I})$.

(h) Now suppose that $\mathbf{X} \in R^n$ is a random vector and that there exists a constant vector $a \in R^n$ such that the distribution of $\mathbf{\Gamma}(\mathbf{X} - \mathbf{a}) + \mathbf{a}$ is the same as the distribution of $\mathbf{X}$. Let $\mathbf{Y} = \mathbf{X} - \mathbf{a}$. Show that the distribution of $\mathbf{\Gamma Y}$ is the same as the distribution of $\mathbf{Y}$ for all orthogonal $\mathbf{\Gamma}$.

(This derivation assumes that $\mathbf{X}$ has a moment generating function. The result can be extended with trivial modifications to the more general case if the moment generating function is replaced by the characteristic function.)

CHAPTER 4

Introduction to Linear Models

In this chapter we define the linear model. Let V be a given p-dimensional subspace of R^n. Suppose that we observe the random vector

$$\mathbf{Y} \sim N_n(\boldsymbol{\mu}, \sigma^2 \mathbf{I}), \qquad \boldsymbol{\mu} \in V, \qquad \sigma^2 > 0. \tag{4.1}$$

This model is called the (ordinary) *linear model.*

Let $\mathbf{X}$ be a basis matrix for V. Then $\mathbf{X}$ is an $n \times p$ matrix of rank p, and there exists a unique $\boldsymbol{\beta} \in R^p$ such that $\boldsymbol{\mu} = \mathbf{X}\boldsymbol{\beta}$. The relationship between $\boldsymbol{\mu}$ and $\boldsymbol{\beta}$ is given by

$$\boldsymbol{\mu} = \mathbf{X}\boldsymbol{\beta}, \qquad \boldsymbol{\beta} = (\mathbf{X}'\mathbf{X})^{-1}\mathbf{X}'\boldsymbol{\mu}. \tag{4.2}$$

Since the transformation from $\boldsymbol{\mu}$ to $\boldsymbol{\beta}$ is an invertible transformation from V to R^p, $\boldsymbol{\beta}$ is just a reparametrization of the problem. Also, note that every subspace V has a basis matrix, and every $n \times p$ matrix of rank p is a basis matrix for the space spanned by its columns. Therefore, an equivalent version of the linear model occurs when we choose an $n \times p$ matrix $\mathbf{X}$ that has rank p, and we observe the random vector

$$\mathbf{Y} \sim N_n(\mathbf{X}\boldsymbol{\beta}, \sigma^2 \mathbf{I}), \qquad \boldsymbol{\beta} \in R^p, \qquad \sigma^2 > 0. \tag{4.3}$$

We call the model defined by (4.3) the *coordinatized* version of the linear model, since in choosing a basis matrix, we have coordinatized the space V. We call the version of the model given in (4.1) the *coordinate-free* version of the linear model.

In this book we use the coordinate-free version, although results for the coordinate-free version are often restated for the coordinatized version. The following is a list of some reasons that the coordinate-free version of the problem is more appealing to study.

(1) It is not necessary to choose a basis matrix $\mathbf{X}$. All that is necessary is to describe a subspace V. This subspace can be described in two rather different ways. We can first pick a basis matrix, and then

define V as the space spanned by its columns. We are then in a coordinatized version of the linear model. However, we can also define a subspace by defining a set of linear constraints that the elements of the subspace must satisfy. The second approach seems more reasonable for analysis of variance problems, since for those problems, there is often no natural basis matrix for V.

(2) Most statistics used in the coordinate-free version of the problem can be expressed as projections or lengths of projections. Since the projections do not depend on the basis matrix chosen, we can often act as though the basis matrix were orthogonal, which greatly simplifies the proofs. (See, for example, the proof of Theorem 3.12.)

(3) The formulas for the statistics are much easier to interpret when expressed as projections than when expressed in matrices. Choosing a particular basis often obscures the intuition.

(4) Many of the formulas for particular analysis of variance problems are easier to derive using the least squares property of projections (see Theorem 2.5b), rather than the equivalent matrix expression. Both approaches are available with the coordinate-free approach.

In the next chapter, complete sufficient statistics are found for the general linear model. In Chapter 6, minimum variance unbiased estimators and maximum likelihood estimators are found for $\boldsymbol{\mu}$ and σ^2 (also for $\boldsymbol{\beta}$). We also prove the Gauss-Markov theorem. Chapter 7 discusses F-tests and t-tests, and establishes that they are UMP invariant.

Chapter 8 covers simultaneous confidence intervals, based on the F-distribution. Chapter 9 is about tests concerning σ^2. In Chapter 10, we establish under fairly general conditions that asymptotically (as $n \to \infty$), the F-tests and t-tests are not sensitive to the assumption of normality nor are the simultaneous confidence intervals. Chapter 11 considers estimators (James-Stein and ridge) that are, in some sense, better than the minimum variance unbiased estimators discussed in Chapter 6. In Chapter 12, alternative methods of finding confidence intervals and tests (based on the studentized range and Bonferroni's inequality) are considered. There is also a discussion of procedures for multiple comparisons. In Chapters 13–16, we consider other models that are closely related to the linear model defined in this section.

4.1. EXAMPLES.

We now look at some examples of linear models.

EXAMPLE 1 (The one-sample model). Let $Y_1, \ldots, Y_n$ be independent with $Y_i \sim N_1(\theta, \sigma^2)$, $-\infty < \theta < \infty$, $\sigma^2 > 0$. We now show that this is a particular case of a linear model. Let

$$\mathbf{Y}=\begin{pmatrix}Y_1\\ \vdots\\ Y_n\end{pmatrix},\qquad \mathbf{u}=\begin{pmatrix}\theta\\ \vdots\\ \theta\end{pmatrix}.$$

Then $\mathbf{Y}\sim N_n(\boldsymbol{\mu},\sigma^2\mathbf{I})$. Now, let V_1 be the one-dimensional subspace of R^n consisting of vectors all of whose components are the same. Then $\boldsymbol{\mu}=(\theta,\ldots,\theta)'$ for some θ if and only if $\boldsymbol{\mu}\in V_1$. Therefore, an equivalent version of this model is one in which we observe

$$\mathbf{Y}\sim N_n(\boldsymbol{\mu},\sigma^2\mathbf{I}),\qquad \boldsymbol{\mu}\in V_1,\qquad \sigma^2>0,$$

so that this model is a particular case of a linear model.

EXAMPLE 2 (The k-sample model). Let Y_{ij} be independent, $Y_{ij}\sim N_1(\theta_i,\sigma^2)$, $-\infty<\theta_i<\infty$, $\sigma^2>0$, for $i=1,\ldots,k$; $j=1,\ldots,n_i$. To show that this is a special case of a linear model, let

$$\mathbf{Y}=\begin{pmatrix}Y_{11}\\ \vdots\\ Y_{1n_1}\\ Y_{21}\\ \vdots\\ Y_{kn_k}\end{pmatrix},\qquad \boldsymbol{\mu}=\begin{pmatrix}\theta_1\\ \vdots\\ \theta_1\\ \theta_2\\ \vdots\\ \theta_k\end{pmatrix},\qquad n=\sum_i n_i.$$

The $\mathbf{Y}\sim N_n(\boldsymbol{\mu},\sigma^2\mathbf{I})$. Let V_2 be the k-dimensional subspace of R^n consisting of vectors whose first n_1 components are the same, whose next n_2 component are the same, . . . , and whose last n_k components are the same. Then $\boldsymbol{\mu}$ has the form given above if and only if $\boldsymbol{\mu}\in V_2$. Therefore, an equivalent version of this model is one in which we observe

$$\mathbf{Y}\sim N_n(\boldsymbol{\mu},\sigma^2\mathbf{I}),\qquad \boldsymbol{\mu}\in V_2,\qquad \sigma^2>0.$$

This model is often called the one-way analysis of variance model.

EXAMPLE 3 (The two-way analysis of variance model, no interaction). Let Y_{ij} be independent, $Y_{ij}\sim N_1(\theta+\alpha_i+\beta_j,\sigma^2)$, $\sum_i\alpha_i=0$, $\sum_j\beta_j=0$,

$-\infty < \theta < \infty$, $\sigma^2 > 0$ for $i = 1, \ldots, r$; $j = 1, \ldots, c$. Now, let

$$\mathbf{Y} = \begin{pmatrix} Y_{11} \\ \vdots \\ Y_{1c} \\ Y_{21} \\ \vdots \\ Y_{rc} \end{pmatrix}, \qquad \boldsymbol{\mu} = \begin{pmatrix} \theta + \alpha_1 + \beta_1 \\ \vdots \\ \theta + \alpha_1 + \beta_c \\ \theta + \alpha_2 + \beta_1 \\ \vdots \\ \theta + \alpha_r + \beta_c \end{pmatrix}, \qquad n = rc.$$

Then $\mathbf{Y} \sim N_n(\boldsymbol{\mu}, \sigma^2\mathbf{I})$. Now let V_3 be the $(r + c - 1)$-dimensional subspace of R^n consisting of vectors $\boldsymbol{\mu}$ having the form given above. (Note that there are $r + c + 1$ parameters for the problem, but that $\sum_i \alpha_i = 0$ and $\sum_j \beta_j = 0$, so that there are only $r + c - 1$ linearly independent parameters.) Then an equivalent version of this model is one in which we observe

$$\mathbf{Y} \sim N_n(\boldsymbol{\mu}, \sigma^2\mathbf{I}), \qquad \boldsymbol{\mu} \in V_3, \qquad \sigma^2 > 0.$$

EXAMPLE 4 (Multiple regression). Let Y_i be independent, $Y_i \sim N_1(\sum_{j=1}^{r} x_{ij}\beta_j, \sigma^2)$, $-\infty < \beta_j < \infty$, $\sigma^2 > 0$, $i = 1, \ldots, n$, where the x_{ij} are known constants. Let

$$\mathbf{Y} = \begin{pmatrix} Y_1 \\ \vdots \\ Y_n \end{pmatrix}, \qquad \mathbf{X} = \begin{pmatrix} X_{11} & \cdots & X_{1r} \\ \vdots & & \\ X_{n1} & \cdots & X_{nr} \end{pmatrix}, \qquad \boldsymbol{\beta} = \begin{pmatrix} \beta_1 \\ \vdots \\ \beta_r \end{pmatrix}, \qquad \boldsymbol{\mu} = \mathbf{X}\boldsymbol{\beta}.$$

Then $\mathbf{Y} \sim N_n(\boldsymbol{\mu}, \sigma^2\mathbf{I})$. Suppose that $\mathbf{X}$ has rank p, and let V_4 be the p-dimensional subspace of R^n spanned by the columns of $\mathbf{X}$. Then $\boldsymbol{\mu} = \mathbf{X}\boldsymbol{\beta}$ if and only if $\boldsymbol{\mu} \in V_4$, so that this model is the one in which we observe

$$\mathbf{Y} \sim N_n(\boldsymbol{\mu}, \sigma^2\mathbf{I}), \qquad \boldsymbol{\mu} \in V_4, \qquad \sigma^2 > 0.$$

If $p = r$, then this model is just a coordinatized linear model. However, if $p < r$, then this model is still a linear model. (In future chapters where we discuss regression models we assume that $p = r$.)

In Chapter 7 we consider additional examples of linear models, including nested analysis of variance, Latin squares, and analysis of covariance models.

4.2. DISCUSSION OF THE ASSUMPTION FOR THE LINEAR MODEL.

As we mentioned above, we study the linear model for the next eight chapters. We now give an elementary discussion of the assumptions that we are making in this model. Let

$$\mathbf{e} = \begin{pmatrix} e_1 \\ \vdots \\ e_n \end{pmatrix} = \mathbf{Y} - \boldsymbol{\mu}.$$

We first discuss the assumptions made about **e**. By Theorem 3.6c, we see that we are assuming that

$$\mathbf{e} \sim N_n(\mathbf{0}, \sigma^2 \mathbf{I}).$$

This statement includes four different assumptions about the e_i:

(1) The e_i are independent.
(2) The $Ee_l = 0$.
(3) The $\text{var}(e_i) = \sigma^2$.
(4) The e_i are normally distributed.

We now discuss each of these assumptions and some of the remedies for situations in which they are not met.

(1) The e_i Are Independent. This is a rather strong assumption, but one that is often met in applications. Observations on different individuals (rats, people, cars, etc.) can often be assumed independent if the experiment is done carefully. In Chapter 14, it is shown that this assumption of independence can be weakened to exchangeability without affecting the optimality of many of the procedures. (The components of **e** are exchangeable if the distribution of **Pe** is the same as the distribution of **e** for all permutation matrices **P**.) One situation in which this assumption is not met is the situation in which we make several observations on each individual. In this book, we consider two models for this situation, the repeated measures model discussed in Chapter 14, and the generalized repeated measures model discussed in Sections 18.7 and 19.8. Another situation in which the assumption of independence cannot be made is in time series data. In this situation, it is often possible to use the autocorrelation model discussed in Chapter 13. Finally, in some situations in which the e_i are not independent, it is possible to assume that the covariance matrix of **e** is $\sigma^2 \mathbf{A}$ for some known matrix **A**. In this case, we have the generalized linear model discussed in Chapter 13.

(2) $Ee_i = 0$. This assumption implies that the mean vector of **Y** actually is in V. In a multiple regression model, this assumption is violated if we do not include enough explanatory variables (x_{ij}) in the model. In the one sample model, it is violated if we have both men and women in the sample when the mean for men is different from the mean for women. In principle, this problem can be corrected by adding more

variables to the model. In practice, of course, this is often rather difficult.

(3) $\text{var}(e_i) = \sigma^2$. This assumption implies that all the observations have the same variance. In the k-sample model, it implies that the variance is the same for all k samples. In the multiple regression model, it implies that the variance of an observation is not dependent on the x_{ij}. This assumption is particularly bothersome, because it is often difficult to argue that the variances should be equal. Furthermore, it is not clear what to do if the variances are unequal. It can sometimes be argued that the $\text{var}(e_i) = \sigma^2 k_i$, where the k_i are known constants. In this case, we can use the generalized linear model discussed in Chapter 13.

(4) The e_i Are Normally Distributed. This assumption is the least bothersome of the four assumptions (at least to me), for the following two reasons. The first reason is that in Chapter 10 the procedures derived under the assumption of normality are shown under fairly general conditions to be valid asymptotically (as $n \to \infty$) even in the presence of nonnormal errors. The second reason is that there are many procedures in the literature that have been suggested for the model with nonnormal errors. We do not discuss these procedures in this book. These procedures come under the heading of "nonparametric" procedures, "robust" procedures, or "resistant" procedures. See, for example, Lehmann (1975), Andrews (1974), Huber (1977), and Hettmansperger and McKean (1977).

There are two additional assumptions about the mean vector $\boldsymbol{\mu}$. We have already discussed the assumption that $\boldsymbol{\mu} \in V$. These two additional assumptions are the following:

(5) The subspace V is a known, nonrandom subspace.
(6) The vector $\boldsymbol{\mu}$ is an unknown, but nonrandom vector.

These two assumptions are somewhat easier to interpret for the coordinatized version of the linear model. Therefore, let $\boldsymbol{\mu} = \mathbf{X}\boldsymbol{\beta}$, where $\mathbf{X}$ is a basis matrix for V. Then these two assumptions can be restated in the following way:

(5′) The matrix $\mathbf{X}$ is a known, nonrandom matrix.
(6′) The vector $\boldsymbol{\beta}$ is an unknown, nonrandom vector.

An example of a model in which the matrix $\mathbf{X}$ has random components is the correlation model discussed in Chapter 16. For that model it is shown that optimal procedures for the linear model are also optimal for the correlation model, so that it is not too serious an error to assume that the model is a linear model when it is actually a correlation model. An example of a model in which $\boldsymbol{\beta}$ has random components is the random effects model discussed in Chapter 15. For this model it is shown that optimal procedures for the linear model may not even be valid for the random effects model, so that it can be a rather severe mistake to assume that the model is a linear model when it is a random effects model.

In the next eight chapters (except Chapter 10), we assume that the above assumptions are satisfied, and that the model is in fact a linear model.

CHAPTER 5

A Sufficient Statistic

In this chapter we find a sufficient statistic for the general linear model defined in Chapter 4. In Section 5.1 this statistic is defined, and its joint distribution is determined. In Section 5.2 it is shown to be sufficient, and it is shown to be complete in Section 5.3. In Sections 5.4 and 5.5, the results are reinterpreted for the coordinatized version of the problem.

5.1. THE STATISTIC.

In the general linear model, we observe

$$\mathbf{Y} \sim N_n(\boldsymbol{\mu}, \sigma^2 \mathbf{I}), \qquad \boldsymbol{\mu} \in V, \qquad \sigma^2 > 0,$$

where V is a given p-dimensional subspace of R^n. Define

$$\hat{\boldsymbol{\mu}} = \mathbf{P}_V \mathbf{Y}, \qquad \hat{\sigma}^2 = \frac{\|\mathbf{P}_{V^\perp}\mathbf{Y}\|^2}{n-p} = \frac{\|\mathbf{Y} - \mathbf{P}_V \mathbf{Y}\|^2}{n-p} = \frac{\|\mathbf{Y} - \hat{\boldsymbol{\mu}}\|^2}{n-p} = \frac{\|\mathbf{Y}\|^2 - \|\hat{\boldsymbol{\mu}}\|^2}{n-p} \tag{5.1}$$

(see Theorem 2.5d). In the next section, we show that $(\hat{\boldsymbol{\mu}}, \hat{\sigma}^2)$ is a sufficient statistic. In this section we find the joint distribution of $\hat{\boldsymbol{\mu}}$ and $\hat{\sigma}^2$.

THEOREM 5.1. $\hat{\boldsymbol{\mu}}$ and $\hat{\sigma}^2$ are independent,

$$\hat{\boldsymbol{\mu}} \sim N_n(\boldsymbol{\mu}, \sigma^2 \mathbf{P}_V), \qquad (n-p)\hat{\sigma}^2 \sim \sigma^2 \chi^2_{n-p}(0).$$

PROOF. V and $V^\perp$ are orthogonal, so that by Theorem 3.11b, $\mathbf{P}_V\mathbf{Y}$ and $\mathbf{P}_{V^\perp}\mathbf{Y}$ are independent, and therefore so are $\hat{\boldsymbol{\mu}}$ and $\hat{\sigma}^2$. By Theorem 3.11a,

$$\mathbf{P}_V \mathbf{Y} \sim N_n(\mathbf{P}_V \boldsymbol{\mu}, \sigma^2 \mathbf{P}_V).$$

However, $\mathbf{P}_V\boldsymbol{\mu} = \boldsymbol{\mu}$, since $\boldsymbol{\mu} \in V$. Since $\dim(V^\perp) = n - \dim(V) = n - p$, by Theorem 3.12,

$$(n-p)\hat{\sigma}^2 = \|\mathbf{P}_{V^\perp}\|^2 \sim \sigma^2\chi^2_{n-p}\left(\frac{\|\mathbf{P}_{V^\perp}\boldsymbol{\mu}\|^2}{\sigma^2}\right).$$

However, since $\boldsymbol{\mu} \in V$, $\mathbf{P}_{V^\perp}\boldsymbol{\mu} = \mathbf{0}$. □

Note that the distribution of $\hat{\boldsymbol{\mu}}$ is a singular normal distribution, since $\mathbf{P}_V$ is an $n \times n$ matrix of rank p.

5.2. SUFFICIENCY.

We now show that $(\hat{\boldsymbol{\mu}}, \hat{\sigma}^2)$ is sufficient.

THEOREM 5.2. $(\hat{\boldsymbol{\mu}}, \hat{\sigma}^2)$ is a sufficient statistic.

PROOF.

$$f(\mathbf{y};\, \boldsymbol{\mu}, \sigma^2) = \frac{1}{(2\pi)^{n/2}\sigma^n}\exp\left(-\frac{1}{2\sigma^2}\|\mathbf{y} - \boldsymbol{\mu}\|^2\right).$$

By the Pythagorean theorem (Theorem 2.5.a),

$$\|\mathbf{y} - \boldsymbol{\mu}\|^2 = \|\mathbf{y} - \hat{\boldsymbol{\mu}}\|^2 + \|\hat{\boldsymbol{\mu}} - \boldsymbol{\mu}\|^2 = (n-p)\hat{\sigma}^2 + \|\hat{\boldsymbol{\mu}} - \boldsymbol{\mu}\|^2.$$

The result follows from the factorization criterion. □

5.3. COMPLETENESS.

The proof of the completeness is a little bit trickier. We present first what would seem to be an obvious approach to showing the completeness, and then show why it is incorrect.

Let

$$h(y) = 1, \qquad k(\boldsymbol{\mu}, \sigma^2) = \frac{1}{(2\pi)^{n/2}\sigma^n}\exp(-\|\boldsymbol{\mu}\|^2/2\sigma^2)$$

$$\mathbf{R}(\boldsymbol{\mu}, \sigma^2) = \begin{bmatrix} -\dfrac{1}{2\sigma^2} \\ \dfrac{1}{\sigma^2}\boldsymbol{\mu} \end{bmatrix}, \qquad \mathbf{S}(\mathbf{y}) = \begin{pmatrix} (n-p)\hat{\sigma}^2 + \|\hat{\boldsymbol{\mu}}\|^2 \\ \hat{\boldsymbol{\mu}} \end{pmatrix}$$

Then

$$f(\mathbf{y};\, \boldsymbol{\mu}, \sigma^2) = h(\mathbf{y})k(\boldsymbol{\mu}, \sigma^2)\exp\left[(\mathbf{R}(\boldsymbol{\mu}, \sigma^2))'\mathbf{S}(\mathbf{y})\right]$$

Therefore, we might argue that $\mathbf{S}(\mathbf{Y})$ is a complete sufficient statistic, by the exponential criterion. Since $(\hat{\boldsymbol{\mu}}, \hat{\sigma}^2)$ is an invertible function of $\mathbf{S}(\mathbf{Y})$ it would also be complete and sufficient. The fallacy with this argument is that the image of $\mathbf{R}(\boldsymbol{\mu}, \boldsymbol{\sigma}^2)$ does not have any interior, as the following argument shows. Let $\mathbf{w} \in V^{\perp}$. Then

$$\left[\mathbf{R}(\boldsymbol{\mu}, \sigma^2)\right]'\begin{pmatrix} 0 \\ \mathbf{w} \end{pmatrix} = 0$$

so that the image of R lies in a proper subspace of R^{n+1} (i.e., a hyperplane), which has no interior.

To clean up the argument suggested above, we must choose an orthonormal basis matrix $\mathbf{X}$ for V. Then let $\boldsymbol{\beta} \in R^p$ be defined by

$$\boldsymbol{\mu} = \mathbf{X}\boldsymbol{\beta}, \qquad \boldsymbol{\beta} = \mathbf{X}'\boldsymbol{\mu}.$$

Define

$$\mathbf{Q}(\boldsymbol{\mu}, \sigma^2) = \begin{bmatrix} -\dfrac{1}{2\sigma^2} \\ \dfrac{1}{\sigma^2}\mathbf{X}'\boldsymbol{\mu} \end{bmatrix} = \begin{bmatrix} -\dfrac{1}{2\sigma^2} \\ \dfrac{1}{\sigma^2}\boldsymbol{\beta} \end{bmatrix},$$

$$\mathbf{T}(\mathbf{y}) = \begin{pmatrix} (n-p)\hat{\sigma}^2 + \|\hat{\boldsymbol{\mu}}\|^2 \\ \mathbf{X}'\hat{\boldsymbol{\mu}} \end{pmatrix} = \begin{pmatrix} T_1(\mathbf{y}) \\ \mathbf{T}_2(\mathbf{y}) \end{pmatrix}$$

Then, as before,

$$f(\mathbf{y};\, \boldsymbol{\mu}, \sigma^2) = h(\mathbf{y})k(\boldsymbol{\mu}, \sigma^2)\exp\left\{\left[\mathbf{Q}(\boldsymbol{\mu}, \sigma^2)\right]'\mathbf{T}(\mathbf{y})\right\}.$$

Now the image of $\mathbf{Q}$ is all those vectors in R^{p+1} whose first coordinate is negative, and this set has interior points. Therefore, $\mathbf{T}$ is complete and sufficient. Also

$$\begin{pmatrix} \hat{\sigma}^2 \\ \hat{\boldsymbol{\mu}} \end{pmatrix} = \begin{bmatrix} \left[T_1(\mathbf{y}) - \|\mathbf{X}\mathbf{T}_2(\mathbf{y})\|^2\right]/(n-p) \\ \mathbf{X}\mathbf{T}_2(\mathbf{y}) \end{bmatrix} \tag{5.2}$$

is an invertible function of $\mathbf{T}$. Therefore, we have proved the following theorem:

THEOREM 5.3. $(\hat{\boldsymbol{\mu}}, \hat{\sigma}^2)$ is a complete sufficient statistic for the general linear model.

At first inspection, it might seem that $(\hat{\boldsymbol{\mu}}, \hat{\sigma}^2)$ could not be complete for the following reason. Let $\mathbf{w} \in V^{\perp}$. Then, $(\boldsymbol{\mu} \in V)$

$$E\mathbf{w}'\hat{\boldsymbol{\mu}} = \mathbf{w}'\boldsymbol{\mu} = 0$$

and we therefore have an unbiased estimator of 0, which would seem to violate the definition of completeness. However, $\hat{\boldsymbol{\mu}} \in V$ also, and therefore

$$\mathbf{w}'\hat{\boldsymbol{\mu}} = \mathbf{0},$$

and this estimator is the 0 function everywhere. The definition is therefore not violated.

5.4. THE COORDINATIZED MODEL.

We now consider the coordinatized version of the general linear model. We observe

$$\mathbf{Y} \sim N_n(\mathbf{X}\boldsymbol{\beta}, \sigma^2\mathbf{I}), \qquad \boldsymbol{\beta} \in R^p, \qquad \sigma^2 > 0,$$

where $\mathbf{X}$ is a known $n \times p$ matrix of rank p (not necessarily orthonormal). Let V be the subspace spanned by the columns of $\mathbf{X}$. Then $\mathbf{X}$ is a basis matrix for V and the dimension of V is p. Let $\boldsymbol{\mu} = \mathbf{X}\boldsymbol{\beta}$. We have now reduced this model to the model considered in previous sections. Define $\hat{\boldsymbol{\beta}} = (\mathbf{X}'\mathbf{X})^{-1}\mathbf{X}'\mathbf{Y}$. Then, it is easily shown (using Theorem 2.4c) that

$$\hat{\boldsymbol{\mu}} = \mathbf{X}\hat{\boldsymbol{\beta}}, \qquad \hat{\boldsymbol{\beta}} = (\mathbf{X}'\mathbf{X})^{-1}\mathbf{X}'\hat{\boldsymbol{\mu}}.$$

Therefore, $\hat{\boldsymbol{\beta}}$ is an invertible function of $\hat{\boldsymbol{\mu}}$. Also, by (5.1),

$$\begin{aligned}(n-p)\hat{\sigma}^2 &= \|\mathbf{Y} - \hat{\boldsymbol{\mu}}\|^2 = \|\mathbf{Y} - \mathbf{X}\hat{\boldsymbol{\beta}}\|^2 = \|\mathbf{Y}\|^2 - \|\hat{\boldsymbol{\mu}}\|^2 \\ &= \|\mathbf{Y}\|^2 - \|\mathbf{X}\hat{\boldsymbol{\beta}}\|^2 = \mathbf{Y}'\big(\mathbf{I} - \mathbf{X}(\mathbf{X}'\mathbf{X})^{-1}\mathbf{X}'\big)\mathbf{Y}.\end{aligned}$$

THEOREM 5.4. $(\hat{\boldsymbol{\beta}}, \hat{\sigma}^2)$ is a complete sufficient statistic for the coordinatized linear model. $\hat{\boldsymbol{\beta}}$ and $\hat{\sigma}^2$ are independent and

$$\hat{\boldsymbol{\beta}} \sim N_p\big(\boldsymbol{\beta}, \sigma^2(\mathbf{X}'\mathbf{X})^{-1}\big), \qquad (n-p)\hat{\sigma}^2 \sim \sigma^2\chi^2_{n-p}(0).$$

PROOF. See Exercise B1. □

We now note one property of the statistic $\hat{\boldsymbol{\beta}}$. Using the fact that $\mathbf{v} \in V$ if and only if $\mathbf{v} = \mathbf{X}\mathbf{b}$ for some $\mathbf{b} \in R^p$, we see from Theorem 2.5b that

$$\|\mathbf{Y} - \mathbf{X}\hat{\boldsymbol{\beta}}\|^2 \leqslant \|\mathbf{Y} - \mathbf{X}\mathbf{b}\|^2 \tag{5.3}$$

for all $\mathbf{b} \in R^p$. For this reason the vector $\hat{\boldsymbol{\beta}}$ is called the *ordinary least squares estimator* of $\boldsymbol{\beta}$.

5.5. REGRESSION WITH AN INTERCEPT TERM.

In most applications of the coordinatized linear model, the regression equation has an intercept term. That is, $\mathbf{Y} \sim N_n(\mathbf{X}\boldsymbol{\beta}, \sigma^2\mathbf{I})$ and

$$\mathbf{X} = \begin{pmatrix} 1 & \mathbf{T}_1 \\ \vdots & \vdots \\ 1 & \mathbf{T}_n \end{pmatrix}, \qquad \boldsymbol{\beta} = \begin{pmatrix} \delta \\ \boldsymbol{\gamma} \end{pmatrix}$$

where $\mathbf{T}_i'$ are known $(p-1)\times 1$ vectors such that $\mathbf{X}$ has rank p, δ is an unknown number, and $\boldsymbol{\gamma}$ is an unknown $(p-1)\times 1$ vector.

We now derive alternative formulas for

$$\hat{\boldsymbol{\beta}} = \begin{pmatrix} \hat{\delta} \\ \hat{\gamma} \end{pmatrix}$$

and $\hat{\sigma}^2$ for this particular case.

Let $\mathbf{Y}' = (Y_1, \ldots, Y_n)$,

$$\overline{\mathbf{T}} = \frac{1}{n}\sum_{i=1}^{n}\mathbf{T}_i, \qquad \overline{Y} = \frac{1}{n}\sum_{i=1}^{n} Y_i, \qquad \tilde{Y}_i = Y_i - \overline{Y},$$

$$\tilde{\mathbf{T}}_i = \mathbf{T}_i - \overline{\mathbf{T}}, \qquad \tilde{\mathbf{Y}} = \begin{bmatrix} \tilde{\mathbf{Y}}_1 \\ \vdots \\ \tilde{\mathbf{Y}}_n \end{bmatrix}, \qquad \tilde{\mathbf{T}} = \begin{bmatrix} \tilde{\mathbf{T}}_1 \\ \vdots \\ \tilde{\mathbf{T}}_n \end{bmatrix} = \mathbf{X}\begin{pmatrix} -\overline{\mathbf{T}} \\ \mathbf{I} \end{pmatrix}. \tag{5.4}$$

LEMMA 5.5. $\hat{\gamma} = (\tilde{\mathbf{T}}'\tilde{\mathbf{T}})^{-1}\tilde{\mathbf{T}}'\tilde{\mathbf{Y}}$, $(n-p)\hat{\sigma}^2 = \|\tilde{\mathbf{Y}} - \tilde{\mathbf{T}}\hat{\gamma}\|^2$, and $\hat{\delta} = \overline{Y} - \overline{\mathbf{T}}\hat{\gamma}$.

PROOF. We first show that $\tilde{\mathbf{T}}$ has rank $p-1$ and hence $\mathbf{T}'\mathbf{T}$ is invertible. Let

$$\tilde{\mathbf{X}} = \begin{bmatrix} 1 & \tilde{\mathbf{T}}_1 \\ \vdots & \vdots \\ 1 & \tilde{\mathbf{T}}_n \end{bmatrix}. \qquad \text{Then } \tilde{\mathbf{X}} = \mathbf{X}\begin{pmatrix} 1 & -\overline{\mathbf{T}} \\ 0 & \mathbf{I} \end{pmatrix}. \qquad \text{Since } \begin{pmatrix} 1 & -\overline{\mathbf{T}} \\ 0 & \mathbf{I} \end{pmatrix}$$

is invertible and $\mathbf{X}$ has rank p, $\tilde{\mathbf{X}}$ has rank p. Therefore, the columns of $\mathbf{X}$ are linearly independent and $\tilde{\mathbf{T}}$ has rank $p-1$. To find $\hat{\delta}$ and $\hat{\gamma}$, we need to minimize

$$Q(\delta, \gamma) = \sum_{i=1}^{n}(Y_i - \delta - \mathbf{T}_i\gamma)^2.$$

[see (5.3)]. Taking the partial derivative of Q with respect to δ and setting it equal to 0, we see that $\hat{\delta} = \overline{Y} - \overline{\mathbf{T}}\hat{\gamma}$. We now need to find γ that minimizes

$$Q(\hat{\delta}, \gamma) = \sum_{i=1}^{n}\left(Y_i - \overline{Y} - (\mathbf{T}_i - \overline{\mathbf{T}})\gamma\right)^2 = \sum_{i=1}^{n}(\tilde{Y}_i - \tilde{\mathbf{T}}_i\gamma)^2 = \|\tilde{\mathbf{Y}} - \tilde{\mathbf{T}}\gamma\|^2.$$

That is, $\hat{\gamma}$ is the ordinary least squares estimator for the model in which we observe $\tilde{\mathbf{Y}}$ instead of $\mathbf{Y}$ with basis matrix $\tilde{\mathbf{T}}$ instead of $\mathbf{X}$. Therefore, $\hat{\gamma} = (\tilde{\mathbf{T}}'\tilde{\mathbf{T}})^{-1}\tilde{\mathbf{T}}'\tilde{\mathbf{Y}}$. Finally,

$$(n-p)\hat{\sigma}^2 = \|\mathbf{Y} - \mathbf{X}\hat{\boldsymbol{\beta}}\|^2 = Q(\hat{\delta}, \hat{\gamma}) = \|\tilde{\mathbf{Y}} - \tilde{\mathbf{T}}\hat{\gamma}\|^2. \quad \square$$

Using Lemma 5.5 to calculate $\hat{\boldsymbol{\beta}}$ and $\hat{\sigma}^2$ is often easier than using the original formulas for two reasons. One reason is that the formulas in Lemma 5.5 involve only deviations from the mean, which are often much smaller than the actual observations. The second reason is that the matrix $\tilde{\mathbf{T}}'\tilde{\mathbf{T}}$ is $(p-1)\times(p-1)$, while $\mathbf{X}'\mathbf{X}$ is $p \times p$, so that $\tilde{\mathbf{T}}'\tilde{\mathbf{T}}$ is smaller and hence easier to invert. These formulas will also be used to derive results for the correlation model studied in Chapter 16.

EXERCISES.

Type A

1. Suppose that we observe the model

$$Y_i = \beta_0 + \beta_1 X_i + \beta_2 X_i^2 + e_i$$

where the e_i are independent and $e_i \sim N_1(0, \sigma^2)$. We observe the following data:

X	1	2	3	4	5
Y	2	8	9	5	6.

Let $\boldsymbol{\beta} = (\beta_0, \beta_1, \beta_2)'$, $\mu_i = EY_i$, $\boldsymbol{\mu} = (\mu_1, \ldots, \mu_5)'$.

(a) Find $\hat{\beta}$ and $\hat{\sigma}^2$ for this model (use Section 5.5).

(b) Find $\hat{\mu}$ for this model.

Type B

1. Prove Theorem 5.4.

2. Verify (5.2).

CHAPTER 6

Estimation

In this chapter, we consider estimating linear functions of the mean vector, $\boldsymbol{\mu}$, and estimating the common variance, σ^2. The estimators discussed in this chapter were first derived in the last century by Gauss (1809) and Legendre (1806). Additional results were derived by Markov (1900).

6.1. MINIMUM VARIANCE UNBIASED ESTIMATORS AND MAXIMUM LIKELIHOOD ESTIMATORS.

We now consider the general linear model in which we observe

$$\mathbf{Y} \sim N_n(\boldsymbol{\mu}, \sigma^2 I), \qquad \boldsymbol{\mu} \in V, \qquad \sigma^2 > 0$$

where V is a specified p-dimensional subspace of R^n. Let $\mathbf{A}$ be an arbitrary $s \times n$ matrix (s is also arbitrary), and let $\hat{\boldsymbol{\mu}}$ and $\hat{\sigma}^2$ be defined by (5.1).

THEOREM 6.1. a. $\mathbf{A}\hat{\boldsymbol{\mu}}$ and $\hat{\sigma}^2$ are the minimum variance unbiased estimators of $\mathbf{A}\boldsymbol{\mu}$ and σ^2.

b. $\hat{\boldsymbol{\mu}}$ and $((n-p)/n)\hat{\sigma}^2$ are the MLEs of $\boldsymbol{\mu}$ and σ^2.

c. $\mathbf{A}\hat{\boldsymbol{\mu}}$ is the MLE of $\mathbf{A}\boldsymbol{\mu}$.

PROOF. a. This result follows directly from Theorems 5.1 and 5.3.

b. The density of $\mathbf{Y}$ is given by

$$f(\mathbf{y}; (\boldsymbol{\mu}, \sigma^2)) = \frac{1}{(2\pi)^{n/2}\sigma^2} \exp \frac{-\|\mathbf{y}-\boldsymbol{\mu}\|^2}{2\sigma^2}.$$

By Theorem 2.5b,

$$\|\mathbf{y}-\hat{\boldsymbol{\mu}}\|^2 \leqslant \|\mathbf{y}-\boldsymbol{\mu}\|^2$$

for all $\boldsymbol{\mu} \in V$. Therefore, for all σ^2, $f(\mathbf{y}; (\boldsymbol{\mu}, \sigma^2))$ is maximized by $\boldsymbol{\mu} = \hat{\boldsymbol{\mu}}$.

Now

$$f(\mathbf{y}; (\hat{\boldsymbol{\mu}}, \sigma^2)) = \frac{1}{(2\pi)^{n/2}\sigma^{n/2}} \exp \frac{-\|\mathbf{y} - \hat{\boldsymbol{\mu}}\|^2}{2\sigma^2}$$

$$= \frac{1}{(2\pi)^{n/2}\sigma^{n/2}} \exp \frac{-(n-p)\hat{\sigma}^2}{2\sigma^2}.$$

Since

$$\frac{\partial}{\partial \sigma} \log f(\mathbf{y}; (\hat{\boldsymbol{\mu}}, \sigma)) = -\frac{n}{\sigma} + (n-p)\frac{\hat{\sigma}^2}{\sigma^3}.$$

The maximum of $f(\mathbf{y}; (\hat{\boldsymbol{\mu}}, \sigma^2))$ occurs for $\sigma^2 = ((\mathrm{n} - \mathrm{p})/\mathrm{n})\hat{\sigma}^2$.

c. This follows from the definition of maximum likelihood estimators (see Section 1.2). □

We now consider the coordinatized version of the problem. Let $\mathbf{X}$ be a basis matrix for V, and let $\boldsymbol{\beta}$ be the unique solution to the equation $\boldsymbol{\mu} = \mathbf{X}\boldsymbol{\beta}$. As in Chapter 5, let $\hat{\boldsymbol{\beta}}$ be the unique solution to the equation $\hat{\boldsymbol{\mu}} = \mathbf{X}\hat{\boldsymbol{\beta}}$ and let $\mathbf{C}$ be an arbitrary $s \times p$ matrix.

COROLLARY . $\mathbf{C}\hat{\boldsymbol{\beta}}$ is the minimum variance unbiased estimator and MLE of $\mathbf{C}\boldsymbol{\beta}$.

PROOF. See Exercise B1. □

6.2. BEST INVARIANT ESTIMATORS.

We now find best invariant estimators for $\mathbf{A}\boldsymbol{\mu}$ and σ^2. We first consider estimating $\mathbf{A}\boldsymbol{\mu}$, where $\mathbf{A}$ is an arbitrary $s \times n$ matrix of rank s. We consider a fairly general class of quadratic loss functions. Let $\mathbf{d}_1 \in R^s$. The loss function we use for this problem is

$$L_1(\mathbf{d}_1; (\boldsymbol{\mu}, \sigma^2)) = \frac{(\mathbf{d}_1 - \mathbf{A}\boldsymbol{\mu})'\mathbf{H}(\mathbf{d}_1 - \mathbf{A}\boldsymbol{\mu})}{\sigma^2}$$

where $\mathbf{H} > 0$ is an arbitrary known matrix. This problem is invariant under the group G_1 of transformations given by

$$g_1(\hat{\boldsymbol{\mu}}, \hat{\sigma}^2) = (c\hat{\boldsymbol{\mu}} + \mathbf{b}, c^2\hat{\sigma}^2), \qquad \bar{g}_1(\boldsymbol{\mu}, \sigma^2) = (c\boldsymbol{\mu} + \mathbf{b}, c^2\sigma^2),$$

$$\tilde{g}_1(\mathbf{d}_1) = c\mathbf{d}_1 + \mathbf{A}\mathbf{b}$$

where $c > 0$ and $\mathbf{b} \in V$. Note that

$$L_1(\tilde{g}_1(\mathbf{d}_1);\bar{g}_1(\mu,\sigma^2)) = \frac{(c\mathbf{d}_1 + \mathbf{Ab} - (c\mathbf{A}\mu + \mathbf{Ab}))'\mathbf{H}(c\mathbf{d}_1 + \mathbf{Ab} - (c\mathbf{A}\mu + \mathbf{Ab}))}{c^2\sigma^2}$$

$$= \frac{(\mathbf{d}_1 - \mathbf{A}\mu)'\mathbf{H}(\mathbf{d}_1 - \mathbf{A}\mu)}{\sigma^2} = L(\mathbf{d}_1;(\mu,\sigma^2)).$$

THEOREM 6.2. $\mathbf{A}\hat{\mu}$ is the best invariant estimator of $\mathbf{A}\mu$ for the loss function L_1 under the group G_1.

PROOF. $T(\hat{\mu},\hat{\sigma}^2)$ is an invariant estimator under G_1 if and only if $\mathbf{T}(g_1(\hat{\mu},\hat{\sigma}^2)) = \tilde{g}_1(T(\hat{\mu},\hat{\sigma}^2))$, or equivalently

$$\mathbf{T}(c\hat{\mu} + \mathbf{b}, c^2\hat{\sigma}^2) = c\mathbf{T}(\hat{\mu},\hat{\sigma}^2) + \mathbf{Ab}$$

for all $c > 0$, $\mathbf{b} \in V$. We note first that $\mathbf{A}\hat{\mu}$ is an invariant estimator of $\mathbf{A}\mu$. Now, let $\mathbf{T}(\hat{\mu},\hat{\sigma}^2)$ be any other invariant estimator of $\mathbf{A}\mu$, and let $c = 1/\hat{\sigma} > 0$, $b = (-1/\hat{\sigma})\hat{\mu} \in V$. Since $\mathbf{T}$ is invariant

$$\mathbf{T}(\mathbf{0},1) = (1/\hat{\sigma})\mathbf{T}(\hat{\mu},\hat{\sigma}) + (1/\hat{\sigma})\mathbf{A}\hat{\mu}$$

or equivalently,

$$\mathbf{T}(\hat{\mu},\hat{\sigma}) = \hat{\sigma}\mathbf{T}(\mathbf{0},1) + \mathbf{A}\hat{\mu}.$$

(Note that $\mathbf{T}$ is completely determined by $\mathbf{T}(\mathbf{0},1)$.) Therefore,

$$R(\mathbf{T};(\mu,\sigma^2)) = \frac{E(\hat{\sigma}\mathbf{T}(\mathbf{0},1) + \mathbf{A}\hat{\mu} - \mathbf{A}\mu)'\mathbf{H}(\hat{\sigma}\mathbf{T}(\mathbf{0},1) + \mathbf{A}\hat{\mu} - \mathbf{A}\mu)}{\sigma^2}$$

$$= E\frac{\hat{\sigma}^2}{\sigma^2}\mathbf{T}'(\mathbf{0},1)\mathbf{HT}(\mathbf{0},1) + \frac{E(\mathbf{A}\hat{\mu} - \mathbf{A}\mu)\mathbf{H}(\mathbf{A}\hat{\mu} - \mathbf{A}\mu)}{\sigma^2}$$

$$+ 2\mathbf{T}(\mathbf{0},1)'\mathbf{HA}E(\hat{\mu} - \mu)E\hat{\sigma}/\sigma^2,$$

since $\hat{\mu}$ and $\hat{\sigma}$ are independent.

However, $E(\hat{\mu} - \mu) = 0$, $R(\mathbf{A}\hat{\mu},(\mu,\sigma^2)) = E(\mathbf{A}\hat{\mu} - \mathbf{A}\mu)'\mathbf{H}(\mathbf{A}\hat{\mu} - \mathbf{A}\mu)/\sigma^2$, and $\mathbf{T}(\mathbf{0},1) \neq 0$ (since $\mathbf{T} \neq \mathbf{A}\hat{\mu}$). Therefore,

$$R(\mathbf{T};(\mu,\sigma^2)) = E\frac{\hat{\sigma}^2}{\sigma^2}\mathbf{T}'(\mathbf{0},1)\mathbf{HT}(\mathbf{0},1) + R(\mathbf{A}\hat{\mu};(\mu,\sigma^2)) > R(\mathbf{A}\hat{\mu};(\mu,\sigma^2))$$

(since $\mathbf{H} > 0$). Hence $\mathbf{A}\hat{\mu}$ has smaller risk than any other invariant estimator and is the best invariant estimator of $\mathbf{A}\mu$. □

COROLLARY. $\hat{\beta}$ is the best invariant estimator of β for the loss function $(\mathbf{d}_1 - \mathbf{B})'\mathbf{H}(\mathbf{d}_1 - \beta)$ for any $\mathbf{H} > 0$.

PROOF. This follows directly from the theorem above with $\mathbf{A} = (\mathbf{X}'\mathbf{X})^{-1}\mathbf{X}'$. □

We now look at the problem of estimating σ^2. Let $d_2 \in R$; $d_2 > 0$. We use the quadratic loss function

$$L_2(d_2;(\mu,\sigma^2)) = \frac{(d_2-\sigma^2)}{\sigma^4} = \left(\frac{d_2}{\sigma^2} - 1\right)^2.$$

This problem is invariant under the group G_2 of transformations

$$g_2(\hat{\mu},\hat{\sigma}^2) = (c\hat{\mu} + \mathbf{b}, c^2\hat{\sigma}^2), \qquad \bar{g}_2(\mu,\sigma^2) = (c\mu + \mathbf{b}, c^2\sigma^2),$$
$$\tilde{g}_2(d_2) = c^2 d_2.$$

Note that

$$L_2(\tilde{g}_2(d_2);\bar{g}(\mu,\sigma^2)) = \frac{(c^2d_2 - c^2\sigma^2)^2}{c^4\sigma^4} = \frac{(d_2-\sigma^2)^2}{\sigma^4} = L_2(d_2;(\mu,\sigma^2)).$$

THEOREM 6.3. $(n-p)\hat{\sigma}^2/(n-p+2)$ is the best invariant estimator of σ^2 under the group G_2 with the loss function L_2.

PROOF. See Exercise B2. □

Note that both the minimum variance unbiased estimator, $\hat{\sigma}^2$, and the MLE, $(n-p)\hat{\sigma}^2/n$ are invariant estimators, and hence the best invariant estimator is better than either of these estimators (at least for the quadratic loss L_2).

6.3. CONFIDENCE, INTERVALS AND PREDICTION INTERVALS.

We can find confidence intervals for linear functions $\mathbf{c}'\boldsymbol{\mu}$ and $\mathbf{d}'\boldsymbol{\beta}$ by using the following lemma.

LEMMA 6.4. Let $\mathbf{c} \in R^n$, $\mathbf{c} \notin V^{\perp}$, and $\mathbf{d} \in R^p$, $d \neq 0$. Then

$$\frac{\mathbf{c}'\hat{\boldsymbol{\mu}} - \mathbf{c}'\boldsymbol{\mu}}{\hat{\sigma}\sqrt{\mathbf{c}'\mathbf{P}_V\mathbf{c}}} \sim t_{n-p}(0) \qquad \frac{\mathbf{d}'\hat{\boldsymbol{\beta}} - \mathbf{d}'\boldsymbol{\beta}}{\hat{\sigma}\sqrt{\mathbf{d}'(\mathbf{X}'\mathbf{X})^{-1}\mathbf{d}}} \sim t_{n-p}(0).$$

PROOF. See Exercise B3. □

(We need to assume that $\mathbf{c} \notin V^{\perp}$ so that

$$\mathbf{c}'\mathbf{P}_V\mathbf{c} = \|\mathbf{P}_V\mathbf{c}\|^2 > 0.$$

If $\mathbf{c} \in V^{\perp}$, then $\mathbf{c}'\boldsymbol{\mu}$ is identically 0 so that confidence intervals for $\mathbf{c}'\boldsymbol{\mu}$ would not be very interesting.)

Using Lemma 6.4, we see that the intervals

$$\mathbf{c}'\hat{\boldsymbol{\mu}} - t_{n-p}^{\alpha/2}\hat{\sigma}\sqrt{\mathbf{c}'\mathbf{P}_V\mathbf{c}} < \mathbf{c}'\boldsymbol{\mu} < \mathbf{c}'\hat{\boldsymbol{\mu}} + t_{n-p}^{\alpha/2}\hat{\sigma}\sqrt{\mathbf{c}'\mathbf{P}_V\mathbf{c}}$$

$$\mathbf{d}'\hat{\boldsymbol{\beta}} - t_{n-p}^{\alpha/2}\hat{\sigma}\sqrt{\mathbf{d}'(\mathbf{X}'\mathbf{X})^{-1}\mathbf{d}} < \mathbf{d}'\boldsymbol{\beta} < \mathbf{d}'\hat{\boldsymbol{\beta}} + t_{n-p}^{\alpha/2}\hat{\sigma}\sqrt{\mathbf{d}'(\mathbf{X}'\mathbf{X})^{-1}\mathbf{d}}$$

are $1 - \alpha$ confidence intervals for $\mathbf{c}'\boldsymbol{\mu}$ and $\mathbf{d}'\boldsymbol{\mu}$, respectively. From Lemma 5.1, we see that

$$(n-p)\frac{\hat{\sigma}^2}{\sigma^2} \sim \chi^2_{n-p}(0)$$

and therefore

$$\frac{(n-p)\hat{\sigma}^2}{\chi_{n-p}^{2(1-\alpha/2)}} > \sigma^2 > \frac{(n-p)\hat{\sigma}^2}{\chi_{n-p}^{2(\alpha/2)}}$$

is a $1 - \alpha$ confidence interval for σ^2.

In a coordinatized linear model, we often want to predict Y_0 where

$$Y_0 \sim N_1(\mathbf{x}_0\boldsymbol{\beta}, \sigma^2)$$

independently of $\mathbf{Y}$ and $\mathbf{x}_0$ is a known $1 \times p$ vector. The obvious estimate of Y_0 is $\mathbf{x}_0\hat{\boldsymbol{\beta}}$, which is the minimum variance unbiased estimator of EY_0. We can also find an interval for Y_0 by using the following lemma.

LEMMA 6.5.

$$\frac{Y_0 - \mathbf{x}_0\hat{\boldsymbol{\beta}}}{\hat{\sigma}\sqrt{1 + \mathbf{x}_0(\mathbf{X}'\mathbf{X})^{-1}\mathbf{x}_0'}} \sim t_{n-p}(0).$$

PROOF. Y_0 and $\mathbf{x}_0\hat{\boldsymbol{\beta}}$ are independent and

$$\mathbf{x}_0\hat{\boldsymbol{\beta}} \sim N_1\left(\mathbf{x}_0\boldsymbol{\beta}, \sigma^2\mathbf{x}_0(\mathbf{X}'\mathbf{X})^{-1}\mathbf{x}_0'\right).$$

Therefore,

$$Y_0 - \mathbf{x}_0\hat{\boldsymbol{\beta}} \sim N_1\left(0, \sigma^2\left(1 + \mathbf{x}_0(\mathbf{X}'\mathbf{X})^{-1}\mathbf{x}_0'\right)\right)$$

independently of $\hat{\sigma}^2$. The result follows directly from the definition of the t-distribution. □

Using this lemma, we see that

$$P\left(\mathbf{x}_0\hat{\boldsymbol{\beta}} + t_{n-p}^{\alpha/2}\hat{\sigma}\sqrt{1 + \mathbf{x}_0(\mathbf{X}'\mathbf{X})^{-1}\mathbf{x}_0'} < Y_0 < \mathbf{x}_0\hat{\boldsymbol{\beta}} + t_{n-p}^{\alpha/2}\hat{\sigma}\sqrt{1 + \mathbf{x}_0(\mathbf{X}'\mathbf{X})^{-1}\mathbf{x}_0}\right)$$

$$= 1 - \alpha.$$

We note that the interval given above is not a confidence interval for Y_0 since Y_0 is not a parameter. The probability is computed over the distribution of both Y_0 and $\mathbf{Y}$. This interval is called a *prediction interval* for Y_0.

6.4. FURTHER DISCUSSION.

We now state some further results about the estimation of $\boldsymbol{\mu}$, $\boldsymbol{\beta}$, and σ^2. We first discuss the estimation of $\boldsymbol{\mu}$. We know that $\hat{\boldsymbol{\mu}}$ is the minimum variance unbiased estimator, the MLE, and the best invariant estimator with respect to the general quadratic loss function L_1 defined in Section 6.1. It is also minimax for this loss function. Since $\hat{\boldsymbol{\mu}}$ has all these nice properties, it is perhaps hard to imagine why we should use any other estimator. Unfortuantely, if $p > 2$, then $\hat{\boldsymbol{\mu}}$ is inadmissible. For the remainder of this paragraph, we restrict attention to the loss function

$$L_0(\mathbf{d}_1; (\boldsymbol{\mu}, \sigma^2)) = \frac{\|\mathbf{d}_1 - \boldsymbol{\mu}\|^2}{\sigma^2}.$$

When $p = 1$, $\hat{\boldsymbol{\mu}}$ has long been known to be admissible for this loss function. Stein (1955) showed that when $p = 2$, $\hat{\boldsymbol{\mu}}$ is admissible, but that when $p > 2$, $\hat{\boldsymbol{\mu}}$ is an inadmissible estimator of $\boldsymbol{\mu}$. Note that any estimator that is better than $\hat{\boldsymbol{\mu}}$ is better than any invariant estimator and better than any unbiased estimator. In particular, such an estimator cannot be unbiased or invariant. James and Stein (1960) derived a class of estimators that are better than $\hat{\boldsymbol{\mu}}$. These estimators are defined in Chapter 11 and shown to be better than $\hat{\boldsymbol{\mu}}$. This class of estimators has the following interesting property. We can first choose a point that we believe $\boldsymbol{\mu}$ should be near. We then shrink $\hat{\boldsymbol{\mu}}$ toward that point in a prescribed fashion. If we are correct and $\boldsymbol{\mu}$ is near the point, then our shrunken estimator is much better than $\hat{\boldsymbol{\mu}}$ (as we would expect). However, even if we are wrong and $\boldsymbol{\mu}$ is far from the point we have shrunk toward, the shrunken estimator is still moderately better than $\hat{\boldsymbol{\mu}}$. Using this type of estimator gives us a free chance. We can guess a value for $\boldsymbol{\mu}$ at no cost.

We now discuss the estimation of $\boldsymbol{\beta}$. We have seen that $\hat{\boldsymbol{\beta}}$ is the MLE, the minimum variance unbiased estimator and the best invariant estimator for general quadratic loss functions. It is also minimax for these loss functions. However, in Chapter 11, we define a class of estimators that are better than $\boldsymbol{\beta}$ for the loss function

$$L_0^*(\mathbf{d}_1^*; (\boldsymbol{\beta}, \sigma^2)) = \frac{(\mathbf{d}_1^* - \boldsymbol{\beta})'\mathbf{X}'\mathbf{X}(\mathbf{d}_1^* - \boldsymbol{\beta})}{\sigma^2}$$

[(the Mahalanobis distance loss function, since $\hat{\boldsymbol{\beta}} \sim N_p(\boldsymbol{\beta}, \sigma^2(\mathbf{X}'\mathbf{X})^{-1})$. See Section 3.3.] We also define another class of estimators (ridge estimators) that simulation studies have indicated may be better than $\boldsymbol{\beta}$ for

$$L^*(\mathbf{d}_1^*; (\boldsymbol{\beta}, \sigma^2)) = \frac{\|\mathbf{d}_1^* - \boldsymbol{\beta}\|^2}{\sigma^2}.$$

Most of the results that have been derived for estimating $\boldsymbol{\mu}$ and $\boldsymbol{\beta}$ have been derived for the loss functions L_0 and L_0^*. However, the estimators $\hat{\boldsymbol{\mu}}$ and $\hat{\boldsymbol{\beta}}$ appear to be inadmissible for general quadratic losses.

For any quadratic loss functions, $\hat{\boldsymbol{\mu}}$ (or $\hat{\boldsymbol{\beta}}$) is minimax and has constant risk. In this situation, another estimator is as good as $\hat{\boldsymbol{\mu}}$ (or $\hat{\boldsymbol{\beta}}$) if and only if it is minimax. Therefore, searching for minimax estimators for $\boldsymbol{\mu}$ is the same as searching for estimators that are as good as $\hat{\boldsymbol{\mu}}$. If another estimator is minimax and has different risk from $\hat{\boldsymbol{\mu}}$ it must be better than $\hat{\boldsymbol{\mu}}$. Although a minimax estimator often has constant risk, if there are several minimax estimators, then one with nonconstant risk is better than one with constant risk. Equivalently, a minimax estimator with constant risk is the worst minimax estimator and is only admissible if all minimax estimators have constant risk.

We now discuss the estimation of σ^2. We note that

$$\frac{(n-p)\hat{\sigma}^2}{(n-p+2)}$$

is the best invariant estimator of σ^2 for the quadratic loss function L_2 defined in Section 6.2. Therefore, the MLE and the minimum variance unbiased estimator are inadmissible for this loss function. Stein (1964) has shown that the best invariant estimator is also inadmissible for this loss function.

6.5. THE GAUSS-MARKOV THEOREM.

We now consider a generalization of the linear model defined in Chapter 4. We drop the assumption of joint normality of the errors. We assume that we observe the $n \times 1$ vector $\mathbf{Y}$ such that

$$E(\mathbf{Y}) = \boldsymbol{\mu}, \qquad \operatorname{cov}(\mathbf{Y}) = \sigma^2 \mathbf{I}, \qquad \boldsymbol{\mu} \in V, \qquad \sigma^2 > 0,$$

where, as before, V is a specified p-dimensional subspace of R^n. It is not possible to find minimum variance unbiased estimators or maximum likelihood estimators for this model. Surprisingly, we can find best linear unbiased estimators. Let $\mathbf{a}$ be an $n \times 1$ vector, and let $T(\mathbf{Y})$ be a one-dimensional statistic. We say that $T(\mathbf{Y})$ is a *linear unbiased estimator* of $\mathbf{a}'\boldsymbol{\mu}$ if

1. $T(\mathbf{Y})$ is a linear function of $\mathbf{Y}$, say $T(\mathbf{Y}) = \mathbf{b}'\mathbf{Y}$.
2. $E(T(\mathbf{Y})) = \mathbf{a}'\boldsymbol{\mu}$.

We say that $T(\mathbf{Y})$ is the *best linear unbiased estimator* of $\mathbf{a}'\boldsymbol{\mu}$ if it is the linear unbiased estimator that has the smallest variance.

LEMMA 6.6. **$\mathbf{b}'\mathbf{Y}$ is a linear unbiased estimator of $\mathbf{a}'\boldsymbol{\mu}$ if and only if $\mathbf{P}_V\mathbf{a} = \mathbf{P}_V\mathbf{b}$.**

PROOF. $E(\mathbf{b}'\mathbf{Y}) = \mathbf{b}'E(\mathbf{Y})$. Therefore, $\mathbf{b}'\mathbf{Y}$ is a linear unbiased estimator of $\mathbf{a}'\boldsymbol{\mu}$ if and only if $\mathbf{b}'\boldsymbol{\mu} = \mathbf{a}'\boldsymbol{\mu}$ for every $\boldsymbol{\mu} \in V$, if and only if $\langle \mathbf{a} - \mathbf{b}, \boldsymbol{\mu} \rangle = 0$ for all $\boldsymbol{\mu} \in V$ if and only if $(\mathbf{a} - \mathbf{b}) \in V^{\perp}$, if and only if $\mathbf{P}_V(\mathbf{a} - \mathbf{b}) = 0$. □

THEOREM 6.7. (Gauss-Markov). $\mathbf{a}'\hat{\boldsymbol{\mu}}$ is the best linear unbiased estimator of $\mathbf{a}'\boldsymbol{\mu}$.

PROOF. Clearly $\mathbf{a}'\hat{\boldsymbol{\mu}} = \mathbf{a}'\mathbf{P}_V\mathbf{Y}$ is linear. Also

$$E(\mathbf{a}'\hat{\boldsymbol{\mu}}) = E(\mathbf{a}'\mathbf{P}_V\mathbf{Y}) = \mathbf{a}'\mathbf{P}_V E(\mathbf{Y}) = \mathbf{a}'\mathbf{P}_V\boldsymbol{\mu} = \mathbf{a}'\boldsymbol{\mu},$$

since $\boldsymbol{\mu} \in V$. Therefore, $\mathbf{a}'\hat{\boldsymbol{\mu}}$ is a linear unbiased estimator of $\mathbf{a}'\boldsymbol{\mu}$. In addition

$$\operatorname{var}(\mathbf{a}'\hat{\boldsymbol{\mu}}) = \operatorname{var}(\mathbf{a}'\mathbf{P}_V\mathbf{Y}) = \sigma^2\mathbf{a}'\mathbf{P}_V\mathbf{a}.$$

Now let $\mathbf{b}'\mathbf{Y}$ be any other linear unbiased estimator of $\mathbf{a}'\boldsymbol{\mu}$. By Lemma 6.6, $\mathbf{P}_V\mathbf{b} = \mathbf{P}_V\mathbf{a}$. Therefore,

$$\begin{aligned}\operatorname{var}(\mathbf{b}'\mathbf{Y}) &= \sigma^2\mathbf{b}'\mathbf{b} = \sigma^2\|\mathbf{b}\|^2 = \sigma^2\|\mathbf{b} - \mathbf{P}_V\mathbf{b}\|^2 + \sigma^2\|\mathbf{P}_V\mathbf{b}\|^2 \\ &= \sigma^2\|\mathbf{b} - \mathbf{P}_V\mathbf{b}\|^2 + \sigma^2\|\mathbf{P}_V\mathbf{a}\|^2 \geqslant \sigma^2\|\mathbf{P}_V\mathbf{a}\|^2 \\ &= \operatorname{var}(\mathbf{a}'\hat{\boldsymbol{\mu}}),\end{aligned}$$

with equality if and only if $b = \mathbf{P}_V\mathbf{b} = \mathbf{P}_V\mathbf{a}$ if and only if $\mathbf{b}'Y = \mathbf{a}'\mathbf{P}_V\mathbf{Y} = \mathbf{a}'\hat{\boldsymbol{\mu}}$. □

Now let $\mathbf{A}$ be a $k \times n$ matrix. Let $\mathbf{T}(\mathbf{Y})$ be a $k \times 1$ statistic. We say that $\mathbf{T}(\mathbf{Y})$ is the *best linear unbiased estimator* of $\mathbf{A}\boldsymbol{\mu}$ if the components of $\mathbf{T}$ are the best linear unbiased estimators of the components of $\mathbf{A}\boldsymbol{\mu}$.

COROLLARY 1. a. $\hat{\boldsymbol{\mu}}$ is the best linear unbiased estimator of $\boldsymbol{\mu}$.
b. $\mathbf{A}\hat{\boldsymbol{\mu}}$ is the best linear unbiased estimator of $\mathbf{A}\boldsymbol{\mu}$.

We now consider the coordinatized version of the problem. Let $\mathbf{X}$ be a basis matrix for V, and let $\boldsymbol{\beta}$ and $\hat{\boldsymbol{\beta}}$ be the unique solutions to $\boldsymbol{\mu} = \mathbf{X}\boldsymbol{\beta}$ and $\hat{\boldsymbol{\mu}} = \mathbf{X}\hat{\boldsymbol{\beta}}$.

COROLLARY 2. $\hat{\boldsymbol{\beta}}$ is the best linear unbiased estimator of $\boldsymbol{\beta}$.

The following corollary shows that we can get a slightly stronger result.

COROLLARY 3. Let $\mathbf{BY}$ be a linear unbiased estimator of $\mathbf{A}\boldsymbol{\mu}$. Then

$$\operatorname{cov}(\mathbf{BY}) - \operatorname{cov}(\mathbf{A}\hat{\boldsymbol{\mu}}) \geqslant 0.$$

[That is, $\operatorname{cov}(\mathbf{BY}) - \operatorname{cov}(\mathbf{A}\hat{\boldsymbol{\mu}})$ is nonnegative definite.]

PROOF. See Exercise B5. □

As a final result in this section, we show that $\hat{\sigma}^2$ is an unbaised estimator of σ^2 in the more general model defined in this section.

THEOREM 6.8. $E\hat{\sigma}^2 = \sigma^2$.

PROOF. By Lemma 3.1,

$$E\hat{\sigma}^2 = \frac{1}{n-p} E\|\mathbf{P}_{V^\perp}\mathbf{Y}\|^2 = \frac{1}{n-p}\left(\operatorname{tr}(\operatorname{cov}(\mathbf{P}_{V^\perp}\mathbf{Y}))\right) + \|E\mathbf{P}_{V^\perp}\mathbf{Y}\|^2$$

$$= \frac{1}{n-p}\left(\sigma^2 \operatorname{tr}\mathbf{P}_{V^\perp} + 0\right) = \frac{1}{n-p}(n-p)\sigma^2 = \sigma^2. \quad \square$$

The results of this section can be interpreted as showing that the estimators derived under the normal assumptions of Section 6.1, are not bad estimators even when that assumption is removed. The estimators for linear functions of the mean vector are still best linear unbiased estimators for this more general model, whereas the estimator of σ^2 is at least unbiased.

EXERCISES.

Type A

1. Use the data of Exercise A1 of Chapter 5 to do the following:

(a) Find 95% confidence intervals for β_0, β_1, and β_2.

(b) Find a 95% confidence interval for σ^2.

(c) Find a 90% confidence interval for $2\beta_0 + 3\beta_1 + \beta_2$.

(d) Find a prediction interval for the Y value that would be observed if X were 3.

Type B

1. Prove the corollary to Theorem 6.1. (Note that $\mathbf{C\beta} = \mathbf{A\mu}$ for some $\mathbf{A}$.)

2. Prove Theorem 6.3. [Note that $T(\hat{\boldsymbol{\mu}}, \hat{\sigma}^2)$ is an invariant estimator of σ^2 if and only if $T(\hat{\boldsymbol{\mu}}, \hat{\sigma}^2) = \hat{\sigma}^2 T(\mathbf{0}, 1)$. Why? Note also that $(n-p)\hat{\sigma}^2 \sim \chi^2_{n-p}$.]

3. Prove Lemma 6.4.

4. Let $\mathbf{A}$ be a $k \times n$ matrix such that $\mathbf{A}P_V\mathbf{A}'$ has rank k. A $(1-\alpha)$ confidence region for $\mathbf{A\mu}$ is a set $S(\hat{\boldsymbol{\mu}}, \hat{\sigma}^2)$ such that

$$P\left(\mathbf{A\mu} \in S(\hat{\boldsymbol{\mu}}, \hat{\sigma}^2)\right) = 1 - \alpha$$

where this probability is computed over the distribution of $(\hat{\boldsymbol{\mu}}, \hat{\sigma}^2)$.

(a) Show that

$$\frac{(\mathbf{A}\hat{\boldsymbol{\mu}} - \mathbf{A\mu})'(\mathbf{AP}_V\mathbf{A})^{-1}(\mathbf{A}\hat{\boldsymbol{\mu}} - \mathbf{A\mu})}{k\hat{\sigma}^2} \sim F_{k,n-p}(0).$$

(Hint: See Theorem 3.10.)

(b) Let

$$S(\hat{\boldsymbol{\mu}}, \hat{\sigma}^2) = \left\{ \mathbf{t}: \frac{(\mathbf{A}\hat{\boldsymbol{\mu}} - \mathbf{t})'(\mathbf{A}\mathbf{P}_V\mathbf{A}')^{-1}(\mathbf{A}\hat{\boldsymbol{\mu}} - \mathbf{t})}{k\hat{\sigma}^2} < F^{\alpha}_{k,n-p} \right\}.$$

Show that $S(\hat{\boldsymbol{\mu}}, \hat{\sigma}^2)$ is a $1 - \alpha$ confidence region for $\mathbf{A}\boldsymbol{\mu}$.

(c) Find a $1 - \alpha$ confidence region for $\mathbf{C}\boldsymbol{\beta}$ where $\mathbf{C}$ is a $k \times p$ matrix of rank k.

5. Prove the corollaries to Theorem 6.7. (Hint: $\mathbf{s}'(\mathrm{cov}(\mathbf{BY}) - \mathrm{cov}(\mathbf{A}\hat{\boldsymbol{\mu}}))\mathbf{s} = \mathrm{var}\,\mathbf{s}'\mathbf{BY} - \mathrm{var}\,\mathbf{s}'\mathbf{A}\hat{\boldsymbol{\mu}}$.)

6. An alternative proof of the Gauss-Markov theorem. Consider the general model of Section 6.5.

(a) Show that mean and variance of any linear function $\mathbf{a}'\mathbf{Y}$ are the same for the general model as for the normal model considered in Sections 6.1–6.4.

(b) Show that an estimator is a linear unbiased estimator for the more general model if and only if it is a linear unbiased estimator for the normal model.

(c) Show that $\mathbf{a}'\hat{\boldsymbol{\mu}}$ is the best linear unbiased estimator of $\mathbf{a}'\boldsymbol{\mu}$ for the general model. (It is the best linear unbiased estimator for the normal model, since it is the minimum variance unbaised estimator for that model.)

Type C

1. Case in which $\mathbf{X}$ has less than full rank. Let $\mathbf{X}$ be an $n \times r$ matrix of rank $p < r$. In this problem we consider the model in which we observe $\mathbf{Y} \sim N_n(\mathbf{X}\boldsymbol{\beta}, \sigma^2\mathbf{I})$, $\boldsymbol{\beta} \in R^r$, $\sigma^2 > 0$. (Let $\boldsymbol{\beta}_1$ and $\boldsymbol{\beta}_2$ be vectors such that $\mathbf{X}\boldsymbol{\beta}_1 = \mathbf{X}\boldsymbol{\beta}_2$. Then the distribution of $\mathbf{Y}$ is the same whether $\boldsymbol{\beta}_1$ or $\boldsymbol{\beta}_2$ is the true parameter value. Therefore, there is no way we can distinguish from the data whether $\boldsymbol{\beta}_1$ or $\boldsymbol{\beta}_2$ is the true parameter. In this situation we can say that the parameter $\boldsymbol{\beta}$ is not identified.) An estimator $\hat{\boldsymbol{\beta}}$ of $\boldsymbol{\beta}$ is an *ordinary least squares* (OLS) estimator of $\boldsymbol{\beta}$ if

$$\|\mathbf{Y} - \mathbf{X}\hat{\boldsymbol{\beta}}\|^2 \leqslant \|\mathbf{Y} - \mathbf{X}\mathbf{b}\|^2$$

for all $\mathbf{b} \in R^r$. Now, let $\boldsymbol{\mu} = \mathbf{X}\boldsymbol{\beta}$ and let V be the p-dimensional subspace spanned by the columns of $\mathbf{X}$. Finally, let $\hat{\boldsymbol{\mu}} = \mathbf{P}_V\mathbf{Y}$, $\hat{\sigma}^2 = \|\mathbf{P}_{V^\perp}\mathbf{Y}\|^2/(n - p)$ and that $(n - p)\hat{\sigma}^2 = \|\mathbf{Y} - \mathbf{X}\hat{\mathbf{B}}\|^2$

(a) Show that $\hat{\boldsymbol{\beta}}$ is an OLS estimator of $\boldsymbol{\beta}$ if and only if $\mathbf{X}\hat{\boldsymbol{\beta}} = \hat{\boldsymbol{\mu}}$.

(b) Show that $\mathbf{X}\hat{\boldsymbol{\beta}} = \hat{\boldsymbol{\mu}}$ if and only if $\mathbf{X}'\mathbf{X}\hat{\boldsymbol{\beta}} = \mathbf{X}'\mathbf{Y}$. (Hint: use Exercise C4 of Chapter 2.)

(c) Let $\mathbf{A}$ be a generalized inverse of $\mathbf{X}'\mathbf{X}$. Show that $\hat{\boldsymbol{\beta}}$ is an OLS estimator of $\boldsymbol{\beta}$ if and only if

$$\hat{\boldsymbol{\beta}} = \mathbf{AX'Y} + (\mathbf{I} - \mathbf{AX'X})\mathbf{b}(\mathbf{Y})$$

for some function $\mathbf{b}(\mathbf{Y})$. In particular, $\mathbf{AX'Y}$ is an OLS estimator of $\boldsymbol{\beta}$. (Hint: If $\mathbf{X'X}\hat{\boldsymbol{\beta}} = \mathbf{X'Y}$, then $\hat{\boldsymbol{\beta}} = \mathbf{AX'Y} + (\mathbf{I} - \mathbf{AX'X})\hat{\boldsymbol{\beta}}$. Why?)

(d) Let $\mathbf{B}$ be the g-inverse of $\mathbf{X'X}$. Show that the OLS estimator of shortest length is $\hat{\boldsymbol{\beta}} = \mathbf{BX'Y}$. [Hint: Show that $\mathbf{X'XB} = \mathbf{BX'X}$ and then show that $\mathbf{BX'Y}$ and $(\mathbf{I} - \mathbf{BX'X})\mathbf{b}(\mathbf{Y})$ are orthogonal.]

(e) Show that there is no unbiased estimator of $\boldsymbol{\beta}$. (Hint: Suppose that $E\hat{\boldsymbol{\beta}} = \boldsymbol{\beta}$. Let $\mathbf{X}\boldsymbol{\beta} = \mathbf{X}\boldsymbol{\beta}^*$. Then the distribution of $\mathbf{Y}$ is the same when $\boldsymbol{\beta}^*$ is true as when $\boldsymbol{\beta}$ is true, and therefore, $E\hat{\boldsymbol{\beta}} = \boldsymbol{\beta}^*$ also, which is a contradiction.)

(f) Let $\hat{\boldsymbol{\beta}}$ be an OLS estimator of $\boldsymbol{\beta}$, and let $\hat{\hat{\sigma}}^2 = \|\mathbf{Y} - \mathbf{X}\hat{\boldsymbol{\beta}}\|^2/n$. Show that $(\hat{\boldsymbol{\beta}}, \hat{\sigma}^2)$ is a MLE of $(\boldsymbol{\beta}, \sigma^2)$. Note that the MLE is not unique for this model.

(g) A linear function $\mathbf{c}'\boldsymbol{\beta}$ is said to be an *estimable* function if there exists a linear function $\mathbf{a}'\mathbf{Y}$ that is an unbiased estimator of $\mathbf{c}'\boldsymbol{\beta}$. Show that $\mathbf{c}'\boldsymbol{\beta}$ is an estimable function if and only if $\mathbf{c} = \mathbf{X}'\mathbf{a}$ for some $\mathbf{a}$, that is, if and only if $\mathbf{c}'\boldsymbol{\beta} = \mathbf{a}'\boldsymbol{\mu}$. [Hint: If $E\mathbf{a}'\mathbf{Y} = \mathbf{c}'\boldsymbol{\beta}$, then $(\mathbf{c}' - \mathbf{a}'\mathbf{X})\boldsymbol{\beta} = 0$ for all $\boldsymbol{\beta} \in R^r$.]

(h) Let $\mathbf{c}'\boldsymbol{\beta} = \mathbf{a}'\boldsymbol{\mu}$ be an estimable function and let $\hat{\boldsymbol{\beta}}$ be any OLS estimator of $\boldsymbol{\beta}$. Show that $\mathbf{c}'\hat{\boldsymbol{\beta}} = \mathbf{a}'\hat{\boldsymbol{\mu}}$ and hence $\mathbf{c}'\hat{\boldsymbol{\beta}}$ is the minimum variance unbiased estimator of $\mathbf{c}'\boldsymbol{\beta}$. (Note that $\mathbf{c}'\hat{\boldsymbol{\beta}}$ is uniquely defined. It does not depend on the choice of OLS estimator.)

CHAPTER 7

Tests about the Mean

In this chapter we consider the testing problem that often is called testing the general linear hypothesis. The problem studied in this chapter contains most analysis of variance testing problems, as well as many regression problems. In Section 7.1, we state the problem, and define the F-statistic. We find its distribution and define the usual F-test. In Sections 7.2–7.5, we give some examples of problems that are of the form considered in Section 7.1. For most of these problems we derive formulas for the F-statistic. In Section 7.2, the examples have to do with regression problems, whereas those in Sections 7.3–7.5 are analysis of variance and analysis of covariance problems. The methods used to derive the F-statistics in these two situations are different. Regression problems have a natural basis matrix, which is then used to find the projections used in the F-statistic. For analysis of variance problems, there is usually no natural basis matrix, so we use a method based on the least-squares property (Theroem 2.5b) to derive the projections for these models. Following these examples, we show in Section 7.6 that the F-test is unbiased and UMP invariant. Other properties of the F-test are also stated. In Sections 7.7–7.11 other aspects of this testing problem are treated.

Many of the F-test discussed in Section 7.2–7.5 were derived by Fisher. (See, for example, Fisher, 1918, Fisher, 1925, and Fisher, 1935.) The UMP invariance of these tests is due to Hunt and Stein (1946).

7.1. THE F-TEST.

We consider the testing problem in which we observe the n-variate random vector $\mathbf{Y}$ such that

$$\mathbf{Y} \sim N_n(\boldsymbol{\mu}, \sigma^2\mathbf{I}), \qquad \boldsymbol{\mu} \in V, \qquad \sigma^2 > 0,$$

and we are testing the null hypothesis that $\boldsymbol{\mu} \in W \subset V$, where V and W are both subspaces, $\dim(V) = p$, and $\dim(W) = k$, $k < p < n$. Many of the most commonly tested hypotheses fall into this framework. Define

$$F = \frac{\|\mathbf{P}_{V|W}\mathbf{Y}\|^2(n-p)}{\|\mathbf{P}_{V^\perp}\mathbf{Y}\|^2(p-k)}. \tag{7.1}$$

THEOREM 7.1.

$$F \sim F_{p-k,n-p}\left(\frac{\|\mathbf{P}_{V|W}\boldsymbol{\mu}\|^2}{\sigma^2}\right).$$

PROOF. $(V|W) \perp V^\perp$. Therefore, by Theorem 3.11b, $\mathbf{P}_{V|W}\mathbf{Y}$ is independent of $\mathbf{P}_{V^\perp}\mathbf{Y}$, and hence $\|\mathbf{P}_{V|W}\mathbf{Y}\|^2$ is independent of $\|\mathbf{P}_{V^\perp}\mathbf{Y}\|^2$. Also, $\dim(V^\perp) = n - p$. By Theorem 3.12, therefore,

$$\|\mathbf{P}_{V^\perp}\mathbf{Y}\|^2 \sim \sigma^2\chi^2_{n-p}\left(\frac{\|\mathbf{P}_{V^\perp}\boldsymbol{\mu}\|^2}{\sigma^2}\right)$$

However, $\boldsymbol{\mu} \in$ V, and therefore $\mathbf{P}_{V^\perp}\boldsymbol{\mu} = \mathbf{0}$. Similarly, $\dim(V|W) = p - k$. Therefore,

$$\|\mathbf{P}_{V|W}\mathbf{Y}\|^2 \sim \sigma^2\chi^2_{p-k}\left(\frac{\|\mathbf{P}_{V|W}\boldsymbol{\mu}\|^2}{\sigma^2}\right)$$

The result follows. □

COROLLARY . If $\boldsymbol{\mu} \in W$, then $F \sim F_{p-k,n-p}(0)$.

PROOF. If $\boldsymbol{\mu} \in W$, then $\mathbf{P}_{V|W}\boldsymbol{\mu} = \mathbf{0}$. □

We now find some alternative expressions for F.

LEMMA 7.2.

$$F = \frac{\|\mathbf{P}_{V|W}\hat{\boldsymbol{\mu}}\|^2}{(p-k)\hat{\sigma}^2},$$

$$\|\mathbf{P}_{V|W}\hat{\boldsymbol{\mu}}\|^2 = \|\hat{\boldsymbol{\mu}} - \mathbf{P}_W\hat{\boldsymbol{\mu}}\|^2 = \|\hat{\boldsymbol{\mu}}\|^2 - \|\mathbf{P}_W\hat{\boldsymbol{\mu}}\|^2.$$

PROOF. See Exercise B1. □

Note that F is a function of the complete sufficient statistic $(\hat{\boldsymbol{\mu}}, \hat{\sigma}^2)$.

By the corollary to Lemma 7.1, we can design a size α test in the following way. Let F be defined by (7.1) and let

$$\phi(F) = \begin{cases} 1 & \text{if } F > F^{\alpha}_{p-k,n-p} \\ 0 & \text{if } F \leqslant F^{\alpha}_{p-k,n-p} \end{cases}. \tag{7.2}$$

In a later section we show that this test is UMP invariant size α and unbiased. In Exercise B2, it is shown that ϕ is the likelihood ratio test. In the next sections, we look at some examples of hypotheses of this form and work out expressions for F.

7.2. MULTIPLE REGRESSION.

We first consider some testing problems in regression analysis. These problems are characterized by the existence of a natural basis matrix $\mathbf{X}$ for the subspace V. As before, let $\boldsymbol{\beta}$ and $\hat{\boldsymbol{\beta}}$ be the unique solutions to the equations $\boldsymbol{\mu} = \mathbf{X}\boldsymbol{\beta}$, $\hat{\boldsymbol{\mu}} = \mathbf{X}\hat{\boldsymbol{\beta}}$.

7.2.1. *Multiple Regression in the Homogeneous Case*

Consider the problem in which we observe

$$\mathbf{Y} \sim N_n(\mathbf{X}\boldsymbol{\beta}, \sigma^2\mathbf{I}), \qquad \boldsymbol{\beta} \in R^p, \qquad \sigma^2 > 0$$

and we want to test the null hypothesis that $\mathbf{A}\boldsymbol{\beta} = \mathbf{0}$ for some known $(p-k) \times p$ matrix $\mathbf{A}$ of rank $p-k$. Let V be the subspace spanned by the columns of $\mathbf{X}$, (i.e., $\mathbf{X}$ is a basis matrix for V). Let W be the subspace of V in which

$$\mathbf{0} = \mathbf{A}\boldsymbol{\beta} = \mathbf{A}(\mathbf{X}'\mathbf{X})^{-1}\mathbf{X}'\mathbf{X}\boldsymbol{\beta} = \mathbf{A}(\mathbf{X}'\mathbf{X})^{-1}\mathbf{X}'\boldsymbol{\mu}.$$

We are therefore testing the null hypothesis that $\boldsymbol{\mu} \in W$ against the alternative that $\boldsymbol{\mu} \in V$. We have a basis matrix for V. We now find a basis matrix for $V|W$.

LEMMA 7.3. $\mathbf{C} = \mathbf{X}(\mathbf{X}'\mathbf{X})^{-1}\mathbf{A}'$ is a basis matrix for $V|W$; $\dim(W) = k$.

PROOF. We first show that the columns of $\mathbf{C}$ are linearly independent. Let $\mathbf{C} = (\mathbf{C}_1, \ldots, \mathbf{C}_{p-k})$ and suppose that $\mathbf{0} = \sum_{i=1}^{p-k} b_i\mathbf{C}_i = \mathbf{C}\mathbf{b}$. Then $\mathbf{0} = (\mathbf{A}\mathbf{A}')^{-1}\mathbf{A}\mathbf{X}'\mathbf{C}\mathbf{b} = \mathbf{b}$. ($\mathbf{A}\mathbf{A}'$ is invertible because $\mathbf{A}$ has rank $p-k$.) Therefore, the columns of $\mathbf{C}$ are linearly independent. Now let U be the subspace spanned by the columns of $\mathbf{C}$, so that $\mathbf{C}$ is a basis matrix for U. Let $\mathbf{u} \in U$. Then $\mathbf{u} = \mathbf{X}(\mathbf{X}'\mathbf{X})^{-1}\mathbf{A}'\mathbf{b} = \mathbf{X}\mathbf{c}$ for some $\mathbf{c} \in R^p$. Hence $u \in V$. Therefore $U \subset V$. W is the subspace in which $\mathbf{C}'\boldsymbol{\mu} = \mathbf{0}$ and hence $W = V|U$. By Lemma 2.9, $V|W = V|V|U) = U$. Since $\mathbf{C}$ is $n \times (p-k)$, $\dim(V|W) = p-k$ and hence $\dim(W) = k$. □

THEOREM 7.4. For testing $\mathbf{A}\boldsymbol{\beta} = \mathbf{0}$, in the coordinatized linear model.

$$F = \frac{\hat{\boldsymbol{\beta}}'\mathbf{A}'\left(\mathbf{A}(\mathbf{X}'\mathbf{X})^{-1}\mathbf{A}'\right)^{-1}\mathbf{A}\hat{\boldsymbol{\beta}}}{(p-k)\hat{\sigma}^2} \sim F_{p-k,\,n-p}\left(\frac{\boldsymbol{\beta}'\mathbf{A}'\left(\mathbf{A}(\mathbf{X}'\mathbf{X})^{-1}\mathbf{A}'\right)^{-1}\mathbf{A}\boldsymbol{\beta}}{\sigma^2}\right)$$

PROOF. See Exercise B3. □

7.2.2. Multiple Regression in the Nonhomogeneous Case

We continue the notation of the last section, except that we now want to test the null hypothesis that $\mathbf{A}\boldsymbol{\beta} = \mathbf{b}$, for some specified vector $\mathbf{b}$. We first make a transformation to reduce this hypothesis to the homogeneous case considered in the last section. Let $\boldsymbol{\beta}_0$ be a solution to the equation $\mathbf{A}\boldsymbol{\beta} = \mathbf{b}$. Let $\mathbf{Y}^* = \mathbf{Y} - \mathbf{X}\boldsymbol{\beta}_0$, $\boldsymbol{\beta}^* = \boldsymbol{\beta} - \boldsymbol{\beta}_0$. Then

$$\mathbf{Y}^* \sim N_n(\mathbf{X}\boldsymbol{\beta}^*, \sigma^2\mathbf{I})$$

and $\mathbf{A}\boldsymbol{\beta} = \mathbf{b}$ if and only if $\mathbf{A}\boldsymbol{\beta}^* = \mathbf{0}$. Therefore the hypothesis tested in the transformed problem is $\mathbf{A}\boldsymbol{\beta}^* = \mathbf{0}$. Since the transformation from $\mathbf{Y}$ to $\mathbf{Y}^*$ is invertible, this transformed problem is really the same as the original problem (see Section 1.6). It is also of the form considered in Section 7.2.1. Therefore, let

$$\hat{\boldsymbol{\beta}}^* = (\mathbf{X}'\mathbf{X})^{-1}\mathbf{X}'\mathbf{Y}^* = \hat{\boldsymbol{\beta}} - \boldsymbol{\beta}_0,$$

$$(n-p)\hat{\sigma}^{*2} = (\mathbf{Y}^* - \mathbf{X}\hat{\boldsymbol{\beta}}^*)(\mathbf{Y} - \mathbf{X}\hat{\boldsymbol{\beta}}^*) = (n-p)\hat{\sigma}^2.$$

By Theorem 7.4,

$$F = \frac{\hat{\boldsymbol{\beta}}^*\mathbf{A}'\left(\mathbf{A}(\mathbf{X}'\mathbf{X})^{-1}\mathbf{A}'\right)^{-1}\mathbf{A}\hat{\boldsymbol{\beta}}^*}{(p-k)\hat{\sigma}^{*2}}$$

$$= \frac{(\mathbf{A}\hat{\boldsymbol{\beta}} - \mathbf{b})'\left(\mathbf{A}(\mathbf{X}'\mathbf{X})^{-1}\mathbf{A}'\right)^{-1}(\mathbf{A}\hat{\boldsymbol{\beta}} - \mathbf{b})}{(p-k)\hat{\sigma}^2}. \tag{7.3}$$

Note that F does not depend on the choice of $\boldsymbol{\beta}_0$.

Now, let $\boldsymbol{\beta} = \binom{\boldsymbol{\beta}_1}{\boldsymbol{\beta}_2}$ where $\boldsymbol{\beta}_1$ is $(p-k)\times 1$. We want to test that $\boldsymbol{\beta}_1 = \mathbf{b}$ against general alternatives. Let $\mathbf{A} = (\mathbf{I}, \mathbf{0})$ where $\mathbf{I}$ is $(p-k)\times(p-k)$, and $\mathbf{0}$ is $(p-k)\times k$. Then $\boldsymbol{\beta}_1 = \mathbf{b}$ if and only if $\mathbf{A}\boldsymbol{\beta} = \mathbf{b}$. This problem is therefore in the above framework. Let

$$\hat{\boldsymbol{\beta}} = \begin{pmatrix} \hat{\boldsymbol{\beta}}_1 \\ \hat{\boldsymbol{\beta}}_2 \end{pmatrix}, \qquad (\mathbf{X}'\mathbf{X})^{-1} = \begin{pmatrix} \mathbf{Q}_{11} & \mathbf{Q}_{12} \\ \mathbf{Q}_{21} & \mathbf{Q}_{22} \end{pmatrix},$$

where $\hat{\boldsymbol{\beta}}_1$ is $(p-k)\times 1$ and $\mathbf{Q}_{11}$ is $(p-k)\times(p-k)$. Then

$$F = \frac{(\hat{\boldsymbol{\beta}}_1 - \mathbf{b})'\mathbf{Q}_{11}^{-1}(\hat{\boldsymbol{\beta}}_1 - \mathbf{b})}{(p-k)\hat{\sigma}^2}. \tag{7.4}$$

7.2.3. *Regression with an Intercept Term*

We now consider the regression model with an intercept defined in Section 5.5, in which

$$\mathbf{X} = \begin{bmatrix} 1 & \mathbf{T}_1 \\ \vdots & \vdots \\ 1 & \mathbf{T}_n \end{bmatrix}, \qquad \boldsymbol{\beta} = \begin{pmatrix} \delta \\ \boldsymbol{\gamma} \end{pmatrix},$$

where the $\mathbf{T}_i'$ are known $(p-1)$-dimensional vectors, δ is an unknown number, and $\boldsymbol{\gamma}$ is an unknown $(p-1)$-dimensional vector. We want to test that $\mathbf{A}\boldsymbol{\gamma} = \mathbf{0}$, where $\mathbf{A}$ is a $(p-k) \times (p-1)$ matrix of rank $p-k$. We now derive the test of this hypothesis. Although this model fits in the framework of the last section, it is easier to derive the results directly, rather than using Theorem 7.4.

Let $\tilde{\mathbf{T}}$ and $\tilde{\mathbf{Y}}$ be defined by (5.4), and let $\mathbf{1}$ be an $\mathbf{n}$-dimensional vector all of whose elements are 1. Then

$$\boldsymbol{\mu} = E\mathbf{Y} = \delta\mathbf{1} + \mathbf{T}\boldsymbol{\gamma} = (\delta + \overline{\mathbf{T}}\boldsymbol{\gamma})\mathbf{1} + \tilde{\mathbf{T}}\boldsymbol{\gamma},$$
$$\hat{\boldsymbol{\mu}} = \hat{\delta}\mathbf{1} + \mathbf{T}\hat{\boldsymbol{\gamma}} = (\hat{\delta} + \overline{\mathbf{T}}\hat{\boldsymbol{\gamma}})\mathbf{1} + \tilde{\mathbf{T}}\hat{\boldsymbol{\gamma}}.$$

Also $\tilde{\mathbf{T}}'\mathbf{1} = \mathbf{0}$. Therefore $\tilde{\mathbf{T}}'\boldsymbol{\mu} = \tilde{\mathbf{T}}'\tilde{\mathbf{T}}\boldsymbol{\gamma}$ and $\tilde{\mathbf{T}}'\hat{\boldsymbol{\mu}} = \tilde{\mathbf{T}}'\tilde{\mathbf{T}}\hat{\boldsymbol{\gamma}}$. Hence $\mathbf{A}(\tilde{\mathbf{T}}'\tilde{\mathbf{T}})^{-1}\tilde{\mathbf{T}}'\boldsymbol{\mu} = \mathbf{A}\boldsymbol{\gamma}$. As before, let V be the p-dimensional subspace spanned by the columns of $\mathbf{X}$, and let W be the k-dimensional subspace in which $\mathbf{A}\boldsymbol{\gamma} = 0$. The following lemma is proved in the same way as Lemma 7.3. The proof is left for an exercise.

LEMMA 7.5. $\mathbf{C} = \tilde{\mathbf{T}}(\tilde{\mathbf{T}}'\tilde{\mathbf{T}})^{-1}\mathbf{A}'$ is a basis matrix for $V \mid W$.

COROLLARY 1. For testing that $\mathbf{A}\boldsymbol{\gamma} = 0$ in the regression model with an intercept,

$$F = \frac{(\mathbf{A}\hat{\boldsymbol{\gamma}})'\left(\mathbf{A}(\tilde{\mathbf{T}}'\tilde{\mathbf{T}})^{-1}\mathbf{A}'\right)^{-1}(\mathbf{A}\hat{\boldsymbol{\gamma}})}{(p-k)\hat{\sigma}^2}.$$

PROOF. The same as Theorem 7.4. □

Let $\boldsymbol{\gamma} = \binom{\boldsymbol{\gamma}_1}{\boldsymbol{\gamma}_2}$ where $\boldsymbol{\gamma}_1$ is a $(p-k)$-dimensional vector. We now consider testing that $\boldsymbol{\gamma}_1 = \mathbf{0}$. As before, let $\mathbf{A} = (\mathbf{I}\ \mathbf{0})$. Then $\mathbf{A}\boldsymbol{\gamma} = \boldsymbol{\gamma}_1$. Partition $\hat{\boldsymbol{\gamma}}$ by $\hat{\boldsymbol{\gamma}} = \binom{\hat{\boldsymbol{\gamma}}_1}{\hat{\boldsymbol{\gamma}}_2}$ where $\hat{\boldsymbol{\gamma}}_1$ is also $(p-k)$-dimensional. Then $\mathbf{A}\hat{\boldsymbol{\gamma}} = \hat{\boldsymbol{\gamma}}_1$. Now, let

$$\tilde{\mathbf{Q}} = (\tilde{\mathbf{T}}'\tilde{\mathbf{T}})^{-1} = \begin{pmatrix} \tilde{\mathbf{Q}}_{11} & \tilde{\mathbf{Q}}_{12} \\ \tilde{\mathbf{Q}}_{21} & \tilde{\mathbf{Q}}_{22} \end{pmatrix}.$$

Then $\mathbf{A}\tilde{\mathbf{Q}}\mathbf{A}' = \tilde{\mathbf{Q}}_{11}$.

Finally, let

$$\tilde{\mathbf{T}} = (\tilde{\mathbf{T}}_1\ \tilde{\mathbf{T}}_2) \qquad (\tilde{\mathbf{T}}'\tilde{\mathbf{T}}) = \begin{pmatrix} \tilde{\mathbf{T}}_1'\tilde{\mathbf{T}}_1 & \tilde{\mathbf{T}}_1'\tilde{\mathbf{T}}_2 \\ \tilde{\mathbf{T}}_2'\tilde{\mathbf{T}}_1 & \tilde{\mathbf{T}}_2'\tilde{\mathbf{T}}_2 \end{pmatrix} = \tilde{\mathbf{Q}}^{-1}$$

where $\tilde{\mathbf{T}}_1$ is $n \times (p-k)$. By Lemma 2 of the appendix,

$$\tilde{\mathbf{Q}}_{11}^{-1} = \tilde{\mathbf{T}}_1'\tilde{\mathbf{T}}_1 - \tilde{\mathbf{T}}_1'\tilde{\mathbf{T}}_2(\tilde{\mathbf{T}}_2'\tilde{\mathbf{T}}_2)^{-1}\tilde{\mathbf{T}}_2'\tilde{\mathbf{T}}_1.$$

We have therefore proved the following corollary:

COROLLARY 2. For testing the $\gamma_1 = \mathbf{0}$ in the regression model with an intercept

$$F = \frac{\hat{\gamma}_1'\left(\tilde{\mathbf{T}}_1'\tilde{\mathbf{T}}_1 - \tilde{\mathbf{T}}_1'\tilde{\mathbf{T}}_2(\tilde{\mathbf{T}}_2'\tilde{\mathbf{T}}_2)^{-1}\tilde{\mathbf{T}}_2'\tilde{\mathbf{T}}_1\right)\hat{\gamma}_1}{(p-k)\hat{\sigma}^2}$$

$$\sim F_{p-k,n-p}\left(\frac{\gamma_1'\left(\tilde{\mathbf{T}}_1'\tilde{\mathbf{T}}_1 - \tilde{\mathbf{T}}_1'\tilde{\mathbf{T}}_2(\tilde{\mathbf{T}}_2'\tilde{\mathbf{T}}_2)^{-1}\tilde{\mathbf{T}}_2'\tilde{\mathbf{T}}_1\right)\gamma_1}{\sigma^2}\right)$$

This result will be especially useful when we discuss the correlation model.

7.3. BALANCED ANALYSIS OF VARIANCE.

We now consider testing problems that are often considered under the heading of analysis of variance problems. One characteristic of all these problems is that there are no natural bases for the spaces involved. One appraoch that is often used for deriving the F statistic is to find bases for the spaces and then use the results of the last section. A second approach is given in Section 7.8. A third approach uses the least squares property of projections to determine them. This approach seems more efficient for generating formulas for the "clean" cases. This approach is based on the fact (see Theorem 2.5b) that $\hat{\boldsymbol{\mu}} = \mathbf{P}_V\mathbf{Y}$ minimizes

$$\|\mathbf{Y} - \mathbf{v}\|^2 = \sum_i (Y_i - v_i)^2 \tag{7.5}$$

among all vectors $\mathbf{v} \in V$. Simlarly $\hat{\hat{\boldsymbol{\mu}}} = \mathbf{P}_W\mathbf{Y}$ minimizes

$$\|\mathbf{Y} - \mathbf{w}\|^2 = \sum_i (Y_i - w_i)^2 \tag{7.6}$$

among all vectors $\mathbf{w} \in W$. Finally

$$F = \frac{\|\mathbf{P}_{V|W}\mathbf{Y}\|^2}{\|\mathbf{Y} - \mathbf{P}_V\mathbf{Y}\|^2}\,\frac{n-p}{p-k} = \frac{\|\mathbf{P}_V\mathbf{Y} - \mathbf{P}_W\mathbf{Y}\|^2}{\|\mathbf{Y} - \mathbf{P}_V Y\|^2}\,\frac{n-p}{p-k} = \frac{\sum_i(\hat{\mu}_i - \hat{\hat{\mu}}_i)^2}{\sum_i(Y_i - \hat{\mu}_i)^2}\,\frac{n-p}{p-k}. \tag{7.7}$$

In the least squares method, we first determine the $\hat{\mu}_i$ and $\hat{\hat{\mu}}_i$ by minimizing (7.5) and (7.6). We then substitute these into (7.7). To find the dim(V), note that dim(V) is the number of linearly independent parameters or is the total number of parameters minus the number of linearly independent constraints. We find dim W similarly. Finally, the numerator of the non-centrality parameter is $\|\mathbf{P}_{V|W}\boldsymbol{\mu}\|$ and is therefore the same function of the μ_i that the numerator sum of squares of F is of the Y_i.

In the following sections we assume that the reader is familiar with the usual notation used in analysis of variance problems, notation in which a dot replacing a subscript indicates that the number is an average over that subscript of the original random variables. For example, $\bar{a}_{i.k}$ is the average of all the a_{ijk} whose first subscript is i and whose last subscript is k. The average is an unweighted average of the original numbers, rather than an unweighted average of averages. For example

$$\bar{a}_{i..} = \frac{\sum_j \sum_k a_{ijk}}{\sum_j \sum_k 1}.$$

7.3.1. *Balanced One-Way Analysis of Variance*

In this model we observe Y_{ij} independent with $Y_{ij} \sim N_1(\mu_i, \sigma^2)$, $i = 1, \ldots, p$; $j = 1, \ldots, m$. This model is also called the p-sample model and is a natural extension of the two-sample model discussed in Chapter 1. The null hypothesis we want to test is that the μ_i are all the same.

We first put this model into a different form. Let

$$\theta = \bar{\mu}_., \alpha_i = \mu_i - \bar{\mu}.$$

Then

$$\mu_i = \theta + \alpha_i, \qquad \sum_i \alpha_i = 0.$$

The constraint that $\sum \alpha_i = 0$ makes θ and the α_i uniquely defined. In addition, the μ_i are all equal if and only if all the α_i are zero. Therefore, an equivalent version of this model is one in which we observe Y_{ij} independent, $Y_{ij} \sim N_1(\theta + \alpha_i, \sigma^2)$, $i = 1, \ldots, p$; $j = 1, \ldots, m$; where $\sum_i \alpha_i = 0$. We wish to test that all the $\alpha_i = 0$. To put this in the form of a linear model, let $\mathbf{Y}$ be an ordering of the Y_{ij}'s into a vector with mp variables. Similarly, let $\boldsymbol{\mu}$ be the same ordering of $\mu_{ij} = \theta + \alpha_i$. Then

$$\mathbf{Y} \sim N_{mp}(\boldsymbol{\mu}, \sigma^2 I).$$

Let V be the set of vectors $\boldsymbol{\mu}$ such that $\mu_{ij} = \theta + \alpha_i$ for some θ and α_i such that $\sum_i \alpha_i = 0$. This is clearly a subspace, by the definition of subspace. Since there are $p + 1$ parameters in V and one constraint, dim(V) = p. Now let W be the set of vectors $\boldsymbol{\mu}$ such that $\mu_{ij} = \theta$ for some θ. Then again by the definition of subspace, W is a subspace and has dimension 1. We are testing the hypothesis that $\boldsymbol{\mu} \in W$ against the alternative that $\boldsymbol{\mu} \in V$. To find $\hat{\boldsymbol{\mu}} = \mathbf{P}_V\mathbf{Y}$, we must minimize

$$\|\mathbf{Y} - \boldsymbol{\mu}\|^2 = \sum_i \sum_j (Y_{ij} - \theta - \alpha_i)^2.$$

However,

$$\begin{aligned}\sum_i \sum_j (Y_{ij} - \theta - \alpha_i)^2 &= \sum_i \sum_j \left[(Y_{ij} - \overline{Y}_{i.}) + (\overline{Y}_{i.} - \overline{Y}_{..} - \alpha_i) + (\overline{Y}_{..} - \theta) \right]^2 \\ &= \sum_i \sum_j (Y_{ij} - \overline{Y}_{i.})^2 + m \sum_i (\overline{Y}_{i.} - \overline{Y}_{..} - \alpha_i)^2 \\ &\quad + mp(\overline{Y}_{..} - \theta)^2, \qquad (7.8)\end{aligned}$$

since the cross product terms drop out. Therefore $\hat{\theta} = \overline{Y}_{..}$, $\hat{\alpha}_i = \overline{Y}_{i.} - \overline{Y}_{..}$, and $\hat{\mu}_{ij} = \hat{\mu} + \hat{\alpha}_i = \overline{Y}_{i.}$. Similarly to find $\hat{\hat{\boldsymbol{\mu}}} = \mathbf{P}_W\mathbf{Y}$, we must minimize $\sum\sum(Y_{ij} - \theta)^2$. Putting $\alpha_i = 0$ into (7.8), we see that $\hat{\hat{\theta}} = \overline{Y}_{..}$, and $\hat{\hat{\mu}}_{ij} = \overline{Y}_{..}$. Hence ($n = mp$, $p = p$, $k = 1$)

$$F = \frac{\sum_i \sum_j (\hat{\mu}_{ij} - \hat{\hat{\mu}}_{ij})^2 p(m-1)}{\sum_i \sum_j (Y_{ij} - \hat{\mu}_{ij})^2 (p-1)} = \frac{m \sum_i (\overline{Y}_{i.} - \overline{Y}_{..})^2 p(m-1)}{\sum_i \sum_j (Y_{ij} - \overline{Y}_{i.})^2 (p-1)},$$

In addition

$$\frac{\|P_{V|W}\mu\|^2}{\sigma^2} = \frac{m\sum_i (\bar{\mu}_{i.} - \bar{\mu}_{..})^2}{\sigma^2} = \frac{m\sum_i \alpha_i^2}{\sigma^2}.$$

Therefore

$$F \sim F_{p-1,\, p(m-1)}\left(\frac{m\sum_i \alpha_i^2}{\sigma^2} \right).$$

7.3.2. *Two-Way Analysis of Variance with No Replication*

In this model we observe Y_{ij} independent with $Y_{ij} \sim N_1(\mu_{ij}, \alpha^2)$, $i = 1, \ldots, r$; $j = 1, \ldots, c$. In order to be able to estimate σ^2, we must make some assumption about the μ_{ij}. The usual assumption that is made is that

$$\mu_{ij} = s_i + t_j$$

for some numbers s_i and t_j. This model is called the *additive* model for the μ_{ij}. (A more general model for this situation is treated in Section 7.9.) We want to test the hypothesis that the μ_{ij} depend only on j.

We put this model into an alternative form. Let

$$\theta = \bar{s}_. + \bar{t}_., \qquad \alpha_i = s_i - \bar{s}_., \qquad \beta_j = t_j - \bar{t}_..$$

Then

$$\mu_{ij} = \theta + \alpha_i + \beta_j, \qquad \sum_i \alpha_i = 0, \qquad \sum_j \beta_j = 0.$$

As in the previous section, the μ_{ij} are an invertible function of θ, the α_i and the β_j. (θ, the α_i and the β_j are uniquely defined with the constraints given.) We call the α_i the *row effects* and the β_j the *column effects*. Testing that the μ_{ij} depend only on j is the same as testing that the $\alpha_i = 0$ in this model. Therefore, an equivalent version of this model is one in which we observe Y_{ij} independent, $Y_{ij} \sim N_1(\theta + \alpha_i + \beta_j, \sigma^2)$; $i = 1, \ldots, r$; $j = 1, \ldots, c$; where $\sum_i \alpha_i = 0$ and $\sum_j \beta_j = 0$. We want to test that all the $\alpha_i = 0$. As before, let $\mathbf{Y}$ be an ordering of the rcY_{ij}'s and let $\boldsymbol{\mu}$ be the same ordering of the $\mu_{ij} = \theta + \alpha_i + \beta_j$. Then

$$\mathbf{Y} \sim N_{rc}(\boldsymbol{\mu}, \sigma^2 \mathbf{I}).$$

Let V be the set of $\boldsymbol{\mu}$ such that $\mu_{ij} = \theta + \alpha_i + \beta_j$ for some θ, α_i and β_j satisfying the above constraints. Then V is a subspace with dimension $p = 1 + r + c - 2 = r + c - 1$. Let W be the set of $\boldsymbol{\mu}$ such that $\mu_{ij} = \theta + \beta_j$ for some θ and β_j satisfying the constraints. Then W is a subspace of dimension $k = 1 + c - 1 = c$. We are testing that $\boldsymbol{\mu} \in W$ versus the alternative that $\boldsymbol{\mu} \in V$. To find $\mathbf{P}_V \mathbf{Y}$, we must minimize

$$\begin{aligned} &\sum_i \sum_j (Y_{ij} - \theta - \alpha_i - \beta_j)^2 \\ &\quad = \sum_i \sum_j \left(Y_{ij} - \bar{Y}_{i.} - \bar{Y}_{.j} + \bar{Y}_{..}\right)^2 + c \sum_i \left(\bar{Y}_{i.} - \bar{Y}_{..} - \alpha_i\right)^2 \qquad (7.9) \\ &\quad\quad + r \sum_j \left(\bar{Y}_{.j} - \bar{Y}_{..} - \beta_j\right)^2 + rc\left(\bar{Y}_{..} - \theta\right)^2. \end{aligned}$$

Therefore we see that $\hat{\alpha}_i = \bar{Y}_{i.} - \bar{Y}_{..}$, $\hat{\beta}_j = \bar{Y}_{.j} - \bar{Y}_{..}$, $\hat{\theta} = \bar{Y}_{..}$, and $\hat{\mu}_{ij} = \hat{\theta} + \hat{\alpha}_i + \hat{\beta}_j = \bar{Y}_{i.} + \bar{Y}_{.j} - \bar{Y}_{..}$. Similarly, using (7.9) with $\alpha_i = 0$, we find that $\hat{\hat{\theta}} = \bar{Y}_{..}$, $\hat{\hat{\beta}}_j = \bar{Y}_{.j} - \bar{Y}_{..}$ and $\hat{\hat{\mu}}_{ij} = \bar{Y}_{.j}$. Hence, ($n = rc$, $p = r + c - 1$, $k = c$)

$$F = \frac{c\sum_i \left(\bar{Y}_{i.} - \bar{Y}_{..}\right)^2}{\sum_i \sum_j \left(Y_{ij} - \bar{Y}_{i.} - \bar{Y}_{.j} + \bar{Y}_{..}\right)^2} \; \frac{(r-1)(c-1)}{(r-1)},$$

and $\|\mathbf{P}_{V|W} \boldsymbol{\mu}\|^2 = \dfrac{c\sum_i (\bar{\mu}_{i.} - \bar{\mu}_{..})^2}{\sigma^2} = \dfrac{c\sum_i \alpha_i^2}{\sigma^2}$. Therefore,

$$F \sim F_{r-1,(r-1)(c-1)}\left(\frac{c\sum_i \alpha_i^2}{\sigma^2}\right).$$

We could test $\beta_j = 0$ similarly.

7.3.3 *Two-Way Analysis of Variance with Balanced Replications*

In this model we have $m > 1$ observations on each (i, j). That is, we observe Y_{ijk} independent such that $Y_{ijk} \sim N_1(\mu_{ij}, \sigma^2)$, $i = 1, \ldots, r$; $j = 1, \ldots, c$; $k = 1, \ldots, m$. In this model it is not necessary to make any assumption about the μ_{ij}. Let

$$\theta = \bar{\mu}_{..}, \qquad \alpha_i = \bar{\mu}_{i.} - \bar{\mu}_{..}, \qquad \beta_j = \bar{\mu}_{.j} - \bar{\mu}_{..}, \qquad \gamma_{ij} = \mu_{ij} - \bar{\mu}_{i.} - \bar{\mu}_{.j} + \mu_{..}.$$

Then

$$\mu_{ij} = \theta + \alpha_i + \beta_j + \gamma_{ij}, \qquad \sum_i \alpha_i = 0, \qquad \sum_j \beta_j = 0,$$

$$\sum_i \gamma_{ij} = 0, \qquad \sum_j \gamma_{ij} = 0.$$

The constraints on α_i, β_j, and γ_{ij} make the new parameters uniquely defined. We call the α_i *row effects*, the β_j *column effects* and the γ_{ij} *interaction effects*. The α_i and the β_j are called *main effects*. The interactions represent what cannot be explained by the additive model used in the last section. Testing that the interactions are 0 is equivalent to testing that the effects are additive.

Therefore, we now consider the model in which we observe Y_{ijk} independent, such that $Y_{ijk} \sim N_1(\theta + \alpha_i + \beta_j + \gamma_{ij}, \sigma^2)$, $i = 1, \ldots, r$; $j = 1, \ldots, c$; $k = 1, \ldots, m$; where $\sum_i \alpha_i = 0$, $\sum_j \beta_j = 0$, $\sum_i \gamma_{ij} = 0$, $\sum_j \gamma_{ij} = 0$. We want to test three different hypotheses:

1. All the $\alpha_i = 0$ ($\bar{\mu}_{i.}$ does not depend on i).
2. All the $\beta_j = 0$ ($\bar{\mu}_{.j}$ does not depend on j).
3. All the $\gamma_{ij} = 0$ (the μ_{ij} are additive).

Let $\mathbf{Y}$ be an ordering of the Y_{ijk} and let $\boldsymbol{\mu}$ be the same ordering of the $\mu_{ijk} = \theta + \alpha_i + \beta_j + \gamma_{ij}$. Then $\mathbf{Y} \sim N_{mrc}(\boldsymbol{\mu}, \sigma^2\mathbf{I})$. Let V be the set of $\boldsymbol{\mu}$ such that $\mu_{ijk} = \theta + \alpha_i + \beta_j + \gamma_{ij}$. Then V is a subspace of dimension $1 + r + c + rc - 1 - 1 - (r + c - 1) = rc$ (since there are only $r + c - 1$ linearly independent constraints on the γ_{ij}). The appropriate equality is

$$\begin{aligned}\sum_i \sum_j \sum_k (Y_{ijk} - \theta - \alpha_i - \beta_j - \gamma_{ij})^2 &= \sum_i \sum_j \sum_k \left(Y_{ijk} - \bar{Y}_{ij.}\right)^2 \\ &+ m \sum_i \sum_j \left(\bar{Y}_{ij.} - \bar{Y}_{i..} - \bar{Y}_{.j.} + \bar{Y}_{...} - \gamma_{ij}\right)^2 + mr \sum_j \left(\bar{Y}_{.j.} - \bar{Y}_{...} - \beta_j\right)^2 \\ &+ mc \sum_i \left(\bar{Y}_{i..} - \bar{Y}_{...} - \alpha_i\right)^2 + rcn\left(\bar{Y}_{...} - \theta\right)^2. \qquad (7.10)\end{aligned}$$

Therefore $\hat{\theta} = \bar{Y}_{...}$, $\hat{\alpha}_i = \bar{Y}_{i..} - \bar{Y}_{...}$, $\hat{\beta}_j = \bar{Y}_{.j.} - \bar{Y}_{...}$, $\hat{\gamma}_{ij} = \bar{Y}_{ij.} - \bar{Y}_{i..} - \bar{Y}_{.j.} + \bar{Y}_{...}$, and $\hat{\mu}_{ijk} = \bar{Y}_{ij.}$. To test hypothesis 1, let W_1 be the set such that μ_{ijk}

$= \theta + \beta_j + \gamma_{ij}$. Using (7.10) with $\alpha_i = 0$, we see that $\hat{\hat{\theta}} = \bar{Y}_{...}$, $\hat{\hat{\beta}}_j = \bar{Y}_{.j.} - \bar{Y}_{...}$, $\hat{\gamma}_{ij} = \bar{Y}_{ij.} - \bar{Y}_{i..} - \bar{Y}_{.j.} + \bar{Y}_{...}$, and $\hat{\hat{\mu}}_{ijk} = \bar{Y}_{ij.} - \bar{Y}_{i..} + \bar{Y}_{...}$. $\dim(W_1) = 1 + c + rc - 1 - (r + c - 1) = rc - r + 1$. Putting these together, we get

$$F_1 = \frac{mc\sum_i\left(\bar{Y}_{i..} - \bar{Y}_{...}\right)^2}{\sum_i\sum_j\sum_k\left(Y_{ijk} - \bar{Y}_{ij.}\right)^2}\,\frac{rc(m-1)}{(r-1)}$$

and

$$F_1 \sim F_{rc(m-1),\,r-1}\left(\frac{mc\sum_i\alpha_i^2}{\sigma^2}\right).$$

Hypothesis 2 is tested similarly. In Exercise B7, you are asked to show that the F-statistic for hypothesis 3 is

$$F_3 = \frac{m\sum_i\sum_j\left(\bar{Y}_{ij.} - \bar{Y}_{i..} - \bar{Y}_{.j.} + \bar{Y}_{...}\right)^2}{\sum_i\sum_j\sum_k\left(Y_{ijk} - \bar{Y}_{ij.}\right)^2}\,\frac{rc(m-1)}{(r-1)(c-1)} \tag{7.11}$$

and

$$F_3 \sim F_{(r-1)(c-1),\,rc(m-1)}\left(\frac{m\sum_i\sum_j\gamma_{ij}^2}{\sigma^2}\right).$$

7.3.4. The Balanced Nested Model

We now consider the same model as in the last section in a different situation. In this situation, we consider the j effect to be nested inside the i effect. This would occur, for example, when the i effect represents species and the j effect represents subspecies. In this section, we let

$$\theta = \bar{\mu}_{..}, \qquad \alpha_i = \bar{\mu}_{i.} - \bar{\mu}_{..}, \qquad \delta_{ij} = \mu_{ij} - \bar{\mu}_{i.}.$$

Then

$$\mu_{ij} = \theta + \alpha_i + \delta_{ij}, \qquad \sum_i \alpha_i = 0, \qquad \sum_j \delta_{ij} = 0.$$

(Note that $\sum_i \delta_{ij} \neq 0$.) As before, these new parameters are uniquely defined with the constraints given. Therefore, we now consider the model in which we observe $Y_{ijk} \sim N_1(\theta + \alpha_i + \delta_{ij}, \sigma^2)$, $i = 1, \ldots, r$; $j = 1, \ldots, c$; $k = 1, \ldots, m$; where $\sum_i \alpha_i = 0$ and $\sum_j \delta_{ij} = 0$. We want to test two hypotheses for this model:

(1) All the $\alpha_i = 0$ (that the $\bar{\mu}_{i.}$ are all the same).
(2) All the $\delta_{ij} = 0$ (that the μ_{ij} depend only on i).

It can be shown that

$$\sum_i \sum_j \sum_k (Y_{ijk} - \theta - \alpha_i - \delta_{ij})^2$$

$$= \sum_i \sum_j \sum_k \left(Y_{ijk} - \overline{Y}_{ij.}\right)^2 + m \sum_i \sum_j \left(\overline{Y}_{ij} - \overline{Y}_{i.} - \delta_{ij}\right)^2$$

$$+ cm \sum_i \left(\overline{Y}_{i.} - \overline{Y}_{..} - \alpha_i\right)^2 + rcm\left(\overline{Y}_{...} - \theta\right)^2. \quad (7.12)$$

To test the hypothesis that $\alpha_i = 0$, $i = 1, \ldots, r$, we let

$$F = \frac{rc(m-1)}{(r-1)} \frac{cm\sum_i\left(\overline{Y}_{i..} - \overline{Y}_{...}\right)^2}{\sum_i\sum_j\sum_k\left(Y_{ijk} - \overline{Y}_{ij.}\right)^2},$$

$$F \sim F_{r-1, rc(m-1)}\left(\frac{mc\sum_i \alpha_i^2}{\sigma^2}\right). \quad (7.13)$$

To test the hypothesis that $\delta_{ij} = 0$, $i = 1, \ldots, r$, $j = 1, \ldots, c$, let

$$F = \frac{rc(m-1)}{r(c-1)} \frac{m\sum_i\sum_j\left(\overline{Y}_{ij.} - \overline{Y}_{i..}\right)^2}{\sum_i\sum_j\sum_k\left(Y_{ijk} - \overline{Y}_{ij.}\right)^2}, \qquad F \sim F_{r(c-1), rc(m-1)}\left(\frac{m\sum_i\sum_j\delta_{ij}^2}{\sigma^2}\right).$$

Note that the hypothesis that all the $\alpha_i = 0$ is tested the same for this model as for the model in the last section, which is not surprising since the α_i are the same for both models. However, $\delta_{ij} = \beta_j + \gamma_{ij}$, and testing that all the $\delta_{ij} = 0$ in this model is equivalent to testing that all the $\beta_j = 0$ and all the $\gamma_{ij} = 0$ for the model of the previous section.

It is clear that we could extend this to more complicated models. For example, we could let Y_{ijkm} be independent,

$$Y_{ijkm} \sim N_1\left(\theta + \alpha_i + \beta_j + \delta_{jk}, \sigma^2\right).$$

Then the k effect would be nested inside the j effect.

7.3.5. *Latin Squares*

In this model, we observe Y_{ij} independent, $i = 1, \ldots, m$; $j = 1, \ldots, m$. Let $k(i, j)$ be a function such that for each fixed i, $k(i, j)$ is a permutation of the integers, $1, \ldots, m$, and similarly for each fixed j, $k(i, j)$ is a permutation of the integers $1, \ldots, m$. The model we assume is that

$$Y_{ij} \sim N_1\left(\theta + \alpha_i + \beta_j + \gamma_{k(i,j)}, \sigma^2\right)$$

where α_i, β_j, and γ_k are constants such that $\sum_i \alpha_i = 0$, $\sum_j \beta_j = 0$, $\sum_k \gamma_k = 0$.

It can be shown that

$$\sum_i \sum_j (Y_{ij} - \theta - \alpha_i - \beta_j - \gamma_{k(i,j)})^2$$
$$= \sum_i \sum_j \left(Y_{ij} - \overline{Y}_{i.} - \overline{Y}_{.j} - \overline{Y}_{..k(i,j)} + 2\overline{Y}_{..}\right)^2 + m\sum_i \left(\overline{Y}_{i.} - \overline{Y}_{..} - \alpha_i\right)^2$$
$$+ m\sum_j \left(\overline{Y}_{.j} - \overline{Y}_{..} - \beta_j\right)^2 + m\sum_k \left(\overline{Y}_{..k} - \overline{Y}_{..} - \gamma_k\right)^2 + m^2\left(\overline{Y}_{..} - \theta\right)^2$$

where, as you would expect, $\overline{Y}_{..k}$ is the average of all Y_{ij} such that $k(i, j) = k$, and hence $Y_{..k(i,j)}$ is the average of all $Y_{i'j'}$ such that $k(i', j') = k(i, j)$. To test that $\alpha_i = 0$, we use

$$F = \frac{(m-1)(m-2)}{m-1} \frac{m\sum_i \left(\overline{Y}_{i.} - \overline{Y}_{..}\right)^2}{\sum_i \sum_j \left(Y_{ij} - \overline{Y}_{i.} - \overline{Y}_{.j} - \overline{Y}_{..k(i,j)} + 2\overline{Y}_{..}\right)^2},$$

$$F \sim F_{m-1,(m-1)(m-2)}\left(\frac{m\sum_i \alpha_i^2}{\sigma^2}\right).$$

Similarly to test that $\gamma_k = 0$,

$$F = \frac{(m-1)(m-2)}{m-1} \frac{m\sum_k \left(\overline{Y}_{..k} - \overline{Y}_{..}\right)^2}{\sum_i \sum_j \left(Y_{ij} - \overline{Y}_{i.} - \overline{Y}_{.j} - \overline{Y}_{..k(i,j)} + 2\overline{Y}_{..}\right)^2},$$

$$F \sim F_{(m-1),(m-1)(m-2)}\left(\frac{m\sum \gamma_k^2}{\sigma^2}\right).$$

The great advantage of this kind of design is that it is possible to test for three effects with only m^2 measurements. For a three-way anova, even with only one observation per cell, there would be m^3 measurements. The disadvantages are that we must assume no interaction and must have the same number of levels for each effect. It turns out that we can also make a design that allows us to study four effects with only a two-way layout. Such designs are called Greco-Latin squares.

7.4. UNBALANCED ANALYSIS OF VARIANCE.

In this section, we discuss unbalanced analysis of variance models, that is, models which have unequal numbers of observations in the various cells. We assume throughout the discussion of these models that there is at least one observation in each cell and that the total number of observations is greater than the number of linearly independent parameters.

For some of these models, there is no simple formula for the F-statistic. In addition, the derivation of a formula is quite messy and not too instructive. For this reason, for several models, we merely state the minimization to be solved, and using the solution to that problem, we state the F-statistic. Not having a complete formula for the F-statistic for these models is not too great a problem in practice, since that statistic is usually found by a computer.

The reader should realize that once we have specified the subspaces V and W and have found their dimensions, the statistical part of the problem is finished. Computing the F-statistic at this point just involves the technical problem of computing the projections on the spaces V and W, which is an algebra or numerical analysis problem, and not a statistical one.

In the next sections, we try to emphasize what the hypotheses of the testing problems are. In particular, we show that some of the hypotheses depend on the constraints necessary to define the parameters. We establish which hypotheses are independent of the constraints and also discuss what constraints to use when the hypothesis does depend on the constraints. The reader should recognize that any assessment of which constraints to use is not a mathematical result, but merely an opinion as to which system seems the most advantageous. The problem of which constraints to use in certain models is still controversial at this time.

We use the following notation. For any numbers a_{ij}, $i = 1, \ldots, p$; $j = 1, \ldots, n_i$; with $N = \sum_i n_i$, we write $\mathbf{a} = (a_{ij})$ to mean that $\mathbf{a}$ is an N-dimensional vector where components are the a_{ij}. Although the order in which the components are arranged is unimportant, it is important that the same order be maintained throughout the section. Similarly, if a_{ijk} are numbers, we write $\mathbf{a} = (a_{ijk})$ to mean that $\mathbf{a}$ is a vector whose components are the a_{ijk}.

7.4.1. *Unbalanced One-Way Analysis of Variance*

We consider the model in which we observe Y_{ij} independent, such that

$$Y_{ij} \sim N_1(\theta + \alpha_i, \sigma^2), \qquad i = 1, \ldots, p; \qquad j = 1, \ldots, n_i, \qquad N = \sum n_i.$$

We want to test the hypothesis all the $\alpha_i = 0$. However, to have θ and α_i defined, we need a constraint on the α_i. Let $w_1, \ldots, w_p$ be nonnegative weights such that $\sum_i w_i > 0$. In this model we use the restraint that

$$\sum_i w_i \alpha_i = 0.$$

We want to test the hypothesis that all the $\alpha_i = 0$. Therefore, let V be the p-dimensional subspace of vectors $\boldsymbol{\mu} = (\mu_{ij})$ such that $\mu_{ij} = \theta + \alpha_i$ for some θ and α_i such that $\sum_i w_i \alpha_i = 0$. Let W be the one-dimensional subspace of vectors $\boldsymbol{\mu} = (\mu_{ij})$ such that $\mu_{ij} = \theta$ for some θ. We are therefore testing that $\boldsymbol{\mu} \in W$ against $\boldsymbol{\mu} \in V$.

LEMMA 7.6. V and W do not depend on the weights w_i.

PROOF. It is clear that W does not depend on the w_i since they are not used in its definition. Let V^* be the subspace of all vectors $\boldsymbol{\mu} = (\mu_{ij})$ such that μ_{ij} does not depend on j. We note that V^* does not depend on the w_i. We now show that $V = V^*$. It is clear that $V \subset V^*$. Now, suppose that $\boldsymbol{\mu} = (\mu_{ij}) \in V^*$. Then μ_{ij} does not depend on j, so that $\mu_{ij} = \mu_i$. Now let

$$\theta = \frac{\sum_i w_i \mu_i}{\sum_i w_i}, \qquad \alpha_i = \mu_i - \theta.$$

Then $\mu_{ij} = \mu_i = \theta + \alpha_i$ and $\sum_i w_i \alpha_i = 0$. Hence $V^* \subset V$. □

Because of Lemma 7.6, we can choose any weights w_i we like. The hypothesis being tested does not depend on the weights w_i, nor does the F-statistic, since it is completely determined by V and W. We choose the weights $w_i = n_i$. Using these weights, we are assuming that $\sum_i n_i \alpha_i = 0$. With this constraint, we can show that

$$\sum_i \sum_j (Y_{ij} - \theta - \alpha_i)^2 = \sum_i \sum_j \left(Y_{ij} - \overline{Y}_{i.}\right)^2 + \sum_i n_i \left(\overline{Y}_{i.} - \overline{Y}_{..} - \alpha_i\right)^2 + N\left(\overline{Y}_{..} - \theta\right)^2.$$

Therefore, we see that $\hat{\mu}_{ij} = \overline{Y}_{i.}$ and $\hat{\hat{\mu}}_{ij} = \overline{Y}_{..}$. Hence, the F-statistic is

$$F = \frac{(N-p)\sum_i n_i \left(\overline{Y}_{i.} - \overline{Y}_{..}\right)^2}{(p-1)\sum_i \sum_j \left(Y_{ij} - \overline{Y}_{i.}\right)^2} \sim F_{p-1,\,n-p}\left(\frac{\sum_i n_i \alpha_i^2}{\sigma^2}\right).$$

7.4.2. *Unbalanced Two-Way Analysis of Variance with No Interaction*

We consider now the model in which we observe Y_{ijk} independent with

$$Y_{ijk} \sim N_1(\theta + \alpha_i + \beta_j, \sigma^2),$$

$$i = 1, \ldots, r; \qquad j = 1, \ldots, c; \qquad k = 1, \ldots, n_{ij} > 0$$

We want to test that all the $\alpha_i = 0$. To have θ, α_i, and β_j defined we need to have some constraints. Let $w_1, \ldots, w_r$ and $v_1, \ldots, v_c$ be nonnegative weights such that $\sum_i w_i > 0$ and $\sum_j v_j > 0$. We use the constraints

$$\sum_i w_i \alpha_i = 0, \qquad \sum_j v_j \beta_j = 0.$$

Let $n_{i.} = \sum_j n_{ij}$, $n_{.j} = \sum_i n_{ij}$, $N = \sum_i \sum_j n_{ij}$. Now, let V be the $(r + c - 1)$-dimensional subspace of vectors $\boldsymbol{\mu} = (\mu_{ijk})$ such that $\mu_{ijk} = \theta + \alpha_i + \beta_j$ for some θ, α_i and β_j such that $\sum_i w_i \alpha_i = 0$ and $\sum_j v_j \beta_j = 0$. Let W be the c-dimensional subspace of vectors $\boldsymbol{\mu} = (\mu_{ijk})$ such that $\mu_{ijk} = \theta + \beta_j$ for some θ and β_j such that $\sum_j v_j \beta_j = 0$. Testing the $\alpha_i = 0$ is the same as testing that $\boldsymbol{\mu} \in W$ against $\boldsymbol{\mu} \in V$.

LEMMA 7.7. V and W do not depend on the weights w_i and v_j.

PROOF. Let V^* be the subspace of all vectors $\boldsymbol{\mu} = (\mu_{ijk})$ such that $\mu_{ijk} = s_i + t_j$ for some numbers s_i and t_j. V^* does not depend on the w_i or the v_j. It is clear that $V \subset V^*$. Now, let $\boldsymbol{\mu} = (\mu_{ijk}) \in V^*$. Then $\mu_{ijk} = s_i + t_j$. Define

$$\theta = \frac{\sum_i w_i s_i}{\sum_i w_i} + \frac{\sum_j v_j t_j}{\sum_j v_j}, \qquad \alpha_i = s_i - \frac{\sum_i w_i s_i}{\sum_i w_i}, \qquad \beta_j = t_j - \frac{\sum_j v_j t_j}{\sum_j v_j}.$$

Then $\mu_{ijk} = \theta + \alpha_j + \beta_j$ and $\sum_i w_i \alpha_i = 0$, $\sum_j v_j \beta_j = 0$. Therefore $V^* = V$, and V does not depend on the weights. Now, let W^* be the subspace of all vectors $\boldsymbol{\mu} = (\mu_{ijk})$ such that μ_{ijk} does not depend on i or k. W^* does not depend on the weights. By a proof similar to Lemma 7.6, $W = W^*$. □

To compute F, we first find $\hat{\theta}$, $\hat{\alpha}_i$, and $\hat{\beta}_j$ by minimizing

$$\sum_i \sum_j \sum_k (Y_{ijk} - \theta - \alpha_i - \beta_j)^2$$

subject to $\sum_i w_i \alpha_i = 0$ and $\sum_j v_j \beta_j = 0$ for some weights w_i and v_j. Then $\hat{\mu}_{ijk} = \hat{\theta} + \hat{\alpha}_i + \hat{\beta}_j$. To find the estimator under the null hypothesis, we minimize the above expression subject to $\alpha_i = 0$ and $\sum_j v_j \beta_j = 0$, finding $\hat{\hat{\theta}}$ and $\hat{\hat{\beta}}_j$. Finally

$$F = \frac{(N - r - c + 1)\sum_i \sum_j n_{ij}\left(\hat{\theta} + \hat{\alpha}_i + \hat{\beta}_j - \hat{\hat{\theta}} - \hat{\hat{\beta}}_j\right)^2}{(r - 1)\sum_i \sum_j \sum_k \left(Y_{ijk} - \hat{\theta} - \hat{\alpha}_i - \hat{\beta}_j\right)^2} \sim F_{r-1, N-r-c+1}(0)$$

under the null hypothesis. One situation in which it is possible to find a simple expression for F is in the case of proportional sampling in which $n_{ij} = n_{i.} n_{.j} / N$. Using (7.14) of the next section with $\gamma_{ij} = 0$, we see that in that case $\hat{\theta} = \hat{\hat{\theta}} = \overline{Y}_{...}$, $\hat{\alpha}_i = (\overline{Y}_{i..} - \overline{Y}_{...})$ and $\hat{\beta}_j = \hat{\hat{\beta}}_j = (Y_{.j.} - \overline{Y}_{...})$ and

$$F = \frac{(N - r - c + 1)\sum_i n_{i.}\left(\overline{Y}_{i..} - \overline{Y}_{...}\right)^2}{(r - 1)\sum_i \sum_j \sum_k \left(Y_{ijk} - \overline{Y}_{i.} - \overline{Y}_{.j} + \overline{Y}_{...}\right)^2} \sim F_{r-1, N-r-c+1}\left(\frac{\sum n_i \alpha_i^2}{\sigma^2}\right)$$

We would, of course, test the hypothesis $\beta_j = 0$ in a similar fashion.

7.4.3. Unbalanced Two-Way Analysis of Variance with Interaction.

We now add an interaction term to the two-way model. We assume that we observe Y_{ijk} independent, with

$$Y_{ijk} \sim N_1(\theta + \alpha_i + \beta_j + \gamma_{ij}, \sigma^2),$$
$$i = 1, \ldots, r; \quad j = 1, \ldots, c; \quad k = 1, \ldots, n_{ij} > 0.$$

As in the previous section, we have nonnegative weights w_i and v_j such that $\sum_i w_i > 0$, $\sum_j v_j > 0$. In order to define θ, α_i, β_j, and γ_{ij} we use the constraints

$$\sum_i w_i \alpha_i = 0, \qquad \sum_j v_j \beta_j = 0, \qquad \sum_i w_i \gamma_{ij} = 0, \qquad \sum_j v_j \gamma_{ij} = 0.$$

Let $n_{i.} = \sum_j n_{ij}$, $n_{.j} = \sum_i n_{ij}$, $N = \sum_i \sum_j n_{ij}$. Now, let V be the subspace consisting of vectors $\boldsymbol{\mu} = (\mu_{ijk})$ such that $\mu_{ijk} = \theta + \alpha_i + \beta_j + \gamma_{ij}$ where α_i, β_j, and γ_{ij} satisfy the constraints given above. We show that V does not depend on the weights w_i and v_j.

LEMMA 7.8. Let V^* be the subspace of $\boldsymbol{\mu} = (\mu_{ijk})$ such that μ_{ijk} does not depend on k. Then $V = V^*$, and hence V does not depend on the weights w_i and v_j.

PROOF. See Exercise B12. □

We note that $\dim(V) = \dim(V^*) = rc$. We now find the estimator $\hat{\mu}_{ijk}$ under the general hypothesis that $\boldsymbol{\mu} \in V$. Using Lemma 7.8, we see that we need to minimize

$$\sum_i \sum_j \sum_k (Y_{ijk} - \mu_{ij})^2 = \sum_i \sum_j \sum_k \left(Y_{ijk} - \overline{Y}_{ij.}\right)^2 + \sum_i \sum_j n_{ij}\left(\overline{Y}_{ij.} - \mu_{ij}\right)^2.$$

Therefore, $\hat{\mu}_{ijk} = \hat{\mu}_{ij} = \overline{Y}_{ij.}$.

We now consider testing the hypothesis that all the $\gamma_{ij} = 0$. Let W_1 be $(r + c - 1)$-dimensional subspace of V in which $\gamma_{ij} = 0$. We note that W_1 defined here is the same as V defined in the last section. Therefore, by Lemma 7.7; we see that W_1 does not depend on the weights chosen. Therefore the F-statistic for testing that $\gamma_{ij} = 0$ is given by

$$F_1 = \frac{(N - rc)\sum_i \sum_j n_{ij}\left(\overline{Y}_{ij.} - \hat{\theta} - \hat{\alpha}_i - \hat{\beta}_j\right)^2}{(r-1)(c-1)\sum_i \sum_j \sum_k \left(Y_{ijk} - \overline{Y}_{ij.}\right)^2}$$

$$\sim F_{(r-1)(c-1), N-rc}(0)$$

under the null hypothesis where $\hat{\theta}$, $\hat{\alpha}_i$ and $\hat{\beta}_j$ are defined in the last section.

We now consider the problem of testing that all the $\alpha_i = 0$. Let W_2 be the subspace in which $\alpha_i = 0$. Unfortunately, the subspace W_2 *does* depend on the weights chosen. To find the F-statistic for this model, we must minimize

$$\sum_i \sum_j \sum_k (Y_{ijk} - \theta - \beta_j - \gamma_{ij})^2$$

subject to the constraints that $\sum_j v_j \beta_j = 0$, $\sum_i w_i \gamma_{ij} = 0$, and $\sum_j v_j \gamma_{ij} = 0$, getting the estimators $\hat{\hat{\theta}}$, $\hat{\hat{\beta}}_j$, and $\hat{\hat{\gamma}}_{ij}$. Then $\hat{\hat{\mu}}_{ijk} = \hat{\hat{\theta}} + \hat{\hat{\beta}}_j + \hat{\hat{\gamma}}_{ij}$, and the F-statistic is given by

$$F_2 = \frac{(N - rc)\sum_i \sum_j n_{ij}\left(\overline{Y}_{ij.} - \hat{\hat{\theta}} - \hat{\hat{\beta}}_j - \hat{\hat{\gamma}}_{ij}\right)^2}{(r-1)\sum_i \sum_j \sum_k \left(Y_{ijk} - \overline{Y}_{ij.}\right)^2}.$$

One case in which it is particularly easy to find the estimators is the case of proportional sampling ($n_{ij} = n_{i.}n_{.j}/N$) with the weights $w_i = n_{i.}$, $v_j = n_{.j}$. In this case,

$$\begin{aligned}\sum_i \sum_j \sum_k &(Y_{ijk} - \theta - \alpha_i - \beta_j - \gamma_{ij})^2 \\ &= \sum_i \sum_j \sum_k \left(Y_{ijk} - \overline{Y}_{ij.}\right)^2 + \sum \sum_j n_{ij}\left(\overline{Y}_{ij.} - \overline{Y}_{i..} - \overline{Y}_{.j.} + \overline{Y}_{...} - \gamma_{ij}\right)^2 \\ &\quad + \sum_i n_{i.}\left(\overline{Y}_{i..} - \overline{Y}_{...} - \alpha_i\right)^2 + \sum_j n_{.j}\left(\overline{Y}_{.j.} - \overline{Y}_{...} - \beta_j\right)^2 + N\left(\overline{Y}_{...} - \theta\right)^2.\end{aligned} \tag{7.14}$$

Therefore,

$$F_1 = \frac{(N - rc)\sum_i \sum_j n_{ij}\left(\overline{Y}_{ij.} - \overline{Y}_{i..} - \overline{Y}_{.j.} + \overline{Y}_{...}\right)^2}{(r-1)(c-1)\sum_i \sum_j \sum_k \left(Y_{ijk} - \overline{Y}_{ij.}\right)^2};$$

$$F_2 = \frac{(N - rc)\sum_i n_{i.}\left(\overline{Y}_{i..} - \overline{Y}_{...}\right)^2}{(r-1)\sum_i \sum_j \sum_k \left(Y_{ijk} - \overline{Y}_{ij.}\right)^2}$$

in the case of proportional sampling with the weights $w_i = n_{i.}$, $v_j = n_{.j}$. Usually, in practice, these weights would not be appropriate weights to choose however (see below). Therefore, the statistic F_2 would be an inappropriate choice for testing that $\alpha_i = 0$, in that it implies an inappropriate choice for the weights, and hence for the α_i. The statistic F_1 would be correct, because the test for no interactions is independent of the weights. We could test $\beta_j = 0$ in a similar fashion.

The problem of testing that $\alpha_i = 0$ in the presence of possible interactions has attracted a considerable amount of interest, and even a fair amount of controversy in recent years. The controversy is not in the derivations given above, but that the hypothesis $\alpha_i = 0$ depends on the weights and in many problems there are no natural weights to use to define the hypothesis. (It should be emphasized at this time that the controversy should also extend to balanced models. In discussing the balanced model in Section 7.3.3, we used the constraints associated with $w_i = v_j = 1$. If these weights are not correct for the unbalanced model, it seems that they would be incorrect also for the balanced model. That is, the assumed structure of the θ, α_i, β_j, and γ_{ij} should not depend on whether the sampling is balanced or unbalanced, but should be determinable before the sampling scheme is defined.) In the next five paragraphs, we present five approaches for dealing with testing that $\alpha_i = 0$ in the presence of possible interactions.

The first approach is not to do it. If we imagine that interactions are possible, we do not test that all the $\alpha_i = 0$, but rather test that $\alpha_i + \gamma_{ij} = 0$ for all i and j. This hypothesis does not involve weights, and can therefore be tested without specification of the weights. This would imply, in particular, in the balanced two-way model that we not test that $\alpha_i = 0$. This is the approach used in Searle (1971) and is a consistent approach to the problem. However, it is hard to imagine a setting when we could conclude a priori that the interactions are zero and this seems like an overly restrictive solution to the problem.

The second solution is a two-step procedure. First test that the interactions are all zero. If this hypothesis is rejected, the procedure is stopped. We conclude that there is an effect due to the subscript i since the interactions are not zero. If we accept the hypothesis that the interactions are 0, we would then proceed to test the hypothesis that $\alpha_i = 0$ under the assumption that the interactions are 0 by the procedures derived in the last section. This method does not involve the weights. This method is used quite often in practice and is the method suggested in Snedecor and Cochran (1967) and Seber (1977). (However, neither of these books suggests this approach for the balanced case.)

One disadvantage of this procedure is that accepting a null hypothesis cannot be construed as proof (even statistical proof) that the null hypothesis is true. If we accept with the 0.05 size test, but would have rejected with the 0.10 size test, we have not established that the null hypothesis is true, but merely have data that suggests that the null hypothesis is false, but is not conclusive enough to prove it. Therefore, accepting the null hypothesis that the interactions are all zero is pretty weak evidence for concluding that they are. Hence, the F-test defined in the last section may be very inappropriate even after the hypothesis of no interaction has been accepted. Furthermore, if we use this procedure, we would be saying that there is an effect due to the rows if we reject the hypothesis of no interaction or if we accept the hypothesis of no interaction but reject the hypothesis that all the $\alpha_i = 0$. The size of this procedure would be considerably greater than α and might be near 2α. (Note that the two F-tests are not independent so that the size of the procedure cannot be immediately computed.)

The third and fourth procedures involve arbitrarily picking weights w_i

and v_j. The third method would choose the weights $w_i = n_{i.}$, $v_j = n_{.j}$. Note that this is the system of weights that leads to a nice solution for the model with proportional sampling. It also reduces to the weights used in the balanced model. These weights are used in Scheffe (1959, pp. 112–119). However, these weights seem particularly unappealing since they depend on the sample sizes n_{ij}. These weights would imply that two different experimenters testing that $\alpha_i = 0$ in a given situation would actually be testing different hypotheses if they had different sample sizes. As mentioned earlier, it seems that the hypotheses should be definable before establishing the sampling scheme. [One situation in which these weights would be sensible occurs when the n_{ij} are themselves random variables, representing the number of observations actually observed in the ijth cell in a random sample from the whole population. Although this situation (random n_{ij}) does not fit into the framework of this chapter, it would seem sensible to use the procedures defined in this chapter with weights $w_i = n_{i.}/N$, $v_j = n_{.j}/N$ since these weights would be estimators of the proportion of the total population in the ith row and the jth column.]

The fourth procedure is to let $w_i = v_j = 1$. This procedure amounts to treating all the rows as equally important and all the columns as equally important. This approach has the great advantage of being a systematic procedure that is determined independently of the sampling scheme. It of course, reduces to the usual weights in the balanced two-way model. The formulas for the F-statistic for the weights $w_i = v_j = 1$ were derived by Yates (1934). These formulas are given in Graybill (1976).

A fifth approach is to make the experimenter choose weights w_i and v_j before performing this experiment. If he does not have weights, then he should not test that all the $\alpha_i = 0$. To help him choose the weights, we note that

$$\theta = \frac{\sum_i \sum_j w_i v_j \mu_{ij}}{\sum_i \sum_j w_i v_j}, \qquad \alpha_i = \frac{\sum_j v_j \mu_{ij}}{\sum_j v_j} - \theta.$$

Therefore, an equivalent formulation of the hypothesis being tested is that

$$\mu_{i.}^{v} = \frac{\sum_j v_j \mu_{ij}}{\sum_j v_j}$$

does not depend on i. We note $\mu_{i.}^{v}$ is just a weighted average of the μ_{ij} averaged over v_j. We think of $\mu_{i.}^{v}$ as representing the "average" of values in the ith row. We want to test that these "averages" are the same. The weight v_j represents the importance we attach to the jth column. For example, v_j might be the proportion of the total population in the jth column. (Note that the above analysis implies that the hypothesis that the $\alpha_i = 0$ does not depend on the w_i, so that it is only necessary to specify the v_j for this approach).

The fourth procedure above implies that we are applying equal weights to all the columns, that is, that we are calling the "average" in the ith class

$$\mu_{i.} = \frac{\sum_j \mu_{ij}}{c} .$$

While this system of weights may often be sensible, we give an example to show that the unthinking application of these weights sometimes leads to inconsistencies. Suppose that a researcher collects data on a model with c columns. Suppose that a second researcher combines the last two columns into one column and suppose that both researchers use weights $w_i = v_j = 1$. Then the "average" of the ith row for the first researcher is

$$\bar{\mu}_{i.} = \frac{\sum_{j=1}^{c} \mu_{ij}}{c}$$

while the "average" for the second researcher is

$$\bar{\mu}_{i.} = \frac{\sum_{j=1}^{c-2} \mu_{ij}}{c-1} + \frac{\sum_{j=c-1}^{c} \mu_{ij}}{2(c-1)}$$

In other words, in assuming that the classes have the same weight, the second researcher is putting half as much weight on the last two columns as on the first $c - 2$ columns. This example emphasizes that we should use the weights $w_i = v_j = 1$ only when we believe the columns have equal weight and the rows have equal weight. (It also implies that when columns are combined, the weights associated with those columns should be added. Note that the third method in which we let $v_j = n_{.j}$ would satisfy this criterion.)

I find the fifth method to be the most appealing one for this problem. Note that the first and fifth method are very similar. Both imply that if we have no specific weights in mind, then the hypothesis of no main effects is not defined and hence should not be tested. However, the first method discourages the researcher from ever testing for main effects, whereas the fifth method encourages him to decide on some weights and then use the appropriate test. It seems that in practice he should often be able to establish the relative importance of the columns. If he cannot, then he probably should not be testing for main effects. They are undefined. (Calling the columns equally important because we cannot decide which ones are most important, as is done in method 4, seems to me unreasonable.)

Use of the fifth approach would necessitate a change in the way analysis of variance is approached. At present, most computer packages offer several options for testing for main effects in the presence of possible interactions. These options depend, among other things, on the weights chosen for the analysis. To use the fifth approach the programs would have to be rewritten so that the researcher inputs his weights along with his data, and would also have to be rewritten to allow general systems of weights.

Perhaps the most important aspect of this discussion is that it emphasizes that the weights should, in general, be chosen before the data is collected. This suggests the possibility of designing the sampling scheme to account for weights. One approach to such a design is given in Section 7.7. We emphasize again that the difficulty encountered in this section is that the hypothesis being tested depends on the weights used to define it. It is not a situation in which there are several possible tests for the same hypothesis, but rather a situation in which there are several possible hypotheses associated with the simple statement that there is "no row effect." Each of these hypotheses represents a different interpretation of that statement and each is tested differently.

As a final comment in this section, we mention again that the previous discussion is really concerned with the applications of the theory. The theory is clear. If we have weights v_j then we should use the test defined earlier. If we do not have weights, then the hypothesis that all the $\alpha_i = 0$ is not defined and hence cannot be tested.

7.4.4. *Unbalanced Nested Analysis of Variance*

We consider the model in which we observe Y_{ijk} independent such that

$$Y_{ijk} \sim N_1\left(\theta + \alpha_i + \delta_{ij}, \sigma^2\right),$$
$$i = 1, \ldots, r; \quad j = 1, \ldots, c_i; \quad k = 1, \ldots, n_{ij} > 0.$$

We want to test the hypothesis that all the $\alpha_i = 0$ and the hypothesis that all the $\delta_{ij} = 0$. In order to make the α_i and δ_{ij} well-defined, we choose nonnegative weights v_i and w_{ij} such that $\sum_i v_i > 0$ and $\sum_j w_{ij} > 0$ and assume that

$$\sum_i v_i \alpha_i = 0, \qquad \sum_j w_{ij} \delta_{ij} = 0.$$

Let $C = \sum_i c_i$, $n_{i.} = \sum_j n_{ij}$, $N = \sum_i \sum_j n_{ij}$. Let V be the C-dimensional subspace of vectors $\boldsymbol{\mu} = (\mu_{ijk})$ such that $\mu_{ijk} = \theta + \alpha_l + \delta_{ij}$ where α_i and δ_{ij} satisfy the above constraints. Let W_1 be r-dimensional subspace of vectors $\boldsymbol{\mu} = (\mu_{ijk})$ such that $\mu_{ijk} = \theta + \alpha_i$ with α_i satisfying the above constraint. Let W_2 be the $(C - r + 1)$-dimensional subspace of vectors $\boldsymbol{\mu} = (\mu_{ijk})$ such that $\mu_{ijk} = \theta + \delta_{ij}$ where δ_{ij} satisfies the above constraint. Then testing that the $\delta_{ij} = 0$ is the same as testing that $\boldsymbol{\mu} \in W_1$ against $\boldsymbol{\mu} \in V$, whereas testing that the $\alpha_i = 0$ is the same as testing that $\boldsymbol{\mu} \in W_2$ against $\boldsymbol{\mu} \in V$.

LEMMA 7.9. a. Let V^* be the subspace of vectors $\boldsymbol{\mu} = (\mu_{ijk})$ such that

μ_{ijk} does not depend on k. Then $V = V^*$ and hence V does not depend on the weights chosen.

b. Let W_1^* be the subspace of vectors $\boldsymbol{\mu} = (\mu_{ijk})$ such that μ_{ijk} does not depend on j or k. Then $W_1 = W_1^*$, and hence W_1 does not depend on the weights chosen.

c. Let W_2^* be the subspace of vectors $\mu = (\mu_{ijk})$ such that $\mu_{ijk} = \mu_{ij}$ does not depend on k and $\sum_j w_{ij}\mu_{ij}/\sum_j w_{ij}$ does not depend on i. Then $W_2 = W_2^*$ and hence W_2 does not depend on v_j. (It does depend on the w_{ij}.)

PROOF. a. Clearly $V \subset V^*$. Now, let $\boldsymbol{\mu} = (\mu_{ijk}) \in V^*$. Then $\mu_{ijk} = \mu_{ij}$. Let

$$\theta = \frac{\sum_i v_i \dfrac{\sum_j w_{ij}\mu_{ij}}{\sum_j w_{ij}}}{\sum_i v_i}, \qquad \alpha_i = \frac{\sum_j w_{ij}\mu_{ij}}{\sum_j w_{ij}} - \theta, \qquad \delta_{ij} = \mu_{ij} - \theta - \alpha_i.$$

Then $\mu_{ijk} = \mu_{ij} = \theta + \alpha_i + \delta_{ij}$, where α_i and δ_{ij} satisfy the above restraints. Therefore, $\mu \in V$, and hence $V = V^*$.

b. See Exercise B13.

c. Suppose $\boldsymbol{\mu} = (\mu_{ijk}) \in W_2$. Then μ_{ijk} does not depend on k and

$$\sum_j w_{ij}\mu_{ijk} \Big/ \sum_j w_{ij} = \theta + \sum_j w_{ij}\delta_{ij} \Big/ \sum_j w_{ij} = \theta.$$

Therefore $\boldsymbol{\mu} \in W_2^*$. Now suppose that $\boldsymbol{\mu} = (\mu_{ijk}) \in W_2^*$. Then $\mu_{ijk} = \mu_{ij}$. Let

$$\theta = \sum_j w_{ij}\mu_{ij} \Big/ \sum_j w_{ij}, \qquad \delta_{ij} = \mu_{ij} - \theta.$$

Then $\mu_{ijk} = \mu_{ij} = \theta + \delta_{ij}$ and δ_{ij} satisfies the above constraints. Hence $\boldsymbol{\mu} \in W_2$ and $W_2 = W_2^*$. □

We now find the estimator $\hat{\mu}_{ijk}$. By Lemma 7.9a, we need to minimize

$$\sum_i \sum_j \sum_k (Y_{ijk} - \mu_{ij})^2 = \sum_i \sum_j \sum_k \left(Y_{ijk} - \overline{Y}_{ij.}\right)^2 + \sum_i \sum_j n_{ij}\left(\overline{Y}_{ij.} - \mu_{ij}\right)^2.$$

Therefore, $\hat{\mu}_{ijk} = \overline{Y}_{ij.}$.

We now find the estimator $\hat{\hat{\mu}}_{ijk}$ under the hypothesis that $\mu \in W_1$. By Lemma 7.9b, we must minimze

$$\sum_i \sum_j \sum_k (Y_{ijk} - \mu_i)^2 = \sum_i \sum_j \sum_k \left(Y_{ijk} - \overline{Y}_{i..}\right)^2 + \sum_i n_{i.}\left(\overline{Y}_{i..} - \mu_i\right)^2.$$

Hence $\hat{\hat{\mu}}_{ijk} = \overline{Y}_{i..}$. Therefore, the F-statistic for testing that $\delta_{ij} = 0$ is

$$F_1 = \frac{(N - C)\sum_i \sum_j n_{ij}\left(\overline{Y}_{ij.} - \overline{Y}_{i..}\right)^2}{(C - r)\sum_i \sum_j \sum_k \left(\overline{Y}_{ijk} - \overline{Y}_{ij.}\right)^2}.$$

We now consider testing that $\alpha_i = 0$. We note that this hypothesis depends on the weights w_{ij}, and hence on the constraints on the δ_{ij}. In order to find the estimator $\hat{\mu}_{ijk}$ under the hypothesis that $\alpha_i = 0$, we must minimize

$$\sum_i \sum_j \sum_k (Y_{ijk} - \theta - \delta_{ij})^2$$

subject to the constraints that $\sum_j w_{ij}\delta_{ij} = 0$ for all i. Let $\hat{\hat{\theta}}$ and $\hat{\hat{\delta}}_{ij}$ be the estimators. Then $\hat{\hat{\mu}}_{ijk} = \hat{\hat{\theta}} + \hat{\hat{\delta}}_{ij}$, and hence the F-statistic for testing that the $\alpha_i = 0$ is given by

$$F_2 = \frac{(N - C)\sum_i \sum_j n_{ij}\left(\bar{Y}_{ij.} - \hat{\hat{\theta}} - \hat{\hat{\delta}}_{ij}\right)^2}{(r-1)\sum_i \sum_j \sum_k \left(Y_{ijk} - \bar{Y}_{ij.}\right)^2}.$$

If we use the weights $w_{ij} = n_{ij}$, then

$$\begin{aligned}\sum_i \sum_j \sum_k (Y_{ijk} - \theta - \delta_{ij})^2 = &\sum_i \sum_j \sum_k \left(Y_{ijk} - \bar{Y}_{ij.}\right)^2 \\ &+ \sum_i \sum_j n_{ij}\left(\bar{Y}_{ij.} - \bar{Y}_{i..} - \delta_{ij}\right)^2 \\ &+ \sum_i n_i\left(Y_{i..} - \bar{Y}_{...}\right)^2 + N(Y_{...} - \theta)^2.\end{aligned}$$

Therefore, in this case, $\hat{\hat{\mu}}_{ijk} = \bar{Y}_{ij.} - \bar{Y}_{i..} + \bar{Y}_{...}$, and,

$$F_2 = \frac{(N - C)\sum_i n_{i.}\left(\bar{Y}_{i..} - \bar{Y}_{...}\right)^2}{(r-1)\sum_i \sum_j \sum_k \left(Y_{ijk} - \bar{Y}_{ij.}\right)^2}.$$

We emphasize again that this F-statistic is only appropriate for testing that $\alpha_i = 0$ if we use the weights $w_{ij} = n_{ij}$. These weights are usually not very appealing, since they depend on the sample sizes n_{ij}. In most problems, we should choose weights w_{ij} before choosing the sampling scheme. These weights would lead to a messier F-statistic, but at least the statistic would be testing a hypothesis that is independent of the sampling scheme. It should be mentioned that when the n_{ij}'s are all the same (and hence the only unbalanced part of the model is in the numbers of subclasses), then using the weights $w_{ij} = n_{ij} = n$ is the same as using the weights $w_{ij} = 1$. Therefore, if the number of observations in each subclass is the same and each subclass has the same importance, then the F-statistic for testing that $\alpha_i = 0$ is given by the simplified formula above, even if the numbers of subclasses c_i are not equal.

7.5. ANALYSIS OF COVARIANCE.

The analysis of covariance model is a cross between analysis of variance and regression models. As an example, let Y_{ij} be independent

$$Y_{ij} \sim N_1(\theta + \alpha_i + \beta x_{ij}, \sigma^2), \quad i = 1, \ldots, r; \quad j = 1, \ldots, n_i;$$

where $\sum_i w_i \alpha_i = 0$. We want to test the hypothesis that all the $\alpha_i = 0$ and the hypothesis that $\beta = 0$. This model represents a setting in which we have r regression lines all with slope β. The y-intercept of the ith line is $\theta + \alpha_i$. Therefore, the regression lines are all parallel with possibly different intercepts. In testing that all the $\alpha_i = 0$, we are really testing that the intercepts of the lines are the same when we assume that the slopes are.

Since the hypotheses being tested do not depend on w_i, (see Exercise B14) we choose the weights $w_i = n_{i.}$. We are therefore assuming that $\sum_i n_i \alpha_i = 0$. To find the estimators under the general alternative, we want to minimze

$$\sum_i \sum_j (Y_{ij} - \theta - \alpha_i - \beta x_{ij})^2.$$

Differentiating with respect to θ, and using the constraint $\sum_i n_i \alpha_i = 0$, we find that $\hat{\theta} = \overline{Y}_{..} - \hat{\beta}\overline{x}_{..}$. Differentiation with respect to α_i yields $\hat{\alpha}_i = \overline{Y}_{i.} - \hat{\theta} - \hat{\beta}\overline{x}_{i.}$. To find $\hat{\beta}$, we therefore have to minimize

$$\sum_i \sum_j \left[Y_{ij} - \overline{Y}_{i.} - \beta(x_{ij} - \overline{x}_{i.}) \right]^2.$$

Differentiation with respect to β yields

$$\hat{\beta} = \frac{\sum_i \sum_j (Y_{ij} - \overline{Y}_{i.})(x_{ij} - \overline{x}_{i.})}{\sum_i \sum_j (x_{ij} - \overline{x}_{i.})^2}.$$

Therefore,

$$\|\mathbf{P}_{V^\perp} Y\|^2 = \sum_i \sum_j (Y_{ij} - \hat{\theta} - \hat{\alpha} - \hat{\beta} x_{ij})^2 = \sum_i \sum_j \left[Y_{ij} - \overline{Y}_{i.} - \hat{\beta}(x_{ij} - \overline{x}_{i.}) \right]^2$$

Similarly, under the null hypothesis that the $\alpha_i = 0$, $\hat{\hat{\theta}} = \overline{Y}_{..} - \hat{\hat{\beta}} x_{..}$, $\hat{\alpha}_i = 0$,

$$\hat{\hat{\beta}} = \frac{\sum_i \sum_j (Y_{ij} - \overline{Y}_{..})(x_{ij} - \overline{x}_{..})}{\sum_i \sum_j (x_{ij} - \overline{x}_{..})^2}.$$

Therefore, $(N = \sum_i n_i)$

$$F_1 = \frac{\sum_i \sum_j \left[\overline{Y}_{i.} - \overline{Y}_{..} + \hat{\beta}(x_{ij} - \overline{x}_{i.}) - \hat{\hat{\beta}}(x_{ij} - \overline{x}_{..}) \right]^2}{\sum_i \sum_j \left[Y_{ij} - \overline{Y}_{i.} - \hat{\beta}(x_{ij} - x_{i.}) \right]^2} \; \frac{N - r - 1}{r - 1}$$

and

$$F_1 \sim F_{r-1, N-r-1}(0)$$

when the $\alpha_i = 0$.

We could also test the hypothesis that $\beta = 0$. The estimates under the alternative hypothesis are the same as before. Under this null hypothesis, $\hat{\theta} = \bar{Y}_{..}$, $\hat{\alpha}_i = \bar{Y}_{i.} - \bar{Y}_{..}$, and therefore,

$$F_2 = \frac{\hat{\beta}^2 \Sigma_i \Sigma_j (x_{ij} - \bar{x}_{i.})^2 (N - r - 1)}{\Sigma_i \Sigma_j \left(Y_{ij} - \bar{Y}_{i.} - \hat{\beta}(x_{ij} - \bar{x}_{i.}) \right)^2} \sim F_{1, N-r-1}(0)$$

when $\beta = 0$. To test that $\beta = c$, simply let $Y_{ij}^* = Y_{ij} - cx_{ij}$, and proceed as above.

This kind of model could obviously be extended in many directions. For example, we could let

$$Y_{ijk} \sim N_1\left(\theta + \alpha_i + \beta_j + \gamma_{ij} + \delta x_{ijk} + \epsilon z_{ijk}, \sigma^2\right).$$

This would be a cross between a two way analysis of variance and a regression on two variables. The variables x_{ij} (or x_{ijk} and z_{ijk}) are called *covariates*.

7.6. OPTIMALITY OF THE F-TEST.

We now return to the general model defined in Section 7.1. We show that the test defined in (7.2) is UMP invariant and is unbiased. However, we first transform the problem into what is called the canonical form.

7.6.1. *The Canonical Form*

In the general problem given in Section 7.1, we observe

$$\mathbf{Y} \sim N_n(\boldsymbol{\mu}, \sigma^2 \mathbf{I}), \qquad \boldsymbol{\mu} \in V, \qquad \sigma^2 > 0$$

where V is a p-dimensional subspace of R^n. Let $W \subset V$ be a k-dimensional subspace of R^n. We want to test the null hypothesis that $\boldsymbol{\mu} \in W$ versus the alternative hypothesis that $\boldsymbol{\mu} \in V$. Let $\mathbf{X}_1$ be an orthonormal basis matrix for W, let $\mathbf{X}_2$ be an orthonormal basis for $V|W$, and let $\mathbf{X}_3$ be an orthonormal basis matrix for $V^{\perp}$. Define

$$\mathbf{X} = (\mathbf{X}_1, \mathbf{X}_2, \mathbf{X}_3).$$

Then the columns of $\mathbf{X}$ are orthonormal, and there are n of them, so $\mathbf{X}$ is an orthogonal matrix. Let

$$\mathbf{Z} = \begin{pmatrix} \mathbf{Z}_1 \\ \mathbf{Z}_2 \\ \mathbf{Z}_3 \end{pmatrix} = \begin{pmatrix} \mathbf{X}_1'\mathbf{Y} \\ \mathbf{X}_2'\mathbf{Y} \\ \mathbf{X}_3'\mathbf{Y} \end{pmatrix} = \mathbf{X}'\mathbf{Y}, \qquad \boldsymbol{\nu} = \begin{pmatrix} \boldsymbol{\nu}_1 \\ \boldsymbol{\nu}_2 \\ \boldsymbol{\nu}_3 \end{pmatrix} = \begin{pmatrix} \mathbf{X}_1'\boldsymbol{\mu} \\ \mathbf{X}_2'\boldsymbol{\mu} \\ \mathbf{X}_3'\boldsymbol{\mu} \end{pmatrix} = \mathbf{X}'\boldsymbol{\mu}$$

Then, since $\mathbf{X}$ is orthogonal,

$$\mathbf{Z} \sim N_n(\boldsymbol{\nu}, \sigma^2\mathbf{I})$$

Therefore $\mathbf{Z}_1$, $\mathbf{Z}_2$, and $\mathbf{Z}_3$ are independent, and

$$\mathbf{Z}_1 \sim \mathbf{N}_k(\boldsymbol{\nu}_1, \sigma^2\mathbf{I}), \qquad \mathbf{Z}_2 \sim N_{p-k}(\boldsymbol{\nu}_2, \sigma^2\mathbf{I}), \qquad \mathbf{Z}_3 \sim N_{n-p}(\boldsymbol{\nu}_3, \sigma^2\mathbf{I}).$$

Since $\mathbf{X}$ is an orthogonal matrix, and hence invertible, observing $\mathbf{Z}$ is equivalent to observing $\mathbf{Y}$ (see Section 1.6). The following lemma shows how the hypotheses transform.

LEMMA 7.10. a. $\boldsymbol{\mu} \in V$ if and only if $\boldsymbol{\nu}_3 = \mathbf{0}$.
b. $\boldsymbol{\mu} \in W$ if and only if $\boldsymbol{\nu}_2 = \mathbf{0}$ and $\boldsymbol{\nu}_3 = \mathbf{0}$.

PROOF. By Lemma 2.6c, $\|\mathbf{P}_{V^\perp}\boldsymbol{\mu}\|^2 = \|\boldsymbol{\nu}_3\|^2$, $\|\mathbf{P}_{V|W}\boldsymbol{\mu}\|^2 = \|\boldsymbol{\nu}_2\|^2$, $\|\mathbf{P}_W\boldsymbol{\mu}\|^2 = \|\boldsymbol{\nu}_1\|^2$

a. See Exercise B16.

b. By Lemma 2.5c, $\boldsymbol{\mu} \in W$ if and only if $\mathbf{P}_{W^\perp}\boldsymbol{\mu} = \mathbf{0}$. However, by Lemma 2.5d and f, $\mathbf{P}_{W^\perp}\boldsymbol{\mu} = \mathbf{P}_{V^\perp}\boldsymbol{\mu} + \mathbf{P}_{V|W}\boldsymbol{\mu}$. In addition $(V|W)\perp(V^\perp)$. Therefore,

$$\|\mathbf{P}_{V^\perp}\boldsymbol{\mu} + \mathbf{P}_{V|W}\boldsymbol{\mu}\|^2 = \|\mathbf{P}_{V^\perp}\boldsymbol{\mu}\|^2 + \|\mathbf{P}_{V|W}\boldsymbol{\mu}\|^2 = \|\boldsymbol{\nu}_3\|^2 + \|\boldsymbol{\nu}_2\|^2.$$

Therefore $\boldsymbol{\mu} \in W$ if and only if $0 = \|\mathbf{P}_{W^\perp}\boldsymbol{\mu}\|^2 = \|\boldsymbol{\nu}_3\|^2 + \|\boldsymbol{\nu}_2\|^2$ if and only if $\boldsymbol{\nu}_2 = \mathbf{0}$, $\boldsymbol{\nu}_3 = \mathbf{0}$. □

We have now transformed the problem given in Section 7.1 to the problem in which we observe $\mathbf{Z}_1$, $\mathbf{Z}_2$, and $\mathbf{Z}_3$ independent, with

$$\mathbf{Z}_1 \sim N_k(\boldsymbol{\nu}_1, \sigma^2\mathbf{I}), \qquad \mathbf{Z}_2 \sim N_{p-k}(\boldsymbol{\nu}_2, \sigma^2\mathbf{I}), \qquad \mathbf{Z}_3 \sim N_{n-p}(\boldsymbol{\nu}_3, \sigma^2\mathbf{I}),$$

and we are testing the null hypothesis that $\boldsymbol{\nu}_2 = \mathbf{0}$ and $\boldsymbol{\nu}_3 = \mathbf{0}$ versus the alternative hypothesis that only $\boldsymbol{\nu}_3 = \mathbf{0}$. This form of the problem is often called the *canonical form for the univariate linear model.* The following lemma expresses F in terms of the transformed variables.

LEMMA 7.11. Let F be defined in (7.1). Then

$$F = \frac{\|\mathbf{Z}_2\|^2}{\|\mathbf{Z}_3\|^2} \frac{n-p}{p-k}.$$

PROOF. $\|\mathbf{Z}_2\|^2 = \|\mathbf{X}_2'\mathbf{Y}\|^2 = \|\mathbf{P}_{V|W}\mathbf{Y}\|^2$ by Theorem 2.6c. Similarly, $\|\mathbf{Z}_3\|^2 = \|\mathbf{P}_{V^\perp}\mathbf{Y}\|^2$. $\square$

We now discuss the relationship between $(\mathbf{Z}_1, \mathbf{Z}_2, \mathbf{Z}_3)$ and the complete sufficient statistic $(\hat{\boldsymbol{\mu}}, \hat{\sigma}^2)$. We first note that

$$\hat{\boldsymbol{\mu}} = \mathbf{P}_V\mathbf{Y} = \mathbf{P}_W\mathbf{Y} + \mathbf{P}_{V|W}\mathbf{Y} = \mathbf{X}_1\mathbf{X}_1'\mathbf{Y} + \mathbf{X}_2\mathbf{X}_2'\mathbf{Y} = \mathbf{X}_1\mathbf{Z}_1 + \mathbf{X}_2\mathbf{Z}_2.$$

Also, $\mathbf{X}_1'\mathbf{X}_1 = \mathbf{I}, \mathbf{X}_2'\mathbf{X}_2 = \mathbf{I}$ and $\mathbf{X}_1'\mathbf{X}_2 = 0$. Therefore,

$$\mathbf{X}_1'\hat{\boldsymbol{\mu}} = \mathbf{Z}_1, \qquad \mathbf{X}_2'\hat{\boldsymbol{\mu}} = \mathbf{Z}_2.$$

Hence $(\mathbf{Z}_1, \mathbf{Z}_2)$ is an invertible function of $\hat{\boldsymbol{\mu}}$. Now,

$$(n-p)\hat{\sigma}^2 = \|\mathbf{P}_{V^\perp}\mathbf{Y}\|^2 = \|\mathbf{X}_3'\mathbf{Y}\|^2 = \|\mathbf{Z}_3\|^2.$$

Therefore, $\|\mathbf{Z}_3\|^2$ is an invertible function of $\hat{\sigma}^2$. Hence $(\mathbf{Z}_1, \mathbf{Z}_2, \|\mathbf{Z}_3\|^2)$ is an invertible function of the complete sufficient statistic $(\hat{\boldsymbol{\mu}}, \hat{\sigma}^2)$ and is also a complete sufficient statistic.

The final topic of this section is a geometric description of the transformation to canonical form. We first review some facts about change of basis. Let $\mathbf{e}_i$ be the n-dimensional vector with 1 in the ith place and 0 elsewhere, and let $\mathbf{u}_1, \ldots, \mathbf{u}_n$ be a set of orthonormal vectors in R^n. Then the $\mathbf{e}_i$ form an orthonormal basis for R^n as do the $\mathbf{u}_i$. Let

$$\mathbf{U} = (\mathbf{u}_1, \ldots, \mathbf{u}_n).$$

Let $\mathbf{y} \in R^n$. The the coordinate vector $\mathbf{y}$ in the basis $\mathbf{u}_i$ is the vector $\mathbf{b} = (b_1, \ldots, b_n)'$ such that

$$\mathbf{y} = \sum_i b_i\mathbf{u}_i = \mathbf{U}\mathbf{b}.$$

In fact

$$\mathbf{b} = \mathbf{U}'\mathbf{U}\mathbf{b} = \mathbf{U}'\mathbf{y}.$$

We note, therefore, that the coordinate vector of $\mathbf{y}$ in the basis $\mathbf{e}_i$ is just the vector $\mathbf{y}$ itself. We also note that

$$\|\mathbf{y}\|^2 = \|\mathbf{U}\mathbf{b}\|^2 = \|\mathbf{b}\|^2,$$

and hence that lengths are preserved under orthonormal change of basis.

The matrix $\mathbf{X}$ defined above in this section is an $n \times n$ matrix whose columns are orthonormal. Hence the columns of $\mathbf{X}$ form an orthonormal basis for R^n. The first k columns of $\mathbf{X}$ lie in W, the next $p-k$ in $V|W$ and the remaining $n-p$ in $V^\perp$. We note that

$$\mathbf{Z} = \mathbf{X}'\mathbf{Y}, \qquad \boldsymbol{\nu} = \mathbf{X}'\boldsymbol{\mu}.$$

Therefore $\mathbf{Z}$ is just the coordinate vector of $\mathbf{Y}$ with respect to the basis $\mathbf{X}$, while $\boldsymbol{\nu}$ is the coordinate vector of $\boldsymbol{\mu}$ with respect to the basis $\mathbf{X}$. The spherical symmetry of the spherical normal distribution implies that one orthonormal basis is the same as another. So we choose an orthonormal basis matrix that is better for this problem than the natural orthonormal basis $\mathbf{e}_1, \ldots, \mathbf{e}_n$. We note that $\boldsymbol{\mu} \in V$ if and only if $\boldsymbol{\mu}$ is orthogonal to all vectors in $V^{\perp}$, if and only if the last $p - k$ components of $\boldsymbol{\nu}$ are $\mathbf{0}$. That is $\boldsymbol{\mu} \in V$ if and only if $\boldsymbol{\nu}_3 = \mathbf{0}$. Similarly $\boldsymbol{\mu} \in W$ if and only if $\boldsymbol{\nu}_2 = \mathbf{0}$, and $\boldsymbol{\nu}_3 = \mathbf{0}$.

It should be emphasized that the reduction to canonical form depends very heavily on the spherical symmetry of the spherical normal distribution. In exercise C4 of Chapter 3, it is shown that the spherical normal distribution is the only spherically symmetric distribution with independent components. If the components of $\mathbf{Y}$ have unequal variances, or have a distribution that is not normal, then the components of $\mathbf{Z}$ will not necessarily be independent, even if the components of $\mathbf{Y}$ are. Therefore, in making the transformation to canonical form, we may be destroying the independence of the observations unless the assumptions of the model are met exactly. If the assumptions of the model are met, then the canonical form is an easier model for deriving results than the original model defined in Section 7.1.

7.6.2. *Invariance*

We now show that ϕ defined in (7.2) is UMP invariant. The problem is invariant under the following groups of transformations:

(1) G_1: $g_1(\mathbf{Z}_1, \mathbf{Z}_2, \mathbf{Z}_3) = (\mathbf{Z}_1 + \mathbf{a}, \mathbf{Z}_2, \mathbf{Z}_3)$, where $\mathbf{a}$ is an arbitrary $k \times 1$ vector.
(2) G_2: $g_2(\mathbf{Z}_1, \mathbf{Z}_2, \mathbf{Z}_3) = (\mathbf{Z}_1, \boldsymbol{\Gamma}\mathbf{Z}_2, \mathbf{Z}_3)$ where $\boldsymbol{\Gamma}$ is an arbitrary $(p - k) \times (p - k)$ orthogonal matrix.
(3) G_3: $g_3(\mathbf{Z}_1, \mathbf{Z}_2, \mathbf{Z}_3) = (c\mathbf{Z}_1, c\mathbf{Z}_2, c\mathbf{Z}_3)$ where c is an arbitrary positive scalar.

(This problem is also invariant under the group G_4 consisting of transformations of the form $g_4(\mathbf{Z}_1, \mathbf{Z}_2, \mathbf{Z}_3) = (\boldsymbol{\Gamma}_1\mathbf{Z}_1, \mathbf{Z}_2, \boldsymbol{\Gamma}_3\mathbf{Z}_3)$ where $\boldsymbol{\Gamma}_1$ and $\boldsymbol{\Gamma}_3$ are $k \times k$ and $(n - p) \times (n - p)$ orthogonal matrices. However this group gives no additional reduction in the problem, since the maximal invariant under G_1, G_2, and G_3 is also invariant under G_4, and would hence also be the maximal invariant under G_1, G_2, G_3 and G_4. We therefore ignore G_4 when reducing the problem by invariance.)

We first reduce the problem by sufficiency, then by invariance under G_1, then by invariance under G_2, and finally by invariance under G_3. Let $U_3 = \|\mathbf{Z}_3\|^2$. Then $(\mathbf{Z}_1, \mathbf{Z}_2, U_3)$ is a sufficient statistic.

(1) To reduce the problem under G_1, let $\mathbf{T}_1(\mathbf{Z}_1, \mathbf{Z}_2, U_3) = (\mathbf{Z}_2, U_3)$. Then $\mathbf{T}_1$ is a maximal invariant under G_1 as the following argument shows.

a. $\mathbf{T}_1(g_1(\mathbf{Z}_1, \mathbf{Z}_2, U_3)) = T_1(\mathbf{Z}_1 + \mathbf{a}, \mathbf{Z}_2, U_3) = (\mathbf{Z}_2, U_3) = \mathbf{T}_1(\mathbf{Z}_1, \mathbf{Z}_2, U_3)$.

b. Now suppose that $\mathbf{T}_1(\mathbf{Z}_1, \mathbf{Z}_2, U_3) = \mathbf{T}_1(\mathbf{Z}_1^*, \mathbf{Z}_2^*, U_3^*)$ Then $\mathbf{Z}_2 = \mathbf{Z}_2^*$, $U_3 = U_3^*$, and $\mathbf{Z}_1 = \mathbf{Z}_1^* + (\mathbf{Z}_1 - \mathbf{Z}_1^*) = \mathbf{Z}_1^* + \mathbf{a}$. Therefore $(\mathbf{Z}_1, \mathbf{Z}_2, U_3) = g_1(\mathbf{Z}_1^*, \mathbf{Z}_2^*, U_3^*)$.

(2) We now reduce the problem by G_2. Define $\mathbf{T}_2(\mathbf{Z}_2, U_3) = (U_2, U_3)$, where $U_2 = \|\mathbf{Z}_2\|^2$. This is a maximal invariant, as the following argument shows.

a. $$\mathbf{T}_2(g_2(\mathbf{Z}_2, U_3)) = \mathbf{T}_2(\boldsymbol{\Gamma}\mathbf{Z}_2, U_3) = (\|\boldsymbol{\Gamma}\mathbf{Z}_2\|^2, U_3) = (\|\mathbf{Z}_2\|^2, U_3) = \mathbf{T}_2(\mathbf{Z}_2, U_3).$$

b. Now suppose that $\mathbf{T}_2(\mathbf{Z}_2^*, U_3^*) = \mathbf{T}_2(\mathbf{Z}_2, U_3)$. Then $\|\mathbf{Z}_2\|^2 = \|\mathbf{Z}_2^*\|^2$, and $U_3 = U_3^*$. Therefore, $\mathbf{Z}_2 = \boldsymbol{\Gamma}\mathbf{Z}_2^*$ for some orthogonal $\boldsymbol{\Gamma}$, and $(\mathbf{Z}_2, U_3) = g_2(\mathbf{Z}_2^*, U_3^*)$. (See Lemma 10 of the appendix.)

(3) We now reduce further by G_3, which is now

$$g_3(U_2, U_3) = (c^2U_2, c^2U_3).$$

A maximal invariant is

$$T_3(U_2, U_3) = \frac{U_2}{U_3}\frac{n-p}{p-k} = F$$

(See Exercise B17.)

We have therefore proved the following theorem.

THEOREM 7.12. *F* is a maximal invariant under the groups G_1, G_2, and G_3.

Therefore, the reduced problem is the problem in which we observe

$$F \sim F_{p-k, n-p}(\delta), \qquad \delta = \frac{\|\boldsymbol{\nu}_2\|^2}{\sigma^2}$$

and we are testing the null hypothesis that $\delta = 0$ versus the alternative that $\delta \geqslant 0$. Let $f(F, \delta)$ be the density of F. By Lemma 1.9, $f(F,\delta)/f(F,0)$ is an increasing function of F. By Theorem 1.7, therefore, the test ϕ defined in (7.2) is UMP size α for the reduced problem and hence is UMP invariant size α for the original problem. We have therefore proved the following theorem.

THEOREM 7.13. The test ϕ defined in (7.2) is UMP invariant size α under G_1, G_2, G_3, for the canonical form of the invariate linear model.

COROLLARY. The test ϕ is unbiased.

PROOF. This follows directly from Theorem 1.13. □

In fact, we can get a result that is slightly stronger than unbiasedness. We use the following lemma.

LEMMA 7.14. Let $U \sim F_{s,t}(\delta)$. Then $P_\delta(U > d)$ is an increasing function of δ.

PROOF. See Exercise C3, part d. □

This lemma implies that the power function of the F-test defined in (7.2) is an increasing function of the noncentrality parameter $\|\mathbf{P}_{V|W}\boldsymbol{\mu}\|^2/\sigma^2$.

7.6.3. *Other Properties of the F-Test.*

We have now shown that the F-test ϕ defined in (7.2) is UMP invariant size α, is unbiased and its power function is an increasing function of $\delta = \|\mathbf{P}_{V|W}\boldsymbol{\mu}\|^2/\sigma^2$. In Chapter 10, we show under fairly general conditions that the size and power function of this test are asymptotically (as $n \to \infty$) independent of the normal assumption used in deriving them.

We now state two more properties of the F-test. First, it is an admissible test. (See Lehmann and Stein (1953) for a proof when $k=0$. See Ghosh (1964) for a proof in a more general setting.) Secondly it is a most stringent test. (see Lehmann, 1959, pp. 326–347 for a careful discussion of this result and other minimaxity properties.)

7.7. ORTHOGONAL DESIGNS.

In Sections 7.3 and 7.4, we considered models in which several hypotheses were being tested. For example, in the two-way model, we discussed tests that $\alpha_i = 0$, $\beta_j = 0$, and $\gamma_{ij} = 0$. In such a setting, it is especially nice to have what is called an orthogonal design, which we now describe.

Suppose that we observe $\mathbf{Y} \sim N_n(\boldsymbol{\mu}, \sigma^2\mathbf{I})$, $\boldsymbol{\mu} \in V$, $\sigma^2 > 0$, and that $\boldsymbol{\mu} = \sum_i \boldsymbol{\delta}_i$ where $\boldsymbol{\delta}_i \in U_i$, U_i is a subspace and $U_i \perp U_j$ for all $i \neq j$. Finally, suppose that we are interested in testing the hypotheses that $\boldsymbol{\delta}_1 = \mathbf{0}$, that $\boldsymbol{\delta}_2 = \mathbf{0}$, and so on. Since the $U_i \perp U_j$, we see that $\boldsymbol{\delta}_i = \mathbf{0}$ if and only if $\boldsymbol{\mu} \in V | U_i = W_i$. We therefore want to test that $\boldsymbol{\mu} \in W_1$, that $\boldsymbol{\mu} \in W_2$, and so on. In this setting, we say that we have an *orthogonal design*.

We now look at the balanced two-way analysis of variance model with equal weights as an example. There $\boldsymbol{\mu} = (\mu_{ijk})$, $\mu_{ijk} = \theta + \alpha_i + \beta_j + \gamma_{ij}$ where $\sum_i \alpha_i = 0$, $\sum_j \beta_j = 0$, $\sum_i \gamma_{ij} = 0$, $\sum_j \gamma_{ij} = 0$. Let V be the subspace of such vectors $\boldsymbol{\mu}$. Let $\boldsymbol{\delta}_1 = (\delta^1_{ijk})$ where $\delta^1_{ijk} = \theta$, $\boldsymbol{\delta}_2 = (\delta^2_{ijk})$, where $\delta^2_{ijk} = \alpha_i$, $\boldsymbol{\delta}_3 = (\delta^3_{ijk})$, where $\delta^3_{ijk} = \beta_j$ and $\boldsymbol{\delta}_4 = (\delta^4_{ijk})$ where $\delta^4_{ijk} = \gamma_{ij}$. Then $\boldsymbol{\mu} = \boldsymbol{\delta}_1 + \boldsymbol{\delta}_2 + \boldsymbol{\delta}_3 + \boldsymbol{\delta}_4$. Note that $\delta_1 \in U_1$ where U_1 is the one-dimensional subspace of vectors all of whose components are the same; $\boldsymbol{\delta}_2 \in U_2$ the $(r-1)$-dimensional subspace of vectors (δ_{ijk}^2), $\delta_{ijk}^2 = \alpha_i$, $\sum_i \alpha_i = 0$; $\boldsymbol{\delta}_3 \in U_3$ the $(c$-$1)$-dimensional subspace of vectors (δ_{ijk}^3), $\delta_{ijk}^3 = \beta_j$, $\sum_j \beta_j = 0$; and $\boldsymbol{\delta}_4 \in U_4$ the $(r-1)(c-1)$-dimensional subspace of all vectors (δ_{ijk}^4), $\delta_{ijk}^4 = \gamma_{ij}$, $\sum_i \gamma_{ij} = 0$, $\sum_j \gamma_{ij} = 0$. We now show that $U_i \perp U_j$. Let $\boldsymbol{\delta}_1 \in U_1$, $\boldsymbol{\delta}_2 \in U_2$, $\boldsymbol{\delta}_3 \in U_3$, $\boldsymbol{\delta}_4 \in U_4$. Then

$$\langle \boldsymbol{\delta}_1, \boldsymbol{\delta}_2 \rangle = nc\theta \sum_i \alpha_i = 0, \qquad \langle \boldsymbol{\delta}_1, \boldsymbol{\delta}_3 \rangle = nr\theta \sum_j \beta_j = 0,$$

$$\langle \boldsymbol{\delta}_1, \boldsymbol{\delta}_4 \rangle = n\theta \sum_i \sum_j \gamma_{ij} = 0, \qquad \langle \boldsymbol{\delta}_2, \boldsymbol{\delta}_3 \rangle = n \sum_i \alpha_i \sum_j \beta_j = 0,$$

$$\langle \boldsymbol{\delta}_2, \boldsymbol{\delta}_4 \rangle = n \sum_i \alpha_i \sum_j \gamma_{ij} = 0, \qquad \langle \boldsymbol{\delta}_3, \boldsymbol{\delta}_4 \rangle = n \sum_j \beta_j \sum_i \gamma_{ij} = 0.$$

Therefore $U_i \perp U_j$. Note that testing that all the $\alpha_i = 0$ is the same as testing that $\boldsymbol{\delta}_2 = \mathbf{0}$, testing that all the $\beta_j = 0$ is the same as testing that $\boldsymbol{\delta}_3 = \mathbf{0}$, and testing that all the $\gamma_{ij} = 0$, is the same as testing that $\boldsymbol{\delta}_4 = \mathbf{0}$. Finally we could also test that $\theta = 0$, that is, that $\boldsymbol{\delta}_1 = \mathbf{0}$. So this model is an orthogonal design.

When we are testing the various hypotheses that $\boldsymbol{\delta}_i = \mathbf{0}$ in an orthogonal design, the hypotheses that we are testing are separate hypotheses. There is no overlap between them. In this paragraph, we give some specific examples of nice properties of orthogonal designs that illustrate this separation. The first aspect of orthogonal designs is that $\|\mathbf{P}_{V|W_i}\mathbf{Y}\|^2 = \|\mathbf{P}_{U_i}\mathbf{Y}\|^2$ are independent, since the U_i are orthogonal. Therefore, the numerator sums of squares are independent χ^2 statistics. Another important property of an orthogonal design is the following equality

$$\|\mathbf{Y} - \boldsymbol{\mu}\|^2 = \|\mathbf{P}_{V^\perp}\mathbf{Y}\|^2 + \sum_i \|\mathbf{P}_{U_i}\mathbf{Y} - \boldsymbol{\delta}_i\|^2$$

(see Exercise C2). Various versions of this equality were used in previous sections for deriving the estimators of $\boldsymbol{\delta}_i$ under the null and alternative hypotheses. This equation implies that estimators $\hat{\boldsymbol{\delta}}_i$ under the alternative are given by $\hat{\boldsymbol{\delta}}_i = \mathbf{P}_{U_i}\mathbf{Y}$. Under the null hypothesis that $\boldsymbol{\delta}_i = \mathbf{0}$, the estimators are given by $\hat{\boldsymbol{\delta}}_i = \mathbf{0}$, $\hat{\boldsymbol{\delta}}_j = \mathbf{P}_{U_j}\mathbf{Y}$. We note that the estimators for $\boldsymbol{\delta}_j$ are the same whether or not we assume that $\boldsymbol{\delta}_i$ is in the model. If the estimators of $\boldsymbol{\delta}_j$ change when we assume $\boldsymbol{\delta}_i = \mathbf{0}$, it is difficult to think of the $\boldsymbol{\delta}_j$ as separate parameters. They must have some overlap. Also

$$\|\mathbf{Y}\|^2 = \|\mathbf{P}_{V^\perp}\mathbf{Y}\|^2 + \sum_i \|\mathbf{P}_{U_i}\mathbf{Y}\|^2.$$

That is, if we have an orthogonal design we can partition the total sum of squares $\|\mathbf{Y}\|^2$ into a sum of terms for each of the effects $(\|\mathbf{P}_{U_i}\mathbf{Y}\|^2)$ plus a term for the variance $(\|\mathbf{P}_{V^\perp}\mathbf{Y}\|^2)$. The various sums of squares add up to the total sum of squares only with an orthogonal design. We also note that the numerator of the F-statistic for testing that $\boldsymbol{\delta}_i = \mathbf{0}$ is just $\|\mathbf{P}_{U_i}\mathbf{Y}\|^2 = \|\hat{\boldsymbol{\delta}}_i\|^2$, that is, the length of the estimator of $\boldsymbol{\delta}_i$ under the alternative hypothesis. Finally, the noncentrality parameter is

$$\frac{\|\mathbf{P}_{U_i}\boldsymbol{\mu}\|^2}{\sigma^2} = \frac{\|\boldsymbol{\delta}_i\|^2}{\sigma^2}.$$

All these nice properties of orthogonal designs indicate that we should try to have an orthogonal design wherever possible.

We now discuss which of the models discussed in previous sections are orthogonal designs. To all the models we add the problem of testing that $\theta = 0$. If we use equal weights, then the balanced one-way, balanced two-way (with or without interaction) Latin square and balanced nested models are all orthogonal designs. The unbalanced one-way model is an orthogonal design if we use the weights $w_i = n_i$. (Although the hypothesis that $\alpha_i = 0$ does not depend on the weights, the hypothesis that $\theta = 0$ does depend on the weights.) The unbalanced two-way model with or without interaction is an orthogonal design in the case of proportional sampling with the weights $w_i = n_i$, $v_j = n_j$. The unbalanced nested model is an orthogonal design with the weights $v_i = n_{i.}$, $w_{ij} = n_{ij}$. The analysis of covariance model is an orthogonal design with the weights $w_i = n_i$ if $x_{i.} = 0$. The above models are not orthogonal designs in other settings and with other weights. The fact that analysis of variance designs are sometimes not orthogonal explains why it is often difficult to find formulas for the estimators of the parameters under certain hypotheses.

The results stated above for orthogonal designs indicate that we should try to use an orthogonal design wherever possible. If the design is not orthogonal, then the various null hypotheses being tested are not really separated. As mentioned in Section 7.4.3, it does not seem reasonable to change the weights for a particular sampling scheme in order to make the design orthogonal. A more reasonable approach is to design the sampling scheme for a given set of weights. For example, suppose that in a two-way model, we have the weights w_i and v_j (which should be determined before the data is collected, and hence can often be determined before the sampling scheme is defined). We then sample with the scheme defined by

$$n_{ij} = K w_i v_j$$

for some K. Then, we see that

$$n_{i.} = K w_i \sum_j v_j, \qquad n_{.j} = K v_j \sum_i w_i, \qquad N = K \sum_i w_i \sum_j v_j.$$

Hence $n_{ij} = n_{i.} n_{.j}/N$, and the sampling scheme is a proportional sampling scheme. Also, $\sum_i n_{i.}\alpha_i = 0$ if and only if $\sum_i w_i\alpha_i = 0$ and $\sum_j n_{.j}\beta_j = 0$ if and only if $\sum_j v_j\beta_j = 0$. Therefore, using the weights w_i and v_j is the same as using the weights $n_{i.}$ and $n_{.j}$. Hence, this design would be orthogonal.

Similarly, in the nested analysis of variance model discussed in Section 7.4.4, if we want to use the weights w_{ij}, we should let

$$n_{ij} = Kw_{ij}.$$

Since neither of the hypotheses of interest depend on the weights v_j, we could let $v_i = n_{i.}$, or equivalently, let $v_i = \sum_j w_{ij}$. This would lead to an orthogonal design for that model.

In some models with a large number of parameters, it is very difficult for the computer to solve the appropriate minimization problem. For these models, it is a particular advantage to have an orthogonal design, since we can find formulas for the F-statistics in these designs, and it is not necessary to solve the minimization problem numerically. This would be the case, for example, in a two-way model with 100 rows and 50 columns.

7.8. ESTIMABLE FUNCTIONS AND TESTABLE HYPOTHESES.

In Sections 7.3 and 7.4, we studied several analysis of variance models and derived F-tests for appropriate hypotheses for those models. In all those models we imposed constraints to make the parameters uniquely defined. There is another approach to these models in which no constraints are assumed. This approach is used in several books on linear models as well as many papers on this subject. In this section we present a very elementary introduction to this method of studying analysis of variance models. For a more detailed treatment of this approach, see Searle (1971) or Graybill (1976).

As an example, consider the two-way analysis of variance model discussed in Section 7.4c in which we observe Y_{ijk} independent, $i = 1, \ldots, r$; $j = 1, \ldots, c$; $k = 1, \ldots, n_{ij}$; such that $Y_{ijk} \sim N_1(\theta + \alpha_i + \beta_j + \gamma_{ij}, \sigma^2)$. In this section, we impose no constraints on the parameters $\theta, \alpha_i, \beta_j, \gamma_{ij}$. We first note that the parameters are not uniquely defined. For example, if we add d to α_i and subtract d from the γ_{ij}, we have not changed the model. In Section 7.4.3, we imposed constraints on the α_i, β_j, and γ_{ij} to make them uniquely defined.

Now, let $N = \sum_i\sum_j n_{ij}$ and let $\mathbf{Y}$ be an N-dimensional vector whose elements are the Y_{ijk}. Let $\mu_{ij} = EY_{ijk}$, $\boldsymbol{\mu} = E\mathbf{Y}$. Then

$$\mu_{ij} = \theta + \alpha_i + \beta_j + \gamma_{ij}$$

is a linear function of θ, α_i, β_j, and γ_{ij}. Therefore

$$\boldsymbol{\mu} = \mathbf{X}\boldsymbol{\beta}, \qquad \boldsymbol{\beta}' = (\theta, \alpha_1, \ldots, \alpha_r, \beta_1, \ldots, \beta_c, \gamma_{11}, \ldots, \gamma_{rc})$$

for some matrix $\mathbf{X}$. Note that $\mathbf{X}$ is an $Nx(rc + r + c + 1)$ matrix of rank rc and hence does not have full rank and that $\boldsymbol{\beta}$ is not uniquely defined.

With this example in mind, we now consider the model in which we observe

$$\mathbf{Y} \sim N_n(\mathbf{X}\boldsymbol{\beta}, \sigma^2\mathbf{I}), \qquad \boldsymbol{\beta} \in R^s, \qquad \sigma^2 > 0$$

where $\mathbf{X}$ is a known $n \times s$ matrix of rank p. Since $\mathbf{X}$ does not have full rank, this model is not a coordinatized linear model as discussed in earlier sections. As mentioned above, the vector $\boldsymbol{\beta}$ is not uniquely defined in the sense that there are many vectors $\boldsymbol{\beta}$ such that $\mathbf{X}\boldsymbol{\beta}$ is the same, and hence many $\boldsymbol{\beta}$ such that the distribution of $\mathbf{Y}$ is the same.

Now, let $\boldsymbol{\mu} = \mathbf{X}\boldsymbol{\beta}$ and let V be the p-dimensional subspace spanned by the columns of $\mathbf{X}$. Note that $\boldsymbol{\mu}$ and V are uniquely defined. Also,

$$\mathbf{Y} \sim N_n(\boldsymbol{\mu}, \sigma^2\mathbf{I}), \qquad \boldsymbol{\mu} \in V, \qquad \sigma^2 > 0$$

so that this model is a linear model of the form we have been studying in previous sections. As before, let

$$\hat{\boldsymbol{\mu}} = \mathbf{P}_V\mathbf{Y}, \qquad \hat{\sigma}^2 = \frac{\|\mathbf{P}_{V^\perp}\mathbf{Y}\|^2}{(n-p)}$$

Let $\mathbf{c} \in R^p$. We say that a function $\mathbf{c}'\boldsymbol{\beta}$ is *estimable* if there is a linear unbiased estimator of $\mathbf{c}'\boldsymbol{\beta}$ or equivalently if there exists $\mathbf{a} \in R^n$ such that $\mathbf{a}'\mathbf{Y}$ is an unbiased estimator of $\mathbf{c}'\boldsymbol{\beta}$. We say that the estimator $\hat{\boldsymbol{\beta}}$ is an *ordinary least-squares* estimator of $\boldsymbol{\beta}$ if

$$\|\mathbf{Y} - \mathbf{X}\hat{\boldsymbol{\beta}}\|^2 \leqslant \|\mathbf{Y} - \mathbf{X}\boldsymbol{\beta}\|^2$$

for all $\boldsymbol{\beta} \in R^p$. The following thereom summarizes the results of Exercise C1 of Chapter 6.

THEOREM 7.15. a. There is no unbiased estimator of $\boldsymbol{\beta}$ since $\boldsymbol{\beta}$ is not uniquely defined.

b. There are many ordinary least-squares estimators of $\boldsymbol{\beta}$.

c. $\hat{\boldsymbol{\beta}}$ is a least squares estimator of $\boldsymbol{\beta}$ if and only if $\hat{\boldsymbol{\mu}} = \mathbf{X}\hat{\boldsymbol{\beta}}$.

d. If $\hat{\boldsymbol{\beta}}$ is a least-squares estimator of $\boldsymbol{\beta}$, then $\hat{\sigma}^2 = \|\mathbf{Y} - \mathbf{X}\hat{\boldsymbol{\beta}}\|^2/(n-p)$.

e. $\mathbf{c}'\boldsymbol{\beta}$ is estimable if and only if there exists $\mathbf{a} \in R^n$ such that $\mathbf{c}'\boldsymbol{\beta} = \mathbf{a}'\boldsymbol{\mu}$.

f. Let $\hat{\boldsymbol{\beta}}$ be an ordinary least squares estimator of $\boldsymbol{\beta}$. If $\mathbf{c}'\boldsymbol{\beta}$ is an estimable function then $\mathbf{c}'\hat{\boldsymbol{\beta}}$ is the minimum variance unbiased estimator of $\mathbf{c}'\boldsymbol{\beta}$.

g. Let $\hat{\boldsymbol{\beta}}$ be an ordinary least squares estimator of β. Then $\hat{\boldsymbol{\beta}}$ and $(n-p)/n)\hat{\sigma}^2$ are MLEs of $\boldsymbol{\beta}$ and σ^2. Note that the MLE of $\boldsymbol{\beta}$ is not uniquely defined.

h. $\hat{\boldsymbol{\beta}}$ is an ordinary least squares estimator of $\boldsymbol{\beta}$ if and only if $\mathbf{X}'\mathbf{X}\hat{\boldsymbol{\beta}} = \mathbf{X}'\mathbf{Y}$.

i. Let $\mathbf{A}$ be a matrix such that $\mathbf{X'XAX'X} = \mathbf{X'X}$ and let $\hat{\boldsymbol{\beta}} = \mathbf{AX'Y}$. Then $\hat{\boldsymbol{\beta}}$ is an ordinary least-squares estimator of $\boldsymbol{\beta}$. (The matrix $\mathbf{A}$ is called a *generalized inverse* of $\mathbf{X'X}$ and exists for all $\mathbf{X}$, but is not unique when $\mathbf{X}$ does not have full rank.)

To use this theorem, we first find an ordinary least-squares estimator $\hat{\boldsymbol{\beta}}$ of $\boldsymbol{\beta}$ using part h or part i. We then find $\hat{\sigma}^2$ using part d. The minimum variance unbiased estimator of any estimable function $\mathbf{c}'\boldsymbol{\beta}$ is $\mathbf{c}'\hat{\boldsymbol{\beta}}$. Note that $\mathbf{c}'\hat{\boldsymbol{\beta}}$ does not depend on the choice of ordinary least-squares estimator $\hat{\boldsymbol{\beta}}$ as long as $\mathbf{c}'\boldsymbol{\beta}$ is estimable.

The least-squares estimator $\hat{\boldsymbol{\beta}}$ is not uniquely defined. To compute a particular estimator $\hat{\boldsymbol{\beta}}$, constraints are often imposed on the estimator to make it uniquely defined. These constraints are very similar to the constraints imposed on the parameters in Sections 7.3 and 7.4. However, the constraints have a different meaning in this version of the model. In the constrained version of the model, the constraints are used to define the parameters and hence define the model. In this version, the constraints are merely used to get a least-squares solution, since any ordinary least-squares solution can be used. In other words, the constraints in Sections 7.3 and 7.4 are used to uniquely define the parameters, whereas the constraints in this version of the model are only used to get a particular least-squares estimator.

Now, let $\mathbf{H}$ be a $(p-k) \times s$ matrix of rank $p-k$. We now consider testing that $\mathbf{H}\boldsymbol{\beta} = \mathbf{0}$. In general, it is not possible to test this hypothesis. We say that hypothesis is *testable* if the components of $\mathbf{H}\boldsymbol{\beta}$ are estimable functions. In that case, by Theorem 7.15e, there exists $\mathbf{T}$ such that

$$\mathbf{H}\boldsymbol{\beta} = \mathbf{T}\boldsymbol{\mu}.$$

Therefore, testing that $\mathbf{H}\boldsymbol{\beta} = \mathbf{0}$ is equivalent to testing that $\mathbf{T}\boldsymbol{\mu} = 0$. Let W be the subspace of V in which $\mathbf{T}\boldsymbol{\mu} = \mathbf{0}$. Testing that $\mathbf{H}\boldsymbol{\beta} = \mathbf{0}$ is equivalent to testing that $\boldsymbol{\mu} \in W$. The dimension of W is k (see Exercise C1, part c). Therefore, the appropriate test for this problem is

$$F = \frac{\|\mathbf{P}_{V|W}\mathbf{Y}\|^2(n-p)}{\|\mathbf{P}_{V^\perp}\mathbf{Y}\|^2(p-k)}, \qquad \phi(F) = \begin{cases} 1 & \text{if } F > F^{\alpha}_{p-k,n-p} \\ 0 & \text{if } F \leqslant F^{\alpha}_{p-k,n-p} \end{cases}$$

From Exercise C1, we see that

$$F = \frac{(\mathbf{H}\hat{\boldsymbol{\beta}})'(\mathbf{HAH})^{-1}(\mathbf{H}\hat{\boldsymbol{\beta}})}{(p-k)\hat{\sigma}^2}$$

where, as before, $\mathbf{A}$ is any matrix such that $\mathbf{X'XAX'X} = \mathbf{X'X}$ and $\hat{\boldsymbol{\beta}}$ is any ordinary least-squares estimator $\boldsymbol{\beta}$.

We now return to the two-way analysis of variance example given at the start of this section. We first consider testing that there is no interaction in the model. In the version of the model considered in this section, we cannot test that $\gamma_{ij} = 0$, since we could add d to all the γ_{ij} and subtract d from θ without changing the model. We also cannot test that $\gamma_{ij} = \gamma_{i'j'}$ since we could add d to the γ_{ij} and subtract d from α_i without changing the model. In other words the hypotheses that $\gamma_{ij} = 0$ and $\gamma_{ij} - \gamma_{ij'} = 0$ are not testable hypotheses. One hypothesis that is testable is the hypothesis that

$$\gamma_{ij} - \gamma_{i'j} - \gamma_{ij'} + \gamma_{i'j'} = 0$$

since

$$\gamma_{ij} - \gamma_{i'j} - \gamma_{ij'} + \gamma_{ij} = \mu_{ij} - \mu_{i'j} - \mu_{ij'} + \mu_{i'j'}.$$

(By Theorem 7.15e, a linear hypothesis is testable if and only if it can be written as a function of the μ_{ij}.) Therefore, we could use the methods of this section to test that $\gamma_{ij} - \gamma_{i'j} - \gamma_{ij'} + \gamma_{ij} = 0$. The test derived using these methods is the same as the test discussed in Section 7.4.3 for testing for no interactions in the two-way model with constraints. (Note that the test for no interactions derived in Section 7.4.3 did not depend on the constraints.)

Now consider testing that there is no row effect. We cannot test that $\alpha_i = 0$, since we could add d to all the α_i and subtract d from θ without changing the model. We also cannot test that $\alpha_i = \alpha_{i'}$, since we could add d to α_1 and subtract d from γ_{1j} without changing the model. In fact, there does not seem to be any natural test of no row effect for the unconstrained model, which is not surprising since the test derived in Section 7.4.3 for no row effects depends on the constraints. Therefore, let $v_j \geqslant 0$ be such that $\sum_j v_j > 0$. The hypothesis of no row effects considered in Section 7.4.3 is the same as the hypothesis that

$$\sum_j v_j(\mu_{ij} - \mu_{i'j}) = 0$$

for all i and i'. Since this hypothesis is in terms of the μ_{ij}, it is a testable hypothesis in the present version. Also

$$\sum_j v_j(\mu_{ij} - \mu_{i'j}) = (\alpha_i - \alpha_{i'})\sum_j v_j + \sum_j v_j(\gamma_{ij} - \gamma_{i'j}).$$

Therefore, the test of no row effects derived in Section 7.4.3 with the constraints suggested there is the same as the test that

$$(\alpha_i - \alpha_{i'})\sum_j v_j + \sum_j v_j(\gamma_{ij} - \gamma_{i'j}) = 0$$

for all i and i' for the unconstrained model.

We have now considered two versions of the same model. It should be emphasized that these two versions are just two different ways of looking at the same model. Testing that the mean vector is in a particular subspace can be given two interpretations depending on whether the parameters are constrained or not, but the F-statistic depends only on the subspace chosen.

One important implication of the material in this section is that we won't get into any difficulties if we state all hypotheses in terms of the means μ_{ij} rather than in terms of the artificial parameters θ, α_i, β_j, and γ_{ij}. For the unconstrained model, hypotheses are testable if and only if they can be expressed in terms of the means. For the constrained model, we have chosen enough restraints so that the artificial parameters θ, α_i, β_j, and γ_{ij} can be determined from the means, and hence any hypotheses about these parameters can be expressed in terms of the means μ_{ij}.

7.9. INTERACTION IN THE TWO-WAY MODEL WITH NO REPLICATION.

We now return to the two-way model in which we observe Y_{ij} independent, $Y_{ij} \sim N_1(\theta + \alpha_i + \beta_j, \sigma^2)$. In many applications, it is hard to justify the assumption that the interactions are all 0. In this section we look at the more general model in which we observe Y_{ij} independent, $Y_{ij} \sim N_1(\theta + \alpha_i + \beta_j + \gamma_{ij}, \sigma^2)$, $i = 1, \ldots, r$; $j = 1, \ldots, c$; where $\sum_i \alpha_i = 0$, $\sum_j \beta_j = 0$, $\sum_i \gamma_{ij} = 0$, $\sum_j \gamma_{ij} = 0$. Let $\mu_{ij} = \theta + \alpha_i + \beta_j + \gamma_{ij}$ and let V^* be the space of all such μ_{ij}. Then $V^* = R^{rc}$ and hence $V^{*\perp} = 0$ and we have no estimator of σ^2 for this model. For this model, however, it is still true that $SS_1 = c\sum_i(\overline{Y}_{i.} - \overline{Y}_{..})^2$ and $SS_2 = \sum_i\sum_j(Y_{ij} - \overline{Y}_{i.} - \overline{Y}_{.j} + \overline{Y}_{..})$ are independent, and that

$$\frac{SS_1}{\sigma^2} \sim \chi^2_{r-1}\left(\frac{c\sum_i \alpha_i^2}{\sigma^2}\right), \qquad \frac{SS_2}{\sigma^2} \sim \chi^2_{(r-1)(c-1)}\left(\frac{\sum_i\sum_j \gamma_{ij}^2}{\sigma^2}\right).$$

Now, let ϕ be the F-test for testing that $\alpha_i = 0$ in the model that occurs when we assume that the interactions are 0. That is,

$$F = (c-1)\frac{SS_1}{SS_2}, \qquad \phi(F) = \begin{cases} 1 & \text{if } F > F^{\alpha}_{r-1,(r-1)(c-1)} \\ 0 & \text{if } F \leqslant F^{\alpha}_{r-1,(r-1)(c-1)} \end{cases}.$$

Let $\delta_1 = c\sum_i \alpha_i^2/\sigma^2$, $\delta_2 = \sum_i\sum_j \gamma_{ij}^2/\sigma^2$. Then the power function of ϕ depends only on δ_1 and δ_2, and is in fact a decreasing funciton of δ_2 for all fixed δ_1 (see Exercise C3f). Therefore, the probability of rejecting the null hypothesis when it is true is always less than or equal to α. Since the probability of rejecting when $\delta_1 = 0$ and $\delta_2 = 0$ is α, we see that this test has size α even for the model with possible interactions. However, if the interactions are large (i.e., if δ_2 is large) then the power of the test would not be very high. The test is also not unbiased.

We now consider the problem of testing that the interactions are 0. We cannot find a sensible test of this hypothesis in the most general possible case. However in the case in which the interactions can be assumed to have the form

$$\gamma_{ij} = \eta\alpha_i\beta_j$$

we can use Tukey's test for nonadditivity, which we now discuss. (We first note that the set of μ_{ij} of the form $\mu_{ij} = \theta + \alpha_i + \beta_j + \eta\alpha_i\beta_j$ is not a subspace. Hence we cannot find an F-test of the form suggested in Section 7.1.) To motivate this test, we first find the ordinary least squares estimator of η, which is given by

$$\tilde{\eta} = \frac{\sum_i \sum_j \tilde{\alpha}_i \tilde{\beta}_j Y_{ij}}{\sum_i \tilde{\alpha}_i^2 \sum_j \tilde{\beta}_j^2}$$

where $\tilde{\alpha}_i$ and $\tilde{\beta}_j$ are the ordinary least squares estimators of α_i and β_j. These estimators are difficult to find, so we substitute the estimators

$$\hat{\alpha}_i = \overline{Y}_{i.} - \overline{Y}_{..}, \qquad \hat{\beta}_j = \overline{Y}_{.j} - \overline{Y}_{..}, \qquad \hat{\eta} = \frac{\sum_i \sum_j \hat{\alpha}_i \hat{\beta}_j Y_{ij}}{\sum_i \hat{\alpha}_i^2 \sum_j \hat{\beta}_j^2}$$

($\hat{\alpha}$ and $\hat{\beta}$ are the ordinary least squares estimators of α_i and β_j under the assumption of no interactions.) Now, let

$$SS_3 = \sum \sum (\hat{\eta}\hat{\alpha}_i \hat{\beta}_j)^2, \qquad SS_4 = SS_2 - SS_3.$$

We think of SS_3 as representing the sum of squares due to η and think of SS_4 as the residual sum of squares due to variance. It can be shown that when $\gamma_{ij} = 0$ (or equivalently $\eta = 0$), then

$$F^* = \left[(r-1)(c-1) - 1 \right] \frac{SS_3}{SS_4} \sim F_{1,(r-1)(c-1)-1}(0)$$

(See the Exercise C4f). Therefore, a sensible test that $\eta = 0$ would be given by

$$\phi^*(F^*) = \begin{cases} 1 & \text{if } F^* > F^{\alpha}_{1,(r-1)(c-1)} \\ 0 & \text{if } F^* \leqslant F^{\alpha}_{1,(r-1)(c-1)} \end{cases}.$$

This test is clearly a size α test. However, its power function is quite difficult to compute. It should be emphasized that this test has not been shown to have the optimality we established for F-tests of the type defined in Section 7.1. However, it does seem to be a sensible test of the hypothesis that the interactions are 0 for the case in which the interactions have the form $\gamma_{ij} = \eta\alpha_i\beta_j$.

In light of the discussion in the previous paragraph, it might seem sensible when testing that $\alpha_i = 0$ to use SS_4 as the denominator sum of squares, since we think of this as the residual sum of squares. [For example, this approach seems to be implied by Rao (1965), p. 209.] However, unless $\gamma_{ij} = 0$, the distribution of SS_4 is not independent of $\hat{\alpha}_i$ (see Exercise C4e) and is therefore not independent of $SS_1 = c\sum\hat{\alpha}_i^2$. In this situation, it is difficult to make any statements about the actual size and power of a test based on the ratio of SS_1 and SS_4. Hence SS_4 should *not* be used as the denominator sum of squares for testing that $\alpha_i = 0$ even in the case in which the interactions cannot be assumed to be 0.

7.10. ONE-SIDED TESTS.

Throughout this section we consider testing that $\boldsymbol{\mu} \in W$ where $\dim(V|W) = 1$. Let $\mathbf{d}$ be a basis vector of $V|W$. We now consider testing that $\langle \mathbf{d}, \boldsymbol{\mu} \rangle = 0$ (or equivalently $\boldsymbol{\mu} \in W$) against $\langle \mathbf{d}, \boldsymbol{\mu} \rangle \geqslant 0$. (Note that we have altered the parameter space for this testing problem to one in which $\boldsymbol{\mu} \in V$ and $\langle \mathbf{d}, \boldsymbol{\mu} \rangle \geqslant 0$. Note also that $\mathbf{d} \in V$ by assumption.) Define

$$t = \frac{\langle \mathbf{d}, \hat{\boldsymbol{\mu}} \rangle}{\|\mathbf{d}\|\hat{\sigma}}\ .$$

Then

$$t \sim t_{n-p}\left(\frac{\langle \mathbf{d}, \boldsymbol{\mu} \rangle}{\|\mathbf{d}\|\sigma} \right)$$

(see Exercise B20). Therefore, we can design a size α test for this problem which has the following form:

$$\phi(t) = \begin{cases} 1 & \text{if } t > t_{n-p}^{\alpha} \\ 0 & \text{if } t \leqslant t_{n-p}^{\alpha} \end{cases}. \tag{7.15}$$

This test is in fact the UMP invariant test for this problem (see Exercise B20).

As an example of a test of this form, we consider a two-way analysis of variance model with no replication or interaction. We consider this model in the case that there are only two rows, and we want to test that there is no row effect. We therefore consider the model in which we observe Y_{ij} independent with $Y_{ij} \sim N_1(\theta + \alpha_i + \beta_j, \sigma^2)$, $i = 1, 2$; $j = 1, \ldots, c$; where $\alpha_1 + \alpha_2 = 0$ and $\sum_j \beta_j = 0$. We want to test that $\alpha_1 = \alpha_2 = 0$ (or equivalently $\alpha_1 - \alpha_2 = 0$) against the alternative that $\alpha_2 \geqslant 0$ (or equivalently $\alpha_2 - \alpha_1 \geqslant 0$). As in previous sections, let $\mathbf{Y}$ be an ordering of the Y_{ij}, let $\boldsymbol{\mu}$ be the same ordering of the $\mu_{ij} = \theta + \alpha_i + \beta_j$. Let $\mathbf{d}$ be the same ordering of $d_{ij} = (-1)^i$. Then

$$\langle \mathbf{d}, \boldsymbol{\mu} \rangle = \sum_i \sum_j d_{ij}(\theta + \alpha_i + \beta_j) = c(\alpha_2 - \alpha_1)$$

Therefore, testing that $\alpha_2 - \alpha_1 = 0$ against $\alpha_2 - \alpha_1 \geqslant 0$ is equivalent to testing that $\langle \mathbf{d}, \boldsymbol{\mu} \rangle = 0$ against $\langle \mathbf{d}, \boldsymbol{\mu} \rangle \geqslant 0$. The appropriate t-statistic for this model is

$$t = \frac{\langle \mathbf{d}, \hat{\boldsymbol{\mu}} \rangle}{\|\mathbf{d}\| \hat{\sigma}} = \frac{\sum_i \sum_j d_{ij} \left(\bar{Y}_{i.} + \bar{Y}_{.j} - \bar{Y}_{..} \right)}{\left(\sum_i \sum_j d_{ij}^2 \right)^{1/2} \hat{\sigma}}$$

$$= \frac{\sqrt{c} \left(\bar{Y}_{2.} - \bar{Y}_{1.} \right)}{\sqrt{2}\, \hat{\sigma}} \sim t_{c-1} \left[\frac{\sqrt{c}\, (\alpha_2 - \alpha_1)}{\sqrt{2}\, \sigma} \right]$$

where, as in Section 7.3.2

$$\hat{\sigma}^2 = \frac{1}{c-1} \sum_i \sum_j \left(Y_{ij} - \bar{Y}_{i.} - \bar{Y}_{.j} + \bar{Y}_{..} \right)^2.$$

We can put this statistic into somewhat simpler form. Let $Z_j = Y_{2j} - Y_{1j} \sim N_1(\alpha_2 - \alpha_1, 2\sigma^2)$. Then

$$\bar{Y}_{2.} - \bar{Y}_{1.} = \bar{Z}, \qquad 2\hat{\sigma}^2 = \frac{1}{c-1} \sum_i \left(Z_i - \bar{Z} \right)^2 \tag{7.16}$$

(see Exercise B21). Therefore, the t-test defined above is just a one-sample t-test performed on the differences Z_i. This test is called the paired comparison test.

One-sided tests of this type can be used any time the dimension of $V \mid W$ is one. If the dimension of $V \mid W$ is greater than one, then there are no one-sided tests.

We could also use two-sided t-tests in this setting. If we want to test that $\langle \mathbf{d}, \boldsymbol{\mu} \rangle = 0$ (or equivalently $\boldsymbol{\mu} \in W$) against $\boldsymbol{\mu} \in V$, we could use the following test

$$\phi(t) = \begin{cases} 1 & \text{if } t > t_{n-p}^{\alpha} \text{ or } t < -t_{n-p}^{\alpha} \\ 0 & \text{if } -t_{n-p} < t < t_{n-p}^{\alpha} \end{cases}$$

However,

$$t^2 = \frac{\langle \mathbf{d}, \hat{\boldsymbol{\mu}} \rangle^2}{\|\mathbf{d}\|^2 \hat{\sigma}^2} = \frac{\hat{\boldsymbol{\mu}}'\mathbf{d}(\mathbf{d}'\mathbf{d})^{-1}\mathbf{d}\hat{\boldsymbol{\mu}}}{\hat{\sigma}^2} = \frac{\hat{\boldsymbol{\mu}}'\mathbf{P}_{V \mid W} \hat{\boldsymbol{\mu}}}{\hat{\sigma}^2} = \frac{\|\mathbf{P}_{V \mid W} \hat{\boldsymbol{\mu}}\|^2}{\hat{\sigma}^2},$$

and therefore, this test is the same as the F-test defined in (7.2) for this problem.

7.11. THE CASE IN WHICH σ^2 IS KNOWN.

We now consider the linear model when we assume that $\sigma^2 = \sigma_0^2$ is a known constant. Although it is hard to imagine a situation in which σ^2 would be known, in Chapter 10 we give some examples of models that approximate linear models with known variance. The model that we consider in this section is one in which we observe

$$\mathbf{Y} \sim N_n(\boldsymbol{\mu}, \sigma_0^2 \mathbf{I}), \qquad \boldsymbol{\mu} \in V,$$

where $V \subset R^n$ is a p-dimensional subspace. It is no longer necessary to assume that $p < n$. That is, we allow the possibility that $V = R^n$. We want to test the hypotheesis $\boldsymbol{\mu} \in W \subset V$ against $\boldsymbol{\mu} \in V$, where W is a k-dimensional subspace of V. (Clearly we need to assume that $p > k$, or the hypothesis is rather uninteresting.) It seems evident that a sensible test would just use the numerator of the F-statistic defined in Section 7.1, since there is no reason to estimate the variance. Therefore, a sensible test would be given by

$$\chi^2 = \frac{\|\mathbf{P}_{V|W}\mathbf{Y}\|^2}{\sigma_0^2} \sim \chi^2_{p-k}\left(\frac{\|\mathbf{P}_{V|W}\boldsymbol{\mu}\|^2}{\sigma_0^2} \right) \tag{7.17}$$

$$\phi(\chi^2) = \begin{cases} 1 & \text{if } \chi^2 > \chi^{2\,\alpha}_{p-k} \\ 0 & \text{if } \chi^2 \leqslant \chi^{2\,\alpha}_{p-k} \end{cases}$$

In fact, this test is the UMP invariant size α test for this problem (see Exercise B22).

EXERCISES.

Type A

All of the following problems are to be tested with size $\alpha = 0.05$.

1. Use the data and model of Exercise A1 of Chapter 5 to test the following hypothesis:

(a) $\beta_0 = 1$ and $\beta_1 = 2$.

(b) $2\beta_0 + \beta_1 = 5$ and $\beta_2 = 3$.

2. Let $Y_{11} = 4$, $Y_{12} = 4$, $Y_{21} = 7$, $Y_{22} = 6$, $Y_{31} = 8$, $Y_{32} = 7$. Consider this data as coming from a balanced one-way model and use the test defined in Section 7.3.1 to test that $\alpha_1 = \alpha_2 = \alpha_3 = 0$.

3. Consider the data of Exercise 2 as a two-way model with no replication (and no interaction). Using the procedure derived in Section 7.3.2 test the hypothesis that there is no row effect.

4. Let $Y_{111} = 2$, $Y_{112} = 4$, $Y_{121} = 0$, $Y_{122} = 6$, $Y_{131} = 6$, $Y_{132} = 6$, $Y_{211} = 8$, $Y_{212} = 12$, $Y_{221} = 10$, $Y_{222} = 18$, $Y_{231} = 8$, $Y_{232} = 16$. Consider this data as coming from a balanced two-way model (with equal weights). Use the procedures derived in Section 7.3.3 to test that there is no row effect, to test that there is no column effect, and to test that there is no interaction effect.

5. Consider the data of Exercise 4 as from a nested model (with j nested inside i). Using the procedures derived in Section 7.3.4, test the hypothesis that the $\alpha_i = 0$ and the hypothesis that the $\delta_{ij} = 0$.

6. Consider the following Latin square $(k(i, j))$ and data Y_{ij}:

	$k(i,j)$			Y_{ij}		
i	$j=1$	$j=2$	$j=3$	$j=1$	$j=2$	$j=3$
1	1	2	3	4	6	8
2	3	1	2	11	4	9
3	2	3	1	9	11	4

(a) Using the procedure derived in Section 7.3.5 test that the $\alpha_i = 0$.
(b) Using the procedure derived in Section 7.3.5 test that the $\gamma_k = 0$.

7. Let $Y_{11} = 6$, $Y_{21} = 8$, $Y_{22} = 10$, $Y_{31} = 8$, $Y_{32} = 6$, $Y_{33} = 4$. Consider this data as coming from an unbalanced one-way model. Use the procedure derived in Section 7.3.7 to test that the $\alpha_i = 0$.

8. Use Tukey's test of nonadditivity (see Section 7.6) to test for interactions with the data in Exercise A2.

Type B

1. Prove Lemma 7.2.

2. (a) Let λ be the likelihood ratio test statistic for testing that $\boldsymbol{\mu} \in W$ against $\boldsymbol{\mu} \in V$. Show that

$$\lambda = \left(\frac{\|\mathbf{P}_{V^\perp}\mathbf{Y}\|^2}{\|\mathbf{P}_{W^\perp}\mathbf{Y}\|^2} \right)^{n/2}.$$

(b) Show that the test ϕ defined in (7.2) is the likelihood ratio test.

3. Prove Theorem 7.4. (By Lemma 7.3, $\mathbf{P}_{V|W}\hat{\boldsymbol{\mu}} = \mathbf{C}(\mathbf{C}'\mathbf{C})^{-1}\mathbf{C}'\hat{\boldsymbol{\mu}}$.)

4. Prove Lemma 7.5.

5. Derive (7.8).

6. Derive (7.9).

7. Use (7.10) to establish that F_3 defined in (7.11) is the appropriate statistic to use for testing hypothesis 3 in the two-way anova model.

8. Derive (7.12) and use it to find the F-statistics for testing the hypotheses considered in Section 7.3.4.

9. Verify the equation in Section 7.3.5 and use it to verify the F-statistics given in that section.

10. Suppose I observe Y_{ijkm} independent, $i = 1, \ldots, r$; $j = 1, \ldots, c$; $k = 1, \ldots, s$; $m = 1, \ldots, t$ such that

$$Y_{ijkm} \sim N_1(\theta + \alpha_i + \beta_j + \gamma_{ij} + \delta_{ijk}, \sigma^2), \qquad \sum_i \alpha_i = 0, \qquad \sum_j \beta_j = 0,$$

$$\sum_i \gamma_{ij} = 0, \qquad \sum_j \gamma_{ij} = 0, \qquad \sum_k \delta_{ijk} = 0.$$

(a) Find the F-statistic for testing that the $\alpha_i = 0$. How is it distributed?
(b) Find the F-statistic for testing that $\delta_{ijk} = 0$ for all i, j, and k. How is it distributed?

11. Let α_i be defined as in the proof of Lemma 7.7. Show that $\sum_i w_i \alpha_i = 0$.

12. Prove Lemma 7.8.

13. Prove Lemma 7.9b.

14. Consider the analysis of covariance model defined in Section 7.5 with general weights w_i.
(a) Show that the alternative hypothesis does not depend on the weights w_i.
(b) Show that the null hypothesis $\beta = 0$ does not depend on the weights w_i. (It is clear that the null hypothesis $\alpha_i = 0$ does not depend on the weights.)

15. Suppose that I observe Y_{ij} independent, $i = 1, \ldots, r$; $j = 1, \ldots, c$; such that

$$Y_{ij} \sim N_1(\theta + \alpha_i + \beta_j + \gamma x_{ij}, \sigma^2) \qquad \sum_i \alpha_i = 0, \qquad \sum_j \beta_j = 0$$

where the x_{ij} are known fixed numbers. Find the F-statistics and their distributions for testing the following hypotheses.
(a) $\gamma = 0$.
(b) $\alpha_i = 0$ for all i.

16. Prove Lemma 7.10a.

17. Show that a maximal invariant under the group of transformations of the form $g_3(W_2, W_3) = (c^2W_2, c^2W_3)$ is

$$F = \frac{W_2}{W_3}\,\frac{n-p}{p-k}.$$

18. Show that the analysis of covariance model is an orthogonal design with the weights $w_i = n_i$ if $x_{i.} = 0$.

19. Show that the unbalanced two-way analysis of variance model with proportional sampling is an orthogonal design with the weights $w_i = n_{i.}$, $v_j = n_{.j}$.

20. One-sided tests. Consider the one-sided model discussed in Section 7.10.

(a) Following the derivation in Section 7.6.1, put this problem in canonical form.

(b) Show that this problem is invariant under the groups G_1 and G_3 defined in Section 7.6.2. (Note that this problem is not invariant under G_2 since -1 is an orthogonal 1×1 matrix.)

(c) Show that t defined in Section 7.10 is a maximal invariant for this problem.

(d) Show that t has the distribution given in Section 7.10.

(e) Show that the test defined in (7.15) is the UMP invariant size α test for this model.

21. Verify (7.16).

22. Case of known σ_0^2. Consider the linear model in which we observe $\mathbf{Y} \sim N_n(\boldsymbol{\mu}, \sigma_0^2\mathbf{I})$ where σ_0^2 is a known constant. Consider testing that $\boldsymbol{\mu} \in W$ against $\boldsymbol{\mu} \in V$, where $W \subset V$ are subspaces, $\dim W = k$, $\dim V = p$. Let $\mathbf{Z}_1, \mathbf{Z}_2$, and $\mathbf{Z}_3$ be defined as in Section 7.6.1.

(a) Show that a sufficient statistic for this model is $(\mathbf{Z}_1, \mathbf{Z}_2)$.

(b) Show that this problem is invariant under the groups G_1 and G_2 defined in Section 7.6.2.

(c) Show that a maximal invariant under G_1 and G_2 (after reducing by sufficiency) is

$$\chi^2 = \frac{\|\mathbf{Z}_2\|^2}{\sigma_0^2} = \frac{\|\mathbf{P}_{V|W}\mathbf{Y}\|^2}{\sigma_0^2}.$$

(d) Show that the UMP invariant size α test for this problem is given by

$$\phi(\chi^2) = \begin{cases} 1 & \text{if } \chi^2 > \chi_{p-k}^{2\,\alpha} \\ 0 & \text{if } \chi^2 \leqslant \chi_{p-k}^{2\,\alpha} \end{cases}.$$

23. Consider a coordinatized linear model with an intercept in which we observe $\mathbf{Y} \sim N_n(\mathbf{X}\boldsymbol{\beta}, \sigma^2\mathbf{I})$ where

$$\mathbf{X} = \begin{bmatrix} 1 & \mathbf{T}_1 \\ \vdots & \vdots \\ 1 & \mathbf{T}_n \end{bmatrix}, \qquad \boldsymbol{\beta} = \begin{pmatrix} \delta \\ \gamma \end{pmatrix}$$

and $\mathbf{T}_i$ and γ are $(p-1) \times 1$. Let $\gamma = \binom{\gamma_1}{\gamma_2}$ where γ_1 is 1×1. We want to test that $\gamma_1 = 0$ against $\gamma_1 \geqslant 0$. Let

$$\tilde{\mathbf{T}} = (\tilde{\mathbf{T}}_1, \tilde{\mathbf{T}}_2) = \begin{bmatrix} \mathbf{T}_1 - \bar{\mathbf{T}} \\ \vdots \\ \mathbf{T}_n - \bar{\mathbf{T}} \end{bmatrix}, \qquad \hat{\gamma} = \begin{pmatrix} \hat{\gamma}_1 \\ \hat{\gamma}_2 \end{pmatrix} = (\tilde{\mathbf{T}}'\tilde{\mathbf{T}})^{-1}\tilde{\mathbf{T}}\tilde{\mathbf{Y}}$$

where $\tilde{\mathbf{T}}_1$ is $n \times 1$ and $\hat{\gamma}_1$ is 1×1.

(a) Show that the t-statistic for this one-sided hypothesis is given by

$$t = \frac{\hat{\gamma}_1 \sqrt{\tilde{\mathbf{T}}_1'\tilde{\mathbf{T}}_1 - \tilde{\mathbf{T}}_1'\tilde{\mathbf{T}}_2(\tilde{\mathbf{T}}_2'\tilde{\mathbf{T}}_2)^{-1}\tilde{\mathbf{T}}_2'\tilde{\mathbf{T}}_1}}{\hat{\sigma}}.$$

(b) Show that

$$t \sim t_{n-p}\left[\frac{\gamma_1 \sqrt{\tilde{\mathbf{T}}_1'\tilde{\mathbf{T}}_1 - \tilde{\mathbf{T}}_1'\tilde{\mathbf{T}}_2(\tilde{\mathbf{T}}_2'\tilde{\mathbf{T}}_2)^{-1}\tilde{\mathbf{T}}_2'\tilde{\mathbf{T}}_1}}{\sigma}\right].$$

Type C

1. Case of $\mathbf{X}$ matrix of less than full rank. In this problem, we consider the model in which we observe $\mathbf{Y} \sim N_n(\mathbf{X}\mathbf{B}, \sigma^2\mathbf{I})$ where $\mathbf{X}$ is a known $n \times r$ matrix of rank $p < r$. Let $\mathbf{H}$ be a $(p-k) \times r$ matrix of rank $p-k$ such that the elements of $\mathbf{H}\boldsymbol{\beta}$ are estimable functions. (See Exercise C1 of Chapter 6.) We now consider the problem of testing that $\mathbf{H}\boldsymbol{\beta} = \mathbf{0}$. Let V be the p-dimensional subspace spanned by the columns of $\mathbf{X}$ and let W be the subspace of V consisting of vectors $\mathbf{X}\boldsymbol{\beta}$ where $\mathbf{H}\boldsymbol{\beta} = \mathbf{0}$. Let $\mathbf{A}$ be a generalized inverse of $\mathbf{X}'\mathbf{X}$.

(a) Show that there exists $\mathbf{B}$ such that $\mathbf{H} = \mathbf{B}\mathbf{X}$.

(b) Show that $\mathbf{HAX'X} = \mathbf{H}$. (Hint: $\mathbf{XAX'X} = \mathbf{X}$. See exercise C4 of Chapter 2.)

(c) Show that $\mathbf{XA'H'}$ is a basis matrix for $V|W$ and hence that $V|W$ has dimension $p-k$.

(d) Let F be defined by (7.1). Show that

$$F = \frac{(\mathbf{H}\hat{\boldsymbol{\beta}})'(\mathbf{HAH}')^{-1}\mathbf{H}\hat{\boldsymbol{\beta}}}{(p-k)\hat{\sigma}^2} \sim F_{p-k,n-p}\left(\frac{(\mathbf{H}\boldsymbol{\beta})'(\mathbf{HAH}')^{-1}\mathbf{H}\boldsymbol{\beta}}{\sigma^2}\right).$$

(where $\hat{\beta}$ and $\hat{\sigma}^2$ are defined in Exercise C1 of Chapter 6).

2. Orthogonal designs. Suppose that we observe $\mathbf{Y} \sim N_n(\boldsymbol{\mu}, \sigma^2\mathbf{I})$, $\boldsymbol{\mu} \in V$, $\sigma^2 > 0$ where V is a p-dimensional subspace of R^n. Suppose that there exist subspaces U_i such that $U_i \perp U_j$ for $i \neq j$ and such that $\boldsymbol{\mu} = \sum_i \boldsymbol{\delta}_i$, $\boldsymbol{\delta}_i \in U_i$. Let $W_i = V \mid U_i$. We want to test that $\boldsymbol{\mu} \in W_i$.

(a) Show that $V = \sum_i U_i$. (See Exercise C1 of Chapter 2.)

(b) Show that $\mathbf{P}_{U_i}\boldsymbol{\mu} = \boldsymbol{\delta}_i$ (so that $\boldsymbol{\delta}_i$ is uniquely defined).

(c) Show that $\boldsymbol{\mu} \in W_i$ if and only if $\boldsymbol{\delta}_i = \mathbf{0}$.

(d) Show that $\|\mathbf{Y} - \boldsymbol{\mu}\|^2 = \|\mathbf{P}_{V^\perp}\mathbf{Y}\|^2 + \sum_i \|\mathbf{P}_{U_i}\mathbf{Y} - \boldsymbol{\delta}_i\|^2$ for all $\boldsymbol{\mu} = \sum_i \boldsymbol{\delta}_i \in U_i$.

(e) Show that the ordinary least squares estimator (and minimum variance unbiased estimator) of $\boldsymbol{\delta}_j$ is $\hat{\boldsymbol{\delta}}_j = \mathbf{P}_{U_j}\mathbf{Y}$ and that the estimator of $\boldsymbol{\delta}_j$ under the hypothesis that $\boldsymbol{\delta}_i = \mathbf{0}$ is $\hat{\hat{\boldsymbol{\delta}}}_j = \mathbf{P}_{U_j}\mathbf{Y}$ for $i \neq j$. (That is, the estimator of $\boldsymbol{\delta}_j$ under the null hypothesis that $\boldsymbol{\delta}_i = \mathbf{0}$ is the same as the estimator of $\boldsymbol{\delta}_j$ under the alternative hypothesis.)

(f) Show that the F-statistic for testing that $\boldsymbol{\delta}_j = \mathbf{0}$ is given by

$$\frac{\|\hat{\boldsymbol{\delta}}_j\|^2}{k_j\hat{\sigma}^2} \sim F_{k_j,n-p}\left(\frac{\|\boldsymbol{\delta}_j\|^2}{\sigma^2}\right)$$

where k_j is the dimension of U_i.

(g) Show that

$$\|\mathbf{Y}\|^2 = \|\mathbf{P}_{V^\perp}\mathbf{Y}\|^2 + \sum \|\hat{\boldsymbol{\delta}}_i\|^2.$$

(This result shows that for orthogonal designs, the total sum of squares is the sum of the sums of squares for each of the effects. Note that this partititon of the total sum of squares is only true for orthogonal designs.)

3. (a) Let U be a random variable whose distribution does not depend on θ. Show that if $h(U, \theta)$ is an increasing function of θ for all U then $Eh(U, \theta)$ is an increasing function of θ.

(b) Let $Z \sim N_1(\mu, 1)$, $\mu > 0$. Show geometrically that $P(|Z| < a)$ is a decreasing function of μ.

(c) Let $U \sim \chi^2_{s-1}(0)$ independently of Z and let $S_1 = U + Z^2$, $\delta = \mu^2$. Then $S_1 \sim \chi^2_{s-1}(\delta)$. Show that $P_\delta(S_1 < b)$ is a decreasing function of δ. [Hint: $P_\delta(S_1 < b) = EP_\mu(|Z| < \sqrt{b - U} \mid U)$.]

(d) Let $F_1 \sim F_{s,t}(\delta)$. Show that $P_\delta(F_1 > c)$ is an increasing function of δ. [Hint: Let $F_1 = (S_1 t / S_2 s)$ where $S_1 \sim \chi_s^2(\delta_1)$, $S_2 \sim \chi_t^2(0)$ and S_1 and S_2 are independent. Then

$$P_\delta(F_1 > c) = EP_\delta(S_1 > (s/t)S_2 \mid S_2).]$$

(e) Show that the power function of the F-test defined in (7.2) is an increasing function of the noncentraility parameter δ.

(f) Let $S_1 \sim \chi_s^2(\delta_1)$, $S_2 \sim \chi_t^2(\delta_2)$ and let

$$F_2 = \frac{S_1 t}{S_2 s}$$

Show that for each δ_1, $P_{\delta_1, \delta_2}(F_2 > c)$ is a decreasing function of δ_2.

4. Tukey's test of nonadditivity. Consider the two-way model with interaction but without replication in which we observe Y_{ij} independent with $Y_{ij} \sim N_1(\theta + \alpha_i + \beta_j + \gamma_{ij}, \sigma^2)$. Let $\mathbf{Y} = (Y_{ij})$, $\boldsymbol{\mu} = (\mu_{ij})$ be orderings of the Y_{ij} and μ_{ij} into rc-dimensional vectors. Let V be the subspace of $\boldsymbol{\mu} = (\mu_{ij})$, $\mu_{ij} = \theta + \alpha_i + \beta_j$ (i.e., the subspace of no interaction). Let SS_3 and SS_4, η, $\hat{\alpha}_i$, $\hat{\beta}_j$ be defined as in Section 7.9.

(a) Show that

$$\hat{\eta} = \frac{\Sigma_i \Sigma_j \left(Y_{ij} - \overline{Y}_{i.} - \overline{Y}_{.j} + \overline{Y}_{..}\right)\hat{\alpha}_i \hat{\beta}_j}{\Sigma_i \hat{\alpha}_i^2 \Sigma_j \hat{\beta}_j^2}$$

(b) Show that the $Y_{ij} - \overline{Y}_{i.} - \overline{Y}_{.j} + \overline{Y}_{..}$ are independent of the $\hat{\alpha}_i$ and the $\hat{\beta}_j$. [These two parts imply that we may treat the $\hat{\alpha}_i$ and $\hat{\beta}_j$ as constants in the conditional distribution of (SS_3, SS_4) given the $\hat{\alpha}_i$ and the $\hat{\beta}_j$. Why?]

(c) Let $\mathbf{X}^* = (\hat{\alpha}_i \hat{\beta}_j)$. (That is $\mathbf{X}^*$ is the vector with $\hat{\alpha}_i \hat{\beta}_j$ in the place that $\mathbf{Y}$ has Y_{ij}.) Let W^* be the one-dimensional subspace spanned by $\mathbf{X}^*$. Show that $W^* \subset V^\perp$.

(d) Show that $SS_3 = \|\mathbf{P}_{W^*}\mathbf{Y}\|^2$, $SS_4 = \|\mathbf{P}_{V_\perp \mid W^*}\mathbf{Y}\|^2$.

(e) Show that conditionally on $\hat{\alpha}_i$ and $\hat{\beta}_j$, SS_3 and SS_4 are independent and

$$SS_3 \sim \chi_1^2\left(\delta_1(\hat{\alpha}_i, \hat{\beta}_j, \gamma_{ij})\right), \qquad SS_4 \sim \chi_{(r-1)(c-1)-1}^2\left(\delta_2(\hat{\alpha}_i, \hat{\beta}_j, \gamma_{ij})\right)$$

where

$$\delta_1 = \frac{\left(\Sigma_i \Sigma_j \hat{\alpha}_i \hat{\beta}_j \gamma_{ij}\right)^2}{\Sigma_i \hat{\alpha}_i^2 \Sigma_j \hat{\beta}_j^2}, \qquad \delta_2 = \sum_i \sum_j \gamma_{ij}^2 - \delta_1.$$

(f) Show that if $\gamma_{ij} = 0$ then

$$F = ((r-1)(c-1)-1)\frac{SS_3}{SS_4} \sim F_{1,(r-1)(c-1)-1}(0).$$

(Note that the alternative distribution of F given $\hat{\alpha}_i$ and $\hat{\beta}_j$ is a doubly noncentral F and that SS_4 is not independent of $\hat{\alpha}_i$ unless $\gamma_{ij} = 0$.)

5. Spherical distributions. Suppose that we observe an n-dimensional random vector $\mathbf{Y}$ whose distribution is spherical about the unknown vector $\boldsymbol{\mu}$. Let F be defined as in Section 7.1. In this problem, we show that under the null hypothesis that $\boldsymbol{\mu} \in W$, F has the same distribution as when $\mathbf{Y}$ is normally distributed. (It is interesting to note that by a proof similar to that in Exercise C3 of Chapter 3, we could show that the numerator and denominator are independent only when the spherical distribution is normal.) Let $\mathbf{Z}, \mathbf{Z}_i$, $\boldsymbol{\nu}$ and $\boldsymbol{\nu}_i$ be defined as in Section 7.6.1.

(a) Show that the distribution of $\mathbf{Z}$ is spherical about $\boldsymbol{\nu}$.

(b) Under the null hypothesis, show that the distribution of $\mathbf{T} = (\mathbf{Z}_2', \mathbf{Z}_3')'$ is spherical about $\mathbf{0}$.

(c) Let $\mathbf{U} = (1/\|\mathbf{T}\|)\mathbf{T}$. Under the null hypothesis show that $\mathbf{U}$ is uniformly distributed on the unit sphere of vectors $\mathbf{u} \in R^{n-k}$ such that $\|\mathbf{u}\| = 1$. (Since $\|\mathbf{U}\| = 1$, it is sufficient to show that the distribution of $\boldsymbol{\Gamma}\mathbf{U}$ is the same as the distribution of $\mathbf{U}$ for all orthogonal $\boldsymbol{\Gamma}$.)

(d) Show that the null distribution of F is the same for all spherical distributions. (Hint: F depends only on $\mathbf{U}$ whose distribution is the same for all spherical distributions.)

CHAPTER 8

Simultaneous Confidence Intervals

In this chapter, we find the simultaneous confidence intervals associated with the hypotheses studied in the last chapter. The simultaneous confidence intervals discussed in this chapter are all based on the F-distribution. They were first derived by Scheffé (1953) and are often called Scheffé-type simultaneous confidence intervals. Other types of simultaneous confidence intervals based on the studentized range distribution and on Bonferroni's inequality are given in Chapter 12.

Let $\mathbf{Z}$ be a random vector whose distribution depends on the parameter $\boldsymbol{\theta}$. Let $\{g_\lambda(\boldsymbol{\theta}); \lambda \in \Lambda\}$ be a collection of real-valued functions of $\boldsymbol{\theta}$. A set of $(1-\alpha)$ *simultaneous confidence intervals* for the g_λ is a set of intervals $a_\lambda(\mathbf{Z}) \leqslant g_\lambda(\boldsymbol{\theta}) \leqslant b_\lambda(\mathbf{Z})$ such that

$$P_{\boldsymbol{\theta}}(a_\lambda(\mathbf{Z}) \leqslant g_\lambda(\boldsymbol{\theta}) \leqslant b_\lambda(\mathbf{Z}) \quad \text{for all } \lambda \in \Lambda) = 1-\alpha, \tag{8.1}$$

where this probability is computed over the distribution of $\mathbf{Z}$. The $\boldsymbol{\theta}$ is not random. Let $S_\lambda(\boldsymbol{\theta})$ be the set of Z such that $a_\lambda(\mathbf{Z}) \leqslant g_\lambda(\boldsymbol{\theta}) \leqslant b_\lambda(\mathbf{Z})$. Then (8.1) is equivalent to

$$P_{\boldsymbol{\theta}}\Big(\mathbf{Z} \in \bigcap_\lambda S_\lambda(\boldsymbol{\theta})\Big) = 1-\alpha.$$

8.1. THE BASIC RESULT.

Before deriving the primary result, we give a simple lemma about linear algebra.

LEMMA 8.1. Let $\mathbf{v} \in R^n$ and let $T \subset R^n$ be a subspace. Then

$$\sup_{\substack{\mathbf{c}\in T\\ \mathbf{c}\neq \mathbf{0}}} \frac{(\langle \mathbf{c}, \mathbf{v}\rangle)^2}{\|\mathbf{c}\|^2} = \|\mathbf{P}_T\mathbf{v}\|^2.$$

PROOF. If $\mathbf{v} \in T^{\perp}$, then the lemma is obvious. We assume therefore that $\mathbf{v} \notin T^{\perp}$. Let $\mathbf{c} = \mathbf{P}_T\mathbf{v} \in T$. Then $\langle \mathbf{c}, \mathbf{v}\rangle^2/\|\mathbf{c}\|^2 = \|\mathbf{P}_T\mathbf{v}\|^2$. Therefore,

$$\sup_{\substack{\mathbf{c}\in T\\ \mathbf{c}\neq \mathbf{0}}} \frac{(\langle \mathbf{c}, \mathbf{v}\rangle)^2}{\|\mathbf{c}\|^2} \geqslant \|\mathbf{P}_T v\|^2.$$

Now let $\mathbf{c} \in T$. Then

$$\langle \mathbf{c}, \mathbf{v}\rangle^2 = \langle \mathbf{P}_T\mathbf{c}, \mathbf{v}\rangle = \langle \mathbf{c}, \mathbf{P}_T\mathbf{v}\rangle$$

Therefore, by the Cauchy-Scwartz inequality,

$$\frac{\langle \mathbf{c}, \mathbf{v}\rangle^2}{\|\mathbf{c}\|^2} = \frac{\langle \mathbf{c}, \mathbf{P}_T\mathbf{v}\rangle^2}{\|\mathbf{c}\|^2} \leqslant \|\mathbf{P}_T\mathbf{v}\|^2.$$

Hence

$$\sup_{\substack{\mathbf{c}\in T\\ \mathbf{c}\neq 0}} \frac{(\langle \mathbf{c}, \mathbf{v}\rangle)^2}{\|\mathbf{c}\|^2} \leqslant \|\mathbf{P}_T\mathbf{v}\|^2. \quad \square$$

We now return to the linar model in which we observe

$$\mathbf{Y} \sim N_n(\boldsymbol{\mu}, \sigma^2\mathbf{I}), \qquad \boldsymbol{\mu} \in V, \qquad \sigma^2 > 0.$$

We want to test that $\boldsymbol{\mu} \in W \subset V$, where V and W are specified subspaces, $\dim(W) = k$ and $\dim(V) = p$. As in Section 7.1, define

$$F = \frac{\|\mathbf{P}_{V|W}\hat{\boldsymbol{\mu}}\|^2}{(p-k)\hat{\sigma}^2}, \qquad \phi(F) = \begin{cases} 1 & \text{if } F > F^{\alpha}_{p-k,n-p} \\ 0 & \text{if } F \leqslant F^{\alpha}_{p-k,n-p} \end{cases}.$$

We note that $\boldsymbol{\mu} \in W$ if and only if $< \mathbf{d}, \boldsymbol{\mu} > = 0$ for all $\mathbf{d} \in V|W$. In this section we find simultaneous confidence intervals for the set of all $< \mathbf{d}, \boldsymbol{\mu} >$, $\mathbf{d} \in V|W$, and we discuss the relationship of these intervals to the F-test defined above. For all $\mathbf{d} \in V|W$, define

$$a_{\mathbf{d}}(\hat{\boldsymbol{\mu}}, \hat{\sigma}^2) = \langle \mathbf{d}, \hat{\boldsymbol{\mu}}\rangle - \left[(p-k)F^{\alpha}_{p-k,n-p}\right]^{1/2}\|\mathbf{d}\|\hat{\sigma}$$
$$b_{\mathbf{d}}(\hat{\boldsymbol{\mu}}, \hat{\sigma}^2) = \langle \mathbf{d}, \hat{\boldsymbol{\mu}}\rangle + \left[(p-k)F^{\alpha}_{p-k,n-p}\right]^{1/2}\|\mathbf{d}\|\hat{\sigma}.$$

THEOREM 8.2. a. The set of intervals $a_d(\hat{\boldsymbol{\mu}}, \hat{\sigma}^2) \leqslant \langle \mathbf{d}, \boldsymbol{\mu}\rangle \leqslant b_d(\hat{\boldsymbol{\mu}}, \hat{\sigma}^2)$ is a set of $(1-\alpha)$ simultaneous confidence intervals for $\langle \mathbf{d}, \boldsymbol{\mu}\rangle$, $\mathbf{d} \in V|W$.

b. The hypothesis $\boldsymbol{\mu} \in W$ is rejected with the F-test defined above if and only if at least one of the simultaneous confidence intervals does not contain 0.

PROOF. a. We first note that the confidence interval for $\mathbf{d} = \mathbf{0}$ is satisfied. Therefore,

$$P\left(a_{\mathbf{d}}(\hat{\boldsymbol{\mu}},\hat{\sigma}^2) \leqslant \langle \mathbf{d}, \boldsymbol{\mu}\rangle \leqslant b_{\mathbf{d}}(\hat{\boldsymbol{\mu}},\hat{\sigma}^2) \quad \text{for all } \mathbf{d} \in V \mid W\right)$$

$$= P\left[\frac{(\langle \mathbf{d}, \hat{\boldsymbol{\mu}} - \boldsymbol{\mu}\rangle)^2}{(p-k)\hat{\sigma}^2\|\mathbf{d}\|^2} \leqslant F^{\alpha}_{p-k,n-p} \quad \text{for all } \mathbf{d} \in V \mid W, \mathbf{d} \neq \mathbf{0}\right]$$

$$= P\left[\sup_{\substack{\mathbf{d} \in V \mid W \\ \mathbf{d} \neq \mathbf{0}}} \frac{(\langle \mathbf{d}, \hat{\boldsymbol{\mu}} - \boldsymbol{\mu}\rangle)^2}{(p-k)\hat{\sigma}^2\|\mathbf{d}\|^2} \leqslant F^{\alpha}_{p-k,n-p}\right]$$

$$= P\left[\frac{\|P_{V \mid W}(\hat{\boldsymbol{\mu}} - \boldsymbol{\mu})\|^2}{(p-k)\hat{\sigma}^2} \leqslant F^{\alpha}_{p-k,n-p}\right]$$

(using Lemma 8.1). However,

$$\frac{\|\mathbf{P}_{V \mid W}(\hat{\boldsymbol{\mu}} - \boldsymbol{\mu})\|^2}{(p-k)\hat{\sigma}^2} \sim F_{p-k,n-p}(0),$$

and the result is proved.

b. We note that 0 is contained in the interval associated with $\mathbf{d} = \mathbf{0}$. Using the definition of the intervals, we see that 0 is contained in all the intervals if and only if

$$\frac{(\langle \mathbf{d}, \hat{\boldsymbol{\mu}}\rangle)^2}{\|\mathbf{d}\|^2(p-k)\hat{\sigma}^2} \leqslant F^{\alpha}_{p-k,n-p}$$

for all $\mathbf{d} \in V \mid W$, $\mathbf{d} \neq \mathbf{0}$. That is, 0 is in all the intervals if and only if

$$F = \frac{\|\mathbf{P}_{V \mid W}\hat{\boldsymbol{\mu}}\|^2}{(p-k)\hat{\sigma}^2} = \sup_{\substack{\mathbf{d} \in V \mid W \\ \mathbf{d} \neq 0}} \frac{(\langle \mathbf{d}, \hat{\boldsymbol{\mu}}\rangle)^2}{\|\mathbf{d}\|^2(p-k)\hat{\sigma}^2} \leqslant F^{\alpha}_{p-k,n-p}.$$

if and only if the hypothesis is accepted. □

We call the function $\langle \mathbf{d}, \boldsymbol{\mu}\rangle$, $\mathbf{d} \in V \mid W$ a *contrast* associated with testing that $\boldsymbol{\mu} \in W$. We say that the contrast $\langle \mathbf{d}, \boldsymbol{\mu}\rangle$ is *significant* if 0 is not contained in the interval associated with $\langle \mathbf{d}, \boldsymbol{\mu}\rangle$. Theorem 8.2 implies that if we reject the hypothesis that $\boldsymbol{\mu} \in W$ with the F-test, then at least one contrast must be significant. Therefore, after a hypothesis is rejected with the F-test, we search for those contrasts that are significant. We think of these contrasts as the ones that are causing the hypothesis to be rejected.

We often use the following notation for the confidence intervals given in Theorem 8.2 part a:

$$\langle \mathbf{d}, \boldsymbol{\mu}\rangle \in \langle \mathbf{d}, \hat{\boldsymbol{\mu}}\rangle \pm \left[(p-k)F^{\alpha}_{p-k,n-p}\right]^{1/2}\hat{\sigma}\|\mathbf{d}\|.$$

8.2. EXAMPLES.

We now find simultaneous confidence intervals for contrasts for some of the testing problems considered in the last chapter. As in that chapter we write $\mathbf{a} = (a_{ij})$ to mean that $\mathbf{a}$ is a vector whose components are the a_{ij}. Although the particular order of the components is unimportant, it is important that the same order be maintained throughout a given section.

8.2.1. *One-Way Analysis of Variance*

We look first at the one-way model (possibly unbalanced) discussed in Sections 7.3.1 and 7.4.1. In that model we observe Y_{ij} independent with $Y_{ij} \sim N_1(\theta + \alpha_i, \sigma^2)$, $i = 1, \ldots, p$; $j = 1, \ldots, n_i$; where $\sum_i n_i \alpha_i = 0$. (Since the weights do not affect the hypothesis of interest, we choose the weights that are most convenient.) In this model, we want to test that all the $\alpha_i = 0$. As in Section 7.4.1, let $N = \sum_i n_i$, $\mathbf{Y} = (Y_{ij})$, $\boldsymbol{\mu} = (\mu_{ij})$. Let V be the subspace of $\boldsymbol{\mu} = (\mu_{ij})$ such that $\mu_{ij} = \theta + \alpha_i$ for some θ and α_i satisfying the above constraint and let W be the subspace of $\boldsymbol{\mu} = (\mu_{ij})$ such that $\mu_{ij} = \theta$ for some θ. Then testing that the $\alpha_i = 0$ is the same as testing that $\boldsymbol{\mu} \in W$ against $\boldsymbol{\mu} \in V$. A constrast associated with this hypothesis is a function $\langle \mathbf{d}, \boldsymbol{\mu} \rangle$ where $\mathbf{d} \in V|W$. We now characterize $V|W$. Let U be subspace of all vectors $\mathbf{d} = (d_{ij})$ such that $d_{ij} = b_i/n_i$ where $\sum_i b_i = 0$. We now show that $U = V|W$. Clearly $U \subset V$ (let $\theta = 0$, $\alpha_i = b_i/n_i$). We now show that $U \perp W$. Let $\mathbf{d} \in U$, $\boldsymbol{\mu} \in W$. Then

$$\langle \mathbf{d}, \boldsymbol{\mu} \rangle = \sum_i \sum_j \frac{b_i}{n_i} \theta = \theta \sum_i b_i = 0.$$

Therefore $U \subset V|W$. However, the dimension of U is $p - 1$ which is the same as the dimension of $V|W$, and hence $U = V|W$.

Let $\mathbf{d} \in U$. Then

$$\langle \mathbf{d}, \boldsymbol{\mu} \rangle = \sum_i \sum_j \frac{b_i}{n_i}(\theta + \alpha_i) = \sum_i b_i \alpha_i,$$

$$\langle \mathbf{d}, \hat{\boldsymbol{\mu}} \rangle = \sum_i \sum_j \frac{b_i}{n_i} \overline{Y}_{i.} = \sum_i b_i \overline{Y}_{i.},$$

$$\|\mathbf{d}\|^2 = \sum_i \sum_j \frac{b_i^2}{n_i^2} = \sum_i \frac{b_i^2}{n_i}.$$

Therefore, contrasts associated with testing that the $\alpha_i = 0$ have the form $\sum b_i \alpha_i$, where $\sum b_i = 0$. A set of $(1 - \alpha)$ confidence intervals for these contrasts is given by $(\hat{\sigma}^2 = \sum\sum(Y_{ij} - \bar{Y}_{i.})/(N-P))$

$$\sum_i b_i \alpha_i \in \sum_i b_i \bar{Y}_{i.} \pm \hat{\sigma}\left[(p-1)F^{\alpha}_{p-1,N-p}\sum_i \frac{b_i^2}{n_i}\right]^{1/2}.$$

Often, the only contrasts of interest are the *comparisons*, that is, contrasts of the form $\alpha_i - \alpha_j$. These are contrasts with $b_i = 1$, $b_j = -1$ and $b_m = 0$ for all other m. Since the set of comparisons is a subset of the set of contrasts, we see that

$$P\left(\alpha_i - \alpha_j \in \bar{Y}_{i.} - \bar{Y}_{j.} \pm \hat{\sigma}\left[(p-1)F^{\alpha}_{p-1,N-p}\left(\frac{1}{n_i} + \frac{1}{n_j}\right)\right]^{1/2}\right.$$

$$\left.\text{for all } i \neq j\right) \geqslant 1 - \alpha.$$

Although the null hypothesis that all the $\alpha_i = 0$ is true if and only if all the comparisons are 0, it is not necessarily true that the F-test rejects if and only if some comparison is significant. When the F-test rejects, at least one contrast must be significant, but there may be no signficiant comparison. In this situation, we do not have enough data to conclude that any comparison is significantly different from 0, but can conclude that some more complicated contrast is significantly different from 0.

8.2.2. *Balanced Two-Way Analysis of Variance*

We now return to the balanced two-way model (with equal weights) discussed in Section 7.3.3. In that model we observe Y_{ijk} independent such that $Y_{ijk} \sim N_1(\theta + \alpha_i + \beta_j + \gamma_{ij}, \sigma^2)$ $i = 1, \ldots, r$; $j = 1, \ldots, c$; $k = 1, \ldots, m$ where $\sum_i \alpha_i = 0$, $\sum_j \beta_j = 0$, $\sum_i \gamma_{ij} = 0$, $\sum_j \gamma_{ij} = 0$. We now find the contrasts associated with testing that all the $\alpha_i = 0$ and the contrasts associated with testing that all the $\gamma_{ij} = 0$. Let $\mathbf{Y} = (Y_{ijk})$, $\boldsymbol{\mu} = (\mu_{ijk})$ and let V be the subspace of $\boldsymbol{\mu} = (\mu_{ijk})$ such that $\mu_{ijk} = \theta + \alpha_i + \beta_j + \gamma_{ij}$ for some θ, α_i, β_j, and γ_{ij} satisfying the above constraints.

We first consider the contrasts associated with testing that all the $\alpha_i = 0$. Let W_1 be the subspace of $\boldsymbol{\mu} = (\mu_{ijk})$ such that $\mu_{ijk} = \theta + \beta_j + \gamma_{ij}$ where β_j and γ_{ij} satisfy the above constraints. Testing that the $\alpha_i = 0$ is the same as testing that $\boldsymbol{\mu} \in W_1$. Let U_1 be the subspace of $\mathbf{d} = (d_{ijk})$ such that $d_{ijk} = b_i/mc$ where $\sum_i b_i = 0$. We now show that $U_1 = V|W_1$. Clearly $U_1 \subset V$. Let $\mathbf{d} = (d_{ijk}) \in U_1$, $\boldsymbol{\mu} = (\mu_{ijk}) \in W_1$. Then

$$\langle \mathbf{d}, \boldsymbol{\mu} \rangle = \sum_i \sum_j \sum_k d_{ijk}\mu_{ijk} = \sum_i \sum_j \sum_k \frac{b_i}{m}(\theta + \beta_j + \gamma_{ij}) = 0.$$

Therefore, $U_1 \perp W_1$ and hence $U_1 \subset V|W_1$. Finally, U_1 and $V|W_1$ have the same dimension and hence $U_1 = V \mid W_1$. Now let $\mathbf{d} \in U_1$, then

$$\langle \mathbf{d}, \boldsymbol{\mu} \rangle = \sum_i \sum_j \sum_k \frac{b_i}{mc}(\theta + \alpha_i + \beta_j + \gamma_{ij}) = \sum_i b_i \alpha_i$$

$$\langle \mathbf{d}, \hat{\boldsymbol{\mu}} \rangle = \sum_i \sum_j \sum_k \frac{b_i}{mc} \overline{Y}_{ij.} = \sum_i b_i \overline{Y}_{i..}, \qquad \|\mathbf{d}\|^2 = \sum_i \sum_j \sum_k \frac{b_i^2}{n^2c^2} = \frac{\sum_i b_i^2}{mc}.$$

Therefore, the contrasts associated with testing that all the $\alpha_i = 0$ have the form $\sum_i b_i \alpha_i$ where $\sum_i b_i = 0$, and a set of $(1 - \alpha)$ simultaneous confidence intervals for the set of all contrasts is given by $(\hat{\sigma}^2 = \sum\sum\sum(Y_{ijk} - \overline{Y}_{ij}) / rc(n-1))$

$$\sum b_i \alpha_i \in \sum b_i \overline{Y}_{i.} \pm \hat{\sigma}\left[(r-1)F^{\alpha}_{r-1,\,rc(m-1)} \sum_i b_i^2 / mc\right]^{1/2}.$$

We now consider contrasts associated with testing that all the $\gamma_{ij} = 0$. Let W_2 be the subspace of $\boldsymbol{\mu} = (\mu_{ijk})$ such that $\mu_{ijk} = \theta + \alpha_i + \beta_j$ for some θ, α_i, and β_j satisfying the above constraints. Testing that all the $\gamma_{ij} = 0$ is the same as testing that $\boldsymbol{\mu} \in W_2$. Let U_2 be the subspace of all vectors $\mathbf{d} = (d_{ijk})$ such that $d_{ijk} = b_{ij}/m$, where $\sum_i b_{ij} = 0$ and $\sum_j b_{ij} = 0$. By arguments similar to those given above, $U_2 = V \mid W_2$. If $\mathbf{d} \in U_2$, then

$$\langle \mathbf{d}, \boldsymbol{\mu} \rangle = \sum_i \sum_j b_{ij}\gamma_{ij}, \qquad \langle \mathbf{d}, \hat{\boldsymbol{\mu}} \rangle = \sum_i \sum_j b_{ij} \overline{Y}_{ij.}, \qquad \|\mathbf{d}\|^2 = \frac{\sum_i \sum_j b_{ij}^2}{m}.$$

Therefore, contrasts associated with testing that all the $\gamma_{ij} = 0$ have the form $\sum_i \sum_j b_{ij}\gamma_{ij}$ where $\sum_i b_{ij} = 0$ and $\sum_j b_{ij} = 0$. A set of $(1 - \alpha)$ simultaneous confidence intervals is given by

$$\sum_i \sum_j b_{ij}\gamma_{ij} \in \sum_i \sum_j b_{ij} \overline{Y}_{ij.} \pm \hat{\sigma}\left[\frac{(r-1)(c-1)F^{\alpha}_{(r-1)(c-1),\,rc(m-1)} \sum_i \sum_j b_{ij}^2}{m}\right]^{1/2}.$$

8.2.3. *Unbalanced Two-Way Analysis of Variance with Interaction*

We return to the unbalanced two-way model discussed in Section 7.4.3. In that model we observe Y_{ijk} independent with

$$Y_{ijk} \sim N_1(\theta + \alpha_i + \beta_j + \gamma_{ij}, \sigma^2),$$
$$i = 1, \ldots, r;\ j = 1, \ldots, c;\ k = 1, \ldots, n_{ij}.$$

To make the parameters uniquely defined, we choose nonnegative weights w_i and v_j such that $\sum_i w_i > 0$ and $\sum_j v_j > 0$. We assume that

$$\sum_i w_i \alpha_i = 0, \qquad \sum_j v_j \beta_j = 0, \qquad \sum_i w_i \gamma_{ij} = 0, \qquad \sum_j v_j \gamma_{ij} = 0.$$

We now find simultaneous confidence intervals for contrasts associated with testing that the $\alpha_i = 0$ and for contrasts associated with testing that the $\gamma_{ij} = 0$. Let $\mathbf{Y} = (Y_{ijk})$, $\boldsymbol{\mu} = (\mu_{ijk})$ and let V be the subspace of $\boldsymbol{\mu} = (\mu_{ijk})$ such that $\mu_{ijk} = \theta + \alpha_i + \beta_j + \gamma_{ij}$ for some $\theta, \alpha_i, \beta_j$, and γ_{ij} satisfying the above constraints. Let $N = \sum_i \sum_j n_{ij}$.

We first consider contrasts associated with testing that the $\alpha_i = 0$. Let W_1 be the subspace of all $\boldsymbol{\mu} = (\mu_{ijk})$ such that $\mu_{ijk} = \theta + \beta_j + \gamma_{ij}$ for some θ, β_j, and γ_{ij} satisfying the given constraints. Testing that the $\alpha_i = 0$ is the same as testing that $\boldsymbol{\mu} \in W_1$. Let U_1 be the subspace consisting of $\mathbf{d} = (d_{ijk})$ such that

$$d_{ijk} = \frac{b_i v_j}{n_{ij} v_.}, \qquad \sum_i b_i = 0 \qquad v_. = \sum_j v_j$$

We now show that $U_1 = V | W_1$. We first note that by Lemma 7.8, $U_1 \subset V$. Let $\boldsymbol{\mu} = (\mu_{ijk}) \in W_1$. Then $\mu_{ijk} = \theta + \beta_j + \gamma_{ij}$, where $\sum_j v_j \beta_j = 0$ and $\sum_j v_j \gamma_{ij} = 0$. Therefore,

$$\begin{aligned}\langle \mathbf{d}, \boldsymbol{\mu} \rangle &= \sum_i \sum_j \sum_k d_{ijk} \mu_{ijk} \\ &= \sum_i \sum_j n_{ij} d_{ij} (\theta + \beta_j + \gamma_{ij}) = \frac{\sum_i \sum_j b_i v_j (\theta + \beta_j + \gamma_{ij})}{v_.} \\ &= \frac{\theta (\sum_i b_i)(\sum_i v_j)}{v_.} + \frac{\sum_i b_i (\sum_j v_j \beta_j)}{v_.} + \frac{\sum_i b_i (\sum_j v_j \gamma_{ij})}{v_.} = 0.\end{aligned}$$

Hence $U_1 \perp W_1$ and $U_1 \subset V | W_1$. Now, $\dim(U_1) = r - 1 = \dim(V | W_1)$. Hence $U_1 = V | W_1$.

Now, let $\mathbf{d} \in U_1 = V | W_1$. Then

$$\langle \mathbf{d}, \boldsymbol{\mu} \rangle = \sum_i \sum_j n_{ij} d_{ij} \mu_{ij} = \sum_i b_i \alpha_i,$$

$$\langle \mathbf{d}, \hat{\boldsymbol{\mu}} \rangle = \sum_i \sum_j n_{ij} d_{ij} \bar{Y}_{ij.} = \sum_i \sum_j n_i \frac{v_j}{v_.} \bar{Y}_{ij.},$$

$$\|\mathbf{d}\|^2 = \sum_i \sum_j n_{ij} d_{ij}^2 = \sum_i \sum_j \frac{b_i^2 v_j^2}{n_{ij} v_.^2}.$$

Therefore, the contrasts for testing that $\alpha_i = 0$ in the two-way analysis of variance model are functions of the form

$$\sum_i b_i \alpha_i, \qquad \sum_i b_i = 0$$

and the simultaneous confidence intervals for the set of all contrasts are given by ($\hat{\sigma}^2 = \sum\sum\sum(Y_{ijk} - \overline{Y}_{ij})/(N - rc)$)

$$\sum_i b_i \alpha_i \in \frac{\sum_i \sum_j b_i v_j \overline{Y}_{ij}}{v_.} + \hat{\sigma}\left[(r-1)F^{\alpha}_{r-1,N-rc} \sum_i \sum_j \frac{b_i^2 v_j^2}{n_{ij} v_.^2}\right]^{1/2}.$$

On the homework, you are asked to show that contrasts associated with testing that $\gamma_{ij} = 0$ are given by

$$\sum_i \sum_j b_{ij}\gamma_{ij}, \qquad \sum_i b_{ij} = 0, \qquad \sum_j b_{ij} = 0$$

and that the set of simultaneous confidence intervals is given by

$$\sum_i \sum_j b_{ij}\gamma_{ij} \in \sum_i \sum_j c_{ij}\overline{Y}_{ij.} \pm \hat{\sigma}\left[(r-1)(c-1)F^{\alpha}_{(r-1)(c-1),N-rc} \sum_i \sum_j \frac{b_{ij}^2}{n_{ij}}\right]^{1/2}.$$

Note that the contrasts associated with the problem of testing that $\alpha_i = 0$ depend on the weights v_j, whereas those associated with the problem of testing that $\gamma_{ij} = 0$ do not.

8.2.4. *Multiple Regression*

We now return to the multiple regression model. We consider testing that $\mathbf{A}\boldsymbol{\beta} = \mathbf{0}$ where $\mathbf{A}$ is $(p-k) \times p$ of rank $p - k$. Let V be the subspace of all $\boldsymbol{\mu} = \mathbf{X}\boldsymbol{\beta}$ and let W be the subspace of all $\boldsymbol{\mu} = \mathbf{X}\boldsymbol{\beta}$ with $\mathbf{A}\boldsymbol{\beta} = 0$. By Lemma 7.3, we see that $\mathbf{X}(\mathbf{X}'\mathbf{X})^{-1}\mathbf{A}'$ is a basis matrix for $V|W$. Therefore, $\mathbf{d} \in V|W$ if and only if $\mathbf{d} = \mathbf{X}(\mathbf{X}'\mathbf{X})^{-1}\mathbf{A}'\mathbf{s}$ for some $\mathbf{s} \in R^{p-k}$. In this case,

$$\langle \mathbf{d}, \boldsymbol{\mu} \rangle = \mathbf{s}'\mathbf{A}(\mathbf{X}'\mathbf{X})^{-1}\mathbf{X}'\mathbf{X}\boldsymbol{\beta} = \mathbf{s}'\mathbf{A}\boldsymbol{\beta},$$

$$\langle \mathbf{d}, \hat{\boldsymbol{\mu}} \rangle = \mathbf{s}'\mathbf{A}\hat{\boldsymbol{\beta}}, \|\mathbf{d}\|^2 = \mathbf{s}'\mathbf{A}(\mathbf{X}'\mathbf{X})^{-1}\mathbf{A}'\mathbf{s}.$$

Therefore, the set of contrasts associated with the problem of testing that $\mathbf{A}\boldsymbol{\beta} = \mathbf{0}$ are the functions $\mathbf{s}'\mathbf{A}\boldsymbol{\beta}$, $\mathbf{s} \in R^{p-k}$ and the set of simultaneous confidence intervals for the set of all contrasts is given by

$$\mathbf{s}'\mathbf{A}\boldsymbol{\beta} \in \mathbf{s}'\mathbf{A}\hat{\boldsymbol{\beta}} + \hat{\sigma}\left[(p-k)F^{\alpha}_{p-k,n-p}\mathbf{s}'\mathbf{A}(\mathbf{X}'\mathbf{X})^{-1}\mathbf{A}'\mathbf{s}\right]^{1/2}.$$

8.2.5. *Confidence Band for the Response Surface*

In multiple regression models, the function

$$g(\mathbf{s}) = \mathbf{s}'\boldsymbol{\beta}$$

is often called the response surface. For each $\mathbf{s} \in R^p$, $g(\mathbf{s})$ is the expected value of a random variable whose independent variables have value $\mathbf{s}$. In this section we find simultaneous confidence intervals for the set of all $g(\mathbf{s}) = \mathbf{s}'\boldsymbol{\beta}$.

Consider testing that $\boldsymbol{\beta} = \mathbf{0}$. This problem fits into the framework of the last section with $\mathbf{A} = \mathbf{I}$. By the results of the last section, we see that the set of contrasts for this problem is the set of all

$$g(\mathbf{s}) = \mathbf{s}'\boldsymbol{\beta}$$

and the set of simultaneous confidence intervals for this whole set is given by

$$\mathbf{s}'\boldsymbol{\beta} \in \mathbf{s}'\hat{\boldsymbol{\beta}} \pm \hat{\sigma}\left[pF^{\alpha}_{p,n-p}\mathbf{s}'(\mathbf{X}'\mathbf{X})^{-1}\mathbf{s}\right]^{1/2}.$$

These intervals are therefore a set of simultaneous confidence intervals for the response surface.

EXERCISES.

Type A

1. Using the data and model of Exercise A2 of Chapter 7, find the Scheffé simultaneous confidence intervals for all possible comparisons.

2. (a) Using the data and model of problem A3 of Chapter 7, find the Scheffé simultaneous confidence intervals for all comparisons $\alpha_i - \alpha_{i'}$. Find also the intervals for $2\alpha_1 - \alpha_2 - \alpha_3$ and $\alpha_1 + 2\alpha_2 - 3\alpha_3$.

(b) Using the same data find the Scheffé simultaneous confidence intervals for the set of all $\gamma_{ij} - \gamma_{ij'} - \gamma_{i'j} + \gamma_{ij}$.

Type B

1. (a) Consider the balanced two-way model of Section 8.2.2. Show that $U_2 = V \mid W_2$.

(b) Verify the simultaneous confidence intervals for the contrasts in the γ_{ij} have the form given in that section.

2. Consider the problem of testing that $\gamma_{ij} = 0$ in the two-way analysis of variance model discussed in Section 7.4.3. Let V and W_2 be defined as in that section, and let U be the subspace of all vectors $\mathbf{d} = (d_{ijk})$ such that $d_{ijk} = c_{ij}/n_{ij}$, $\sum_i c_{ij} = 0$, $\sum_j c_{ij} = 0$.

(a) Show that U has dimension $(r-1)(c-1)$.

(b) Show that $U = V \mid W_2$.

(c) Derive the contrasts associated with testing that $\gamma_{ij} = 0$ and the set of simultaneous confidence intervals for the contrasts.

3. Consider the nested analysis of variance model discussed in Section 7.4.4.

(a) Find the contrasts associated with testing that $\beta_{ij} = 0$ and find the set of simultaneous confidence intervals for these contrasts.

(b) Find the contrasts associated with testing that $\alpha_i = 0$ and the set of simultaneous confidence intervals for these contrasts.

CHAPTER 9

Tests about the Variance

We now consider methods for doing tests involving σ^2. Those methods depend on the fact that

$$(n-p)\hat{\sigma}^2 \sim \sigma^2 \chi^2_{n-p}(0). \tag{9.1}$$

In the next chapter, we show that the procedures discussed in this chapter for drawing inference about the variance are much more sensitive to the assumption of normality than those derived in previous chapters for drawing inference about the mean.

We consider three testing problems concerning σ^2. Let σ_0^2 be a known constant. We consider testing that $\sigma^2 = \sigma_0^2$ against each of the following alternatives: (a) $\sigma^2 \geqslant \sigma_0^2$; (b) $\sigma^2 \leqslant \sigma_0^2$; and (c) $\sigma^2 \geqslant 0$ (i.e., against both one-sided alternatives and against the two-sided alternative). All three testing problems are invariant under the group G consisting of transformations of the form

$$g(\hat{\boldsymbol{\mu}}, \hat{\sigma}^2) = (\hat{\boldsymbol{\mu}} + \mathbf{a}, \hat{\sigma}^2), \qquad \mathbf{a} \in V.$$

A maximal invariant for this group is $\hat{\sigma}^2$. Define

$$\phi_1(\hat{\sigma}^2) = \begin{cases} 1 & \text{if } (n-p)\hat{\sigma}^2 > \sigma_0^2 \chi_{n-p}^{2\alpha} \\ 0 & \text{if } (n-p)\hat{\sigma}^2 \leqslant \sigma_0^2 \chi_{n-p}^{2\alpha} \end{cases},$$

$$\phi_2(\hat{\sigma}^2) = \begin{cases} 1 & \text{if } (n-p)\hat{\sigma}^2 < \sigma_0^2 \chi_{n-p}^{2(1-\alpha)} \\ 0 & \text{if } (n-p)\hat{\sigma}^2 \geqslant \sigma_0^2 \chi_{n-p}^{2(1-\alpha)} \end{cases}.$$

THEOREM 9.1. ϕ_1 is UMP invariant size α under G for testing that $\sigma^2 = \sigma_0^2$ against $\sigma^2 \geqslant \sigma_0^2$, and ϕ_2 is UMP invariant size α under G for testing $\sigma^2 = \sigma_0^2$ against $\sigma^2 \leqslant \sigma_0^2$.

PROOF. See Exercise B1. □

ϕ_1 and ϕ_2 are also likelihood ratio tests (see exercise B2).

A parameter maximal invariant for all these problems is σ^2. The problem of testing that $\sigma^2 = \sigma_0^2$ against $\sigma^2 \geqslant 0$ is a two-sided problem, even after being reduced by invariance. Therefore, there is no UMP invariant size α test for this problem. The likelihood ratio test is given by

$$\lambda = k\left(\frac{\hat{\sigma}^2}{\sigma_0^2}\right)^{n/2} \exp -\tfrac{1}{2}(n-p)\frac{\hat{\sigma}^2}{\sigma_0^2},$$

$$\phi_3(\lambda) = \begin{cases} 1 & \text{if } \lambda < c \\ 0 & \text{if } \lambda \geqslant c \end{cases}$$

(see Exercise B2). However this test is not even unbiased. For this reason, we often use the following modified likelihood ratio test, which is unbiased.

$$\lambda^* = k^*\left(\frac{\hat{\sigma}^2}{\sigma_0^2}\right)^{(n-p)/2} \exp -\tfrac{1}{2}(n-p)\frac{\hat{\sigma}^2}{\sigma_0^2},$$

$$\phi_3^*(\lambda^*) = \begin{cases} 1 & \text{if } \lambda^* < c^* \\ 0 & \text{if } \lambda^* \geqslant c^* \end{cases}.$$

On the homework, you are asked to show that both ϕ_3 and ϕ_3^* have the following form:

$$\phi_{a,b}(\hat{\sigma}^2) = \begin{cases} 1 & \text{if } (n-p)\dfrac{\hat{\sigma}^2}{\sigma_0^2} > a \text{ or } (n-p)\dfrac{\hat{\sigma}^2}{\sigma_0^2} < b \\ 0 & \text{if } b \leqslant (n-p)\dfrac{\hat{\sigma}^2}{\sigma_0^2} \leqslant a \end{cases}.$$

That is, for both tests, we reject if $\hat{\sigma}^2$ is too large or too small, as we would expect. Putting the test in this form makes it easier to calculate the size of ϕ or ϕ^*. If $n-p$ is reasonably large, then the equal tails test given by

$$a = \chi_{n-p}^{2\,(\alpha/2)}, \qquad b = \chi_{n-p}^{2\,[1-(\alpha/2)]}$$

is a fairly good approximation to the unbiased size α test, and this is the test that is usually used.

The UMP invariant tests ϕ_1 and ϕ_2 are also UMP among unbiased size α tests, as is the modified likelihood ratio test ϕ_3^*. See Lehmann (1959, pp. 163–168) for a derivation of these facts in a similar setting.

EXERCISES.

Type B

1. Prove Theorem 9.1.

2. Show that ϕ_1, ϕ_2, and ϕ_3 are the likelihood ratio tests for the three testing problems considered in Section 9.2.

3. Show graphically that ϕ_3 and $\phi_3{}^*$ have the form of $\phi_{a,b}$ for some a and b. {Hint: Let $h(\hat{\sigma}^2/\sigma_0^2) = \log[\lambda(\hat{\sigma}^2/\sigma_0^2)]$. Then $h'' \leqslant 0$ everywhere and $\lim_{t\to 0} h(t) = \lim_{t\to\infty} h(t) = -\infty$}.

4. Let $\mathbf{Y}_1$ and $\mathbf{Y}_2$ be independent with $\mathbf{Y}_i \sim N_{n_i}(\boldsymbol{\mu}_i, \sigma_i^2\mathbf{I})$, $\boldsymbol{\mu}_i \in V_i$, $\sigma_i^2 > 0$, where V_i is a known p_i-dimensional subspace.

(a) Find a sufficient statistic for this situation.

(b) Find the UMP invariant size α test that $\sigma_1^2 = \sigma_2^2$ against $\sigma_1^2 \geqslant \sigma_2^2$.

(c) Find the likelihood ratio test that $\sigma_1^2 = \sigma_2^2$ against $\sigma_1^2 > 0$, $\sigma_2^2 > 0$.

(d) Show that the likelihood ratio rejects if $F = \hat{\sigma}_1^2/\hat{\sigma}_2^2$ is too large or too small.

CHAPTER 10

Asymptotic Validity of Procedures under Nonnormal Distributions

In this chapter we derive results that indicate, under fairly general conditions, that the size of the test derived in Chapter 7 is asymptotically (as $n \to \infty$) independent of the normal assumption used in deriving it. Similarly, the confidence coefficients for the simultaneous confidence intervals derived in Chapter 8 are asymptotically insensitive to the normal assumption. Finally, we show that the sizes and confidence coefficients of the procedures derived in Chapters 6 and 9 for drawing inferences about the variance σ^2 are *not* independent of the normal assumption used in deriving them. Therefore, if n is large, and p is moderate, the procedures for drawing inference about the mean vector μ are much less sensitive to the normal assumption than are the procedures for drawing inference about the variance σ^2.

The theorems studied in this chapter are based on results that would be derived in a course on probability theory. These results from probability theory are stated in Section 10.1. In Section 10.2, the model studied in this chapter is defined, and the results are summarized. In addition to some assumptions on the errors that are very similar to those used to derive the Gauss-Markov theorem, there is an assumption due to Huber about the mean vector, which is discussed in more detail in Section 10.3. In Section 10.4, we state Theorem 10.3, and use it to derive the asymptotic results. Even a student who has had no probability theory should be able to understand the derivations in this section (at least if he is willing to accept the results of Section 10.2). The proof of Theorem 10.3 is somewhat more difficult and is deferred to Section 10.7. It is a rather interesting proof, because it uses the full generality of the Lindeberg-Feller theorem.

The basic result of this chapter is due to Huber (1973). The application of that result to the present situation is from Arnold (1980a).

10.1. DEFINITIONS AND THEOREMS FROM PROBABILITY THEORY.

We now give some definitions and summarize some results that would be discussed in a course in probability theory. Proofs of these results can be found, for example, in Chung (1974).

Let $\{\mathbf{X}_n\}$ be a sequence of p-dimensional random vectors with $\mathbf{X}_n$ having distribution function F_n. Let F be a p-dimensional distribution function. We say that $\mathbf{X}_n$ converges in distribution to F, and write $\mathbf{X}_n \xrightarrow{d} F$, if

$$\lim_{n\to\infty} F_n(\mathbf{x}) = F(\mathbf{x})$$

at all points $\mathbf{x}$ where F is continuous. Let $\mathbf{a}$ be a p-dimensional (constant) vector. We say that $\mathbf{X}_n$ converges in probability to $\mathbf{a}$, and write $\mathbf{X}_n \xrightarrow{p} \mathbf{a}$ if for all $\epsilon > 0$

$$\lim_{n\to\infty} P(\|\mathbf{X}_n - \mathbf{a}\| > \epsilon) = 0$$

The first theorem gives two of the most important theorems in probability theory. Let $\{U_n\}$ be a sequence of independently, identically distributed random variables. Let

$$\overline{X}_n = \frac{1}{n}\sum_{i=1}^{n} U_i$$

THEOREM 10.1. a (Weak law of large numbers). Let $EU_i = \mu$ (μ finite). Then $\overline{X}_n \xrightarrow{p} \mu$.

b. (Central limit theorem). Suppose in addition that $\text{var}(U_i) = \sigma^2 < \infty$. Then $\sqrt{n}(\overline{X}_n - \mu) \xrightarrow{d} N(0, \sigma^2)$

We now state some other elementary results from probability theory.

THEOREM 10.2. a. Let $U_1, U_2, \ldots; V_1, V_2, \ldots,$ be two sequences of random variables such that $U_n \xrightarrow{p} a$ and $V_n \xrightarrow{p} b$. Then $U_n + V_n \xrightarrow{p} a + b$ and $U_n V_n \xrightarrow{p} ab$.

b. Let $X_1, X_2, \ldots,; U_1, U_2, \ldots;$ and $V_1, V_2, \ldots;$ be sequences of random variables such that $X_n \xrightarrow{d} F$, $U_n \xrightarrow{p} 0$ and $V_n \xrightarrow{p} 1$. Then $X_n + U_n \xrightarrow{d} F$, $X_n U_n \xrightarrow{p} 0$ and $X_n / V_n \xrightarrow{d} F$.

c. Let $\mathbf{X}_1, \mathbf{X}_2, \ldots,$ be a sequence of random vectors such that $\mathbf{X}_n \xrightarrow{p} \mathbf{a}$. If g is continuous at $\mathbf{a}$ then $g(\mathbf{X}_n) \xrightarrow{p} g(\mathbf{a})$.

d. Let $\mathbf{X}_1, \mathbf{X}_2, \ldots,$ be a sequence of random vectors such that $\mathbf{X}_n \xrightarrow{d} F$. If g is a continuous function then $g(\mathbf{X}_n) \xrightarrow{d} F_g$ where F_g is the distribution function of $g(\mathbf{X})$ when $\mathbf{X}$ has distribution F.

e. Cramer-Wold. $\mathbf{X}_n \xrightarrow{d} N_p(\boldsymbol{\theta}, \boldsymbol{\Sigma})$ if and only if $\mathbf{c}'\mathbf{X}_n \xrightarrow{d} N_1(\mathbf{c}'\boldsymbol{\theta}, \mathbf{c}'\boldsymbol{\Sigma}\mathbf{c})$ for all $\mathbf{c} \in R^p$ such that $\|\mathbf{c}\| = 1$.

If $d_1, d_2, \ldots,$ is a sequence of constants such that $\lim d_n = d$, then the d_n are degenerate random variables and $d_n \xrightarrow{p} c$ from the definition of convergence in probability. Therefore in Theorem 10.2, parts a and b, U_n and V_n can be constants that converge to the appropriate limits ($a, b, 0$, or 1).

10.2. DEFINING THE MODEL.

In this chapter we study a sequence of models (over n) in which we observe the n-dimensional vector $\mathbf{Y}_n$ such that

$$\mathbf{Y}_n = \boldsymbol{\mu}_n + \mathbf{e}_n, \qquad \boldsymbol{\mu}_n \in V_n, \qquad \mathbf{e}'_n = (f_1, \ldots, f_n) \tag{10.1}$$

where $\boldsymbol{\mu}_n$ is a sequence of unknown constant vectors, and V_n is a sequence of known p-dimensional subspaces, $V_n \subset R^n$. In addition, we assume that the f_i are independently, indentically distributed with

$$Ef_i = 0, \qquad \operatorname{var}(f_i) = \sigma^2. \tag{10.2}$$

(Note that if we add the assumption that the f_i are normally distributed, this becomes a sequence of linear models of the type discussed in previous chapters.) In this chapter, we study this model as $n \to \infty$, but p remains fixed. (For a more general approach in which p can also go to ∞ as long as $p/n \to 0$ see Arnold, 1980a.) The results in this chapter are therefore relevant when n, the number of observations, is much larger than p, the number of parameters in the mean.

We also need one more assumption to establish the results. This assumption was suggested by Huber (1973).

Huber's Condition: The largest diagonal element of $\mathbf{P}_{V_n}$ goes to 0 as $n \to \infty$.

Now, let $\mathbf{A}$ be a matrix whose i, j element is a_{ij}. Define

$$m(\mathbf{A}) = \max_{i,j} |a_{ij}|.$$

If $\mathbf{A} \geqslant 0$, then $\mathbf{A}$ is a covariance matrix. Hence, by the Cauchy-Schwartz inequality, we see that

$$|a_{ij}| \leqslant (a_{ii}a_{jj})^{1/2} \leqslant \max a_{ii}.$$

Therefore, if $\mathbf{A} \geqslant 0$, then $m(\mathbf{A})$ is just the largest diagonal element of $\mathbf{A}$. Hence, we could restate Huber's condition in the following way:

$$m(\mathbf{P}_{V_n}) \to 0 \qquad \text{as } n \to \infty.$$

Under the assumptions given above, we will show in Section 10.4 that the procedures defined in Chapters 7 and 8 for drawing inferences about $\boldsymbol{\mu}$ are insensitive, asymptotically, to the normal assumption used in deriving them. However, we show that the procedures discussed in Chapter 9 for drawing inference about σ^2 are only asymptotically valid if

$$Ef_i^4 = 3\sigma^4.$$

Therefore, if the actual distribution of the errors, f_i, does not satisfy this assumption, then the procedures derived in Chapter 9 are not even asymptotically valid. It is for this reason that we said in Chapter 9 that those procedures are very sensitive to the normal assumption. The quantity

$$\delta = \frac{Ef_i^4}{\left[\mathrm{var}(f_i)\right]^2} - 3$$

is called the *kurtosis* of the distribution of the f_i, and is a measure of how severely nonnormal that distribution is, if we are concerned with inference about σ^2. We emphasize again, that the kurtosis only affects inference about σ^2. It has no asymptotic affect on inference about $\boldsymbol{\mu}$.

We now present a discussion of Huber's condition, together with some examples.

10.3. DISCUSSION OF HUBER'S CONDITION WITH EXAMPLES.

The ith element of $\hat{\boldsymbol{\mu}}_n = \mathbf{P}_{V_n}\mathbf{Y}_n$ is the estimator of the mean of the ith element of $\mathbf{Y}_n$. Therefore, the ith diagonal element of $\mathbf{P}_{V_n}$ represents the coefficient of the ith element of $\mathbf{Y}_n$ in the estimator of its own mean. Hence, Huber's condition can be restated as saying that the effect of any particular element of $\mathbf{Y}_n$ on its estimated mean must go to 0.

We now give two simple examples in which Huber's condition is not satisfied. In both examples $p = 1$.

EXAMPLE 1. Let $\mathbf{X}_n' = (\sqrt{2}, \sqrt{4}, \sqrt{8}, \ldots, \sqrt{2^n})$. The largest diagonal element of $\mathbf{P}_{V_n} = (1/\mathbf{X}_n'\mathbf{X}_n)(\mathbf{X}_n\mathbf{X}_n')$ is the nth diagonal element which is $2^n/(2^n - 2)$, and the limit of this expression is $\frac{1}{2}$.

EXAMPLE 2. Let $\mathbf{X}_n' = (\sqrt{\frac{1}{2}}, \sqrt{\frac{1}{4}}, \sqrt{\frac{1}{8}}, \ldots, \sqrt{2^{-n}})$. Then the largest diagonal element of $\mathbf{P}_{V_n}$ is the first one, which is $(\frac{1}{2})/(1 - 2^{-n})$ and the limit is again $\frac{1}{2}$.

These two examples are rather different. In the first one, the x's are growing so fast that the effect of the last observation on its estimated mean dominates the effect of all the previous observations, and so Huber's condition is not satisfied. In the second example, the x's are going to 0 fast enough that the first observation has a nontrivial effect, even in the limit. Therefore, Huber's condition is not satisfied for this example either.

We now return to some examples considered in Chapter 7.

EXAMPLE 3. (Multiple regression). Suppose that we observe $\mathbf{Y}_n$ such that $E\mathbf{Y}_n = \mathbf{X}_n\boldsymbol{\beta}$, $\operatorname{cov}\mathbf{Y}_n = \sigma^2\mathbf{I}$. Then Huber's condition is satisfied if

$$m(\mathbf{P}_{V_n}) = m\big(\mathbf{X}_n(\mathbf{X}_n'\mathbf{X}_n)^{-1}\mathbf{X}_n'\big) \to 0.$$

Now, let $\hat{\boldsymbol{\beta}}_n = (\mathbf{X}_n'\mathbf{X}_n)^{-1}\mathbf{X}_n'\mathbf{Y}_n$. Then $E\hat{\boldsymbol{\beta}}_n = \boldsymbol{\beta}$ and $\operatorname{cov}(\hat{\boldsymbol{\beta}}_n) = \sigma^2(\mathbf{X}_n'\mathbf{X}_n)^{-1}$. Therefore, $\hat{\boldsymbol{\beta}}_n \xrightarrow{P} \boldsymbol{\beta}$ if $m[(\mathbf{X}_n'\mathbf{X}_n)^{-1}] \to 0$. In Example 1, $m(\mathbf{X}_n'\mathbf{X}_n)^{-1}) \to 0$ but $m(\mathbf{X}_n(\mathbf{X}_n'\mathbf{X}_n)^{-1}\mathbf{X}_n')$ does not go to 0. Therefore the condition that $m[(\mathbf{X}_n'\mathbf{X}_n)^{-1}] \to 0$ is not sufficient to guarantee Huber's condition.

We now look at several examples of analysis of variance models. In the following examples, we work out the coefficient of each observation in its estimated mean and find conditions that guarantee that all these coefficients go to 0. To keep the notation within reasonable limits, we suppress the subscript n used in the previous examples.

EXAMPLE 4. (One-way analysis of variance). In this model, we observe Y_{ij}, independent, $i = 1, \ldots, p$; $j = 1, \ldots, n_i$; such that $EY_{ij} = \theta + \alpha_i$, $\operatorname{var} Y_{ij} = \sigma^2$. By Section 7.3.7, we see that $\hat{\mu}_{ij} = \overline{Y}_{i.}$, so that he coefficient of Y_{ij} in $\hat{\mu}_{ij}$ is $1/n_i$. Therefore, Huber's condition is satisfied if and only if $n_i \to \infty$. It is not enough that the total number of observations go to ∞. We need to assume that the number of observations in each class goes to ∞. Therefore, the results of this chapter would apply to the one-way model in a situation where every class had a large number of observations.

EXAMPLE 5. (Two-way analysis of variance with interaction). In this model, we observe Y_{ijk} independent, $i = 1, \ldots, r$; $j = 1, \ldots, c$; $k = 1, \ldots, n_{ij}$, such that $EY_{ijk} = \theta + \alpha_i + \beta_j + \gamma_{ij}$, $\operatorname{var}(Y_{ijk}) = \sigma^2$. By the results in Section 7.4.3, we see that $\hat{\mu}_{ijk} = \overline{Y}_{ij.}$ for this model. Therefore, the coefficient of Y_{ijk} in μ_{ijk} is $1/n_{ij}$. Hence Huber's condition is satisfied for this model if and only if all the $n_{ij} \to \infty$.

EXAMPLE 6. (Two-way analysis of variance with no interaction). We have not derived any explicit formula for $\hat{\mu}_{ijk}$ for this model (except in the case of proportional sampling) and therefore cannot give an explicit interpretation for Huber's condition for this model. However, the alternative subspace for this model is contained in the alternative subspace for the model with interaction, and therefore (by exercise B3) the condition that $n_{ij} \to \infty$ for all i and j is a sufficient condition to guarantee that Huber's condition is satisfied for this model.

EXAMPLE 7. (Analysis of covariance). In this model, we observe Y_{ij} independent, $i = 1, \ldots, r$; $j = 1, \ldots, n_i$, such that $EY_{ij} = \theta + \alpha_i + \beta x_{ij}$, $\text{var}(Y_{ijk}) = \sigma^2$. On the homework, you are asked to show that the coefficient of Y_{ij} in $\hat{\mu}_{ij}$ is given by

$$\frac{1}{n_i} + \frac{(x_{ij} - \bar{x}_{i.})^2}{\sum_i \sum_j (x_{ij} - \bar{x}_{i.})^2}.$$

Therefore, Huber's condition is satisfied for this model if and only if $n_i \to \infty$ and $(x_{ij} - \bar{x}_{i.})^2 / \sum_i \sum_j (x_{ij} - \bar{x}_{i.})^2 \to 0$ as $N = \sum_i n_i \to \infty$.

The nested analysis of variance model discussed in Section 7.4.4 has the same alternative space as the two-way model with interactions. Therefore, Huber's condition for this model is satisfied if and only if $n_{ij} \to \infty$. Finally, consider the Latin square model discussed in Section 7.3.5. In this model $p = 3m - 2$ and $n = m^2$. Therefore, we cannot let $n \to \infty$ and keep p fixed for this model. Hence, Huber's condition is not relevant for this model.

We can interpret the results of this chapter as saying that if n is "large" and the diagonal elements of P_{V_n} are all "small" then the size of the F-test for the general linear model is not too sensitive to the normal assumption used in deriving it nor is the confidence coefficient for Scheffe simultaneous confidence intervals. How large n has to be and how small the diagonal elements of P_{V_n} must be for these asymptotic results to be relevant depend on the actual distribution of the errors f_i.

The average diagonal element of $\mathbf{P}_{V_n}$ is $\text{tr}\mathbf{P}_{V_n}/n = p/n$ (see Theorem 2.5i). Therefore the average diagonal element of $\mathbf{P}_{V_n}$ is sure to go to zero as n goes to ∞. For a particular n and p, we can minimize the largest diagonal element by making all the diagonal elements the same (since the average is completely determined by n and p). Therefore, the results of this chapter suggest that, when possible, we should try to design experiments so that the diagonal elements of $\mathbf{P}_{V_n}$ are nearly the same. In many analysis of variance models, the diagonal elements of $\mathbf{P}_{V_n}$ are the sample sizes in the various cells (see Examples 4 and 5 above). For these models, the results of this chapter suggest we should use designs that have nearly the same number of observations in each cell.

Huber's condition is also a necessary condition for the results in this chapter. If Huber's condition is not satisfied, then there is a subspace W_n such that for any symmetric nonnormal distribution of the errors f_i, the asymptotic distribution of the F-statistic for testing that $\boldsymbol{\mu} \in W_n$ would not be the same as the asymptotic distribution for a normal distribution of the f_i (see Arnold, 1980a). Therefore, if Huber's condition is not satisfied, then the asymptotic distribution of at least one such F-statistic would be dependent on the normal assumption. The necessity of Huber's condition indicates that it is not enough to collect a large amount of data for these asymptotic results to be relevant. We must also be sure that the diagonal elements of $\mathbf{P}_{V_n}$ are all small.

10.4. DERIVATIONS.

We now derive the results mentioned in previous sections. Throughout this section and the next, we assume the model defined in Section 10.2 by (10.1) and (10.2). We keep p fixed, and let $n(> p)$ go to ∞. Define

$$\hat{\mu}_n = \mathbf{P}_{V_n}\mathbf{Y}_n, \qquad \hat{\sigma}_n^2 = \frac{\|\mathbf{P}_{V_{n\perp}}\mathbf{Y}_n\|^2}{n-p}. \tag{10.3}$$

We first state the basic result of this section. Its proof involves some techniques from measure theory, and is postponed to Section 10.7.

THEOREM 10.3. Let $\{\mathbf{a}_n\}$ be a sequence of vectors such that $\mathbf{a}_n \in R^n$ and $\|\mathbf{a}_n\| = 1$. If $m(\mathbf{a}_n) \to 0$, then

$$\mathbf{a}_n'\mathbf{e}_n \xrightarrow{d} N_1(0, \sigma^2). \tag{10.4}$$

PROOF. See Section 10.7. □

Before proceeding to derive more interesting results for the model considered in this chapter, we give two elementary facts about $m(\mathbf{A})$.

$$m(\mathbf{AB}) \leqslant km(\mathbf{A})m(\mathbf{B}), \qquad (m(\mathbf{A}))^2 \leqslant m(\mathbf{AA}'), \tag{10.5}$$

where k is the common dimension of A and B.

THEOREM 10.4. Let $\{W_n\}$ be a sequence of k dimensional subspaces such that $W_n \subset R^n$ and $m(\mathbf{P}_{W_n}) \to 0$. Then

$$\frac{\|\mathbf{P}_{W_n}\mathbf{e}_n\|^2}{\sigma^2} \xrightarrow{d} \chi_k^2(0).$$

PROOF. Let $\mathbf{X}_n$ be an orthonormal basis matrix for W_n. We first show that $\mathbf{X}_n'\mathbf{e}_n \xrightarrow{d} N_k(\mathbf{0}, \sigma^2\mathbf{I})$. By Theorem 10.2e this is equivalent to showing that $\mathbf{c}'\mathbf{X}_n'\mathbf{e}_n \to N_1(0, \sigma^2)$ for all k-dimensional vectors $\mathbf{c}$ such that $\|\mathbf{c}\| = 1$. Let $\mathbf{a}_n = \mathbf{X}_n\mathbf{c}$. Then

$$\mathbf{c}'\mathbf{X}_n'\mathbf{e}_n = \mathbf{a}_n'\mathbf{e}_n, \qquad \|\mathbf{a}_n\|^2 = \mathbf{c}'\mathbf{X}_n'\mathbf{X}_n\mathbf{c} = \mathbf{c}'\mathbf{c} = 1.$$

Since $\mathbf{P}_{V_n} = \mathbf{X}_n\mathbf{X}_n'$, and $m(\mathbf{c}) \leqslant 1$,

$$m(\mathbf{a}_n) \leqslant km(\mathbf{X}_n) \leqslant k(m(\mathbf{P}_{V_n}))^{1/2}.$$

Therefore, $m(\mathbf{a}_n) \to 0$ and by Theorem 10.3,

$$\mathbf{c}'\mathbf{X}_n'\mathbf{e}_n = \mathbf{a}_n'\mathbf{e}_n \xrightarrow{d} N_1(0, \sigma^2).$$

Hence by Theorem 10.2e, $\mathbf{X}_n'\mathbf{e}_n \xrightarrow{d} N_k(0, \sigma^2\mathbf{I})$ and by Theorem 10.2d,

$$\frac{\|\mathbf{P}_{W_n}\mathbf{e}_n\|^2}{\sigma^2} = \frac{\|\mathbf{X}_n'\mathbf{e}_n\|^2}{\sigma^2} \xrightarrow{d} \chi_k^2(0). \quad \square$$

We need one final preliminary result. Note that we do not need Huber's assumption for this result.

THEOREM 10.5. If $n \to \infty$ then $\hat{\sigma}_n^2 \xrightarrow{p} \sigma^2$.

PROOF.

$$\frac{n-p}{n}\hat{\sigma}_n^2 = \frac{\|\mathbf{P}_{V_n^\perp}\mathbf{Y}_n\|^2}{n} = \frac{\|\mathbf{P}_{V_n^\perp}\mathbf{e}_n\|^2}{n} = \frac{\|\mathbf{e}_n\|^2}{n} - \frac{\|\mathbf{P}_{V_n}\mathbf{e}_n\|^2}{n}.$$

Now,

$$E\mathbf{P}_{V_n}\mathbf{e}_n = \mathbf{P}_{V_n}E\mathbf{e}_n = \mathbf{0}, \qquad \operatorname{cov}\mathbf{P}_{V_n}\mathbf{e}_n = \sigma^2\mathbf{P}_{V_n}\mathbf{I}\mathbf{P}_{V_n} = \sigma^2\mathbf{P}_{V_n}.$$

Therefore,

$$\frac{E\|\mathbf{P}_{V_n}\mathbf{e}_n\|^2}{n} = \sigma^2\frac{\operatorname{tr}\mathbf{P}_{V_n}}{n} = \sigma^2\frac{p}{n}$$

(see Lemmas 3.1b and 2.5i). Now $\|\mathbf{P}_{V_n}\mathbf{e}_n\|^2/n > 0$ and $E\|\mathbf{P}_{V_n}\mathbf{e}_n\|^2/n \to 0$. This implies that

$$\frac{\|\mathbf{P}_{V_n}\mathbf{e}_n\|^2}{n} \xrightarrow{p} 0$$

(see Exercise B5). Let $U_i = f_i^2$. The f_i are independently, identically distributed, and therefore the U_i are also. $EU_i = Ef_i^2 = \sigma^2$ and

$$\frac{\|\mathbf{e}_n\|^2}{n} = \frac{1}{n}\sum_{i=1}^{n} U_i$$

Therefore, by the weak law of large numbers (Theorem 10.1a), we have

$$\frac{\|\mathbf{e}_n\|^2}{n} \xrightarrow{p} \sigma^2.$$

Finally, using Theorem 10.2a and b

$$\hat{\sigma}^2 = \frac{n}{n-p}\left(\frac{\|\mathbf{e}_n\|^2}{n} - \frac{\|\mathbf{P}_{V_n}\mathbf{e}_n\|^2}{n}\right) \xrightarrow{p} \sigma^2. \quad \square$$

We now derive the main results of this chapter.

THEOREM 10.6. Let $W_n \subset V_n$, and let $\dim W_n = k$. If Huber's condition is satisfied then

$$F_n = \frac{\|\mathbf{P}_{V_n \mid W_n}(\mathbf{Y}_n - \boldsymbol{\mu}_n)\|^2}{(p-k)\hat{\sigma}_n^2} \xrightarrow{d} \frac{\chi^2_{p-k}(0)}{p-k}.$$

PROOF. $W_n \subset V_n$. Therefore

$$m(\mathbf{P}_{V_n \mid W_n}) = m(\mathbf{P}_{V_n} - \mathbf{P}_{W_n}) \leqslant m(\mathbf{P}_{V_n})$$

(since the diagonal elements of $\mathbf{P}_{W_n}$ are nonnegative) and $m(\mathbf{P}_{V_n \mid W_n}) \to 0$. Therefore,

$$\frac{\|\mathbf{P}_{V_n \mid W_n}(\mathbf{Y}_n - \boldsymbol{\mu}_n)\|^2}{\sigma^2} = \frac{\|\mathbf{P}_{V_n \mid W_n}\mathbf{e}_n\|^2}{\sigma^2} \xrightarrow{d} \chi^2_{p-k}(0)$$

by Theorem 10.4. Also, by Theorems 10.5 and 10.2,

$$\frac{\hat{\sigma}_n^2}{\sigma^2} \xrightarrow{P} 1.$$

Therefore, by Theorem 10.2b,

$$F_n = \frac{\|\mathbf{P}_{V_n \mid W_n}(\mathbf{Y}_n - \boldsymbol{\mu}_n)\|^2}{(p-k)\sigma^2} \Big/ \frac{\hat{\sigma}_n^2}{\sigma^2} \xrightarrow{d} \frac{\chi^2_{p-k}(0)}{p-k}.$$

COROLLARY. If $\boldsymbol{\mu}_n \in W_n$, then

$$F_n^* = \frac{\|\mathbf{P}_{V_n \mid W_n} Y_n\|^2}{(p-k)\hat{\sigma}_n^2} \xrightarrow{d} \frac{\chi^2_{p-k}(0)}{p-k}.$$

PROOF. If $\boldsymbol{\mu}_n \in W_n$, then $\mathbf{P}_{V_n \mid W_n}\boldsymbol{\mu}_n = 0$, and hence $F_n^* = F_n$. □

Now consider the testing problem discussed in Chapter 7 in which we test that $\boldsymbol{\mu}_n \in W_n$ against $\boldsymbol{\mu}_n \in V_n$. The statistic that we use is F_n^* defined in the corollary above. By that corollary, the asymptotic distribution of F_n^* under the null hypothesis does not depend on the distribution of the f_i (at least under the conditions given in Section 10.2). Therefore, the size of the F-test is asymptotically unaffected by the underlying distribution and hence is the same as if the f_i were normally distributed. This result establishes that the F-test derived in Chapter 7 under the assumption of normality is also asymptotically size α even for nonnormal distributions, as long as the assumptions of Section 10.2 (including Huber's condition) are met. In Exercise C1, you are asked to show that if $\lim_{n\to\infty}\|\mathbf{P}_{V_n \mid W_n}\boldsymbol{\mu}_n\| = \gamma$,

then

$$F_n^* \xrightarrow{d} \frac{\chi^2_{p-k}\left(\dfrac{\gamma^2}{\sigma^2}\right)}{p-k}.$$

This result implies that the power of the F-test is also unaffected asymptotically by the distribution of the f_i.

Now, consider the simultaneous confidence intervals discussed in Chapter 8. These confidence intervals depend only on the distribution of F_n derived in Theorem 10.6. Therefore, the simultaneous confidence intervals which were derived in Chapter 8 under the normal assumption are asymptotically simultaneous $(1-\alpha)$ confidence intervals even without the normal assumption.

By arguments similar to those in this section, we can show that the t-test derived in Chapter 7 and the confidence intervals for a function of $\boldsymbol{\mu}$ derived in Chapter 6 are also asymptotically valid (see Exercise B7).

We now look at the procedures defined in Chapters 6 and 9 for finding confidence intervals for σ^2 and testing hypotheses about σ^2. These procedures depend only on the distribution of $\hat{\sigma}_n^2$. We therefore find its asymptotic distribution. We need to make the additional assumption that the errors have a finite fourth moment. Therefore, suppose that

$$Ef_i^4 = \delta\sigma^4, \qquad \delta < \infty. \tag{10.6}$$

(Note that $Ef_i^4 \geqslant (Ef_i^2)^2$ and therefore $\delta \geqslant 1$.)

THEOREM 10.7. $\sqrt{n}\,(\hat{\sigma}_n^2 - \sigma^2) \xrightarrow{d} N_1(0, \sigma^4(\delta - 1))$.

PROOF. See Exercise B7. □

Note that the limiting distribution $\hat{\sigma}_n^2$ *does* depend on the underlying distribution of the errors. The procedures for drawing inference about σ^2 defined in Chapters 6 and 10 depend only on the distribution of $\hat{\sigma}_n^2$. Therefore, unless the distribution of the errors is such that $\delta = 3$, the value for the normal distribution, the procedures defined in Chapters 6 and 9 will be asymptotically invalid, in the sense that the tests discussed in that chapter are no longer even asymptotically size α, nor do the confidence coefficients for the confidence intervals agree with those calculated for the normal model. Therefore, the procedures for making inferences about σ^2 are much more sensitive to the normal assumption made in deriving them than those for making inferences about $\boldsymbol{\mu}$. One way of interpreting this additional sensitivity to the normal assumption is to note that the procedures based on $\hat{\sigma}_n^2$ have not been "studentized" with an estimator of the variance of $\hat{\sigma}_n^2$, and therefore depend heavily on what that variance would be for the normal distribution.

10.5. FURTHER DISCUSSION.

Theorem 10.6 implies, under fairly general conditions, that the F-tests and Scheffé type simultaneous confidence intervals derived in Chapters 7 and 8 under the assumption of normal errors are asymptotically valid when the errors are not normally distributed. This result indicates that the procedures derived in Chapters 7 and 8 under the normal assumption are not too senstive to that assumption. However, this robustness against the normal assumption must be interpreted in the light of the following three comments.

The first comment is that we have made no discussion of the rate of convergence. For some distributions with extremely heavy tails, the convergence can be very slow. In such a case, it is necessary to have an unreasonably large number of observations in order for these asymptotic results to be relevant.

We have shown that the F-test is UMP invariant size α for the linear model with normal errors, and that its size and power are unaffected asymptotically by nonnormal errors (see Exercise C1). This might lead us to conclude that the F-test is the "best" test to use, asymptotically, for these nonnormal models. However, there exist nonparametric size α tests whose power is not quite as high as the F-test if the errors are normally distributed (where the F-test is "best"), but is substantially higher for other distributions. Using this sort of procedure in place of the F-test, we face a small loss of power if the errors really are normally distributed, but possibly a large gain in power if they are not. Therefore, if the errors are known to have a distribution that is far from normal, we should not use the F-test. However, Theorem 10.6 indicates that if the errors are not dramatically nonnormal, then the F-test should be a good test.

The third comment that should be made here is that we have only dealt with departures from the assumption of normality. We are still assuming independence and equal variances for the errors, as well as the other assumptions discussed in Chapter 4. The F-test seems to be very sensitive to the assumption of independence of errors. (For an interesting situation in which it is not sensitive to this dependence, see Section 14.7.) It is also often sensitive to the assumption of equal variances. However, in a balanced analysis of variance model with equal weights, the F-test for main effects does not seem too sensitive to different variances in different classes. See Scheffé (1959), Chapter 10, for a more detailed discussion of the effects of violations of these assumptions on the F-test.

10.6. VARIANCE STABILIZING TRANSFORMATIONS.

In many problems we have random variables that are not normally distributed and we want to test that equality of means of these observations. Often the variance of these random variables depends on the mean, so that we cannot use the procedures discussed in previous chapters. In this setting we often transform the observations so the variance is independent of the mean at least asymptotically. The basic result that we need for such transformations is the following.

THEOREM 10.8. If $\sqrt{n}(T_n - a) \xrightarrow{d} N_1(0, b^2)$ and $g(t)$ is differentiable at a, then

$$\sqrt{n}\,(g(T_n) - g(a)) \xrightarrow{d} N_1\big(0, (g'(a))^2 b^2\big).$$

PROOF. We first note that $T_n - a = (1/\sqrt{n})(\sqrt{n}(T_n - a)) \to 0$ by Theorem 10.2b. Now let

$$h(t) = \begin{cases} \dfrac{g(t) - g(a)}{t - a} - g'(a) & \text{if } t \neq a \\ 0 & \text{if } t = a \end{cases}.$$

By the definition of a derivative, we see that $h(t)$ is continuous at a. Therefore, by Theorem 10.2c, $h(T_n) \xrightarrow{p} h(a) = 0$. Now

$$\sqrt{n}\,(g(T_n) - g(a)) = g'(a)\big(\sqrt{n}(T_n - a)\big) + \sqrt{n}(T_n - a)h(T_n).$$

By Theorem 10.2b, we see that $\sqrt{n}(T_n - a)h(T_n) \xrightarrow{p} 0$. By Theorem 10.2d we see that $g'(a)(\sqrt{n}(T_n - a)) \xrightarrow{d} N_1(0, (g'(a))^2 b^2)$. The result follows from Theorem 10.2b. □

We now consider several models that have the following form [after reduction by sufficiency and invariance (where applicable)]. We observe U_i independent, $i = 1, \ldots, k$, such that

$$\sqrt{n_i}\,(U_i - \lambda_i) \xrightarrow{d} N_1\big(0, (h(\lambda_i))^2\big) \qquad \text{as } n_i \to \infty$$

for some known function h. We want to test the hypothesis that the λ_i are equal. We cannot use the results for linear models in this setting because the variances of the U_i depend on the λ_i. (We cannot assume that the variances are all equal and then test whether the λ_i are equal.) We therefore transform the model to one that has constant variance, at least asymptotically. Let g be a function such that

$$g'(s) = \frac{1}{h(s)}\,. \tag{10.7}$$

By Theorem 10.8, we see that

$$\sqrt{n_i}\,(g(U_i) - g(\lambda_i)) \xrightarrow{d} N_1(0,1). \tag{10.8}$$

The function g has changed the variables U_i to variables $g(U_i)$ that have asymptotically constant variance, and hence g is called a *variance stabilizing transformation*. In addition, if $h(s) > 0$ (i.e., the variances are always positive), then the function $g(s)$ is strictly increasing and hence invertible. Therefore, the $g(U_i)$ are a sufficient statistic for the reduced model defined above.

Now, let $T_i = g(U_i)$, $\theta_i = g(\lambda_i)$, $\mathbf{Y}' = (\sqrt{n_1}\,T_1, \ldots, \sqrt{n_k}\,T_k)$, $\boldsymbol{\mu}' = (\sqrt{n_1}\,\theta_1, \ldots, \sqrt{n_k}\,\theta_k)$. Since the U_i are independent, the T_i are independent. Therefore, if the n_i are all large, then

$$\mathbf{Y} \dot{\sim} N_k(\boldsymbol{\mu}, \mathbf{I})$$

(where the $\cdot$ above the $\sim$ indicates that the distribution is only approximate). Let W be the one-dimensional subspace of R^k consisting of vectors $\boldsymbol{\mu}' = (\sqrt{n_1}\,\theta, \ldots, \sqrt{n_k}\,\theta)$ for some θ. Then we are testing that $\boldsymbol{\mu} \in W$ against $\boldsymbol{\mu} \in R^p$. Therefore, this model is approximately a linear model with known variance as discussed in Section 7.11. From that section, we see that the UMP invariant size α test for testing that $\boldsymbol{\mu} \in W$ for the linear model depends on $\chi^2 = \|\mathbf{P}_{W^\perp}\mathbf{Y}\|^2 = \sum(Y_i - \sqrt{n_i}\,\hat{\hat{\theta}})^2$ where $\hat{\hat{\theta}}$ minimizes $(Y_i - \sqrt{n_i}\,\theta)^2$. That is, the UMP invariant size α test for the linear model is given by

$$\chi^2 = \sum_i \left(Y_i - \sqrt{n_i}\,\hat{\hat{\theta}}\right)^2 = \sum_i n_i\left(T_i - \hat{\hat{\theta}}\right)^2, \hat{\hat{\theta}} = \frac{\sum_i \sqrt{n_i}\,Y_i}{\sum_i n_i} = \frac{\sum_i n_i T_i}{\sum_i n_i}$$

$$\phi(\chi^2) = \begin{cases} 1 & \text{if } \chi^2 > \chi^{2\alpha}_{p-1} \\ 0 & \text{if } \chi^2 \leqslant \chi^{2\alpha}_{p-1} \end{cases}. \tag{10.9}$$

Since this test is UMP invariant size α for the linear model that approximates the original model, the test ϕ should have size near α for the original model and should be a fairly powerful test for the original model, at least if the n_i are all large.

We now look at some example of models having the form given above, and work out the variance stabilizing transformations for these models.

EXAMPLE 1. (Poisson distribution). Consider the model in which we observe U_{ij} independent, $i = 1, \ldots, k$; $j = 1, \ldots, n_i$; such that U_{ij} has a Poisson distribution with mean λ_i. We want to test that the λ_i are all equal. We assume that the n_i are all large. A sufficient statistic for this model is $(\bar{U}_{1.}, \ldots, \bar{U}_{k.})$. Since $EU_{ij} = \lambda_i$, $\text{var}(U_{ij}) = \lambda_i$, we see from the central limit theorem that

$$\sqrt{n_i}\,(U_{i.}-\lambda_i) \xrightarrow{d} N_1(0,\lambda_i).$$

A variance stabilizing transformation for this distribution must satisfy $g'(s) = s^{-1/2}$. We therefore let $g(s) = 2s^{1/2}$. Hence, we define $T_i = 2(\overline{U}_{i.})^{1/2}$, and use the test ϕ defined in (10.9) for this model.

EXAMPLE 2. (Exponential distribution). Now let U_{ij} be independent, $i = 1, \ldots, k$; $j = 1, \ldots, n_i$; such that U_{ij} is exponentially distributed with mean λ_i. That is, U_{ij} has continuous density function

$$f(u_{ij};\lambda_i) = \frac{1}{\lambda_i}\exp -\left(\frac{u_{ij}}{\lambda_i}\right), \qquad u_{ij} > 0.$$

We want to test that the λ_i are all the same. We note again that $(\overline{U}_{1.}, \ldots, \overline{U}_{k.})$ is a complete sufficient statistic for this model. Since $EU_{ij} = \lambda_i$, $\operatorname{var}(U_{ij}) = \lambda_i^2$, we see that

$$\sqrt{n_i}\,(U_{i.}-\lambda_i) \xrightarrow{d} N_1(0,\lambda_i^2).$$

We therefore want a transformation that satisfies $g'(s) = 1/s$. We let $g(s) = \log s$. Therefore, let $T_i = \log(\overline{U}_{i.})$ and let ϕ be defined by (10.9). Then ϕ should be a good approximate test for this problem.

EXAMPLE 3. (Equality of variances in linear models). Suppose we observe k linear models and want to test that the variances are the same for all the models. We observe Q_i, independent random vectors, $i = 1, \ldots, k$, such that

$$\mathbf{Q}_i \sim N_{n_i}(\boldsymbol{\mu}_i, \sigma_i^2\mathbf{I}), \qquad \boldsymbol{\mu}_i \in V_i, \qquad \sigma_i^2 > 0$$

where the V_i are known p_i-dimensional subspaces. We want to test that the σ_i^2 are all equal. A complete sufficient statistic for this model is $(\hat{\boldsymbol{\mu}}_1, \hat{\sigma}_1^2, \ldots, \hat{\boldsymbol{\mu}}_k, \hat{\sigma}_k^2)$ where $\hat{\boldsymbol{\mu}}_i = \mathbf{P}_{V_i}\mathbf{Q}_i, (n_i - p_i)\hat{\sigma}_i^2 = \|\mathbf{P}_{V_i^\perp}\mathbf{Q}_i\|^2$. This problem is invariant under transformations that take $\hat{\boldsymbol{\mu}}_i$ to $\hat{\boldsymbol{\mu}}_i + \mathbf{a}_i$ where $\mathbf{a}_i \in V_i$. A maximal invariant under this group of transformations is $(\hat{\sigma}_1^2, \ldots, \hat{\sigma}_k^2)$. By Thereom 10.7, we see that

$$\sqrt{n_i}\,(\hat{\sigma}_i^2 - \sigma_i^2) \xrightarrow{d} N_1(0, 2\sigma_i^4)$$

(since $\delta = 3$ for normal distributions). Therefore, we need a transformation $g(s)$ such that $g'(s) = 1\sqrt{2}\,s$. We therefore let $g(s) = (\log s)/\sqrt{2}$. Hence, we let $T_i = (\log \hat{\sigma}_i^2)/2$ and use ϕ defined by equation (10.9). In the case $k = 2$, ϕ has the form

$$F = \frac{\hat{\sigma}_1^2}{\hat{\sigma}_2^2}, \qquad \phi(F) = \begin{cases} 1 & \text{if } F > a \text{ or } F < b \\ 0 & \text{if } b \leqslant F \leqslant a \end{cases}$$

(see Exercise B8). That is, in this case ϕ reduces to an F-test for the equality of two variances.

The test derived in Example 3 is called Bartlett's test of the equality of variances. It is very sensitive to the normal assumption used in deriving it. If the errors are not normally distributed, then the kurtosis may not be 0, or equivalently δ need not be 3. In fact, it need not even be constant. Therefore, this test should be used with great caution.

One situation in which Bartlett's test is sometimes used is in analysis of variance problems before testing hypotheses about the means. However, that procedure is *not* appropriate for the following three reasons. The first reason is the sensitivity to the normal assumption. The hypothesis of equal variance may be rejected bacause the data is not normal, even when the variances are equal. In earlier sections, we showed that the F-test is not too sensitive to the normal assumption. Therefore, we might avoid using the F-test in a situation in which it is appropriate, if we first perform a test for equal variances. The second reason for not performing the test of equal variances first is that accepting the hypothesis of equal variances is certainly no proof that they are equal. It may be that they are unequal, but we do not have enough data to prove them so, or it may be that they are unequal, but that the underlying distribution is nonnormal in such a way as to mask this inequailty. The third reason is that the F-test for main effects in analysis of variance models is not too sensitive to the assumption of equal variances, at least if the cell sizes are nearly equal and the weights used are nearly equal (see Scheffé, 1959, pp. 351–358). For these reasons, it is not recommended that Bartlett's test be used before the F-tests in analysis of variance.

10.7. PROOF OF THEOREM 10.3.

Before proving the basic result of this chapter, we state two more theorems that are proved in most textbooks on probability theory. The first theorem is a generalization of the central limit theorem, while the second result is a sufficient condition for bringing a limit inside an expectation. We use the following notation. For any $a \geqslant 0$, and $x \in R$ we define

$$I_a(x) = \begin{cases} 1 & \text{if } |x| > a \\ 0 & \text{if } |x| \leqslant a \end{cases}.$$

THEOREM 10.9. a. Lindeberg-Feller theorem. Let W_{ni} be a collection of random variables, $n = 1, 2, \ldots,$; $i = 1, \ldots, n$ with $EW_{ni} = 0$, $\text{var}(W_{ni}) = \sigma_{ni}^2$. Suppose that $\sum_{i=1}^n \sigma_{ni}^2 = a$, $\lim_{n\to\infty} \max_{1 \leqslant i \leqslant n} \sigma_{ni}^2 = 0$ and for all $\epsilon > 0$

$$\lim_{n\to\infty} \sum_{i=1}^n EW_{ni}^2 I_\epsilon(W_{ni}) = 0.$$

Then

$$\sum_{i=1}^n W_{ni} \xrightarrow{d} N_1(0, a).$$

b. Bounded convergence theorem. Let X be a random variable and let $g_n(x)$ be a sequence of functions such that $\lim g_n(x)$ exists and $|g_n(x)| \leqslant h(x)$ where $Eh(X)$ exists. Then

$$\lim Eg_n(X) = E(\lim g_n(X)).$$

We now return to the proof of Theorem 10.3. Let $f_1, f_2, \ldots,$ be a sequence of independently identically distributed random variables such that $Ef_1 = 0$, $\text{var}\, f_1 = \sigma^2 < \infty$, and let $e_n' = (f_1, \ldots, f_n)$.

THEOREM 10.3. Let $\{a_n\}$ be a sequence of vectors such that a_n is $n \times 1$, $\|a_n\| = 1$ and $m(a_n) \to 0$. Then

$$a_n' e_n \xrightarrow{d} N_1(0, \sigma^2).$$

PROOF. Let $a_n' = (a_{n1}, \ldots, a_{nn})$ and let $W_{ni} = a_{ni} f_i$. Then

$$a_n' e_n = \sum_{i=1}^n a_{ni} f_i = \sum_{i=1}^n W_{ni}.$$

Since the f_i are independent, for each n the W_{ni} are independent. Also

$$EW_{ni} = 0, \qquad \text{var}(W_{ni}) = a_{ni}^2 \sigma^2, \qquad \sum_i \text{var}(W_{ni}) = \sigma^2 \|a_n\|^2 = \sigma^2,$$

$$\max_{1 \leqslant i \leqslant n} \text{var}(W_{ni}) = \sigma^2 \max_{1 \leqslant i \leqslant n} a_{ni}^2 = \sigma^2 (m(a_n))^2 \to 0.$$

By the Lindeberg-Feller theorem, we will be finished when we show that for all $\epsilon > 0$

$$C_n = \sum_{i=1}^n EW_{ni}^2 I_\epsilon(W_{ni}) \to 0.$$

Let $m_n = m(a_n)$. Then

$$C_n = \sum_{i=1}^n EW_{ni}^2 I_\epsilon(W_{ni}) = \sum_{i=1}^n a_{ni}^2 Ef_i^2 I_{\epsilon/|a_{ni}|}(f_i) \leqslant \sum_{i=1}^n a_{ni}^2 Ef_i^2 I_{\epsilon/m_n}(f_i)$$

$$= \sum_{i=1}^n a_{ni}^2 Ef_1^2 I_{\epsilon/m_n}(f_1) = Ef_1^2 I_{\epsilon/m_n}(f_1) \sum_{i=1}^n a_{ni}^2 = Ef_1^2 I_{\epsilon/m_n}(f_1).$$

(If $a_{ni} = 0$, then $I_{\epsilon/|a_{ni}|}(x) = 0$.) Now, $|f_1^2 I_{\epsilon/m_n}(f_1)| \leqslant f_1^2$ and $Ef_1^2 = \sigma^2 < \infty$. Therefore, by the bounded convergence theorem we see that

$$\lim_{n\to\infty}\left(Ef_1^2 I_{\epsilon/m_n}(f_1)\right) = E\lim_{n\to\infty}\left(f_1^2 I_{\epsilon/m_n}(f_1)\right) = 0.$$

Therefore, C_n is trapped between 0 and a sequence that is converging to 0, and must converge to 0. □

EXERCISES.

Type B

1. Show that if $f_i \sim N_1(0, \sigma^2)$, then $Ef_i^4 = 3\sigma^4$.

2. (a) Show that $m(\mathbf{AB}) \leqslant km(\mathbf{A})m(\mathbf{B})$, where k is the common dimension of $\mathbf{A}$ and $\mathbf{B}$.

(b) Show that $m(\mathbf{AA}') \geqslant (m(\mathbf{A}))^2$.

3. Show that if $V_n^* \subset V_n$, then $m(\mathbf{P}_{V_n^*}) \leqslant m(\mathbf{P}_{V_n})$.

4. Show that the coefficient of Y_{ij} in μ_{ij} for the analysis of covariance model is given by

$$\frac{1}{n_i} + \frac{\left(X_{ij} - \overline{X}_{i.}\right)^2}{\Sigma\Sigma\left(X_{ij} - \overline{X}_{i.}\right)^2}.$$

5. (a) Let $U > 0$ be a random variable such that $EU < \infty$. Show that $P(|U| > \epsilon) < EU/\epsilon$.

(b) Let $U_1, U_2, \ldots,$ be a sequence of random variables such that $U_n \geqslant 0$ and $EU_n \to 0$. Show that $U_n \xrightarrow{P} 0$.

6 Consider the model defined by (10.1) and (10.2). Let Huber's condition be satisfied.

(a) If $\mathbf{c}_n \in V_n$, show that

$$\frac{<\mathbf{c}_n, \boldsymbol{\mu}_n>}{\hat{\sigma}_n \|\mathbf{c}_n\|} \xrightarrow{d} N_1(0, 1).$$

(b) Use this result to establish the asymptotic validity of the size of the t-test defined in Section 7.9.

(c) Use part a to establish the asymptotic validity of the confidence coefficient for the confidence interval for functions of $\boldsymbol{\mu}$ given in Section 6.2. [Note that $\mathbf{P}_{V_n}\mathbf{c}_n \in V_n$ and $\mathbf{c}_n'\boldsymbol{\mu}_n = (\mathbf{P}_{V_n}\mathbf{c}_n)'\boldsymbol{\mu}_n$.]

7. Prove Theorem 10.7. [Hint: Use the central limit theorem in place of the weak law of large numbers in the proof of Theorem 10.5.]

8. Show that when $k = 2$, Bartlett's test of the equality of variances reduces to an F-test.

Type C

1 Asymptotic power of F-test. Let $W_n \mid V_n$ be a k-dimensional subspace. Suppose that $\lim_{n\to\infty} \|\mathbf{P}_{V_n \mid W_n} \boldsymbol{\mu}_n\| = \gamma \neq 0$ and that $\mathbf{P}_{V_n \mid W_n} \boldsymbol{\mu}_n \neq 0$ for any n. Let X_n be an orthonormal basis matrix for $V_n \mid W_n$ whose first column is given by

$$\frac{1}{\|\mathbf{P}_{V_n \mid W_n} \boldsymbol{\mu}_n\|} \mathbf{P}_{V_n \mid W_n} \boldsymbol{\mu}_n .$$

(a) Show that

$$\mathbf{X}_n' \mathbf{Y}_n \xrightarrow{d} N_{p-k}\left[\begin{pmatrix} \gamma \\ 0 \\ \vdots \\ 0 \end{pmatrix}, \sigma^2 \mathbf{I}\right].$$

[Hint: $c'\mathbf{X}_n{}'\mathbf{Y}_n = c'\mathbf{X}_n'\mathbf{e}_n + c_1\gamma + c_1(\|\mathbf{P}_{V_n \mid W_n} \boldsymbol{\mu}_n\| - \gamma)$. Why? Use the proof of Theorem 10.4, and Theorem 10.2, parts b and d.]

(b) Show that

$$\frac{\|\mathbf{P}_{V_n \mid W_n} \mathbf{Y}_n\|^2}{\sigma^2} \xrightarrow{d} \chi^2_{p-k}\left(\frac{\gamma^2}{\sigma^2}\right).$$

(c) Show that

$$\frac{\|\mathbf{P}_{V_n \mid W_n} \mathbf{Y}_n\|^2}{(p-k)\hat{\sigma}_n^2} \xrightarrow{d} \frac{\chi^2_{p-k}\left(\dfrac{\gamma^2}{\sigma^2}\right)}{(p-k)} .$$

2. Multivariate central limit theorem. Let $\mathbf{U}_1, \mathbf{U}_2, \ldots,$ be a sequence of independently, identically distributed random p-dimensional vectors with $E\mathbf{U}_i = \mathbf{a}$, $\operatorname{cov}(\mathbf{U}_i) = \mathbf{B}$. Show that $\sqrt{n}\,(\bar{\mathbf{U}} - \mathbf{a}) \xrightarrow{d} N_p(\mathbf{0}, \mathbf{B})$, (Hint: Let $\mathbf{t} \subset R^p$ and let $v_i = \mathbf{t}'\mathbf{U}_i$. Then $\bar{v} = \mathbf{t}'\mathbf{U}$. Use Theorems 10.1b and 10.2e.)

3. Multivariate Theorem 10.8. Suppose that $\sqrt{n}\,(\mathbf{T}_n - \mathbf{a}) \xrightarrow{d} N_p(\mathbf{0}, \mathbf{B})$. Let $g(\mathbf{t})$ be a differentiable function from R^p to R^1 with gradient

$$\Delta(\mathbf{t}) = \left(\frac{\partial g_1}{\partial t_1}, \ldots, \frac{\partial g_p}{\partial t_p}\right)'.$$

Show that $\sqrt{n}\,(g(T_n) - g(a)) \xrightarrow{d} N_1(0, \Delta'(a) B \Delta(a))$. {You may assume that $[g(t) - g(a) - \Delta'(a)(t-a)]/\|t-a\|$ goes to 0 as t goes to a.}

CHAPTER 11

James-Stein and Ridge Estimators

In this chapter we derive some alternative estimators for $\boldsymbol{\mu}$ and $\boldsymbol{\beta}$ that are in some ways better than the ordinary least squares estimators defined in Chapter 6. As in previous chapters we study the model in which we observe

$$\mathbf{Y} \sim N_n(\boldsymbol{\mu}, \sigma^2 \mathbf{I}), \qquad \boldsymbol{\mu} \in V, \qquad \sigma^2 > 0.$$

As before, let

$$\hat{\boldsymbol{\mu}} = \mathbf{P}_V \mathbf{Y}, \qquad \hat{\sigma}^2 = \frac{\|\mathbf{P}_{V^\perp} \mathbf{Y}\|^2}{n - p}.$$

The estimators considered in this chapter are basically shrinking estimators in that they modify $\hat{\boldsymbol{\mu}}$ (or $\hat{\boldsymbol{\beta}}$) to make it into a shorter vector. The following lemma indicates why this could be a sensible procedure.

LEMMA 11.1. $E\|\hat{\boldsymbol{\mu}}\|^2 = \|\boldsymbol{\mu}\|^2 + p\sigma^2$.

PROOF. By Lemma 3.1b and Theorem 2.5i, we see that

$$E\|\hat{\boldsymbol{\mu}}\|^2 = \|\boldsymbol{\mu}\|^2 + \sigma^2 \operatorname{tr} \mathbf{P}_V = \|\boldsymbol{\mu}\|^2 + p\sigma^2. \quad \square$$

This lemma indicates that if we estimate $\boldsymbol{\mu}$ by $\hat{\boldsymbol{\mu}}$, we should be substantially overestimating the length of $\boldsymbol{\mu}$. (For any estimator $\hat{\hat{\boldsymbol{\mu}}}$, it is possible to find functions $h(\hat{\hat{\boldsymbol{\mu}}})$ such that $Eh(\hat{\hat{\boldsymbol{\mu}}}) \geqslant h(\boldsymbol{\mu})$. What makes Lemma 11.1 interesting is that if p is large and $\|\boldsymbol{\mu}\|/\sigma^2$ is small, then we would be overestimating $\boldsymbol{\mu}$ by a very large amount if we use $\hat{\boldsymbol{\mu}}$.)

This lemma suggests that we might improve the ordinary least-squares estimator $\hat{\boldsymbol{\mu}}$ by shrinking it (i.e., multiplying it by a scalar $a(\hat{\boldsymbol{\mu}}, \hat{\sigma}^2)$ where $0 \leqslant a(\hat{\boldsymbol{\mu}}, \hat{\sigma}^2) \leqslant 1$). It also suggests that when $\|\hat{\boldsymbol{\mu}}\|^2/\hat{\sigma}^2$ is small (and hence $\|\boldsymbol{\mu}\|^2/\sigma^2$ is probably small), then $a(\hat{\boldsymbol{\mu}}, \hat{\sigma}^2)$ should also be small, but when $\|\hat{\boldsymbol{\mu}}\|^2/\hat{\sigma}^2$ is large, then $a(\hat{\boldsymbol{\mu}}, \hat{\sigma}^2)$ should be near 1. In this chapter, we look at estimators of this form with

$$a(\hat{\boldsymbol{\mu}}, \hat{\sigma}^2) = 1 - c\hat{\sigma}^2/\|\hat{\boldsymbol{\mu}}\|^2$$

where c is constant. In Section 11.1, we find the optimal choice for c, which leads to the James-Stein estimator. In Section 11.2, it is noted that $a(\hat{\boldsymbol{\mu}}, \hat{\sigma}^2)$ as given above is not necessarily nonnegative. It is replaced with

$$a^*(\hat{\boldsymbol{\mu}}, \hat{\sigma}^2) = \max\left(a(\hat{\boldsymbol{\mu}}, \hat{\sigma}^2), 0\right)$$

which leads to the modified James-Stein estimator.

As a standard for comparing different estimators, we use the loss function

$$L\left(\mathbf{d}; (\boldsymbol{\mu}, \sigma^2)\right) = \frac{\|\boldsymbol{\mu} - \mathbf{d}\|^2}{\sigma^2}.$$

When $p = 1$, the estimator $\hat{\boldsymbol{\mu}}$ has been known to be admissible for some time. Stein (1955) showed that for $p = 2$, $\hat{\boldsymbol{\mu}}$ is an admissible estimator, but that for $p > 2$, $\hat{\boldsymbol{\mu}}$ is inadmissible. Then James and Stein (1960) exhibited an estimator that is better than $\hat{\boldsymbol{\mu}}$ when $p > 2$. In Section 11.1, the James-Stein estimator of $\boldsymbol{\mu}$ is defined and shown to be better than $\hat{\boldsymbol{\mu}}$. The modified James-Stein estimator is discussed in Section 11.2. In Section 11.3, another shrinking estimator, the ridge estimator, is defined. In Sections 11.4–11.7 there is a discussion of some other properties of these estimators.

Let $\mathbf{d}(\mathbf{Y})$ be an unbiased estimator of $\boldsymbol{\mu}$. In Chapter 6, we showed that $\hat{\boldsymbol{\mu}}$ is the minimum varianced unbiased estimator of $\boldsymbol{\mu}$. By Theorem 1.5, $\hat{\boldsymbol{\mu}}$ is as good as any unbiased estimator. Hence, to find an estimator that is better than $\hat{\boldsymbol{\mu}}$, we must look outside the set of unbiased estimators. None of the estimators considered in this chapter is unbiased. Since $\hat{\boldsymbol{\mu}}$ is inadmissible, all unbiased estimators are inadmissible. Similarly, in Chapter 6 we showed that $\hat{\boldsymbol{\mu}}$ is the best invariant estimator. Therefore, all invariant estimators must be inadmissible when $p > 2$, and an estimator that is better than $\hat{\boldsymbol{\mu}}$ could not be invariant.

For the remainder of this chapter we assume that $p > 2$.

11.1. THE JAMES-STEIN ESTIMATOR FOR $\boldsymbol{\mu}$.

The first result in this section is the determination of the risk function for the ordinary least-squares estimator $\hat{\boldsymbol{\mu}}$.

LEMMA 11.2. $R(\hat{\boldsymbol{\mu}}; (\boldsymbol{\mu}, \sigma^2)) = p$.

PROOF. See Exercise B1. □

In this section we find another estimator whose risk is everywhere smaller than p and whose risk when $\boldsymbol{\mu} = \mathbf{0}$ is about 2. Therefore, if p is much larger than 2, this estimator is much better than $\hat{\boldsymbol{\mu}}$ for $\boldsymbol{\mu}$ near $\mathbf{0}$, and moderately better than $\hat{\boldsymbol{\mu}}$ even when $\|\boldsymbol{\mu}\|/\sigma$ is large.

We first state the following lemma.

LEMMA 11.3. Let $\mathbf{U} \sim N_p(\boldsymbol{\theta}, \mathbf{I})$ and K have a Poisson distribution with mean $\|\boldsymbol{\theta}\|^2/2$. Then

a.

$$E\frac{1}{\|\mathbf{U}\|^2} = E\frac{1}{p-2+2K}.$$

b.

$$E\frac{\mathbf{U}'(\mathbf{U}-\boldsymbol{\theta})}{\|\mathbf{U}\|^2} = E\frac{p-2}{p-2+2K}.$$

PROOF. a. Let $V|K \sim \chi^2_{p+2K}(0)$. By Lemma 1.8, $V \sim \chi^2_p(\|\boldsymbol{\theta}\|^2)$. Also, by the definition of a χ^2 distribution, $\|\mathbf{U}\|^2 \sim \chi^2_p(\|\boldsymbol{\theta}\|^2)$. Therefore

$$E\frac{1}{\|\mathbf{U}\|^2} = E\frac{1}{V} = E\left(E\frac{1}{V}\,|K\right) = E\frac{1}{p-2+2K}.$$

(See Exercise B2.)

b. Let $\mathbf{U}' = (U_1, \ldots, U_p)$, $\boldsymbol{\theta}' = (\theta_1, \ldots, \theta_p)$. We note first that

$$\frac{\partial}{\partial\theta_i}E\frac{1}{\|\mathbf{U}\|^2} = E\frac{U_i-\theta_i}{\|\mathbf{U}\|^2}, \qquad \frac{\partial}{\partial\theta_i}E\frac{1}{p-2+2K} = \frac{\theta_i}{\|\boldsymbol{\theta}\|^2}E\frac{2K-\|\boldsymbol{\theta}\|^2}{p-2+2K}$$

(see Exercise C1). By part a,

$$E\frac{1}{\|\mathbf{U}\|^2} = E\frac{1}{p-2+2K}.$$

Therefore,

$$E\frac{1}{\|\mathbf{U}\|^2}(U_i-\theta_i) = \frac{\delta}{\delta\theta_i}E\frac{1}{\|\mathbf{U}\|^2} = \frac{\delta}{\delta\theta_i}E\frac{1}{p-2+2K} = \frac{\theta_i}{\|\boldsymbol{\theta}\|^2}E\frac{2K-\|\boldsymbol{\theta}\|}{p-2+2K}$$

and

$$E\frac{1}{\|\mathbf{U}\|^2}(\mathbf{U}-\boldsymbol{\theta}) = \frac{1}{\|\boldsymbol{\theta}\|^2}\left(E\frac{2K-\|\boldsymbol{\theta}\|^2}{p-2+2K}\right)\boldsymbol{\theta}.$$

Hence

$$E\frac{\mathbf{U}'(\mathbf{U}-\boldsymbol{\theta})}{\|\mathbf{U}\|^2} = E\frac{\mathbf{U}'\mathbf{U}-\boldsymbol{\theta}'\boldsymbol{\theta}-\boldsymbol{\theta}'(\mathbf{U}-\boldsymbol{\theta})}{\|\mathbf{U}\|^2}$$

$$= 1 - \|\boldsymbol{\theta}\|^2 E\frac{1}{\|\mathbf{U}\|^2} - \boldsymbol{\theta}' E\frac{1}{\|\mathbf{U}\|^2}(\mathbf{U}-\boldsymbol{\theta})$$

$$= 1 - \|\boldsymbol{\theta}\|^2 E\frac{1}{p-2+2K} - E\frac{2K-\|\boldsymbol{\theta}\|^2}{p-2+2K}$$

$$= E\frac{p-2}{p-2+2K}. \quad \square$$

The proof of part b is from Baranchik (1973). Although it is rather tricky, it is much easier than the original proof in James and Stein (1960).

We are now ready to derive an estimator that is better than $\hat{\boldsymbol{\mu}}$. Define

$$\mathbf{d}_c(\mathbf{Y}) = \left(1 - \frac{c\hat{\sigma}^2}{\|\hat{\boldsymbol{\mu}}\|^2}\right)\hat{\boldsymbol{\mu}}.$$

Note that if $c = 0$, then $\mathbf{d}_c = \hat{\boldsymbol{\mu}}$. Since $\mathbf{d}_c$ is a (random) scalar times $\hat{\boldsymbol{\mu}}$, $\mathbf{d}_c(\mathbf{Y}) \in V$. We now find the risk function of $\mathbf{d}_c$.

LEMMA 11.4

$$R(\mathbf{d}_c;(\boldsymbol{\mu},\sigma^2)) = p - \left[2c(p-2) - c^2\frac{n-p+2}{n-p}\right]E\frac{1}{p-2+2K}$$

where K has a Poisson distribution with mean $\|\boldsymbol{\mu}\|^2/2\sigma^2$.

PROOF.

$$R(\mathbf{d}_c;(\boldsymbol{\mu},\sigma^2)) = E\frac{\|\mathbf{d}_c-\boldsymbol{\mu}\|^2}{\sigma^2} = \frac{E\left\|\hat{\boldsymbol{\mu}}-\boldsymbol{\mu}-\dfrac{c\hat{\sigma}^2}{\|\hat{\boldsymbol{\mu}}\|^2}\hat{\boldsymbol{\mu}}\right\|^2}{\sigma^2}$$

$$= \frac{E\|\hat{\boldsymbol{\mu}}-\boldsymbol{\mu}\|^2}{\sigma^2} - \frac{2c}{\sigma^2}E\hat{\sigma}^2 E\frac{(\hat{\boldsymbol{\mu}}-\boldsymbol{\mu})'\hat{\boldsymbol{\mu}}}{\|\hat{\boldsymbol{\mu}}\|^2} + \frac{c^2}{\sigma^2}E\hat{\sigma}^4 E\frac{1}{\|\hat{\boldsymbol{\mu}}\|^2},$$

using the independence of $\hat{\boldsymbol{\mu}}$ and $\hat{\sigma}^2$. By Lemma 11.2, we see that $E(\|\hat{\boldsymbol{\mu}}-\boldsymbol{\mu}\|^2/\hat{\sigma}^2) = p$. Also $(n-p)\hat{\sigma}^2/\sigma^2 \sim \chi^2_{n-p}(0)$. Therefore, $E\hat{\sigma}^2 = \sigma^2$ and $E\hat{\sigma}^4 = (n-p+2)\sigma^4/(n-p)$. Now let $\mathbf{X}$ be an orthonormal basis matrix for V. Define $\mathbf{U} = (1/\sigma)\mathbf{X}'\mathbf{Y}$, $\boldsymbol{\theta} = (1/\sigma)\mathbf{X}'\boldsymbol{\mu}$. Then $\mathbf{U} \sim N_p(\boldsymbol{\theta}, \mathbf{I})$. Since $\boldsymbol{\mu} \in V$, $\boldsymbol{\mu} = \mathbf{P}_V\boldsymbol{\mu} = \mathbf{X}\mathbf{X}'\boldsymbol{\mu} = \sigma\mathbf{X}\boldsymbol{\theta}$, $\hat{\boldsymbol{\mu}} = \mathbf{P}_V\mathbf{Y} = \mathbf{X}\mathbf{X}'\mathbf{Y} = \sigma\mathbf{X}\mathbf{U}$, and

$$\frac{\|\boldsymbol{\mu}\|^2}{2\sigma^2} = \frac{\|\boldsymbol{\theta}\|^2}{2}, \qquad \frac{1}{\|\hat{\boldsymbol{\mu}}\|^2} = \frac{1}{\sigma^2\|\mathbf{U}\|^2}, \qquad \frac{(\hat{\boldsymbol{\mu}}-\boldsymbol{\mu})'\hat{\boldsymbol{\mu}}}{\|\hat{\boldsymbol{\mu}}\|^2} = \frac{(\mathbf{U}-\boldsymbol{\theta})'\mathbf{U}}{\|\mathbf{U}\|^2}.$$

Therefore, by Lemma 11.3, we see that

$$E\frac{(\hat{\boldsymbol{\mu}}-\boldsymbol{\mu})'\hat{\boldsymbol{\mu}}}{\|\hat{\boldsymbol{\mu}}\|^2}=E\frac{(\mathbf{U}-\boldsymbol{\theta})'\mathbf{U}}{\|\mathbf{U}\|^2}=E\frac{p-2}{p-2+2K},$$

$$E\frac{1}{\|\hat{\boldsymbol{\mu}}\|^2}=\frac{1}{\sigma^2}E\frac{1}{\|\mathbf{U}\|^2}=\frac{1}{\sigma^2}E\frac{1}{p-2+2K}.$$

Hence

$$R(\mathbf{d}_c;(\boldsymbol{\mu},\sigma^2))=p-2cE\frac{p-2}{p-2+2K}+c^2\left(\frac{n-p+2}{n-p}\right)E\frac{1}{p-2+2K}$$

$$=p-\left(2c(p-2)-c^2\frac{n-p+2}{n-p}\right)E\frac{1}{p-2+2K}. \quad \square$$

Now we need to find the c that minimizes $R(\mathbf{d}_c;(\boldsymbol{\mu},\sigma^2))$. Since $p>2$, $K\geqslant 0$, we see that $E(p-2+2K)^{-1}>0$. Hence $R(\mathbf{d}_c;(\boldsymbol{\mu},\sigma^2))$ is minimized by

$$c=\frac{(p-2)(n-p)}{n-p+2}.$$

Therefore, let

$$\hat{\hat{\boldsymbol{\mu}}}=\left(1-\frac{(p-2)(n-p)}{n-p+2}\,\frac{\hat{\sigma}^2}{\|\hat{\boldsymbol{\mu}}\|^2}\right)\hat{\boldsymbol{\mu}}. \tag{11.1}$$

$\hat{\hat{\boldsymbol{\mu}}}$ is called the *James-Stein estimator* of $\boldsymbol{\mu}$.

THEOREM 11.5. The estimator $\hat{\hat{\boldsymbol{\mu}}}$ is better than $\hat{\boldsymbol{\mu}}$. Its risk function is

$$R(\hat{\hat{\boldsymbol{\mu}}};(\boldsymbol{\mu},\sigma^2))=p-\frac{(p-2)^2(n-p)}{n-p+2}E\frac{1}{p-2+2K}<p$$

where K has a Poisson distribution with mean $\|\boldsymbol{\mu}\|^2/2\sigma^2$. In particular,

$$R(\hat{\hat{\boldsymbol{\mu}}};(\mathbf{0},\sigma^2))=\frac{2n}{n-p+2}$$

PROOF. These results follow directly from substituting $c=(p-2)\cdot(n-p)/(n-p+2)$ into Lemma 11.4. (Note that if $\boldsymbol{\mu}=\mathbf{0}$, then K has a Poisson distribution with mean 0, i.e., $K\equiv 0$, and hence $E(p-2+2K)^{-1}=(p-2)^{-1}$.) $\square$

This theorem establishes that $\hat{\hat{\boldsymbol{\mu}}}$ is much better than $\hat{\boldsymbol{\mu}}$ if $\boldsymbol{\mu}$ is near $\mathbf{0}$, and somewhat better for all $\boldsymbol{\mu}$. Note that $R(\hat{\hat{\boldsymbol{\mu}}};(\boldsymbol{\mu},\sigma^2))$ depends on $(\boldsymbol{\mu},\sigma^2)$ only through $\|\boldsymbol{\mu}\|^2/\sigma^2$.

In the previous discussion, it is not necessary to shrink $\hat{\boldsymbol{\mu}}$ toward $\mathbf{0}$. Let

$\mathbf{v} \in V$, and define

$$\hat{\hat{\boldsymbol{\mu}}}_{\mathbf{v}} = \left[1 - \frac{(p-2)(n-p)\hat{\sigma}^2}{(n-p+2)\|\hat{\boldsymbol{\mu}} - \mathbf{v}\|^2}\right](\hat{\boldsymbol{\mu}} - \mathbf{v}) + \mathbf{v}. \tag{11.2}$$

Then

$$R(\hat{\hat{\boldsymbol{\mu}}}_{\mathbf{v}}; (\boldsymbol{\mu}, \sigma^2)) = R(\hat{\boldsymbol{\mu}}; (\boldsymbol{\mu} - \mathbf{v}, \sigma^2)) = p - \frac{(p-2)^2(n-p)}{n-p+2} E \frac{1}{p-2+2K} \tag{11.3}$$

where K has a Poisson distribution with mean $\|\boldsymbol{\mu} - \mathbf{v}\|^2/2\sigma^2$ (see Exercise B5). Therefore, $\hat{\hat{\boldsymbol{\mu}}}_v$ does very well if $\boldsymbol{\mu}$ is near $\mathbf{v}$, and better than $\hat{\boldsymbol{\mu}}$ for all $\boldsymbol{\mu}$. This fact furnishes an interesting aspect of James-Stein estimation. A researcher can choose a value $\mathbf{v}$ that he thinks that $\boldsymbol{\mu}$ should be near. He can then compute $\hat{\hat{\boldsymbol{\mu}}}_{\mathbf{v}}$. If he is correct, and $\boldsymbol{\mu}$ is near $\mathbf{v}$, his estimator is much better than $\hat{\boldsymbol{\mu}}$, but even if he is wrong, his estimator will be better than the minimum variance unbiased estimator, $\hat{\boldsymbol{\mu}}$. He must, of course choose $\mathbf{v}$ before he looks at his data. This aspect is also a disadvantage to James-Stein estimation in that two different experimentors would find different estimators from the same data if they choose to shrink to different points. The James-Stein estimator has sacrificed the "objectivity" of the ordinary least-squares estimator. In addition, on a particular experiment, a researcher may have no idea where $\boldsymbol{\mu}$ is. In that case, there does not seem to be any obvious way to determine a point to shrink toward.

If n is large, then $R(\hat{\hat{\boldsymbol{\mu}}}_{\mathbf{v}}; (\boldsymbol{\mu}, \sigma^2))$ is approximately equal to

$$h_p(\delta) = p - (p-2)E(p-2+2k)^{-1}$$

where K has a Poisson distribution with $\delta/2 = \|\boldsymbol{\mu} - \mathbf{v}\|^2/2\sigma^2$. In Figure 11.1, there are graphs of $h_p(\delta)/p$ for various choices of p. (Note that p is the risk of the ordinary least-squares estimator.) For example, we see that if $p = 12$, and $\|\boldsymbol{\mu} - \mathbf{v}\|^2 \leqslant 20\sigma^2$, then $h_p(\delta)/p < 0.7$. This implies that if our choice for $\mathbf{v}$ satisfies $\|\boldsymbol{\mu} - \mathbf{v}\|^2 \leqslant 20\sigma^2$, then the risk of $\hat{\hat{\boldsymbol{\mu}}}_v$ is only about 70% of the risk of the ordinary least-squares estimator. This indicates the tremendous savings that are possible using shrinking estimators.

Let $\mathbf{X}$ be a basis matrix for V and let $\boldsymbol{\beta}$ be the unique solution to $\boldsymbol{\mu} = \mathbf{X}\boldsymbol{\beta}$. Let $\mathbf{b} \in R^p$ and let $\hat{\hat{\boldsymbol{\mu}}}_{\mathbf{Xb}}$ be as defined in (11.2). Let $\hat{\hat{\boldsymbol{\beta}}}_{\mathbf{b}}$ be the unique solution to

$$\hat{\hat{\boldsymbol{\mu}}}_{\mathbf{Xb}} = \mathbf{X}\hat{\hat{\boldsymbol{\beta}}}_b.$$

Then $\hat{\hat{\boldsymbol{\beta}}}_{\mathbf{b}}$ is called a *James-Stein estimator* of $\boldsymbol{\beta}$. In Exercise B6, it is shown

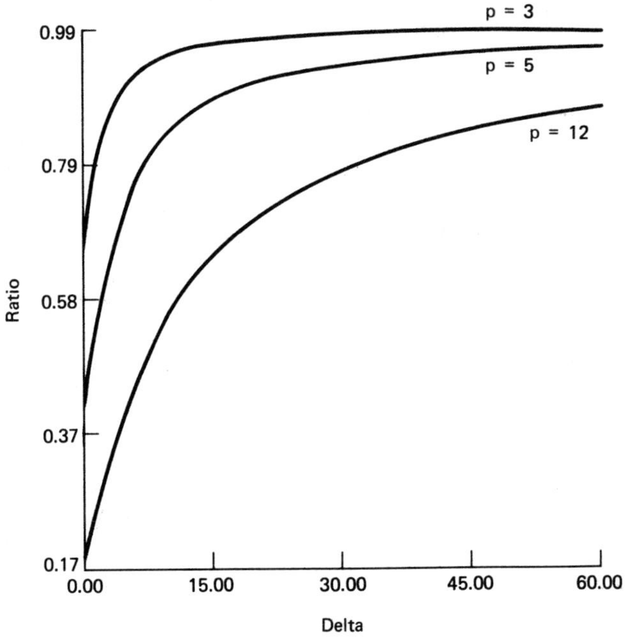

Figure 11.1

that

$$\hat{\hat{\boldsymbol{\beta}}}_{\mathbf{b}} = \left[1 - \frac{(p-1)(n-p)\hat{\sigma}^2}{(n-p+2)\|\mathbf{X}\hat{\boldsymbol{\beta}} - \mathbf{X}\mathbf{b}\|^2}\right](\hat{\boldsymbol{\beta}} - \mathbf{b}) + \mathbf{b}. \tag{11.4}$$

11.2. THE MODIFIED JAMES-STEIN ESTIMATOR.

Let

$$a_{\mathbf{v}}(\hat{\boldsymbol{\mu}}, \hat{\sigma}^2) = 1 - \frac{(p-2)(n-p)\hat{\sigma}^2}{(n-p+2)\|\hat{\boldsymbol{\mu}} - \mathbf{v}\|^2}.$$

Then

$$\hat{\hat{\boldsymbol{\mu}}}_{\mathbf{v}} = a_{\mathbf{v}}(\hat{\boldsymbol{\mu}}, \hat{\sigma}^2)(\hat{\boldsymbol{\mu}} - \mathbf{v}) + \mathbf{v}.$$

It is apparent that $a_{\mathbf{v}}(\hat{\boldsymbol{\mu}}, \hat{\sigma}^2) < 1$. If $a_{\mathbf{v}}(\hat{\boldsymbol{\mu}}, \hat{\sigma}^2) \geqslant 0$, then $\hat{\hat{\boldsymbol{\mu}}}_{\mathbf{v}}$ shrinks $\boldsymbol{\mu}$ toward $\mathbf{v}$. However, if $\mathbf{u}_{\mathbf{v}}(\hat{\boldsymbol{\mu}}, \hat{\sigma}^2) < 0$, then $\hat{\hat{\boldsymbol{\mu}}}_{\mathbf{v}}$ shrinks $\hat{\boldsymbol{\mu}}$ past $\mathbf{v}$. In this case we can improve $\hat{\hat{\mu}}_v$ by letting it be $\mathbf{0}$. That is, we define

$$\hat{\hat{\hat{\mu}}}_v = \max\bigl(0, a_v(\hat{\mu}, \hat{\sigma}^2)\bigr)(\hat{\mu} - v) + v.$$

Then $\hat{\hat{\hat{\mu}}}_v$ is better than $\hat{\hat{\mu}}_v$ (see Stein, 1966). $\hat{\hat{\hat{\mu}}}_v$ is called a *modified James-Stein estimator* of $\boldsymbol{\mu}$. Similar modifications could be made for $\hat{\hat{\beta}}_b$.

One of the nicer aspects of the James-Stein estimator is the simplicity of the risk function. The modified James-Stein estimator does not have such a simple risk function.

11.3. THE RIDGE ESTIMATOR.

We now give another shrinking estimator which numerical studies have indicated is better than the ordinary least-squares estimator, and possibly better than the James-Stein estimator (see, for example, Dempster et al., 1977). A *ridge estimator* for $\boldsymbol{\beta}$ in the coordinatized linear model is an estimator of the form

$$\tilde{\beta} = \left[\mathbf{X}'\mathbf{X} + k(\mathbf{Y})\mathbf{I}\right]^{-1}\mathbf{X}'\mathbf{Y} = \left[\mathbf{X}'\mathbf{X} + k(\mathbf{Y})\mathbf{I}\right]^{-1}\mathbf{X}'\mathbf{X}\hat{\beta}$$

where $k(\mathbf{Y})$ is a real-valued function of the data. At present, it is not clear how one actually chooses k. This type of estimator was originally suggested by Hoerl and Kennard (1970a and b) for problems in which $|\mathbf{X}'\mathbf{X}|$ is near 0, and hence $(\mathbf{X}'\mathbf{X})^{-1}$ is unstable [i.e., small changes in $\mathbf{X}$ can make large changes in $(\mathbf{X}'\mathbf{X})^{-1}$]. However, they also suggested applying it to other settings, and it seems to do quite well, judging from the numerical results that have been presented.

We could shrink to an arbitrary point $\mathbf{b} \in R^p$, by letting

$$\tilde{\beta}_{\mathbf{b}} = \left[\mathbf{X}'\mathbf{X} + k^*(\mathbf{Y})\mathbf{I}\right]^{-1}\mathbf{X}'\mathbf{X}(\hat{\beta} - \mathbf{b}) + \mathbf{b}.$$

We could also find a ridge estimator for $\boldsymbol{\mu}$ by letting $\mathbf{v} = \mathbf{X}\mathbf{b}$

$$\tilde{\mu}_v = \mathbf{X}\tilde{\beta}_{\mathbf{b}} = \mathbf{X}\left[\mathbf{X}'\mathbf{X} + k^*(\mathbf{Y})\mathbf{I}\right]^{-1}\mathbf{X}'(\mathbf{Y} - \mathbf{v}) + \mathbf{v}$$

It should be apparent that the estimator $\tilde{\mu}_v$ depends on the basis matrix chosen for V. The James-Stein estimator, however, depends only on $(\hat{\mu}, \hat{\sigma}^2)$ and is therefore independent of the basis matrix.

If $\mathbf{X}$ is an orthonormal basis matrix for V, then

$$\tilde{\mu} = \left(1 - \frac{k^*(\mathbf{Y})}{1 + k^*(\mathbf{Y})}\right)(\hat{\mu} - \mathbf{v}) + \mathbf{v}$$

and hence in this case the James-Stein estimator is a ridge estimator with

$$\frac{k^*(\mathbf{Y})}{1+k^*(\mathbf{Y})}=\frac{(p-2)(n-p)\hat{\sigma}^2}{(n-p+2)\|\hat{\boldsymbol{\mu}}-\mathbf{v}\|^2}.$$

11.4. AN EMPIRICAL BAYES PERSPECTIVE.

Let $\pi(\boldsymbol{\mu},\sigma^2)$ be a prior distribution for the parameters $\boldsymbol{\mu}$ and σ^2. The Bayes estimator $\mathbf{d}^*$ of $\boldsymbol{\mu}$ is defined by

$$d^*(\mathbf{Y})=E\boldsymbol{\mu}\,|\,\mathbf{Y}.$$

If the prior distribution itself has some unknown constants and we substitute some "sensible" estimators for those constants, we get an estimator $\mathbf{d}^{**}(\mathbf{Y})$, which is called an *empirical Bayes estimator* of $\boldsymbol{\mu}$. When we use an empirical Bayes estimator, we are letting the observations choose which prior in a given class to use. Such procedures can be considered a compromise between the classical frequentist approach to inference and the Bayesian approach.

In this section, we show that both the James-Stein and ridge estimators are empirical Bayes estimators. The prior distribution that we use is one in which σ^2 is degenerate at δ^2 [i.e., $P(\sigma^2=\delta^2)=1$], and

$$\boldsymbol{\mu}\sim N_n(\mathbf{v},\tau^2\mathbf{A}).$$

We consider $\mathbf{v}$ and $\mathbf{A}$ as known, and then estimate δ^2 and τ^2 from the data. One choice for $\mathbf{A}$ leads to the James-Stein estimator, whereas another leads to the ridge estimator. To guarantee that $\boldsymbol{\mu}\in V$, it is necessary to assume that $\mathbf{c}'\mathbf{v}=0$ and $\mathbf{c}'\mathbf{A}\mathbf{c}=0$ for all $\mathbf{c}\in V^\perp$ (i.e., that $\mathbf{v}\in V$ and the rows and columns of $\mathbf{A}$ are in V) (see Exercise B8).

We now find the Bayes estimators of $\boldsymbol{\mu}$.

LEMMA 11.6. The Bayes estimator of $\boldsymbol{\mu}$ is given by

$$d^*(\mathbf{Y})=\mathbf{v}+\tau^2\mathbf{A}(\delta^2\mathbf{I}+\tau^2\mathbf{A})^{-1}(\mathbf{Y}-\mathbf{v}).$$

PROOF. Since $\sigma^2=\delta^2$, $\mathbf{Y}\,|\,\boldsymbol{\mu}\sim N_n(\boldsymbol{\mu},\delta^2\mathbf{I})$. From Exercise B9, we see that

$$\begin{pmatrix}\mathbf{Y}\\ \boldsymbol{\mu}\end{pmatrix}\sim N_{2n}\left(\begin{pmatrix}\mathbf{v}\\ \mathbf{v}\end{pmatrix},\begin{pmatrix}\delta^2\mathbf{I}+\tau^2\mathbf{A} & \tau^2\mathbf{A}\\ \tau^2\mathbf{A} & \tau^2\mathbf{A}\end{pmatrix}\right)$$

Therefore, by Lemma 3.7c, we see that

$$E\boldsymbol{\mu}\,|\,\mathbf{Y}=\mathbf{v}+\tau^2\mathbf{A}(\delta^2\mathbf{I}+\tau^2\mathbf{A})^{-1}(\mathbf{Y}-\mathbf{v}).\quad\square$$

To derive the James-Stein estimator as an empirical Bayes estimator, we let $\mathbf{A} = \mathbf{P}_V$. Clearly $\mathbf{c}'\mathbf{P}_V\mathbf{c} = 0$ for all $\mathbf{c} \in V^{\perp}$, so this is an acceptable choice. Also,

$$\mathbf{P}_V(\delta^2\mathbf{I} + \tau^2\mathbf{P}_V)^{-1} = \frac{1}{\delta^2 + \tau^2}\mathbf{P}_V$$

(see Exercise B10). Therefore the Bayes estimator of $\boldsymbol{\mu}$ for $\mathbf{A} = \mathbf{P}_V$ is given by

$$\mathbf{d}_1^*(\mathbf{Y}) = \mathbf{v} + \frac{\tau^2}{\delta^2 + \tau^2}\mathbf{P}_V(\mathbf{Y} - \mathbf{v}) = \left(1 - \frac{\delta^2}{\delta^2 + \tau^2}\right)(\hat{\boldsymbol{\mu}} - \mathbf{v}) + \mathbf{v}$$

(since $\mathbf{v} \in V$). To find the James-Stein estimator $\hat{\hat{\boldsymbol{\mu}}}_{\mathbf{v}}$, we merely substitute $k(\hat{\boldsymbol{\mu}}, \hat{\sigma}^2) = (n-p)(p-2)\hat{\sigma}^2/((n-p+2)\|\hat{\boldsymbol{\mu}} - \mathbf{v}\|^2)$ for $\delta^2/(\delta^2 + \tau^2)$. In Exercise C3 we indicate that $k(\hat{\boldsymbol{\mu}}, \hat{\sigma}^2)$ is a "sensible" estimator of $\delta^2/(\delta^2 + \tau^2)$.

Hence the James-Stein estimator $\hat{\hat{\boldsymbol{\mu}}}_{\mathbf{v}}$ is an empirical Bayes estimator of $\boldsymbol{\mu}$ for the prior distribution given with $\mathbf{A} = \mathbf{P}_V$. Let $\mathbf{X}$ be an orthonormal basis matrix for V. Then this prior implies that $\mathbf{X}'\boldsymbol{\mu} \sim N_p(\mathbf{X}'\mathbf{v}, \tau^2\mathbf{I})$. Therefore, this prior distribution is just a spherical normal distribution over the p-dimensional subspace V, centered at the vector $\mathbf{v} \in V$.

We now show that the ridge estimator is an empirical Bayes estimator. Let $\mathbf{X}$ be a basis matrix for V (not necessarily an orthonormal basis matrix), and let $\mathbf{A} = \mathbf{X}\mathbf{X}'$. It is clear that $\mathbf{c}'\mathbf{X}\mathbf{X}'\mathbf{c} = 0$ for all $\mathbf{c} \in V^{\perp}$, so that this is an acceptable choice for $\mathbf{A}$. Also,

$$\mathbf{X}'(\delta^2\mathbf{I} + \tau^2\mathbf{X}\mathbf{X}')^{-1} = (\delta^2\mathbf{I} + \tau^2\mathbf{X}'\mathbf{X})^{-1}\mathbf{X}'$$

(see Exercise B10). Then the Bayes estimator for this choice for $\mathbf{A}$ is given by

$$\mathbf{d}_2^*(\mathbf{Y}) = \mathbf{v} + \tau^2\mathbf{X}\mathbf{X}'(\delta^2\mathbf{I} + \tau^2\mathbf{X}\mathbf{X}')^{-1}(\mathbf{Y} - \mathbf{v}) = \mathbf{X}\left(\mathbf{X}'\mathbf{X} + \frac{\delta^2}{\tau^2}\mathbf{I}\right)^{-1}\mathbf{X}'(\mathbf{Y} - \mathbf{v}) + \mathbf{v}$$

which is just the ridge estimator $\tilde{\boldsymbol{\mu}}_{\mathbf{v}}$ with $k = \delta^2/\tau^2$. We can therefore think of $\mathbf{k}^*(\mathbf{Y})$ in the ridge estimator as an estimator of δ^2/τ^2.

A ridge estimator is an empirical Bayes estimator with respect to the family in which $\boldsymbol{\mu} \sim N_n(\mathbf{v}, \tau^2\mathbf{X}'\mathbf{X})$. Now let $\mathbf{X}\boldsymbol{\beta} = \boldsymbol{\mu}$ and $\mathbf{X}\mathbf{b} = \mathbf{v}$. Then $\boldsymbol{\beta} \sim N_n(\mathbf{b}, \tau^2\mathbf{I})$. Therefore the prior distribution for the ridge estimator is a spherical normal distribution in the p-dimensional space of values for $\boldsymbol{\beta}$, while the prior distribution for the James-Stein estimator is a spherical normal distribution in the p-dimensional space of values for $\boldsymbol{\mu}$. If $\mathbf{X}$ is an orthonormal basis matrix the $\mathbf{X}'\mathbf{X} = \mathbf{P}_V$ and the priors are the same.

One of the difficulties in using either the James-Stein or ridge estimators is choosing the vector $\mathbf{v}$ to shrink toward. One approach to this difficulty is suggested by the empirical Bayes derivations given above. We could let $\mathbf{v}$ be an unknown in the prior distribution and estimate it from the data. Note that

$$E\hat{\mu} = E\frac{E\hat{\mu}}{\mu} = E\mu = \mathbf{v}.$$

Therefore $\hat{\mu}$ would be a "sensible" estimator of $\mathbf{v}$. However, if we substitute $\hat{\mu}$ for $\mathbf{v}$ in the Bayes estimators derived in Lemma 11.6, we just get $\hat{\mu}$ as an empirical Bayes estimator. This fact illustrates an important aspect of empirical Bayes estimation. The empirical Bayes approach is only effective when the number of unknown constants to be estimated in the prior distribution is much smaller than the number of parameters to be estimated in the original problem. In the derivations in this section, there were only two unknown constants in the prior distribution (δ^2 and τ^2), whereas there were p independent parameters to be estimated in the original problem. If we also let the vector $\mathbf{v}$ be unknown, there are more unknown constants in the prior distribution ($p + 2$) than there are parameters in the prior distribution. In this situation the empirical Bayes approach leads to the same estimator as the classical approach does.

11.5. SENSITIVITY TO UNITS OF MEASUREMENT.

We now consider the effects of change of units (in both $\mathbf{Y}$ and $\mathbf{X}$) on the three estimators: $\hat{\mu}$ (the ordinary least-squares estimator), $\hat{\hat{\mu}}_{\mathbf{v}}$ (the James-Stein estimator shrunk to $\mathbf{v}$) and $\tilde{\mu}_{\mathbf{v}}$ (a ridge estimator shrunk to $\mathbf{v}$). We assume throughout this section that $\mathbf{1} = (1, \ldots, 1)' \in V$, as it is in most practical models. (It is in the examples in Chapter 7.)

Let $\mathbf{U}$ be a vector of observations in some units (e.g., degrees Fahrenheit). Suppose that we change the units of the elements of $\mathbf{U}$ (e.g., to degrees Centigrade). Let $\mathbf{T}$ be the vector of observations in the new units. Then

$$\mathbf{T} = c\mathbf{U} + a\mathbf{1} \tag{11.6}$$

for some scalars $c > 0$ and a.

We first consider a change of units in columns of the $\mathbf{X}$ matrix. This change will not affect $\mu = \mathbf{X}\beta$ since μ does not depend on the units of $\mathbf{X}$. Therefore, we would hope that an estimator of μ would be unaffected by this change of units. A change of units in the ith column of $\mathbf{X}$ would replace $\mathbf{X}_i$ with $c\mathbf{X}_i + a\mathbf{1}$. The subspace spanned by $\mathbf{X}_1, \ldots, \mathbf{X}_{i-1}, c\mathbf{X}_i + a\mathbf{1}, \mathbf{X}_{i+1}, \ldots, \mathbf{X}_p$ is the same as the subspace spanned by the columns of $\mathbf{X}$ (since $c \neq 0$ and $\mathbf{1} \in V$). Therefore $\hat{\mu}$ and $\hat{\sigma}^2$ would not be changed by this change of units in the $\mathbf{X}$ matrix. In addition, $\hat{\hat{\mu}}_{\mathbf{v}}$ depends on $\mathbf{X}$ only through $\hat{\mu}$ and $\hat{\sigma}^2$ and is also unchanged by this change of units. However, $\tilde{\mu}_{\mathbf{v}}$, the ridge estimator would be affected by changes in units for the $\mathbf{X}$ matrix. For this reason, it has been suggested that before computing the ridge estimator we replace all the variables in $\mathbf{X}$ by standardized variables (i.e., replace X_{ij} by $(X_{ij} - \overline{X}_{.j})/S_j$ where $\overline{X}_{.j}$ and S_j are the sample mean and standard deviations of the observations in the jth column.) This still leaves the problem of how to scale the indicator variables such as $\mathbf{1}$ in the regression model with an intercept term.

Now, consider a change in units in the $\mathbf{Y}$ vector. Let $\mathbf{Y}^* = c\mathbf{Y} + a\mathbf{1}$. Then $\mathbf{Y}^* \sim N_n(c\boldsymbol{\mu} + a\mathbf{1}, \sigma^2\mathbf{I})$. Since $\boldsymbol{\mu} \in V$, $\mathbf{1} \in V$, we see that $c\boldsymbol{\mu} + a\mathbf{1} \in V$. Therefore we might hope that if $d(\mathbf{Y})$ is a sensible estimator of $\boldsymbol{\mu}$, then

$$\mathbf{d}(c\mathbf{Y} + a\mathbf{1}) = c\mathbf{d}(\mathbf{Y}) + a\mathbf{1}. \tag{11.7}$$

Consider first $d(\mathbf{Y}) = \hat{\boldsymbol{\mu}}$. Then

$$\hat{\boldsymbol{\mu}}(c\mathbf{Y} + a\mathbf{1}) = \mathbf{P}_V(c\mathbf{Y} + a\mathbf{1}) = c\mathbf{P}_V\mathbf{Y} + a\mathbf{P}_V\mathbf{1} = c\hat{\boldsymbol{\mu}} + a\mathbf{1},$$

since $\mathbf{1} \in V$. Therefore, $\hat{\boldsymbol{\mu}}$ satisfies (11.7). However, the James-Stein estimator $\hat{\hat{\boldsymbol{\mu}}}_{\mathbf{v}}$ does not (unless $a = 0$, $\mathbf{v} = \mathbf{0}$). In fact,

$$c\hat{\hat{\boldsymbol{\mu}}}_{\mathbf{v}}(\mathbf{Y}) + a\mathbf{1} = \hat{\hat{\boldsymbol{\mu}}}_{c\mathbf{v}+a\mathbf{1}}(c\mathbf{Y} + a\mathbf{1}). \tag{11.8}$$

This equation implies that if we are shrinking toward the vector $\mathbf{v}$ in the units u, then we should shrink toward the vector $c\mathbf{v} + a\mathbf{1}$ in the units $cu + a$. That is, we should also change the units of the vector that we are shrinking toward. If we always express $\mathbf{v}$ in the new units, then the James-Stein estimator does in fact act sensibly under unit changes. It is impossible to tell how a ridge estimator behaves without knowing more about $k^*(\mathbf{Y})$.

In summary, both the ordinary least-squares estimator and the James-Stein estimator react sensibly to change of units in either $\mathbf{X}$ or $\mathbf{Y}$, whereas ridge estimators seem to be much more sensitive to the units used.

11.6. OTHER COMMENTS.

At the present time, the theory of these shrinking estimators is not very well understood. We do know, as has been shown, that the ordinary least-squares estimators are inadmissible, (at least for the loss function L) and that there is a great potential gain in using some form of shrinking estimator. However, it is not clear what type to use with a particular data set. It may be that neither the James-Stein nor ridge estimator is as good as some other estimator. We know that the James-Stein estimator is inadmissible, since the modified James-Stein estimator is better. Baranchik (1971) gives a class of estimators that are better than $\hat{\boldsymbol{\mu}}$ for the case in which σ^2 is assumed known. Strawderman (1973) extends this class to the case of unknown σ^2 and also derives a class of admissible Bayes estimators that are better than $\hat{\boldsymbol{\mu}}$ for the case $p \geqslant 5$ and $n + p \geqslant 9$. Efron and Morris (1973) give a class of empirical Bayes estimators that are better than $\hat{\boldsymbol{\mu}}$. In another paper Efron and Morris (1976) derive an unbiased estimator for the risk functions for a large class of estimators. They then show that this estimated risk function is less than or equal to p (for all $\hat{\boldsymbol{\mu}}$ and $\hat{\sigma}$) for many estimators. Therefore, the risk function (which is the expectation of the unbiased estimator) must also be less than or equal to p and hence these estimators are better than $\hat{\boldsymbol{\mu}}$.

The loss function $L(\mathbf{d}; (\boldsymbol{\mu}, \sigma^2)) = \|\boldsymbol{\mu} - \mathbf{d}\|^2/\sigma^2$ is not the only possible loss function we could consider for this problem. A *quadratic loss function* is a loss function of the form

$$\tilde{L}_{\mathbf{B}}\big(\mathbf{d}; (\boldsymbol{\mu}, \sigma^2)\big) = \frac{(\mathbf{d} - \boldsymbol{\mu})'\mathbf{B}(\mathbf{d} - \boldsymbol{\mu})}{\sigma^2}$$

where $\mathbf{B} \geqslant 0$ is a known matrix. [Note that $L(\mathbf{d}; (\boldsymbol{\mu}, \sigma^2)) = \tilde{L}_{\mathbf{I}}(\mathbf{d}; (\boldsymbol{\mu}, \sigma^2))$.] The derivation in Section 11.1 showing that the James-Stein estimator is better than the ordinary least-squares estimator is only valid for the loss function L. In fact the James-Stein estimator is not better than the ordinary least-squares estimator for arbitrary quadratic loss functions. However, the results of Brown (1966) indicate that the ordinary least-squares estimator is inadmissible for all quadratic loss functions (and in fact for most convex loss functions). The results in Brown (1973) indicate that there exist estimators that are better than the ordinary least-squares estimator for broad classes of quadratic loss functions. (That is, there are estimators that are better than $\hat{\boldsymbol{\mu}}$ for all the loss functions in the class simultaneously.) Unfortunately, these papers do not suggest what these estimators would be.

The ordinary least squares estimator is a minimax estimator of $\boldsymbol{\mu}$ for all quadratic loss functions and it has constant risk equal to the $\operatorname{tr} \mathbf{B}\mathbf{P}_V$ for the loss function $\tilde{L}_B$. (See Exercise B11.) In Exercise B12 it is shown that an estimator is as good as a constant risk minimax estimator if and only if it is a minimax estimator. Therefore, searching for estimators of $\boldsymbol{\mu}$ that are minimax for a particular quadratic loss function is equivalent to looking for estimators that are as good as the ordinary least-squares estimator for that loss function.

We now define a class of estimators that includes both the James-Stein and ridge estimators as special cases. Let $\mathbf{A} > 0$ be a specified $p \times p$ matrix and let $\mathbf{v} \in V$ be a specified vector. Let $k(\mathbf{Y})$ be a real-valued function of $\mathbf{Y}$. A *generalized ridge estimator* is an estimator $\tilde{\boldsymbol{\mu}}_{\mathbf{v},\mathbf{A}}$ of the form

$$\tilde{\boldsymbol{\mu}}_{\mathbf{v},\mathbf{A}} = \mathbf{X}\big[\mathbf{X}'\mathbf{X} + k(\mathbf{Y})\mathbf{A}\big]^{-1}\mathbf{X}'(\mathbf{Y} - \mathbf{v}) + \mathbf{v}.$$

If $\mathbf{A} = \mathbf{I}$, then this estimator is just a ridge estimator. If $\mathbf{A} = \mathbf{X}'\mathbf{X}$ and $[1 + k(\mathbf{Y})]^{-1} = 1 - (p-2)(n-p)\hat{\sigma}^2/(n-p+2)\|\hat{\boldsymbol{\mu}} - \mathbf{v}\|^2$, then this estimator is a James-Stein estimator. The generalized ridge estimator $\tilde{\boldsymbol{\mu}}_{\mathbf{v},A}$ is an empirical Bayes estimator with respect to the prior distribution $\boldsymbol{\mu} \sim N_n(\mathbf{v}, \tau^2\mathbf{X}\mathbf{A}^{-1}\mathbf{X}')$ (see Exercise B15).

One reason that we might suspect that generalized ridge estimators are sensible is given in the following lemma.

THEOREM 11.7. Let $\mathbf{A} > 0$ and $\mathbf{B} \geqslant 0$ be $p \times p$ and $n \times n$ matrices such that $\mathbf{BX} \neq \mathbf{0}$. Let $m \in R$ and define

$$\tilde{\boldsymbol{\mu}}_{\mathbf{v},\mathbf{A},m} = \mathbf{X}(\mathbf{X}'\mathbf{X} + m\mathbf{A})^{-1}\mathbf{X}'(\mathbf{Y} - \mathbf{v}) + \mathbf{v}.$$

For all $(\boldsymbol{\mu}, \sigma^2)$ there exists $m(\boldsymbol{\mu}, \sigma^2)$ such that

$$E_{\boldsymbol{\mu},\sigma^2}(\tilde{\boldsymbol{\mu}}_{\mathbf{v},\mathbf{A},m(\boldsymbol{\mu},\sigma^2)} - \boldsymbol{\mu})'\mathbf{B}(\tilde{\boldsymbol{\mu}}_{\mathbf{v},\mathbf{A},m(\boldsymbol{\mu},\sigma^2)} - \boldsymbol{\mu}) < E_{\boldsymbol{\mu},\sigma^2}(\hat{\boldsymbol{\mu}} - \boldsymbol{\mu})'\mathbf{B}(\hat{\boldsymbol{\mu}} - \boldsymbol{\mu}).$$

PROOF. Let

$$h(m) = E_{\boldsymbol{\mu},\sigma^2}(\tilde{\boldsymbol{\mu}}_{\mathbf{v},\mathbf{A},m} - \boldsymbol{\mu})'\mathbf{B}(\tilde{\boldsymbol{\mu}}_{\mathbf{v},\mathbf{A},m} - \boldsymbol{\mu}).$$

Then $h(0) = E_{\boldsymbol{\mu},\sigma^2}(\hat{\boldsymbol{\mu}} - \boldsymbol{\mu})'\mathbf{B}(\hat{\boldsymbol{\mu}} - \boldsymbol{\mu})$. We will therefore be finished when we show that h has a negative derivative at 0. In Exercise C5, it is shown that

$$h'(0) = -2\,\mathrm{tr}\left[(\mathbf{X}'\mathbf{X})^{-1}\mathbf{X}'\mathbf{B}\mathbf{X}(\mathbf{X}'\mathbf{X})^{-1}\mathbf{A}\right] \tag{11.9}$$

However,

$$\mathrm{tr}\left[(\mathbf{X}'\mathbf{X})^{-1}\mathbf{X}'\mathbf{B}\mathbf{X}(\mathbf{X}'\mathbf{X})^{-1}\mathbf{A}\right]$$
$$= \mathrm{tr}\left[\mathbf{A}^{1/2}(\mathbf{X}'\mathbf{X})^{-1}\mathbf{X}'\mathbf{B}^{1/2}\right]\left[\mathbf{A}^{1/2}(\mathbf{X}'\mathbf{X})^{-1}\mathbf{X}'\mathbf{B}^{1/2}\right]' > 0.$$

(Note that $\mathbf{0} \neq \mathbf{X}'\mathbf{B} = \mathbf{X}'\mathbf{B}^{1/2}\mathbf{B}^{1/2}$ and hence $\mathbf{X}'\mathbf{B}^{1/2} \neq \mathbf{0}$. Also $\mathbf{A}^{1/2}$ and $(\mathbf{X}'\mathbf{X})^{-1}$ are invertible matrices. Therefore $\mathbf{A}^{1/2}(\mathbf{X}'\mathbf{X})^{-1}\mathbf{X}'\mathbf{B}^{1/2} \neq \mathbf{0}$.) Hence $h'(0) < 0$ and the result is proved. □

Now suppose that we are interested in estimating $\boldsymbol{\mu}$ with the quadratic loss function $\tilde{L}_B$. We first choose an $\mathbf{A} > 0$ and a $\mathbf{v} \in V$. For all $(\boldsymbol{\mu}, \sigma^2)$, there exists $m(\boldsymbol{\mu}, \sigma^2)$ such that $\tilde{\boldsymbol{\mu}}_{\mathbf{v},\mathbf{A},m}$ is better than $\hat{\boldsymbol{\mu}}$. We now find a sensible estimator $k(\mathbf{Y})$ for $m(\boldsymbol{\mu}, \sigma^2)$. Then the generalized ridge estimator $\tilde{\boldsymbol{\mu}}_{\mathbf{v},\mathbf{A}}$ for that choice of k may be better than the ordinary least-squares estimator. However, this lemma must be interpreted with some caution, because it is true even in the cases $p = 1$ and $p = 2$, when we know that it is not possible to improve $\hat{\boldsymbol{\mu}}$. As one final comment about this lemma, we note that the calculation of $h(m)$ depends only on the first two moments of the random variables defined and is therefore valid in the more general Gauss-Markov setting of Section 6.5. Furthermore, the appropriate $m(\boldsymbol{\mu}, \sigma^2)$ does not depend on any normal assumption, but would be the same for any distribution satisfying the assumptions of that section.

The estimators defined in this chapter are not compatible with the confidence intervals and prediction intervals discussed in Chapter 6, nor with the simultaneous confidence intervals discussed in Chapter 8. It is not necessarily true that $\mathbf{c}'\hat{\hat{\boldsymbol{\mu}}}_v$ is even in the confidence interval defined in Chapter 6 for $\mathbf{c}'\boldsymbol{\mu}$. It seems much more difficult to find sensible confidence intervals that are compatible with the estimators discussed here. The primary difficulty is that the distribution of $\hat{\hat{\boldsymbol{\mu}}}_v - \boldsymbol{\mu}$ depends on $\boldsymbol{\mu}$, (the distribution of $\hat{\boldsymbol{\mu}} - \boldsymbol{\mu}$ does not). It is therefore very difficult to find a sensible function of $(\boldsymbol{\mu}, \hat{\hat{\boldsymbol{\mu}}}_v, \hat{\sigma}^2)$ that does not depend on any unknown parameters to use for a pivotal quantity. For this reason, any interval estimators of $\boldsymbol{\mu}$ based on these shrunken estimators would have to be much more complicated than those based on the ordinary least squares estimator $\hat{\boldsymbol{\mu}}$. In many practical problems, however, it seems that confidence intervals and prediction intervals are computed because they are easily available, but the point estimator is the object of interest. In such a situation, the estimators discussed in this chapter offer the possibility of great improvement over the ordinary least-squares estimator.

11.7. ESTIMATING $\boldsymbol{\beta}$.

In the previous sections, we have primarily discussed the estimation of $\boldsymbol{\mu}$. We now show how the results of previous sections can be applied to the problem of estimating $\boldsymbol{\beta}$. Let $\mathbf{d}^* \in R^p$ be an estimate of $\boldsymbol{\beta}$. A *quadratic loss function* for estimating $\boldsymbol{\beta}$ is a loss function of the form

$$L_{\mathbf{C}}^*\big(\mathbf{d}^*; (\boldsymbol{\beta}, \sigma^2)\big) = \frac{(\mathbf{d}^* - \boldsymbol{\beta})'\mathbf{C}(\mathbf{d}^* - \boldsymbol{\beta})}{\sigma^2}$$

where $\mathbf{C} \geqslant 0$ is a known matrix. Now, let $\mathbf{d}^*(\mathbf{Y})$ be an estimator of $\boldsymbol{\beta}$ and let $\mathbf{d}(\mathbf{Y}) = \mathbf{X}\mathbf{d}^*(\mathbf{Y})$ be the associated estimator of $\boldsymbol{\mu}$. Then

$$\begin{aligned} R_{\mathbf{C}}^*\big(\mathbf{d}^*; (\boldsymbol{\beta}, \sigma^2)\big) &= EL_{\mathbf{C}}^*\big(\mathbf{d}^*(\mathbf{Y}); (\boldsymbol{\beta}, \sigma^2)\big) \\ &= E\tilde{L}_{\mathbf{B}}\big(\mathbf{d}(\mathbf{Y}); (\boldsymbol{\mu}, \sigma^2)\big) = \tilde{R}_{\mathbf{B}}\big(\mathbf{d}; (\boldsymbol{\mu}, \sigma^2)\big) \end{aligned} \tag{11.10}$$

where $\tilde{L}_{\mathbf{B}}$ is defined in the last section and

$$\mathbf{B} = \mathbf{X}(\mathbf{X}'\mathbf{X})^{-1}\mathbf{C}(\mathbf{X}'\mathbf{X})^{-1}\mathbf{X}'$$

(see Exercise B13). Therefore, the risk function for using $\mathbf{d}^*(\mathbf{Y})$ to estimate $\boldsymbol{\beta}$ with the loss function $L_{\mathbf{C}}^*$ is the same as the risk function for using $\mathbf{d}(\mathbf{Y})$ to estimate $\boldsymbol{\mu}$ with the loss function $\tilde{L}_{\mathbf{B}}$. Hence, any results derived in previous sections for estimating $\boldsymbol{\mu}$ with the loss function $\tilde{L}_{\mathbf{B}}$ can be restated as results for estimating $\boldsymbol{\beta}$ with the loss function $L_{\mathbf{C}}$. In particular, the ordinary least-squares estimator $\hat{\boldsymbol{\beta}}$ is a minimax estimator of $\boldsymbol{\beta}$ for any quadratic loss function, and $\hat{\boldsymbol{\beta}}$ has constant risk $\operatorname{tr}\mathbf{P}_V\mathbf{B} = \operatorname{tr}\mathbf{C}(\mathbf{X}'\mathbf{X})^{-1}$ for the loss function

L_C^*. Hence an estimator of $\boldsymbol{\beta}$ is as good as $\hat{\boldsymbol{\beta}}$ for the loss function L_C^* if and only if it is minimax for that loss function.

A *generalized ridge estimator* of $\boldsymbol{\beta}$ is an estimator of the form

$$\tilde{\boldsymbol{\beta}}_{\mathbf{b},\mathbf{A}} = \left[\mathbf{X}'\mathbf{X} + k(\mathbf{Y})\mathbf{A}\right]^{-1}\mathbf{X}'(\mathbf{Y} - \mathbf{X}\mathbf{b}) + \mathbf{b}$$

where $\mathbf{b} \in R^p$ is a specified vector, $\mathbf{A} > 0$ is a specified $p \times p$ matrix and $k(\mathbf{Y})$ is a specified real-valued function of $\mathbf{Y}$. Note that $\mathbf{X}\tilde{\boldsymbol{\beta}}_{\mathbf{b},\mathbf{A}} = \tilde{\boldsymbol{\mu}}_{\mathbf{X}\mathbf{b},\mathbf{A}}$. By Theorem 11.7, reinterpreted for estimating $\boldsymbol{\beta}$ we may expect to find generalized ridge estimators that are better than $\hat{\boldsymbol{\beta}}$.

There have been at least two quadratic loss functions considered for the problem of estimating $\boldsymbol{\beta}$:

$$L_1^*(\mathbf{d}^*;(\boldsymbol{\beta},\sigma^2)) = L_{\mathbf{X}'\mathbf{X}}^*(\mathbf{d}^*;(\boldsymbol{\beta},\sigma^2)) = \frac{(\mathbf{d}^* - \boldsymbol{\beta})'\mathbf{X}'\mathbf{X}(\mathbf{d}^* - \boldsymbol{\beta})}{\sigma^2}$$

$$= \frac{\|\mathbf{X}\mathbf{d}^* - \mathbf{X}\boldsymbol{\beta}\|^2}{\sigma^2} = \frac{\|\mathbf{d} - \boldsymbol{\mu}\|^2}{\sigma^2} = \tilde{L}_{\mathbf{I}}(\mathbf{d};(\boldsymbol{\mu},\sigma^2)),$$

and

$$L_2^*(\mathbf{d}^*;(\boldsymbol{\beta},\sigma^2)) = L_{\mathbf{I}}^*(\mathbf{d}^*;(\boldsymbol{\beta},\sigma^2)) = \frac{\|\mathbf{d} - \boldsymbol{\beta}\|^2}{\sigma^2} = \tilde{L}_{\mathbf{B}}(\mathbf{d};(\boldsymbol{\mu},\sigma^2)),$$

where

$$\mathbf{B} = \mathbf{X}(\mathbf{X}'\mathbf{X})^{-2}\mathbf{X}'.$$

We now discuss each of these loss functions.

We consider first L_1^*. One important advantage of L_1^* is that it is unit-free. It does not depend on the units either of $\mathbf{Y}$ or of the columns of the $\mathbf{X}$ matrix. It is the Mahalanobis distance loss function for this problem. [Note that $\hat{\boldsymbol{\beta}} \sim N_p(\boldsymbol{\beta},\sigma^2(\mathbf{X}'\mathbf{X})^{-1})$.] Another advantage of this loss function is that results are more easily derived for this loss function than for other loss functions. Note that using this loss function for the problem of estimating $\boldsymbol{\beta}$ is equivalent to using the loss function $L_{\mathbf{I}} = L$ for the problem of estimating $\boldsymbol{\mu}$. Therefore, the James-Stein estimator of $\boldsymbol{\beta}$ is better than the ordinary least-squares estimator for this loss function. The disadvantage of this loss function is that it depends on the values of the independent variables (x_{ij}) that are actually included in the model. Two different experimenters fitting the same model would have different loss functions if they chose different values for their independent variables. In addition, using this loss function implies that a good estimator $\mathbf{d}^*(\mathbf{Y})$ is one such that $\mathbf{X}\mathbf{d}^*(\mathbf{Y})$ is near $\mathbf{X}\boldsymbol{\beta} = \boldsymbol{\mu}$.

The loss function L_2^* seems much more appealing, at least on first impression. It implies that a good estimator $\mathbf{d}^*(\mathbf{Y})$ of $\boldsymbol{\beta}$ is one which is close to $\boldsymbol{\beta}$. However, this loss function is very dependent on the units of the $\mathbf{X}$ matrix. Note that if the ith column of the $\mathbf{X}$ matrix has units a_i, then

$(d_i^* - \beta_i)^2/\sigma^2$ has units $1/a_i^2$. This fact implies that in using the loss function L_2^* we are emphasizing those β_i associated with columns that have large units. In particular, two experiments fitting the same model would have different loss functions if they used different units.

Because of the severe unit problems associated with the loss function L_2^*, it is often suggested that the columns of the $\mathbf{X}$ matrix be centered and standardized before applying this loss function. Although this approach leads to a unit-free loss function, in Exercise B14 several disadvantages of this loss function are derived. The most important disadvantage is that this loss function again depends on the actual values for the independent variables included in the experiment. In fact, this loss function gives heaviest weight to those coefficients associated with columns of $\mathbf{X}$ that have large sample variances. In addition it is shown that using this approach is very close to using the Mahalanobis distance loss function L_1^*.

The above discussion illustrates the oversimplification that results when we pick a loss function for an estimation problem. In this problem, we want an estimator $\mathbf{d}^*(\mathbf{Y})$ that is "close" to $\boldsymbol{\beta}$. However, there does not seem to be any single loss function that measures how "far" an estimate is from $\boldsymbol{\beta}$. Any loss function that we have considered is either quite sensitive to the units used in measuring the independent variable or is sensitive to the choices for the values of those variables to be included in the experiment.

EXERCISES.

Type A

1. Use the data of Exercise A1 of Chapter 5 to find the James-Stein estimator shrunk to $\mathbf{0}$ and shrunk to $(2, 8, 0)'$.

Type B

1. Prove Lemma 11.2 ($\|\hat{\boldsymbol{\mu}} - \boldsymbol{\mu}\|^2/\sigma^2 \sim \chi_p^2$. Why?)

2. Show that if $V\,|\,K \sim \chi^2_{p+2K}(0)$, then $EV^{-1}\,|\,K = (p-2+2K)^{-1}$.

3. Show that $E\hat{\sigma}^4 = (n-p+2)\sigma^4/(n-p)$.

4. Fill in the details of the proof of Theorem 11.5.

5. Derive (11.3).

6. Show that if $\mathbf{X}\hat{\hat{\boldsymbol{\beta}}}_{\mathbf{b}} = \hat{\hat{\boldsymbol{\mu}}}_{\mathbf{X}\mathbf{b}}$, then $\hat{\hat{\boldsymbol{\beta}}}_{\mathbf{b}}$ satisfies (11.4).

7. Let $\mathbf{d}(\mathbf{Y})$ be an estimator of $\boldsymbol{\mu}$ and let $\pi(\boldsymbol{\mu}, \sigma^2)$ be a prior distribution. Define $r(\mathbf{d}) = ER(\mathbf{d}; (\boldsymbol{\mu}, \sigma^2))$, where this expectation is taken over the distribution of $(\boldsymbol{\mu}, \sigma^2)$. Let $\mathbf{d}^*(\mathbf{Y}) = E\boldsymbol{\mu} \mid \mathbf{Y}$. Show that

$$r(\mathbf{d}) = E\|\mathbf{d}(\mathbf{Y}) - \mathbf{d}^*(\mathbf{Y})\|^2 + r(\mathbf{d}^*),$$

and hence that $\mathbf{d}^*$ minimizes the Bayes risk.

8. Let $\boldsymbol{\mu} \sim N_n(\mathbf{v}, \tau^2\mathbf{A})$. Show that $P(\boldsymbol{\mu} \in V) = 1$ if and only if $\mathbf{c}'\mathbf{v} = 0$ and $\mathbf{c}'\mathbf{A}\mathbf{c} = 0$ for all $\mathbf{c} \in V$. (Hint: Let $\mathbf{X}$ be a basis matrix for V. How is $\mathbf{X}'\boldsymbol{\mu}$ distributed?)

9. Suppose that

$$\begin{pmatrix} \mathbf{Y} \\ \boldsymbol{\mu} \end{pmatrix} \sim N_{2n}\left(\begin{pmatrix} \mathbf{v} \\ \mathbf{v} \end{pmatrix}, \begin{pmatrix} \delta^2\mathbf{I} + \tau^2\mathbf{A} & \tau^2\mathbf{A} \\ \tau^2\mathbf{A} & \tau^2\mathbf{A} \end{pmatrix} \right).$$

(a) Show that $\mathbf{Y} - \boldsymbol{\mu}$ and $\boldsymbol{\mu}$ are independent and $\mathbf{Y} - \boldsymbol{\mu} \sim N_n(\mathbf{0}, \delta^2\mathbf{I})$, $\boldsymbol{\mu} \sim N_n(\mathbf{v}, \tau^2\mathbf{A})$.

(b) Show that $\mathbf{Y} \mid \boldsymbol{\mu} \sim N_n(\boldsymbol{\mu}, \delta^2\mathbf{I})$ and that $\boldsymbol{\mu} \sim N_n(\mathbf{v}, \tau^2\mathbf{A})$. (Note that you cannot use Lemma 3.7 since $\mathbf{A}$ is not invertible.)

10. (a) Show that $\mathbf{P}_V(\delta^2\mathbf{I} + \tau^2\mathbf{P}_V)^{-1} = (\delta^2 + \tau^2)^{-1}\mathbf{P}_V$.

(b) Show that $\mathbf{X}'(\delta^2\mathbf{I} + \tau^2\mathbf{X}\mathbf{X}')^{-1} = (\delta^2\mathbf{I} + \tau^2\mathbf{X}'\mathbf{X})^{-1}\mathbf{X}'$.

11. (a) Let $\mathbf{U}$ be a random vector with $E\mathbf{U} = \mathbf{a}$, $\text{cov}(\mathbf{U}) = \mathbf{C}$. Show that $E\mathbf{U}'\mathbf{B}\mathbf{U} = \mathbf{a}'\mathbf{B}\mathbf{a} + \text{tr}\,\mathbf{B}\mathbf{C}$.

(b) Show that the risk of the ordinary least-squares estimator $\hat{\boldsymbol{\mu}}$ for the loss function $L_{\mathbf{B}}$ (defined in Section 11.6) is given by $\text{tr}\,\mathbf{B}\mathbf{P}_V$.

(c) Show that the risk of estimator $\hat{\boldsymbol{\beta}}$ for the loss function L_C^* (defined in Section 11.7) is $\text{tr}\,\mathbf{C}(\mathbf{X}'\mathbf{X})^{-1}$.

12. Let $\mathbf{d}(\mathbf{Y})$ be a minimax estimator of the function $\mathbf{v}(\boldsymbol{\theta})$ for the loss function $L(\mathbf{d}; \boldsymbol{\theta})$. Suppose that $\mathbf{d}(\mathbf{Y})$ has constant risk. Show that another estimator $\mathbf{d}^*(\mathbf{Y})$ is as good as $\mathbf{d}(\mathbf{Y})$ for the loss function L if and only if $\mathbf{d}^*(\mathbf{Y})$ is also minimax. (This result implies that a constant risk minimax estimator is the worst minimax estimator.)

13. Verify (11.10).

14. Let

$$\mathbf{X} = \begin{bmatrix} 1 & X_{12} & \cdots & X_{1p} \\ \vdots & \vdots & & \vdots \\ 1 & X_{n2} & \cdots & X_{np} \end{bmatrix}, \qquad \boldsymbol{\beta} = \begin{bmatrix} \alpha \\ \gamma_2 \\ \vdots \\ \gamma_p \end{bmatrix}$$

Define

$$\mathbf{X}^* = \begin{bmatrix} 1/\sqrt{n} & X_{12}^* & \cdots & X_{1p}^* \\ \vdots & \vdots & & \vdots \\ 1/\sqrt{n} & X_{n1}^* & \cdots & X_{2p}^* \end{bmatrix}, \qquad \boldsymbol{\beta}^* = \begin{bmatrix} \alpha^* \\ \gamma_2^* \\ \vdots \\ \gamma_p^* \end{bmatrix}$$

where

$$X_{ij}^* = (X_{ij} - \overline{X}_j)/S_j, \qquad \alpha^* = \sqrt{n}\left(\alpha + \sum_{j=2}^{p} \overline{X}_j \gamma_j\right), \qquad \gamma_j^* = \gamma_j S_j,$$

$$\overline{X}_j = \frac{1}{n}\sum_{i=1}^{n} X_{ij}, \qquad S_j^2 = \sum_{i=1}^{n} X_{ij}^2 - n\overline{X}_j^2.$$

(a) Show that $\mathbf{X}\boldsymbol{\beta} = \mathbf{X}^*\boldsymbol{\beta}^*$.

(b) Now, let $\mathbf{d} = (d_1, \ldots, d_p)'$ be an estimate of $\boldsymbol{\beta}$, and let

$$\mathbf{d}^* = (d_1^*, \ldots, d_p^*)' \text{ where } d_1^* = d_1 + \sum_{j=2}^{p} \overline{X}_j d_j,\ d_j^* = d_j S_j$$

for $j > 1$. Show that

$$\|\mathbf{d}^* - \boldsymbol{\beta}^*\|^2 = n\left(d_1 - \alpha + \sum_{j=1}^{p} \overline{X}_j(d_j - \gamma_j)\right)^2 + \sum_{j=2}^{p} S_j^2(d_j - \gamma_j)^2.$$

(c) Show that $\|\mathbf{d}^* - \boldsymbol{\beta}^*\|^2 = (\mathbf{d} - \boldsymbol{\beta})'C(\mathbf{d} - \boldsymbol{\beta})$, where

$$\mathbf{C} = \begin{bmatrix} n & n\overline{X}_2 & n\overline{X}_3 & \cdots & x\overline{N}_p \\ n\overline{X}_2 & \Sigma X_{i2}^2 & n\overline{X}_2\overline{X}_3 & \cdots & n\overline{X}_2\overline{X}_p \\ n\overline{X}_3 & n\overline{X}_2\overline{X}_3 & \Sigma X_{i3}^2 & \cdots & n\overline{X}_3\overline{X}_p \\ \vdots & \vdots & \vdots & \ddots & \vdots \\ n\overline{X}_p & n\overline{X}_2\overline{X}_p & n\overline{X}_3\overline{X}_p & \cdots & \Sigma X_{ip}^2 \end{bmatrix}.$$

(d) Show that

$$\mathbf{X}'\mathbf{X} = \mathbf{C} + \begin{bmatrix} 0 & 0 & 0 & \cdots & 0 \\ 0 & 0 & S_{23} & \cdots & S_{2p} \\ 0 & S_{23} & 0 & \cdots & S_{3p} \\ \vdots & \vdots & \vdots & \ddots & \vdots \\ 0 & S_{2p} & S_{3p} & \cdots & 0 \end{bmatrix} \qquad S_{ij} = \sum_{i=1}^{n} X_{ij}X_{ik} - n\overline{X}_j\overline{X}_k.$$

15. (a) Show that

$$\tau^2\mathbf{X}\mathbf{A}^{-1}\mathbf{X}'(\delta^2\mathbf{I} + \tau^2\mathbf{X}\mathbf{A}^{-1}\mathbf{X}')^{-1} = (\mathbf{X}'\mathbf{X} + \delta^2/\tau^2\mathbf{A})^{-1}\mathbf{X}'.$$

(b) Show that the generalized ridge estimator $\mu_{\mathbf{v},\mathbf{A}}$ is an empirical Bayes estimator with respect to the prior distribution $\boldsymbol{\mu} \sim N_n(\mathbf{v}, \tau^2\mathbf{X}\mathbf{A}^{-1}\mathbf{X}')$.

Type C

1. We follow the notation of the proof of Theorem 11.3.

(a) Using the normal density function show that

$$\frac{\partial}{\partial\theta_i} E\frac{1}{\|\mathbf{U}\|^2} = E\frac{U_i - \theta}{\|\mathbf{U}\|^2}.$$

(b) Using the Poisson density function, show that

$$\frac{\partial}{\partial\theta_i} E\frac{1}{p-2-2K} = \frac{\theta_i}{\|\boldsymbol{\theta}\|^2} E\frac{2K - \|\boldsymbol{\theta}\|^2}{p-2+2K}.$$

(You may pull derivatives inside the integrals and sums because both the Poisson and normal distributions are exponential families.)

2. Suppose that $\mathbf{Z} \sim N_p(\boldsymbol{\mu}, \mathbf{I})$, where $\boldsymbol{\mu} \in R^p$.

(a) Show that $\mathbf{Z}$ is the minimum variance unbiased estimator and the MLE of $\boldsymbol{\mu}$.

(b) Show that $E\|\mathbf{Z}\|^2 = \|\boldsymbol{\mu}\|^2 + p$.

(c) Consider the class of estimators of the form $d_c = (1 - c/\|\mathbf{Z}\|^2)\mathbf{Z}$. Find c such that $E\|\mathbf{d}_c - \boldsymbol{\mu}\|^2$ is minimized.

3. We follow the notation of Section 11.4 with $\mathbf{A} = \mathbf{P}_V$.

(a) Show that $E\hat{\sigma}^2 = \delta^2$.

(b) Show that the marginal distribution of $\hat{\boldsymbol{\mu}}$ is given by

$$\hat{\boldsymbol{\mu}} \sim N_n\big(\mathbf{v}, (\delta^2 + \tau^2)\mathbf{P}_V\big).$$

(c) Show that $\|\hat{\boldsymbol{\mu}} - \mathbf{v}\|^2 \sim (\delta^2 + \tau^2)\chi^2_p$. (Hint: Let $\mathbf{X}$ be an orthonormal basis matrix for V. Then $\|\hat{\boldsymbol{\mu}} - \mathbf{v}\|^2 = \|\mathbf{X}'(\hat{\boldsymbol{\mu}} - \mathbf{v})\|^2$. How is $\mathbf{X}'(\hat{\boldsymbol{\mu}} - \mathbf{v})$ distributed?)

(d) Show that

$$E\frac{p-2}{\|\hat{\boldsymbol{\mu}} - \mathbf{v}\|^2} = \frac{1}{\delta^2 + \tau^2}.$$

(e) Show that

$$E\frac{(n-p)(p-2)\hat{\sigma}^2}{(n-p+2)\|\hat{\boldsymbol{\mu}}-\mathbf{v}\|^2}=\frac{(n-p)\sigma^2}{(n-p+2)(\delta^2+\tau^2)}\approx\frac{\sigma^2}{(\delta^2+\tau^2)}.$$

4. Let $\mathbf{U}(m)$ be a matrix-valued function of the real variable m. Let $(d/dm)\mathbf{U}(m)$ be the matrix whose (i, j) element is the derivative of the (i, j) element of $\mathbf{U}$. Verify the following identities:

(a) $(d/dm)[\mathbf{U}(m)\mathbf{V}(m)] = \mathbf{U}(m)[(d/dm)\mathbf{V}(m)] + [(d/dm)\mathbf{U}(m)]\mathbf{V}(m)$. Note that $[(d/dm)\mathbf{U}(m)]\mathbf{V}(m) \neq \mathbf{V}(m)[(d/dm)\mathbf{U}(m)]$. [Hint: Find the derivative of the (i, j) component of $\mathbf{U}(m)\mathbf{V}(m)$.]

(b) $(d/dm)(\mathbf{A}\mathbf{U}(m)\mathbf{B}) = \mathbf{A}[(d/dm)\mathbf{U}(m)]\mathbf{B}$.

(c) $(d/dm)\{[\mathbf{U}(m)]^{-1}\} = -[\mathbf{U}(m)]^{-1}[(d/dm)\mathbf{U}(m)][\mathbf{U}(m)]^{-1}$. {Hint: differentiate both sides of $\mathbf{U}(m)[\mathbf{U}(m)]^{-1} = \mathbf{I}$.}

(d) $(d/dm)[\operatorname{tr}\mathbf{U}(m)] = \operatorname{tr}[(d/dm)\mathbf{U}(m)]$.

5. We follow the notation of Theorem 11.7. Let

$$\mathbf{U}(m) = \mathbf{X}(\mathbf{X}'\mathbf{X} + m\mathbf{A})^{-1}\mathbf{X}',$$

and let $\mathbf{V}(m) = \mathbf{U}(m)\mathbf{Y} - \boldsymbol{\mu}$.

(a) Show that $E\mathbf{V}(m) = [\mathbf{U}(m) - \mathbf{I}]\boldsymbol{\mu}$, $\operatorname{cov}[\mathbf{V}(m)] = \sigma^2[\mathbf{U}(m)]^2$.

(b) Show that

$$h(m)=\frac{\boldsymbol{\mu}'[\mathbf{U}(m)-\mathbf{I}]\mathbf{B}[\mathbf{U}(m)-\mathbf{I}]\boldsymbol{\mu}}{\sigma^2}+\operatorname{tr}[\mathbf{U}(m)]^2\mathbf{B}$$

(See Exercise B11a.)

(c) Show that

$$h'(m)=\frac{2\boldsymbol{\mu}'[\mathbf{U}(m)-\mathbf{I}]\mathbf{B}\left(\frac{d}{dm}\mathbf{U}(m)\right)\boldsymbol{\mu}}{\sigma^2}+2\operatorname{tr}\mathbf{U}(m)\left[\frac{d}{dm}\mathbf{U}(m)\right]\mathbf{B}.$$

(d) Show that $(d/dm)\mathbf{U}(m) = -\mathbf{X}(\mathbf{X}'\mathbf{X} + m\mathbf{A})^{-1}\mathbf{A}(\mathbf{X}'\mathbf{X} + m\mathbf{A})^{-1}\mathbf{X}'$.

(e) Verify (11.9).

CHAPTER 12

Inference Based on the Studentized Range Distribution and Bonferroni's Inequality

In this chapter we consider some alternative methods for drawing inference about means. For most of the chapter, we limit discussion to the balanced one way analysis of variance model, although the procedures that we suggest are applicable for drawing inference about main effects in any balanced analysis of variance model with equal weights. In Section 12.1, we define the studentized range distribution. In Section 12.2, we use this distribution in the design of the studentized range test, and in Section 12.3, we apply the distribution to derive Tukey-type simultaneous confidence intervals for the set of all contrasts. In Section 12.4, we give an introduction to multiple comparisons. We establish the asymptotic validity of procedure based on the studentized range in Section 12.5. In Section 12.6, we derive Bonferroni's inequality and show how it can be used to find appropriate simultaneous confidence intervals.

12.1. THE STUDENTIZED RANGE DISTRIBUTION.

Let $Z_1, \ldots, Z_k$ and U be independent random variables with $Z_i \sim N_1(0,1)$, and $U \sim \chi_m^2(0)$. Define

$$q = \max_{i \neq j} \frac{|Z_i - Z_j|}{\sqrt{U/m}}$$

We say that q has a *studentized range distribution* and write that

$$q \sim q_{k,m}$$

Tables of critical values for this distribution can be found in most textbooks in applied linear models, (for example, Neter and Wasserman, 1974, pp. 824–825). Throughout this chapter, let $q^{\alpha}_{k,m}$ be the upper α point of this studentized range distribution.

LEMMA 12.1. Let $Y_1, \ldots, Y_k$ and S^2 be independent with

$$Y_i \sim N(\mu, a\sigma^2), \qquad mS^2 \sim \sigma^2\chi^2_m(0).$$

Then

$$\max_{i \neq j} \frac{|Y_i - Y_j|}{\sqrt{a}\, S} \sim q_{k,m}$$

PROOF. See Exercise B1. □

Note that in Lemma 12.1, the Y_i must all have the same variance. This restriction limits the exact application of the studentized range distribution to balanced models.

12.2. THE STUDENTIZED RANGE TEST.

The balanced one-way analysis of variance model is one in which we observe Y_{ij} independent, $i = 1, \ldots, k$; $j = 1, \ldots, n$; with

$$Y_{ij} \sim N_1(\theta + \alpha_i, \sigma^2), \qquad \sum_i \alpha_i = 0$$

(Note that we are assuming that each class has the same number, n, of observations.) In this section we derive a size α test that the α_i are 0. The UMP invariant size α test for this problem is derived in Section 7.3.1.

Let

$$\overline{Y}_{i\cdot} = \frac{1}{n} \sum_{j=1}^{n} Y_{ij}, \qquad \hat{\sigma}^2 = \frac{1}{k(n-1)} \sum_{i=1}^{k} \sum_{j=1}^{n} \left(Y_{ij} - \overline{Y}_{i.}\right)^2 \qquad (12.1)$$

THEOREM 12.2. a. $(\overline{Y}_{1.}, \ldots, \overline{Y}_{k.}, \hat{\sigma}^2)$ is a complete sufficient statistic for the one-way model.

b. $\overline{Y}_{1.}, \ldots, \overline{Y}_{k.}$ and $\hat{\sigma}^2$ are independent.

$$\overline{Y}_{i.} \sim N_1\left(\theta + \alpha_i, (1/n)\sigma^2\right), \qquad k(n-1)\hat{\sigma}^2 \sim \sigma^2\chi^2_{k(n-1)}(0).$$

PROOF. a. By the discussion in Section 7.3.1, we see that $\hat{\boldsymbol{\mu}}$ is a vector whose first n components are $\overline{Y}_{1.}$, next n components are $\overline{Y}_{2.}$, and so on, and that $\hat{\sigma}^2$ is given by (12.1). Therefore, $(\overline{Y}_{1.}, \ldots, \overline{Y}_{k.}, \hat{\sigma}^2)$ is an invertible function of the complete sufficient statistic $(\hat{\mu}, \hat{\sigma}^2)$ and hence is also a sufficient statistic.

b. See Exercise B1. □

From part b and Lemma 12.1, we have the following corollary.

COROLLARY.

$$\max_{i \neq j} \frac{\sqrt{n}\,|\overline{Y}_{i.} - \overline{Y}_{j.} - (\alpha_i - \alpha_j)|}{\hat{\sigma}} \sim q_{k,k(n-1)}.$$

Now, let

$$q = \max_{i \neq j} \sqrt{n}\, \frac{|\overline{Y}_{i.} - \overline{Y}_{j.}|}{\hat{\sigma}}. \tag{12.2}$$

Under the null hypothesis that the α_i are all 0, we see that

$$q \sim q_{k,k(n-1)}$$

and therefore, a size α test for this problem is given by

$$\phi(q) = \begin{cases} 1 & \text{if } q > q^{\alpha}_{k,k(n-1)} \\ 0 & \text{if } q \leqslant q^{\alpha}_{k,k(n-1)} \end{cases}. \tag{12.3}$$

The test ϕ defined by (12.2) and (12.3) is called the *studentized range* test for this problem. Although the F-test derived in Section 7.3.1 is UMP invariant size α, it is not necessarily more powerful than the studentized range test, since the studentized range test is not invariant.

One drawback of the studentized range test is that the alternative distribution of q has not been studied as much as the alternative distribution of the F-statistic (i.e., as the noncentral F-distribution). One nice property of the alternative distribution of the F-statistic is that it depends on only one parameter, the noncentrality parameter. The alternative distribution of q does not seem to share this property.

12.3. SIMULTANEOUS CONFIDENCE INTERVALS—TUKEY TYPE.

We continue with the balanced one-way analysis of the variance model described in the last section. We now find a set of simultaneous confidence intervals for the set of all contrasts associated with testing that the $\alpha_i = 0$. These simultaneous confidence intervals were suggested by Tukey (1953) and are called Tukey simultaneous confidence intervals. We first find simultaneous confidence intervals for the set of all comparisons, $\alpha_i - \alpha_j$.

THEOREM 12.3. $P(\alpha_i - \alpha_j \in \overline{Y}_{i\cdot} - \overline{Y}_{j\cdot} \pm (\hat{\sigma}/\sqrt{n}) q^{\alpha}_{k,k(n-1)}$ for all $i \neq j) = 1 - \alpha$.

PROOF.

$$P\left(\alpha_i - \alpha_j \in \overline{Y}_{i\cdot} - \overline{Y}_{j\cdot} \pm \left(\hat{\sigma}/\sqrt{n}\right) q^{\alpha}_{k,k(n-1)} \qquad \text{for all } i \neq j\right)$$

$$= P\left(\frac{\sqrt{n}\,|\overline{Y}_{i\cdot} - \overline{Y}_{j\cdot} - (\alpha_i - \alpha_j)|}{\hat{\sigma}} \leqslant q^{\alpha}_{k,k(n-1)} \qquad \text{for all } i \neq j\right)$$

$$= P\left(\max_{i \neq j} \frac{\sqrt{n}\,|\overline{Y}_{i\cdot} - \overline{Y}_{j\cdot} - (\alpha_i - \alpha_j)|}{\hat{\sigma}} \leqslant q^{\alpha}_{k,k(n-1)}\right) = 1 - \alpha. \quad \square$$

Therefore, the intervals

$$\alpha_i - \alpha_j \in \overline{Y}_{i\cdot} - \overline{Y}_{j\cdot} \pm \frac{\hat{\sigma}}{\sqrt{n}} q^{\alpha}_{k,k(n-1)}$$

are a set of $(1 - \alpha)$ simultaneous confidence intervals for the set of all comparisons. They are shorter than the Scheffé intervals derived in Section 8.2.1 (see Exercise B2) which is not surprising, since the Tukey intervals are exact, whereas the Scheffé intervals are conservative (i.e., the confidence coefficient for the Scheffé type intervals is greater than $1 - \alpha$). These intervals can be used after the hypothesis of no mean differences has been rejected, either with the F-test derived in Section 7.3.1 or with the studentized range test discussed in the last section. If the hypothesis of equal means has been rejected, then we would define a pair of means $\mu_i = \theta + \alpha_i$ and $\mu_j = \theta + \alpha_j$ to be significantly different if the Tukey interval for $\alpha_i - \alpha_j$ does not contain 0. The studentized range test rejects the null hypothesis if and only if at least one of these intervals does not contain 0, so that if we reject with the studentized range test, we are sure to find at least one pair of means to be different with the Tukey intervals. If we use the F-test first, then we may find no significant contrasts even when the F-test is rejected. The Tukey intervals are shorter than the Scheffé intervals, and therefore more comparisons are significant with the Tukey intervals than the Scheffé intervals.

A contrast in the means is a function $\sum_i c_i \alpha_i$ where $\sum_i c_i = 0$. We now extend the Tukey confidence intervals to the set of all contrasts by using the following lemma.

LEMMA 12.4. Let $a_1, \ldots, a_k$ be numbers. Then

$$|a_i - a_j| \leqslant b \qquad \text{for all } i \text{ and } j$$

if and only if

$$\left|\sum_i c_i a_i\right| \leqslant b \sum_i \frac{|c_i|}{2} \qquad \text{for all } c_i \text{ such that } \sum_i c_i = 0.$$

PROOF. If

$$\left|\sum_i c_i a_i\right| \leqslant b \sum_i \frac{|c_i|}{2},$$

then

$$|a_i - a_j| = |a_i + (-a_j)| \leqslant b\frac{1+1}{2} = b.$$

Conversely, suppose that $|a_i - a_j| \leqslant b$ for all i and j. Let c_i be such that $\sum_i c_i = 0$. If $c_i = 0$ for all i, then $|\sum_i c_i a_i| = b\sum_i |c_i|/2$ and the condition is satisfied. Now let c_i be such that some $c_i \neq 0$. Let P be the set of subscripts in which $c_i > 0$ and N be the set in which $c_i < 0$. Let $g = \sum_i |c_i|/2$. Then

$$0 = \sum_{i \in P} c_i + \sum_{i \in N} c_i, \qquad 2g = \sum_{i \in P} c_i - \sum_{i \in N} c_i = 2\sum_{i \in P} c_i$$

and therefore

$$g = \sum_{i \in P} c_i = -\sum_{i \in N} c_i > 0.$$

Now

$$g\sum_i c_i a_i = g\left(\sum_{i \in P} c_i a_i + \sum_{j \in N} c_j a_j\right) = \sum_{i \in P}\sum_{j \in N} c_i(-c_j)a_i + \sum_{i \in P}\sum_{j \in N} c_i c_j a_j$$

$$= \sum_{i \in P}\sum_{j \in N} -c_i c_j (a_i - a_j).$$

However, if $i \in P, j \in N$, then

$$|-c_i c_j (a_i - a_j)| = -c_i c_j |a_i - a_j| \leqslant -c_i c_j b,$$

and

$$\left|g\sum_i c_i a_i\right| = \left|\sum_{i \in P}\sum_{j \in N} -c_i c_j (a_i - a_j)\right| \leqslant \sum_{i \in P}\sum_{j \in N} |-c_i c_j (a_i - a_j)|$$

$$\leqslant b\sum_{i \in P}\sum_{j \in N} -c_i c_j = bg^2.$$

Therefore

$$\left|\sum_i c_i a_i\right| \leqslant bg. \quad \square$$

COROLLARY.

$$P\left(\sum_i c_i\alpha_i \in \sum_i c_i \bar{Y}_{i.} \pm \frac{\hat{\sigma}}{\sqrt{n}} q^{\alpha}_{k,k(n-1)} \frac{\sum_i |c_i|}{2} \quad \text{for all } c_i \text{ such that } \sum_i c_i = 0\right)$$
$$= 1 - \alpha.$$

PROOF.

$$P\left(\sum_i c_i\alpha_i \in \sum_i c_i \bar{Y}_{i.} \pm \frac{\hat{\sigma}}{\sqrt{n}} q^{\alpha}_{k,k(n-1)} \frac{\sum_i |c_i|}{2} \quad \text{for all } c_i \text{ such that } \sum_i c_i = 0\right)$$
$$= P\left(\left|\sum_i c_i(\bar{Y}_{i.} - \mu_i)\right| \leqslant \frac{\hat{\sigma}}{\sqrt{n}} q^{a}_{k,k(n-1)} \frac{\sum_i |c_i|}{2} \quad \text{for all } c_i \text{ such that } \sum_i c_i = 0\right)$$
$$= P\left(|Y_{i.} - \bar{Y}_{j.} - (\mu_i - \mu_j)| \leqslant \frac{\hat{\sigma}}{\sqrt{n}} q^{\alpha}_{k,k(n-1)} \quad \text{for all } i \neq j\right) = 1 - \alpha$$

where the next to last inequality follows from Lemma 12.4, and the last inequality follows from the proof of Theorem 12.3 Therefore, the intervals

$$\sum_i c_i\alpha_i \in \sum_i c_i \bar{Y}_{i.} \pm \frac{\hat{\sigma}}{\sqrt{n}} q^{\alpha}_{k,k(n-1)} \sum_i \frac{|c_i|}{2}$$

are a set of $1 - \alpha$ confidence intervals for the set of all contrasts. Note that the intervals for the comparisons (i.e., contrasts in which $c_i = 1$, $c_j = -1$ and $c_k = 0$ for $k \neq i$, $k \neq j$) reduce to the intervals derived in Theorem 12.3.

We now compare these Tukey intervals with the Scheffé intervals derived in Section 8.3.1. The midpoints of the two sets of intervals are the same. Therefore, the better set of intervals is the set with shorter intervals. The lengths of the Tukey and Scheffé intervals are

$$\frac{\hat{\sigma}}{\sqrt{n}} q^{\alpha}_{k,k(n-1)} \sum_i |c_i|, \qquad \frac{2\hat{\sigma}}{\sqrt{n}} \left[(k-1) F^{\alpha}_{k-1,k(n-1)} \sum_i c_i^2\right]^{1/2}$$

respectively. The Tukey intervals are shorter for some choices of c_i and the Scheffé intervals are shorter for others. In general, the Tukey intervals are shorter for simple contrasts in which most of the c_i are 0, while the Scheffé intervals are shorter for more complicated contrasts. See Scheffê (1959) for some numerical comparisons of the lengths of the intervals.

It should again be emphasized that Tukey simultaneous confidence intervals are only exact for contrasts in main effects in balanced analysis of variance models.

12.4. MULTIPLE COMPARISONS.

We continue with the balanced one-way analysis of variance model discussed in Sections 12.2 and 12.3. In this section we consider the problem of testing whether the $\mu_i = \theta + \alpha_i$ are different and, if so, then deciding which ones are different. This problem is called the *multiple comparisons* problem. A procedure for this problem is a rule that for each i and j determines whether μ_i and μ_j are significantly different. We say that a procedure makes a *type I error* any time it decides two means are significantly different when those means are in fact the same. The procedure makes a *type II error* when it decides that two means are the same when they are different. It is possible for a procedure to make both a type I and type II error with the same data. For example, if $k = 3$, $\mu_1 = \mu_2 \neq \mu_3$, and the procedure decided that μ_1 and μ_2 are significantly different, but that μ_2 and μ_3 are not, then this procedure would be making both a type I and a type II error.

Consider a procedure d. For any $\mu_1, \ldots, \mu_k, \sigma^2$, define

$$\alpha(\mu_1, \ldots, \mu_k, \sigma^2)$$

to be the probability that the procedure d makes at least one type I error. If the μ_i are all different, then this probability would be zero. Traditionally it has been felt that the most serious type I error occurs when we find at least one significant difference when all the means are the same (see Miller (1966)). The *size* α^* of a procedure d is defined to be the supremum of this probability. That is,

$$\alpha^* = \sup_{\mu_1 = \cdots = \mu_k} \alpha(\mu_1, \ldots, \mu_k, \sigma^2).$$

More recent researchers have felt that any type I error is serious (see Gabriel, 1969). We define the *generalized size* α of a procedure d to be the supremum of the probability of any type I error. That is,

$$\alpha = \sup \alpha(\mu_1, \ldots, \mu_k, \sigma^2).$$

In this section, we look at procedures that have generalized size α. In this class of procedures, we search for procedures that make few type II errors. Unfortunately, there is no natural power function that can be used to compare different procedures. [For two different definitions of power, see Einot and Gabriel (1975) and Ramsey (1978). The conclusions of these papers are somewhat different due to different definitions of power.] Now let d_1 and d_2 be two different procedures. There is one situation in which it is clear that d_1 is more powerful than d_2 for any sensible definition of power. That situation occurs when any significant difference for d_2 is also a significant difference for d_1. We therefore say that d_1 is *more powerful than*

d_2 if d_1 and d_2 are not identical and any significant difference with d_2 is also a significant difference with d_1.

We now look at some procedures that have been suggested for multiple comparisons. The first procedure is the *Scheffé* procedure which declares μ_i and μ_j to be significantly different if the $(1 - \alpha)$ Scheffé simultaneous confidence interval for $\mu_i - \mu_j = \alpha_i - \alpha_j$ does not contain 0. This procedure makes a type I error only if 0 is not in some Scheffé confidence interval when $\mu_i - \mu_j = 0$. Since these confidence intervals are $(1 - \alpha)$ simultaneous intervals, the probability of such an event much be at most α. Hence the Scheffé procedure has generalized size at most α. (Actually the generalized size of this procedure is less than α, since the Scheffé confidence intervals are not exact for the set of all comparisons.) The second procedure that we consider is the *Tukey* procedure, which declares μ_i and μ_j to be significantly different if the $(1 - \alpha)$ Tukey-type confidence interval for $\mu_i - \mu_j = \alpha_i - \alpha_j$ does not contain 0. By the same argument as given above for the Scheffé procedure, the Tukey procedure has generalized size at most α. However, this procedure finds at least one significant difference if and only if the studentized range rejects the hypothesis of equal means. Therefore, the size of the Tukey procedure is exactly α. Since the generalized size must be at least as large as the actual size, we see that the generalized size of the Tukey procedure is exactly α. Both the Tukey and Scheffé confidence intervals have the same midpoints $(\overline{Y}_{i\cdot} - \overline{Y}_{j\cdot})$, and the Tukey intervals are shorter. Therefore, if 0 is not in the Scheffé intervals, then 0 is not in the Tukey intervals. Hence the Tukey procedure is more powerful than the Scheffé procedure.

The third procedure is presented primarily for historical interest. It was suggested in Fisher (1935) and was probably the first multiple comparison procedure suggested. The *Fisher least significant difference* procedure first uses a size α F-test to test the equality of the means. If this test is accepted, then the means are declared to be insignificantly different. If the test is rejected, then individual size α two sample t-tests are used to determine which means are different. Clearly, we find at least one significant difference only if the hypothesis is rejected with the F-test. Therefore, the size of this procedure is at most α. [It is possible (but not probable) that the F-test is rejected when none of the individual t-tests is significant. Therefore, the size of this procedure is less than α.] However, the generalized size of this procedure is quite large. To illustrate the reason for the large generalized size, let $k = 10$, $\mu_1 = \mu_2 = \cdots = \mu_9$, and let $\mu_{10} - \mu_9 = 1000\sigma$. Then we would be almost sure to reject the F-test. The probability that at least one of the 36 pairs of the μ_i, $i = 1, \ldots, 9$ would be found significant with individual size α t-tests is considerably larger than α.

We now define two classes of multiple comparison procedures, the multiple range procedures and multiple F procedures. Let S be a subset of $(1, \ldots, k)$ with s elements. We say that S is *F-significant* at size α_s if

$$F_S = \frac{\sum_{i \in S}\left(\bar{Y}_{i\cdot} - \bar{Y}^S_{\cdot\cdot}\right)^2}{(s-1)\hat{\sigma}^2} > F^{\alpha_s}_{s-1,k(n-1)}, \qquad \bar{Y}^S_{\cdot\cdot} = \frac{1}{s}\sum_{i \in S} \bar{Y}_{i\cdot} \quad (12.4)$$

That is, S if F-significant if the hypothesis that $\mu_i = \mu_j$ for all $i, j \in S$ would have been rejected with a size α_s F-test. (Note that the denominator is $\hat{\sigma}^2$ which is computed from all classes.) Similarly, we say that S is *q-significant* at size α_s if

$$q_S = \max_{(i,j) \in S} \frac{\sqrt{n}\,|\bar{Y}_{i\cdot} - \bar{Y}_{j\cdot}|}{\hat{\sigma}} > q^{\alpha_s}_{s,k(n-1)}. \qquad (12.5)$$

That is, S is q-significant if the hypothesis that $\mu_i = \mu_j$ for all $i, j \in S$ is rejected with a size α_s studentized range test. A procedure that declares μ_i and μ_j to be significantly different if and only if every subset S containing i and j is F-significant at size α_s is called a *multiple F-test*. The choice of α_s depends on the particular multiple F test chosen. Similarly, a procedure that declares μ_i and μ_j to be significantly different if and only if every subset S is q-significant at size α_s is called a *multiple range test*. We now present several choices for α_s that have been suggested.

The *Newman-Keuls multiple F-test* and *multiple range test* choose $\alpha_s = \alpha$. The Newman-Keuls multiple range test was the first test of this sort suggested (see Newman, 1938, and Keuls, 1952). Since either of these procedures find at least one significant difference only if the whole set $(1, \ldots, k)$ is found to be significant, both these procedures have size at most α. (Actually, the multiple range test has size exactly α.) However, neither of these procedures has generalized size α, unless $k \leqslant 3$. To illustrate why not, suppose that $k = 4$, that $\mu_1 = \mu_2$, $\mu_3 = \mu_4$, and that $\mu_2 - \mu_3 = 1000\sigma$. With either procedure, the whole set is found to be significant with probability very near one, as are the subsets with three elements. Therefore, $\alpha(\mu_1, \mu_2, \mu_3, \mu_4, \sigma^2)$ would be essentially the probability of finding either $(1,2)$ or $(3,4)$ (q- or F-) significant. Since the probability of finding either (q- or F-) significant is α, $\alpha(\mu_1, \mu_2, \mu_3, \mu_4, \sigma^2)$ is considerably greater than α, and probably near 2α. Hence the generalized size is considerably greater than α.

The *Ryan multiple F-test* and *multiple range test* chooses $\alpha_s = (s\alpha)/k$ (see Ryan, 1960). We now show that both these procedures have generalized size at most α, that is, that $\alpha(\mu_1, \ldots, \mu_k, \sigma^2) \leqslant \alpha$. Fix $\mu_1, \ldots, \mu_k$ and σ^2. If the μ_i are all different, this result is trivially true. If the μ_i are not all different, partition the set $(1, \ldots, k)$ into disjoint sets $S_0, S_1, \ldots, S_t$ such that S_0 consists of all i such that μ_i is not equal to any other mean, if i and

i' are in S_j, $j \geqslant 1$, then $\mu_i = \mu_{i'}$, and if $i \in S_j$, $i' \in S_{j'}$, $j \neq j'$, then $\mu_i \neq \mu_{i'}$. That is S_1 is a cluster of equal means, S_2 is a cluster of equal means, and so on. Let s_j be the number of elements in S_j. The Ryan multiple F or multiple range test makes a type I error if and only if at least one of the S_i, $i \geqslant 1$ is found to be (F- or q-) significant. Therefore,

$$\alpha(\mu_1, \ldots, \mu_k, \sigma^2) = P \text{ (at least one } S_i \text{ significant)}$$
$$\leqslant \sum_i P\,(S_i \text{ significant}) \leqslant \sum_i \frac{s_i \alpha}{k} \leqslant \alpha.$$

Hence, the Ryan multiple F- and multiple range tests both have generalized size at most α. If $q^{s\alpha/k}_{s,k(n-1)} \leqslant q^{\alpha}_{k,k(n-1)}$ for all $s \leqslant k$, then the Ryan multiple range test has generalized size exactly α (see Exercise B9). We now show in addition that if $q^{s\alpha/k}_{s,k(n-1)} \leqslant q^{\alpha}_{k,k(n-1)}$, then the Ryan multiple range test is more powerful than the Tukey procedure. Suppose that the Tukey procedure declares that μ_i and μ_j are significantly different. Then $\sqrt{n}\,|Y_{i\cdot} - Y_{j\cdot}|\hat{\sigma} > q^{\alpha}_{k,k(n-1)}$. Now, let S be a set containing i and j. Then

$$\max_{i',j' \in S} \frac{\sqrt{n}\,|Y_{i'\cdot} - Y_{j'\cdot}|}{\hat{\sigma}} \geqslant \frac{\sqrt{n}\,|\overline{Y}_{i\cdot} - \overline{Y}_{j\cdot}|}{\hat{\sigma}} > q^{\alpha}_{k,k(n-1)} \geqslant q^{s\alpha/k}_{s,k(n-1)}.$$

Hence, under these conditions, every subset S containing i and j would be q-significant at size $\alpha = s\alpha/k$. Therefore, if $q^{s\alpha/k}_{s,k(n-1)} \leqslant q^{\alpha}_{k,k(n-1)}$, then the Ryan multiple range test is more powerful than the Tukey procedure.

The *Duncan multiple F-test* and *multiple range test* choose

$$\alpha_s = 1 - (1 - \alpha)^{(s-1)/(k-1)}$$

(see Duncan, 1955). These procedures also both have generalized size at most α (see Exercise C2). If $q^{\alpha_s}_{s,k(n-1)} \leqslant q^{\alpha}_{k,k(n-1)}$ for all $s \leqslant k$, then the Duncan multiple range test has generalized size exactly α and is more powerful than the Tukey procedure. Duncan suggests using $\alpha = 1 - (0.95)^{k-1}$ for this procedure (and labels it $\alpha = 0.05$). This choice of α can be quite large. For example, for $k = 10$, $\alpha = 0.37$ and for $k = 25$, $\alpha = 0.71$. Naturally, with this choice of α, the Duncan multiple range and multiple F procedures are quite powerful. However, they would also make many type I errors. Although it is not too difficult to imagine situations in which it might be sensible to let α get large as k gets large, this particular choice of α seems to get large too quickly.

The *modified Ryan multiple F-test* and *multiple range test* choose

$$\alpha_s = 1 - (1 - \alpha)^{s/k}$$

(see Einot and Gabriel, 1975). This procedure also has generalized size α

(see Exercise C2). Since

$$1-(1-\alpha)^{s/k}>1-(1-\alpha)^{(s-1)/(k-1)}, \qquad 1-(1-\alpha)^{s/k}>\frac{s}{k}\alpha,$$

(see Exercise B4), the modified Ryan multiple range test is more powerful than either the Ryan multiple range test or the Duncan multiple range test, and the modified Ryan multiple F-test is more powerful than the Ryan or Duncan multiple F-tests.

One advantage of the multiple range tests over the multiple F-tests is that there is a fairly simple algorithm for performing them which we now describe. Consider a multiple range test with a general α_s. Define

$$c_s = \max_{2 \leqslant i \leqslant s} q_{i,k(n-1)}^{\alpha_i}$$

[often $c_s = q_{s,k(n-1)}^{\alpha_s}$]. We perform the multiple range test in the following fashion. Let $\overline{Y}_{[1]} \leqslant \cdots \leqslant \overline{Y}_{[k]}$ be the order statistics computed from the $\overline{Y}_{i\cdot}$, and let $\mu_{[i]}$ be the mean of the class from which $\overline{Y}_{[i]}$ comes. (Note that $\mu_{[i]}$ is not necessarily the ith ordered mean, since the $\overline{Y}_{i\cdot}$ may be in "incorrect" order.) At the first stage of the multiple range test, we see whether $\overline{Y}_{[k]} - \overline{Y}_{[1]} > c_k\hat{\sigma}/\sqrt{n}$. If not, all the means would be declared the same and the procedure would stop. On the other hand, if $\overline{Y}_{[k]} - \overline{Y}_{[1]} > c_k\hat{\sigma}/\sqrt{n}$, then we would declare $\mu_{[1]}$ and $\mu_{[k]}$ significantly different. We would now proceed to stage 2, in which we check whether $\overline{Y}_{[k-1]} - \overline{Y}_{[1]} > c_{k-1}\hat{\sigma}/\sqrt{n}$ and $\overline{Y}_{[k]} - \overline{Y}_{[2]} > c_{k-1}\hat{\sigma}/\sqrt{n}$. If neither of these is true, we stop and declare the only significantly different means to be $\mu_{[1]}$ and $\mu_{[k]}$. If $\overline{Y}_{[k-1]} - \overline{Y}_{[1]} \leqslant c_{k-1}\hat{\sigma}/\sqrt{n}$, then the means $\mu_{[1]}, \ldots, \mu_{[k-1]}$ would be declared to be insignificantly different. (As a rather simple method of keeping track of which means are insignificantly different, it is helpful to underline those means that are insignificantly different. It is not necessary to check whether any subsets of an underlined set of means is significant.) On the other hand, if $\overline{Y}_{[k-1]} - \overline{Y}_{[1]} > c_{k-1}\hat{\sigma}/\sqrt{n}$, then the means $\mu_{[k-1]}$ and $\mu_{[1]}$ would be declared significantly different. We make a similar decision for $\mu_{[k]}$ and $\mu_{[2]}$. If at least one pair of means is declared significantly different at stage 2, we proceed to stage 3. If $\mu_{[k-1]}$ and $\mu_{[1]}$ are declared significantly different at stage 2, we look at whether $\overline{Y}_{[k-2]} - \overline{Y}_{[1]} > c_{k-2}\hat{\sigma}/\sqrt{n}$. If $\mu_{[k]}$ and $\mu_{[2]}$ are declared significantly different at stage 2, we look at whether $\overline{Y}_{[k]} - \overline{Y}_{[3]} > c_{k-2}\hat{\sigma}/\sqrt{n}$. If $\mu_{[k-1]}$ and $\mu_{[1]}$ are declared significantly different and $\mu_{[k]}$ and $\mu_{[2]}$ are declared significantly different at stage 2, we look at whether $\overline{Y}_{[k-2]} - \overline{Y}_{[1]} > c_{k-2}\hat{\sigma}/\sqrt{n}$, $\overline{Y}_{[k-1]} - \overline{Y}_{[2]} > c_{k-2}\hat{\sigma}/\sqrt{n}$ and $\overline{Y}_{[k]} - \overline{Y}_{[3]} > c_{k-2}\hat{\sigma}/\sqrt{n}$. On the basis of these inequalities, we decide whether $\mu_{[k-2]}$ and $\mu_{[1]}$ are significantly different in the first case, whether $\mu_{[k]}$ and $\mu_{[3]}$ are significantly different in the second case, and which of $\mu_{[k]}$ and $\mu_{[3]}$, $\mu_{[k-1]}$ and $\mu_{[2]}$, or $\mu_{[k-2]}$ and $\mu_{[1]}$ are significantly different in the third case. We proceed in this fashion until the procedure is stopped, either by having found all possible significant differences or by arriving at the $(k-1)$ stage.

As an example, suppose that $k = 6$ and the $\overline{Y}_{i\cdot}$ are 1, 3, 9, 11, 16, 21 and that the $c_s\hat{\sigma}/\sqrt{n}$ are given by $c_6\hat{\sigma}/\sqrt{n} = 18$, $c_5\hat{\sigma}/\sqrt{n} = 14$, $c_4\hat{\sigma}/\sqrt{n} = 11$, $c_3\hat{\sigma}/\sqrt{n} = 8$ and $c_2\hat{\sigma}/\sqrt{n} = 6$. Then the analysis would look like the following

$$\underline{1\quad 3\quad 9\quad 11}\quad 16\quad 21. \tag{12.6}$$

Two means are declared significantly different if and only if there is no underline connecting them. In this example, the means associated with the following pairs would be declared significantly different: (1, 16), (1, 21), (3, 16), (3, 21) (9, 21), and (11, 21). From this example, we see that this algorithm for the multiple range tests is a fairly quick and easy way to determine which means, if any, are significantly different.

We now show why the algorithm discussed above is equivalent to the multiple range test defined earlier. Let i_s be the i such that $c_s = q^{\alpha_i}_{i,k(n-1)}$. If $\overline{Y}_{[k]} - \overline{Y}_{[1]} \leqslant c_k\hat{\sigma}/\sqrt{n}$, then no subset with i_k elements could be significant, and hence all means would be declared insignificantly different. On the other hand, suppose that $\overline{Y}_{[k]} - \overline{Y}_{[1]} > c_k\hat{\sigma}/\sqrt{n} \geqslant q^{\alpha_i}_{i,k(n-1)}/\hat{\sigma}/\sqrt{n}$ (for all i). Then any subset containing [1] and [k] would be significant and we would declare $\mu_{[1]}$ and $\mu_{[k]}$ to be significantly different. [Note that it is not enough that $\overline{Y}_{[k]} - \overline{Y}_{[1]} > q^{\alpha_k}_{k,k(n-1)}\hat{\sigma}/\sqrt{n}$ unless $q^{\alpha_k}_{k,k(n-1)} = c_k$.] If $\mu_{[1]}$ and $\mu_{[k]}$ are declared significantly different, we proceed to stage 2. If $\overline{Y}_{[k-1]} - \overline{Y}_{[1]} \leqslant c_{k-1}\hat{\sigma}/\sqrt{n}$, then no subset with i_{k-1} elements that did not contain $\mu_{[k]}$ could be significant and hence $\mu_{[1]}$ and $\mu_{[k-1]}$ would be declared insignificantly different. On the other hand, if $\overline{Y}_{[k-1]} - \overline{Y}_{[1]} > c_{k-1}\hat{\sigma}/\sqrt{n} \geqslant q^{\alpha_i}_{i,k(n-1)}\hat{\sigma}/\sqrt{n}$ for $i = 2, 3, \ldots, k-1$. Therefore any subset of at most $k-1$ elements that contains [$k-1$] and [1] would be significant. Since the whole set has already been found to be significant (or we would not have proceeded to stage 2), all subsets containing [$k-1$] and [1] would be significant and hence $\mu_{[1]}$ and $\mu_{[k-1]}$ would be declared significantly different. The argument for $\mu_{[2]}$ and $\mu_{[k]}$ is similar as is the argument for later stages.

There does not seem to be such an easy algorithm for multiple F-tests. In performing those tests we would first look at the whole set $(1, \ldots, k)$. If this set is accepted, we stop and declare all means insignificantly different. However, if that set is rejected, we cannot conclude that any pairs are necessarily significantly different. At stage 2, we would need to compute the F-statistic for all possible subsets with $k-1$ elements. We continue in this fashion. At the jth stage, we would need to check every subset with $k - j + 1$ elements that is not a subset of a set that has already been declared insignificant. (Once a set has been declared insignificant, every pair of means in that subset is insignificant, so that it is not necessary to check subsets of insignificant sets.) Because we need to check so many subsets at each stage, the computations for the multiple F-test are more complicated than those for the multiple range test. In addition, the computation of the numerators of the F-statistics is more complicated than the computations for the range statistics.

The additional computations necessary for the multiple F tests can be a serious disadvantage when k is moderately large. For the multiple range test we test the significance of one set at stage 1, at most 2 sets at stage 2, at most 3 sets at stage 3, and so on. Therefore the total number of sets tested is at most

$$a_k = \sum_{i-1}^{k-1} i = \frac{k(k-1)}{2}. \tag{12.7}$$

On the other hand, for the multiple F-test, we may have to check every subset of $1, \ldots, k$ that has at least 2 elements. Therefore, we may have to check as many as

$$b_k = 2^k - k - 1 \tag{12.8}$$

such subsets. As an example, $a_{20} = 190$, whereas b_{20} is over one million. Therefore, if we have 20 well-separated means, we may have to check over one million subsets before finishing the multiple F-test, but would need to check at most 190 subsets with the multiple range test.

Because of the number of computations for the multiple F-test, it has been suggested that we check only adjacent subsets (as we do with the algorithm for the multiple range tests). That is, if we find the whole sets, $1, \ldots, k$ to be F-significant at size α_k, we declare $\mu_{[1]}$ and $\mu_{[k]}$ to be significantly different. We now look at the sets $[1], \ldots, [k-1]$ and $[2], \ldots, [k]$. If we find $[1], \ldots, [k-1]$ to be F-significant at size α_{k-1}, we declare $\mu_{[1]}$ and $\mu_{[k-1]}$ to be significantly different and so on. As soon as we find any subset is not F-signficant at the appropriate size, we declare all means in the set to be insignificantly different. It is apparent that this multiple F procedure would check at most a_k subsets. (One way of viewing this procedure is as a multiple F-test in which we are declaring all nonadjacent subsets to be significant.) This procedure would have size α_k and would be more powerful than the analagous multiple F-test. However, it seems difficult to make any statement about the generalized size of such a procedure. (An equivalent procedure for multiple-range tests would substitute $q_{s,k(n-1)}^{\alpha_s}$ for c_s in the algorithm suggested above.)

One problem with the multiple range and multiple F-tests (except the Newman-Keuls tests) is the lack of adequate tables. To use these tables we need the upper percentage points of the central F and studentized range distribution for percentages that have not been adequately tabled. Until more tables are provided, these procedures are not possible for many practical problems.

There have been several simulation studies comparing the power of various multiple comparison procedures. Some of the earlier studies compared procedures with quite different generalized sizes, and these studies confirmed that procedures with high generalized size were more powerful than those with low generalized size. More recent studies by Einot and Gabriel (1975) and Ramsey (1978) have restricted the procedures to have generalized size at most α for some α and then compared the power of different procedures. The results of these papers were somewhat different because different definitions of power were used in the two papers. However, both papers indicated the multiple F-tests are "more powerful" than the multiple range tests.

The approach taken in this section has been rather an oversimplification of the multiple comparison problem. For example, we have controlled the type I errors by using the generalized size. When we use this measure of type I errors, we are implying that making several type I errors is no worse than making one such error. A more reasonable measure of the size of a procedure might be the expected number of type I errors. In addition, the definition we have used for more powerful is quite restrictive. If d_1 is more powerful than d_2, then d_1 must find more differences than d_2 on any data set, and therefore must make more type I errors also.

As should be apparent from the previous discussion, the multiple comparison problem is still not very well understood. It is not even clear what a good procedure should do. Should it maximize the probability of no type II errors, minimize the expected number of type II errors, minimize the probability of particular type II errors, or what? Therefore, in this section, we have discussed only some of the more elementary procedures. (For a more technical discussion see Gabriel, 1969.) Of these procedures, the modified Ryan multiple range test is more powerful than any other multiple range test, and if $q^{\alpha_s}_{s,k(n-1)} \leqslant q^{\alpha}_{k,k(n-1)}$ for $\alpha_s = 1 - (1-\alpha)^{s/k}$, $s \leqslant k$, then it is more powerful than the Tukey and Scheffé procedures. Similarly, the modified Ryan multiple F-test is more powerful than any other multiple F-test. Therefore, of the procedures suggested here, the modified Ryan multiple range and multiple F-tests seem to be the best. However, there are other more complicated procedures that are more powerful than these two. See Einot and Gabriel (1975), Welsch (1977), and Ramsey (1978) for more details on these procedures.

The procedures defined in this section can be extended in an obvious way to other balanced analyses of variance models (with equal weights) to determine which main effects, if any, are significantly different. For unbalanced designs, the procedures based on the studentized range distributions would not be exact, but those based on the F-distribution would still be available (as would those discussed in Section 12.6 using Bonferroni's inequality).

12.5. ASYMPTOTIC VALIDITY OF STUDENTIZED RANGE PROCEDURES.

We now show that as $n \to \infty$ (but k remains fixed) the asymptotic distribution of

$$q_{n,k} = \max_{i \neq j} \frac{\sqrt{n}\,|\bar{Y}_{i\cdot} - \alpha_i - (\bar{Y}_{j\cdot} - \alpha_j)|}{\hat{\sigma}}$$

does not depend on the normal assumption used in deriving it. The model we assume is one in which we observe

$$Y_{ij} = \theta + \alpha_i + e_{ij}, \qquad i = 1, \ldots, k; \qquad j = 1, \ldots, n.$$

We assume that the e_{ij} are independently identically distributed with $Ee_{ij} = 0$, $\operatorname{var}(e_{ij}) = \sigma^2$. We define the range distribution r_k to be tbe distribution of the range of k independently identically distributed normal random variables with mean 0 and variance 1.

THEOREM 12.5. As $n \to \infty$

$$q_{n,k} \xrightarrow{d} r_k.$$

PROOF. By the central limit theorem, we see that $\sqrt{n}\,\bar{e}_{i\cdot}/\sigma \xrightarrow{d} N_1(0,1)$. Also the $\bar{e}_{i\cdot}$ are independent. Therefore,

$$\frac{\sqrt{n}}{\sigma}\begin{pmatrix} \bar{e}_{1\cdot} \\ \vdots \\ \bar{e}_{k\cdot} \end{pmatrix} \xrightarrow{d} N_k(0, I).$$

By Theorem 10.2d, we see that

$$\max_{i \neq j} \frac{\sqrt{n}\,|\bar{e}_{i\cdot} - \bar{e}_{j\cdot}|}{\sigma} \xrightarrow{d} r_k.$$

By Theorem 10.5 $\hat{\sigma}/\sigma \xrightarrow{p} 1$. By Theorem 10.2a,

$$q_{n,k} = \max \frac{\left(\sqrt{n}\,|\bar{e}_{i.} - \bar{e}_{j.}|\right)/\sigma}{\hat{\sigma}/\sigma} \xrightarrow{d} r_k. \quad \square$$

This result implies that the studentized range test discussed in Section 12.2 has asymptotic size α even without the normal assumption used in deriving that test. Similarly, the simultaneous confidence intervals derived in Section 12.3 are asymptotically $(1-\alpha)$ simultaneous confidence intervals. Finally, the Tukey procedure and the multiple range procedures (except Student-Newman-Keuls) asymptotically have generalized size α for the multiple comparisons problem discussed in Section 12.4. (The Scheffé and multiple F-test procedures have asymptotic generalized size α by the results of Chapter 10.) Therefore, the procedures based on the studentized range distribution have the same asymptotic validity as those based on the F-distribution.

12.6. BONFERRONI'S INEQUALITY.

We now return to the linear model in which we observe $\mathbf{Y} \sim N_j(\boldsymbol{\mu}, \sigma^2)$, $\boldsymbol{\mu} \in V$, $\sigma^2 > 0$. Bonferroni's (Boole's) inequality is the following

$$P(\cap A_i) \geqslant 1 - \sum_i P(A_i^c), \tag{12.9}$$

which is proved by noting that $(\cap A_i)^c = \cup(A_i^c)$ and therefore

$$P(\cap A_i) = 1 - P\left[(\cap A_i)^c\right] = 1 - P(\cup A_i^c) \geqslant 1 - \sum P(A_i^c).$$

THEOREM 12.6. Let D be a finite set of functions $\mathbf{d}'\boldsymbol{\mu}$, $\mathbf{d} \in V$, and let s be the number of functions in D. Then

$$P\left(\boldsymbol{d}'\boldsymbol{\mu} \in \boldsymbol{d}'\hat{\boldsymbol{\mu}} \pm t_{n-p}^{\alpha/2s}\,\hat{\sigma}\|\boldsymbol{d}\| \qquad \text{for all } \boldsymbol{d} \in D\right) \geqslant 1 - \alpha.$$

PROOF. See Exercise B7. $\square$

We can use this theorem in the obvious way to find simultaneous confidence intervals for any finite set of functions. These intervals are called Bonferroni intervals. Since D is finite, it cannot be a subspace, as in the previous examples, and in particular cannot be the space of all contrasts. Note also that if s is large, then $t_{n-p}^{\alpha/2s}$ will be large and the intervals will be wide.

To compare the Bonferroni intervals with the Scheffé intervals, let W be the smallest subspace containing D, and let $\dim(W) = k \leqslant s$. Then by Theorem 8.2,

$$P\left(\boldsymbol{d}'\boldsymbol{\mu} \in \boldsymbol{d}'\hat{\boldsymbol{\mu}} \pm \sqrt{kF^{\alpha}_{k,n-p}}\,\hat{\sigma}\|\boldsymbol{d}\| \quad \text{for all } \boldsymbol{d} \in D\right) \geqslant 1-\alpha$$

(with the inequality because we only ask that the probability be true for $\mathbf{d} \in D$, not all $\mathbf{d} \in W$). Both sets of intervals have the same midpoints, $\mathbf{d}'\boldsymbol{\mu}$. To see which is shorter, we need to compare $t_{n-p}^{\alpha/2s}$ with $\sqrt{kF^{\alpha}_{k,n-p}}$. If s and k are at all comparable, the Bonferroni intervals will be shorter. The Scheffé intervals are better than the Bonferroni intervals only if s is much larger than k.

If the Tukey method is applicable (i.e., if the independence and equal variance conditions are met), then it will give shorter simultaneous confidence intervals for the set of all pairwise comparisons than the Bonferroni method (see Exercise B8). However, Bonferroni may do better if only some of the comparisons are of interest (before the data is collected).

Bonferroni's inequality can be used to define multiple Bonferroni procedures that have generalized size at most α analogous to the multiple range and multiple F-procedures defined in the last section. Although the multiple Bonferroni procedures are less powerful than the analogous multiple range procedures, they can be extended to unbalanced situations in a very simple way, while the multiple range procedures cannot. See Exercise C3 for the definition of multiple Bonferroni procedures, as well as some simple properties.

12.7. FURTHER COMMENTS.

The problem of finding which means, if any, are significantly different is an example of a problem of simultaneous statistical inference. For more details on the procedures considered in this chapter and other procedures, see Miller (1966).

EXERCISES.

Type A

1. Find Tukey simultaneous confidence intervals for the set of all comparisons in Exercise A2 of Chapter 7.

2. Verify the underlines in (12.6).

3. Verify that if $k = 10$, then $1 - (1 - 0.05)^{k-1} = 0.37$.

4. Let a_k and b_k be defined by (12.7) and (12.8). Verify that $a_{20} = 190$ and b_{20} is over one million.

Type B

1. Prove Lemma 12.1 and Theorem 12.2b.

2. Show that the Tukey confidence intervals for comparisons are shorter than the Scheffé confidence intervals for comparisons. (Hint: If not, then the probability of simultaneously covering for Tukey intervals would be as great as for Scheffé intervals. Why? Note that the two sets of intervals have the same midpoints and the lengths of all Tukey intervals are the same as are the lengths of all Scheffé intervals.)

3. Show that $1 - (1 - \alpha)^{s/k} > 1 - (1 - \alpha)^{(s-1)/(k-1)}$ and that $1 - (1 - \alpha)^{s/k} > (s/k)\alpha$.

4. Show that if $q_{s,k(n-1)}^{s\alpha/k} \leqslant q_{k,k(n-1)}^{\alpha}$ then the Ryan multiple range test has generalized size exactly α.

5. Let $Z_1, \ldots, Z_k$, U be independent, $Z_i \sim N_1(0, 1)$, $U \sim \chi_m^2(0)$. Let

$$m = \max \frac{|Z_i|}{\sqrt{U/m}} .$$

We say that m has a *studentized maximum modulus* distribution and write $m \sim m_{k,m}$.

(a) Use the studentized maximum modulus distribution to find simultaneous confidence intervals for the set of all $\mu_i = \theta + \alpha_i$ in a balanced one way analysis of variance model.

(b) Let a_i be a set of numbers, $i = 1, \ldots, k$. Show that $\max|a_i| \leqslant c$ if and only if $|\sum_i d_i a_i| \leqslant c \sum_i |d_i|$ for all $d_1, \ldots, d_k$.

(c) Use part b and the studentized maximum modulus distribution to find simultaneous for the set of all $\sum_i d_i \mu_i$.

(d) Find the Scheffé type confidence intervals for the set of all $\sum d_i \mu_i$. (Hint: These are the contrasts associated with testing that the μ_i are all 0.)

(e) Give an elementary comparison of the lengths of the Scheffé intervals with those derived in this problem.

6. Consider a balanced two-way model with interaction and equal weights (see Section 7.3.3). Find the Tukey simultaneous confidence intervals for the set of all $\sum_i c_i \alpha_i$ where $\sum_i c_i = 0$. [Hint:

$$\overline{Y}_{i..} \sim N_1(\theta + \alpha_i, \sigma^2/nc). \qquad \text{Why?}]$$

7. Prove Theorem 12.6.

8. Show that the Tukey $(1 - \alpha)$ simultaneous confidence intervals for the set of all comparisons in a one-way model are shorter than the Bonferroni intervals. (See Exercise B2 above.)

Type C

1. Generalized Bonferroni inequality. Let $g_i(X)$ nondecreasing functions from R^1 to R^1, $i = 1, \ldots, m$. Let X be a random variable such that $Eg_i(X) = \mu_i < \infty$.

(a) Show that $E(g_1(X) - \mu_1)(g_2(X) - \mu_2) \geqslant 0$. {Hint: Let a be such that $g_1(x) \geqslant \mu_1$ if $x > a$ and $g_1(x) \leqslant \mu_1$ if $x < a$. Then $[g_1(x) - \mu_1][g_2(x) - g_2(a)] \geqslant 0$. (Why?) Also $E[g_1(X) - \mu_1][g_2(X) - \mu_2] = E[g_1(X) - \mu_1][g_2(X) - g_2(a)]$. Why?}

(b) Show that $E\prod_{i=1}^m g_i(X) \geqslant \prod_{i=1}^m Eg_i(X)$.

2. We return to the multiple range tests of Section 12.4. Let S_j be disjoint subsets of $(1, \ldots, k)$, and let q_j and F_j be the studentized range statistic and F statistic associated with S_j [see (12.4) and (12.5)].

(a) Let $h_j(\hat{\sigma}) = EP(q_j \leqslant a_j) | \hat{\sigma})$, $g_j(\hat{\sigma}) = EP(F_j \leqslant b_j) | \hat{\sigma})$. Show that h_j and g_j are nondecreasing functions.

(b) Show that $EP(q_j \leqslant a_j$ for all $j | \hat{\sigma}) = \prod_j h_j(\hat{\sigma})$ and that

$$EP(F_j \leqslant b_j \quad \text{for all} \quad j | \hat{\sigma}) = \prod_j g_j(\hat{\sigma}).$$

(c) Show that $P(q_j \leqslant a_j$ for all $j) \geqslant \prod_j P(q_j \leqslant a_j)$ and that

$$P(F_j \leqslant b_j \quad \text{for all} \quad j) \geqslant \prod_j P(F_j \leqslant b_j).$$

(d) Use part c to show that the modified Ryan multiple range and multiple F-test have generalized size α, as do the Duncan multiple range and multiple F-tests.

3. Multiple Bonferroni tests. Consider the model of Section 12.4. Let S be a subset of $1, \ldots, k$ with s elements.

(a) Show that a set of $1 - \alpha$ Bonferroni simultaneous confidence intervals for the set of all comparisons $i, j \in S$ is given by

$$\mu_i - \mu_j \in \overline{Y}_{i.} - \overline{Y}_{j.} \pm t_{k(n-1)}^{\alpha/[s(s-1)]} \hat{\sigma} (2/n)^{1/2}$$

[Note that there are $\binom{s}{2}$ different comparisons in S.] Define the set S to be *Bonferroni-significant* at size α_s if

$$\max_{i,j \in S} \sqrt{n} \, | Y_{i.} - Y_{j.} | > t_{k(n-1)}^{\alpha_s/[s(s-1)]} \sqrt{2} \, \hat{\sigma}$$

(b) Suppose that $\mu_i = \mu_j$ for all $i, j \in S$. Show that the probability that S is Bonferroni-significant at size α_s is at most α_s. A multiple Bonferroni procedure is one in which μ_i and μ_j are declared significantly different if and only if every set containing i and j is declared Bonferroni-significant at size α_s.

(c) Ryan multiple Bonferroni test. Show that if $\alpha_s = s\alpha/k$, then the multiple Bonferroni procedure has generalized size at most α.

(d) Modified Ryan multiple Bonferroni test. Show that if $\alpha_s = 1 - (1 - \alpha)^{s/k}$ then the multiple Bonferroni test has generalized size at most α. (By a similar argument we could show that the Duncan multiple Bonferroni test has generalized size at most α.)

(e) Show that the modified Ryan multiple Bonferroni test is more powerful than the Ryan multiple Bonferroni test.

(f) Show that a multiple Bonferroni test can be performed with the algorithm described for the multiple range test with c_s replaced by

$$2t_{k(n-1)}^{\alpha_s/[s(s-1)]}$$

(g) Show that a multiple range test with a particular α_s is more powerful than the multiple Bonferroni test with the same α_s. (See Exercise B8.)

(h) Define a multiple Bonferroni procedure for the case of unequal sample sizes. (Hint: Start first with the Bonferroni simultaneous confidence intervals for this case.)

CHAPTER 13

The Generalized Linear Model

With this chapter, we begin the study of other models that are closely related to the ordinary linear model discussed in the last nine chapters. The model considered in this chapter is the *generalized linear model*, in which we observe the $n \times 1$ random vector

$$\mathbf{Y} \sim N_n(\boldsymbol{\mu}, \sigma^2\mathbf{A}), \qquad \boldsymbol{\mu} \in V, \qquad \sigma^2 > 0, \tag{13.1}$$

where V is a specified p-dimensional subspace of R^n (as before) and $\mathbf{A}$ is a known $n \times n$ positive definite matrix. If $\mathbf{A} = \mathbf{I}$, then the model is, of course, an ordinary linear model. It is somewhat difficult to imagine a situation in which $\mathbf{A}$ would be known, but this model is useful in applications for the following reason. In applications, it is often assumed that $\mathbf{A}$ has a certain form. Then the unknown parameters in $\mathbf{A}$ are estimated from the data. We then assume that we had known $\mathbf{A}$ to have that value, and use the results derived in this section to find approximate procedures for drawing inference in this setting. If the estimated $\mathbf{A}$ is close to the true $\mathbf{A}$, these approximate procedures will often be satisfactory. This method is used, for example, to analyze models with autocorrelation.

For the generalized linear model, we find a complete sufficient statistic, maximum likelihood estimators, minimum variance unbiased estimators, UMP invariant tests and simultaneous confidence intervals for contrasts. A James-Stein estimator is derived in Exercise C1. Finally, we find necessary and sufficient conditions for the estimators to be the same for the generalized linear model as for the ordinary linear model.

13.1. THE BASIC RESULTS.

Since $\mathbf{A} > 0$, there is a unique matrix $\mathbf{A}^{1/2} > 0$ such that

$$\mathbf{A} = (\mathbf{A}^{1/2})^2, \qquad (\mathbf{A}^{1/2})^{-1}\mathbf{A}(\mathbf{A}^{1/2})^{-1} = \mathbf{I}$$

(see Section 3.2). We write $\mathbf{A}^{-1/2} = (\mathbf{A}^{1/2})^{-1}$. Now let

$$\mathbf{Y}^* = \mathbf{A}^{-1/2}\mathbf{Y}, \qquad \boldsymbol{\mu}^* = \mathbf{A}^{-1/2}\boldsymbol{\mu}, \qquad V^* = \{\mathbf{A}^{-1/2}\mathbf{v};\, \mathbf{v} \in V\}. \quad (13.2)$$

Since V is a subspace, so is V^*. If $\mathbf{X}$ is a basis matrix for V, then $\mathbf{A}^{-1/2}\mathbf{X}$ is a basis matrix for V^*. Therefore the dimension of V^* is the same as the dimension of V. Also

$$\mathbf{Y}^* \sim N_n(\boldsymbol{\mu}^*, \sigma^2\mathbf{I}), \qquad \boldsymbol{\mu}^* \in V^*, \qquad \sigma^2 > 0.$$

Since $\mathbf{A}^{-1/2}$ is an invertible matrix, the transformation from $\mathbf{Y}$ to $\mathbf{Y}^*$ is invertible. The transformed model in which we observe $\mathbf{Y}^*$, is just an ordinary linear model. Therefore, let

$$\hat{\boldsymbol{\mu}}^* = \mathbf{P}_{V^*}\mathbf{Y}^*, \qquad \hat{\hat{\sigma}}^2 = \|\mathbf{P}_{V^{*\perp}}\mathbf{Y}^*\|^2/(n-p) \qquad (13.3)$$

Then $(\hat{\boldsymbol{\mu}}^*, \hat{\hat{\sigma}}^2)$ is a complete sufficient statistic for the transformed problem, and hence for the original problem in which we observe $\mathbf{Y}$. Now let

$$\hat{\hat{\boldsymbol{\mu}}} = \mathbf{A}^{1/2}\hat{\boldsymbol{\mu}}^* \qquad (13.4)$$

Then $(\hat{\hat{\boldsymbol{\mu}}}, \hat{\hat{\sigma}}^2)$ is an invertible function of $(\hat{\boldsymbol{\mu}}^*, \hat{\hat{\sigma}}^2)$, and hence $(\hat{\hat{\boldsymbol{\mu}}}, \hat{\hat{\sigma}}^2)$ is a complete sufficient statistic also. Also

$$E\hat{\hat{\boldsymbol{\mu}}} = E\mathbf{A}^{1/2}\hat{\boldsymbol{\mu}}^* = \mathbf{A}^{1/2}\boldsymbol{\mu}^* = \boldsymbol{\mu}, \qquad E\hat{\hat{\sigma}}^2 = \sigma^2,$$

using known facts about the ordinary linear model involving $\mathbf{Y}^*$. Therefore, $\hat{\hat{\boldsymbol{\mu}}}$ and $\hat{\hat{\sigma}}^2$ are the minimum variance unbiased estimators of $\boldsymbol{\mu}$ and σ^2 for the generalized linear model. Finally, $\hat{\boldsymbol{\mu}}^*$ and $[(n-p)/n]\hat{\hat{\sigma}}^2$ are the MLE's for $\boldsymbol{\mu}^*$ and σ^2, and therefore $\hat{\hat{\boldsymbol{\mu}}} = \mathbf{A}^{1/2}\hat{\boldsymbol{\mu}}^*$ is the MLE for $\boldsymbol{\mu} = \mathbf{A}^{1/2}\boldsymbol{\mu}^*$. In Exercises B8 and B9, it is shown that $\hat{\hat{\mu}}$ and $(n-p)\hat{\hat{\sigma}}^2/(n-p+2)$ are the best invariant estimators of μ and σ^2 for the loss functions

$$L_1(\mathbf{d}_1; (\boldsymbol{\mu}, \sigma^2)) = \frac{(\mathbf{d}_1 - \boldsymbol{\mu})'\mathbf{H}(\mathbf{d}_1 - \boldsymbol{\mu})}{\sigma^2},$$

$$L_2(d_2; (\boldsymbol{\mu}, \sigma^2)) = \frac{(d_2 - \sigma^2)^2}{\sigma^4} = \left(\frac{d_2}{\sigma^2} - 1\right)^2$$

where $\mathbf{H} > 0$ is known. If $\mathbf{H} = \mathbf{A}^{-1}$, then L_1 is the Mahalanobis distance loss function. We have therefore proved the following theorem.

THEOREM 13.1. Let $\hat{\hat{\boldsymbol{\mu}}}$ and $\hat{\hat{\boldsymbol{\sigma}}}^2$ be defined by (13.3) and (13.4).

a. Then $(\hat{\hat{\boldsymbol{\mu}}}, \hat{\hat{\sigma}}^2)$ is a complete sufficient statistic for the generalized linear model.

b. $\hat{\hat{\boldsymbol{\mu}}}$ and $\hat{\hat{\sigma}}^2$ are the minimum variance unbiased estimators of $\boldsymbol{\mu}$ and σ^2.

c. $\hat{\hat{\boldsymbol{\mu}}}$ and $[(n-p)/n]\hat{\hat{\sigma}}^2$ are the MLE's of $\boldsymbol{\mu}$ and σ^2.

d. $\hat{\hat{\boldsymbol{\mu}}}$ is the best invariant estimator of $\boldsymbol{\mu}$ with respect to the loss L_1 and $(n-p)\hat{\hat{\sigma}}^2/(n-p+2)$ is the best invariant estimator of σ^2 for the loss function L_2.

In the exercise, a James-Stein estimator is derived and $\hat{\mu}$ is shown to be inadmissible for the Mahalanobis distance loss function.

The following theorem gives the joint distribution of $\hat{\hat{\mu}}$ and $\hat{\hat{\sigma}}^2$. It follows directly from known results for the ordinary linear model involving $\mathbf{Y}^*$, and its proof is asked for in Exercise B1.

THEOREM 13.2. $\hat{\hat{\mu}}$ and $\hat{\hat{\sigma}}^2$ are independent and

$$\hat{\hat{\mu}} \sim N_n(\mu, \sigma^2 \mathbf{A}^{1/2} \mathbf{P}_{V^*} \mathbf{A}^{1/2\prime}), \qquad (n-p)\hat{\hat{\sigma}}^2 \sim \sigma^2 \chi^2_{n-p}(0).$$

We now dicuss two methods for finding $\hat{\hat{\mu}}$ and $\hat{\hat{\sigma}}^2$. These methods correspond to the two methods used to find statistics for the ordinary linear model, namely, (a) finding a basis matrix and (b) using the least-squares property. Suppose, first that $\mathbf{X}$ is a basis matrix for V. Then $\mathbf{A}^{-1/2}\mathbf{X}$ is a basis matrix for V^*. Therefore

$$\begin{aligned}\hat{\mu}^* &= \mathbf{P}_{V^*}\mathbf{Y}^* = (\mathbf{A}^{-1/2}\mathbf{X})\left[(\mathbf{A}^{-1/2}\mathbf{X})'\mathbf{A}^{-1/2}\mathbf{X}\right]^{-1}(\mathbf{A}^{-1/2}\mathbf{X})'\mathbf{A}^{-1/2}\mathbf{Y}\\ &= \mathbf{A}^{-1/2}\mathbf{X}(\mathbf{X}'\mathbf{A}^{-1}\mathbf{X})^{-1}\mathbf{X}'\mathbf{A}^{-1}\mathbf{Y}\end{aligned}$$

Hence

$$\hat{\hat{\mu}} = \mathbf{A}^{1/2}\hat{\mu}^* = \mathbf{X}(\mathbf{X}'\mathbf{A}^{-1}\mathbf{X})^{-1}\mathbf{X}'\mathbf{A}^{-1}\mathbf{Y}, \tag{13.5}$$

$$(n-p)\hat{\hat{\sigma}}^2 = \|\mathbf{Y}^*\|^2 - \|\hat{\mu}^*\|^2 = \mathbf{Y}'\mathbf{A}^{-1}\mathbf{Y} - \hat{\hat{\mu}}'\mathbf{A}^{-1}\hat{\hat{\mu}} = (\mathbf{Y} - \hat{\hat{\mu}})\mathbf{A}^{-1}(\mathbf{Y} - \hat{\hat{\mu}}).$$

Equation (13.5), therefore gives formulas for $\hat{\hat{\mu}}$ and $\hat{\hat{\sigma}}^2$ for a given basis matrix $\mathbf{X}$.

The second method for finding $\hat{\hat{\mu}}$ and $\hat{\hat{\sigma}}^2$ is to use the fact that

$$(\mathbf{Y} - \hat{\hat{\mu}})'\mathbf{A}^{-1}(\mathbf{Y} - \hat{\hat{\mu}}) \leqslant (\mathbf{Y} - \mathbf{v})'\mathbf{A}^{-1}(\mathbf{Y} - \mathbf{v}) \tag{13.6}$$

for all $\mathbf{v} \in V$ (see Exercise B2). The second approach, therefore, would find $\hat{\hat{\mu}}$ by minimizing $(\mathbf{Y} - \mathbf{v})'\mathbf{A}^{-1}(\mathbf{Y} - \mathbf{v})$ over $\mathbf{v}$ in V, and then find $\hat{\hat{\sigma}}^2$ by using (13.5). Because the estimator $\hat{\hat{\mu}}$ satisfies (13.6), it is often called the *generalized least-squares estimator*. (It is also called the Aitken estimator since it was first derived by Aitken, 1935).

We now consider testing the null hypothesis $\mu \in W$, a particular k-dimensional subspace of V against the alternative that $\mu \in V$. Let

$$W^* = \{\mathbf{A}^{-1/2}\mathbf{w};\ \mathbf{w} \in W\}$$

Then $\mu \in W$ if and only if $\mu^* \in W^*$. Therefore, we are testing $\mu^* \in W^*$ against $\mu^* \in V^*$. (Note that W^* is a k-dimensional subspace of V^*.) Let

$$F = \frac{\|\mathbf{P}_{V^*|W^*}\mathbf{Y}^*\|^2}{\|\mathbf{P}_{V^{*\perp}}\mathbf{Y}^*\|^2}\,\frac{n-p}{p-k}, \qquad \phi(F) = \begin{cases} 1 & \text{if } F > F^{\alpha}_{p-k,n-p} \\ 0 & \text{if } F \leqslant F^{\alpha}_{p-k,n-p}\end{cases}. \tag{13.7}$$

By Theorem 7.13 applied to the ordinary linear model involving $\mathbf{Y}^*$, we have the following result.

THEOREM 13.3. Let $\phi(F)$ be defined by (13.7). Then ϕ is UMP invariant size α and unbiased for testing $\mu \in W$ for the generalized linear model.

We now discuss the computation of F. Let

$$\hat{\hat{\boldsymbol{\mu}}} = \mathbf{A}^{1/2}\mathbf{P}_{W*}\mathbf{Y}^*.$$

Then $\hat{\hat{\boldsymbol{\mu}}}$ is the minimum variance unbiased estimator of $\boldsymbol{\mu}$ under the null hypothesis, and may be computed using (13.5) or (13.6) with W in place of V. In Exercise B3 you are asked to show that

$$F = \frac{\hat{\boldsymbol{\mu}}'A^{-1}\hat{\boldsymbol{\mu}} - \hat{\hat{\boldsymbol{\mu}}}'\mathbf{A}^{-1}\hat{\hat{\boldsymbol{\mu}}}}{\mathbf{Y}'\mathbf{A}^{-1}\mathbf{Y} - \hat{\boldsymbol{\mu}}'\mathbf{A}^{-1}\hat{\boldsymbol{\mu}}} \frac{p-k}{n-p} \tag{13.8}$$

Therefore, we can compute $\hat{\boldsymbol{\mu}}$ and $\hat{\hat{\boldsymbol{\mu}}}$ using (13.5) or (13.6), and then we can compute F using (13.8). We do not need to find $\mathbf{A}^{-1/2}$ for any of the procedures discussed in this section.

We now consider the contrasts associated with testing that $\boldsymbol{\mu} \in W$, and find simultaneous confidence intervals for them. We first look at the model after it has been transformed to an ordinary linear model. There, the contrasts associated with testing that $\boldsymbol{\mu} \in W$, or equivalently $\boldsymbol{\mu}^* \in W^*$ have the form $\mathbf{d}^{*\prime}\boldsymbol{\mu}^*$, $\mathbf{d}^* \in V^* \mid W^*$, and the set of simultaneous confidence intervals for the set of contrasts is given by

$$\mathbf{d}^{*\prime}\boldsymbol{\mu}' \in \mathbf{d}^{*\prime}\hat{\boldsymbol{\mu}}^* \pm \left[(p-k)F^{\alpha}_{p-k,n-p}\right]^{1/2}\|\mathbf{d}^*\|\hat{\hat{\sigma}}$$

(see Theorem 8.2). Now, let $\mathbf{d}^* = \mathbf{A}^{-1/2}\mathbf{d}$ (i.e., let $\mathbf{d} = \mathbf{A}^{1/2}\mathbf{d}^*$). Then

$$\mathbf{d}^{*\prime}\hat{\boldsymbol{\mu}}^* = \mathbf{d}'\mathbf{A}^{-1}\hat{\boldsymbol{\mu}}, \qquad \mathbf{d}^{*\prime}\boldsymbol{\mu}^* = \mathbf{d}'\mathbf{A}^{-1}\boldsymbol{\mu}, \qquad \|\mathbf{d}^*\|^2 = \mathbf{d}'\mathbf{A}^{-1}\mathbf{d}.$$

and, $\mathbf{d}^* \in V^* \mid W^*$ if and only if $\mathbf{d} \in V$ and $\mathbf{d}'\mathbf{A}^{-1}\mathbf{w} = 0$ for all $\mathbf{w} \in W$ (see Exercise B7). Therefore, let $V \| W$ be the set of all $\mathbf{d} \in V$ such that $\mathbf{d}'\mathbf{A}^{-1}\mathbf{w} = 0$ for all $\mathbf{w} \in W$. (Note that if $\mathbf{A} = \mathbf{I}$, then $V \| W = V \mid W$.) The *contrasts* associated with testing that $\boldsymbol{\mu} \in W$ in the generalized linear model are defined to be the functions $\mathbf{d}'\mathbf{A}^{-1}\boldsymbol{\mu}$, $\mathbf{d} \in V \| W$. The set of Scheffé simultaneous $(1-\alpha)$ confidence intervals for the set of all contrasts is therefore given by

$$\mathbf{d}'\mathbf{A}^{-1}\boldsymbol{\mu} \in \mathbf{d}'\mathbf{A}^{-1}\hat{\boldsymbol{\mu}} \pm \left[(p-k)F^{\alpha}_{p-k,n-p}\mathbf{d}'\mathbf{A}^{-1}\mathbf{d}\right]^{1/2}\hat{\sigma}. \tag{13.9}$$

As a final topic in this section, we consider the coordinatized general linear model. Let $\mathbf{X}$ be a basis matrix for V and let $\boldsymbol{\beta}$ and $\hat{\hat{\boldsymbol{\beta}}}$ be the unique solutions to $\boldsymbol{\mu} = \mathbf{X}\boldsymbol{\beta}$, $\hat{\hat{\boldsymbol{\mu}}} = \mathbf{X}\hat{\hat{\boldsymbol{\beta}}}$. Then

$$\hat{\hat{\boldsymbol{\beta}}} = (\mathbf{X}'\mathbf{A}^{-1}\mathbf{X})^{-1}\mathbf{X}'\mathbf{A}^{-1}\mathbf{Y}, \qquad (n-p)\hat{\hat{\sigma}}^2 = (\mathbf{Y} - \mathbf{X}\hat{\hat{\boldsymbol{\beta}}})'\mathbf{A}^{-1}(\mathbf{Y} - \mathbf{X}\hat{\hat{\boldsymbol{\beta}}}) \tag{13.10}$$

(see Exercise B6). The estimator $\hat{\hat{\boldsymbol{\beta}}}$ is called the generalized least-squares estimator or Aitken estimator of $\boldsymbol{\beta}$. Since $(\hat{\hat{\boldsymbol{\beta}}}, \hat{\hat{\sigma}}^2)$ is an invertible function of $(\hat{\hat{\boldsymbol{\mu}}}, \hat{\hat{\sigma}}^2)$, $(\hat{\hat{\boldsymbol{\beta}}}, \hat{\hat{\boldsymbol{\sigma}}}^2)$ is also a complete sufficient statistic for the generalized linear model. Now, consider testing that $\mathbf{C}\boldsymbol{\beta} = \mathbf{0}$ where $\mathbf{C}$ is a $(p-k) \times r$ matrix of rank $p-k$. This testing problem falls in the framework of the testing problem treated earlier in this section. By arguments similar to those in Section 7.2.1, the appropriate F statistic for this model is

$$\frac{\hat{\boldsymbol{\beta}}'\mathbf{C}'(\mathbf{C}\mathbf{X}'\mathbf{A}^{-1}\mathbf{X}\mathbf{C}')^{-1}\mathbf{C}\hat{\boldsymbol{\beta}}}{(p-k)\hat{\sigma}^2}. \tag{13.11}$$

13.2. AUTOCORRELATION.

In most econometric data sets, the observations $Y_1, \ldots, Y_n$ are arriving over time (i.e., they are components of a time series). In this section, we present a model that is often used to analyze such data and give an elementary discussion of how the generalized linear model can be used to find approximate procedures for analyzing this model.

The *autocorrelation model* is one in which we observe $\mathbf{Y}' = (Y_1, \ldots, Y_n)$ having the distribution

$$\mathbf{Y} \sim N_n(\boldsymbol{\mu}, \sigma^2\mathbf{A}(\rho)), \qquad \boldsymbol{\mu} \in V, \qquad \sigma^2 > 0,$$

$$\mathbf{A}(\rho) = \begin{pmatrix} 1 & \rho & \rho^2 & \cdots & \rho^n \\ \rho & 1 & \rho & \cdots & \rho^{n-1} \\ \rho^2 & \rho & 1 & \cdots & \rho^{n-2} \\ \vdots & \vdots & \vdots & & \vdots \\ \rho^n & \rho^{n-1} & \rho^{n-2} & \cdots & 1 \end{pmatrix}$$

(i.e., $\operatorname{cov}(Y_i, Y_j) = \rho^{|i-j|}\sigma^2$). ρ is called the *autocorrelation coefficient*. One motivation for this model is provided by the fact that if i and j are close (and hence Y_i and Y_j are not widely separated in time), then the correlation between Y_i and Y_j is high, whereas, if i and j are far apart (and hence Y_i and Y_j occur at far different times), then the correlation between Y_i and Y_j is small. This aspect of the autocorrelation model agreees well with our intuition in such times series.

Another motivation for the autocorrelation model is provided by the following analysis. Let U_i be independent random variables,

$$U_i \sim N_1\big(\theta_i, \sigma^2(1-\rho^2)\big).$$

Let Y_0 by an unobserved initial random variable independent of the U_i, with $Y_0 \sim N_1(\mu_0, \sigma^2)$. Define Y_i and μ_i recursively by

$$Y_i = \rho Y_{i-1} + U_i, \qquad \mu_i = \rho\mu_{i-1} + \theta_i, \qquad i = 1, 2, \ldots, n.$$

Then, using induction, we can see that

$$\operatorname{var}(Y_i) = \sigma^2, \qquad \operatorname{cov}(Y_i, Y_j) = \rho^{|i-j|}\sigma^2$$

and therefore, the observations $Y' = (Y_1, \ldots, Y_n)$ form an autocorrelation model. Hence, we can think of the autocorrelation model as one in which the ith observation, Y_i is equal to ρ times the $(i-1)$th observation plus an independent observation U_i that represents the contribution at the ith time period. This sort of model is felt to be a reasonable model for many econometric data sets.

If ρ is known, then this model is just a particular case of the generalized linear model discussed in Section 13.1, and should be analyzed by the procedures suggested there. However, it is hard to imagine a practical application of this model in which ρ would be known. If ρ is not known, then it is very difficult to find exact procedures for this model. For example, it is not possible to give formulas for the MLE's. They must be computed iteratively.

One method that is often used to provide an approximate analysis for this model is to find a "sensible" estimator $\hat{\rho}$ for the autocorrelation coefficient ρ, and then use generalized least squares assuming ρ to be $\hat{\rho}$. The approximation in this analysis is obviously only as good as the approximation in using $\hat{\rho}$ for ρ. If $\hat{\rho}$ is close to ρ, then the approximations should be quite good. Therefore, to use this method, it is important to find a good estimator for ρ. Many estimators have been suggested for this model.

The purpose of this section is to provide an example of how the generalized linear model can be used to give approximate procedures in practical situations. A careful discussion of the autocorrelation model would take us far beyond the intent of this chapter. For a more detailed discussion of this model, see Johnston (1972).

13.3. OTHER RESULTS FOR THE GENERALIZED LINEAR MODEL.

It is clear that many results for the ordinary linear model involving $\mathbf{Y}^*$ can be restated as results for the generalized linear model involving $\mathbf{Y}$. For example, we could find one-sided tests as in Chapter 7 or make inference about σ^2 as in Chapter 9. The other optimality results stated in previous chapters, but not proved (most stringent, likelihood ratio test, etc.) carry over to the generalized linear model in a straightforward way. The asymptotic results in Chapter 10 do not carry over quite so easily. If $\mathbf{Y}$ is not normally distributed, the components of $\mathbf{Y}^*$ will be uncorrelated, but not necessarily independent. The derivations in Chapter 10 depend on the independence of the components. (One situation in which the results do carry over occurs when the components of $\mathbf{Y}$ are independent and hence $\mathbf{A}$ is diagonal.)

We finish this chapter with a result due to Kruskal (1968), which relates the generalized least-squares and ordinary least-squares estimators.

THEOREM 13.4. Let $V^* = (A^{-1}v; v \in V)$. The ordinary least squares and generalized least squares estimators are the same if and only if $V = V^*$.

PROOF. See Exercise C2. □

EXERCISES.

Type B

1. Prove Theorem 13.2.

2. Prove (13.6).

3. Derive (13.8).

4. (a) Let $\mathbf{X}$ be a basis matrix for V. Show that $\mathbf{A}^{1/2}\mathbf{P}_{V^*}\mathbf{A}^{1/2} = \mathbf{X}(\mathbf{X}'\mathbf{A}^{-1}\mathbf{X})^{-1}\mathbf{X}'$.
   (b) Let $\mathbf{d} \in V$. Show that the variance of $\mathbf{d}'\mathbf{A}^{-1}\hat{\boldsymbol{\mu}}$ is $\mathbf{d}'\mathbf{A}^{-1}\mathbf{d}$. (Note that $\mathbf{d} = \mathbf{X}\mathbf{c}$ for some $\mathbf{c} \in R^p$. Why?)
   (c) Find a $(1 - \alpha)$ confidence interval for $\mathbf{d}'\mathbf{A}^{-1}\boldsymbol{\mu}$, $\mathbf{d} \in V$.
   (d) Find a $(1 - \alpha)$ confidence interval for σ^2.

5. Let $\mathbf{d} \in V$. Find the UMP invariant size α test that $\mathbf{d}'\mathbf{A}^{-1}\boldsymbol{\mu} = 0$ against $\mathbf{d}'\mathbf{A}^{-1}\boldsymbol{\mu} \geqslant 0$.

6. Verify (13.10) and (13.11).

7. Let $\mathbf{d} \in V$, $\mathbf{d}^* = \mathbf{A}^{-1/2}\mathbf{d}$.
   (a) Show that $\mathbf{d}^{*\prime}\boldsymbol{\mu}^* = \mathbf{d}'\mathbf{A}^{-1}\boldsymbol{\mu}$, $\mathbf{d}^{*\prime}\hat{\boldsymbol{\mu}}^* = \mathbf{d}'\mathbf{A}^{-1}\hat{\boldsymbol{\mu}}$, $\|\mathbf{d}^*\|^2 = \mathbf{d}'\mathbf{A}^{-1}\mathbf{d}$.
   (b) Show that $\mathbf{d}^* \in V | W^*$ if and only if $\mathbf{d} \in V$ and $\mathbf{d}'\mathbf{A}^{-1}\mathbf{w} = 0$ for all $\mathbf{w} \in W$.
   (c) Show that confidence intervals defined in (13.9) are simultaneous $(1 - \alpha)$ confidence intervals.

(d) Show that the hypothesis $\boldsymbol{\mu} \in W$ is rejected with the test defined in (13.7) if and only if at least one of the Scheffé simultaneous confidence intervals associated with that hypothesis does not contain 0.

8. Best invariant estimator of $\boldsymbol{\mu}$. Consider estimating $\boldsymbol{\mu}$ with the loss function L_1 defined in Section 13.1.

(a) Show that this problem is invariant under the group G_1 given by $g_1(\hat{\boldsymbol{\mu}}, \hat{\sigma}^2) = (c\hat{\boldsymbol{\mu}} + \mathbf{b}, c^2\hat{\sigma}^2)$, $g_1(\boldsymbol{\mu}, \sigma^2) = (c\boldsymbol{\mu} + \mathbf{b}, c^2\sigma^2)$, $g_1(\mathbf{d}_1) = c\mathbf{d}_1 + \mathbf{b}$, where $c > 0$, $\mathbf{b} \in V$.

(b) Show that $\mathbf{T}(\hat{\boldsymbol{\mu}}, \hat{\sigma}^2)$ is an invariant estimator if and only if $T(\hat{\boldsymbol{\mu}}, \hat{\sigma}^2) = \hat{\sigma}\mathbf{T}(\mathbf{0}, 1) + \hat{\boldsymbol{\mu}}$.

(c) Show that $\hat{\boldsymbol{\mu}}$ is the best invariant estimator of $\boldsymbol{\mu}$.

9 Best invariant estimator of σ^2. Consider estimating σ^2 with the loss function L_2 defined in Section 13.1.

(a) Under what group is this problem invariant?

(b) Show that the best invariant estimator of σ^2 is $[(n-p)\hat{\sigma}^2]/(n-p+2)$.

10. Generalized linear model with known σ^2. Consider the model in which we observe $\mathbf{Y} \sim N_n(\boldsymbol{\mu}, \mathbf{A})$, $\boldsymbol{\mu} \in V$, where $\mathbf{A}$ is a known positive definite matrix.

(a) Let $\hat{\boldsymbol{\mu}}$ be defined by (13.5). Show that $\hat{\boldsymbol{\mu}}$ is a complete sufficient statistic for this model.

(b) Show that $\hat{\boldsymbol{\mu}}$ is the MLE and minimum variance unbiased estimator of $\boldsymbol{\mu}$.

(c) Consider testing that $\boldsymbol{\mu} \in W$ against $\boldsymbol{\mu} \in V$. Show that the UMP invariant size α test rejects if $\|\mathbf{P}_{V^* | W^*}\mathbf{Y}^*\|^2 \geqslant \chi^{2\alpha}_{p-k}$.

(d) Let $\mathbf{X}$ be a basis matrix for V, let $\boldsymbol{\mu} = \mathbf{X}\boldsymbol{\beta}$ and let $\hat{\boldsymbol{\beta}}$ be defined by (13.10). Show that $\hat{\boldsymbol{\beta}}$ is a complete sufficient statistic for this model and is the MLE and the minimum variance unbiased estimator for $\boldsymbol{\beta}$.

Type C

1. A James-Stein estimator. We consider estimating $\boldsymbol{\mu}$ with the Mahalanobis distance loss function

$$L\big(\mathbf{d}; (\boldsymbol{\mu}, \sigma^2)\big) = \frac{(\mathbf{d} - \boldsymbol{\mu})'\mathbf{A}^{-1}(\mathbf{d} - \boldsymbol{\mu})}{\sigma^2}$$

and risk function

$$R\big(\mathbf{d}(\mathbf{Y}); (\boldsymbol{\mu}, \sigma^2)\big) = \frac{E(\mathbf{d}(\mathbf{Y}) - \boldsymbol{\mu})'\mathbf{A}^{-1}(\mathbf{d}(\mathbf{Y}) - \boldsymbol{\mu})}{\sigma^2}.$$

Find the risk of the estimator

$$\hat{\hat{\boldsymbol{\mu}}}_{\mathbf{v}} = \left[1 - \frac{(p-2)(n-p)\hat{\hat{\sigma}}^2}{(n-p+2)(\hat{\boldsymbol{\mu}} - \mathbf{v})'\mathbf{A}^{-1}(\hat{\boldsymbol{\mu}} - \mathbf{v})}(\hat{\boldsymbol{\mu}} - \mathbf{v}) + \mathbf{v}\right]$$

And show that $\hat{\hat{\boldsymbol{\mu}}}_{\mathbf{v}}$ is better than $\hat{\boldsymbol{\mu}}$. [Hint: Let $\hat{\hat{\boldsymbol{\mu}}}^*_{\mathbf{v}}(\mathbf{Y}) = \mathbf{A}^{-1/2}\hat{\hat{\boldsymbol{\mu}}}_{\mathbf{v}}(\mathbf{A}^{1/2}\mathbf{Y})$. Then

$$R\left(\hat{\hat{\boldsymbol{\mu}}}_{\mathbf{v}}; (\boldsymbol{\mu}, \sigma^2)\right) = \frac{E\|\hat{\hat{\boldsymbol{\mu}}}^*_{\mathbf{v}}(\mathbf{Y}) - \boldsymbol{\mu}^*\|^2}{\sigma^2}.$$

Why? Find an expression for $\hat{\hat{\boldsymbol{\mu}}}^*$ and use (11.3).]

2. Let V^* and V^{**} be defined as in (13.2) and Theorem 13.4.

(a) Show that the ordinary least-squares and generalized least-squares estimators are equal if and only if

$$\mathbf{P}_{V^*}\mathbf{A}^{-1/2}\mathbf{y} = \mathbf{A}^{-1/2}\mathbf{P}_V\mathbf{y}$$

for all $\mathbf{y} \in R^n$.

(b) Show that $\mathbf{P}_{V^*}\mathbf{A}^{-1/2}\mathbf{y} = \mathbf{A}^{-1/2}\mathbf{P}_V\mathbf{y}$ if and only if $\mathbf{A}^{-1/2}\mathbf{P}_{V^\perp}\mathbf{y} \in V^{*\perp}$. (Hint: Use the definition of $\mathbf{P}_{V^*}$.)

(c) Show that $\mathbf{A}^{-1/2}\mathbf{P}_{V^\perp}\mathbf{y} \in V^{*\perp}$ if and only if $\mathbf{P}_{V^\perp}\mathbf{y} \in V^{**\perp}$. (Hint: $\mathbf{A}^{-1/2}\mathbf{P}_{V^\perp}\mathbf{y} \in V^{*\perp}$ if and only if $\mathbf{A}^{-1/2}\mathbf{P}_{V^\perp}\mathbf{y} \perp \mathbf{A}^{-1/2}\mathbf{y}$ for all $\mathbf{v} \in V$ if and only if $\mathbf{P}_{V^\perp}y \perp A^{-1}\mathbf{v}$ for all $\mathbf{v} \in V$.)

(d) Show that $\mathbf{P}_{V^\perp}\mathbf{y} \in V^{**\perp}$ for all $\mathbf{y} \in R^n$ if and only if $V \supset V^{**}$.

(e) Show that V and V^{**} have the same dimensions.

(f) Derive Theorem 13.4.

CHAPTER 14

The Repeated Measures Model

We now consider a model for experiments in which each individual receives several treatments, and hence the observations cannot be assumed independent. We assume throughout this chapter that each individual receives the same number, r, of treatments and that the measurements on each individual have an r-dimensional multivariate normal distribution with common covariance matrix $\mathbf{\Sigma} > 0$. If we make no additional assumptions about $\mathbf{\Sigma}$, this model is called the generalized repeated measures model and is studied in Sections 18.7 and 19.8. Often, however, we do not have enough individuals to get a good estimator for $\mathbf{\Sigma}$, and in this case the procedures discussed in those sections are not very powerful. In this chapter, we assume that

$$\mathbf{\Sigma} = \sigma^2 \mathbf{A}(\rho), \qquad \mathbf{A}(\rho) = \begin{pmatrix} 1 & \rho & \cdots & \rho \\ \rho & 1 & \cdots & \rho \\ \vdots & & \ddots & \vdots \\ \rho & \rho & \cdots & 1 \end{pmatrix} \tag{14.1}$$

that is, that all the measurements have the same variance and all the pairs of measurements on the same individual have the same covariance. By making this stronger assumption about $\mathbf{\Sigma}$, we can get more powerful procedures. However, these procedures are only valid if $\mathbf{\Sigma}$ satisfies (14.1). For example, it seems unlikely that $\mathbf{\Sigma}$ would satisfy (14.1) if the measurements had different units.

Let $\mathbf{Y}_k$ be the r-dimensional vector of observations on the kth individual. In this chapter, we assume that the $\mathbf{Y}_k$ are independent and

$$\mathbf{Y}_k \sim N_r(\boldsymbol{\mu}_k, \sigma^2 \mathbf{A}(\rho)), \qquad k = 1, \ldots, m \tag{14.2}$$

where m is the number of individuals. Let $\mathbf{1}$ be the r-dimensional vector all of whose elements are 1 and let $\mathbf{U}$ be the one-dimensional subspace of R^r spanned by $\mathbf{1}$. Define

$$\delta_k \mathbf{1} = \mathbf{P}_U \boldsymbol{\mu}_k, \qquad \gamma_k = \mathbf{P}_{U^\perp} \boldsymbol{\mu}_k, \qquad \boldsymbol{\delta} = \begin{bmatrix} \delta_1 \\ \vdots \\ \delta_m \end{bmatrix}, \qquad \boldsymbol{\gamma} = \begin{bmatrix} \gamma_1 \\ \vdots \\ \gamma_m \end{bmatrix} \tag{14.3}$$

The parameter space that we consider for this model is given by

$$\boldsymbol{\delta} \in S, \qquad \boldsymbol{\gamma} \in T, \qquad \sigma^2 > 0, \qquad \mathbf{A}(\rho) > 0, \tag{14.4}$$

where S and T are s- and t-dimensional subspaces of R^m and R^{mr}. (Additional assumptions about S and T are given in Section 14.4.) The model defined by (14.2)–(14.4) is called the *repeated measures model.*

The analysis of the repeated measures model is somewhat more complicated than the analysis of previous models. For that reason, in Section 14.1, we state the results for this model. In Section 14.2 we look at several examples of repeated measures models. In Sections 14.3–14.7 we derive the results stated in Section 14.1. In Section 14.8 we summarize other results for this model. In Section 14.9 we look at the case $m = 1$, which is called the exchangeable linear model.

Particular cases of repeated measures models have been analyzed as mixed models as studied in the next chapter. (See Winer, 1971, for an elementary example of this approach.) The general approach used in this chapter is from Arnold (1979b).

14.1. STATEMENT OF THE RESULTS.

We now state the main results for the repeated measures model. The proofs are given in Sections 14.4–14.7. We continue with the notation of the last section. Let $\mathbf{Y}$, $\boldsymbol{\mu}$ and $\mathbf{B}(\rho)$ be defined by

$$\mathbf{Y} = \begin{bmatrix} \mathbf{Y}_1 \\ \vdots \\ \mathbf{Y}_m \end{bmatrix}, \qquad \mathbf{H} = \begin{bmatrix} \mathbf{1} & \mathbf{0} & \cdots & \mathbf{0} \\ \mathbf{0} & \mathbf{1} & \cdots & \mathbf{0} \\ \vdots & \vdots & & \vdots \\ \mathbf{0} & \mathbf{0} & \cdots & \mathbf{1} \end{bmatrix} \qquad \boldsymbol{\mu} = \begin{bmatrix} \boldsymbol{\mu}_1 \\ \vdots \\ \boldsymbol{\mu}_m \end{bmatrix} = \mathbf{H}\boldsymbol{\delta} + \gamma_1$$

$$\mathbf{B}(\rho) = \begin{bmatrix} \mathbf{A}(\rho) & & \mathbf{0} \\ \vdots & \ddots & \vdots \\ \mathbf{0} & \cdots & \mathbf{A}(\rho) \end{bmatrix}. \tag{14.5}$$

(Where $\mathbf{A}(\rho)$ is repeated m times so that $\mathbf{B}(\rho)$ is $mr \times mr$.) Finally let V be the subspace of R^{mr} given by

$$V = \{ \boldsymbol{\mu} = \mathbf{H}\boldsymbol{\delta} + \boldsymbol{\gamma}, \boldsymbol{\delta} \in S, \boldsymbol{\gamma} \in T \}. \tag{14.6}$$

An equivalent version of the repeated measures model is one in which we observe the mr-dimensional vector $\mathbf{Y}$ such that

$$\mathbf{Y} \sim N_{mr}(\boldsymbol{\mu}, \sigma^2 \mathbf{B}(\rho)), \qquad \boldsymbol{\mu} \in V, \qquad \sigma^2 > 0, \qquad \mathbf{B}(\rho) > 0.$$

If $\rho = 0$, then $\mathbf{B}(\rho) = \mathbf{I}$ and this model is an ordinary linear model. In this section we state results that show how results for the ordinary linear model that occurs when $\rho = 0$ in the repeated measures model can be used to give results for the repeated measures model.

We note first that the dimension of V is $s + t$ [see (14.16)]. Therefore, in the notation of Chapters 4–12, $n = mr$ and $p = s + t$. Let

$$\hat{\boldsymbol{\mu}} = \mathbf{P}_V \mathbf{Y}, \qquad \hat{\sigma}^2 = \frac{\|\mathbf{P}_{V^\perp} \mathbf{Y}\|^2}{(mr - s - t)}$$

be the usual unbiased estimators for $\boldsymbol{\mu}$ and σ^2 in the ordinary linear model that occurs when $\rho = 0$. ($\hat{\sigma}^2$ is not an unbiased estimator of σ^2 for the repeated measures model.) Now, let $\mathbf{Y}_k = (Y_{k1}, \ldots, Y_{kr})'$, $k = 1, \ldots, m$ and define

$$\mathbf{Z} = \begin{pmatrix} \bar{Y}_{1\cdot} \\ \vdots \\ \bar{Y}_{m\cdot} \end{pmatrix}, \qquad U_1^2 = \frac{r\|\mathbf{P}_{S^\perp} \mathbf{Z}\|^2}{m - s}, \qquad U_2^2 = \frac{(mr - t - s)\hat{\sigma}^2 - (m - s)U_1^2}{m(r - 1) - t}. \tag{14.7}$$

(That is, the kth component of $\mathbf{Z}$ is just the average of the measurements on the kth individual.)

The formula for U_2^2 can be remembered in the following way. The degrees of freedom for $\hat{\sigma}^2$ is just $mr - t - s$ $(= n - p)$ and the degrees of freedom for U_1^2 and U_2^2 are $m - s$ and $m(r - 1) - t$ (see Theorem A below). Therefore, the degrees of freedom for $\hat{\sigma}^2$ is just the sum of the degrees of freedom for U_1^2 and U_2^2. Now let $SS_1 = (m - s)U_1^2$, $SS_2 = [m(r - 1) - t]U_2^2$ and $SS_3 = (mr - s - t)\hat{\sigma}^2$ be the sum of squares associated with U_1^2, U_2^2, and $\hat{\sigma}^2$. Then $SS_3 = SS_1 + SS_2$. That is, the sum of squares for $\hat{\sigma}^2$ is just the sum of the sums of squares of U_1^2 and U_2^2.

THEOREM A. a. $(\hat{\boldsymbol{\mu}}, U_1^2, U_2^2)$ is a complete sufficient statistic for the repeated measures model.

b. $\hat{\boldsymbol{\mu}}$, U_1^2, and U_2^2 are independent,

$$\hat{\boldsymbol{\mu}} \sim N_{mr}(\boldsymbol{\mu}, \sigma^2 \mathbf{P}_V \mathbf{B}(\rho) \mathbf{P}_V), \qquad (m-s)U_1^2 \sim \sigma^2(1+(r-1)\rho)\chi^2_{m-s}(0),$$
$$[m(r-1)-t]U_2^2 \sim \sigma^2(1-\rho)\chi^2_{m(r-1)-t}(0).$$

c. The minimum variance unbiased estimators of $\boldsymbol{\mu}$, σ^2 and $\rho\sigma^2$ are $\hat{\boldsymbol{\mu}}$,

$$\frac{U_1^2 + (r-1)U_2^2}{r} \quad \text{and} \quad \frac{U_1^2 - U_2^2}{r}.$$

d. The MLE's of $\boldsymbol{\mu}$, σ^2 and $\rho\sigma^2$ are $\hat{\boldsymbol{\mu}}$, $\{(m-s)U_1^2 + [m(r-1)-t]U_2^2\}/mr$ and $\{(m-s)(r-1)U_1^2 - [m(r-1)-t]U_2^2\}/mr(r-1)$.

PROOF. See Theorems 14.7 and 14.8. □

Note that the MLE of σ^2 is the same for the repeated measures model as for the ordinary linear model.

We now consider several problems of hypothesis testing. Let Q_A be a q_A-dimensional subspace of S, $q_A < s$. We say that testing that $\boldsymbol{\delta} \in Q_A$ is testing a *type A hypothesis*. Similarly, if Q_B is a q_B-dimensional subspace of T, $q_B < t$, we say that testing that $\boldsymbol{\gamma} \in Q_B$ is a *type B hypothesis*. Let

$$W_A = \{\boldsymbol{\mu} = \mathbf{H}\boldsymbol{\delta} + \boldsymbol{\gamma}, \boldsymbol{\delta} \in Q_A, \boldsymbol{\gamma} \in T\}$$
$$W_B = \{\boldsymbol{\mu} = \mathbf{H}\boldsymbol{\delta} + \boldsymbol{\gamma}, \boldsymbol{\delta} \in S, \boldsymbol{\gamma} \in Q_B\} \tag{14.8}$$

Since we know that $\boldsymbol{\mu} \in V$ (and hence $\boldsymbol{\delta} \in S$, $\boldsymbol{\gamma} \in T$) testing that $\boldsymbol{\delta} \in Q_A$ is the same as testing that $\boldsymbol{\mu} \in W_A$ and testing that $\boldsymbol{\gamma} \in Q_B$ is the same as testing that $\boldsymbol{\mu} \in W_B$. Also

$$\dim V | W_A = \dim S | Q_A = s - q_A, \qquad \dim V | W_B = \dim T | Q_B = t - q_B$$

(see Lemma 14.10).

THEOREM B. a. The UMP invariant size α test that $\boldsymbol{\delta} \in Q_A$ (or equivalently $\boldsymbol{\mu} \in W_A$) is given by

$$F_A = \frac{\|\mathbf{P}_{V|W_A}\mathbf{Y}\|^2}{(s-q_A)U_1^2}, \qquad \phi_A(F_A) = \begin{cases} 1 & \text{if } F_A > F^{\alpha}_{s-q_A,m-s} \\ 0 & \text{if } F_A \leqslant F^{\alpha}_{s-q_A,m-s} \end{cases}.$$

b. The UMP invariant size α test that $\boldsymbol{\gamma} \in Q_B$ (or equivalently $\boldsymbol{\mu} \in W_B$) is given by

$$F_B = \frac{\|\mathbf{P}_{V|W_B}\mathbf{Y}\|^2}{(t-q_B)U_2^2}, \qquad \phi_B(F_B) = \begin{cases} 1 & \text{if } F_B > F^{\alpha}_{t-q_B,m(r-1)-t} \\ 0 & \text{if } F_B \leqslant F^{\alpha}_{t-q_B,m(r-1)-t} \end{cases}.$$

c.

$$F_A \sim F_{s-q_A,m-s}\left(\frac{\|\mathbf{P}_{V|W_A}\boldsymbol{\mu}\|^2}{\sigma^2(1+(r-1)\rho)}\right) \qquad F_B \sim F_{t-q_B,m(r-1)-t}\left(\frac{\|\mathbf{P}_{V|W_B}\boldsymbol{\mu}\|^2}{\sigma^2(1-\rho)}\right).$$

PROOF. See Theorems 14.9 and 14.11 and Lemma 14.10. □

Note that for both F_A and F_B the numerator sums of squares ($\|\mathbf{P}_{V|W_A}\mathbf{Y}\|^2$ and $\|\mathbf{P}_{V|W_B}\mathbf{Y}\|^2$) are the same as for the ordinary linear model and the numerator degrees of freedom ($s - q_A$ and and $s - q_B$) are also the same as for the ordinary linear model. Therefore, to find the appropriate F-statistic for the repeated measures model (for a type A or type B hypothesis), we merely take the appropriate F-statistic for the ordinary linear model and replace $\hat{\sigma}^2$ and $mr - t - s$ ($= n - p$) with $U_1^2$and $m - s$ or U_2^2 and $m(r-1) - t$ depending on whether the hypothesis is of type A or type B. Similarly to find the noncentrality parameter we merely take the noncentrality parameter in the orinary linear model and replace σ^2 with $\sigma^2(1 + (r-1)\rho)$ or $\sigma^2(1 - p)$ depending on whether the hypothesis is type A or type B.

We now look at simultaneous confidence intervals for contrasts associated with type A and type B hypotheses. As in Chapter 8, a contrast associated with testing the type A hypothesis that $\boldsymbol{\mu} \in W_A$ is a linear function of the form $\langle \mathbf{d}, \boldsymbol{\mu} \rangle$, $\mathbf{d} \in V|W_A$ and a contrast associated with testing the type B hypothesis that $\boldsymbol{\mu} \in W_B$ is a linear function of the form $\langle \mathbf{d}, \boldsymbol{\mu} \rangle$, $\mathbf{d} \in V|W_B$.

THEOREM C. a. A set of simultaneous $(1 - \alpha)$ confidence intervals for the set of all contrasts associated with testing the type A hypothesis that $\boldsymbol{\mu} \in W_A$ is given by

$$\langle \mathbf{d}, \boldsymbol{\mu} > \in \langle \mathbf{d}, \hat{\boldsymbol{\mu}} \rangle \pm U_1 \|\mathbf{d}\| \left[(s - q_A) F^{\alpha}_{s-q_A, m-s} \right]^{1/2}.$$

The hypothesis $\boldsymbol{\mu} \in W_A$ is rejected with the F-test ϕ_A defined in Theorem B if and only if at least one of the confidence intervals does not contain 0.

b. A set of simultaneous $(1 - \alpha)$ confidence intervals associated with testing the type B hypothesis that $\boldsymbol{\mu} \in W_B$ is given by

$$\langle \mathbf{d}, \boldsymbol{\mu} \rangle \in \langle \mathbf{d}, \hat{\boldsymbol{\mu}} \rangle \pm U_2 \|\mathbf{d}\| \left[(t - q_B) F^{\alpha}_{t-q_B, m(r-1)-t} \right]^{1/2}.$$

The hypothesis $\boldsymbol{\mu} \in W_B$ is rejected with the F-test ϕ_B defined in Theorem B if and only if at least one of the confidence intervals does not contains 0.

PROOF. See Theorem 14.12. □

We see again that simultaneous confidence intervals for contrasts associated with type A and type B hypotheses are easy to find. We merely replace $\hat{\sigma}^2$ and $mr - s - t$ in the confidence intervals for the ordinary linear model with U_1 and $m - s$ or U_2 and $m(r-1) - t$ depending on whether the hypothesis is of type A or type B.

We could also test the hypothesis that $\rho = 0$ (i.e., that the measurements are independent). Although there is no UMP invariant size α test for this problem, the following test is a sensible size α test:

$$F_c = \frac{U_1^2}{U_2^2} \sim \frac{1+(r-1)\rho}{1-\rho} F_{m-s,m(r-1)-t}(0),$$

$$\phi(F_c) = \begin{cases} 1 & \text{if } F_c > F^a_{m-s,m(r-1)-t} \text{ or } F_c < F^{1-b}_{m-s,m(r-1)-t} \\ 0 & \text{if } F^{1-b}_{m-s,m(r-1)-t} \leqslant F_c \leqslant F^a_{m-s,m(r-1)-t} \end{cases} \tag{14.9}$$

where $a + b = \alpha$. We could also use (14.9) to find a confidence interval for ρ (see Exercise B2). At this time there is no way to find $1 - \alpha$ confidence intervals for σ^2 or size α tests of hypotheses about σ^2.

14.2. EXAMPLES.

We now look at some examples of repeated measures models. In these examples, the goal is to test certain hypotheses, so we derive the appropriate tests and simultaneous confidence intervals for these hypotheses.

To determine $\boldsymbol{\delta}$ for each of these models, we use the fact that

$$\delta_k \mathbf{1} = \mathbf{1}(\mathbf{1}'\mathbf{1})^{-1}\mathbf{1}'\boldsymbol{\mu}_k = \frac{1}{r}(\mathbf{1}'\boldsymbol{\mu}_k)\mathbf{1}$$

and hence $\delta_k = (1/r)\mathbf{1}'\boldsymbol{\mu}_k$, the average of the elements of $\boldsymbol{\mu}_k$. Let $\mathbf{Z}$ be defined as in the last section and let $\hat{\boldsymbol{\delta}} = (\hat{\delta}_1, \ldots, \hat{\delta}_m)' = \mathbf{P}_S\mathbf{Z}$. Then $\hat{\boldsymbol{\delta}}$ minimizes $\|\mathbf{Z} - \boldsymbol{\delta}\|^2 = \sum_i (Z_i - \delta_i)^2$ out of all $\boldsymbol{\delta} \in S$. Also

$$(m-s)U_1^2 = r\|\mathbf{P}_{S^\perp}\mathbf{Z}\|^2 = r\|\mathbf{Z} - \hat{\boldsymbol{\delta}}\|^2 = r\sum_i (Z_i - \hat{\delta}_i)^2.$$

EXAMPLE 1. Consider a one-way model with d treament levels in which each of n individuals receives each treatment level. Let $\mathbf{Y}_k = (Y_{1k}, \ldots, Y_{dk})'$ be the vector of observations on the kth individual, $k = 1, \ldots, n$. Then the model would be given by

$$Y_{ik} = \theta + \alpha_i + e_{ik}, \qquad \sum_i \alpha_i = 0, \qquad \mathbf{e}_k = (e_{1k}, \ldots, e_{dk})' \sim N_d(\mathbf{0}, \sigma^2\mathbf{A}(\rho))$$

and the $\mathbf{e}_k$ are independent. Hence $\boldsymbol{\mu}_k = E\mathbf{Y}_k = (\theta + \alpha_1, \ldots, \theta + \alpha_d)'$. Note that $\boldsymbol{\mu}_k$ does not depend on k since each individual receives the same treaments. Now, δ_k is the average of the components of $\boldsymbol{\mu}_k$, and hence $\delta_k = \theta$, $\gamma_k = \boldsymbol{\mu}_k - \delta_k\mathbf{1} = (\alpha_1, \ldots, \alpha_d)'$. $\boldsymbol{\delta} = (\delta_1, \ldots, \delta_n)' = (\theta, \ldots, \theta)'$. Therefore, $s = \dim S = 1$, and $t = \dim T = \dim V - \dim S = d - 1$. Note also that m is the number of individuals so that $m = n$, whereas r is the number of measurements on each individual so that $r = d$. From Section

7.3.1, we see that

$$\hat{\sigma}^2 = \frac{\Sigma_i \Sigma_j \left(Y_{ij} - \overline{Y}_{i\cdot} \right)^2}{d(n-1)}.$$

We now find U_1^2 and U_2^2. Note first that $\mathbf{Z}' = (\overline{Y}_{\cdot 1}, \ldots, \overline{Y}_{\cdot m})$. To find $\hat{\theta}$ (and hence $\hat{\delta}$) we must minimize $\Sigma_k (\overline{Y}_{\cdot k} - \theta)^2$, and hence $\hat{\theta} = \overline{Y}_{\cdots}$. Therefore, for this model

$$U_1^2 = \frac{d \Sigma_k \left(\overline{Y}_{\cdot k} - \overline{Y}_{\cdot\cdot} \right)^2}{n-1}, \qquad U_2^2 = \frac{d(n-1)\hat{\sigma}^2 - (n-1)U_1^2}{d(n-1) - (n-1)} = \frac{d\hat{\sigma}^2 - U_1}{d-1}.$$

Now consider testing that the $\alpha_i = 0$. We note that the α_i are in γ so that this is a type B hypothesis. Therefore, using Theorem B and results from Section 7.3.1, we see that the appropriate F-statistic is

$$\frac{n \Sigma_i \left(\overline{Y}_{i\cdot} - \overline{Y}_{\cdot\cdot} \right)^2}{(d-1)U_2^2} \sim F_{d-1,(d-1)(n-1)} \left(\frac{p \Sigma_i \alpha_i^2}{\sigma^2 (1-\rho)} \right).$$

As a final result for this example, we find simultaneous confidence intervals for the set of all contrasts associated with testing that the $\alpha_i = 0$. Using Theorem C and results from Section 8.2.1, we see that

$$\sum_i b_i \alpha_i \in \sum_i b_i \overline{Y}_{i\cdot} \pm U_2 \left[(d-1) F_{d-1,(d-1)(n-1)} \sum_i b_i^2 / n \right]^{1/2}$$

for all b_i such that $\sum_i b_i = 0$.

EXAMPLE 2. Consider a two-way model with d rows and c columns in which each of n individuals receives every pair of treatment levels. Let $\mathbf{Y}_k = (Y_{11k}, \ldots, Y_{d1k}, \ldots, Y_{dck})'$ be the vector of observations on the kth individual, $k = 1, \ldots, n$. Then the model is given by

$$Y_{ijk} = \theta + \alpha_i + \beta_j + \gamma_{ij} + e_{ijk}, \qquad \sum_i \alpha_i = 0, \qquad \sum_j \beta_j = 0,$$

$$\sum_i \gamma_{ij} = 0, \qquad \sum_j \gamma_{ij} = 0 \qquad \mathbf{e}_k = (e_{11k}, \ldots, e_{dck})' \sim N_{dc}\left(\mathbf{0}, \sigma^2 \mathbf{A}(\rho)\right)$$

and the $\mathbf{e}_k$ are independent. Hence $\boldsymbol{\mu}_k = (\alpha + \alpha_1 + \beta_1 + \gamma_{11}, \ldots, \theta + \alpha_d + \beta_c + \gamma_{dc})'$ which again does not depend on k. As before $\delta_k = \theta$. Therefore, $\gamma_k = (\alpha_1 + \beta_1 + \gamma_{11}, \ldots, \alpha_d + \beta_c + \gamma_{dc})'$. Again, $\dim S = 1$. Also, $\dim T = \dim V - \dim S = dc - 1$. Finally, m is the number of individuals and so $m = n$, where r is the number of measurements on each individual and $r = dc$. We now find U_1^2 and U_2^2 for this model. From Section 7.3.3, we see that

$$\hat{\sigma}^2 = \frac{\sum_i \sum_j \sum_k \left(Y_{ijk} - \bar{Y}_{ij.}\right)^2}{dc(n-1)}.$$

In Exercise B3, you are asked to show that

$$U_1^2 = \frac{dc\sum_k\left(\bar{Y}_{..k} - \bar{Y}_{...}\right)^2}{n-1}, \qquad U_2^2 = \frac{dc\hat{\sigma}^2 - U_1}{dc-1}.$$

Note also that $m - s = n - 1$, $m(r-1) - t = n(cd-1) - (cd-1) = (n-1)(cd-1)$. We now consider testing that the $\alpha_i = 0$. We note that this hypothesis is again a type B hypothesis since the α_i are in $\boldsymbol{\gamma}$. Using the results of Section 7.3.3, together with Theorem B, we see that the appropriate F-statistic is

$$F = \frac{cn\sum_i\left(Y_{i..} - \bar{Y}_{...}\right)^2}{(d-1)U_2^2} \sim F_{d-1,(n-1)(dc-1)}\left(\frac{cn\sum_i \alpha_i^2}{\sigma^2(1-\rho)}\right).$$

Using Theorem C together with the results of Section 8.2.2, we see that a set of simultaneous confidence intervals for the set of all contrasts associated with testing that the $\alpha_i = 0$ is given by

$$\sum_i b_i \alpha_i \in \sum_i b_i \bar{Y}_{i.} \pm U_2\left[\frac{(d-1)F^{\alpha}_{d-1,(n-1)(dc-1)}\sum_i b_i^2}{cn}\right]^{1/2}.$$

where $\Sigma b_i = 0$. We could consider testing that $\beta_j = 0$ or that $\gamma_{ij} = 0$. Either of these hypotheses is also of type B and we could use results for the ordinary linear model to find the appropriate statistics and confidence intervals in this model.

We now consider a model that has an interesting type A hypothesis.

EXAMPLE 3. Consider now a two-way model with d rows and c columns in which each individual receives each row treatment but only one column treatment. (This would occur, for example, if columns represented race, sex, degree of illness, etc.). We assume that there are n individuals who receive each column treatment. Let $\mathbf{Y}_{jk} = (Y_{ijk}, \ldots, Y_{djk})'$ be the vector of observations on the kth person who receives the jth column treatment, $j = 1, \ldots, c$; $k = 1, \ldots, n$. The model would then be

$$Y_{ijk} = \theta + \alpha_i + \beta_j + \gamma_{ij} + e_{ijk}, \qquad \sum_i \alpha_i = 0, \qquad \sum_j \beta_j = 0,$$

$$\sum_i \gamma_{ij} = 0, \qquad \sum_j \gamma_{ij} = 0 \qquad \mathbf{e}_{jk} = (e_{ijk}, \ldots, e_{dkj})' \sim N_d\left(\mathbf{0}, \sigma^2\mathbf{A}(\rho)\right)$$

and the $\mathbf{e}_{jk}$ are independent. Then $\boldsymbol{\mu}_{jk} = E\mathbf{Y}_{jk} = (\theta + \alpha_1 + \beta_j + \gamma_{1j}, \ldots, \theta + \alpha_d + \beta_j + \gamma_{dj})'$. Hence δ_{jk}, the average of the components of μ_{jk} is $\theta + \beta_j$, and $\gamma_{jk} = (\alpha_1 + \gamma_{ij}, \ldots, \alpha_d + \gamma_{dj})'$. Then $\boldsymbol{\delta} = (\delta_{11}, \ldots, \delta_{cn})' = (\theta + \beta_1, \ldots, \theta + \beta_1, \ldots, \theta + \beta_c)'$. Therefore, $s = \dim S = c$, $t = \dim T = \dim V - \dim S = dc - c = c(d-1)$. Also m is the number of individuals and hence $m = cn$, whereas r is the number of measurements on each individual and hence $r = c$. Now, $Z_{jk} = \overline{Y}_{\cdot jk}$. To find U_1^2 we must minimize $\Sigma_j\Sigma_k(\overline{Y}_{\cdot jk} - \theta - \beta_j)^2$ and hence $\hat{\theta} = \overline{Y}_{\ldots}$, $\hat{\beta}_j = \overline{Y}_{\cdot j\cdot} - \overline{Y}_{\ldots}$. Therefore,

$$U_1^2 = \frac{d\Sigma_j\Sigma_k\left(\overline{Y}_{\cdot jk} - \overline{Y}_{\cdot j\cdot}\right)}{c(n-1)}, \qquad U_2^2 = \frac{dc(n-1)\hat{\sigma}^2 - c(n-1)U_1}{dc(n-1) - (n-1)c} = \frac{d\hat{\sigma}^2 - U_1}{d-1},$$

where as before

$$\hat{\sigma}^2 = \frac{\Sigma_i\Sigma_j\Sigma_k\left(Y_{ijk} - \overline{Y}_{ij\cdot}\right)^2}{dc(n-1)}.$$

Now, consider testing that the $\alpha_i = 0$. This hypothesis is a type B hypothesis and therefore the appropriate F-statistic is

$$F = \frac{nc\Sigma_i\left(\overline{Y}_{i\ldots} - \overline{Y}_{\ldots}\right)^2}{(d-1)U_2^2} \sim F_{d-1,c(d-1)(n-1)}\left(\frac{nc\Sigma_i\alpha_i^2}{\sigma^2(1-\rho)}\right)$$

and the simultaneous confidence intervals for contrasts are

$$\sum_i b_i\alpha_i \in \sum_i b_i\overline{Y}_{i\ldots} \pm U_2\left[\frac{(d-1)F^{\alpha}_{d-1,c(d-1)(n-1)}\sum b_i^2}{cn}\right]^{1/2}$$

where $\Sigma_i b_i = 0$. Now, consider testing that $\beta_j = 0$. This is a type A hypothesis, since the β_j are in $\boldsymbol{\delta}$. Therefore, the appropriate F-statistic for this hypothesis is

$$\frac{md\Sigma_j\left(\overline{Y}_{\cdot j\cdot} - \overline{Y}_{\ldots}\right)^2}{(c-1)U_1^2} \sim F_{c-1,c(n-1)}\left[\frac{dm\Sigma_j\beta_j^2}{\sigma^2(1+(d-1)\rho)}\right]$$

and the simultaneous confidence intervals for contrasts are

$$\sum_j b_j\beta_j \in \sum_j b_j\overline{Y}_{\cdot j\cdot} \pm U_1\left[\frac{(c-1)F^{\alpha}_{c-1,c(n-1)}\Sigma_j b_j^2}{dn}\right]^{1/2}$$

where $\Sigma b_j = 0$. We could also test that the $\gamma_{ij} = 0$. Note that this hypothesis is a type B hypothesis.

Example 3 illustrates the following two simple rules for determining whether a particular hypothesis has type A, type B, or neither. If the hypothesis involves only effects that are constant for each individual (such as the β_j in Example 3), then the hypothesis is of type A. If the hypothesis involves only effects whose average for each individual is 0 (such as the α_i in Example 3), then the hypothesis is of type B. If the hypothesis involves effects of both types (such as testing for $\beta_j = 0$ and $\gamma_{ij} = 0$) or if it involves effects of neither type (such as the example below), then the hypothesis is neither type A nor type B and cannot be tested by the methods of this chapter.

All the examples above are balanced. Examples 1 and 2 must be balanced since each individual receives every possible treatment in these models. Example 3 need not be balanced and in Exercise B4, an unbalanced version of this model is treated. The only balancing assumption made in this chapter is that the same number of observations is made on each individual.

We finish this section with an example in which an interesting hypothesis is neither type A or type B.

EXAMPLE 4. Consider an analysis of covariance model in which each individual receives each treatment level, and each measurement has a covariate. Let $\mathbf{Y}_k = (Y_{1k}, \ldots, Y_{rk})'$ be the measurements on the kth individual, $k = 1, \ldots, m$. Then the model would be given by

$$Y_{ik} = \theta + \alpha_i + \beta X_{ik}, \qquad \sum \alpha_i = 0, \qquad \mathbf{e}_k = (e_{1k}, \ldots, e_{rk})' \sim N_r(\mathbf{0}, \sigma^2 \mathbf{A}(\rho))$$

and the $\mathbf{e}_k$ are independent. Then $\boldsymbol{\mu}_k = E\mathbf{Y}_k = (\theta + \alpha_1 + \beta X_{1k}, \ldots, \theta + \alpha_r + \beta X_{rk})'$. Therefore $\delta_k = \theta + \beta \bar{X}_{.k}$, and $\gamma_k = [\alpha_1 + \beta(X_{1k} - \bar{X}_{.k}), \ldots, \alpha_r + \beta(X_{rk} - \bar{X}_{.k})]'$. Now consider testing that $\beta = 0$. If $\bar{X}_{.k} = 0$, then this hypothesis is of type B. If X_{ik} depends only on k (i.e., depends only on the individual and not the treatment level) then this hypothesis is of type A. Otherwise, it is not of either type.

14.3. SOME MORE LINEAR ALGEBRA.

In this section we derive some additional elementary results about linear algebra that are useful in this chapter.

LEMMA 14.1. a. Let $V \subset R^n$ be a p-dimensional subspace, and let $\boldsymbol{\Gamma}$ be an $n \times n$ orthogonal matrix. Define $V^* = \{\boldsymbol{\Gamma}\mathbf{v}, \mathbf{v} \in V\}$. Then V^* is a p-dimensional subspace of R^n and $\mathbf{P}_{V^*} = \boldsymbol{\Gamma}\mathbf{P}_V\boldsymbol{\Gamma}'$.

b. Let W be a k-dimensional subspace of V, where V is a p-dimensional subspace of R^n. Let $\mathbf{C}'$ be an orthonormal basis matrix for V. Define $W^* = \{\mathbf{C}\mathbf{w}; \mathbf{w} \in W\}$. The W^* is a k-dimensional subspace of R^p and $\mathbf{P}_{\mathbf{W}^*} = \mathbf{C}\mathbf{P}_W\mathbf{C}'$.

PROOF. a. V^* is a subspace by the definition of subspace. Let $\mathbf{X}$ be an orthonormal basis matrix for V. Let $\mathbf{X}^* = \mathbf{\Gamma X}$. We now show that $\mathbf{X}^*$ is an orthonormal basis matrix for V^*. First, note that

$$(\mathbf{X}^*)'(\mathbf{X}^*) = \mathbf{X}'\mathbf{\Gamma}'\mathbf{\Gamma X} = \mathbf{X}'\mathbf{X} = \mathbf{I},$$

since $\mathbf{X}$ is an orthonormal basis matrix. Therefore, the columns of $\mathbf{X}^*$ are linearly independent. We need to show that they span V^*. Let $\mathbf{v}^* \in V^*$. Then there exists $\mathbf{v} \in V$ such that $\mathbf{v}^* = \mathbf{\Gamma v}$. Since $\mathbf{X}$ is a basis matrix for V, there exists $\mathbf{b} \in R^p$ such that $\mathbf{v} = \mathbf{Xb}$. Therefore $\mathbf{v}^* = \mathbf{\Gamma Xb} = \mathbf{X}^*\mathbf{b}$. Hence the columns of $\mathbf{X}^*$ span V^*, and $\mathbf{X}^*$ is an orthonormal basis matrix for V. Now $\mathbf{X}^*$ has dimension $n \times p$. Therefore, $\dim(V^*) = p$. Also

$$\mathbf{P}_{V^*} = (\mathbf{X}^*)(\mathbf{X}^*)' = \mathbf{\Gamma XX}'\mathbf{\Gamma}' = \mathbf{\Gamma P}_V\mathbf{\Gamma}',$$

since $\mathbf{X}^*$ and $\mathbf{X}$ are orthonormal basis matrices.

b. See Exercise B8. □

We now define a concept that is quite useful in the repeated measures model. Let $V_1 \subset R^n$ and $V_2 \subset R^m$ be p and q dimensional subspaces. The *product* V of V_1 and V_2, written $V = V_1 \times V_2$ is the subspace of R^{n+m} given by

$$V = V_1 \times V_2 = \left\{ \begin{pmatrix} \mathbf{v}_1 \\ \mathbf{v}_2 \end{pmatrix}; \mathbf{v}_1 \in V_1, \mathbf{v}_2 \in V_2 \right\}.$$

LEMMA 14.2. V is a $(p+q)$-dimensional subspace of R^{n+m} and

$$\mathbf{P}_V = \begin{pmatrix} \mathbf{P}_{V_1} & \mathbf{0} \\ \mathbf{0} & \mathbf{P}_{V_2} \end{pmatrix}.$$

PROOF. V is a subspace, by the definition of subspace. Let $\mathbf{X}_i$ be an orthonormal basis matrix for V_i, and let

$$\mathbf{X} = \begin{pmatrix} \mathbf{X}_1 & 0 \\ 0 & \mathbf{X}_2 \end{pmatrix}.$$

Then

$$\mathbf{X}'\mathbf{X} = \begin{pmatrix} \mathbf{X}_1'\mathbf{X}_1 & 0 \\ 0 & \mathbf{X}_2'\mathbf{X}_2 \end{pmatrix} = \mathbf{I}$$

so that the columns of $\mathbf{X}$ are linearly independent. Now, let $\mathbf{V} = \binom{\mathbf{v}_1}{\mathbf{v}_2} \in V$. Then there exist $\mathbf{b}_1 \in R^p$, $\mathbf{b}_2 \in R^q$ such that $\mathbf{v}_i = \mathbf{X}_i\mathbf{b}_i$. Let $\mathbf{b} = \binom{\mathbf{b}_1}{\mathbf{b}_2}$. Then $\mathbf{v} = \mathbf{Xb}$. Therefore, the columns of $\mathbf{X}$ span V, and $\mathbf{X}$ is an orthonormal basis matrix for V. $\mathbf{X}$ is an $(n+m) \times (p+q)$ matrix, so that $\dim(V) = p + q$.

Finally,

$$\mathbf{P}_V = \mathbf{X}\mathbf{X}' = \begin{pmatrix} \mathbf{X}_1\mathbf{X}_1' & 0 \\ 0 & \mathbf{X}_2\mathbf{X}_2' \end{pmatrix} = \begin{pmatrix} \mathbf{P}_{V_1} & 0 \\ 0 & \mathbf{P}_{V_2} \end{pmatrix}. \quad \square$$

If V is a vector space, we define V^n recursively by

$$V^1 = V, \qquad V^n = V^{n-1} \times V.$$

14.4. THE BASIC RESULT.

We now return to the repeated measures model in which we observe $\mathbf{Y}_k$, independent random vectors, such that

$$\mathbf{Y}_k \sim N_r(\boldsymbol{\mu}_k, \sigma^2 \mathbf{A}(\rho)), \qquad \boldsymbol{\mu}_k = \delta_k \mathbf{1} + \boldsymbol{\gamma}_k, \qquad \boldsymbol{\gamma}_k \perp \mathbf{1}, \qquad k = 1, \ldots, m.$$

Let

$$\boldsymbol{\delta} = \begin{bmatrix} \delta_1 \\ \vdots \\ \delta_m \end{bmatrix}, \qquad \boldsymbol{\gamma} = \begin{bmatrix} \gamma_1 \\ \vdots \\ \gamma_m \end{bmatrix}.$$

We note that $\boldsymbol{\gamma} \in (U^{\perp})^m$. The parameter space is given by

$$\boldsymbol{\delta} \in S, \qquad \boldsymbol{\gamma} \in T, \qquad \sigma^2 > 0, \qquad \mathbf{A}(\rho) > 0,$$

where S is an s-dimensional subspace of R^m, $s < m$, and T is a t-dimensional subspace of $(U^{\perp})^m$, $t < m(r-1)$ $(= \dim[(U^{\perp})^m])$. That is, the subspaces S and T are proper subspaces of R^m and $(U^{\perp})^m$.

As in Section 14.1, let

$$\mathbf{Y} = \begin{bmatrix} \mathbf{Y}_1 \\ \vdots \\ \mathbf{Y}_m \end{bmatrix}, \qquad \boldsymbol{\mu} = \begin{bmatrix} \boldsymbol{\mu}_1 \\ \vdots \\ \boldsymbol{\mu}_m \end{bmatrix}, \qquad \mathbf{H} = \begin{bmatrix} \mathbf{1} & \mathbf{0} & \cdots & \mathbf{0} \\ \mathbf{0} & \mathbf{1} & \cdots & \mathbf{0} \\ \vdots & \vdots & & \vdots \\ \mathbf{0} & \mathbf{0} & \cdots & \mathbf{1} \end{bmatrix}$$

$$V = \{ \boldsymbol{\mu} = \mathbf{H}\boldsymbol{\delta} + \boldsymbol{\gamma}, \boldsymbol{\delta} \in S, \boldsymbol{\gamma} \in T \}$$

$$\mathbf{B}(\rho) = \begin{bmatrix} \mathbf{A}(\rho) & 0 & \cdots & 0 \\ 0 & \mathbf{A}(\rho) & \cdots & 0 \\ \vdots & & \ddots & \\ 0 & 0 & \cdots & \mathbf{A}(\rho) \end{bmatrix}.$$

That is, V is the set of possible values of $\boldsymbol{\mu}$. Then V is an $(s+t)$-

dimensional subspace of R^{mr} [see (14.16)]. We note that

$$\mathbf{Y} \sim N_{mr}\big(\boldsymbol{\mu}, \sigma^2 \mathbf{B}(\rho)\big), \qquad \boldsymbol{\mu} \in V.$$

If $\rho = 0$, then $\mathbf{B}(\rho) = \mathbf{I}$, and this model is just an ordinary linear model as studied in Chapter 4–12.

Let $\mathbf{C}$ be an orthonormal basis matrix for $U^{\perp}$, and define

$$\mathbf{F} = \begin{bmatrix} r^{-1/2}\mathbf{1}' & 0 & \cdots & 0 \\ 0 & r^{-1/2}\mathbf{1}' & & 0 \\ \vdots & & \ddots & \vdots \\ 0 & 0 & \cdots & r^{-1/2}\mathbf{1}' \end{bmatrix}, \tag{14.10}$$

$$\mathbf{D} = \begin{bmatrix} \mathbf{C}' & 0 & \cdots & 0 \\ 0 & \mathbf{C}' & & 0 \\ \vdots & & \ddots & \vdots \\ 0 & 0 & \cdots & \mathbf{C}' \end{bmatrix}, \qquad \boldsymbol{\Gamma} = \begin{pmatrix} \mathbf{F} \\ \mathbf{D} \end{pmatrix}$$

where $\mathbf{F}$ is $m \times mr$, $\mathbf{D}$ is $m(r-1) \times mr$. Then $\boldsymbol{\Gamma}$ is an $(mr) \times (mr)$ orthogonal matrix. Now, let

$$\mathbf{Y}^* = \begin{pmatrix} \mathbf{Y}_1^* \\ \mathbf{Y}_2^* \end{pmatrix} = \begin{pmatrix} \mathbf{FY} \\ \mathbf{DY} \end{pmatrix} = \boldsymbol{\Gamma}\mathbf{Y}, \tag{14.11}$$

where $\mathbf{Y}_1^*$ is $m \times 1$. Since $\boldsymbol{\Gamma}$ is orthogonal, this transformation is invertible. We now find the joint distribution of $\mathbf{Y}_1^*$ and $\mathbf{Y}_2^*$.

LEMMA 14.3. $\mathbf{Y}_1^*$ and $\mathbf{Y}_2^*$ are independent and

$$\mathbf{Y}_1^* \sim N_m\big(r^{1/2}\boldsymbol{\delta}, \sigma^2(1 + (r-1)\rho)\mathbf{I}\big), \qquad \mathbf{Y}_2^* \sim N_{m(r-1)}\big(\mathbf{D}\boldsymbol{\gamma}, \sigma^2(1-\rho)\mathbf{I}\big).$$

$\mathbf{A}(\rho) > 0$ if and only if $[1 + (r-1)\rho] > 0$, and $(1-\rho) > 0$.

PROOF. First, let

$$\mathbf{Z}_k = \begin{pmatrix} \mathbf{Z}_{k1} \\ \mathbf{Z}_{k2} \end{pmatrix} = \begin{pmatrix} r^{-1/2}\mathbf{1}' \\ \mathbf{C}' \end{pmatrix} \mathbf{Y}_k$$

where Z_{k1} is 1×1. Then the $\mathbf{Z}_k$ are independent. Also

$$\mathbf{Z}_k \sim N_r\left(\begin{pmatrix} r^{-1/2}\mathbf{1}' \\ \mathbf{C}' \end{pmatrix} \boldsymbol{\mu}_k, \sigma^2 \begin{pmatrix} r^{-1/2}\mathbf{1}' \\ \mathbf{C}' \end{pmatrix} \mathbf{A}(\rho)(r^{-1/2}\mathbf{1}\;\mathbf{C}) \right).$$

Now $\mathbf{A}(\rho) = (1-\rho)\mathbf{I} + \rho\mathbf{1}\mathbf{1}'$. Hence

$$\begin{pmatrix} r^{-1/2}\mathbf{1}' \\ \mathbf{C}' \end{pmatrix} \mathbf{A}(\rho)(r^{-1/2}\mathbf{1}\;\mathbf{C}) = (1-\rho)\mathbf{I} + \begin{pmatrix} r\rho & \mathbf{0} \\ \mathbf{0} & \mathbf{0} \end{pmatrix} = \begin{pmatrix} 1 + (r-1)\rho & \mathbf{0} \\ \mathbf{0} & (1-\rho)\mathbf{I} \end{pmatrix}$$

(since $\mathbf{C}'\mathbf{1} = 0$). Also $\boldsymbol{\mu}_k = \delta_k \mathbf{1} + \gamma_k$. Therefore,

$$\begin{pmatrix} r^{-1/2}\mathbf{1}' \\ \mathbf{C}' \end{pmatrix} \boldsymbol{\mu}_k = \begin{pmatrix} r^{1/2}\delta_k \\ \mathbf{C}'\gamma_k \end{pmatrix}.$$

Hence Z_{k1} and Z_{k2} are independent and

$$Z_{k1} \sim N_1\left(r^{1/2}\delta_k, \sigma^2(1+(r-1)\rho)\right), \qquad Z_{k2} \sim N_{r-1}\left(\mathbf{C}'\gamma_k, \sigma^2(1-\rho)\mathbf{I}\right).$$

Now

$$\mathbf{Y}_1^* = \begin{bmatrix} r^{-1/2}\mathbf{1}' & \mathbf{0} & \cdots & \mathbf{0} \\ \mathbf{0} & r^{-1/2}\mathbf{1}' & \cdots & \mathbf{0} \\ \vdots & & \ddots & \vdots \\ \mathbf{0} & \mathbf{0} & \cdots & r^{-1/2}\mathbf{1}' \end{bmatrix} \begin{bmatrix} \mathbf{Y}_1 \\ \vdots \\ \mathbf{Y}_m \end{bmatrix} = \begin{bmatrix} Z_{11} \\ \vdots \\ Z_{m1} \end{bmatrix}, \qquad E\mathbf{Y}_1^* = r^{1/2}\boldsymbol{\delta}$$

$$\mathbf{Y}_2^* = \begin{bmatrix} \mathbf{C}' & \mathbf{0} & \cdots & \mathbf{0} \\ \mathbf{0} & \mathbf{C}' & \cdots & \mathbf{0} \\ \vdots & & \ddots & \vdots \\ \mathbf{0} & \mathbf{0} & \cdots & \mathbf{C}' \end{bmatrix} \begin{bmatrix} \mathbf{Y}_1 \\ \vdots \\ \mathbf{Y}_m \end{bmatrix} = \begin{bmatrix} \mathbf{Z}_{12} \\ \vdots \\ \mathbf{Z}_{m2} \end{bmatrix}, \qquad E\mathbf{Y}_2^* = \mathbf{D}\boldsymbol{\gamma}.$$

Therefore, $\mathbf{Y}_1^*$ and $\mathbf{Y}_2^*$ are independent, and

$$\mathbf{Y}_1^* \sim N_m\left(r^{1/2}\boldsymbol{\delta}, \sigma^2(1+(r-1)\rho)\mathbf{I}\right)$$
$$\mathbf{Y}_2^* \sim N_{m(r-1)}\left(\mathbf{D}\boldsymbol{\gamma}, \sigma^2(1-\rho)\mathbf{I}\right).$$

Finally, $\mathbf{A}(\rho) > 0$ if and only if

$$\begin{pmatrix} r^{-1/2}\mathbf{1} \\ \mathbf{C}' \end{pmatrix} \mathbf{A}(\rho)(r^{-1/2}\mathbf{1}\ \mathbf{C}) = \begin{pmatrix} (1+(r-1)\rho) & \mathbf{0} \\ \mathbf{0} & (1-\rho)\mathbf{I} \end{pmatrix} > 0$$

if and only if $(1+(r-1)\rho) > 0$ and $(1-\rho) > 0$. □

We now reparametrize the model. Let

$$\boldsymbol{\delta}^* = r^{1/2}\boldsymbol{\delta}, \qquad \boldsymbol{\gamma}^* = \mathbf{D}\boldsymbol{\gamma}, \qquad \boldsymbol{\mu}^* = \boldsymbol{\Gamma}\boldsymbol{\mu} = \begin{pmatrix} \boldsymbol{\delta}^* \\ \boldsymbol{\gamma}^* \end{pmatrix},$$

$$\tau_1^2 = \sigma^2(1+(r-1)\rho), \qquad \tau_2^2 = \sigma^2(1-\rho), \tag{14.12}$$
$$\mathbf{T}^* = \{\mathbf{D}\boldsymbol{\gamma}; \boldsymbol{\gamma} \in T\}, \qquad V^* = \{\boldsymbol{\Gamma}\boldsymbol{\mu}; \boldsymbol{\mu} \in V\}.$$

Then $\boldsymbol{\delta} \in S$ if and only if $\boldsymbol{\delta}^* \in S$, $\boldsymbol{\gamma} \in T$ if and only if $\boldsymbol{\gamma}^* \in T^*$, and $\boldsymbol{\mu} \in V$ if and only if $\boldsymbol{\mu}^* \in V^*$. Also $\sigma^2 > 0$ and $\mathbf{A}(\rho) > 0$ if and only if $\tau_1^2 > 0$ and $\tau_2^2 > 0$. Therefore we have transformed the repeated measures model to a model in which we observe $\mathbf{Y}_1^*$ and $\mathbf{Y}_2^*$ independent, such that

$$\mathbf{Y}_1^* \sim N_m(\boldsymbol{\delta}^*, \tau_1^2\mathbf{I}), \qquad \mathbf{Y}_2^* \sim N_{m(r-1)}(\boldsymbol{\gamma}^*, \tau_2^2\mathbf{I}) \tag{14.13}$$

and the parameter space is given by

$$\boldsymbol{\delta}^* \in S, \qquad \gamma^* \in T^*, \qquad \tau_1^2 > 0, \qquad \tau_2^2 > 0. \tag{14.14}$$

Thus we have transformed the repeated model to a problem that is really two separate ordinary linear models, one involving $\mathbf{Y}_1^*$ and one involving $\mathbf{Y}_2^*$. We use results derived for these two ordinary linear models to find results for the repeated measures model. We also discuss the ordinary linear model that occurs if we assume that $\rho = 0$ in the repeated measures model, or equivalently that $\tau_1^2 = \tau_2^2$ in the transformed version of the repeated measures model. It is quite important to keep these three versions of the ordinary linear model separate.

We now discuss S, T, T^*, V and V^*. We note first that $\mathbf{D}'$ is an orthonormal basis matrix for $(U^{\perp})^m$, and that $T \subset (U^{\perp})^m$. Therefore, by Lemma 14.1,

$$\begin{aligned} \dim(V^*) &= \dim(V), &\qquad \mathbf{P}_{V^*} &= \boldsymbol{\Gamma}\mathbf{P}_V\boldsymbol{\Gamma}', \\ \dim(T^*) &= \dim(T) = t, &\qquad \mathbf{P}_{T^*} &= \mathbf{D}\mathbf{P}_T\mathbf{D}'. \end{aligned} \tag{14.15}$$

Finally, we note that $\boldsymbol{\mu}^* \in V^*$ if and only if $\boldsymbol{\delta}^* \in S$, $\gamma^* \in T^*$. Therefore, by Lemma 14.2,

$$V^* = S \times T^*, \qquad \mathbf{P}_{V^*} = \begin{pmatrix} \mathbf{P}_S & \mathbf{0} \\ \mathbf{0} & \mathbf{P}_{T^*} \end{pmatrix}, \qquad \dim V = \dim V^* = s + t. \tag{14.16}$$

14.5. SUFFICIENCY AND ESTIMATION.

We are now ready to derive a complete sufficient statistic for the repeated measures model. We look first at the transformed version of this problem. Let

$$\hat{\boldsymbol{\delta}}^* = \mathbf{P}_S\mathbf{Y}_1^*, \qquad \hat{\tau}_1^2 = \frac{1}{m-s}\|\mathbf{P}_{S^\perp}\mathbf{Y}_1^*\|^2, \qquad \hat{\gamma}^* = \mathbf{P}_{T^*}\mathbf{Y}_2^*,$$

$$\hat{\tau}_2^2 = \frac{1}{m(r-1)-t}\|\mathbf{P}_{T^{*\perp}}\mathbf{Y}_2^*\|^2. \tag{14.17}$$

$(\hat{\boldsymbol{\delta}}^*, \hat{\tau}_1^2)$ is a complete sufficient statistic for the ordinary linear model involving only $\mathbf{Y}_1^*$, and $(\hat{\gamma}^*, \hat{\tau}_2^2)$ is a complete sufficient statistic for the ordinary linear model involving only $\mathbf{Y}_2^*$.

LEMMA 14.4. $(\hat{\boldsymbol{\delta}}^*, \hat{\gamma}^*, \hat{\tau}_1^2, \hat{\tau}_2^2)$ is a complete sufficient statistic for the repeated measures model.

PROOF. We follow the proof of Theorem 5.3, in which we derived the complete sufficient statistic for the ordinary linear model. Let $\mathbf{X}_1$ and $\mathbf{X}_2$ be orthonormal basis matrices for S and T^*. Define

$$\mathbf{Q}_1(\boldsymbol{\delta}^*, \tau_1^2) = \begin{bmatrix} -\dfrac{1}{2\tau_1^2} \\ \dfrac{1}{\tau_1^2}\mathbf{X}_1'\boldsymbol{\delta}^* \end{bmatrix}, \qquad \mathbf{Q}_2(\boldsymbol{\gamma}^*, \tau_2^2) = \begin{bmatrix} \dfrac{-1}{2\tau_2^2} \\ \dfrac{1}{\tau_2^2}\mathbf{X}_2'\boldsymbol{\gamma}^* \end{bmatrix},$$

$$\mathbf{T}_1(\mathbf{Y}_1^*) = \begin{pmatrix} (m-s)\hat{\tau}_1^2 + \|\hat{\boldsymbol{\delta}}^*\|^2 \\ \mathbf{X}_1'\hat{\boldsymbol{\delta}}^* \end{pmatrix},$$

$$\mathbf{T}_2(\mathbf{Y}_2^*) = \begin{pmatrix} (m(r-1)-t)\hat{\tau}_2^2 + \|\hat{\boldsymbol{\gamma}}^*\|^2 \\ \mathbf{X}_2'\hat{\boldsymbol{\gamma}}^* \end{pmatrix}$$

$$\mathbf{Q}(\boldsymbol{\delta}^*, \boldsymbol{\gamma}^*, \tau_1^2, \tau_2^2) = \begin{bmatrix} \mathbf{Q}_1(\boldsymbol{\delta}^*, \tau_1^2) \\ \mathbf{Q}_2(\boldsymbol{\gamma}^*, \tau_2^2) \end{bmatrix}, \qquad \mathbf{T}(\mathbf{Y}^*) = \begin{pmatrix} \mathbf{T}_1(\mathbf{Y}_1^*) \\ \mathbf{T}_2(\mathbf{Y}_2^*) \end{pmatrix}.$$

As in Theorem 5.3, the joint density of $\mathbf{Y}_1^*$ and $\mathbf{Y}_2^*$ is

$$\begin{aligned} f(\mathbf{y}_1^*, \mathbf{y}_2^*; \boldsymbol{\delta}^*, \tau_1^2, \boldsymbol{\gamma}^*, \tau_2^2) &= f_1(\mathbf{y}_1^*; \boldsymbol{\delta}^*, \tau_1^2) f_2(\mathbf{y}_2^*; \boldsymbol{\gamma}^*, \tau_2^2) \\ &= h_1(\mathbf{y}_1^*)h_2(\mathbf{y}_2^*)k_1(\boldsymbol{\delta}^*, \tau_1^2)k_2(\boldsymbol{\gamma}^*, \tau_2^2) \\ &\quad \times \exp\left[\mathbf{Q}_1(\boldsymbol{\delta}^*, \tau_1^2)\right]'\mathbf{T}_1(\mathbf{y}_1^*)\exp\left[\mathbf{Q}_2(\boldsymbol{\gamma}^*, \tau_2^2)\right]'\mathbf{T}_2(\mathbf{y}_2^*) \\ &= h(\mathbf{y}_1^*, \mathbf{y}_2^*)k(\boldsymbol{\delta}^*, \boldsymbol{\gamma}^*, \tau_1^2, \tau_2^2)\exp\left[\left(\mathbf{Q}(\boldsymbol{\delta}^*, \boldsymbol{\gamma}^*, \tau_1^2, \tau_2^2)\right)'\mathbf{T}(\mathbf{y}^*)\right]. \end{aligned}$$

The image of $\mathbf{Q}$ is all points in R^{s+t+2} whose first and $(s+1)$th coordinates are negative. This set contains an open rectangle. Therefore $\mathbf{T}(\mathbf{Y}^*)$ is a complete sufficient statistic. Finally, $(\hat{\boldsymbol{\delta}}^*, \hat{\boldsymbol{\gamma}}^*, \hat{\tau}_1^2, \hat{\tau}_2^2)$ is an invertible function of $\mathbf{T}$, and hence is also a complete sufficient statistic. □

The following lemma follows directly from results previously derived for the ordinary linear model, applied to the ordinary linear models involving $\mathbf{Y}_1^*$ and $\mathbf{Y}_2^*$.

LEMMA 14.5. $\hat{\boldsymbol{\delta}}^*$, $\hat{\boldsymbol{\gamma}}^*$, $\hat{\tau}_1^2$, and $\hat{\tau}_2^2$ are independent, and

$$\hat{\boldsymbol{\delta}}^* \sim N_m(\boldsymbol{\delta}^*, \tau_1^2\mathbf{P}_S), \qquad \hat{\boldsymbol{\gamma}}^* \sim N_{m(r-1)}(\boldsymbol{\gamma}^*, \tau_2^2\mathbf{P}_{T^*}),$$

$$(m-s)\hat{\tau}_1^2 \sim \tau_1^2\chi^2_{m-s}(0), \qquad \left[m(r-1)-t\right]\hat{\tau}_2^2 \sim \tau_2^2\chi^2_{m(r-1)-t}(0).$$

As in Section 14.1, let $\mathbf{Y}_k = (Y_{k1}, \ldots, Y_{kr})'$, $k = 1, \ldots, n$; and let $\mathbf{Z}$

$= (\bar{Y}_{1\cdot}, \ldots, \bar{Y}_{m\cdot})'$. Define

$$\hat{\mu} = \mathbf{P}_V\mathbf{Y}, \hat{\sigma}^2 = \frac{\|\mathbf{P}_{V^\perp}\mathbf{Y}\|^2}{(mr - t - s)}, \qquad U_1^2 = \frac{r\|\mathbf{P}_{S^\perp}\mathbf{Z}\|^2}{(m - s)}$$

$$U_2^2 = \frac{(mr - t - s)\hat{\sigma}^2 - (m - s)U_1^2}{m(r - 1) - t}.$$

LEMMA 14.6.

$$\hat{\mu} = \Gamma'\begin{pmatrix}\hat{\delta}^* \\ \hat{\gamma}^*\end{pmatrix}, \qquad U_1^2 = \hat{\tau}_1^2, \qquad U_2^2 = \hat{\tau}_2^2.$$

PROOF.

$$\hat{\mu} = \mathbf{P}_V\mathbf{Y} = \Gamma'\mathbf{P}_{V^*}\mathbf{Y}^* = \Gamma'\begin{pmatrix}\mathbf{P}_S\mathbf{Y}_1^* \\ \mathbf{P}_T^*\mathbf{Y}_2^*\end{pmatrix} = \Gamma'\begin{pmatrix}\hat{\delta}^* \\ \hat{\gamma}^*\end{pmatrix}$$

[using (14.15) and (14.16)]. In Exercise B9 you are asked to show that $U_1^2 = \hat{\tau}_1^2$. To verify that $U_2^2 = \hat{\tau}_2^2$, we note that

$$\begin{aligned}(mr - t - s)\hat{\sigma}^2 &= \|\mathbf{P}_{V^\perp}\mathbf{Y}\|^2 = \|\mathbf{Y}\|^2 - \|\mathbf{P}_V\mathbf{Y}\|^2 = \|\mathbf{Y}^*\|^2 - \|\mathbf{P}_{V^*}\mathbf{Y}^*\|^2 \\ &= \|\mathbf{Y}_1^*\|^2 + \|\mathbf{Y}_2^*\|^2 - \|\mathbf{P}_{S^*}\mathbf{Y}_1^*\|^2 - \|\mathbf{P}_{T^*}\mathbf{Y}_2^*\|^2 \\ &= \|\mathbf{P}_{S^\perp}\mathbf{Y}_1^*\|^2 + \|\mathbf{P}_{T^{*\perp}}\mathbf{Y}_2^*\|^2 = (m - s)\hat{\tau}_1^2 + [m(r - 1) - t]\hat{\tau}_2^2,\end{aligned}$$

[using (14.15) and (14.16) again]. Therefore, (since $U_1^2 = \hat{\tau}_1^2$),

$$\hat{\tau}_2^2 = \frac{(mr - t - s)\hat{\sigma}^2 - (m - s)U_1^2}{m(r - 1) - t} = U_2^2. \quad \square$$

We now prove parts a and b of Theorem A.

THEOREM 14.7. a. $(\hat{\mu}, U_1^2, U_2^2)$ is a complete sufficient statistic for the repeated measures model.

b. $\hat{\mu}$, U_1^2 and U_2^2 are independent and

$$\hat{\mu} \sim N_m[\mu, \sigma^2\mathbf{P}_V\mathbf{B}(\rho)\mathbf{P}_V], \qquad (m - s)U_1^2 \sim \sigma^2[1 + (r - 1)\rho]\chi^2_{m-s}(0)$$

$$[m(r - 1) - t]U_2^2 \sim \sigma^2(1 - \rho)\chi^2_{m(r-s)-t}(0).$$

PROOF. a. From Lemma 14.6, $(\hat{\mu}, U_1^2, U_2^2)$ is an invertible function of the complete sufficient statistic $(\hat{\delta}^*, \hat{\gamma}^*, \hat{\tau}_1^2, \hat{\tau}_2^2)$ and hence is complete and sufficient.

b. The independence follows from Lemma 14.5 as do the marginal distributions of U_1^2 and U_2^2. Since $\mathbf{Y} \sim N_n[\boldsymbol{\mu}, \sigma^2\mathbf{B}(\rho))]$, we see that $\hat{\boldsymbol{\mu}} = \mathbf{P}_V\mathbf{Y} \sim N_n[\mathbf{P}_V\boldsymbol{\mu}, \sigma^2\mathbf{P}_V\mathbf{B}(\rho)\mathbf{P}_V]$. However $\mathbf{P}_V\boldsymbol{\mu} = \boldsymbol{\mu}$ since $\boldsymbol{\mu} \in V$. □

We now prove the remaining two parts of Theorem A.

THEOREM 14.8. a. The minimum variance unbiased estimators of $\boldsymbol{\mu}$, σ^2 and $\rho\sigma^2$ are $\hat{\boldsymbol{\mu}}$, $[U_1^2 + (r-1)U_2^2]/r$ and $(U_1^2 - U_2^2)/r$.
b. The MLEs of $\boldsymbol{\mu}$, σ^2 and $\rho\sigma^2$ are $\hat{\boldsymbol{\mu}}$, $\{m-s)U_1^2 + [m(r-1)-t]U_2^2\}/mr$ and $\{(m-s)(r-1)U_1^2 - [m(r-1)-t]U_2^2\}/mr(r-1)$.

PROOF. a. It is easily verified that the estimators are unbiased. Since they are functions of the complete sufficient statistic, they are minimum variance unbiased estimators.
b. We first find the MLE's for the transformed version of the model. Let $\hat{\hat{\tau}}_1^2 = [(m-s)/m]\hat{\tau}_1^2$ and $\hat{\hat{\tau}}_2^2 = \{[m(r-1)-t]/[m(r-1)]\}\hat{\tau}_1^2$. Then $\hat{\boldsymbol{\delta}}^*$ and $\hat{\hat{\tau}}_1^2$ are the MLE's for the ordinary linear model consisting of $\mathbf{Y}_1^*$, while $\hat{\boldsymbol{\gamma}}^*$ and $\hat{\hat{\tau}}_2^2$ are the MLE's for the linear model involving $\mathbf{Y}_2^*$. Therefore

$$f_1(\mathbf{y}_1^*; \hat{\boldsymbol{\delta}}^*, \hat{\hat{\tau}}_1^2) \geqslant f_1(\mathbf{y}_1^*; \boldsymbol{\delta}^*, \tau_1^2), \qquad f_2(\mathbf{y}_2^*; \hat{\boldsymbol{\gamma}}^*, \hat{\hat{\tau}}_2^2) \geqslant f_2(\mathbf{y}_2^*; \boldsymbol{\gamma}^*, \tau_2^2).$$

Hence

$$f_1(\mathbf{y}_1^*; \hat{\boldsymbol{\delta}}^*, \hat{\hat{\tau}}_1^2) f_2(\mathbf{y}_2^*; \hat{\boldsymbol{\gamma}}^*, \hat{\hat{\tau}}_2^2) \geqslant f_1(\mathbf{y}_1^* : \boldsymbol{\delta}^*, \tau_1^2) f_2(\mathbf{y}_2^*; \boldsymbol{\gamma}^*, \tau_2^2)$$

and therefore, $\hat{\boldsymbol{\delta}}^*$, $\hat{\hat{\tau}}_1^2$, $\hat{\boldsymbol{\gamma}}^*$ and $\hat{\hat{\tau}}_2^2$ are the MLE's for $\boldsymbol{\delta}^*$, τ_1^2, $\boldsymbol{\gamma}^*$ and τ_2^2 for the repeated measures model. Now, $\boldsymbol{\mu} = \boldsymbol{\Gamma}'\binom{\boldsymbol{\delta}^*}{\boldsymbol{\gamma}^*}$ so that the MLE for $\boldsymbol{\mu}$ is $\boldsymbol{\Gamma}'\binom{\hat{\boldsymbol{\delta}}^*}{\hat{\boldsymbol{\gamma}}^*} = \hat{\boldsymbol{\mu}} \cdot \sigma^2 = [\tau_1^2 + (r-1)\tau_2^2]/r$, so that the MLE for σ^2 is

$$\hat{\hat{\sigma}}^2 = \left[\hat{\hat{\tau}}_1^2 + (r-1)\hat{\hat{\tau}}_2^2\right]/r = \{(m-s)U_1^2 + [m(r-1)-t]U_2^2\}/mr.$$

The result for $\rho\sigma^2$ follows similarly. □

14.6. HYPOTHESIS TESTING.

We now consider the testing problems mentioned in Section 14.1. We first look at problems of type A, in which we are testing that $\boldsymbol{\delta} \in Q_A$ against $\boldsymbol{\delta} \in S$, where Q_A is a q_A-dimensional subspace of S, $q_A < s$. We first note that $\boldsymbol{\delta} \in Q_A$ if and only if $\boldsymbol{\delta}^* \in Q_A$, so that in the transformed model, we are testing that $\boldsymbol{\delta}^* \in Q_A$ against $\boldsymbol{\delta}^* \in S$. Define

$$F_A = \frac{\|\mathbf{P}_{S|Q_A}\mathbf{Y}_1^*\|^2}{(s-q_A)U_1^2} \qquad \phi_A(F_A) = \begin{cases} 1 & \text{if } F_A > F^{\alpha}_{s-q_A,m-s} \\ 0 & \text{if } F_A \leqslant F^{\alpha}_{s-q_A,m-s} \end{cases}. \tag{14.18}$$

We note that ϕ_A would be the UMP invariant size α test for testing that $\boldsymbol{\delta}^* \in Q_A$ against $\boldsymbol{\delta}^* \in S$ for the ordinary linear model consisting only of

$\mathbf{Y}_1^*$. We now show that it is also UMP invariant size α for the repeated measures model.

THEOREM 14.9.

$$F_A \sim F_{s-q_A, m-s}\left\{\frac{\|\mathbf{P}_{S \mid Q_A}\boldsymbol{\delta}^*\|^2}{\sigma^2[1+(r-1)\rho]}\right\}.$$

The test ϕ_A defined in (14.18) is UMP invariant size α for testing that $\boldsymbol{\delta} \in Q_A$ against $\boldsymbol{\delta} \in S$ for the repeated measures model.

PROOF. The distribution of F_A follows from results for the ordinary linear model involving Y_1^*. To find a maximal invariant for this problem, we first reduce by the group G_1 of transformations of the form

$$g_1(\hat{\boldsymbol{\delta}}^*, \hat{\tau}_1^2, \hat{\boldsymbol{\gamma}}^*, \hat{\tau}_2^2) = (\hat{\boldsymbol{\delta}}^*, \hat{\tau}_1^2, c\hat{\boldsymbol{\gamma}}^* + b, c^2\hat{\tau}_2^2)$$

where $b \in T^*$, $c > 0$. A maximal invariant under G_1 is

$$(\hat{\boldsymbol{\delta}}^*, \hat{\tau}_1^2).$$

Now the model involving $(\hat{\boldsymbol{\delta}}^*, \hat{\tau}_1^2)$ and $(\boldsymbol{\delta}^*, \tau_1^2)$ is just an ordinary linear model. Therefore, by Theorem 7.12, the maximal invariant for testing that $\boldsymbol{\delta}^* \in Q_A$ is just F_A. By Theorem 1.7 and Lemma 1.9, ϕ_A is the UMP invariant test that $\boldsymbol{\delta}^* \in Q_A$. □

In this theorem, we have broken from our usual practice of formally stating the groups for which the test is UMP invariant. It should be clear that these groups would include G_1 mentioned above plus the three groups mentioned in Section 7.4, acting on $(\hat{\boldsymbol{\delta}}^*, \hat{\tau}_1^2)$ (or equivalently, on $\mathbf{Y}_1^*$). Stating those last three groups formally would necessitate putting the ordinary linear model involving Y_1^* into canonical form.

We now express the numerator of F_A in terms of the original observations $\mathbf{Y}$. As in Section 14.1, let W_A be the following subspace of V

$$W_A = \{\boldsymbol{\mu} = \mathbf{H}\boldsymbol{\delta} + \boldsymbol{\gamma}, \boldsymbol{\delta} \in Q_A, \boldsymbol{\gamma} \in T\}.$$

By an argument similar to that in (14.16), we can show that $\dim W_A = q_A + t$. Also $\boldsymbol{\delta} \in Q_A$ if and only if $\boldsymbol{\mu} \in W_A$ (since we are assuming that $\boldsymbol{\gamma} \in T$).

LEMMA 14.10. $\|\mathbf{P}_{V \mid W_A}\mathbf{Y}\|^2 = \|\mathbf{P}_{S \mid Q_A}\mathbf{Y}_1^*\|^2$, $\|\mathbf{P}_{V \mid W_A}\boldsymbol{\mu}\|^2 = \|\mathbf{P}_{S \mid Q_A}\boldsymbol{\delta}^*\|^2$ and $\dim \mathrm{V} \mid W_A = \dim S \mid Q_A$.

PROOF. By (14.15) and (14.16)

$$\|\mathbf{P}_V\mathbf{Y}\|^2 = \|\mathbf{P}_{V^*}\mathbf{Y}^*\|^2 = \|\mathbf{P}_S\mathbf{Y}_1^*\|^2 + \|\mathbf{P}_{T^*}\mathbf{Y}_2^*\|^2.$$

Similarly

$$\|\mathbf{P}_{W_A}\mathbf{Y}\|^2 = \|\mathbf{P}_{Q_A}\mathbf{Y}_1^*\|^2 + \|\mathbf{P}_{T^*}\mathbf{Y}_2^*\|^2.$$

Therefore

$$\|\mathbf{P}_{V|W_A}\mathbf{Y}\|^2 = \|\mathbf{P}_V\mathbf{Y}\|^2 - \|\mathbf{P}_{W_A}\mathbf{Y}\|^2 = \|\mathbf{P}_S\mathbf{Y}_1^*\|^2 - \|\mathbf{P}_{Q_A}\mathbf{Y}_1^*\|^2 = \|\mathbf{P}_{S|Q_A}\mathbf{Y}_1^*\|^2.$$

Similarly $\|\mathbf{P}_{V|W_A}\boldsymbol{\mu}\|^2 = \|\mathbf{P}_{S|Q_A}\boldsymbol{\delta}^*\|^2$. Finally,

$\dim V|W_A = \dim V - \dim W_A = s + t - (q_A + t) = s - q_A = \dim S|Q_A$. □

We have therefore shown that the numerator sum of squares and degrees of freedom are the same for the repeated measures model as they are for the ordinary linear model that occurs when we assume that $\rho = 0$ in the repeated measures model. That is, they are the same as they would be if we treated the repeated measures as independent observations. The denominator sum of squares and degrees of freedom are, however, different for the two models.

We now consider testing hypotheses of type B. Let Q_B be a q_B-dimensional subspace of T, $q_B < t$. We consider testing that $\gamma \in Q_B$ against $\gamma \in T$. Let

$$Q_B^* = \{\mathbf{D}\gamma; \gamma \in Q_B\}.$$

Then Q_B^* is a q_B-dimensional subspace of T^*, and $\gamma \in Q_B$ if and only if $\gamma^* \in Q_B^*$. As in Section 14.1, let W_B be the subspace of V in which $\gamma \in Q_B$. Define

$$F_B = \frac{\|\mathbf{P}_{V|W_B}\mathbf{Y}\|^2}{(t - q_B)U_2^2} \tag{14.19}$$

$$\phi_B(F_B) = \begin{cases} 1 & \text{if } F_B > F^{\alpha}_{t-q_B,\, m(r-1)-t} \\ 0 & \text{if } F_B \leqslant F^{\alpha}_{t-q_B,\, m(r-1)-t} \end{cases}.$$

By proofs similar to those for Theorem 14.9 and Lemma 14.10, we have the following theorem. (See Exercise B11.)

THEOREM 14.11. The UMP invariant size α test for testing $\gamma \in Q_B$ against $\gamma \in T$ for the repeated measures model is given by ϕ_B defined in (14.19).

$$F_B \sim F_{t-q_B,\, m(r-1)-t}\left(\frac{\|\mathbf{P}_{V|W_B}\boldsymbol{\mu}\|^2}{\sigma^2(1-\rho)}\right), \qquad \dim V|W_B = t - q_B.$$

We see again that the numerator sum of squares and degrees of freedom are the same for the repeated measures model as for the associated ordinary linear model.

The final testing problem for the repeated measures model is to test that $\rho = 0$, that is, that the repeated measures model is actually an ordinary linear model. In the transformed model, that becomes the problem of testing that $\tau_1^2 = \tau_2^2$. Therefore, this problem transforms to the problem of testing the equality of variances in two different ordinary linear models. Let

$$F_C = \frac{\hat{\tau}_1^2}{\hat{\tau}_2^2} = \frac{U_1^2}{U_2^2},$$

$$\phi_{a,b}(F_C) = \begin{cases} 1 & \text{if } F_C > F^b_{m-s,m(r-1)-t} \text{ or} \\ & \text{if } F_C < F^{1-a}_{m-s,m(r-1)-t} \\ 0 & \text{if } F^b_{m-s,m(r-1)-t} \geqslant F_C \geqslant F^{1-a}_{m-s,m(r-1)-t} \end{cases} \tag{14.20}$$

where $a + b = \alpha$. Then $\phi_{a,b}$ is a size α test. There is no UMP invariant test for this model, but there exist a and b such that $\phi_{a,b}$ is UMP umbiased, and other a and b such that $\phi_{a,b}$ is a likelihood ratio test. The choice $a = b = \alpha/2$ is the choice used most often.

14.7. SIMULTANEOUS CONFIDENCE INTERVALS FOR CONTRASTS.

In this section, we find simultaneous confidence intervals for the set of all contrasts associated with either a type A or type B hypothesis. A contrast associated with the type A hypothesis that $\boldsymbol{\mu} \in W_A$ is a linear function $\langle \mathbf{d}, \boldsymbol{\mu} \rangle, \mathbf{d} \in V | W_A$, while a contrast associated with the type B hypothesis that $\boldsymbol{\mu} \in W_B$ is a linear function $\langle \mathbf{d}, \boldsymbol{\mu} \rangle, \mathbf{d} \in V | W_B$.

THEOREM 14.12. a. A set of simultaneous $(1 - \alpha)$ confidence intervals for the set of all contrasts associated with testing the type A hypothesis that $\boldsymbol{\mu} \in W_A$ is given by

$$\langle \mathbf{d}, \boldsymbol{\mu} \rangle \in \langle \mathbf{d}, \hat{\boldsymbol{\mu}} \rangle \pm U_1 \|\mathbf{d}\| \left[(s - q_A) F^{\alpha}_{s-q_A, m-s} \right]^{1/2}, \qquad \mathbf{d} \in V | W_A.$$

The hypothesis $\boldsymbol{\mu} \in W_A$ is rejected with the test defined in (14.18) if and only if at least one of the confidence intervals does not contain 0.

b. A set of simultaneous $(1 - \alpha)$ confidence intervals for the set of all contrasts associated with testing the B hypothesis that $\boldsymbol{\mu} \in W_B$ is given by

$$\langle \mathbf{d}, \boldsymbol{\mu} \rangle \in \langle \mathbf{d}, \hat{\boldsymbol{\mu}} \rangle \pm U_2 \|\mathbf{d}\| \left[(t - q_B) F_{t-q_B, m(r-1)-t} \right]^{1/2}, \qquad \mathbf{d} \in V | W_B.$$

The hypothesis $\boldsymbol{\mu} \in W_B$ is rejected with the test given in (14.19) if and only if at least one of the confidence intervals does not contain 0.

PROOF. a. By Lemmas 8.1 and 14.10, we see that

$$\sup_{\substack{\mathbf{d}\in V|W_A\\ \mathbf{d}\neq \mathbf{0}}} \frac{(\mathbf{d}'(\hat{\boldsymbol{\mu}}-\boldsymbol{\mu}))^2}{(s-q_A)U_1^2\|\mathbf{d}\|^2} = \frac{\|\mathbf{P}_{V|W_A}(\hat{\boldsymbol{\mu}}-\boldsymbol{\mu})\|^2}{(s-q_A)U_1^2}$$

$$= \frac{\|\mathbf{P}_{S|Q_A}(\hat{\boldsymbol{\gamma}}^*-\boldsymbol{\gamma}^*)\|^2}{(s-q_A)U_1^2} \sim F^{\alpha}_{s-q_A,t-s}(0).$$

The interval associated with $\mathbf{d}=\mathbf{0}$ is trivially satisfied. Therefore

$$P\Big(\langle \mathbf{d},\boldsymbol{\mu}\rangle \in \langle \mathbf{d},\hat{\boldsymbol{\mu}}\rangle \pm U_1\|\mathbf{d}\|\big[(s-q_A)F^{\alpha}_{s-q_A,m-s}\big]^{1/2} \qquad \text{for all } \mathbf{d}\in V|W_A\Big)$$

$$= P\left[\sup_{\substack{\mathbf{d}\in V|W_A\\ \mathbf{d}\neq \mathbf{0}}} \frac{(\mathbf{d}'(\hat{\boldsymbol{\mu}}-\boldsymbol{\mu}))^2}{(s-q_A)U_1^2\|\mathbf{d}\|^2} \leqslant F^{\alpha}_{s-q_A,m-s}\right] = 1-\alpha.$$

Furthermore, 0 is in all the intervals if and only if

$$\sup_{\substack{\mathbf{d}\in V|W_A\\ \mathbf{d}\neq 0}} \frac{(\mathbf{d}'\hat{\boldsymbol{\mu}})^2}{(s-q_A)\|\mathbf{d}\|^2U_1^2} = \frac{\|\mathbf{P}_{V|W_A}\mathbf{Y}\|^2}{(s-q_A)U_1^2} < F^{\alpha}_{s-q_A,m-s}.$$

That is, 0 is in all the intervals if and only if the F test accepts.
b. See Exercise B12. □

14.8. OTHER RESULTS.

It is clear that we could find UMP invariant tests analogous to those in Chapter 7 for one-sided hypotheses involving linear functions of either $\boldsymbol{\delta}$ or $\boldsymbol{\gamma}$ (but not both). We could also find confidence intervals and test hypotheses about $\sigma^2(1+(r-1)\rho)$ and $\sigma^2(1-\rho)$. We could find James-Stein estimators for this model. (In Exercise C1, a James-Stein estimator is derived for the simpler exchangeable linear model discussed in the next section. A James-Stein estimator for the repeated measures model would be derived similarly.)

The tests ϕ_A and ϕ_B derived in Section 14.5 are most stringent, admissible, unbiased, and likelihood ratio tests (see Arnold, 1979b). Therefore, these tests have the same optimality for the repeated measures model as for the ordinary linear model.

We now discuss the asymptotic validity of procedures for this model. Consider the model in which we observe the r-dimensional random vectors

$\mathbf{Y}_k$, $k = 1, \ldots, m$ such that the $\mathbf{Y}_k$ are independent and

$$E\mathbf{Y}_k = \delta_k \mathbf{1} + \gamma_k, \quad \operatorname{cov}(\mathbf{Y}_k) = \sigma^2 \mathbf{A}(\rho), \quad \boldsymbol{\delta} = \begin{bmatrix} \delta_1 \\ \vdots \\ \delta_m \end{bmatrix} \in S, \quad \gamma = \begin{bmatrix} \gamma_1 \\ \vdots \\ \gamma_m \end{bmatrix} \in T$$

(where S and T are defined in Section 14.1). We consider this model (without the normal assumption) as the number of individuals, m, goes to ∞, but the number of measurements, r, on each individual is fixed. Let the subspace V be defined as in Section 14.1. In Arnold (1980b), it is shown that if the largest diagonal element of $\mathbf{P}_V$ goes to 0 (i.e., if Huber's condition is satisfied), then the sizes of ϕ_A and ϕ_B defined in (14.18) and (14.19) for testing type A and type B hypotheses are asymptotically independent of the normal assumption used in deriving them as are the confidence coefficients for the simultaneous confidence intervals given in Theorem 14.12. Therefore, the sizes of tests and confidence coefficients of simultaneous confidence intervals for type A and type B hypotheses have the same asymptotic validity for the repeated measures model as they would have for the ordinary linear model. The test $\phi_{a,b}$ defined in (14.20) for testing that $\rho = 0$ is really a test of the equality of two variances and its size is quite dependent on the normal assumption used in its derivation.

We now consider an interesting aspect of testing type A hypotheses. We return to the repeated measures model in which the $\mathbf{Y}_k$ are normally distributed. Let Z_k and $\mathbf{Z}$ be defined as in Section 14.1. Then the Z_k are independent and $Z_k \sim N_1(\delta_k, \sigma^2(1 + (r-1)\rho)/r)$. Therefore

$$\mathbf{Z} \sim N_n\left(\boldsymbol{\delta}, \frac{\sigma^2(1 + (r-1)\rho)}{r}\mathbf{I}\right)$$

and $\boldsymbol{\delta} \in S$. Hence, the model in which we observe $\mathbf{Z}$ is just an ordinary linear model. Also the statistic F_A for testing that $\boldsymbol{\delta} \in Q_A$ can be rewritten as

$$F_A = \frac{\|\mathbf{P}_{S \mid Q_A} \mathbf{Z}\|^2}{\|\mathbf{P}_{S^\perp} \mathbf{Z}\|^2} \frac{(m - s)}{s - q_A}, \tag{14.21}$$

(see Exercise B13). We can therefore look at a test of a type A hypothesis in the following way. To gain independence, we first replace the vector of observations on each person with the average of the components of that vector. We then treat these averages as an ordinary linear model. It is somewhat surprising that this type of procedure does not throw away useful information about the type A hypothesis. We have shown, however, that this procedure is UMP invariant size α for testing type A hypotheses. Now suppose that the covariance matrix of $\mathbf{Y}_k$ does not have the nice form assumed in this chapter, but is the same for all individual. That is, suppose that $\mathbf{Y}_k \sim N_r(\delta_k \mathbf{1} + \gamma_k, \boldsymbol{\Sigma})$. Then $Z_k \sim N_1(\delta_k, \mathbf{1}'\boldsymbol{\Sigma}\mathbf{1}/r^2)$ and

$$\mathbf{Z} \sim N_m\left(\boldsymbol{\delta}, \frac{\mathbf{1}'\boldsymbol{\Sigma}\mathbf{1}}{r^2}\mathbf{I}\right).$$

Therefore, the model in which we observe only $\mathbf{Z}$ is still an ordinary linear model and hence

$$F_A \sim F_{s-q_A, n-s}\left(\frac{r^2\|\mathbf{P}_{S\mid Q_A}\boldsymbol{\delta}\|^2}{\mathbf{1}'\boldsymbol{\Sigma}\mathbf{1}}\right).$$

Therefore, this test is still a sensible size α test for this more general model. Procedures for type A hypotheses are therefore not sensitive to the assumption that $\operatorname{cov}(\mathbf{Y}_k) = \sigma^2 \mathbf{A}(\rho)$.

The model considered in this chapter does not include all possible models for which repeated measures models would be appropriate. In particular, V is defined in such a way that V^* is a product of two subspaces. If V does not have this form, then there are no complete sufficient statistic and no minimum variance unbiased estimators for the model. The maximum likelihood estimators are quite messy and optimal tests do not seem possible. One example of such a model is the analysis of covariance model given in Example 4. Another example is a one-way model in which each individual receives b treatments, where b is less than the total number of treatments. This sort of model is discussed in papers dealing with the recovery of interblock information. One approach to models of this sort that do not fall into the framework considered in this chapter is to find a "sensible" estimator $\hat{\rho}$ of ρ, and use the results for the generalized linear model derived in Chapter 13 applied to the generalized linear model that occurs when we assume that $\mathbf{A}(\rho) = \mathbf{A}(\hat{\rho})$. This generates approximate procedures for this generalization of the repeated measures model, which should be sensible procedures provided $\hat{\rho}$ is near ρ.

In addition, it is sometimes desirable to test hypotheses involving both $\boldsymbol{\delta}$ and γ in a repeated measures model. In Example 3 we might want to test that both $\beta_j = 0$ and $\gamma_{ij} = 0$. There is no obvious test for this sort of problem. One approach to this problem is the same as suggested in the last paragraph. Estimate ρ and then use the procedure that would have been optimal if we had known that $\rho = \hat{\rho}$.

14.9. THE EXCHANGEABLE LINEAR MODEL.

We now study a model that is closely related to the repeated measures model discussed earlier in this chapter. The random variables $e_1, \ldots, e_r$ are said to be *exchangeably distributed* if the joint distribution of $(e_{\pi 1}, \ldots, e_{\pi r})$

is the same as the joint distribution of $(e_1, \ldots, e_r)$ for any permutation π of the integers $1, \ldots, r$. If the e_i are independently, identically distributed, then they are exchangeably distributed, but exchangeably distributed random variables need not be independent. The model we study in this section occurs when we replace the assumption in the ordinary linear model of independently, identically, normally distributed errors with the weaker assumption of exchangeably normally distributed errors. This model is therefore the model in which we observe $\mathbf{Y} = \boldsymbol{\mu} + \mathbf{e}$, where $\boldsymbol{\mu}$ is an unobserved constant that is assumed to lie in a p-dimensional subspace V of R^r, and $\mathbf{e} = (e_1, \ldots, e_r)'$ is an unobserved random vector with expectation $\mathbf{0}$ such that the e_i are exchangeably, jointly normally distributed. We call this model the *exchangeable linear model.*

The exchangeable linear model would be applicable in many problems of dependent sampling, particularly those involving sampling from a finite population. In addition, in this section, we show that there is no sensible test of the hypothesis that the errors are independently, identically distributed against the alternative that they are exchangeably distributed.

There is therefore no way to tell from the data whether the ordinary linear model or the exchangeable linear model is correct for a particular problem. We may therefore be using the ordinary linear model in situations in which the exchangeable linear model is appropriate.

The exchangeable linear model has been studied by Halperin (1951) and McElroy (1967). The approach given here follows Arnold (1979a).

Let $\boldsymbol{\Sigma}$ be the covariance matrix of $\mathbf{e}$. Since $\mathbf{e}$ has an r-variate normal distribution with expectation $\mathbf{0}$, we see that the e_i are exchangeably distributed if and only if

$$\boldsymbol{\Sigma} = \sigma^2 \mathbf{A}(\rho)$$

for some $\sigma^2 > 0$, $\mathbf{A}(\rho) > 0$. Therefore, an equivalent statement of the exchangeable linear model is that it is the model in which we observe

$$\mathbf{Y} \sim N_r(\boldsymbol{\mu}, \sigma^2 \mathbf{A}(\rho)), \qquad \boldsymbol{\mu} \in V, \qquad \sigma^2 > 0, \qquad \mathbf{A}(\rho) > 0. \quad (14.22)$$

For the remainder of this section, we assume that $\mathbf{1} \in V$, and hence that $U \subset V$, where U is the subspace spanned by $\mathbf{1}$. Now, let $\delta \mathbf{1} = \mathbf{P}_U \boldsymbol{\mu}$, $\boldsymbol{\gamma} = \mathbf{P}_{V|U} \boldsymbol{\mu}$. Then

$$\boldsymbol{\mu} = \delta \mathbf{1} + \boldsymbol{\gamma} \qquad \boldsymbol{\gamma} \perp \mathbf{1}. \quad (14.23)$$

We see, therefore, that the exchangeable linear model is really just a repeated measures model with $m = 1$, that is, with only one individual.

As in Section 14.4, let $\mathbf{C}$ be an orthonormal basis matrix for $U^{\perp}$, and

define

$$\mathbf{Y}^* = \begin{pmatrix} Y_1^* \\ \mathbf{Y}_2^* \end{pmatrix} = \begin{pmatrix} r^{-1/2}\mathbf{1}' \\ \mathbf{C}' \end{pmatrix}\mathbf{Y}, \qquad \delta^* = r^{1/2}\delta, \qquad \gamma^* = \mathbf{C}'\gamma, \tag{14.24}$$

$$\tau_1^2 = \left[1 + (r-1)\rho\right]\sigma^2, \qquad \tau_2^2 = \sigma^2(1-\rho), \qquad T^* = \{\mathbf{C}'\gamma; \gamma \in V \mid U\}$$

where Y_1^* is 1×1. Then, by Lemma 14.3 (which is true for $m = 1$), we see that Y_1^* and $\mathbf{Y}_2^*$ are independent and

$$\begin{gathered} Y_1^* \sim N_1(\delta^*, \tau_1^2), \qquad \mathbf{Y}_2^* \sim N_{r-1}(\gamma^*, \tau_2^2\mathbf{I}), \\ \delta^* \in R^1, \qquad \gamma^* \in T^* \qquad \tau_1^2 > 0, \qquad \tau_2^2 > 0. \end{gathered} \tag{14.25}$$

We note that the part of this model involving $\mathbf{Y}_2^*$ is just an ordinary linear model, whereas the part involving Y_1^* is a simple model in which we have one observation from a normal distribution with unknown mean and unknown variance. There is, therefore, no unbiased estimator of τ_1^2.

Define

$$\hat{\gamma}^* = \mathbf{P}_{T^*}\mathbf{Y}_2^*, \qquad \hat{\tau}_2^2 = \frac{1}{r-p}\|\mathbf{P}_{T^{*\perp}}\mathbf{Y}_2^*\|^2. \tag{14.26}$$

(Note that $\dim T^* = p - 1$ and hence the natural denominator of $\hat{\tau}_2$ is $(r-1)-(p-1) = r - p$.)

LEMMA 14.13. $(Y_1^*, \hat{\gamma}^*, \hat{\tau}_2^2)$ is a complete sufficient statistic for the exchangeable linear model.

PROOF. Let $\mathbf{X}_2$ be an orthonormal basis matrix for T^*. Using the exponential criterion, we see that $(Y_1^{*2}, Y_1^*, (n-p)\hat{\tau}_2^2 + \|\hat{\gamma}_2^*\|^2, \mathbf{X}_2'\hat{\gamma}_2^*)$ is a complete sufficient statistic for this model. Since $(Y_1^*, \hat{\gamma}_2^*, \hat{\tau}_2^2)$ is an invertible function of that statistic it is also complete and sufficient. □

From (14.25), we see that $Y_1^*, \hat{\gamma}^*$, and $\hat{\tau}_2^2$ are independent and

$$Y_1^* \sim N_1(\delta^*, \tau_1^2), \hat{\gamma}^* \sim N_{r-1}(\gamma^*, \tau_2^2\mathbf{P}_{T^*}), \qquad (r-p)\frac{\hat{\tau}_2^2}{\tau^2} \sim \chi^2_{r-p}(0). \tag{14.27}$$

Now, let

$$\hat{\mu} = \mathbf{P}_V\mathbf{Y}, \qquad \hat{\sigma}^2 = \frac{1}{r-p}\|\mathbf{P}_{V^\perp}\mathbf{Y}\|^2$$

be the usual sufficient statistic for the ordinary linear model that occurs if we assume that $\rho = 0$ in exchangeable linear model. Then

$$\hat{\mu} = (r^{-1/2}\mathbf{1}, \mathbf{C})\begin{pmatrix} Y_1^* \\ \hat{\gamma} \end{pmatrix}, \qquad \hat{\sigma}^2 = \hat{\tau}_2^2 \tag{14.28}$$

(see Exercise B14).

THEOREM 14.14. a. $(\hat{\boldsymbol{\mu}}, \hat{\sigma}^2)$ is a complete sufficient statistic for the exchangeable linear model.

b. $\hat{\boldsymbol{\mu}}$ and $\hat{\sigma}^2$ are the minimum variance unbiased estimators of $\hat{\boldsymbol{\mu}}$ and $\sigma^2(1-\rho)$. There are no unbiased estimators of σ^2 or $\rho\sigma^2$.

c. There are no maximum likelihood estimators for the exchangeable linear model.

d. $\hat{\boldsymbol{\mu}}$ is the best invariant estimator of $\boldsymbol{\mu}$ with respect to quadratic loss functions of the form $L_1(\mathbf{d}_1, (\boldsymbol{\mu}, \sigma^2, \rho)) = (\mathbf{d}_1 - \boldsymbol{\mu})'\mathbf{H}(\rho)(\mathbf{d}_1 - \boldsymbol{\mu})/\sigma^2$ where $\mathbf{H}(\rho) > 0$. There is no best invariant estimator of σ^2 for the quadratic loss function $L_2(d_2; (\boldsymbol{\mu}, \sigma^2, \rho)) = (d_2 - \sigma^2)^2/\sigma^4$.

PROOF. a. From (14.28) we see that $(\hat{\boldsymbol{\mu}}, \hat{\sigma}^2)$ is an invertible function of the complete sufficient statistic $(Y_1^*, \hat{\gamma}^*, \hat{\tau}_2^2)$ and is therefore complete and sufficient.

b. $E\hat{\boldsymbol{\mu}} = \mathbf{P}_V E\mathbf{Y} = \mathbf{P}_V \boldsymbol{\mu} = \boldsymbol{\mu}$, and $E\hat{\sigma}^2 = E\hat{\tau}_2^2 = \tau_2^2 = \sigma^2(1-\rho)$. If there were an unbiased estimator σ^2 or $\rho\sigma^2$ there would be an unbiased estimator of τ_1^2.

c. We consider the transformed version of the model. For all τ_1^2, the likelihood is maximized for $\delta^* = Y_1^*$, $\gamma^* = \hat{\gamma}^*$, $\tau_2^2 = [(r-p)/(r-1)]\hat{\tau}_2^2$. The likelihood can be made arbitrarily large by letting τ_1^2 go to 0. Therefore there is no maximum for the likelihood function, and hence are no maximum likelihood estimators.

d. See Exercises B19 and B20. □

One important consequence of this theorem is that the complete sufficient statistic for the exchangeable linear model is the same as the complete sufficient statistic for the ordinary linear model that occurs when $\rho = 0$. We have the additional parameter ρ in the exchangeable linear model, and so it is not surprising that some parameters cannot be estimated and some hypotheses cannot be sensibly tested.

We now consider the problem of testing that $\rho = 0$ and show that there is no sensible test. Since this problem is the same as testing that $\tau_1^2 = \tau_2^2$, and there is no sensible estimator of τ_1^2, it seems unlikely that there would be any sensible test of this hypothesis. In the next two theorems, we show that the only unbiased or invariant tests ignore the data.

THEOREM 14.15. Let $\phi(\hat{\boldsymbol{\mu}}, \hat{\sigma}^2)$ be an unbiased size α test that $\rho = 0$ in the exchangeable linear model. Then $\phi(\hat{\boldsymbol{\mu}}, \hat{\sigma}^2) = \alpha$ with probability 1.

PROOF. The density of $\mathbf{Y}$ comes from an exponential family, and hence the power function of any critical function must be a continuous function of $\boldsymbol{\mu}$, σ^2, and ρ. (See Lehmann, 1959, p. 52.) This fact, together with the definition of an unbiased size α test implies that

$$E\left[\phi(\hat{\boldsymbol{\mu}}, \hat{\sigma}^2) - \alpha\right] = 0$$

when $\rho = 0$. However, the statistic $(\hat{\boldsymbol{\mu}}, \hat{\sigma}^2)$ has a complete family of densities for the family in which $\rho = 0$ (i.e., for the ordinary linear model). Hence $\phi(\hat{\boldsymbol{\mu}}, \hat{\sigma}^2) = \alpha$ with probability 1. □

This theorem implies that any unbiased test that is based only on $(\hat{\boldsymbol{\mu}}, \hat{\sigma}^2)$ must ignore the data with probability 1. We now show that any unbiased test for this problem has the same power function as one that ignores the data.

COROLLARY. Let $\phi(\mathbf{Y})$ be an unbiased size α test that $\rho = 0$. Then $E\phi(\mathbf{Y}) = \alpha$ for all $\boldsymbol{\mu}$, σ^2, and ρ.

PROOF. See Exercise B15. □

We now consider invariance for this problem. Let $c > 0$, $\mathbf{d} \in V$. Then

$$c\mathbf{Y} + \mathbf{d} \sim N_r\left(c\boldsymbol{\mu} + \mathbf{d}, c^2\sigma^2\mathbf{A}(\rho)\right).$$

Also $c\boldsymbol{\mu} + \mathbf{d} \in V$ if and only if $\boldsymbol{\mu} \in V$. Therefore, this problem is invariant under the group G of transformations of the form

$$g(\mathbf{Y}) = \mathbf{cY} + \mathbf{d} \qquad \bar{g}(\boldsymbol{\mu}, \sigma^2, \rho) = (c\boldsymbol{\mu} + \mathbf{d}, c^2\sigma^2, \rho)$$

where $c > 0$, $\mathbf{d} \in V$.

THEOREM 14.16. Let $\phi(\hat{\boldsymbol{\mu}}, \hat{\sigma}^2)$ be an invariant size α test. Then $\phi(\hat{\boldsymbol{\mu}}, \hat{\sigma}^2) = \alpha$.

PROOF. In terms of the sufficient statistic g becomes

$$g_{c,d}(\hat{\boldsymbol{\mu}}, \hat{\sigma}^2) = (c\hat{\boldsymbol{\mu}} + \mathbf{d}, c^2\hat{\sigma}^2).$$

Since

$$(\hat{\boldsymbol{\mu}}, \hat{\sigma}^2) = g_{\hat{\sigma}^2, \hat{\boldsymbol{\mu}}}(\mathbf{0}, 1),$$

we see that

$$\phi(\hat{\boldsymbol{\mu}}, \hat{\sigma}^2) = \phi(g_{\hat{\sigma}^2, \hat{\boldsymbol{\mu}}}(\mathbf{0}, 1) = \phi(\mathbf{0}, 1) = c.$$

Since ϕ has size α, $c = \alpha$. □

These two results indicate that there is no sensible way to tell from the data whether the correct model is an ordinary linear model or merely an exchangeable linear model.

We now consider the problem of testing that $\delta = 0$ in the exchangeable linear model. Since any information about δ would have to come from Y_1^*, and we cannot estimate the variance of Y_1^*, it seems unlikely that we can find a sensible test for this problem either. In fact, in Exercise B17, you are asked to show that the only invariant test that $\delta = 0$ for the exchangeable linear model ignores the data. Further, in Exercise B18 you are asked to show that the UMP invariant size α test that $\delta = 0$ for the ordinary linear model (that occurs when we assume that $\rho = 0$ in the exchangeable linear model) has size 1 for the exchangeable linear model. Therefore, this test would be completely inappropriate for the exchangeable linear model. This shows that if we are interested in making inference about δ it can be a rather severe mistake to assume that the model is an ordinary linear model when it is, in fact, an exchangeable linear model.

The final testing problem that we discuss in this section is one in which we test that $\boldsymbol{\mu} \in W$ where W is a k-dimensional subspace of V such that $\mathbf{1} \in W$. Since $\mathbf{1} \in W$, we see that $\boldsymbol{\mu} \in W$ if and only if $\boldsymbol{\gamma} \in W | U$. Therefore, we could also think of this problem as testing that $\boldsymbol{\gamma} \in W | U$ against $\boldsymbol{\gamma} \in V | U$.

THEOREM 14.17. The UMP invariant size α test that $\boldsymbol{\mu} \in W$ against $\boldsymbol{\mu} \in V$ where $\mathbf{1} \in W \subset V$ is the same for the exchangeable linear model as for the ordinary linear model that occurs when we assume that $\rho = 0$ in the exchangeable linear model.

PROOF. See Exercise B16. □

This result implies that if we are only interested in inference about $\boldsymbol{\gamma}$, it does not matter whether we assume an ordinary linear model or merely an exchangeable linear model. The optimal procedure is the same for both models. In fact, Arnold (1979a) shows that the UMP invariant test for the ordinary linear model is also most stringent, Bayes, and admissible for the exchangeable linear model. In short, it has the same properties for the exchangeable linear model as for the ordinary linear model, with the exception of the property of being a likelihood ratio test. There can be no likelihood ratio tests for the exchangeable linear model since there are no maximum likelihood estimators for this model.

This result also indicates that as long as we are only interested in inference about $\boldsymbol{\gamma}$, the procedures developed in Chapter 7 are not sensitive to the independence assumption used in deriving them. They are still optimal for the (rather limited) sort of dependence presented in this section. In addition, we could extend the result to show that the t-test developed in Chapter 7 for linear functions of $\boldsymbol{\gamma}$ is optimal for the exchangeable linear model as well as for the ordinary linear model, and that the techniques developed in Chapter 8 for finding simultaneous confidence intervals in contrasts associated with hypotheses involving $\boldsymbol{\gamma}$ are also valid for the exchangeable linear model.

We now present some examples and show that for most examples, the parameters of interest are in $\boldsymbol{\gamma}$ not in δ.

EXAMPLE 5. Regression. Let $\mathbf{Y} = \theta\mathbf{1} + \mathbf{T}\boldsymbol{\beta} + e$. Let $\boldsymbol{\mu} = \theta\mathbf{1} + \mathbf{T}\boldsymbol{\beta}$. Then

$\delta = \theta 1 + \bar{\mathbf{T}}\beta, \gamma = \tilde{\mathbf{T}}\boldsymbol{\beta}$ (see Section 6.2). Since θ is unrestricted in this model, $\boldsymbol{\beta} \in Q$ if and only if $\gamma \in \{\mathbf{T}\boldsymbol{\beta};\ \boldsymbol{\beta} \in Q\}$ (i.e., we get no information about $\boldsymbol{\beta}$ from δ). Therefore, any hypothesis about $\boldsymbol{\beta}$ can be tested, and the UMP invariant test is the same as for the ordinary linear model. However, hypotheses involving θ cannot be tested. Fortunately, θ is usually the least interesting parameter in a regression of this form.

EXAMPLE 6. The one-sample problem. Let $Y_i = \theta + e_i$, and let $\mathbf{Y}' = (Y_1, \ldots, Y_n)$, $\boldsymbol{\mu} = E\mathbf{Y}$, and $\mathbf{e}' = (e, \ldots, e_n)$. Then $\boldsymbol{\mu} = \theta\mathbf{1}$, so that $\delta = \theta$, $\gamma = \mathbf{0}$. Therefore we *cannot* test hypotheses about θ.

EXAMPLE 7. The two-sample problem. Let $\mathbf{Y}_{1i} = \theta_1 + e_{1i}$, $i = 1, \ldots, n$, and $Y_{2i} = \theta_2 + e_{2i}$, $i = 1, \ldots, m$. Let $\mathbf{Y}' = (Y_{11}, \ldots, Y_{1m}, Y_{21}, \ldots, Y_{2m})$, and let $\mu = EY$. Then

$$\delta = \frac{n\theta_1 + m\theta_2}{n+m}, \qquad \gamma' = (\theta_1 - \delta, \ldots, \theta_1 - \delta, \theta_2 - \delta, \ldots, \theta_2 - \delta).$$

We can therefore test the hypothesis that $\theta_1 = \theta_2$, and the UMP invariant test is just the usual two-sample t-test.

EXAMPLE 8. One-way analysis of variance. Suppose that $Y_{ij} = \theta + \alpha_i + e_{ij}$, $i = 1, \ldots, k$; $j = 1, \ldots, n_i$, where $\sum n_i\alpha_i = 0$. Let $\mathbf{Y}' = (Y_{11}, \ldots, Y_{1n_1}, \ldots, Y_{k1}, \ldots, Y_{kn_k})$. Let $\boldsymbol{\mu} = E\mathbf{Y}$. Using the restraint on the α_i we see that

$$\delta = \theta, \qquad \gamma' = (\alpha_1, \ldots, \alpha_1, \ldots, \alpha_k, \ldots, \alpha_k).$$

We can therefore test the hypothesis that the $\alpha_i = 0$, and the test is just the usual F-test.

EXAMPLE 9. Two-way analysis of variance. Let $Y_{ij} = \theta + \alpha_i + \beta_j + e_{ij}$, $i = 1, \ldots, r$; $j = 1, \ldots, c$, where $\sum_i \alpha_i = 0$, $\sum_j \beta_j = 0$. Let $\mathbf{Y}$ be defined as in Example 8. Using the restraints we see that

$$\delta = \theta \qquad \gamma' = (\alpha_1 + \beta_1, \ldots, \alpha_1 + \beta_c, \ldots, \alpha_r + \beta_1, \ldots, \alpha_r + \beta_c).$$

Therefore, we can test that $\alpha_i = 0$ or that the $\beta_j = 0$, and the UMP invariant test is just the F-test derived in Chapter 7.

We could also look at more complicated examples of analysis of variance. For all these examples, the hypotheses of interest can be tested with the usual F-tests used in the ordinary linear model.

EXERCISES.

Type A

1. Test the hypothesis that $\alpha_i = 0$ using the model of Example 1 on the data of Exercise A2 of Chapter 7.

2. Test the hypothesis that the $\alpha_i = 0$ and that the $\beta_j = 0$ using the model of Example 2 on the data of Exercise A4 of Chapter 7.

3. Test the hypothesis that the $\alpha_i = 0$ and that the $\beta_j = 0$ using the model of Example 3 on the data of Exercise A4 of Chapter 7.

Type B

1. Use Theorem A to find the MLE of ρ.

2. Use (14.9) to find a $(1 - \alpha)$ confidence interval for ρ.

3. In Example 2 verify the expressions for U_1^2 and U_2^2.

4. Consider the model in which we observe $Y_{ijk} = \theta + \alpha_i + \beta_j + \gamma_{ij} + e_{ijk}$, $i = 1, \ldots, d$; $j = 1, \ldots, c$; $k = 1, \ldots, p_j$, where $\sum_i \alpha_i = 0$, $\sum_j p_j \beta_j = 0$, $\sum_i \gamma_{ij} = 0$, $\sum_j p_j \gamma_{ij} = 0$, and $\mathbf{e}_{jk} = (e_{1jk}, \ldots, e_{djk})'$ are independent and $\mathbf{e}_{jk} \sim N_d(\mathbf{0}, \sigma^2 \mathbf{A}(\rho))$. Find the UMP invariant size α tests for testing that $\alpha_i = 0$ and for testing that $\beta_j = 0$. Find the $(1 - \alpha)$ confidence intervals associated with these testing problems. (Note that this unbalanced model is a case of proportional sampling with the "nice" weights. Therefore formulas for estimating the parameters for the ordinary linear model can be obtained from Section 7.4.3.)

5. Consider the model described in Example 4 under the condition that $\bar{x}_{.k} = 0$. Find the UMP invariant size α tests for testing that $\alpha_i = 0$ and for testing that $\beta = 0$.

6. Consider the model described in Example 4 under the condition that $x_{ik} = \bar{x}_{.k}$. Find the UMP invariant size α tests for testing that $\alpha_i = 0$ and for testing that $\beta = 0$.

7. Consider a nested repeated measures model with d classes and c subclasses in each class. Assume each individual is in only one class but is in each subclass. Suppose that there are n individuals in each class.

(a) Set up a repeated measures model for this situation.

(b) Find the F-statistic for testing that there is no class effect.

(c) Find the F-statistic for testing that there is no subclass effect.

8. Prove Lemma 14.1b. (If $\mathbf{X}$ is an orthonormal basis matrix for W then $\mathbf{CX}$ is an orthonormal basis matrix for W^*. Why?)

9. Verify that $U_1 = \hat{\tau}_1^2$ in Section 14.5. ($\mathbf{Y}_1^* = r^{1/2}\mathbf{Z}$. Why?)

10. Show that ϕ_A defined in (14.18) is the likelihood ratio test for testing that $\boldsymbol{\delta} \in Q_A$ against $\boldsymbol{\delta} \in S$.

11. Prove Theorem 14.11.

12. Prove Theorem 14.12b.

13. Verify (14.21)

14. Verify (14.28).

15. Prove the corollary to Lemma 14.15. [Hint: Let $\phi^*(\hat{\boldsymbol{\mu}}, \hat{\sigma}^2) = E\phi(\mathbf{Y}) \mid (\hat{\boldsymbol{\mu}}, \hat{\sigma}^2)$.]

16. Prove Theorem 14.17.

17. Consider the problem of testing that $\delta = 0$ (or equivalently $\delta^* = 0$) in the exchangeable linear model.

(a) Show that this problem is invariant under the group G of transformations

$$g(Y_1^*, \mathbf{Y}_2^*) = (bY_1^*, c\mathbf{Y}_2^* + \mathbf{d}) \qquad b \neq 0, \qquad c > 0, \qquad \mathbf{d} \in T^*,$$

or equivalently

$$g_{b,c,d}\left(Y_1^*, \hat{\boldsymbol{\gamma}}^*, \hat{\tau}_2^2\right) = \left(bY_1^*, c\hat{\boldsymbol{\gamma}}^* + \mathbf{d}, c^2\hat{\tau}_2^2\right).$$

(b) Show that $(Y_1^*, \hat{\boldsymbol{\gamma}}^*, \hat{\tau}_1^2) = g_{Y_1^*, \hat{\boldsymbol{\gamma}}^*, \hat{\tau}_2^2}(1, \mathbf{0}, 1)$.

(c) Show that any test that is invariant under G must be constant.

18. (a) Find the UMP invariant size α test for testing that $\delta = 0$ in the ordinary linear model that occurs when we assume that $\rho = 0$ in the exchangeable linear model.

Answer:

$$F = \frac{(Y_1^*)^2}{\hat{\sigma}^2}, \qquad \phi(F) = \begin{cases} 1 & \text{if } F > F_{1,n-p}^{\alpha} \\ 0 & \text{if } F \leqslant F_{1,n-p}^{\alpha} \end{cases}.$$

(b) Show that this test has size 1 for the exchangeable linear model. [Hint: $F^* = (\tau_2^2/\tau_1^2)F \sim F_{1,n-p}(0)$ under the null hypothesis and $E\phi = P(F^* > (\tau_2^2/\tau_1^2)F_{1,n-p})$. Let $\tau_2^2 \to 0$.]

19. Consider estimating $\boldsymbol{\mu}$ with the loss function $L(\mathbf{d}; (\boldsymbol{\mu}, \sigma^2, \rho) = (\mathbf{d} - \boldsymbol{\mu})'\mathbf{H}(\rho)(\mathbf{d} - \boldsymbol{\mu})$, where $\mathbf{H}(\rho)$ is a known positive definite matrix function of ρ.

(a) Show that this problem is invariant under the group G of transformations $g(\hat{\boldsymbol{\mu}}, \hat{\sigma}^2) = (c\hat{\boldsymbol{\mu}} + \mathbf{b}, c^2\hat{\sigma}^2)$ where $c > 0$, $\mathbf{b} \in V$. What are the induced transformations $\bar{G}$ and $\tilde{G}$?

(b) Show that an estimator $\mathbf{T}(\hat{\boldsymbol{\mu}}, \hat{\sigma}^2)$ is invariant if and only if

$$\mathbf{T}(c\hat{\boldsymbol{\mu}} + \mathbf{b}, c^2\hat{\sigma}^2) = c\mathbf{T}(\hat{\boldsymbol{\mu}}, \hat{\sigma}^2) + \mathbf{b}$$

if and only if $\mathbf{T}(\hat{\boldsymbol{\mu}}, \hat{\sigma}^2) = \hat{\sigma}\mathbf{T}(\mathbf{0}, 1) + \hat{\boldsymbol{\mu}}$. [Let $c = 1/\hat{\sigma}$, $\mathbf{b} = (-1/\hat{\sigma})\hat{\boldsymbol{\mu}}$.]

(c) Show that $\hat{\boldsymbol{\mu}}$ is an invariant estimator of $\boldsymbol{\mu}$.

(d) Now, let $T(\hat{\boldsymbol{\mu}}, \hat{\sigma}^2)$ be any other invariant estimator and let $\mathbf{M} = \mathbf{T}(\mathbf{0}, 1) \neq \mathbf{0}$. Show that

$$R(\mathbf{T}; (\boldsymbol{\mu}, \sigma^2, \rho)) = \mathbf{M}'\mathbf{H}(\rho)\mathbf{M} + R(\hat{\boldsymbol{\mu}}; (\boldsymbol{\mu}, \sigma^2, \rho)) > R(\hat{\boldsymbol{\mu}}; (\boldsymbol{\mu}, \sigma^2, \rho))$$

and hence that $\hat{\boldsymbol{\mu}}$ is the best invariant estimator of $\boldsymbol{\mu}$.

20. Now, consider estimating σ^2 with the loss function $L(d; (\boldsymbol{\mu}, \sigma^2, \rho) = (d - \sigma^2)^2/\sigma^4$.

(a) Show that this estimation problem is invariant under the group G given in Exercise B19 above. What are $\overline{G}$ and $\tilde{G}$?

(b) Show that $T(\hat{\boldsymbol{\mu}}, \hat{\sigma}^2)$ is an invariant estimator if and only if $T(\hat{\boldsymbol{\mu}}, \hat{\sigma}^2) = \hat{\sigma}^2 T(\mathbf{0}, 1)$.

(c) Let $T(\hat{\boldsymbol{\mu}}, \hat{\sigma}^2)$ be an invariant estimator of $\hat{\sigma}^2$ and let $M = T(\mathbf{0}, 1)$. Show that $R(T; (\boldsymbol{\mu}, \sigma^2, \rho)) = M^2(n - p + 2)(1 - \rho)^2/(n - p) - 2M(1 - \rho) + 1$.

(d) Show that there is no best invariant estimator of σ^2. [Note that the risk is minimized for $M = (n - p)/(n - p + 2)(1 - \rho)$ which depends on an unknown parameter.]

Type C

1. (A James-Stein estimator for the exchangeable linear model.) Consider estimating $\boldsymbol{\mu}$ in the exchangeable linear model with the loss function $L(\mathbf{d}; (\boldsymbol{\mu}, \sigma^2, \rho) = (\mathbf{d} - \boldsymbol{\mu})'(\mathbf{A}(\rho))^{-1}(\mathbf{d} - \boldsymbol{\mu})/\sigma^2$. Note that this is just the Mahalanobis distance for this model. Let $\mathbf{d}(\mathbf{Y})$ be an estimator of $\boldsymbol{\mu}$ and define

$$\mathbf{d}^*(\mathbf{Y}^*) = \boldsymbol{\Gamma}\mathbf{d}(\boldsymbol{\Gamma}'\mathbf{Y}) = \begin{pmatrix} d_1^*(\mathbf{Y}^*) \\ \mathbf{d}^*{}_2(\mathbf{Y}^*) \end{pmatrix}$$

where $d_1^*(\mathbf{Y}^*) \in R^1$.

(a) Show that $L(\mathbf{d}; (\boldsymbol{\mu}, \sigma^2, \rho)) = (d_1^*(\mathbf{Y}^*) - \delta^*)^2/\tau_1^2 + \|\mathbf{d}_2^*(\mathbf{Y}^*) - \gamma^*\|^2/\tau_2^2$.

(b) Find the risk of the estimator $\hat{\boldsymbol{\mu}}$.

(c) If $p \geqslant 4$, find an estimator $\hat{\hat{\gamma}}^*$ that has lower risk than $\hat{\gamma}^*$ for the ordinary linear model involving only $\mathbf{Y}_2^*$.

(d) Let

$$\hat{\hat{\boldsymbol{\mu}}} = \boldsymbol{\Gamma}'\begin{pmatrix} Y_1^* \\ \hat{\hat{\gamma}}^* \end{pmatrix}.$$

Show that $\hat{\hat{\boldsymbol{\mu}}}$ has smaller risk than $\hat{\boldsymbol{\mu}}$.

CHAPTER 15

Random Effects and Mixed Models

In this chapter, we consider models used in analysis of of variance settings in which the levels of a treatment (or treatments) are not fixed in advance of the experiment, but are a sample from some larger class of levels. To illustrate these models consider a cattle breeding experiment in which r bulls and c cows are crossbred n times. Let Y_{ijk} be the weight at 9 months of the kth calf born from the ith bull and jth cow. We want to determine the effects of the bulls and the cows on this weight. We could analyze this data with the balanced two-way model discussed in Section 7.3.3. In that model, we would assume that

$$Y_{ijk} = \theta + \alpha_i + \beta_j + \delta_{ij} + e_{ijk}$$

where θ, α_i, β_j, and δ_{ij} are unknown constants and the e_{ijk} are unobserved random variables such that $e_{ijk} \sim N_1(0, \sigma_e^2)$. The parameter space for this model is

$$-\infty < \theta < \infty, \qquad \sum_i \alpha_i = 0, \qquad \sum_j \beta_j = 0,$$

$$\sum_i \delta_{ij} = 0, \qquad \sum_j \delta_{ij} = 0, \qquad \sigma_e^2 > 0.$$

This model is called the *fixed effects model* for this situation, and it is appropriate as long as we are only interested in inference for those bulls and cows contained in the experiment. The α_i, β_j, and δ_{ij} are called *fixed effects* for this model.

Now, suppose that the cows and bulls are randomly selected from some larger collection of bulls and cows. Then the effects due to the bulls and cows would no longer be constants, but would be random variables that would depend on which bulls and cows were randomly selected. We would therefore replace the unknown constants, α_i, β_j, and δ_{ij} with unobserved random variables a_i, b_j, and d_{ij}. The *random effects model* for this situation is one in which we observe

$$Y_{ijk} = \theta + a_i + b_j + d_{ij} + e_{ijk}$$

where the a_i, b_j, d_{ij}, and e_{ijk} are unobserved random variables that are all independent and

$$a_i \sim N_1(0, \sigma_a^2), \qquad b_j \sim N_1(0, \sigma_b^2), \qquad d_{ij} \sim N_1(0, \sigma_d^2), \qquad e_{ijk} \sim N_1(0, \sigma_e^2).$$

The parameter space is given by

$$-\infty < \theta < \infty, \qquad \sigma_a^2 \geqslant 0, \qquad \sigma_b^2 \geqslant 0, \qquad \sigma_d^2 \geqslant 0, \qquad \sigma_e^2 > 0.$$

(Note that we assume as usual that $\sigma_e^2 > 0$, but only assume that $\sigma_a^2 \geqslant 0$. If there is no effect due to the bulls then $\sigma_a^2 = 0$). Throughout this chapter, we follow the notation of using Greek letters for unobserved constants and Roman letters for random variables. Note that

$$\operatorname{var}(Y_{ijk}) = \sigma_a^2 + \sigma_b^2 + \sigma_d^2 + \sigma_e^2.$$

For this reason, σ_a^2, σ_b^2, σ_d^2, and σ_e^2 are called the *components of variance* for this model. The random variables a_i, b_j, and d_{ij} are called *random effects*. We are interested in estimating the parameters θ, σ_a^2, σ_b^2, σ_d^2, and σ_e^2 for this model as well as finding confidence intervals for them. We are also interested in testing that $\sigma_a^2 = 0$ (i.e., that the bulls have no effect on the weight), that $\sigma_b^2 = 0$ and that $\sigma_d^2 = 0$. Note that we are not interested in simultaneous confidence intervals for contrasts in the a_i, since these are random variables concerning the particular bulls selected for the study. If we are interested in the particular bulls, we should treat the bulls as a fixed effect.

In fact, in many cattle breeding experiments, we are interested in the particular bulls in the experiment, but not in the particular cows. The cows have to give birth every year in order to give milk. However, often the only function of the bulls is to service the cows, so there may not be too many bulls in the herd and all the bulls would be used. In this case, we could not use a random effects model, since the bulls are not selected randomly from a larger population. We instead use a *mixed model* for this situation in which we assume

$$Y_{ijk} = \theta + \alpha_i + b_j + d_{ij} + e_{ijk},$$

where θ and α_i are unknown constants, b_j, d_{ij}, and e_{ijk} are unobserved random variables. (Note that d_{ij} is random since it depends on the cow chosen.) Two different models have been considered for this situation, depending on the assumptions about the interactions d_{ij}. In the first version, we assume that the b_j, d_{ij}, and e_{ijk} are all independent with $b_j \sim N_1(0, \sigma_b^2)$, $d_{ij} \sim N_1(0, \sigma_d^2)$, $e_{ijk} \sim N_1(0, \sigma_e^2)$. The parameter space for this model is given by

$$-\infty < \theta < \infty, \qquad \sum_i \alpha_i = 0, \qquad \sigma_b^2 \geqslant 0, \qquad \sigma_d^2 \geqslant 0, \qquad \sigma_e^2 > 0.$$

In the second version we assume that $\sum_i d_{ij} = 0$. Clearly, therefore, the d_{ij} cannot be independent. Let $\mathbf{d}_j' = (d_{1j}, \ldots, d_{rj})$. In this version, we assume that b_j, $\mathbf{d}_j$, and e_{ijk} are independent, with $b_j \sim N_1(0, \sigma_b^2)$, $e_{ijk} \sim N_1(0, \sigma_e^2)$. We assume that the $\mathbf{d}_j$ have an r-dimensional normal distribution with $E\,d_{ij} = 0$, $\text{var}(d_{ij}) = \sigma_d^2$, $\text{cov}(d_{ij}, d_{i'j}) = -\sigma_d^2/(r-1)$. [If $\text{cov}(d_{ij}, d_{i'j.})$ does not depend on i or i', then it must be $-\sigma_d^2/(r-1)$. See Exercise B1.] The parameter space is again given by

$$-\infty < \theta < \infty, \qquad \sum_i \alpha_i = 0, \qquad \sigma_b^2 \geqslant 0, \qquad \sigma_d^2 \geqslant 0, \qquad \sigma_e^2 > 0.$$

A heuristic derivation leading to this version of the mixed model is given in Section 15.4. The constants α_i are called *fixed effects* for this model, while the random variables b_j and d_{ij} are called *random effects*. We are interested in estimating the parameters θ, α_i, σ_b^2, σ_d^2, σ_e^2 as well as finding confidence intervals for them. We are also interested in simultaneous confidence intervals for the set of contrasts in the α_i. Finally, we are interested in testing that $\alpha_i = 0$ (i.e., that there is no bull effect), testing that $\sigma_b^2 = 0$ (i.e., that there is no cow effect) and testing that $\sigma_d^2 = 0$ (i.e., that there is no interaction effect).

It should be apparent that we could generate much more complicated random effects and mixed models. In the above example we could add an effect due to diet and have a three-way mixed model in which the diet is a fixed effect. The cows are a random effect and the bulls are a fixed or random effect depending on whether we wanted to limit the inference to the bulls in the experiment or to extend to a larger herd from which the bulls are sampled. In a practical experiment, it is quite easy to tell whether a main effect is a fixed effect or a random effect. If we want to limit the inference to the levels of that main effect that are present in the experiment, then the effect is a fixed effect. If we want to extend the inference to a larger population from which those levels are selected then the effect is a random effect. If an interaction effect has any subscript that is associated with random main effects then the interaction must be a random effect (since it depends on which levels of the random main effect have been selected). If all the subscripts of an interaction term are associated with fixed main effects, then the interaction is a fixed effect.

Unlike the ordinary linear model, generalized linear model, repeated measures model and exchangeable linear model treated in earlier chapters, it does not appear to be possible to give a general treatment of random effects and mixed models. In addition, very little is known about optimal procedures (or even exact ones) for unbalanced models. In this chapter, we therefore look at three particular examples of balanced random effects and mixed models. In Sections 15.1 and 15.2 we look at the balanced one-way and balanced two-way random effects models. In Section 15.3 we look at both versions of the mixed model discussed in previous paragraphs. In the exercises a balanced nested random effects model is studied. In Section 15.4, we give heuristic derivations of the models considered in this chapter from more elementary assumptions. In Section 15.5 we look at the relation between the repeated measures model and certain mixed models. In Section 15.6 we give some other results for random effects and mixed models. Finally in Section 15.7, we derive a general result that is necessary to show completeness of sufficient statistics for the balanced two-way random effects model and other more complicated random effects and mixed models.

The random effects and fixed effects models were treated as one model for many years. In fact, if often appears that Fisher viewed analysis of variance from the perspective of random effects models. See, for example, Fisher (1925, Chapter 7). Eisenhart (1947) seems to be the first paper to make a clear distinction between these two models and even he implied that the analysis would be the same for the two models.

15.1. THE ONE-WAY RANDOM EFFECTS MODEL.

In the balanced one-way fixed effects model, discussed in Section 7.3.1, we observe

$$Y_{ij} = \theta + \alpha_i + e_{ij}, \qquad i = 1, \ldots, k; \qquad j = 1, \ldots, n; \qquad \sum_{i=1}^{k} \alpha_i = 0$$

where θ and α_i are unknown constants and the e_{ij} are unobserved random variables that are independently identically distributed with $e_{ij} \sim N_1(0, \sigma_e^2)$ and $\sigma_e^2 > 0$ is another unknown constant. In the random effects model, we replace the unknown constants α_i by unobserved random variables a_i. In the random effects model, we observe

$$Y_{ij} = \theta + a_i + e_{ij}, \qquad i = 1, \ldots, k; \qquad j = 1, \ldots, n, \tag{15.1}$$

where the a_i and e_{ij} are independent unobserved random variables such that

$$a_i \sim N_1(0, \sigma_a^2), \qquad e_{ij} \sim N_1(0.\sigma_e^2). \tag{15.2}$$

$$-\infty < \theta < \infty, \qquad \sigma_a^2 \geqslant 0, \qquad \sigma_e^2 > 0. \tag{15.3}$$

(While we assume, as usual, that $\sigma_e^2 > 0$, we assume $\sigma_a^2 \geqslant 0$. If $\sigma_a^2 = 0$, this would mean that the a_i's are all 0.)

We note that Y_{ij} and $Y_{ij'}$ are not independent for the random effects model as they are for the fixed effects model. In fact, if $j \neq j'$, then

$$\operatorname{cov}(Y_{ij}, Y_{ij'}) = \operatorname{cov}(\theta + a_i + e_{ij}, \theta + a_i + e_{ij'}) = \operatorname{var}(a_i) = \sigma_a^2$$

$$\operatorname{var}(Y_{ij}) = \operatorname{var}(\theta + a_i + e_{ij}) = \operatorname{var}(a_i) + \operatorname{var}(e_{ij}) = \sigma_a^2 + \sigma_e^2$$

(using the independence of a_i, e_{ij}, and $e_{ij'}$). Therefore, the correlation ρ between Y_{ij} and $Y_{ij'}$ is

$$\rho = \frac{\sigma_a^2}{\sigma_a^2 + \sigma_e^2}$$

ρ is often called the *intraclass correlation coefficient*. (Note that Y_{ij} and $Y_{i'j'}$ are independent as long as $i \neq i'$ even when $j = j'$.)

15.1.1. *Transforming the Problem*

For any integer s, let $\mathbf{1}_s = (1, \ldots, 1)'$ be the $s \times 1$ vector of 1's, let $\mathbf{C}_s$ be an orthonormal basis matrix for the space orthogonal to $\mathbf{1}_s$ and let $\boldsymbol{\Gamma}_s$ be the $s \times s$ orthogonal matrix given by

$$\boldsymbol{\Gamma}_s = \begin{pmatrix} s^{-1/2}\mathbf{1}_s \\ \mathbf{C}_s \end{pmatrix}.$$

Then

$$\boldsymbol{\Gamma}_s \mathbf{1}_s = \begin{pmatrix} s^{1/2} \\ \mathbf{0} \end{pmatrix}. \tag{15.4}$$

Now let

$$\mathbf{Y} = \begin{bmatrix} Y_{11} & \cdots & Y_{1n} \\ \vdots & & \vdots \\ Y_{k1} & \cdots & Y_{kn} \end{bmatrix}, \quad \mathbf{a} = \begin{bmatrix} a_1 \\ \vdots \\ a_k \end{bmatrix}, \quad \mathbf{e} = \begin{bmatrix} e_{11} & \cdots & e_{1n} \\ \vdots & & \vdots \\ e_{k1} & \cdots & e_{kn} \end{bmatrix}.$$

Then an equivalent version of (15.1) is

$$\mathbf{Y} = \theta \mathbf{1}_k \mathbf{1}'_n + \mathbf{a}\mathbf{1}'_n + \mathbf{e}.$$

Let

$$\mathbf{Z} = \begin{bmatrix} Z_{11} & \cdots & Z_{1n} \\ \vdots & & \vdots \\ Z_{k1} & \cdots & Z_{kn} \end{bmatrix} = \boldsymbol{\Gamma}_k \mathbf{Y} \boldsymbol{\Gamma}'_n, \qquad \mathbf{a}^* = \begin{bmatrix} \mathbf{a}_1^* \\ \vdots \\ \mathbf{a}_k^* \end{bmatrix} = \boldsymbol{\Gamma}_k \mathbf{a},$$

$$\mathbf{e}^* = \begin{bmatrix} e_{11}^* & \cdots & e_{1n}^* \\ \vdots & & \vdots \\ e_{k1}^* & \cdots & e_{kn}^* \end{bmatrix} = \boldsymbol{\Gamma}_k \mathbf{e} \boldsymbol{\Gamma}'_n.$$

Since $\boldsymbol{\Gamma}_k$ and $\boldsymbol{\Gamma}_n$ are invertible, observing $\mathbf{Y}$ is equivalent to observing $\mathbf{Z}$. By (15.4) we see that

$$\mathbf{Z} = \begin{pmatrix} \sqrt{nk}\,\theta & \mathbf{0} \\ \mathbf{0} & \mathbf{0} \end{pmatrix} + \left(\sqrt{n}\,\mathbf{a}^*\,\mathbf{0}\right) + \mathbf{e}^*,$$

that is, that

$$Z_{11} = (nk)^{1/2}\theta + n^{1/2}a_1^* + e_{11}^*, \qquad Z_{i1} = n^{1/2}a_i^* + e_{i1}^*, \qquad \text{for } i > 1,$$
$$Z_{ij} = e_{ij}^* \qquad \text{for } j > 1. \tag{15.5}$$

We now find the joint distribution of a^* and e^*.

LEMMA 15.1. The a_i^* and e_{ij}^* are all independent and $a_i^* \sim N_1(0, \sigma_a^2)$, $e_{ij}^* \sim N_1(0, \sigma_e^2)$.

PROOF. Since $\mathbf{a}$ and $\mathbf{e}$ are independent, so are $\mathbf{a}^*$ and $\mathbf{e}^*$. Now, $\mathbf{a} \sim N_k(0, \sigma_{\mathbf{a}}^2\mathbf{I})$, and hence $\mathbf{a}^* = \boldsymbol{\Gamma}_k\mathbf{a} \sim N_k(0, \sigma_{\mathrm{a}}^2\mathbf{I})$, and therefore the a_i^* are independent and $a_i^* \sim N_1(0, \sigma_a^2)$.

We now show that the distribution of e^* is the same as the distribution of e. We use moment generating functions together with two facts about the trace of a product: $\operatorname{tr}\mathbf{AB}' = \sum_i\sum_j a_{ij}b_{ij}$ (where a_{ij} and b_{ij} are the components of $\mathbf{A}$ and $\mathbf{B}$ and $\operatorname{tr}\mathbf{AB} = \operatorname{tr}\mathbf{BA}$. Let $\mathbf{T}$ be a $k \times n$ matrix with components t_{ij}. Then the joint moment generating function of e is

$$\begin{aligned} M_{\mathbf{e}}(\mathbf{t}) &= E\exp\sum_i\sum_j e_{ij}t_{ij} = E\exp\operatorname{tr}\mathbf{eT}' = \prod_i\prod_j \exp\frac{\sigma^2 t_{ij}^2}{2} \\ &= \exp\sigma^2\sum_i\sum_j\frac{t_{ij}^2}{2} = \exp\sigma^2(\operatorname{tr}\mathbf{TT}')/2. \end{aligned}$$

The joint moment generating function of $\mathbf{e}^*$ is

$$\begin{aligned} M_{\mathbf{e}^*}(\mathbf{T}) &= E\exp\operatorname{tr}\mathbf{e}^*\mathbf{T}' = E\exp\operatorname{tr}\boldsymbol{\Gamma}_k\mathbf{e}\boldsymbol{\Gamma}_n\mathbf{T}' = E\exp\operatorname{tr}\mathbf{e}(\boldsymbol{\Gamma}_k\mathbf{T}\boldsymbol{\Gamma}_n)' \\ &= M_{\mathbf{e}}(\boldsymbol{\Gamma}_k\mathbf{T}\boldsymbol{\Gamma}_n) = \exp\sigma^2\left[\operatorname{tr}(\boldsymbol{\Gamma}_k\mathbf{T}\boldsymbol{\Gamma}_n)(\boldsymbol{\Gamma}_k\mathbf{T}\boldsymbol{\Gamma}_n)'\right]/2 \\ &= \exp\sigma^2\left[\operatorname{tr}\mathbf{T}\boldsymbol{\Gamma}_n\boldsymbol{\Gamma}_n'\mathbf{T}'\boldsymbol{\Gamma}_k'\boldsymbol{\Gamma}_k\right]/2 \\ &= \exp\sigma^2\left[\operatorname{tr}\mathbf{TT}'\right]/2 = M_{\mathbf{e}}(\mathbf{T}). \end{aligned}$$

Therefore $\mathbf{e}$ and $\mathbf{e}^*$ have the same moment generating function and hence the same distribution. □

We are now ready for the basic result of this section.

THEOREM 15.2. The Z_{ij} are all independent, $Z_{11} \sim N_1(\sqrt{nk}\,\theta,\ n\sigma_a^2 + \sigma_e^2)$, $Z_{i1} \sim N_1(0, n\sigma_a^2 + \sigma_e^2)$ for $i > 1$, and $Z_{ij} \sim N_1(0, \sigma_e^2)$ for $j > 1$.

PROOF. This follows directly from (15.5) and Lemma 15.1. □

We have now transformed the one-way random effects model to a simpler model in which all the ransom variables Z_{ij} are independent. Note that the Y_{ij} are dependent unless $\sigma_a^2 = 0$.

15.1.2. A Sufficient Statistic.

In this section we find a complete sufficient statistic for the one-way random effects model, find its distribution and express it in terms of the original Y_{ij}. Let

$$U = (nk)^{-1/2} Z_{11}, \qquad S_1^2 = \sum_{i=2}^{k} Z_{i1}^2, \qquad S_2^2 = \sum_{i=1}^{k} \sum_{j=2}^{n} Z_{ij}^2$$

THEOREM 15.3. a. (U, S_1^2, S_2^2) is a complete sufficient statistic for the one-way random effects model.

b. U, S_1^2, S_2^2 are independent and

$$U \sim N_1\left(\theta, \frac{n\sigma_a^2 + \sigma_e^2}{nk}\right), \qquad S_1^2 \sim (n\sigma_a^2 + \sigma_e^2)\chi_{k-1}^2(0), \qquad S_2^2 \sim \sigma_e^2 \chi_{k(n-1)}^2(0).$$

PROOF. a. The joint density function of the Z_{ij} is

$$\frac{1}{(2\pi)^{nk/2}(n\sigma_a^2 + \sigma_e^2)^{k/2}(\sigma_e^2)^{k(n-1)/2}}$$

$$\times \exp -\frac{1}{2}\left[\frac{(Z_{11} - \sqrt{nk}\,\theta)^2}{n\sigma_a^2 + \sigma_e^2} + \frac{S_1^2}{n\sigma_a^2 + \sigma_e^2} + \frac{S_2^2}{\sigma_e^2}\right]$$

$$= \frac{\exp -\dfrac{1}{2}\dfrac{nk\theta^2}{n\sigma_a^2 + \sigma_e^2}}{(2\pi)^{nk/2}(n\sigma_a^2 + \sigma_e^2)^{k/2}(\sigma_e^2)^{k(n-1)/2}}$$

$$\times \exp\left[\frac{Z_{11}\theta\sqrt{nk}}{n\sigma_a^2 + \sigma_e^2} - \frac{Z_{11}^2 + S_1^2}{2(n\sigma_a^2 + \sigma_e^2)} - \frac{S_2^2}{2\sigma_e^2}\right].$$

Since the range of

$$\left[\frac{\theta\sqrt{nk}}{n\sigma_a^2 + \sigma_e^2}, -\frac{1}{2(n\sigma_a^2 + \sigma_e^2)}, \frac{-1}{2\sigma_e^2}\right]'$$

contains an open three-dimensional rectangle, $(Z_{11}, Z_{11}^2 + S_1^2, S_2^2)$ is a complete sufficient statistic, by the exponential criterion. However, $(U, S_1^2,$

S_2^2) is an invertible function of (Z_{11}, $Z_{11}^2 + S_1^2$, S_2^2) and is therefore also complete and sufficient.

b. See Exercise B2. □

We now express U, S_1^2, and S_2^2 in terms of the Y_{ij}. We need the following lemma.

LEMMA 15.4.

$$\sum_{i=1}^{k}\sum_{j=1}^{n} Y_{ij}^2 = \sum_{i=1}^{k}\sum_{j=1}^{n} Z_{ij}^2.$$

PROOF. This is most easily proved using the trace of a matrix again. Since $\operatorname{tr}\mathbf{Y}'\mathbf{Y} = \sum_i\sum_j Y_{ij}^2$, $\operatorname{tr}\mathbf{Z}'\mathbf{Z} = \sum_i\sum_j Z_{ij}^2$, we see that

$$\sum_{i=1}^{k}\sum_{j=1}^{n} Y_{ij}^2 = \operatorname{tr}\mathbf{Y}\mathbf{Y}' = \operatorname{tr}(\mathbf{Y}\boldsymbol{\Gamma}_n')(\boldsymbol{\Gamma}_n\mathbf{Y}') = \operatorname{tr}\boldsymbol{\Gamma}_n\mathbf{Y}'\mathbf{Y}\boldsymbol{\Gamma}_n' = \operatorname{tr}\boldsymbol{\Gamma}_n\mathbf{Y}'\boldsymbol{\Gamma}_k'\boldsymbol{\Gamma}_k\mathbf{Y}\boldsymbol{\Gamma}_n'$$

$$= \operatorname{tr}\mathbf{Z}'\mathbf{Z} = \sum_{i=1}^{k}\sum_{j=1}^{n} Z_{ij}^2. \quad \square$$

We are now ready to express U, S_1^2, and S_2^2 in terms of the Y_{ij}.

LEMMA 15.5.

$$U = \bar{Y}_{..}, \qquad S_1^2 = n\sum_{i=1}^{k}\left(\bar{Y}_{i.} - \bar{Y}_{..}\right)^2, \qquad S_2^2 = \sum_{i=1}^{k}\sum_{j=1}^{n}\left(Y_{ij} - \bar{Y}_{i.}\right)^2.$$

PROOF. By the definition of Γ_s, we see that

$$Z_{11} = (nk)^{-1/2}\mathbf{1}_k\mathbf{Y}\mathbf{1}_n = (nk)^{1/2}\bar{Y}_{...}$$

Therefore, $U = (nk)^{-1/2}Z_{11} = \bar{Y}_{...}$ Now, let

$$\mathbf{Z}_1' = (Z_{11}, \ldots, Z_{k1}), \qquad \bar{\mathbf{Y}}' = \left(\bar{Y}_{1.}, \ldots, \bar{Y}_{k.}\right).$$

Then

$$\mathbf{Z}_1 = n^{-1/2}\boldsymbol{\Gamma}_k\mathbf{Y}\mathbf{1}_n = n^{1/2}\boldsymbol{\Gamma}_k\bar{\mathbf{Y}}.$$

Therefore,

$$S_1^2 + Z_{11}^2 = \sum_{i=1}^{k} Z_{i1}^2 = \|\mathbf{Z}_1\|^2 = n\|\bar{\mathbf{Y}}\|^2 = n\sum_{i=1}^{k}\bar{Y}_{i.}^2.$$

Hence

$$S_1^2 = n\sum_{i=1}^{k}\bar{Y}_{i.}^2 - Z_{11}^2 = n\sum_{i=1}^{k}\bar{Y}_{i.}^2 - nk\bar{Y}_{..}^2 = n\sum_{i=1}^{k}\left(\bar{Y}_{i.} - \bar{Y}_{..}\right)^2.$$

Finally,

$$Z_{11}^2 + S_1^2 + S_2^2 = \sum_{i=1}^{k} \sum_{j=1}^{n} Z_{ij}^2 = \sum_{i=1}^{k} \sum_{j=1}^{n} Y_{ij}^2$$

by Lemma 15.4. Therefore,

$$S_2^2 = \sum_{i=1}^{k} \sum_{j=1}^{n} Y_{ij}^2 - (S_1^2 + Z_{11}^2) = \sum_{i=1}^{k} \sum_{j=1}^{n} Y_{ij}^2 - n \sum_{i=1}^{k} \overline{Y}_{i\cdot}^2$$

$$= \sum_{i=1}^{k} \sum_{j=1}^{n} (Y_{ij} - \overline{Y}_{i\cdot})^2. \quad \square$$

15.1.3. *Estimation*

In this section we consider estimation of the parameters θ, σ_a^2, and σ_e^2. The estimation of these parameters is more difficult than with other models. The reason for this difficulty is that the estimators should take values in the parameter space. In particular any sensible estimator of σ_a^2 must be nonnegative.

We first consider unbiased estimators.

THEOREM 15.6. U and $S_2^2/[k(n-1)]$ are the minimum variance unbiased estimators of θ and σ_e^2. There is no nonnegative unbiased estimator of σ_a^2.

PROOF. We note that

$$E \frac{S_2^2}{k(n-1)} = \sigma_e^2, \qquad EU = \theta.$$

Since (U, S_1^2, S_2^2) is a complete sufficient statistic, $S_2^2/[k(n-1)]$ and U are the minimum variance unbiased estimators of σ_e^2 and θ. Now, let

$$T = \frac{S_1^2}{n(k-1)} - \frac{S_2^2}{nk(n-1)}.$$

Then $ET = \sigma_a^2$. However, T is negative with positive probability. Now suppose that W were a nonnegative unbiased estimator of σ_a^2 (i.e., $EW = \sigma_a^2$ and $W \geqslant 0$). Let $V = EW|(U, S_1^2, S_2^2)$. Then $EV = ET$. Since (U, S_1^2, S_2^2) is a complete sufficient statistic, we see that $V = T$ with probability 1. However, since $W \geqslant 0$, we see that $V \geqslant 0$ and hence $P(T \geqslant 0) = 1$. This is a contradiction, since T is negative with positive probability. $\square$

We now consider maximum likelihood estimators.

THEOREM 15.7. The maximum likelihood estimators of θ, σ_a^2, and σ_e^2 are

$$\hat{\theta} = U, \qquad \hat{\sigma}_a^2 = \max\left(\frac{S_1^2}{nk} - \frac{S_2^2}{nk(n-1)}, 0 \right),$$

$$\hat{\sigma}_e^2 = \min\left[\frac{S_2^2}{k(n-1)}, \frac{S_1^2 + S_2^2}{kn} \right]$$

PROOF. Let $\tau^2 = n\sigma_a^2 + \sigma_e^2$. We first find the maximum likelihood estimators of θ, τ^2, and σ_e^2. The likelihood function (in terms of the Z_{ij}) is

$$L(\theta, \tau^2, \sigma_e^2) = \frac{1}{(2\pi)^{nk/2} \tau^k \sigma_e^{k(n-1)}}$$

$$\times \exp\left[-\frac{1}{2} \frac{\left(Z_{11} - \sqrt{nk}\,\theta\right)^2}{\tau^2} - \frac{1}{2} \frac{S_1^2}{\tau^2} - \frac{1}{2} \frac{S_2^2}{\sigma_e^2} \right].$$

The parameter space is given by

$$-\infty < \theta < \infty, \qquad \tau^2 \geqslant \sigma_e^2 > 0.$$

Direct differentiation shows that the likelihood is maximized by

$$\theta = U, \qquad \tau^2 = \frac{S_1^2}{k}, \qquad \sigma_e^2 = \frac{S_2^2}{k(n-1)}. \tag{15.6}$$

Therefore, if $S_1^2 \geqslant S_2^2/(n-1)$, these are the maximum likelihood estimators. However, if $S_1^2 < S_2^2/(n-1)$, then this point is outside the parameter space. In this case, since the point given by (15.6) is the only critical point for L, the maximum must occur on the boundary (i.e., it must occur when $\sigma_e^2 = \tau^2$). When $\sigma_e^2 = \tau^2$, the likelihood is given by

$$L(\theta, \tau^2, \tau^2) = \frac{1}{(2\pi)^{nk/2} \tau^{nk}} \exp\left[-\frac{1}{2} \frac{\left(Z_{11} - \sqrt{nk}\,\theta\right)^2}{\tau^2} - \frac{1}{2} \frac{S_1^2 + S_2^2}{\tau^2} \right].$$

On this set, the likelihood is maximized by

$$\theta = U, \qquad \tau^2 = \sigma_e^2 = \frac{S_1^2 + S_2^2}{nk}. \tag{15.7}$$

Therefore, if $S_1^2 < S_2^2/(n-1)$, the maximum likelihood estimators are given by (15.7). Note that the maximum likelihood estimator of θ is U in either case. Now $\sigma_a^2 = (\tau^2 - \sigma_e^2)/n$. Therefore, the maximum likelihood estimator of σ_a^2 is given by $(S_1^2 - S_2^2/(n-1))/nk$ if $S_1^2 - S_2^2/(n-1) \geqslant 0$ and is given by 0 if $S_1^2 - S_2^2/(n-1) < 0$, that is, the maximum likelihood estimator of σ_a^2 is given by the maximum of 0 and $(S_1^2 - S_2^2/(n-1))/nk$.

Since $S_1^2 < S_2^2/(n-1)$ if and only if $(S_1^2 + S_2^2)/nk < S_2^2/[k(n-1)]$, we see that the maximum likelihood estimator of σ_e^2 is the minimum of $(S_1^2 + S_2^2)/kn$ and $S_2^2/[k(n-1)]$. □

We note that neither the maximum likelihood estimator for σ_a^2 nor for σ_e^2 is unbiased. The maximum likelihood estimator for σ_e^2 seems preferable to the unbiased estimator for the following reason. If $S_1^2 < S_2^2/(n-1)$, this is fairly strong evidence that $\sigma_a^2 = 0$. If $\sigma_a^2 = 0$, then we should use the pooled estimator, which is essentially the maximum likelihood estimator. The maximum likelihood estimator makes some use of the fact that $\sigma_a^2 \geqslant 0$, and hence $\tau^2 \geqslant \sigma_e^2$, a fact that is ignored by the unbiased estimator.

15.1.4. *Testing that $\sigma_a^2 = 0$*

We now consider the problem of testing that $\sigma_a^2 = 0$. This problem is invariant under the following two groups:

(1) G_1 consists of the transformations $g_1(Y_{ij}) = Y_{ij} + s$, $s \in R^1$.
(2) G_2 consists of the transformations $g_2(Y_{ij}) = tY_{ij}$, $t > 0$. [It is not invariant under $g_3(Y_{ij}) = Y_{ij} + b_i$. Why not?] Let

$$F = \frac{k(n-1)}{k-1}\,\frac{S_1^2}{S_2^2}, \qquad \delta = \frac{n\sigma_a^2 + \sigma_e^2}{\sigma_e^2}$$

$$\phi(F) = \begin{cases} 1 & \text{if} > F^{\alpha}_{k-1,k(n-1)} \\ 0 & \text{if } F \leqslant F^{\alpha}_{k-1,k(n-1)} \end{cases} \tag{15.8}$$

THEOREM 15.8. A maximal invariant under G_1 and G_2 for testing that $\sigma_a^2 = 0$ against $\sigma_a^2 \geqslant 0$ is F and a parameter maximal invariant is δ. $F/\delta \sim F_{k-1,k(n-1)}(0)$. $\phi(F)$ is the UMP invariant size α test.

PROOF. See Exercise B3. □

This test is the same as the UMP invariant size α test for testing that the $\alpha_i = 0$ in the one-way fixed effects model (see Section 7.3.1). However, the groups used to reduce the problems are quite different. Also, the alternative distribution of F is a noncentral F distribution for the fixed effects model and a constant times a central F distribution for the random effects model.

15.1.5. Confidence Intervals

We can easily find confidence intervals for θ, σ_e^2, (σ_a^2/σ_e^2) or $n\sigma_a^2 + \sigma_e^2$, by using the following facts:

$$\sqrt{nk}\,\frac{(\bar{Y}-\theta)}{\sqrt{S_1^2/(k-1)}} \sim t_{k-1}(0), \qquad S_2^2 \sim \sigma_e^2 \chi^2_{k(n-1)}(0),$$

$$\frac{F}{(n\sigma_a^2/\sigma_e^2)+1} \sim F_{k-1,k(n-1)}(0), \qquad S_1^2 \sim (n\sigma_a^2+\sigma_e^2)\chi^2_{k-1}(0),$$

where F is defined in (15.8). However, there does not seem to be any way to find exact confidence intervals for σ_a^2. Satterthwaite (1946) suggested an approximate confidence interval for σ_a^2. See Boardman (1974) for a description of this and other procedures as well as a Monte Carlo comparison of them.

15.2. THE BALANCED TWO-WAY RANDOM EFFECTS MODEL.

In this section we consider the balanced two-way random effects model in which we observe

$$Y_{ijk} = \theta + a_i + b_j + d_{ij} + e_{ijk},$$
$$i = 1, \ldots, r; \qquad j = 1, \ldots, c; \qquad k = 1, \ldots, n;$$

where the a_i, b_j, d_{ij}, and e_{ijk} are unobserved independent random variables such that

$$a_i \sim N_1(0, \sigma_a^2), \qquad b_j \sim N_1(0, \sigma_b^2), \qquad d_{ij} \sim N_1(0, \sigma_d^2), \qquad e_{ijk} \sim N_1(0, \sigma_e^2).$$

The parameter space is given by

$$-\infty < \theta < \infty, \qquad \sigma_a^2 \geqslant 0, \qquad \sigma_b^2 \geqslant 0, \qquad \sigma_d^2 \geqslant 0, \qquad \sigma_e^2 > 0.$$

Now, let

$$U = \bar{Y}_{\cdots}, \qquad S_1^2 = nc\sum_{i=1}^{r}\left(\bar{Y}_{i\cdot\cdot} - \bar{Y}_{\cdots}\right)^2, \qquad S_2^2 = nr\sum_{j=1}^{c}\left(\bar{Y}_{\cdot j\cdot} - \bar{Y}_{\cdots}\right)^2$$

$$S_3^2 = n\sum_{i=1}^{r}\sum_{j=1}^{c}\left(\bar{Y}_{ij\cdot} - \bar{Y}_{i\cdot\cdot} - \bar{Y}_{\cdot j\cdot} + \bar{Y}_{\cdots}\right)^2, \tag{15.9}$$

$$S_4^2 = \sum_{i=1}^{r}\sum_{j=1}^{c}\sum_{k=1}^{n}\left(Y_{ijk} - \bar{Y}_{ij\cdot}\right)^2.$$

The following theorem is the main result of this section. Its proof is just a more complicated version of the results of the last section, and therefore, the proof is merely outlined. The details are left for homework (see Exercise B5).

THEOREM 15.9. $(U, S_1^2, S_2^2, S_3^2, S_4^2)$ is a sufficient statistic for the balanced two-way random effects model. U, S_1^2, S_2^2, S_3^2, and S_4^2 are independent and

$$U \sim N_1\left(\theta, \frac{1}{nrc}\left(nc\sigma_a^2 + nr\sigma_b^2 + n\sigma_d^2 + \sigma_e^2\right)\right), \qquad S_1^2 \sim \left(nc\sigma_a^2 + n\sigma_d^2 + \sigma_e^2\right)\chi_{r-1}^2(0),$$

$$S_2^2 \sim \left(nr\sigma_b^2 + n\sigma_d^2 + \sigma_e^2\right)\chi_{c-1}^2(0), \qquad S_3^2 \sim \left(n\sigma_d^2 + \sigma_e^2\right)\chi_{(r-1)(c-1)}^2(0),$$

$$S_4^2 \sim \sigma_e^2 \chi_{rc(n-1)}^2(0).$$

PROOF. We first define new random variables Z_{ijk} that are independent. Note that the Y_{ijk} are not. Let $\mathbf{\Gamma}_s$ be defined as in the last section and let

$$\mathbf{Y}_{ij} = \begin{pmatrix} Y_{ij1} \\ \vdots \\ Y_{ijn} \end{pmatrix}, \qquad \mathbf{Y}_{ij}^* = \begin{pmatrix} Y_{ij1}^* \\ \vdots \\ Y_{ijn}^* \end{pmatrix} = \mathbf{\Gamma}_n \mathbf{Y}_{ij}.$$

Now, let

$$\mathbf{Y}_1^* = \begin{pmatrix} Y_{111}^* & \cdots & Y_{1c1}^* \\ \vdots & & \vdots \\ Y_{r11}^* & \cdots & Y_{rc1}^* \end{pmatrix}, \qquad \mathbf{Z}_1 = \begin{pmatrix} Z_{111} & \cdots & Z_{1c1} \\ \vdots & & \vdots \\ Z_{r11} & \cdots & Z_{rc1} \end{pmatrix} = \mathbf{\Gamma}_r \mathbf{Y}_1^* \mathbf{\Gamma}_c'.$$

(Note that $(rc)^{-1/2} Y_1^*$ is just the matrix whose components are $\overline{Y}_{ij\cdot}$.) Finally, let

$$Z_{ijk} = Y_{ijk}^* \qquad \text{for } k > 1.$$

By proofs similar to those in the last section, we find that the Z_{ijk} are independent,

$$\begin{aligned}
&Z_{111} \sim N_1\left(\sqrt{ncr}\,\theta, nc\sigma_a^2 + nr\sigma_b^2 + n\sigma_d^2 + \sigma_e^2\right), && && (1 \text{ of these}),\\
&Z_{i11} \sim N_1\left(0, nc\sigma_a^2 + n\sigma_d^2 + \sigma_e^2\right), && i > 1 && (r-1 \text{ of these}),\\
&Z_{1j1} \sim N_1\left(0, nr\sigma_b^2 + n\sigma_d^2 + \sigma_e^2\right), && j > 1 && (c-1 \text{ of these}),\\
&Z_{ij1} \sim N_1\left(0, n\sigma_d^2 + \sigma_e^2\right), && i > 1, j > 1 && \left[(r-1)(c-1) \text{ of these}\right],\\
&Z_{ijk} \sim N_1\left(0, \sigma_e^2\right), && k > 1 && \left[rc(n-1) \text{ of these}\right].
\end{aligned} \tag{15.10}$$

In addition, by proofs similar to the last section again, we see that

$$U = (ncr)^{-1/2} Z_{111}, \qquad S_1^2 = \sum_{i=2}^{r} Z_{i11}^2, \qquad S_2^2 = \sum_{j=2}^{c} Z_{1j1}^2,$$
$$S_3^2 = \sum_{i=2}^{r} \sum_{j=2}^{c} Z_{ij1}^2, \qquad S_4^2 = \sum_{i=1}^{r} \sum_{j=1}^{c} \sum_{k=2}^{n} Z_{ijk}^2. \tag{15.11}$$

The joint density of U, S_1^2, S_2^2, S_3^2, and S_4^2 follows directly from (15.10) and (15.11). In addition the Z_{ijk} are an invertible function of the Y_{ijk}, so that observing the Z_{ijk} is equivalent to observing the Y_{ijk}. The density of the Z_{ijk} has the following form

$$d \exp \frac{-1}{2}\left(\frac{ncr(\theta + U)^2}{nc\sigma_a^2 + nr\sigma_b^2 + n\sigma_d^2 + \sigma_e^2} + \frac{S_1^2}{nc\sigma_a^2 + n\sigma_d^2 + \sigma_e^2} \right.$$
$$\left. + \frac{S_2^2}{nr\sigma_b^2 + n\sigma_d^2 + \sigma_e^2} + \frac{S_3^2}{n\sigma_d^2 + \sigma_e^2} + \frac{S_4^2}{\sigma_e^2} \right)$$

and, therefore, from the factorization criterion, we see that $(U, S_1^2, S_2^2, S_3^2, S_4^2)$ is a sufficient statistic. □

We now discuss the completeness of the statistic $(U, S_1^2, S_2^2, S_3^2, S_4^2)$. We first show why the obvious approach does not work. Let

$$\mathbf{T}(\mathbf{Z}) = (U^2, U, S_1^2, S_2^2, S_3^2, S_4^2)$$

and

$$\mathbf{Q}(\theta, \sigma_a^2, \sigma_b^2, \sigma_d^2, \sigma_e^2)' = (Q_1, Q_2, Q_3, Q_4, Q_5, Q_6)$$
$$= -\frac{1}{2}\left(\frac{ncr}{nc\sigma_a^2 + nr\sigma_b^2 + n\sigma_d^2 + \sigma_e^2}, \frac{-2ncr\theta}{nc\sigma_a^2 + nr\sigma_b^2 + n\sigma_d^2 + \sigma_e^2}, \right.$$
$$\left. \frac{1}{nc\sigma_a^2 + n\sigma_d^2 + \sigma_e^2}, \frac{1}{nr\sigma_b^2 + n\sigma_d^2 + \sigma_e^2}, \frac{1}{n\sigma_d^2 + \sigma_e^2}, \frac{1}{\sigma_e^2} \right).$$

Then it is easily verified that the joint distribution of the Z_{ijk} has the form

$$k(\mathbf{Z}) h(\theta, \sigma_a^2, \sigma_b^2, \sigma_d^2, \sigma_e^2) \exp\left[\mathbf{T}'(\mathbf{Z}) \mathbf{Q}(\theta, \sigma_a^2, \sigma_b^2, \sigma_c^2, \sigma_d^2)\right]$$

and hence the density forms an exponential family. However,

$$\frac{1}{Q_1} = \frac{1}{Q_3} + \frac{1}{Q_4} - \frac{1}{Q_5}$$

and hence the image of **Q** does not contain a six-dimensional rectangle. (Another perspective is **Q** is six-dimensional while there are only five parameters.) For this reason we cannot apply Theorem 1.2 directly to establish the completeness.

The statistic is complete. However, the proof of the completeness uses techniques more advanced than those used in other proofs in this book. The theorem is therefore stated here and proved in Section 15.7.

THEOREM 15.10. $(U, S_1^2, S_2^2, S_3^2, S_4^2)$ is a complete sufficient statistic for the balanced two-way random effects model.

PROOF. See Section 15.7. □

Using these theorems we can find minimum variance unbiased estimators of θ and σ_e^2. However, by an argument similar to one in the last section, there are no nonnegative unbiased estimators of σ_a^2, σ_b^2, or σ_d^2. (See Exercise B7.) Furthermore, MLE's for this model are quite difficult to obtain. However, sensible estimators for this model would seem to be the following

$$\theta = U, \qquad \hat{\sigma}_a^2 = \max\left[\frac{S_1^2}{nc(r-1)} - \frac{S_3^2}{nc(r-1)(c-1)}, 0\right],$$

$$\hat{\sigma}_b^2 = \max\left(\frac{S_2^2}{nr(c-1)} - \frac{S_3^2}{nr(c-1)(r-1)}, 0\right)$$

$$\hat{\sigma}_d^2 = \max\left[\frac{S_3^2}{n(c-1)(r-1)} - \frac{S_4^2}{rc(n-1)}, 0\right] \qquad \hat{\sigma}_e^2 = \frac{S_4^2}{rc(n-1)}.$$

We now consider the problem of testing that $\sigma_a^2 = 0$ against $\sigma_a^2 \geqslant 0$. This problem is invariant under the group of transformations

$$g(Y_{ijk}) = sY_{ijk} + t, \qquad s > 0.$$

A maximal invariant under this group is

$$F_1 = \frac{(c-1)S_1^2}{S_3^2}, \qquad F_2 = \frac{(r-1)S_2^2}{S_3^2}, \qquad F_3 = \frac{rc(n-1)S_3^2}{(r-1)(c-1)S_4^2} \tag{15.12}$$

and a parameter maximal invariant is

$$\delta_1 = \frac{nc\sigma_a^2 + n\sigma_d^2 + \sigma_e^2}{n\sigma_d^2 + \sigma_e^2}, \qquad \delta_2 = \frac{nr\sigma_b^2 + n\sigma_d^2 + \sigma_e^2}{n\sigma_d^2 + \sigma_e^2}, \qquad \delta_3 = \frac{n\sigma_d^2 + \sigma_e^2}{\sigma_e^2}. \tag{15.13}$$

The joint distribution of (F_1, F_2, F_3) is derived in Exercise C3, but is not too helpful. There is no UMP invariant size α test for this problem. The marginal distributions of F_1, F_2, and F_3 are given by

$$\frac{F_1}{\delta_1} \sim F_{r-1,(r-1)(c-1)}(0), \qquad \frac{F_2}{\delta_2} \sim F_{c-1,(r-1)(c-1)}(0),$$
$$\frac{F_3}{\delta_3} \sim F_{(r-1)(c-1),\,rc(n-1)}(0). \tag{15.14}$$

We note that δ_2 and δ_3 do not involve σ_a^2, whereas $\delta_1 \geqslant 1$ and $\sigma_a^2 = 0$ if and only if $\delta_1 = 1$. Therefore, a sensible test for this hypothesis is to reject if $F_1 > F^{\alpha}_{r-1,(r-1)(c-1)}$.

The problems of testing that $\sigma_b^2 = 0$ and $\sigma_d^2 = 0$ are also both invariant under the group given above. Since $\sigma_b^2 = 0$ if and only if $\delta_2 = 1$, a logical test of this hypothesis is to reject if $F_2 > F^{\alpha}_{c-1,(r-1)(c-1)}$. Similarly $\sigma_d^2 = 0$ if and only if $\delta_3 = 1$, so that a sensible test for testing that $\sigma_d^2 = 0$ is to reject if $F_3 > F^{\alpha}_{(r-1)(c-1),rc(n-1)}$. We could also test hypotheses like $\sigma_a^2 = 0$ and $\sigma_d^2 = 0$ (i.e., that there is no difference in rows). We note that $\sigma_a^2 = 0$ and $\sigma_d^2 = 0$ if and only if $\delta_1/\delta_3 = 1$, that $\delta_1/\delta_3 \geqslant 1$ and that $\delta_3 F_1/\delta_1 F_3 \sim F_{r-1,rc(n-1)}(0)$. Therefore a sensible test for this hypothesis is to reject if $F_1/F_3 > F^{\alpha}_{r-1,rc(n-1)}$.

The problem of testing that $\sigma_a^2 = 0$ in the random effects model is analagous to the problem of testing for no row effects in the fixed effects model. However, we note that the tests are different in the two models. In the random effects model we use the mean square for interactions as the denominator of the F-statistic, whereas in the fixed effects model we use the mean square for error. Similarly, testing that $\sigma_b^2 = 0$ in the random effects model is analagous to testing that there are no column effects in the fixed effects model and the hypotheses are tested differently in the two models. The hypothesis that $\sigma_d^2 = 0$ in the random effects model is analagous to testing that there is no interaction in the fixed effects model and is tested in the same way as we test for no interactions in the fixed effects model. Finally, testing that $\sigma_a^2 = 0$ and $\sigma_d^2 = 0$ is analagous to testing for no row effects and no interaction effects in the fixed effects model and is tested the same way.

It is easy to find confidence intervals for a fair number of functions of the parameters in this model [e.g., θ, σ_a^2, $nc\sigma_a^2 + n\sigma_d^2 + \sigma_e^2, \sigma_d^2/\sigma_e^2, \sigma_a^2/(n\sigma_d^2 + \sigma_e^2)$]. However, there are no exact confidence intervals for σ_a^2, σ_b^2, or σ_d^2, which are often the parameters of interest. Approximate confidence intervals of the type mentioned in the last section have been suggested.

Now, consider the case in which there is only one observation in each cell (i.e., $n = 1$). In that case S_4^2 is identically 0, and (U, S_1^2, S_2^2, S_3^2) is a sufficient statistic. We clearly cannot test that $\sigma_d^2 = 0$. However, we can test that $\sigma_a^2 = 0$ in this model, and the test is the one defined earlier in this section for that hypothesis. Note that this is the same test that was derived in Section 7.3.2 for the fixed effects model under the assumption of no interaction. For the random effects model, it is not necessary to assume that there are no interactions (that is, we do not need to assume that $\sigma_d^2 = 0$).

15.3. BALANCED TWO-WAY MIXED MODELS.

There are two versions of the two-way mixed model that are discussed in the literature. In this section we present one version and derive procedures for that model. We then show how the other version can be derived from the first version, and we reinterpret some of the results for this second version.

This first version of the mixed model is one in which we observe

$$Y_{ijk} = \theta + \alpha_i + b_j + d_{ij} + e_{ijk}, \qquad i = 1, \ldots, r; j = 1, \ldots, c; k = 1, \ldots, n; \tag{15.15}$$

where θ and α_i are unknown parameters such that $\sum_i \alpha_i = 0$, the b_j, d_{ij}, and e_{ijk} are independent random variables such that

$$b_j \sim N_1(0, \sigma_b^2), \qquad d_{ij} \sim N_1(0, \sigma_d^2), \qquad e_{ijk} \sim N_1(0, \sigma_e^2). \tag{15.16}$$

The parameter space for this model is given by

$$-\infty < \theta < \infty. \qquad \sum_i \alpha_i = 0, \qquad \sigma_b^2 \geqslant 0, \qquad \sigma_d^2 \geqslant 0, \qquad \sigma_e^2 > 0. \tag{15.17}$$

We first find a complete sufficient statistic for this model. Let U, S_2^2, S_3^2, S_4^2 be defined as in (15.9) for the two-way random effects model. Let $\mathbf{C}_r$ be defined as in Section 15.1.1. Let

$$\mathbf{V} = \begin{bmatrix} V_1 \\ \vdots \\ V_{r-1} \end{bmatrix} = \mathbf{C}_r' \begin{bmatrix} \bar{Y}_{1\cdot\cdot} \\ \vdots \\ \bar{Y}_{r\cdot\cdot} \end{bmatrix}, \qquad \boldsymbol{\gamma} = \begin{bmatrix} \gamma_1 \\ \vdots \\ \gamma_{r-1} \end{bmatrix} = \mathbf{C}_r' \begin{bmatrix} \alpha_i \\ \vdots \\ \alpha_r \end{bmatrix} \tag{15.18}$$

THEOREM 15.11. A complete sufficient statistic for the first version of the balanced two-way mixed model is $(U, \mathbf{V}, S_2^2, S_3^2, S_4^2)$. U, $\mathbf{V}$, S_2^2, S_3^2, S_4^2 are independent,

$$U \sim N_1\left(\theta, \frac{1}{ncr} nr\sigma_b^2 + n\sigma_d^2 + \sigma_e^2\right), \qquad \mathbf{V} \sim N_{r-1}\left(\boldsymbol{\gamma}, \frac{1}{nc}(n\sigma_d^2 + \sigma_e^2)\mathbf{I}\right)$$

$$S_2^2 \sim (nr\sigma_b^2 + n\sigma_d^2 + \sigma_e^2)\chi_{c-1}^2(0), \qquad S_3^2 \sim (n\sigma_d^2 + \sigma_e^2)\chi_{(r-1)(c-1)}^2(0)$$

$$S_4^2 \sim \sigma_e^2 \chi_{rc(n-1)}^2(0).$$

PROOF. See Exercise B11. □

Note that $(\overline{Y}_{1..}, \dots, \overline{Y}_{r..}, S_2^2, S_3^2, S_4^2)$ is an invertible function of $(U, V, S_2^2, S_3^2, S_4^2)$ and is also a complete sufficient statistic. Therefore, the minimum variance unbiased estimators of θ and α_i are $\overline{Y}_{...}$ and $\overline{Y}_{i..} - \overline{Y}_{...}$. The minimum variance unbiased estimator of σ_e^2 is $S_4^2/rc(n-1)$. There are no nonnegative unbiased estimators of σ_b^2 or σ_d^2. Sensible estimators would seem to be

$$\hat{\sigma}_b^2 = \max\left(\frac{1}{nr}\left(\frac{S_2^2}{c-1} - \frac{S_3^2}{(r-1)(c-1)}\right), 0\right),$$

$$\hat{\sigma}_d^2 = \max\left(\frac{S_3^2}{(r-1)(c-1)} - \frac{S_4^2}{rc(n-1)}, 0\right).$$

We now consider testing that $\sigma_b^2 = 0$. This problem is invariant under the groups G_1 and G_2 of transformations

$$g_1(U, \mathbf{V}, S_2^2, S_3^2, S_4^2) = (U + s, \mathbf{V} + \mathbf{t}, S_2^2, S_3^2, S_4^2), \qquad s \in R^1, \qquad \mathbf{t} \in R^{r-1};$$

$$g_2(U, \mathbf{V}, S_2^2, S_3^2, S_4^2) = (hU, H\mathbf{V}, h^2S_2^2, h^2S_3^2, h^2S_4^2), \qquad h > 0.$$

A maximal invariant under G_1 is (S_2^2, S_3^2, S_4^2). Let F_2, F_3, δ_2, and δ_3 be defined as in (15.12) and (15.13) in the last section. Then (F_2, F_3) is a maximal invariant under G_1 and G_2 for this problem and (δ_2, δ_3) is a parameter maximal invariant. As before the joint distribution of F_2 and F_3 is not useful and there is no UMP invariant size α test. However, the marginal distributions of F_2 and F_3 are the same for the mixed model as for the random effects model, that is,

$$F_2/\delta_2 \sim F_{c-1,(r-1)(c-1)}(0), \qquad F_3/\delta_3 \sim F_{(r-1)(c-1),rc(n-1)}(0).$$

Also $\sigma_b^2 = 0$ if and only if $\delta_2 = 1$. Therefore, a sensible test fot this hypothesis is to reject if $F_2 > F^{\alpha}_{c-1,(r-1)(c-1)}$. Similarly, a sensible test that $\sigma_d^2 = 0$ is to reject if $F_3 > F^{\alpha}_{(r-1)(c-1),rc(n-1)}$. These two tests are the same as for the two-way random effects model.

Now consider testing that the fixed effects $\alpha_i = 0$. This is equivalent to testing that $\gamma = \mathbf{0}$. This problem is invariant under the following three groups G_1, G_2, and G_3,

$$g_1(U, \mathbf{V}, S_2^2, S_3^2, S_4^2) = (U + s, \mathbf{V}, S_2^2, S_3^2, S_4^2), \qquad s \in R^1;$$

$$g_2(U, \mathbf{V}, S_2^2, S_3^2, S_4^2) = (U, \boldsymbol{\Delta}\mathbf{V}, S_2^2, S_3^2, S_4^2), \qquad \boldsymbol{\Delta} \text{ orthogonal};$$

$$g_2(U, \mathbf{V}, S_2^2, S_3^2, S_4^2) = (tU, t\mathbf{V}, t^2S_2^2, t^2S_3^2, t^2S_4^2), \qquad t > 0.$$

A maximal invariant under the first group is $(\mathbf{V}, S_2^2, S_3^2, S_4^2)$. Let

$$S_1^2 = nc\|\mathbf{V}\|^2 = nc\sum_{i=1}^{r}(\overline{Y}_{i...} - \overline{Y}_{...})^2 \tag{15.19}$$

(See Exercise B12. Note S_1^2 is the same as for the two-way random effects model.) Then a maximal invariant under the second group is $(S_1^2, S_2^2, S_3^2, S_4^2)$. Finally, let F_1, F_2, and F_3 be as defined in (15.12) for the random effects model. Then (F_1, F_2, F_3) is a maximal invariant for testing that $\alpha_i = 0$ in the mixed model. Let

$$\delta_1^* = \frac{cn\sum \alpha_i^2}{n\sigma_d^2 + \sigma_e^2} .$$

(Note that $\delta_1^* \neq \delta_1$ defined for the random effects model.) As before, the joint distribution of (F_1, F_2, F_3) is not useful and there is no UMP invariant size α test for this hypothesis. However, the marginal distributions of F_1, F_2, F_3 are given by

$$F_1 \sim F_{r-1,(r-1)(c-1)}(\delta_1^*), \qquad F_2/\delta_2 \sim F_{c-1,(r-1)(c-1)}(0),$$
$$F_3/\delta_3 \sim F_{(r-1)(c-1),\, rc(n-1)}(0).$$

(Note that F_1 has a noncentral F-distribution for this model and a constant times a central F-distribution for the random effects model.) Since $\alpha_i = 0$ if and only if $\delta_1^* = 0$, a sensible test for this problem would be to reject if $F_1 > F^{\alpha}_{r-1,(r-1)(c-1)}$. This test is again the same as the test used for the random effects model.

It is easy to find confidence intervals for many functions of the parameters (e.g., θ, σ_e^2, $\sigma_b^2/(n\sigma_d^2 + \sigma_e^2), \sigma_d^2/\sigma_e^2$) but not for σ_b^2 or σ_d^2, which are often the parameters of interest. We can also find simultaneous confidence intervals for the set of all contrasts in the α_i. In Exercise C1 you are asked to show that the confidence intervals

$$\sum_{i=1}^{r} b_i \alpha_i \in \sum_{i=1}^{r} b_i \bar{Y}_{i..} \pm \left(F^{\alpha}_{(r-1),(r-1)(c-1)} \frac{\sum_{i=1}^{r} b_i^2 S_3^2}{nc(c-1)} \right)^{1/2} \tag{15.20}$$

are a set of simultaneous confidence intervals for the set of all contrasts in the α (i.e., all functions $\sum_{i=1}^{r} b_i \alpha_i$ such that $\sum_{i=1}^{r} b_i = 0$.) Note that these confidence intervals can be computed from those for the fixed effects model by substituting $S_3^2/(r-1)(c-1)$ for $\hat{\sigma}^2$ and $(r-1)(c-1)$ for $rc(n-1)$.

We now consider the second version of the mixed model that is suggested in the introduction to this chapter in which we observe

$$Y_{ijk} = \theta^* + \alpha_i^* + b_j^* + d_{ij}^* + e_{ijk}^*;$$
$$j = 1, \ldots, r; \qquad j = 1, \ldots, c; \qquad k = 1, \ldots, n;$$

where θ^* and α_i^* are unknown parameters such that $\sum_i \alpha_i^* = 0$ and b_j^*, d_{ij}^*, and e_{ijk}^* are unobserved random variables such that $\sum_i d_{ij}^* = 0$. Let $\mathbf{d}_j^*$

$= (d_{1j}^*, \ldots, d_{rj}^*)'$. We assume that b_j^*, $\mathbf{d}_j^*$ and e_{ijk}^* are all independent, that

$$b_j^* \sim N_1(0, \tau_b^2), \qquad e_{ijk}^* \sim N_1(0, \tau_e^2),$$

$$\mathbf{d}_j^* \sim N_r\left[\mathbf{0}, \tau_d^2\begin{bmatrix} 1 & -\frac{1}{r-1} & \cdots & -\frac{1}{r-1} \\ -\frac{1}{r-1} & 1 & & -\frac{1}{r-1} \\ -\frac{1}{r-1} & -\frac{1}{r-1} & \cdots & 1 \end{bmatrix}\right].$$

The parameter space for this version of the model is given by

$$-\infty < \theta < \infty, \qquad \sum_i \alpha_i = 0, \qquad \tau_b^2 \geqslant 0, \qquad \tau_d^2 \geqslant 0, \qquad \tau_e^2 > 0.$$

We now show how the first version of the mixed model defined in equations (15.15) – (15.17) can be transformed to this version of the mixed model. In that model

$$Y_{ijk} = \theta + \alpha_i + \left(b_j + \overline{d_{\cdot j}}\right) + \left(d_{ij} - \overline{d_{\cdot j}}\right) + e_{ijk}.$$

Therefore, let

$$\theta^* = \theta, \qquad \alpha_i^* = \alpha_i, \qquad b_j^* = b_j + \bar{d}_{\cdot j}, \qquad d_{ij}^* = d_{ij} - \overline{d_{\cdot j}}\,, e_{ijk}^* = e_{ijk}.$$

It is easily verified that $\sum_i d_{ij}^* = 0$. Since $d_{ij} - \overline{d_{\cdot j}}$ and $\overline{d_{\cdot j}}$ are uncorrelated, they are independent. Therefore, b_j^*, e_{ijk}^*, and $\mathbf{d}_j^* = (d_{1j}^*, \ldots, d_{rj}^*)'$ are independent. Also

$$b_j^* \sim N_1\left(0, \sigma_b^2 + \frac{1}{r}\sigma_d^2\right), \qquad e_{ijk}^* \sim N_1(0, \sigma_e^2),$$

$$\mathbf{d}_j^* \sim N_r\left[0, \sigma_d^2 \frac{r-1}{r}\begin{bmatrix} 1 & -\frac{1}{r-1} & \cdots & -\frac{1}{r-1} \\ -\frac{1}{r-1} & 1 & \cdots & -\frac{1}{r-1} \\ \ddots & \vdots & \ddots & \vdots \\ -\frac{1}{r-1} & -\frac{1}{r-1} & \cdots & 1 \end{bmatrix}\right] \tag{15.21}$$

Therefore the first version of the model has been transformed to the second version with

$$\tau_b^2 = \sigma_b^2 + \frac{1}{r}\sigma_d^2, \qquad \tau_d^2 = \frac{r-1}{r}\sigma_d^2, \qquad \tau_e^2 = \sigma_e^2.$$

Note that the transformed version of the first model has a slightly different parameter space from the second version in that $\sigma_b^2 = \tau_b^2 - (1/(r-1))\tau_d^2 \geqslant 0$ for the transformed version of the first model. Therefore, the second version of the model is slightly more general than the first version. However, this slight additional generality would not affect the complete sufficient statistic nor its distribution. Therefore, we can translate Theorem 15.11 to the following theorem. Let U, V, S_2^2, S_3^2, and S_4^2 be defined, as before, by (15.9) and (15.18).

THEOREM 15.12. A complete sufficient statistic for the second version of the balanced two-way mixed model is $(U, \mathbf{V}, S_2^2, S_3^2, S_4^2)$. U, $\mathbf{V}$, S_2^2, S_3^2, and S_4^2 are independent and

$$U \sim N_1\left(\theta, \frac{1}{ncr}\left(nr\tau_b + \tau_e^2\right)\right), \qquad \mathbf{V} \sim N_n\left(\gamma, \frac{1}{rc}\left(n\left(\frac{r}{r-1}\right)\tau_d^2 + \tau_e^2\right)\mathbf{I}\right)$$

$$S_2^2 \sim \left(nr\tau_b^2 + \tau_e^2\right)\chi_{r-1}^2, \qquad S_3^2 \sim \left(n\left(\frac{r}{r-1}\right)\tau_d^2 + \tau_e^2\right)\chi_{(r-1)(c-1)}^2$$

$$S_4^2 \sim \tau_e^2 \chi_{rc(n-1)}^2,$$

where γ is defined in (15.18).

PROOF. This is just a restatement of Theorem 15.11. □

As before, the minimum variance unbiased estimators of θ, α_i, and σ_e^2 are given by $\overline{Y}_{...}, \overline{Y}_{i..} - \overline{Y}_{...}$ and $S_4^2/rc(n-1)$. There are no nonnegative unbiased estimators of τ_b^2 or τ_d^2. Sensible estimators for these parameters would be the following

$$\hat{\tau}_b^2 = \max\left\{\frac{1}{nr}\left[\frac{S_2^2}{r-1} - \frac{S_4^2}{rc(n-1)}\right], 0\right\},$$

$$\hat{\tau}_d^2 = \max\left\{\frac{r}{n(r-1)}\left[\frac{S_3^2}{(r-1)(c-1)} - \frac{S_4^2}{rc(n-1)}\right], 0\right\}$$

Now consider testing that $\alpha_i^* = 0$. Since $\alpha_i^* = \alpha_i$, this would be tested in the same way as for the first version of the model. Also, simultaneous confidence intervals for contrasts in the α_i^* would be the same as those for the α_i in the first model Similarly, $\tau_d^2 = 0$ if and only if $\sigma_d^2 = 0$, so that we would test $\tau_d^2 = 0$ in the second version in exactly the same way as we tested $\sigma_d^2 = 0$ in the first version. Finally, consider testing that $\tau_b^2 = 0$. Let

$$F_4 = \frac{rc(n-1)S_2^2}{(c-1)S_4^2}, \qquad \delta_4 = \frac{nr\tau_b^2 + \tau_e^2}{\tau_e^2}.$$

Then $F_4/\delta_4 \sim F_{c-1, rc(n-1)}(0)$. Also $\delta_4 = 1$ if and only if $\tau_b^2 = 0$. Therefore, a sensible test that $\tau_b^2 = 0$ in the second version of the model would be to reject if $F_1 > F^{\alpha}_{c-1,rc(n-1)}$. Note that this test is *not* the same as the test that $\sigma_b^2 = 0$ for the first version of the model.

We now compare these two versions of the mixed model. As long as we are interested in inference about the fixed main effect, it does not matter which version of the model is used. The estimators for the fixed effects are the same for both models as are the simultaneous confidence intervals for the set of all contrasts in the fixed effects. The F-test that the fixed effects are 0 is also the same for both models. Now consider inference about the interaction. We note that τ_d^2 is just a rescaling of σ_d^2 (and even the rescaling constant is near 1). Therefore, inference about the interaction is not very sensitive to the choice of version of the model. Finally, consider inference about the random effect in the model. As we have seen above, inference about the random effect is sensitive to the version of the model used. Testing that the main random effects are 0 has two different meanings in the two versions of the model.

The comments in the last paragraph indicate that if we are interested in making inference about the main random effect in a mixed model, it is quite important which model is selected. At present it is not clear which version to select in practice. The first version has the advantage of being easier to work with and leading to somewhat simpler formulas. These advantages become more important for more complicated mixed models. However, in the next section, we give a heuristic motivation for the second version of the mixed model (as well as the one-way and two-way random effect models). This derivation allows us to interpret b_j^*. I know of no such intepretation for b_j, and I have a great deal of difficulty understanding what b_j represents in the first version of the model. Therefore, I would select the second version of the model. However, the question of which model to use is a very difficult question, and one for which there is no definite answer at this time.

15.4. DERIVING THE RANDOM EFFECTS AND MIXED MODELS.

In this section, we derive heuristically the one-way and two-way random effects model and the second version of the mixed model from some assumptions that are more elementary. These heuristic derivations are helpful in understanding what the random effects represent as well as understanding why it is reasonable to assume that the random effects are independent.

We look first at the balanced one-way random effects model. Let Y_{ij} be the jth observation in the ith class and let m_i be the expected value of an observation in the ith class. Note that m_i is a random variable since the classes in a random effects model are selected at random from a larger population of classes. We now postulate two assumptions about the distribution of Y_{ij} and m_i and then show that these assumptions guarantee that the model is the same as the one-way random effects model discussed in Section 15.1. The assumptions are as follows:

(1) The m_i are independent and $m_i \sim N_1(\theta, \sigma_a^2)$.

(2) Given the m_i, the Y_{ij} are independent and $Y_{ij} \mid (m_1, \ldots, m_k) \sim N_1(m_i, \sigma_e^2)$.

THEOREM 15.13. Let $a_i = m_i - \theta$, $e_{ij} = Y_{ij} - m_i$. Then $Y_{ij} = \theta + a_i + e_{ij}$. Under the assumptions given above, the a_i and e_{ij} are all independent, and $a_i \sim N(0, \sigma_a^2)$, $e_{ij} \sim N_1(0, \sigma_e^2)$.

PROOF. It is clear that $Y_{ij} = \theta + a_i + e_{ij}$ and that $a_i \sim N(0, \sigma_a^2)$, $e_{ij} \sim N_1(0, \sigma_e^2)$, and that the a_i are independent. We now show that the a_i and e_{ij} are all independent. Since they have a joint normal distribution, it is enough to establish pairwise independence (see Exercise C1 of Chapter 3). The e_{ij} are independent, since the Y_{ij} given m_i are independent. Finally, the conditional distribution of the e_{ij} given the m_i does not depend on the m_i. Hence the e_{ij} are independent of m_i and therefore of a_i. □

This theorem implies that as long as we can consider the classes in a one-way model as being sampled from an infinite population of possible classes, that the means of the classes have a normal distribution, and that the actual observations have a normal distribution about the mean for the class, then the random effects model is the correct model to use. These assumptions are often quite reasonable in practical situations. We note that σ_a^2 is the variance of the class means m_i, and therefore, testing that $\sigma_a^2 = 0$ is the same as testing that the classes all have the same mean.

We now consider the two-way random effects model. Let Y_{ijk} be the kth observation in the ith row and the jth column. Let m_{ij} be the expected value of an observation in the ith row and the jth column. Note that m_{ij} is random since the rows and columns are randomly selected from larger populations. Define s_i to be the expected value of an observation in the ith row and a randomly selected column, where this column is randomly selected from the set of all possible columns, not just those in the experiment. In the example with bulls and cows, s_i would be expected weight of calf fathered by the ith bull averaged over the populations of all possible cows. Similarly, let t_j be the expected value of an observation in the jth column. Let θ be the expected value of an observation from a randomly selected row and randomly selected column. Note that θ is not a random variable. In the example of the cows and bulls, θ would be the average weight of a calf averaged over all bulls and cows. Note that $Et_i = Es_j = Em_{ij} = \theta$. We now make some assumptions about the joint distribution of Y_{ijk}, m_{ij}, s_i, and t_j and show that these assumptions guarantee that the model is a random effects model. The assumptions are as follows:

(1) The t_i and s_j are all independent, $s_i \sim N_1(\theta, \sigma_a^2)$, $t_j \sim N_1(\theta, \sigma_b^2)$.
(2) If $i \neq i'$, $j \neq j'$, then m_{ij} and $m_{i'j'}$ are independent, Given s_i, m_{ij} and $m_{ij'}$ are independent for $j \neq j'$. Given t_j, m_{ij} and $m_{i'j}$ are independent for $i \neq i'$. The m_{ij} have a joint normal distribution and var $(m_{ij}) = \tau^2$.
(3) Given the m_{ij}, the Y_{ijk} are independent, $Y_{ijk}|(m_{11}, \ldots, m_{rc}) \sim N_1(m_{ij}, \sigma_e^2)$.

In addition to normal assumptions, the first two assumptions are equivalent to assuming that the rows are a sample from a large class of possible rows, the columns are a sample from a larger class of possible columns, and that the rows and columns are sampled independently. Note that assumption 2 does not guarantee that the m_{ij} are independent.

THEOREM 15.14. Let $a_i = s_i - \theta$, $b_j = t_j - \theta$, $d_{ij} = m_{ij} - s_i - t_j + \theta$, $e_{ijk} = Y_{ijk} - m_{ij}$. Then $Y_{ijk} = \theta + a_i + b_j + d_{ij} + e_{ijk}$. Under the above assumptions, the a_i, b_j, d_{ij}, and e_{ijk} are all independent, $a_i \sim N_1(0, \sigma_a^2)$, $b_j \sim N_1(0, \sigma_b^2)$, $d_{ij} \sim N_1(0, \sigma_d^2)$, $e_{ijk} \sim N_1(0, \sigma_e^2)$ where $\sigma_d^2 = \tau^2 - \sigma_a^2 - \sigma_b^2$.

PROOF. Clearly $Y_{ijk} = \theta + a_i + b_j + d_{ij} + e_{ijk}$ and $Ea_i = 0$, $Eb_j = 0$, $Ed_{ij} = 0$, $Ee_{ijk} = 0$. We now show that the a_i, b_j, d_{ij}, e_{ijk} are all independent. Since they have a joint normal distribution it is enough to show that they are uncorrelated. We first note that the e_{ijk} are all independent and are also independent of the m_{ij} by assumption 3. They are therefore all uncorrelated and also uncorrelated with the a_i, b_j, and d_{ij}. The a_i and b_j are all independent since the s_i and t_j are. Hence they are all uncorrelated. If $i \neq i'$, $j \neq j'$, then d_{ij} and $d_{i'j'}$ are independent since (m_{ij}, s_i, t_j) and $(m_{i'j'}, s_{i'}, t_{j'})$ are independent. We therefore need only show that a_i and d_{ij} are uncorrelated, that b_j and d_{ij} are uncorrelated, that d_{ij} and $d_{ij'}$ are uncorrelated if $j \neq j'$ and that d_{ij} and $d_{i'j}$ are uncorrelated if $i \neq i'$. We look first at a_i and d_{ij}. Since $Em_{ij} | s_i = s_i$, $Ed_{ij} | s_i = 0$. Therefore

$$\operatorname{cov}(a_i, d_{ij}) = Ea_i d_{ij} = E(a_i E(d_{ij} \,|\, s_i)) = 0.$$

Hence a_i and d_{ij} are uncorrelated. By a similar argument, b_j and d_{ij} are uncorrelated. We now look at d_{ij} and $d_{ij'}$. We note that assumptions 1 and 2 imply that given s_i, d_{ij}, and $d_{ij'}$ are independent. Therefore,

$$\operatorname{cov}(d_{ij}, d_{ij'}) = Ed_{ij}d_{ij'} = E(E(d_{ij}d_{ij'}|s_i)) = E(E(d_{ij} \,|\, s_i)E(d_{ij'} \,|\, s_i)) = 0.$$

Hence d_{ij} and $d_{ij'}$ are uncorrelated. By a similar argument, d_{ij} and $d_{i'j}$ are uncorrelated. We have therefore shown that the a_i, b_j, d_{ij}, and e_{ijk} are all uncorrelated and hence independent. Since they have a joint normal distribution, they must have marginal normal distributions. (They are linear functions of the Y_{ijk}, m_{ij}, s_i, and t_j which have a joint normal distribution.) It is clear that $a_i \sim N_1(0, \sigma_a^2)$, $b_j \sim N_1(0, \sigma_b^2)$, $e_{ijk} \sim N_1(0, \sigma_e^2)$. To find the variance of d_{ij}, we note that the independence of a_i, b_j, and d_{ij} implies that

$$\tau^2 = \text{var}(m_{ij}) = \text{var}(a_i) + \text{var}(b_j) + \text{var}(d_{ij}) = \sigma_a^2 + \sigma_b^2 + \text{var}(d_{ij}).$$

Hence, $d_{ij} \sim N_1(0, \sigma_d^2)$. □

We note that this theorem implies that assumptions 1–3 guarantee that the model is a random effects model. We see that testing that $\sigma_a^2 = 0$ is equivalent to testing that the s_i are all equal.

We now look at the mixed model from this more elementary perspective. Let Y_{ijk} be the kth observation in the ith row and the jth column. We suppose that the rows are fixed in advance of the experiment but that the columns are randomly selected from a larger class of columns. Let m_{ij} be the expected value of an observation in the ith row and the jth column. Note that m_{ij} is a random variable since the columns are selected at random. Let s_i be the expected value of an observation in the ith row and a randomly selected column (selected from the set of all columns, not just those in the experiment). Note that the s_i are constants since the rows are selected in advance. We need now to make an analagous definition for t_j, the "mean value" for an observation in the jth column. It seems that t_j should not depend on any rows but those included in the experiment (i.e., in the bulls and cows example, it should only depend on those bulls in the experiment, not on any larger class). This leads us to let t_j be the average of the expected values of an observation in the jth column and the ith row averaged over the rows, i.e., $t_j = \overline{m}_{.j}$. Let θ be the average expected value of a measurement in the ith row and a randomly selected column, averaged over the rows. That is $\theta = Et_i = \bar{s}_{..}$ We now suggest some reasonable assumptions for this model and then give a theorem stating that these assumptions lead to the second version of the mixed model. The assumptions are as follows:

(1) The t_j are independent, $t_j \sim N_1(\theta, \sigma_b^2)$.
(2) The conditional distribution of the m_{ij} given the t_j is a multivariate normal distribution.

$$E[m_{ij} \mid (t_1, \ldots, t_c)] = t_j$$

$$\text{var}[m_{ij} \mid (t_1, \ldots, t_c)] = \sigma_d^2$$

$$\text{cov}[m_{ij}, m_{i'j} \mid (t_1, \ldots, t_c)] = -\frac{1}{r-1}\sigma_d^2 \qquad \text{for } i \neq i'$$

$$\text{cov}[m_{ij}, m_{ij'} \mid (t_1, \ldots, t_c)] = 0 \qquad \text{for } j \neq j'.$$

(3) Given the m_{ij}, the Y_{ijk} are independent and $Y_{ijk}|(m_{11}, \ldots, m_{rc}) \sim N_1(m_{ij}, \sigma_e^2)$.

THEOREM 15.15. Let $\alpha_i = s_i - \theta$, $b_j = t_j - \theta$, $d_{ij} = m_{ij} - s_i - t_j + \theta$, $e_{ijk} = Y_{ijk} - m_{ij}$. Then the α_i are constants such that $\sum_i \alpha_i = 0$, the d_{ij} are random variables such that $\sum_i d_{ij} = 0$ and $Y_{ijk} = \theta + \alpha_i + b_j + d_{ij} + e_{ijk}$. If the above assumptions are satisfied, then the b_j, $\mathbf{d}_j' = (d_{1j}, \ldots, d_{rj})$ and e_{ijk} are all independent, and $b_j \sim N_1(0, \sigma_b^2)$, $e_{ijk} \sim N_1(0, \sigma_e^2)$, and

$$\mathbf{d}_j \sim N_n\left(\mathbf{0}, \sigma_d^2 \begin{bmatrix} 1 & -\dfrac{1}{r-1} & \cdots & -\dfrac{1}{r-1} \\ -\dfrac{1}{r-1} & 1 & \cdots & -\dfrac{1}{r-1} \\ \vdots & \vdots & \ddots & \vdots \\ -\dfrac{1}{r-1} & -\dfrac{1}{r-1} & \cdots & 1 \end{bmatrix}\right)$$

PROOF. See Exercise B14. □

Note that if $\operatorname{cov}[m_{ij}, m_{i'j}|(t_1, \ldots, t_c)] = \tau^2$ then $\tau^2 = -\sigma^2/(r-1)$ since $\sum_i m_{ij} = rt_j$ (see Exercise B1). Note also that assuming that $\operatorname{cov}[m_{ij}, m_{i'j'}|(t_1, \ldots, t_c)] = 0$ for $j \neq j'$ we are assuming that values from different cows are independent. This theorem therefore implies that if the three assumptions above are met then the second version of the mixed model is the correct one to use. We note also that $\sum_i d_{ij} = 0$ as long as we define θ, α_i, b_j, and d_{ij} as we have done above. As long as these definitions are correct for the effects, the first version of the mixed model could not be correct. The definitions for θ and α_i seem evident, as does the definition for d_{ij} given the definitions for θ, α_i and b_j. Therefore, the only question in this definition of the model is whether the b_j are defined sensibly. It is hard for me to imagine any other definition for the b_j. The b_j should not depend on any larger class from which the rows are drawn, since those particular rows are the ones of interest or we would not be using a mixed model. In addition there often is no larger class from which these rows are randomly chosen. (An example of this would occur when the fixed effect represents level of a hormone treatment or diet.) For this reason, the second version of the model seems preferable to me. Another reason is that testing that $\sigma_b^2 = 0$ for this model is equivalent to testing that the $m_{.j}$ are all the same, that is, that there is no difference between the average performances of the cows, averaged over the bulls. It seems to me that this is what is meant by testing that there is no "effect due to columns." Therefore, the second version of the model is preferable for testing for no column effect. As mentioned earlier, it does not matter which model is used when testing for row effects or for interaction effects.

It should be clear that the derivations in this section are merely heuristic. It is somewhat difficult to define carefully what is meant by statements like "the expected value of an observation from the ith row and a randomly chosen column." However, I find this development quite helpful in my understanding of random effects and mixed models, helpful in understanding why the various independence assumptions are reasonable and helpful in understanding which mixed model is more reasonable.

15.5. THE RELATIONSHIP BETWEEN THE REPEATED MEASURES MODEL AND CERTAIN MIXED MODELS.

We now illustrate the relationship between the repeated measures model discussed in the last chapter and mixed models with only one random effect and no interaction between the random effect and any fixed effect. Consider first a balanced two-way mixed model with no interaction and no replication. In such a model, we observe

$$Y_{ij} = \theta + a_i + \beta_j + e_{ij}, \qquad \sum_j \beta_j = 0$$

where θ and β_j are unknown constants and the a_i and e_{ij} are unobserved random variables such that $a_i \sim N_1(0, \sigma_a^2)$, $e_{ij} \sim N_1(0, \sigma_e^2)$. This model is fairly simple to analyze, as we shall see, since it has no interaction term. We can put this model into the form of models studied in Chapter 14. Let $\mathbf{Y}_i = (Y_{i1}, \ldots, Y_{ic})'$. Then the $\mathbf{Y}_i$ are independent (since they involve separate a_i's and e_{ij}'s). Furthermore,

$$\mathbf{Y}_i \sim N_c\left[\begin{pmatrix} \theta + \beta_1 \\ \vdots \\ \theta + \beta_c \end{pmatrix}, \quad \sigma^2 \begin{pmatrix} 1 & \rho \cdots & \rho \\ \rho & 1 \cdots & \rho \\ \vdots & \ddots & \vdots \\ \rho & \cdots & 1 \end{pmatrix}\right], \qquad \rho = \frac{\sigma_a^2}{\sigma_a^2 + \sigma_e^2} \tag{15.22}$$

(see Exercise B15). This model is quite similar to the repeated measures model discussed in Chapter 14. The only real differences is that ρ can be negative for the repeated measures model, whereas it must be nonnegative for the mixed model. However, the null distribution of the F-statistic for testing that the $\alpha_i = 0$ in the repeated measures model does not depend on ρ, so that the size α F-test is also a size α procedure for the random effects model. We also note that $\sigma_a^2 = 0$ if and only if $\rho = 0$. Therefore, to test $\sigma_a^2 = 0$, we test that $\rho = 0$ against $\rho \geqslant 0$, so we use the one-sided analogue of the two-sided test derived in Section 15.6. These procedures are in fact the same as would be derived using the mixed model approach.

Any balanced mixed model that has only one random effect, and has no interaction between the random effect and any fixed effect can be analyzed in the fashion suggested above, (i.e., looking at the model as a repeated measures model).

15.6. OTHER RESULTS.

The class of random effects and mixed models for which optimal procedures (of the type considered in this book) exist seems to be a very small class. We have already seen that it is not possible to find exact confidence intervals or nonnegative unbiased estimators for even such simple random effects models as the balanced one-way and two-way models. There are no UMP invariant tests for any random effects models except the balanced one-way model. In any random effects model with more than two crossed effects, there is no nontrivial exact size α test that main effects are 0. However, for balanced nested models, there is a size α F-test which is UMP among unbiased rules (see Lehmann, 1959, pp. 286–293). We have seen that the MLE's for the balanaced two-way model are quite messy, and so likelihood ratio tests must be also. This has lead to the development of restricted MLEs. For a discussion of these estimators, see Harville (1977). For unbalanced models the situation is even worse. Typically, there is no complete sufficient statistic for the model, and no minimum variance unbiased estimators. The dimension of the smallest sufficient statistic is often the same as the dimension of the original data for these models, so there is very little reduction of the data possible. An F-test for testing that the main effects are zero for the unbalanced one-way random effects model is derived in Exercise C2. It should be emphasized that the alternative distribution of this F-statistic is quite messy. It is not a noncentral F-distribution, nor a constant times a central F-distribution. There are no exact tests for other more complicated unbalanced models. Finally, we have seen that for mixed models, it is not even clear what model to use. In summary, we know very little about optimal procedures for these models. However, they do arise in practice, and many procedures for analyzing them have been suggested (see Searle, 1971, pp. 376–514 and Harville, 1977, for a description of some of these procedures.)

For most of the models studied in this book, the procedures defined are not too sensitive (at least asymptotically) to the normal assumption used in deriving them. However, the procedures defined in this chapter are quite sensitive to those assumptions, particularly those procedures for drawing inference about the random effects. Since the parameters involved in the random effects part of the model are all variances, when we attempt to draw inference about these random effects, we are drawing inference about variances. In Chapter 10, we indicated that such inferences are quite dependent on the normal assumption. If the actual distribution does not have the same kurtosis as the normal distribution, then these inferences are hard to interpret and may be meaningless.

In this chapter, we have also relied on the normal assumption at two other points. The first is in the transformation from the Y_{ijk} to the Z_{ijk}. The Z_{ijk} would be uncorrelated for nonnormal distributions, but need not be independent. Similarly, in the heuristic derivations of the two-way random effects and mixed models, we relied on the fact that uncorrelated normal random variables are independent. In fact, it is not obvious that the random effects should be assumed independent unless they are normally distributed.

These comments indicate that the procedures discussed in this chapter are quite sensitive to the normal assumptions used in deriving them. In principle, then, they should only be used when we are fairly confident that this normal assumption is satisfied. However, at this time, there do not seem to be any realistic alternative procedures for this situation, so these procedures are often used, even when the normal assumption is suspect. (For a procedure for one-way analysis of variance that is less sensitive to the assumed normality, see Arveson and Layard, 1975.)

As a final topic in this section we present a fairly general version of the mixed model and show how we can use results for the generalized linear model to design approximate procedures for the fixed effects in this mixed model. Most mixed models can be put into the following form. We observe the n-dimensional vector $\mathbf{Y}$ such that

$$\mathbf{Y} = \mathbf{X}\boldsymbol{\beta} + \mathbf{Z}\mathbf{b} + \mathbf{e}$$

where $\mathbf{X}$ and $\mathbf{Z}$ are known $n \times p$ and $n \times s$ matrices, $\boldsymbol{\beta}$ is an unobserved $p \times 1$ constant vector (representing the fixed effects) and $\mathbf{b}$ and $\mathbf{e}$ are unobserved random vectors (with $\mathbf{b}$ representing the random effects). We assume that $\mathbf{b}$ and $\mathbf{e}$ are independent with $\mathbf{b} \sim N_s(\mathbf{0}, \sigma_e^2 \boldsymbol{\Xi})$, $\mathbf{e} \sim N_n(\mathbf{0}, \sigma_e^2 \mathbf{I})$ where $\boldsymbol{\Xi}$ is an unknown matrix that has a particular structure determined by the mixed model used. (For example, in the first version of the mixed model discussed in Section 15.3, we could let

$$\boldsymbol{\beta} = \begin{pmatrix} \theta + \alpha_1 \\ \vdots \\ \theta + \alpha_r \end{pmatrix}, \qquad \mathbf{b} = \begin{pmatrix} b_1 \\ \vdots \\ b_2 \\ d_{11} \\ \vdots \\ d_{rc} \end{pmatrix}, \qquad \boldsymbol{\Xi} = \frac{1}{\sigma_e^2} \begin{pmatrix} \sigma_b^2 & \cdots & 0 & 0 & \cdots & 0 \\ \vdots & & \vdots & \vdots & & \vdots \\ 0 & \cdots & \sigma_b^2 & 0 & \cdots & 0 \\ 0 & & 0 & \sigma_d^2 & \cdots & 0 \\ \vdots & & \vdots & \vdots & & \vdots \\ 0 & \cdots & 0 & 0 & \cdots & \sigma_d^2 \end{pmatrix}$$

Then that version of the mixed model would be in the framework given here.)

Now, let $\mathbf{A} = \mathbf{Z}\boldsymbol{\Xi}\mathbf{Z}' + \mathbf{I}$. Then an equivalent version of this model is one in which we observe

$$\mathbf{Y} \sim N_n(\mathbf{X}\boldsymbol{\beta}, \sigma_e^2\mathbf{A}).$$

If we knew $\mathbf{A}$ then this model would just be a generalized linear model as discussed in Chapter 13. This fact suggests the following method for designing procedures for drawing inferences about the fixed effects $\boldsymbol{\beta}$ for this model. We first find an estimator $\hat{\boldsymbol{\Xi}}$ for $\boldsymbol{\Xi}$. (In the example above, this means finding estimators for σ_b^2/σ_e^2 and σ_d^2/σ_e^2.) We then let $\hat{\mathbf{A}} = \mathbf{Z}\hat{\boldsymbol{\Xi}}\mathbf{Z}' + \mathbf{I}$, and use the appropriate procedures for the generalized linear model that occurs when we assume that $\mathbf{A} = \hat{\mathbf{A}}$. While it is true that these procedures are only approximate procedures, in most mixed models $\boldsymbol{\Xi}$ does not contain many different parameters so that for reasonable sizes we can find a fairly good estimator for $\boldsymbol{\Xi}$ and hence the approximations should be fairly good. It should be emphasized that this method of designing procedures is only useful for drawing inferences for the fixed effects in the mixed model.

15.7. PROOF OF THEOREM 15.10.

We now prove a generalization of Theorem 15.10. The generalization is useful for establishing completeness for more complicated random effects and mixed models.

THEOREM 15.16. Let $\mathbf{U}, S_1^2, \ldots, S_k^2$ be independent such that

$$\mathbf{U} \sim N_p(\boldsymbol{\gamma}, \mathbf{A}(\tau_1, \ldots, \tau_k)), \qquad \frac{S_i^2}{\tau_i} \sim \chi_{r_i}^2(0)$$

where $\mathbf{A} > 0$ is a known function of $(\tau_1, \ldots, \tau_k)$. If $\boldsymbol{\gamma}$ and $(\tau_1, \ldots, \tau_k)$ are unrelated and the ranges of $\boldsymbol{\gamma}$ and $(\tau_1, \ldots, \tau_k)$ contain open rectangles, then $(\mathbf{U}, S_1^2, \ldots, S_k^2)$ has a complete family of distributions.

PROOF. Let $\mathbf{T}' = (S_1^2, \ldots, S_k^2)$, $\boldsymbol{\tau}' = (\tau_1, \ldots, \tau_k)$. We can use the exponential criterion on the marginal distribution of $\mathbf{T}$ to see that $\mathbf{T}$ has a complete family of distributions. Similarly, we can use the exponential criterion on the marginal distribution of $\mathbf{U}$ to see that $\mathbf{U}$ has a complete family of distributions for each fixed $\boldsymbol{\tau}$. Now, suppose that

$$Eh(\mathbf{U}, \mathbf{T}) = 0.$$

Define

$$g(\mathbf{U}, \boldsymbol{\tau}) = Eh(\mathbf{U}, \mathbf{T}) \mid \mathbf{U}.$$

[Note that g does not depend on $\boldsymbol{\gamma}$, since the conditional distribution of

$\mathbf{T}|\mathbf{U}$, (which is the same as the marginal distribution of $\mathbf{T}$) does not depend on γ.] Now,

$$Eg(\mathbf{U},\boldsymbol{\tau}) = Eh(\mathbf{U},\mathbf{T}) = \mathbf{0}.$$

Since $\mathbf{U}$ has a complete family of distributions for each fixed $\boldsymbol{\tau}$, we see that

$$P[g(\mathbf{U},\boldsymbol{\tau}) = 0] = 1.$$

Let

$$N(\boldsymbol{\tau}) = [u : g(\mathbf{u},\boldsymbol{\tau}) \neq 0], \qquad N = \bigcup_{\boldsymbol{\tau}} N(\boldsymbol{\tau}).$$

If $\mathbf{u} \in N^c$, than $g(\mathbf{u},\boldsymbol{\tau}) = 0$ for all $\boldsymbol{\tau}$. We now show that N has probability 0. (Note that it is an uncountable union of sets of probability 0.) Let Q be the set of all rational $\boldsymbol{\tau}$. Define

$$M = \bigcup_{\boldsymbol{\tau} \in Q} N(\boldsymbol{\tau}).$$

Then M is a countable union of sets of probability 0, and hence has probability 0. We now show that $N = M$. Clearly, $M \subset N$. Let $\mathbf{u} \in M^c$. Then $g(\mathbf{u},\boldsymbol{\tau}) = 0$ for all rational $\boldsymbol{\tau}$. The distribution of $\mathbf{T}|\mathbf{U}$ (i.e., that of $\mathbf{T}$) comes from an exponential family. Hence $g(\mathbf{u},\boldsymbol{\tau})$ is a continuous function of $\boldsymbol{\tau}$. This implies that $g(\mathbf{u},\boldsymbol{\tau}) = 0$ for all $\boldsymbol{\tau}$. Hence $\mathbf{u} \in N^c$. Therefore, $M^c \subset N^c$ and $M = N$. Therefore N has probability 0. Now let $\mathbf{u} \in N^c$. Then

$$g(\mathbf{u},\boldsymbol{\tau}) = E[h(\mathbf{U},\mathbf{T})\,|\,\mathbf{U} = \mathbf{u}] = 0.$$

Since the conditional distribution of $\mathbf{T}|\mathbf{U}$ (i.e., the marginal distribution of $\mathbf{T}$) has a complete family of distributions, we see that

$$P[h(\mathbf{U},\mathbf{T}) = 0\,|\,\mathbf{U} = \mathbf{u}] = 1.$$

Therefore,

$$P[h(\mathbf{U},\mathbf{T}) \neq 0] \leqslant P(\mathbf{U} \in N) + P\{\mathbf{U} \in N^c \text{ and } [h(\mathbf{U},\mathbf{T}) \neq 0]\}$$

$$0 + \int_{\mathbf{u} \in N^c} P[h(\mathbf{U},\mathbf{T}) \neq 0\,|\,\mathbf{U} = \mathbf{u}] f(\mathbf{u})\, d\mathbf{u} = 0$$

where $f(\mathbf{u})$ is the density of $\mathbf{U}$. Therefore, $(\mathbf{U},\mathbf{T})$ has a complete family of distributions. □

COROLLARY. Let $\mathbf{U}$, S_1^2, S_2^2, S_3^2, S_4^2 be defined as in Section 15.2. Then ($\mathbf{U}$, S_1^2, S_2^2, S_3^2, S_4^2) is a complete sufficient statistic for the balanced two-way random effects model.

PROOF. The sufficiency follows from Theorem 15.9. The completeness follows from Theorem 15.16 with $p = 1$, $k = 4$, $\theta = \gamma$, $\tau_1 = nc\sigma_a^2 + n\sigma_d^2 + \sigma_e^2$, $\tau_2 = nr\sigma_b^2 + n\sigma_d^2 + \sigma_e^2$, $\tau_3 = n\sigma_d^2 + \sigma_e^2$, $\tau_4 = \sigma_e^2$, $A(\boldsymbol{\tau}) = (\tau_1 + \tau_2 + \tau_3)/nrc \in c$. □

Theorem 15.16 can be extended to the case in which the range of $(\boldsymbol{\gamma}', \boldsymbol{\tau}')$ contains a $(p + k)$-dimensional rectangle, since that rectangle must contain a set on which $\boldsymbol{\tau}$ and $\boldsymbol{\gamma}$ are unrelated and the ranges of $\boldsymbol{\gamma}$ and $\boldsymbol{\tau}$ contain rectangles.

EXERCISES.

Type A

1. Consider the data of Exercise A4 of Chapter 7 as coming from a two-way random effects model.

(a) Using $\alpha = 0.05$, test the hypothesis that the row effects are all 0.

(b) Using $\alpha = 0.05$, test the hypothesis that the interactions are all 0.

(c) Find 95% confidence intervals for $\sigma_a^2/(n\sigma_d^2 + \sigma_e^2)$ and σ_d^2/σ_e^2.

2. Consider the data of Exercise A4 of Chapter 7 as coming from a two-way mixed model (with the row effects as the fixed effects and the column and interaction effects as random effects). Use the second version of the mixed model as discussed in Section 15.3.

(a) Using $\alpha = 0.05$, test the hypothesis that there is no row effect.

(b) Using $\alpha = 0.05$, test the hypothesis that there is no column effect.

(c) Using $\alpha = 0.05$, test the hypothesis that there is no interaction effect.

(d) Find the 95% simultaneous confidence intervals for the comparisons $\alpha_i - \alpha_j$.

(e) Find 95% confidence intervals for $\tau_b^2/(n\tau_d^2 + \tau_e^2)$ and τ_d^2/τ_e^2.

Type B

1. Suppose that U_j are random variables such that $\sum_j U_j = C$, a constant, $\mathrm{Var}(U_j) = \sigma^2$, and $\mathrm{cov}(U_j, U_k) = \tau^2$ for all j and k. Show that $\tau^2 = \dfrac{-1}{r-1}\sigma^2$. [Hint: $\mathrm{Var}(\sum_j U_j) = 0$.]

2. Prove Theorem 15.3b.

3. Derive Theorem 15.8.

4. Find confidence intervals for θ, σ_a^2, $\sigma_a^2 | \sigma_e^2$, $n\sigma_a^2 + \sigma_e^2$, in the one-way random effects model.

5. Derive (15.10) and (15.11).

6. Find the correlations between the Y_{ijk} in the two-way random effects model.

7. Show that there are no nonnegative unbiased estimators σ_a^2, σ_b^2, or σ_d^2 in the two-way random effects model.

8. Show that (F_1, F_2, F_3) defined in (15.12) is a maximal invariant for testing that $\sigma_a^2 = 0$ in the balanced two-way random effects model. Verify (15.14).

9. In the balanced two-way random effects model, find confidence intervals for θ, σ_e^2, $nc\sigma_a^2 + n\sigma_d^2 + \sigma_e^2$, $n\sigma_d^2 + \sigma_e^2$, $\sigma_a^2/(n\sigma_d^2 + \sigma_e^2)$, σ_d^2/σ_e^2.

10. Consider the balanced nested random effects model in which we observe $Y_{ijk} = \theta + a_i + d_{ij} + e_{ijk}$, $i = 1, \ldots, r$; $j = 1, \ldots, c$; $k = 1, \ldots, n$; where a_i, d_{ij}, and e_{ijk} are unobserved independent random variables such that $a_i \sim N_1(0, \sigma_a^2)$, $d_{ij} \sim N_1(0, \sigma_d^2)$, $e_{ijk} \sim N_1(0, \sigma_e^2)$. Find a complete sufficient statistic for this model and find sensible estimators of the parameters θ, σ_a^2, σ_d^2, and σ_e^2. Find a sensible test that $\sigma_a^2 = 0$ and a sensible test that $\sigma_d^2 = 0$. (Note that this model can be thought of as a two-way model in which $\sigma_b^2 = 0$.)

11. Verify Theorem 15.11. (Hint: Use the Z_{ijk} defined in the proof of Theorem 15.9.)

12. Verify (15.19) and show that $F_1 \sim F_{r-1,(r-1)(c-1)}(\delta_1^*)$ for the mixed model.

13. Verify (15.21).

14. Prove Theorem 15.15.

15. Verify (15.22).

Type C

1. In this problem, we derive the simultaneous confidence intervals for the mixed model. We follow the notation of Section 15.3.

(a) Show that

$$\|V - \gamma\|^2 = \sup_{s \neq 0} \frac{[s'(V - \gamma)]^2}{\|s\|^2} = \sup_{\substack{\sum b_i = 0 \\ \sum b_i^2 \neq 0}} \frac{\left[\sum b_i(\bar{Y}_{i\cdot\cdot} - \alpha_i)\right]^2}{\sum b_i^2}$$

[Hint: See Lemma 8.1. Let $b = (b_1, \ldots, b_n)' = C_r s$. As s ranges over R^{r-1}, b ranges over vectors such that $\sum b_i = 0$. Why? Also $C_r' C_r = I$. Therefore $\sum b_i^2 = \|b\|^2 = \|s\|^2$.]

(b) Show that

$$nc(c-1)\frac{\|\mathbf{V} - \gamma\|^2}{S_3^2} \sim F_{r-1,(r-1)(c-1)}(0).$$

(c) Derive the simultaneous confidence intervals given in (15.20).

2. Consider the unbalanced one-way random effects model in which we observe

$$Y_{ij} = \theta + a_i + e_{ij}, \qquad i = 1, \ldots, r; \qquad j = 1, \ldots, n_i;$$

where a_i and e_{ij} are independent random variables such that $a_i \sim N_1(0, \sigma_a^2)$, $e_{ij} \sim N_1(0, \sigma_e^2)$. Let

$$F = \frac{(N-r)\sum_i n_i\left(Y_{i\cdot} - \overline{Y}_{\cdot\cdot}\right)^2}{(r-1)\sum_i\sum_j\left(Y_{ij} - \overline{Y}_{i\cdot}\right)^2}, \qquad \phi(F) = \begin{cases} 1 & \text{if } F > F^{\alpha}_{N-r,r-1} \\ 0 & \text{if } F \leqslant F^{\alpha}_{N-r,r-1} \end{cases}$$

(a) Show that ϕ is a size α test for testing that $\sigma_a^2 = 0$. (Hint: If $\sigma_a^2 = 0$, then the a_i are all 0, and the model is an ordinary linear model.)

(b) Show that ϕ is unbiased. [Hint: Suppose that $\sigma_a^2 \geqslant 0$. Then the a_i are not all equal. Consider the model conditionally on the a_i. Then this model is a one-way fixed effects model. Since the a_i are not all equal, by the corollary to Theorem 7.13, $E\phi\,|\,(a_1, \ldots, a_r) \geqslant \alpha$.]

3. Joint distribution of F_1, F_2, and F_3 for random effects model. We follow the notation of Section 15.2.

(a) Let F_1, F_2, and F_3 be defined by (15.12) and let $F_4 = S_4^2$. Using Theorem 15.9, find the joint density of F_1, F_2, F_3, and F_4.

(b) Find the joint density of F_1, F_2, and F_3.

CHAPTER 16

The Correlation Model

In this chapter we study a variation on the coordinatized version of the linear model in which most of the columns of the basis matrix are random vectors. In particular, we study the model in which we observe $(Y_1, \mathbf{T}_1), \ldots, (Y_n, \mathbf{T}_n)$ independent random vectors, where Y_i is 1×1 and $\mathbf{T}_i$ is $1 \times s$. We assume that $n > s + 1$ and

$$Y_i \mid \mathbf{T}_i \sim N_1(\delta + T_i\gamma, \sigma^2), \qquad \mathbf{T}_i' \sim N_s(\boldsymbol{\nu}, \boldsymbol{\Xi}), \qquad i = 1, \ldots, n, \quad (16.1)$$

(where $\delta \in R^1$, $\gamma \in R^s$, $\boldsymbol{\nu} \in R^s$). The parameter space we consider is given by

$$-\infty < \delta < \infty, \qquad -\infty < \gamma < \infty, \qquad -\infty < \boldsymbol{\nu} < \infty,$$
$$\sigma^2 > 0, \qquad \boldsymbol{\Xi} > 0. \quad (16.2)$$

The model defined by (16.1) and (16.2) is called the *correlation model*. Let

$$\mathbf{Y} = \begin{pmatrix} Y_i \\ \vdots \\ Y_n \end{pmatrix}, \qquad \mathbf{T} = \begin{pmatrix} \mathbf{T}_1 \\ \vdots \\ \mathbf{T}_n \end{pmatrix}, \qquad \mathbf{X} = \begin{pmatrix} 1 & \mathbf{T}_1 \\ \vdots & \vdots \\ 1 & \mathbf{T}_n \end{pmatrix}, \qquad \boldsymbol{\beta} = \begin{pmatrix} \delta \\ \gamma \end{pmatrix}. \quad (16.3)$$

Since $(Y_1, \mathbf{T}_1), \ldots, (Y_n, \mathbf{T}_n)$ are independent,

$$\mathbf{Y} \mid \mathbf{T} \sim N_n(\mathbf{X}\boldsymbol{\beta}, \sigma^2\mathbf{I}), \quad (16.4)$$

and $\mathbf{X}$ is an $n \times (s + 1)$ matrix. In Theorem 17.8b, in the next chapter, we show that

$$P(\mathbf{X} \text{ has rank } s + 1) = 1. \quad (16.5)$$

Throughout this chapter, we assume therefore the $\mathbf{X}$ has rank $s + 1$. Conditional on $\mathbf{T}$, this model is a coordinatized linear model as defined in Chapter 4 (with $p = s + 1$). In this model, our primary goal is to draw inference about γ. In this chapter we show that optimal procedures for this purpose for the ordinary linear model are also optimal for the correlation model.

For a simple example of this model, let Y_i be the income of the ith person, and let $\mathbf{T}_i$ be a vector consisting of such variables as his father's income, his education, his age, and so on. The correlation model would seem to be an appropriate model for this setting. However, if one of the variables in $\mathbf{T}_1$ were the person's sex or race, this model would not be appropriate, since a variable of this type could not have a normal distribution.

In Section 16.1 a complete sufficient statistic for the correlation model is given, as are minimum variance unbiased estimators, and MLE's of the parameters. The minimum variance unbiased estimators and MLE's for $\boldsymbol{\beta}$ and σ^2 are the same in the correlation model as in the coordinatized ordinary linear model that occurs when the correlation model is conditioned on $\mathbf{T}$.

Let $\boldsymbol{\gamma} = \binom{\gamma_1}{\gamma_2}$ where γ_1 is $q \times 1$. In Section 16.2, we consider the problem of testing that $\boldsymbol{\gamma} = \mathbf{0}$ and in Section 16.3, we consider testing that $\gamma_1 = 0$. In Section 16.4, other testing problems concerning $\boldsymbol{\gamma}$ and σ^2 are discussed. For all these problems it is shown that the tests that are UMP invariant size α for testing the problems in the ordinary linear model (that occurs when we assume that $\mathbf{T}$ is fixed) are also UMP invariant size α tests for the correlation model. Although the UMP invariant size α tests are the same for the two models, the groups used to reduce the problems are quite different as are the power functions of the tests. In Section 16.5 we discuss simultaneous confidence intervals for the correlation model. The asymptotic validity of the procedures derived in Sections 16.2–16.5 is discussed in Section 16.6. In Section 16.7 we find a best invariant estimator for $\boldsymbol{\gamma}$. In Section 16.8, other results for the correlation model are discussed.

Throughout this chapter, the goal is to show that optimal procedures for the linear model that occurs when we condition on $\mathbf{T}$ are also optimal procedures for the correlation model. As will be seen, the arguments for this model are much more complicated than those for the linear model. This is not too surprising since the correlation model has additional random vectors (the $\mathbf{T}_i$) and additional parameters ($\boldsymbol{\nu}$ and $\boldsymbol{\Xi}$).

The procedures discussed in this chapter are often discussed from a perspective that is somewhat different from the one presented here. That perspective uses various sorts of correlation coefficients. In Section 16.9 these correlation coefficients are defined, and the results in Sections 16.2 and 16.3 are reinterpreted in terms of these coefficients.

Fisher (1925) was certainly aware of the relationship between the regression approach to this model given in Sections 16.2–16.8 and the correlation approach given in 16.9. He also used multiple and partial condition coefficients and their distributions. The UMP invariance of the test procedures is due to Hunt and Stein (1946). For an interesting expository paper on the relationship between the ordinary linear model and the correlation model, see Sampson (1974).

16.1. SUFFICIENCY AND ESTIMATION.

In this section we find a complete sufficient statistic and estimators for the various parameters in the correlation model. We first find the joint distribution of the random variables involved. Let

$$\mathbf{Z}_i = \begin{pmatrix} Y_i \\ \mathbf{T}_i' \end{pmatrix}. \tag{16.6}$$

LEMMA 16.1. The $\mathbf{Z}_i$ are independent, and

$$\mathbf{Z}_i \sim N_{s+1}\left(\begin{pmatrix} \delta + \boldsymbol{\nu}'\boldsymbol{\gamma} \\ \boldsymbol{\nu} \end{pmatrix}, \begin{pmatrix} \sigma^2 + \boldsymbol{\gamma}'\boldsymbol{\Xi}\boldsymbol{\gamma} & \boldsymbol{\gamma}'\boldsymbol{\Xi} \\ \boldsymbol{\Xi}\boldsymbol{\gamma} & \boldsymbol{\Xi} \end{pmatrix} \right).$$

PROOF. See Exercise B1. □

Therefore, let

$$\boldsymbol{\mu} = \begin{pmatrix} \mu_1 \\ \boldsymbol{\mu}_2 \end{pmatrix} = \begin{pmatrix} \delta + \boldsymbol{\nu}'\boldsymbol{\gamma} \\ \boldsymbol{\nu} \end{pmatrix}, \qquad \boldsymbol{\Sigma} = \begin{pmatrix} \Sigma_{11} & \boldsymbol{\Sigma}_{12} \\ \boldsymbol{\Sigma}_{21} & \boldsymbol{\Sigma}_{22} \end{pmatrix} = \begin{pmatrix} \sigma^2 + \boldsymbol{\gamma}'\boldsymbol{\Xi}\boldsymbol{\gamma} & \boldsymbol{\gamma}'\boldsymbol{\Xi} \\ \boldsymbol{\Xi}\boldsymbol{\gamma} & \boldsymbol{\Xi} \end{pmatrix}, \tag{16.7}$$

where μ_1 and Σ_{11} are 1×1. Then

$$\boldsymbol{\nu} = \boldsymbol{\mu}_2, \qquad \boldsymbol{\gamma} = \boldsymbol{\Sigma}_{22}^{-1}\boldsymbol{\Sigma}_{21}, \qquad \delta = \mu_1 - \boldsymbol{\mu}_2'\boldsymbol{\Sigma}_{22}^{-1}\boldsymbol{\Sigma}_{21}, \qquad \boldsymbol{\Xi} = \boldsymbol{\Sigma}_{22},$$
$$\sigma^2 = \Sigma_{11} - \boldsymbol{\Sigma}_{12}\boldsymbol{\Sigma}_{22}^{-1}\boldsymbol{\Sigma}_{21}. \tag{16.8}$$

Therefore, $(\boldsymbol{\mu}, \boldsymbol{\Sigma})$ is an invertible function of $(\delta, \boldsymbol{\gamma}, \boldsymbol{\nu}, \sigma^2, \boldsymbol{\Xi})$. Hence an equivalent version of the correlation model is one in which we observe $\mathbf{Z}_1, \ldots, \mathbf{Z}_n$ independent, with $\mathbf{Z}_i \sim N_{s+1}(\boldsymbol{\mu}, \boldsymbol{\Sigma})$. This model is the one-sample model discussed in Chapter 18. The following lemma summarizes some results from that chapter that will be used in this chapter. Let

$$\hat{\boldsymbol{\mu}} = \begin{pmatrix} \hat{\mu}_1 \\ \hat{\boldsymbol{\mu}}_1 \end{pmatrix} = \frac{1}{n}\sum_{i=1}^{n} \mathbf{Z}_i = \overline{\mathbf{Z}},$$

$$\hat{\boldsymbol{\Sigma}} = \frac{1}{n-1}\sum_{i=1}^{n} (\mathbf{Z}_i - \overline{\mathbf{Z}})(\mathbf{Z}_i - \overline{\mathbf{Z}})' = \begin{pmatrix} \hat{\Sigma}_{11} & \hat{\boldsymbol{\Sigma}}_{12} \\ \hat{\boldsymbol{\Sigma}}_{21} & \hat{\boldsymbol{\Sigma}}_{22} \end{pmatrix} \tag{16.9}$$

where $\hat{\mu}_1$ and $\hat{\Sigma}_{11}$ are 1×1.

LEMMA 16.2. $(\hat{\boldsymbol{\mu}}, \hat{\boldsymbol{\Sigma}})$ is a complete sufficient statistic for the one sample model, and hence for the correlation model. $\hat{\boldsymbol{\mu}}$ and $\hat{\boldsymbol{\Sigma}}$ are minimum variance unbiased estimators of $\boldsymbol{\mu}$ and $\boldsymbol{\Sigma}$. $\hat{\boldsymbol{\mu}}$ and $[(n-1)/n]\hat{\boldsymbol{\Sigma}}$ are maximum likelihood estimators of $\boldsymbol{\mu}$ and $\boldsymbol{\Sigma}$.

PROOF. See Theorems 18.1, 18.3, and 18.5. □

We now find a complete sufficient statistic that is more natural for the correlation model. Let

$$\hat{\boldsymbol{\beta}} = \begin{pmatrix} \hat{\delta} \\ \hat{\boldsymbol{\gamma}} \end{pmatrix} = (\mathbf{X}'\mathbf{X})^{-1}\mathbf{X}'\mathbf{Y}, \qquad \hat{\sigma}^2 = \frac{1}{n-s-1}\|\mathbf{Y} - \mathbf{X}\hat{\boldsymbol{\beta}}\|^2$$

$$\hat{\boldsymbol{\nu}} = \bar{\mathbf{T}}' \qquad \hat{\boldsymbol{\Xi}} = \frac{1}{n-1}\sum_{i=1}^{n}(\mathbf{T}_i - \bar{\mathbf{T}})'(\mathbf{T}_i - \bar{\mathbf{T}}) \tag{16.10}$$

where $\hat{\delta}$ is 1×1. Note that $\hat{\boldsymbol{\beta}}$ and $\hat{\sigma}^2$ are the usual estimators in the ordinary linear model that occurs when we condition on $\mathbf{T}$. We now use the results of Section 5.5 to give alternative formulas for these statistics. As in that chapter, let $\tilde{\mathbf{Y}}$ and $\tilde{\mathbf{T}}$ be the $n \times 1$ and $n \times s$ matrices given by

$$\tilde{\mathbf{Y}} = \begin{bmatrix} Y_1 - \bar{Y} \\ \vdots \\ Y_n - \bar{Y} \end{bmatrix}, \qquad \tilde{\mathbf{T}} = \begin{bmatrix} \mathbf{T}_1 - \bar{\mathbf{T}} \\ \vdots \\ \mathbf{T}_n - \bar{\mathbf{T}} \end{bmatrix}.$$

Then

$$\hat{\Sigma}_{11} = \frac{1}{n-1}\tilde{\mathbf{Y}}'\tilde{\mathbf{Y}}, \qquad \hat{\boldsymbol{\Sigma}}_{12} = \frac{1}{n-1}\tilde{\mathbf{Y}}'\tilde{\mathbf{T}},$$

$$\hat{\boldsymbol{\Sigma}}_{21} = \frac{1}{n-1}\tilde{\mathbf{T}}'\tilde{\mathbf{Y}}, \qquad \hat{\boldsymbol{\Sigma}}_{22} = \frac{1}{n-1}\tilde{\mathbf{T}}'\tilde{\mathbf{T}}.$$

By Lemma 5.5 we see that

$$\hat{\boldsymbol{\gamma}} = (\tilde{\mathbf{T}}'\tilde{\mathbf{T}})^{-1}\tilde{\mathbf{T}}'\tilde{\mathbf{Y}} = \hat{\boldsymbol{\Sigma}}_{22}^{-1}\hat{\boldsymbol{\Sigma}}_{21}, \qquad \hat{\delta} = \hat{\mu}_1 - \hat{\boldsymbol{\mu}}_2'\hat{\boldsymbol{\gamma}},$$

$$\hat{\sigma}^2 = \left(\frac{n-1}{n-s-1}\right)(\hat{\Sigma}_{11} - \hat{\boldsymbol{\Sigma}}_{12}\hat{\boldsymbol{\Sigma}}_{22}^{-1}\hat{\boldsymbol{\Sigma}}_{21}) \tag{16.11}$$

$$\hat{\boldsymbol{\Xi}} = \hat{\boldsymbol{\Sigma}}_{22}, \qquad \hat{\boldsymbol{\nu}} = \hat{\boldsymbol{\mu}}_2.$$

We are now ready for the main result of this section.

THEOREM 16.3. a. $(\hat{\boldsymbol{\beta}}, \hat{\boldsymbol{\nu}}, \hat{\sigma}^2, \hat{\boldsymbol{\Xi}})$ is a complete sufficient statistic for the correlation model.

b. $\hat{\boldsymbol{\beta}}$, $\hat{\boldsymbol{\nu}}$, $\hat{\sigma}^2$, and $\hat{\boldsymbol{\Xi}}$ are the minimum variance unbiased estimators of $\boldsymbol{\beta}$, $\boldsymbol{\nu}$, σ^2 and $\boldsymbol{\Xi}$.

c. $\hat{\boldsymbol{\beta}}$, $[(n-s-1)/n]\hat{\sigma}^2$, $\hat{\boldsymbol{\nu}}$ and $[(n-1)/n]\hat{\boldsymbol{\Xi}}$ are the MLE's of $\boldsymbol{\beta}$, σ^2, $\boldsymbol{\nu}$, and $\boldsymbol{\Xi}$ for the correlation model.

PROOF. a. By (16.11), $(\hat{\boldsymbol{\beta}}, \hat{\boldsymbol{\nu}}, \hat{\sigma}^2, \hat{\boldsymbol{\Xi}})$ is an invertible function of the complete sufficient statistic $(\hat{\boldsymbol{\mu}}, \hat{\boldsymbol{\Sigma}})$ and is therefore complete and sufficient.

b. By part a, we need only show that the estimators are unbiased. By Lemma 16.2 $\hat{\boldsymbol{\mu}}$ and $\hat{\boldsymbol{\Sigma}}$ are unbiased estimators of $\boldsymbol{\mu}$ and $\boldsymbol{\Sigma}$. Hence $\hat{\boldsymbol{\nu}}$ and $\hat{\boldsymbol{\Xi}}$ are unbiased estimators of $\boldsymbol{\nu}$ and $\boldsymbol{\Xi}$. To show that $\hat{\boldsymbol{\beta}}$ and $\hat{\sigma}^2$ are unbiased, we use the fact that they are unbiased for the ordinary linear model that occurs when the correlation model is conditioned on $\mathbf{T}$. Therefore,

$$E\hat{\boldsymbol{\beta}} = E(E\hat{\boldsymbol{\beta}} \mid \mathbf{T}) = \boldsymbol{\beta}. \qquad E\hat{\sigma}^2 = E(E\hat{\sigma}^2 \mid \mathbf{T}) = \sigma^2.$$

c. By Lemma 16.2 $\hat{\boldsymbol{\mu}}$ and $[(n-1)/n]\hat{\boldsymbol{\Sigma}}$ are maximum likelihood estimators for $\boldsymbol{\mu}$ and $\boldsymbol{\Sigma}$. Therefore, $\hat{\boldsymbol{\nu}}$ and $[(n-1)/n]\hat{\boldsymbol{\Xi}}$ are maximum likelihood estimators of $\boldsymbol{\nu}$ and $\boldsymbol{\Xi}$. By the definition of maximum likelihood estimators, the maximum likelihood estimator of $\boldsymbol{\gamma} = \boldsymbol{\Sigma}_{22}^{-1}\boldsymbol{\Sigma}_{21}$ is

$$\{[(n-1)/n]\hat{\boldsymbol{\Sigma}}_{22}\}^{-1}\{[(n-1)/n]\hat{\boldsymbol{\Sigma}}_{12}\} = \hat{\boldsymbol{\gamma}}.$$

The results for δ and σ^2 follow similarly. □

Note that the minimum variance unbiased estimators and MLE's of $\boldsymbol{\beta}$ and σ^2 are the same for the ordinary linear model and the correlation model.

16.2 TESTING THAT $\boldsymbol{\gamma} = \mathbf{0}$.

We now look at the problem of testing that $\boldsymbol{\gamma} = \mathbf{0}$, against $-\infty < \boldsymbol{\gamma} < \infty$ or equivalently that the Y_i are independent of the $\mathbf{T}_i$. From Corollary 1 to Lemma 7.5, we see that the UMP invariant size α test that $\boldsymbol{\gamma} = \mathbf{0}$ in the ordinary linear model (that occurs when we assume that the $\mathbf{T}_i$ are constant) is given by

$$F_1 = \frac{(n-1)\hat{\boldsymbol{\gamma}}'\hat{\boldsymbol{\Xi}}\hat{\boldsymbol{\gamma}}}{s\hat{\sigma}^2}, \phi_1(F_1) = \begin{cases} 1 & \text{if } F_1 > F^{\alpha}_{s,n-s-1} \\ 0 & \text{if } F_1 \leqslant F^{\alpha}_{s,n-s-1} \end{cases} \tag{16.12}$$

We also note that

$$F_1 \mid \mathbf{T} \sim F_{s,n-s-1}\left[(n-1)\boldsymbol{\gamma}'\hat{\boldsymbol{\Xi}}\boldsymbol{\gamma}/\sigma^2\right] \qquad \text{where } \hat{\boldsymbol{\Xi}}(\mathbf{T}) = \frac{1}{n-1}\tilde{\mathbf{T}}'\tilde{\mathbf{T}} \tag{16.13}$$

and therefore under the null hypothesis $F_1 \sim F_{s,n-s-1}(0)$ and ϕ_1 is hence a size α test for the correlation model. (The alternative distribution of F_1 is not an F-distribution. It is derived in Exercise C1.)

We now show that ϕ_1 is also the UMP invariant size α test for the correlation model. This problem is invariant under the following three groups.

(1) The first group, G_1, consists of transformations

$$g_1(Y_i, \mathbf{T}_i) = (Y_i, \mathbf{T}_i) + (a_1, \mathbf{a}_2)$$

where a_1 and $\mathbf{a}_2$ are arbitrary 1×1 and $1 \times s$ vectors. Let $Y_i^* = Y_i + a_1$, $\mathbf{T}_i^* = \mathbf{T}_i + \mathbf{a}_2$. Then

$$Y_i^* \mid \mathbf{T}_i^* \sim N_1(\delta + a_1 - \mathbf{a}_2\boldsymbol{\gamma} + \mathbf{T}_i\boldsymbol{\gamma}, \sigma^2), \qquad \mathbf{T}^{*\prime}_i \sim N_s(\boldsymbol{\nu} + \mathbf{a}_2', \boldsymbol{\Xi})$$

Therefore,

$$\bar{g}_1(\delta, \boldsymbol{\gamma}, \sigma^2, \boldsymbol{\nu}, \boldsymbol{\Xi}) = (\delta + a_1 - \mathbf{a}_2\boldsymbol{\gamma}, \boldsymbol{\gamma}, \sigma^2, \boldsymbol{\nu} + \mathbf{a}_2', \boldsymbol{\Xi})$$

and hence the problem of testing that $\boldsymbol{\gamma} = \mathbf{0}$ is invariant under G_1.

(2) The second group G_2 consists of the transformations

$$g_2(Y_i, \mathbf{T}_i) = (Y_i, \mathbf{T}_i\mathbf{B})$$

where $\mathbf{B}$ is an arbitrary $s \times s$ invertible matrix. The induced transformation is given by

$$\bar{g}_2(\delta, \boldsymbol{\gamma}, \sigma^2, \boldsymbol{\nu}, \boldsymbol{\Xi}) = (\delta, \mathbf{B}^{-1}\boldsymbol{\gamma}, \sigma^2, \mathbf{B}\boldsymbol{\nu}, \mathbf{B}'\boldsymbol{\Xi}\mathbf{B})$$

and hence leaves the set where $\boldsymbol{\gamma} = \mathbf{0}$ unchanged.

(3) The third group G_3 is given by

$$g_3(y_i, \mathbf{T}_i) = (fy_i, \mathbf{T}_i)$$

where $f > 0$ is a scalar. The induced transformation is given by

$$g_3(\delta, \boldsymbol{\gamma}, \sigma^2, \boldsymbol{\nu}, \boldsymbol{\Xi}) = (f\delta, f\boldsymbol{\gamma}, f^2\sigma^2, \boldsymbol{\nu}, \boldsymbol{\Xi})$$

and hence leaves the set where $\boldsymbol{\gamma} = \mathbf{0}$ unchanged.

We now find a maximal invariant under G_1, G_2, and G_3. Let

$$\theta_1 = \frac{\boldsymbol{\gamma}'\boldsymbol{\Xi}\boldsymbol{\gamma}}{\sigma^2}. \tag{16.14}$$

LEMMA 16.4. A maximal invariant for testing that $\boldsymbol{\gamma} = \mathbf{0}$ is F_1 defined in (16.12) and a parameter maximal invariant is θ_1 defined in (16.14)

PROOF. As in Chapter 7, we first reduce by sufficiency, then by invariance under G_1, then by invariance under G_2 and finally by invariance under G_3. By Lemma 16.2, a sufficient statistic is given by $(\hat{\delta}, \hat{\boldsymbol{\gamma}}, \hat{\sigma}^2, \hat{\boldsymbol{\nu}}, \hat{\boldsymbol{\Xi}})$.

(1) We need to reinterpret G_1 in terms of the sufficient statistic. We note that g_1 does not change $\tilde{Y}$ or $\tilde{\mathbf{T}}$, and hence leaves $\hat{\boldsymbol{\gamma}}$, $\hat{\sigma}^2$ and $\hat{\boldsymbol{\Xi}}$ unchanged. Also g_1 maps $\hat{\delta}$ onto $\hat{\delta} + a_1 - \mathbf{a}_2\hat{\boldsymbol{\gamma}}$ and $\boldsymbol{\nu}_2$ onto $\hat{\boldsymbol{\nu}}_2 + \mathbf{a}_2'$. Therefore g_1 becomes

$$g_1(\hat{\delta}, \hat{\boldsymbol{\gamma}}, \hat{\sigma}^2, \hat{\boldsymbol{\nu}}, \hat{\boldsymbol{\Xi}}) = (\hat{\delta} + a_1 - \mathbf{a}_2\hat{\boldsymbol{\gamma}}, \hat{\boldsymbol{\gamma}}, \hat{\sigma}^2, \hat{\boldsymbol{\nu}} + \mathbf{a}_2', \hat{\boldsymbol{\Xi}}).$$

A maximal invariant under this group is

$$Q_1 = (\hat{\gamma}, \hat{\sigma}^2, \hat{\Xi}).$$

(2) We note that g_2 maps $\tilde{\mathbf{T}}$ onto $\mathbf{B}\tilde{\mathbf{T}}$ and hence maps $\hat{\gamma}$ onto $\mathbf{B}^{-1}\hat{\gamma}$ and maps $\hat{\Xi}$ onto $\mathbf{B}'\hat{\Xi}\mathbf{B}$, while leaving $\hat{\sigma}^2$ unchanged. Therefore g_2 becomes

$$g_2(\hat{\gamma}, \hat{\sigma}^2, \hat{\Xi}) = (\mathbf{B}^{-1}\hat{\gamma}, \hat{\sigma}^2, \mathbf{B}'\hat{\Xi}\mathbf{B}).$$

A maximal invariant under this group is

$$Q_2(\hat{\gamma}\hat{\Xi}\hat{\gamma}, \hat{\sigma}^2)$$

as we now show.

a. It is easily seen that Q_2 is invariant.
b. Suppose that $Q_2(\hat{\gamma}, \hat{\sigma}^2, \hat{\Xi}) = Q_2(\hat{\gamma}^*, \hat{\sigma}^{*2}, \hat{\Xi}')$. Then $\hat{\sigma}^{*2} = \hat{\sigma}^2$ and $\hat{\gamma}'\hat{\Xi}\hat{\gamma} = \hat{\gamma}^{*\prime}\hat{\Xi}^*\hat{\gamma}^*$. By Theorem 12 of the appendix there exists $\mathbf{B}$ invertible such that

$$\hat{\gamma}^* = \mathbf{B}^{-1}\hat{\delta}, (\hat{\Xi}^*)^{-1} = \mathbf{B}^{-1}\hat{\Xi}^{-1}(\mathbf{B}')^{-1}$$

or equivalently, $\hat{\Xi}^* = \mathbf{B}'\hat{\Xi}\mathbf{B}$. Therefore, Q_2 is maximal.

(3) Now g_3 maps $\hat{\gamma}$ onto $f\hat{\gamma}$, $\hat{\Xi}$ onto $\hat{\Xi}$ and $\hat{\sigma}^2$ into $f^2\hat{\sigma}^2$. Therefore, g_3 becomes

$$g_3(\hat{\gamma}'\hat{\Xi}\hat{\gamma}, \hat{\sigma}^2) = (f^2\hat{\gamma}'\hat{\Xi}\hat{\gamma}, f^2\hat{\sigma}^2)$$

and a maximal invariant is F_1.

The derivation for the parameter maximal invariant follows similarly. □

We need one final result before we show that ϕ_1 is UMP invariant size α. Let $f(F_1; \theta_1)$ be the density function of F_1 (which is not an F density function).

LEMMA 16.5. If $\theta_1 > 0$, then $f(F_1; \theta_1)/f(F_1; 0)$ is an increasing function of F_1.

PROOF. Let $g(F_1; a)$ be the density function of a noncentral F-distribution with s and $n - s - 1$ degrees of freedom and noncentrality parameter a. Then

$$f(F_1; \theta_1) = \int \cdots \int g\left(F_1; \frac{(n-1)}{\sigma^2}(\gamma'\hat{\Xi}(\mathbf{T})\gamma)\right) h(\mathbf{T})\, dT_{1,1}\, dT_{1,2} \cdots dT_{n,s}$$

where $h(\mathbf{T})$ is the joint density of $\mathbf{T}$ (see 16.13). Now, let

$$k(F_1, \mathbf{T}) = \frac{g\left(F_1; ((n-1)/\sigma^2)\gamma'\hat{\Xi}(\mathbf{T})\gamma\right)}{f(F_1; 0)}$$

If $\theta_1 > 0$, then $\boldsymbol{\gamma} \neq 0$ and hence $(n-1)\boldsymbol{\gamma}'\hat{\boldsymbol{\Xi}}\boldsymbol{\gamma}/\sigma^2 > 0$ (since $\hat{\boldsymbol{\Xi}} > 0$). Therefore, by Lemma 1.9, $k(F_1, \mathbf{T})$ is an increasing function of F_1 for all $\mathbf{T}$. Hence

$$\frac{f(F_1;\theta_1)}{f(F_1;0)} = \int \cdots \int k(F_1,\mathbf{T})h(\mathbf{T})\,dT_{1,1}, dT_{1,2} \ldots dT_{n,s}$$

is an increasing function of F_1 when $\theta_1 > 0$. □

THEOREM 16.6. The test ϕ_1 defined in (16.12) is the UMP invariant size α test that $\boldsymbol{\gamma} = \mathbf{0}$ against $-\infty < \boldsymbol{\gamma} < \infty$ in the correlation model.

PROOF. We are testing that $\theta_1 = 0$ against $\theta_1 \geqslant 0$. Therefore this follows directly from Lemmas 16.4 and 16.5 and Theorem 1.7. □

We have therefore shown that the UMP invariant size α test that $\boldsymbol{\gamma} = \mathbf{0}$ in the ordinary linear model is also UMP invariant size α for the correlation model. Note, however, that the groups used to reduce the testing problems are different in the two models.

We now consider the special case in which $s = 1$. In that case γ is 1×1 and we consider the one-sided testing problem of testing that $\gamma = 0$ against $\gamma \geqslant 0$. Let

$$t_1 = \frac{\sqrt{n-1}\,\hat{\gamma}\sqrt{\hat{\Xi}}}{\hat{\sigma}}, \qquad \phi_1^*(t_1) = \begin{cases} 1 & \text{if } t_1 > t_{n-2} \\ 0 & \text{if } t_1 \leqslant t_{n-2} \end{cases} \tag{16.15}$$

Then t_1 is the maximal invariant and ϕ_1^* is the UMP invariant size α test for testing that $\gamma = 0$ against $\gamma \geqslant 0$ in the ordinary linear model that occurs when we assume that $\mathbf{T}$ is known (see Exercise B23 of Chapter 7). Since

$$E\phi_1^*(t_1) = E(E\phi_1^*(t_1)\,|\,\mathbf{T})$$

we see that ϕ_1^* is a size α test for the correlation model. We now show that it is UMP invariant size α. We first note that this problem is invariant under the groups G_1 and G_3 defined above Lemma 16.4. It is not invariant under G_2, since -1 is an invertible 1×1 matrix and multiplying γ by -1 changes the alternative set to one on which $\gamma \leqslant 0$. It is however, invariant under the subgroup G_2' of transformations

$$g_2'(Y_i, T_i) = (Y_i, bT_i), \qquad b > 0.$$

Let

$$\xi_1 = \frac{\gamma\sqrt{\Xi}}{\sigma}. \tag{16.16}$$

LEMMA 16.7. A maximal invariant under the groups G_1, G_2', and G_3 is t_1 defined in (16.15) and a parameter maximal invariant is ξ defined in (16.16).

PROOF. After reducing by sufficiency, a maximal invariant under G_1 is $(\hat{\gamma}, \hat{\sigma}^2, \hat{\Xi})$. In terms of this maximal invariant G_2' becomes transformations $g_2'(\hat{\gamma}, \hat{\sigma}^2, \hat{\Xi}) = (b^{-1}\hat{\gamma}, \hat{\sigma}^2, b^2\hat{\Xi})$. Therefore, a maximal invariant under G_2' is $(\gamma\sqrt{\bar{\Xi}}, \sigma^2)$. In terms of this maximal invariant, G_3 becomes transformations of the form $g_3(\hat{\gamma}\sqrt{\hat{\bar{\Xi}}}, \hat{\sigma}^2) = (f\hat{\gamma}\sqrt{\hat{\bar{\Xi}}}, f^2\hat{\sigma}^2)$ and a maximal invariant is t_1. The derivation for the parameter maximal invariant follows similarly. □

LEMMA 16.8. Let $m(t_1; \xi_1)$ be the density of t_1. If $\xi_1 > 0$, then

$$\frac{m(t_1; \xi_1)}{m(t_1; 0)}$$

is an increasing function of t_1.

PROOF. See Exercise B3. □

THEOREM 16.9. The test ϕ_1^* defined in (16.15) is the UMP invariant size α test for testing that $\gamma = 0$ against $\gamma \geqslant 0$ in the correlation model (with $s = 1$).

PROOF. This follows directly from Lemmas 16.7 and 16.8 and Theorem 1.7. □

16.3. TESTING THAT $\gamma_1 = 0$.

Let

$$\boldsymbol{\gamma} = \begin{pmatrix} \boldsymbol{\gamma}_1 \\ \boldsymbol{\gamma}_2 \end{pmatrix}, \qquad \hat{\boldsymbol{\gamma}} = \begin{pmatrix} \hat{\boldsymbol{\gamma}}_1 \\ \hat{\boldsymbol{\gamma}}_2 \end{pmatrix}$$

where $\boldsymbol{\gamma}_1$ and $\hat{\boldsymbol{\gamma}}_1$ are $q \times 1$. We now study the problem of testing that $\boldsymbol{\gamma}_1 = \mathbf{0}$ against the alternative that $\boldsymbol{\gamma}_1$ is unrestricted. We first give the UMP invariant size α test that $\boldsymbol{\gamma}_1 = \mathbf{0}$ for the ordinary linear model that occurs when $\mathbf{T}$ is assumed fixed. Let

$$\mathbf{T}_i = (\mathbf{T}_{i1}, \mathbf{T}_{i2}), \qquad \tilde{\mathbf{T}} = (\tilde{\mathbf{T}}_1, \tilde{\mathbf{T}}_2),$$

$$\hat{\boldsymbol{\Xi}} = \begin{pmatrix} \hat{\boldsymbol{\Xi}}_{11} & \hat{\boldsymbol{\Xi}}_{12} \\ \hat{\boldsymbol{\Xi}}_{21} & \hat{\boldsymbol{\Xi}}_{22} \end{pmatrix} = \frac{1}{n-1} \begin{pmatrix} \tilde{\mathbf{T}}_1'\tilde{\mathbf{T}}_1 & \tilde{\mathbf{T}}_1'\tilde{\mathbf{T}}_2 \\ \tilde{\mathbf{T}}_2'\tilde{\mathbf{T}}_1 & \tilde{\mathbf{T}}_2'\tilde{\mathbf{T}}_2 \end{pmatrix} \tag{16.17}$$

where $\mathbf{T}_{i1}$ is $1 \times q$, $\tilde{\mathbf{T}}_1$ is $n \times q$ and $\hat{\boldsymbol{\Xi}}_{11}$ is $q \times q$. Define

$$F_2 = \frac{(n-1)\hat{\boldsymbol{\gamma}}_1'(\hat{\boldsymbol{\Xi}}_{11} - \hat{\boldsymbol{\Xi}}_{12}\hat{\boldsymbol{\Xi}}_{22}^{-1}\hat{\boldsymbol{\Xi}}_{21})\hat{\boldsymbol{\gamma}}_1}{q\hat{\sigma}^2},$$

$$\phi_2(F_2) = \begin{cases} 1 & \text{if } F_2 > F^{\alpha}_{q,n-s-1} \\ 0 & \text{if } F_2 \leqslant F^{\alpha}_{q,n-s-1} \end{cases}. \tag{16.18}$$

By Corollary 2 to Lemma 7.5, (with $p = s + 1$, $k = s - q + 1$) ϕ_2 is the UMP invariant size α test for testing that $\gamma_1 = \mathbf{0}$ for the ordinary linear model. Also

$$F_2 | \mathbf{T} \sim F_{q,n-s-1}\left[\frac{(n-1)\gamma_1'(\hat{\boldsymbol{\Xi}}_{11} - \hat{\boldsymbol{\Xi}}_{12}\hat{\boldsymbol{\Xi}}_{22}^{-1}\hat{\boldsymbol{\Xi}}_{22})\gamma_1}{\sigma^2}\right] \tag{16.19}$$

Therefore, if $\gamma_1 = 0$, $F_2 \sim F_{q,n-s-1}(0)$, and hence ϕ_2 is a size α test for the correlation model. The alternative distribution of F is not, however, a noncentral F distribution. We now show that ϕ_2 is also the UMP invariant size α test for testing that $\gamma_1 = \mathbf{0}$ for the correlation model.

We define six groups that leave this problem invariant. For each group, the reader should verify that the parameters transform as stated. It is easier to use the joint distribution of Y_i and $\mathbf{T}_i$ defined in (16.1) than to use the distribution given in Lemma 16.1.

(1) The first group G_1 of the transformations, g_1, is given by

$$g_1(Y_i, \mathbf{T}_{i1}, \mathbf{T}_{i2}) = (Y_i, \mathbf{T}_{i1}, \mathbf{T}_{i2}) + (a_1, \mathbf{a}_2)$$

where a_1 and $\mathbf{a}_2$ are arbitrary 1×1 and $1 \times s$ vectors. Then,

$$\bar{g}_1(\delta, \gamma, \sigma^2, \boldsymbol{\nu}, \boldsymbol{\Xi}) = (\delta + a_1 - \mathbf{a}_2\gamma, \gamma, \sigma^2, \boldsymbol{\nu} + \mathbf{a}_2', \boldsymbol{\Xi}),$$

and hence the set where $\gamma_1 = \mathbf{0}$ is unchanged by this transformation.

(2) The group G_2 consists of transformations of the form

$$g_2(Y_i, \mathbf{T}_{i1}, \mathbf{T}_{i2}) = (Y_i + \mathbf{T}_{i2}\mathbf{B}, \mathbf{T}_{i1}, \mathbf{T}_{i2})$$

where $\mathbf{B}$ is an arbitrary $(s - q) \times 1$ vector. To see how the parameters transform, let $Y_i^* = Y_i + \mathbf{T}_{i2}\mathbf{B}, \mathbf{T}_{i1}^* = \mathbf{T}_{i1}, \mathbf{T}_{i2}^* = \mathbf{T}_{i2}$. Then

$$Y_i^* | (\mathbf{T}_{i1}^*, \mathbf{T}_{i2}^*) \sim N_1(\delta + \mathbf{T}_{i1}^*\gamma_1 + \mathbf{T}_{i2}^*(\gamma_2 + \mathbf{B}), \sigma^2),$$

$$\begin{pmatrix} \mathbf{T}_{i1}^{*\prime} \\ \mathbf{T}_{i2}^{*\prime} \end{pmatrix} \sim N_{p-1}(\boldsymbol{\nu}, \boldsymbol{\Xi}).$$

Hence, $\bar{g}_2(\delta, \gamma_1, \gamma_2, \sigma^2, \boldsymbol{\nu}, \boldsymbol{\Xi}) = (\delta, \gamma_1, \gamma_2 + \mathbf{B}, \sigma^2, \boldsymbol{\nu}, \boldsymbol{\Xi})$ and the set where $\gamma_1 = \mathbf{0}$ is unchanged.

(3) G_3 consists of transformations of the form

$$g_3(Y_i, \mathbf{T}_{i1}, \mathbf{T}_{i2}) = (Y_i, \mathbf{T}_{i1} + \mathbf{T}_{i2}\mathbf{C}, \mathbf{T}_{i2})$$

where $\mathbf{C}$ is an arbitrary $(s - q) \times q$ matrix. Let

$$Y_i^* = Y_i, \mathbf{T}_{i1}^* = \mathbf{T}_{i1} + \mathbf{T}_{i2}\mathbf{C}, \mathbf{T}_{i2}^* = \mathbf{T}_{i2}.$$

Then

$$Y_i^* | (\mathbf{T}_{i1}^*, \mathbf{T}_{i2}^*) \sim N_1[\delta + \mathbf{T}_{i1}^*\gamma_1 + \mathbf{T}_{i2}^*(\gamma_2 - \mathbf{C}\gamma_1), \sigma^2],$$

$$\begin{pmatrix} \mathbf{T}_{i1}^{*\prime} \\ \mathbf{T}_{i2}^{*\prime} \end{pmatrix} \sim N_s\left[\begin{pmatrix} \mathbf{I} & \mathbf{C}' \\ \mathbf{0} & \mathbf{I} \end{pmatrix}\boldsymbol{\nu}, \quad \begin{pmatrix} \mathbf{I} & \mathbf{C}' \\ \mathbf{0} & \mathbf{I} \end{pmatrix}\boldsymbol{\Xi}\begin{pmatrix} \mathbf{I} & \mathbf{0} \\ \mathbf{C} & \mathbf{I} \end{pmatrix}\right].$$

Hence

$$\bar{g}_3(\delta, \gamma_1, \gamma_2, \sigma^2, \boldsymbol{\nu}, \boldsymbol{\Xi}) = \left[\delta, \gamma_1, \gamma_2 - \mathbf{C}\gamma_1, \begin{pmatrix} \mathbf{I} & \mathbf{C}' \\ \mathbf{0} & \mathbf{I} \end{pmatrix}\boldsymbol{\nu}, \begin{pmatrix} \mathbf{I} & \mathbf{C}' \\ \mathbf{0} & \mathbf{I} \end{pmatrix}\boldsymbol{\Xi}\begin{pmatrix} \mathbf{I} & \mathbf{0} \\ \mathbf{C} & \mathbf{I} \end{pmatrix}\right].$$

which again leaves the set in which $\gamma_1 = 0$ unchanged.

(4) G_4 consists of transformations of the form

$$g_4(Y_i, \mathbf{T}_{i1}, \mathbf{T}_{i2}) = (Y_i, \mathbf{T}_{i1}, \mathbf{T}_{i2}\mathbf{D})$$

where $\mathbf{D}$ is an arbitrary $(s-q) \times (s-q)$ invertible matrix, $\bar{g}_4$ maps γ_2 onto $\mathbf{D}^{-1}\gamma_2$, $\boldsymbol{\nu}$ onto $(\begin{smallmatrix} \mathbf{I} & \mathbf{0}' \\ \mathbf{0} & \mathbf{D} \end{smallmatrix})\boldsymbol{\nu}$ and $\boldsymbol{\Xi}$ onto $(\begin{smallmatrix} \mathbf{I} & \mathbf{0}' \\ \mathbf{0} & \mathbf{D} \end{smallmatrix})\boldsymbol{\Xi}(\begin{smallmatrix} \mathbf{I} & \mathbf{0} \\ \mathbf{0} & \mathbf{D} \end{smallmatrix})$ and hence leaves the problem invariant.

(5) G_5 consists of transformations of the form

$$g_5(Y_i, \mathbf{T}_{i1}, \mathbf{T}_{i2}) = (Y_i, \mathbf{T}_{i1}\mathbf{E}, \mathbf{T}_{i2})$$

where $\mathbf{E}$ is an arbitrary $q \times q$ invertible matrix. Then $\overline{g_5}$ maps γ_1 onto $\mathbf{E}^{-1}\gamma_1, \boldsymbol{\nu}$ onto $(\begin{smallmatrix} \mathbf{E}' & \mathbf{0} \\ \mathbf{0} & \mathbf{I} \end{smallmatrix})\boldsymbol{\nu}$ and $\boldsymbol{\Xi}$ onto $(\begin{smallmatrix} \mathbf{E}' & \mathbf{0} \\ \mathbf{0} & \mathbf{I} \end{smallmatrix})\boldsymbol{\Xi}(\begin{smallmatrix} \mathbf{E} & \mathbf{0} \\ \mathbf{0} & \mathbf{I} \end{smallmatrix})$. Note that this is the first transformation that changes γ_1. However, it leaves the set in which $\gamma_1 = \mathbf{0}$ unchanged, and hence leaves the problem invariant.

(6) The last group, G_6, consists of transformations of the form

$$g_6(Y_i, \mathbf{T}_{i1}, \mathbf{T}_{i2}) = (fY_i, \mathbf{T}_{i1}, \mathbf{T}_{i2}), \qquad f > 0.$$

then $\overline{g_6}$ maps δ onto $f\delta$, γ onto $f\gamma$, σ^2 onto $f^2\sigma^2$, and hence leaves the problem invariant.

Therefore, the problem of testing that $\gamma_1 = \mathbf{0}$ for the correlation model is invariant under each of the six groups given above. [Note that is would not, for example, be invariant under the group $g_7(Y_i, T_{i1}, T_{i2}) = (Y_i + \mathbf{T}_{i1}\mathbf{H}, \mathbf{T}_{i1}, \mathbf{T}_{i2})$, because the induced transformations $\bar{g}_7$ would map γ_1 onto $\gamma_1 + \mathbf{H}$, and would therefore change the null set.] The groups for this problem are more complicated than for the ordinary linear model, which is not surprising, since the correlation model has more random variables and more parameters. Invariance under the six groups given above is equivalent to invariance under the single group G of transformations of the following form

$$g(Y_i, \mathbf{T}_{i1}, \mathbf{T}_{i2}) = (Y_i, \mathbf{T}_{i1}, \mathbf{T}_{i2})\begin{bmatrix} f & \mathbf{0} & \mathbf{0} \\ \mathbf{0} & \mathbf{E} & \mathbf{0} \\ \mathbf{B} & \mathbf{C} & \mathbf{D} \end{bmatrix} + (a_1, \mathbf{a}_2)$$

It is much easier, however, to reduce by each of the groups $G_1, \ldots, G_6$ separately than to reduce by the whole group G. Now let

$$\theta_2 = \frac{\gamma_1'(\boldsymbol{\Xi}_{11} - \boldsymbol{\Xi}_{12}\boldsymbol{\Xi}_{22}^{-1}\boldsymbol{\Xi}_{21})\gamma_1}{\sigma^2}. \qquad (16.20)$$

LEMMA 16.10. F_2 defined by (16.18) is a maximal invariant and θ_2 defined by (16.20) is a parameter maximal invariant under the groups $G_1, \ldots, G_6$.

PROOF. As before, we first reduce the problem by sufficiency, and then by each of the groups, getting a smaller maximal invariant at each stage, until we get to F after reducing by G_6. By Lemma 16.2, $(\hat{\delta}, \hat{\gamma}, \hat{\sigma}^2, \hat{\boldsymbol{\nu}}, \hat{\boldsymbol{\Xi}})$ is a sufficient statistic.

(1) As before, g_1 becomes

$$g_1(\hat{\delta}, \hat{\gamma}, \hat{\sigma}^2, \hat{\boldsymbol{\nu}}, \boldsymbol{\Xi}) = (\hat{\delta} + a_1 - \mathbf{a}_2\hat{\gamma}, \hat{\gamma}, \hat{\sigma}^2, \hat{\nu} + \mathbf{a}_2', \hat{\boldsymbol{\Xi}})$$

where a_1 and $\mathbf{a}_2$ are arbitrary. A maximal invariant for this group is

$$Q_1 = (\hat{\gamma}, \hat{\sigma}^2, \hat{\boldsymbol{\Xi}}).$$

(2) We note that $g_2(\tilde{\mathbf{Y}}, \tilde{\mathbf{T}}) = (\tilde{\mathbf{Y}} + \tilde{\mathbf{T}}_2\mathbf{B}, \tilde{\mathbf{T}})$. Therefore, $\hat{\gamma} = (\tilde{\mathbf{T}}', \tilde{\mathbf{T}})^{-1}\tilde{\mathbf{T}}'\tilde{\mathbf{Y}}$ is mapped onto $(\tilde{\mathbf{T}}'\tilde{\mathbf{T}})^{-1}\tilde{\mathbf{T}}'(\tilde{\mathbf{Y}} + \tilde{\mathbf{T}}_2\mathbf{B}) = \hat{\gamma} + (\tilde{\mathbf{T}}'\tilde{\mathbf{T}})^{-1}\tilde{\mathbf{T}}'\tilde{\mathbf{T}}_2\mathbf{B}$. Also $(\tilde{\mathbf{T}}'\tilde{\mathbf{T}})^{-1}\tilde{\mathbf{T}}'\tilde{\mathbf{T}} = \mathbf{I}$, and hence $(\tilde{\mathbf{T}}'\tilde{\mathbf{T}})^{-1}\tilde{\mathbf{T}}'\tilde{\mathbf{T}}_2 = (\mathbf{0}, \mathbf{I})$. Therefore, $\hat{\gamma}_1$ maps onto $\hat{\gamma}_1$ and $\hat{\gamma}_2$ map onto $\hat{\gamma}_2 + \mathbf{B}$. Also $(n - s - 1)\hat{\sigma}^2 = \| \tilde{Y} - \tilde{\mathbf{T}}\hat{\gamma}\|^2$ maps onto $\|\tilde{\mathbf{Y}} + \tilde{\mathbf{T}}_2\mathbf{B} - (\tilde{\mathbf{T}}\hat{\gamma} + \tilde{\mathbf{T}}_2\mathbf{B})\|^2 = (n - s - 1)\hat{\sigma}^2$, and hence $\hat{\sigma}^2$ is mapped onto $\hat{\sigma}^2$. Clearly $\hat{\boldsymbol{\Xi}}$ is mapped on $\hat{\boldsymbol{\Xi}}$. Therefore,

$$g_2(Q_1) = g_2(\hat{\gamma}_1, \hat{\gamma}_2, \hat{\sigma}^2, \hat{\boldsymbol{\Xi}}) = (\hat{\gamma}_1, \hat{\gamma}_2 + \mathbf{B}, \hat{\sigma}^2, \hat{\boldsymbol{\Xi}}).$$

A maximal invariant under this group is

$$Q_2 = (\hat{\gamma}_1, \hat{\sigma}^2, \hat{\boldsymbol{\Xi}}).$$

(3) This is the most interesting reduction. It can be verified g_3 has the form

$$g_3(Q_2) = g_3(\hat{\gamma}_1, \hat{\sigma}^2, \hat{\boldsymbol{\Xi}}) = \left[\hat{\gamma}_1, \hat{\sigma}^2, \begin{pmatrix} \mathbf{I} & \mathbf{C}' \\ \mathbf{0} & \mathbf{I} \end{pmatrix} \hat{\boldsymbol{\Xi}} \begin{pmatrix} \mathbf{I} & \mathbf{0} \\ \mathbf{C} & \mathbf{I} \end{pmatrix}\right].$$

(See Exercise B5.) Now

$$\begin{pmatrix} \mathbf{I} & \mathbf{C}' \\ \mathbf{0} & \mathbf{I} \end{pmatrix} \begin{pmatrix} \hat{\boldsymbol{\Xi}}_{11} & \hat{\boldsymbol{\Xi}}_{12} \\ \hat{\boldsymbol{\Xi}}_{21} & \hat{\boldsymbol{\Xi}}_{22} \end{pmatrix} \begin{pmatrix} \mathbf{I} & \mathbf{0} \\ \mathbf{C} & \mathbf{I} \end{pmatrix}$$

$$= \begin{pmatrix} \hat{\boldsymbol{\Xi}}_{11} + \mathbf{C}'\hat{\boldsymbol{\Xi}}_{21} + \hat{\boldsymbol{\Xi}}_{12}\mathbf{C} + \mathbf{C}'\hat{\boldsymbol{\Xi}}_{22}\mathbf{C} & \hat{\boldsymbol{\Xi}}_{12} + C'\hat{\boldsymbol{\Xi}}_{22} \\ \hat{\boldsymbol{\Xi}}_{21} + \hat{\boldsymbol{\Xi}}_{22}\mathbf{C} & \hat{\boldsymbol{\Xi}}_{22} \end{pmatrix}.$$

Let $\mathbf{U} = \hat{\boldsymbol{\Xi}}_{11} - \hat{\boldsymbol{\Xi}}_{12}\hat{\boldsymbol{\Xi}}_{22}^{-1}\hat{\boldsymbol{\Xi}}_{21}$ and

$$Q_3 = (\hat{\gamma}_1, \hat{\sigma}^2, \mathbf{U}, \hat{\boldsymbol{\Xi}}_{22}).$$

We now show that Q_3 is a maximal invariant under G_3.

a. It is easily verified that Q_3 is invariant.

b. To establish maximality, suppose that

$$\hat{\Xi}^* = \begin{pmatrix} \hat{\Xi}_{11}^* & \hat{\Xi}_{12}^* \\ \hat{\Xi}_{21}^* & \hat{\Xi}_{22}^* \end{pmatrix}, \qquad \mathbf{U}^* = \hat{\Xi}_{11}^* - \hat{\Xi}_{12}^*(\hat{\Xi}_{22}^*)^{-1}\hat{\Xi}_{21}^*$$

and that

$$Q_3(\hat{\gamma}_1, \hat{\sigma}^2, \hat{\Xi}) = Q_3(\hat{\gamma}_1^*, \hat{\sigma}^{*2}, \hat{\Xi}^*).$$

Then $\hat{\gamma}_1^* = \hat{\gamma}_1$, $\hat{\sigma}^{*2} = \hat{\sigma}^2$, $\hat{\Xi}_{22}^* = \hat{\Xi}_{22}$. Let $\mathbf{C}$ satisfy

$$\hat{\Xi}_{21}^* = \hat{\Xi}_{21} + \hat{\Xi}_{22}\mathbf{C} \qquad \left[\text{i.e., let } \mathbf{C} = \hat{\Xi}_{22}^{-1}(\hat{\Xi}_{21}^* - \hat{\Xi}_{21})\right].$$

Then (since $\mathbf{U} = \mathbf{U}^*$, and $\hat{\Xi}_{12}^* = \hat{\Xi}_{21}^{*\prime}$)

$$\begin{aligned} \hat{\Xi}_{11}^* &= \hat{\Xi}_{11} - \hat{\Xi}_{12}\hat{\Xi}_{22}^{-1}\hat{\Xi}_{21} + \hat{\Xi}_{12}^*(\hat{\Xi}_{22}^*)^{-1}\hat{\Xi}_{21}^* \\ &= \hat{\Xi}_{11} - \hat{\Xi}_{12}\hat{\Xi}_{22}^{-1}\hat{\Xi}_{21} + (\hat{\Xi}_{12} + \mathbf{C}'\hat{\Xi}_{22})\hat{\Xi}_{22}^{-1}(\hat{\Xi}_{21} + \hat{\Xi}_{22}\mathbf{C}) \\ &= \hat{\Xi}_{11} + \mathbf{C}'\hat{\Xi}_{21} + \hat{\Xi}_{12}\mathbf{C} + \mathbf{C}'\hat{\Xi}_{22}\mathbf{C}. \end{aligned}$$

Therefore $(\hat{\gamma}_1^*, \hat{\sigma}^{*2}, \hat{\Xi}^*) = g_3(\hat{\gamma}_1, \hat{\sigma}^2, \hat{\Xi})$, and Q_3 is maximal.

(4) G_4 becomes the set of transformations of the form

$$g_4(\hat{\gamma}_1, \hat{\sigma}^2, \hat{\Xi}) = \left(\hat{\gamma}_1, \hat{\sigma}^2, \begin{pmatrix} \mathbf{I} & \mathbf{0} \\ \mathbf{0} & \mathbf{D}' \end{pmatrix}\hat{\Xi}\begin{pmatrix} \mathbf{I} & \mathbf{0} \\ \mathbf{0} & \mathbf{D} \end{pmatrix}\right).$$

Hence

$$g_4(Q_3) = g_4(\hat{\gamma}_1, \hat{\sigma}^2, \mathbf{U}, \hat{\Xi}_{22}) = (\hat{\gamma}_1, \hat{\sigma}^2, \mathbf{U}, \mathbf{D}'\hat{\Xi}_{22}\mathbf{D}).$$

Let

$$Q_4 = (\hat{\gamma}_1, \hat{\sigma}^2, \mathbf{U}).$$

We now show that Q_4 is a maximal invariant under G_1.

a. Q_4 is clearly invariant.

b. Suppose that $Q_4(\hat{\gamma}_1, \hat{\sigma}^2, \mathbf{U}, \hat{\Xi}_{22}) = Q_4(\hat{\gamma}_1^*, \hat{\sigma}^{*2}, \mathbf{U}^*, \hat{\Xi}_{22}^*)$. Then $\hat{\gamma}_1^* = \hat{\gamma}_1$, $\hat{\sigma}^{*2} = \hat{\sigma}^2$, $\mathbf{U}^* = \mathbf{U}$. Since $\hat{\Xi}_{22}$ and $\hat{\Xi}_{22}^*$ are positive definite, there exist positive definite matrices $\mathbf{R}$ and $\mathbf{R}^*$ such that $\hat{\Xi}_{22} = \mathbf{R}^2 = \mathbf{R}\mathbf{R}'$ and $\hat{\Xi}_{22}^* = (\mathbf{R}^*)^2 = \mathbf{R}^*\mathbf{R}^{*\prime}$. Hence $\hat{\Xi}_{22}^* = \mathbf{R}^*\mathbf{R}^{-1}\hat{\Xi}_{22}(\mathbf{R}^{-1})'\mathbf{R}^{*\prime}$. Therefore, $(\hat{\gamma}_1^*, \hat{\sigma}^{*2}, \mathbf{U}^*, \hat{\Xi}_{22}^*) = g_4(\hat{\gamma}_1, \hat{\sigma}^2, \mathbf{U}, \hat{\Xi}_{22})$, and Q_4 is maximal.

(5) G_5 becomes transformations of the form

$$g_5(\hat{\gamma}_1, \hat{\sigma}^2, \mathbf{U}) = (\mathbf{E}^{-1}\hat{\gamma}_1, \hat{\sigma}^2, \mathbf{E}'\mathbf{U}\mathbf{E}).$$

(See Exercise B5.) Let

$$Q_5 = (\hat{\gamma}_1'\mathbf{U}\hat{\gamma}_1, \hat{\sigma}^2).$$

We now show Q_5 is a maximal invariant under G_5.

a. Q_5 is invariant.

b. Suppose that $Q_5(\hat{\gamma}_1, \hat{\sigma}^2, \mathbf{U}) = Q_5(\gamma_1^*, \hat{\sigma}^{*2}, \mathbf{U}^*)$. Then $\hat{\sigma}^2 = \hat{\sigma}^{*2}$ and

$$\hat{\gamma}_1'(\mathbf{U}^{-1})^{-1}\hat{\gamma}_1 = \hat{\gamma}^{*\prime}_1(\mathbf{U}^{*-1})^{-1}\hat{\gamma}_1^*.$$

$\mathbf{U} > 0$ and $\mathbf{U}^* > 0$ and hence $\mathbf{U}^{-1} > 0$ and $\mathbf{U}^{*-1} > 0$. (See Theorems 5 and 6 of the appendix). Therefore, by Theorem 12 of the appendix, there exists an invertible matrix $\mathbf{A}$ such that $\hat{\gamma}_1^* = \mathbf{A}\hat{\gamma}_1, \mathbf{U}^{*-1} = \mathbf{A}\mathbf{U}^{-1}\mathbf{A}'$. Now, let $\mathbf{E} = \mathbf{A}^{-1}$. Then $\hat{\gamma}_1^* = \mathbf{E}^{-1}\hat{\gamma}_1$ and $\mathbf{U}^* = \mathbf{E}'\mathbf{U}\mathbf{E}$. Hence Q_5 is maximal.

(6) G_6 becomes the group of transformations

$$g_6(\hat{\gamma}_1'\mathbf{U}\hat{\gamma}_1, \hat{\sigma}^2) = (f^2\hat{\gamma}_1\mathbf{U}'\hat{\gamma}_1, f^2\hat{\sigma}^2).$$

A maximal invariant under this group is

$$F_2 = \frac{(n-1)\hat{\gamma}_1'\mathbf{U}\hat{\gamma}_1}{q\hat{\sigma}^2}.$$

We have now shown that F_2 is a maximal invariant. The proof that θ_2 is a parameter maximal invariant follows similarly. □

We are now ready for the main result of this section.

THEOREM 16.11. The test ϕ_2 defined in (16.19) is the UMP invariant size α test for testing that $\gamma_1 = \mathbf{0}$ for the correlation model.

PROOF. Let $f(F_2; \theta_2)$ be the density of F_2. By a proof similar to that of Lemma 16.5, we can show that $f(F_2; \theta)/f(F_2; 0)$ is an increasing function of F_2 when $\theta_2 > 0$. Therefore, by Lemma 16.10 and Theorem 1.7 the test ϕ_2 is UMP among all invariant size α tests. □

Now suppose that $q = 1$, that is, that γ_1 is 1×1. We now consider testing that $\gamma_1 = 0$ against the one-sided alternative that $\gamma_1 \geqslant 0$. Let

$$t_2 = \frac{\sqrt{n-1}\,\hat{\gamma}_1\sqrt{\hat{\Xi}_{11} - \hat{\Xi}_{12}\hat{\Xi}_{22}^{-1}\hat{\Xi}_{21}}}{\hat{\sigma}}, \qquad \phi_2^*(t_2) = \begin{cases} 1 & \text{if } t_2 > t_{n-s-1}^{\alpha} \\ 0 & \text{if } t_2 \leqslant t_{n-s-1}^{\alpha} \end{cases}. \tag{16.21}$$

ϕ_2^* is the UMP invariant size α test for testing that $\gamma_1 = 0$ against $\gamma_1 > 0$ for the ordinary linear model (see Exercise B23 of Chapter 7). By arguments analogous to those in previous sections, ϕ_2^* has size α for the correlation model. We now show that ϕ_2^* is UMP invariant size α for the correlation model.

We note first that the one-sided problem is invariant under the groups $G_1, \ldots, G_4$ defined in the last section, since for those groups the induced transformations on the parameter space do not involve γ_1. Also, the one-sided problem is invariant under G_6, since $\gamma_1 \geqslant 0$ if and only if $f\gamma_1 \geqslant 0$.

The one-sided problem is *not* invariant under G_5 (since if $E < 0$, then the set in which $\gamma_1 \geqslant 0$ is transformed to the set in which $\gamma_1 \leqslant 0$). It is however invariant under the subgroup G_5' consisting of transformations of the form

$$g_5'(Y_i, T_{i1}, T_{i2}) = (Y_i, eT_{i1}, \mathbf{T}_{i2})$$

where $e > 0$. (Note that $e \in R^1$.) Let

$$\xi_2 = \frac{\gamma_1\sqrt{\Xi_{11} - \Xi_{12}\Xi_{22}^{-1}\Xi_{21}}}{\sigma}. \tag{16.22}$$

LEMMA 16.12. A maximal invariant under $G_1, \ldots, G_4, G_5', G_6$ is t defined in (16.21). A parameter maximal invariant is ξ defined in (16.22).

PROOF. From the proof of Lemma 16.10 we see that a maximal invariant under G_1, G_2, G_3, and G_4 is

$$Q_4 = (\hat{\gamma}_1, \hat{\sigma}^2, U).$$

Now G_5' operates in the following manner:

$$g_5'(\hat{\gamma}_1, \hat{\sigma}^2, U) = (e^{-1}\hat{\gamma}_1, \hat{\sigma}^2, e^2U)$$

and a maximal invariant is

$$Q_5' = (\hat{\gamma}_1\sqrt{U}, \hat{\sigma}^2).$$

(Note that $e > 0$.) G_6 operates in the following fashion:

$$g_6(\hat{\gamma}_1\sqrt{U}, \hat{\sigma}^2) = (f\hat{\gamma}_1\sqrt{U}, f^2\hat{\sigma}^2)$$

and a maximal invariant is

$$t_2 = \frac{\sqrt{n-1}\,\hat{\gamma}_1\sqrt{U}}{\hat{\sigma}}.$$

The proof for the parameter maximal invariant follows similarly. □

THEOREM 16.13. The test ϕ_2^* defined by (16.21) is UMP invariant size α for testing that $\gamma_1 = 0$ against $\gamma_1 \geqslant 0$ (when $q = 1$) under the groups $G_1, \ldots, G_4, G_5', G_6$.

PROOF. Let $m(t_2; \xi_2)$ be the density of t_2. By a proof similar to that of Lemma 16.8, $m(t_2; \xi_2)/m(t_2; 0)$ is an increasing function of t_2 for all $\xi_2 > 0$. The result therefore follows from Lemma 16.12 and Theorem 1.7. □

16.4. TESTING OTHER HYPOTHESES.

We now consider testing that $\mathbf{A}\boldsymbol{\gamma} = \mathbf{0}$ against $\boldsymbol{\gamma}$ unrestricted, where $\mathbf{A}$ is a known $q \times s$ matrix of rank q. We show for this problem as for previous problems, the UMP invariant size α test for the ordinary linear model is also UMP invariant size α for the correlation model. We accomplish this by transforming the correlation model to another correlation model in which we are testing a hypotheses of the form considered in Section 16.3.

Let $\mathbf{B}$ be an $(s-q)\times s$ dimensional matrix such that

$$\mathbf{C}=\begin{pmatrix}\mathbf{A}\\ \mathbf{B}\end{pmatrix}$$

is invertible. This is possible because $\mathbf{A}$ has full rank. Define

$$\mathbf{T}_i^*=\mathbf{T}_i\mathbf{C}^{-1},\qquad \boldsymbol{\gamma}^*=\begin{pmatrix}\boldsymbol{\gamma}_1^*\\ \boldsymbol{\gamma}_2^*\end{pmatrix}=\mathbf{C}\boldsymbol{\gamma}=\begin{pmatrix}\mathbf{A}\boldsymbol{\gamma}\\ \mathbf{B}\boldsymbol{\gamma}\end{pmatrix},\qquad \boldsymbol{\nu}^*=(\mathbf{C}^{-1})'\boldsymbol{\nu},$$

$$\boldsymbol{\Xi}^*=(\mathbf{C}^{-1})'\boldsymbol{\Xi}\mathbf{C}^{-1}.$$

Then the mapping from $(Y_i,\mathbf{T}_i)$ to $(Y_i,\mathbf{T}_i^*)$ is invertible, so that the model in which we observe $(Y_i,\mathbf{T}_i^*)$ is just a transformation of the model in which we observe (Y_i,T_i). In addition, $(Y_1,\mathbf{T}_1^*),\ldots,(Y_n,\mathbf{T}_n^*)$ are independent and

$$Y_i\mid\mathbf{T}_i^*\sim N_1(\delta+\mathbf{T}_i^*\boldsymbol{\gamma}^*,\sigma^2),\qquad \mathbf{T}_i^{*\prime}\sim N_s(\boldsymbol{\nu}^*,\boldsymbol{\Xi}^*).$$

Therefore, the model in which we observe $(Y_i,\mathbf{T}_i^*)$ is also a correlation model. Furthermore, testing that $\mathbf{A}\boldsymbol{\gamma}=\mathbf{0}$ in the original correlation model is equivalent to testing that $\boldsymbol{\gamma}_1^*=\mathbf{0}$ in the transformed correlation model. Define

$$\tilde{\mathbf{T}}^*=(\tilde{\mathbf{T}}_1^*,\tilde{\mathbf{T}}_2^*)=\begin{bmatrix}\mathbf{T}_1^*-\bar{\mathbf{T}}^*\\ \vdots\\ \mathbf{T}_n^*-\bar{\mathbf{T}}^*\end{bmatrix},\qquad \hat{\boldsymbol{\gamma}}^*=\begin{pmatrix}\hat{\boldsymbol{\gamma}}_1^*\\ \hat{\boldsymbol{\gamma}}_2^*\end{pmatrix}=(\tilde{\mathbf{T}}^{*\prime}\tilde{\mathbf{T}}^*)^{-1}\tilde{\mathbf{T}}^{*\prime}\tilde{\mathbf{Y}},$$

$$\hat{\sigma}^{*2}=\frac{1}{n-s-1}\|\tilde{\mathbf{Y}}-\tilde{\mathbf{T}}^*\hat{\boldsymbol{\gamma}}^*\|^2.$$

By Theorem 16.11, we see that the UMP invariant size α test that $\boldsymbol{\gamma}_1^*=\mathbf{0}$ for this transformed correlation model is given by

$$F^*=\frac{\hat{\boldsymbol{\gamma}}^{*\prime}_1\left[\tilde{\mathbf{T}}_1^{*\prime}\tilde{\mathbf{T}}_1^*-\tilde{\mathbf{T}}_1^{*\prime}\tilde{\mathbf{T}}_2^*(\tilde{\mathbf{T}}_2^{*\prime}\hat{\mathbf{T}}_2^*)^{-1}\tilde{\mathbf{T}}_2^{*\prime}\tilde{\mathbf{T}}_1^*\right]\hat{\boldsymbol{\gamma}}_1^*}{q\hat{\sigma}^{*2}}$$

$$\phi_2^*(F)=\begin{cases}1 & \text{if } F>F^{\alpha}_{q,n-s-1}\\ 0 & \text{if } F\leqslant F^{\alpha}_{q,n-s-1}\end{cases}.$$

In Exercise B6, you are asked to show that

$$F^*=\frac{(\mathbf{A}\hat{\boldsymbol{\gamma}})'\left[\mathbf{A}(\tilde{\mathbf{T}}'\tilde{\mathbf{T}})^{-1}\mathbf{A}'\right]^{-1}\mathbf{A}\hat{\boldsymbol{\gamma}}}{q\hat{\sigma}^2}\tag{16.23}$$

which is the F-statistic used for the ordinary linear model (see Section 7.2.3).

In a similar way, we could show that the UMP invariant size α test for testing that $\mathbf{a}'\gamma = 0$ against $\mathbf{a}'\gamma \geqslant 0$ (where $a \neq 0$ is $s \times 1$) for the ordinary linear model is also UMP invariant size α for the correlation model.

Finally, we consider testing hypotheses about σ^2. It should be apparent from previous sections that any size α test for a hypothesis about σ^2 for the ordinary linear model would also be a size α for the correlation model. In Exercise B7, you are asked to show that the UMP invariant size α test for testing that $\sigma^2 = c$ against $\sigma^2 \geqslant c$ for the ordinary linear model is also UMP invariant size α for the correlation model, although the groups used to reduce the problem are again quite different.

16.5. SIMULTANEOUS CONFIDENCE INTERVALS.

Let $a_\lambda(\mathbf{Y},\mathbf{T}) \leqslant h_\lambda(\beta) \leqslant b_\lambda(\mathbf{Y},\mathbf{T})$, $\lambda \in \Lambda$, be a set of $(1-\alpha)$ simultaneous confidence intervals for the ordinary linear model that occurs when the $\mathbf{T}_i$ are assumed constant. Then these confidence intervals are also simultaneous $(1-\alpha)$ confidence intervals for the correlation model as the following calculation shows:

$$P\left[a_\lambda(\mathbf{Y},\mathbf{T}) \leqslant h_\lambda(\boldsymbol{\beta}) \leqslant b_\lambda(\mathbf{Y},\mathbf{T}) \quad \text{for all } \lambda \in \Lambda\right]$$
$$= E\left\{P\left[a_\lambda(\mathbf{Y},\mathbf{T}) \leqslant h_\lambda(\boldsymbol{\beta}) \leqslant b_\lambda(\mathbf{Y},\mathbf{T}), \quad \text{for all } \lambda \in \Lambda \mid \mathbf{T}\right]\right\} = 1 - \alpha.$$

In particular, the $(1-\alpha)$ confidence band for the response surface is also valid for the correlation model. Similarly, confidence intervals for σ^2 in the ordinary linear model are also valid for the correlation model.

16.6. ASYMPTOTIC VALIDITY OF THE PROCEDURES.

In this section, we discuss the asymptotic validity (as n goes to ∞) of the tests and confidence intervals discussed in Section 16.2–16.6 in the presence of nonnormality (in both $Y_i \mid \mathbf{T}_i$ and in $\mathbf{T}_i$). We consider the model in which we observe $(Y_1, \mathbf{T}_1), \ldots, (Y_n, \mathbf{T}_n)$ independently, identically distributed p-dimensional random vectors such that

$$Y_i = \delta + \mathbf{T}_i\gamma + e_i$$

where Y_i is 1×1 and $\mathbf{T}_i$ is $1 \times s$, δ and γ are unobserved constants, and the e_i are unobserved random variables that are independent of the T_i. In addition, we assume that the e_i are independently, identically distributed with

$$Ee_i = 0, \qquad \text{var}(e_i) = \sigma^2 < \infty$$

(If we assume that e_i and $\mathbf{T}_i$ are normally distributed, this model would be a correlation model. We do not assume that normality in this section.) Now let

$$\mathbf{Y}^n = \begin{bmatrix} Y_1 \\ \vdots \\ Y_n \end{bmatrix}, \qquad \mathbf{X}^n = \begin{bmatrix} 1 & \mathbf{T}_1 \\ \vdots & \vdots \\ 1 & \mathbf{T}_n \end{bmatrix}, \qquad \boldsymbol{\beta} = \begin{pmatrix} \delta \\ \gamma \end{pmatrix}, \qquad e^n = \begin{bmatrix} e_1 \\ \vdots \\ e_n \end{bmatrix}, \qquad \boldsymbol{\mu}^n = \mathbf{X}^n\boldsymbol{\beta}.$$

Then

$$\mathbf{Y}^n = \boldsymbol{\mu}^n + \mathbf{e}^n$$

and therefore, conditionally on the $\mathbf{T}_i$, this model is in the form considered in Chapter 10. If $\operatorname{cov}(T_i) < \infty$, it can be shown that Huber's condition is satisfied with probability 1 (i.e., the largest diagonal element of $\mathbf{X}^n(\mathbf{X}^{n\prime}\mathbf{X}^n)^{-1}\mathbf{X}^{n\prime}$ goes to 0 with probability 1). Therefore if $\operatorname{cov}(\mathbf{T}_i) < \infty$, then the F-tests and t-tests defined in Section 16.2–16.4 are asymptotically valid for the more general model discussed in this section, as are the simultaneous confidence intervals discussed in Section 16.5. Therefore, these procedures are asymptotically insensitive to the normal assumptions used in deriving them. Of course, tests and confidence intervals for σ^2 are not asymptotically valid for the correlation model as they are not in the ordinary linear model. See Arnold (1980a) for more details and derivations.

16.7. THE BEST INVARIANT ESTIMATOR OF γ.

We now find the best invariant estimator of γ. We first need a loss function. Let $\mathbf{d} \in R^s$. The following loss function was suggssted by Stein (1960).

$$L\big(\mathbf{d}; (\delta, \gamma, \sigma^2, \boldsymbol{\nu}, \boldsymbol{\Xi})\big) = \frac{(n-1)(\mathbf{d}-\gamma)'\boldsymbol{\Xi}(\mathbf{d}-\gamma)}{\sigma^2}$$

We present an elementary argument to motivate this loss function. We note first that $\hat{\gamma}\,|\,\mathbf{T} \sim N_s[\gamma, \sigma^2(\tilde{\mathbf{T}}'\tilde{\mathbf{T}})^{-1}]$. Therefore, the Mahalanobis distance for this problem would be $(\mathbf{d}-\gamma)'\tilde{\mathbf{T}}'\tilde{\mathbf{T}}(\mathbf{d}-\gamma)/\sigma^2$. However, this function is not a loss function, since it depends on the random matrix $\tilde{\mathbf{T}}$. We note that $\tilde{\mathbf{T}}'\tilde{\mathbf{T}} = (n-1)\hat{\boldsymbol{\Xi}}$ and hence $E\tilde{\mathbf{T}}'\tilde{\mathbf{T}} = (n-1)\boldsymbol{\Xi}$. We get the loss function L by replacing $\tilde{\mathbf{T}}'\tilde{\mathbf{T}}$ with its expectation in the Mahalanobis distance.

THEOREM 16.14. $\hat{\gamma}$ is the best invariant estimator of γ for the loss function L.

PROOF. The group leaving this problem invariant is a rather messy group. We therefore borrow an idea from invariant testing and list four smaller groups that each leave the problem invariant. The groups we use are the following:

(1) $g_1(Y_i, \mathbf{T}_i) = (Y_i, \mathbf{T}_i) + (a_1, \mathbf{a}_2)$ for arbitrary a_1 and $\mathbf{a}_2$;
(2) $g_2(Y_i, \mathbf{T}_i) = (Y_i, \mathbf{T}_i\mathbf{B})$ for arbitrary invertible $\mathbf{B}$;
(3) $g_3(Y_i, \mathbf{T}_i) = (fY_i, \mathbf{T}_i)$ for $f > 0$;
(4) $g_4(Y_i, \mathbf{T}_i) = (Y_i + \mathbf{T}_i\mathbf{D}, \mathbf{T}_i)$ where $\mathbf{D}$ is an arbitrary $s \times 1$ vector.

For each of these four groups we find the induced transformations on the sufficient statistic $(\hat{\delta}, \hat{\gamma}, \hat{\sigma}^2, \hat{\boldsymbol{\nu}}, \hat{\boldsymbol{\Xi}})$. The induced transformations $\bar{g}$ on the parameter space are the same functions of $(\delta, \gamma, \sigma^2, \boldsymbol{\nu}, \boldsymbol{\Xi})$. (The reader should check this.) We also find transformations $\tilde{g}(\mathbf{d})$. (The reader should also check that the loss function satisfies the appropriate invariance condition.) We then state the condition that an estimator must satisfy to be invariant under each group. We note that the groups G_1, G_2, and G_3 are the same as used in the beginning of Section 16.3 so that the induced groups for the sufficient statistic and the parameter have already been derived.

(1) $g_1(\hat{\delta}, \hat{\gamma}, \hat{\sigma}^2, \hat{\boldsymbol{\nu}}, \hat{\boldsymbol{\Xi}}) = (\hat{\delta} + a_1 - \mathbf{a}_2\hat{\gamma}, \hat{\gamma}, \hat{\sigma}^2, \hat{\boldsymbol{\nu}} + \mathbf{a}_2, \hat{\boldsymbol{\Xi}})$; $\tilde{g}_1(\mathbf{d}) = \mathbf{d}$. Therefore, if $\mathbf{T}$ is invariant under G_1, it must satisfy,

$$\mathbf{T}(\hat{\delta} + a_1 - \mathbf{a}_2\hat{\gamma}, \hat{\gamma}, \hat{\sigma}^2, \hat{\boldsymbol{\nu}} + \mathbf{a}_2, \hat{\boldsymbol{\Xi}}) = \mathbf{T}(\hat{\delta}, \hat{\gamma}, \hat{\sigma}^2, \hat{\boldsymbol{\nu}}, \hat{\boldsymbol{\Xi}}).$$

Therefore, if $\mathbf{T}$ is invariant under G_1, it must not depend on $\hat{\delta}$ or $\hat{\boldsymbol{\nu}}$.

(2) $g_2(\hat{\delta}, \hat{\gamma}, \hat{\sigma}^2, \hat{\boldsymbol{\nu}}, \hat{\boldsymbol{\Xi}}) = (\hat{\delta}, \mathbf{B}^{-1}\hat{\gamma}, \hat{\sigma}^2, \hat{\boldsymbol{\nu}}, \mathbf{B}'\hat{\boldsymbol{\Xi}}\mathbf{B})$, $\tilde{g}_2(\mathbf{d}) = \mathbf{B}^{-1}\mathbf{d}$. Therefore, if $\mathbf{T}$ is invariant under G_2, it must satisfy

$$\mathbf{T}(\hat{\delta}, \mathbf{B}^{-1}\hat{\gamma}, \hat{\sigma}^2, \hat{\boldsymbol{\nu}}, \mathbf{B}'\hat{\boldsymbol{\Xi}}\mathbf{B}) = \mathbf{B}^{-1}\mathbf{T}(\hat{\delta}, \hat{\gamma}, \hat{\sigma}^2, \hat{\boldsymbol{\nu}}, \hat{\boldsymbol{\Xi}}).$$

(3) $g_3(\hat{\delta}, \hat{\gamma}, \hat{\sigma}^2, \hat{\boldsymbol{\nu}}, \hat{\boldsymbol{\Xi}}) = (f\hat{\delta}, f\hat{\gamma}, f^2\hat{\sigma}^2, \hat{\boldsymbol{\nu}}, \hat{\boldsymbol{\Xi}})$, $\tilde{g}_3(\mathbf{d}) = f\mathbf{d}$. Therefore, if $\mathbf{T}$ is invariant under G_3, it must satisfy

$$\mathbf{T}(f\hat{\delta}, f\hat{\gamma}, f^2\hat{\sigma}^2, \hat{\boldsymbol{\nu}}, \hat{\boldsymbol{\Xi}}) = f\mathbf{T}(\hat{\delta}, \hat{\gamma}, \hat{\sigma}^2, \hat{\boldsymbol{\nu}}, \hat{\boldsymbol{\Xi}}).$$

(4) It is easily verified that G_4 becomes $g_4(\hat{\delta}, \hat{\gamma}, \hat{\sigma}^2, \hat{\boldsymbol{\nu}}, \hat{\boldsymbol{\Xi}}) = (\hat{\delta}, \hat{\gamma} + \mathbf{D}, \hat{\sigma}^2, \hat{\boldsymbol{\nu}}, \hat{\boldsymbol{\Xi}})$, and that $\tilde{g}_4(\mathbf{d}) = \mathbf{d} + \mathbf{D}$. Therefore, if $\mathbf{T}$ is invariant under G_4 it must satisfy

$$\mathbf{T}(\hat{\delta}, \hat{\gamma} + \mathbf{D}, \hat{\sigma}^2, \hat{\boldsymbol{\nu}}, \hat{\boldsymbol{\Xi}}) = \mathbf{T}(\hat{\delta}, \hat{\gamma}, \hat{\sigma}^2, \hat{\boldsymbol{\nu}}, \hat{\boldsymbol{\Xi}}) + \mathbf{D}.$$

Now, let G be the smallest group containing G_1, G_2, G_3, and G_4. Then $\mathbf{T}$ is invariant under G if and only if it is invariant under G_1, G_2, G_3, and G_4. Note that $\hat{\gamma}$ is invariant under G. Now let $\mathbf{T}$ be any other invariant estimator. Since it is invariant under G_1, $\mathbf{T}$ does not involve $\hat{\delta}$ or $\hat{\boldsymbol{\nu}}$. In addition, $\mathbf{T}(\hat{\gamma}, \hat{\sigma}^2, \hat{\boldsymbol{\Xi}})$ must satisfy

$$\mathbf{T}(f\mathbf{B}^{-1}\hat{\gamma} + \mathbf{D}, f^2\hat{\sigma}^2, \mathbf{B}'\hat{\boldsymbol{\Xi}}\mathbf{B}) = f\mathbf{B}^{-1}\mathbf{T}(\hat{\gamma}, \hat{\sigma}^2, \hat{\boldsymbol{\Xi}}) + \mathbf{D}$$

since it must be invariant under G_2, G_3, and G_4. Now, let $f = 1/\hat{\sigma}$,

$\mathbf{B} = \hat{\boldsymbol{\Xi}}^{-1/2}$, $\mathbf{D} = (-1/\hat{\sigma})\hat{\boldsymbol{\Xi}}^{-1/2}\hat{\boldsymbol{\gamma}}$. Then $\mathbf{T}$ must satisfy

$$\mathbf{T}(\mathbf{0}, 1, \mathbf{I}) = \frac{1}{\hat{\sigma}}\hat{\boldsymbol{\Xi}}^{-1/2}\mathbf{T}(\hat{\boldsymbol{\gamma}}, \hat{\sigma}^2, \hat{\boldsymbol{\Xi}}) - \frac{1}{\hat{\sigma}}\hat{\boldsymbol{\Xi}}^{-1}\hat{\boldsymbol{\gamma}}$$

or equivalently

$$\mathbf{T}(\hat{\boldsymbol{\gamma}}, \hat{\sigma}^2, \hat{\boldsymbol{\Xi}}) = \hat{\sigma}\hat{\boldsymbol{\Xi}}^{-1/2}\mathbf{T}(\mathbf{0}, 1, \mathbf{I}) + \hat{\boldsymbol{\gamma}}.$$

Now, let $\mathbf{T}(\mathbf{0}, 1, \mathbf{I}) = \mathbf{M}$. Then the risk of $\mathbf{T}$ is

$$\begin{aligned} R\big(\mathbf{T}; (\delta, \boldsymbol{\gamma}, \sigma^2, \boldsymbol{\nu}, \boldsymbol{\Xi})\big) &= (n-1)E(\hat{\sigma}\hat{\boldsymbol{\Xi}}^{-1/2}\mathbf{M} + \hat{\boldsymbol{\gamma}} - \boldsymbol{\gamma})\boldsymbol{\Xi}(\hat{\sigma}\hat{\boldsymbol{\Xi}}^{-1/2}\mathbf{M} + \hat{\boldsymbol{\gamma}} - \boldsymbol{\gamma}) \\ &= \frac{(n-1)E\hat{\sigma}^2\mathbf{M}'\hat{\boldsymbol{\Xi}}^{-1/2}\boldsymbol{\Xi}\hat{\boldsymbol{\Xi}}^{-1/2}\mathbf{M}}{\sigma^2} \\ &\quad + \frac{(n-1)E(\hat{\boldsymbol{\gamma}} - \boldsymbol{\gamma})\boldsymbol{\Xi}(\hat{\boldsymbol{\gamma}} - \boldsymbol{\gamma})}{\sigma^2} \\ &\quad + \frac{2(n-1)E\hat{\sigma}\mathbf{M}'\hat{\boldsymbol{\Xi}}^{-1/2}\boldsymbol{\Xi}(\hat{\boldsymbol{\gamma}} - \boldsymbol{\gamma})}{\sigma^2} \end{aligned}$$

Now, $(n-p)\hat{\sigma}^2/\sigma^2 \,|\, (\mathbf{T}, \hat{\boldsymbol{\gamma}}) \sim \chi^2_{n-p}$ so that $\hat{\sigma}^2$ is independent of $\tilde{\mathbf{T}}$ and $\hat{\boldsymbol{\gamma}}$. Also, $\hat{\boldsymbol{\gamma}} \,|\, \tilde{\mathbf{T}}' \sim N_s(\boldsymbol{\gamma}, \sigma^2(\tilde{\mathbf{T}}'\tilde{\mathbf{T}})^{-1})$. Since $\tilde{\mathbf{T}}'\tilde{\mathbf{T}} = (n-1)\hat{\boldsymbol{\Xi}}$,

$$\hat{\boldsymbol{\gamma}} \,|\, \hat{\boldsymbol{\Xi}} \sim N_s\big(\boldsymbol{\gamma}, \sigma^2, (n-1)\hat{\boldsymbol{\Xi}}^{-1}\big).$$

Therefore,

$$E\hat{\sigma}^2\mathbf{M}'\hat{\boldsymbol{\Xi}}^{-1/2}\boldsymbol{\Xi}(\hat{\boldsymbol{\gamma}} - \boldsymbol{\gamma}) = E\hat{\sigma}^2\big(\mathbf{M}'\hat{\boldsymbol{\Xi}}^{-1/2}\boldsymbol{\Xi}E(\hat{\boldsymbol{\gamma}} - \boldsymbol{\gamma}) \,|\, \hat{\boldsymbol{\Xi}}\big) = 0$$

Hence

$$R(\mathbf{T}), (\delta, \boldsymbol{\gamma}, \sigma^2, \boldsymbol{\nu}, \boldsymbol{\Xi}) = (n-1)\frac{E\hat{\sigma}^2\mathbf{M}'\hat{\boldsymbol{\Xi}}^{-1/2}\boldsymbol{\Xi}\hat{\boldsymbol{\Xi}}^{-1/2}\mathbf{M}}{\sigma^2} + R\big(\hat{\boldsymbol{\delta}}, (\delta, \hat{\boldsymbol{\gamma}}, \sigma^2, \boldsymbol{\nu}, \boldsymbol{\Xi})\big).$$

Now, $\hat{\boldsymbol{\Xi}} > 0$, $\mathbf{M} \neq 0$ (since $\mathbf{T} \neq \hat{\boldsymbol{\gamma}}$), and $\hat{\boldsymbol{\Xi}}^{-1/2}$ is an invertible matrix. Therefore, $\hat{\sigma}^2\mathbf{M}'\hat{\boldsymbol{\Xi}}^{-1/2}\boldsymbol{\Xi}\hat{\boldsymbol{\Xi}}^{1/2}\mathbf{M} > 0$, and hence $\hat{\boldsymbol{\gamma}}$ is better than any other invariant estimator. □

16.8. OTHER OPTIMALITY RESULTS.

In this section we discuss other properties of the procedures derived in earlier sections.

We first discuss the estimation of $\boldsymbol{\gamma}$. We have already shown that $\hat{\boldsymbol{\gamma}}$ is the MLE and the minimum variance unbiased estimator. We have also shown that it is the best invariant estimator with respect to the loss function L defined in the last section. Stein (1960) showed that it is also a minimax estimator for this loss function. Unfortunately, he showed that it is inadmissible when $s > 2$. (Note that we are assuming throughout this chapter that $n > s + 1$.) Baranchik (1973) exhibited a collection of shrinking estimators that are better than $\hat{\gamma}$ for the loss function L.

Stein (1960) also considers estimating $\boldsymbol{\beta} = \binom{\delta}{\gamma}$. Let $d_1 \in R^1$, $\mathbf{d}_2 \in R^s$. The loss function he considers for this problem is

$$L\big((d_1, \mathbf{d}_2); (\delta, \gamma, \sigma^2, \boldsymbol{\nu}, \boldsymbol{\Xi})\big) = \frac{n(d_1 - \delta_1 + \boldsymbol{\nu}'(\mathbf{d}_2 - \gamma))^2 + n(\mathbf{d}_2 - \gamma)'\boldsymbol{\Xi}(\mathbf{d}_2 - \gamma)}{\sigma^2}$$

This loss function can be derived by replacing $\mathbf{X}'\mathbf{X}$ with its expectation in the Mahalanobis distance loss function $L^* = (\mathbf{d} - \boldsymbol{\beta})\mathbf{X}'\mathbf{X}(\mathbf{d} - \boldsymbol{\beta})/\sigma^2$, where $\mathbf{d} = \binom{d_1}{\mathbf{d}_2}$. (Stein (1960) gives another interesting motivation for this loss function in terms of prediction.) By arguments similar to those in the last section, we can show that $\hat{\boldsymbol{\beta}}$ is the best invariant estimator for this loss function, and Stein (1960) shows that it is a minimax estimator, but that it is inadmissible when $s \geqslant 3$. Baranchik (1973) gives a class of shrinking estimators that are better than $\hat{\beta}$ for this loss function. Note that in the linear model we only need $s = (p - 1) \geqslant 2$ to get inadmissibility of $\hat{\beta}$. Stein (1960) showed that $\hat{\beta}$ is admissible when $s = 2$ (and $n \geqslant 6$) for the correlation model (with the loss function L), yet $\hat{\beta}$ is inadmissible for the linear model with the conditional Mahalanobis distance loss function L^* when $s = 2$.

We now discuss the tests about γ derived in earlier sections. These tests have been shown to be admissible tests (see Kiefer and Schwartz 1965). It is now known whether they are most stringent.

16.9. MULTIPLE AND PARTIAL CORRELATION COEFFICIENTS.

The correlation model studied in previous sections is often approached from a different perspective, using various forms of correlation coefficients. In this section, we discuss this approach to the model and show that it leads to the same tests as those derived in Sections 16.2 and 16.3.

We define various sorts of correlation coefficients. Their sample versions are defined to be their MLE's. To find these MLEs, we recall that the MLE of $h(\delta, \gamma, \sigma^2, \boldsymbol{\nu}, \boldsymbol{\Xi})$ is just $h(\hat{\delta}, \hat{\gamma}, (n - s - 1)\hat{\sigma}^2/n, \hat{\boldsymbol{\nu}}, ((n-1)/n)\hat{\boldsymbol{\Xi}}$, that is, the same function of the MLEs of δ, γ, σ^2, $\boldsymbol{\nu}$ and $\boldsymbol{\Xi}$.

16.9.1. The Correlation Coefficient

We now return to the correlation model in the case in which $s = 1$ (i.e., in which Y_i and T_i are both 1×1. In that case we observe $(Y_1, T_1), \ldots,$

(Y_n, T_n) independent two-dimensional vectors such that

$$\begin{pmatrix} Y_i \\ T_i \end{pmatrix} \sim N_2(\boldsymbol{\mu}, \boldsymbol{\Sigma}), \qquad \boldsymbol{\mu} = \begin{pmatrix} \mu_1 \\ \mu_2 \end{pmatrix} = \begin{pmatrix} \delta + \nu\gamma \\ \nu \end{pmatrix},$$

$$\boldsymbol{\Sigma} = \begin{pmatrix} \Sigma_{11} & \Sigma_{12} \\ \Sigma_{21} & \Sigma_{22} \end{pmatrix} = \begin{pmatrix} \sigma^2 + \gamma^2 \Xi & \delta \Xi \\ \gamma \Xi & \Xi \end{pmatrix}$$

The *correlation coefficient* ρ between Y_i and T_i is defined to be

$$\rho = \frac{\operatorname{cov}(Y_i, T_i)}{\sqrt{\operatorname{var}(Y_i)\operatorname{var}(T_i)}} = \frac{\Sigma_{12}}{\sqrt{\Sigma_{11}\Sigma_{22}}} = \frac{\gamma\sqrt{\Xi}}{\sqrt{\sigma^2 + \gamma^2 \Xi}}$$

Note that the Cauchy-Schwartz inequality implies that $-1 \leqslant \rho \leqslant 1$. (In fact, if $\Sigma > 0$, then $-1 < \rho < 1$.) The *sample correlation coefficient* r between Y_i and T_i is defined to be the MLE of ρ, namely

$$r = \frac{\hat{\Sigma}_{12}}{\sqrt{\hat{\Sigma}_{11}\hat{\Sigma}_{22}}} = \frac{\hat{\gamma}\sqrt{\hat{\Xi}}}{\sqrt{\frac{n-2}{n-1}\hat{\sigma}^2 + \hat{\gamma}^2 \hat{\Xi}}} \tag{16.24}$$

We first consider testing that $\rho = 0$ against $\rho \geqslant 0$. Since $\rho = \gamma\sqrt{\Xi} / \sqrt{\sigma^2 + \gamma^2 \Xi}$ this problem is the same as testing that $\gamma = 0$ against $\gamma \geqslant 0$. A maximal invariant, parameter maximal invariant and UMP invariant size α test for this problem are given by t_1, ξ_1, and ϕ_1^* defined in (16.15) and (16.16). We now find the relationship between (t_1, ξ_1) and (r, ρ).

LEMMA 16.15.

$$r = \frac{t_1}{\sqrt{n - 2 + t_1^2}}, \qquad \rho = \frac{\xi}{\sqrt{1 + \xi^2}}, \qquad t_1 = \frac{\sqrt{n-2}\, r}{\sqrt{1 - r^2}}$$

PROOF.

$$r = \frac{\hat{\gamma}\sqrt{\hat{\Xi}}}{\sqrt{\frac{n-2}{n-1}\hat{\sigma}^2 + \hat{\gamma}^2 \hat{\Xi}}} = \frac{\sqrt{n-1}\,\dfrac{\hat{\gamma}\sqrt{\hat{\Xi}}}{\hat{\sigma}}}{\sqrt{n - 2 + (n-1)\hat{\gamma}^2 \hat{\Xi}/\hat{\sigma}^2}} = \frac{t_1}{\sqrt{n - 2 + t_1^2}}$$

The other results follow similarly. □

We note therefore that r is an increasing function of t_1 and ρ is an increasing fnnction of ξ_1. Therefore, r is invariant under the groups G_1, G_2 and G_3 defined in Section 16.2. Equivalently r is invariant under the group of transformations

$$g(Y_i, T_i) = (b_1 Y_i + a_1, b_2 T_i + a_2)$$

for $b_1 > 0$, $b_2 > 0$.

COROLLARY 1. The distribution of r depends only on ρ (and n).

PROOF. Since r is an invertible function of t_1 it is also a maximal invariant for this problem. Similarly ρ is a parameter maximal invariant. Therefore, the distribution of r depends only on ρ (and the sample size). □

COROLLARY 2. The UMP invariant size α test for testing that $\rho = 0$ against $\rho \geqslant 0$ is given by

$$\phi_1^*(r) = \begin{cases} 1 & \text{if } \dfrac{\sqrt{n-2}\, r}{\sqrt{1-r^2}} > t_{n-2}^{\alpha} \\ 0 & \text{if } \dfrac{\sqrt{n-2}\, r}{\sqrt{1-r^2}} \leqslant t_{n-2}^{\alpha} \end{cases}$$

PROOF. This follows directly from Lemma 16.15. □

Now, consider testing that $\rho = 0$ against $-1 < \rho < 1$. This is the same as testing that $\gamma = 0$ against $-\infty < \gamma < \infty$. A maximal invariant, parameter maximal and UMP invariant size α test are given by F_1, θ_1 and ϕ_1 defined in (16.12) and (16.14). By arguments similar to those given above, we can show that

$$r^2 = \frac{F_1}{\sqrt{n-2+F_1}}, \qquad \rho^2 = \frac{\theta_1}{1+\theta_1}, \qquad F_1 = \frac{(n-2)r^2}{1-r^2}$$

Therefore, r^2 is an increasing function of F_1 and ρ^2 is an increasing function of θ_1. Hence the distribution of r^2 depends only on ρ^2 and the UMP invariant size α test can be written as

$$\phi_1(r) = \begin{cases} 1 & \text{if } \dfrac{(n-2)r^2}{1-r^2} > F_{1,n-2}^{\alpha} \\ 0 & \text{if } \dfrac{(n-2)r^2}{1-r^2} \leqslant F_{1,n-2}^{\alpha} \end{cases}$$

It is occasionally desirable to find confidence intervals for ρ or to test hypotheses about ρ other than testing $\rho = 0$. For this reason, the following result is quite useful. Let

$$h(u) = \tfrac{1}{2}\log(1+u)/(1-u)$$

and let r_n be the sample correlation coefficient computed from a sample of size n from a bivariate normal distribution with correlation coefficient ρ. Then

$$\sqrt{n}\,(h(r_n) - h(\rho)) \xrightarrow{d} N_1(0, 1) \tag{16.25}$$

(See Exercise C2). This result was first derived by Fisher (1915) and is called Fisher's Z transformation. Actually, the approximation can be improved by replacing $h(\rho)$ with $h(\rho) + \rho/2(n-1)$ and $\sqrt{n}$ by $\sqrt{n-3}$. David (1938) has made a study of the accuracy of the above approximation. She suggests that it be used as long as $n > 25$. She has also prepared tables of the distribution function of r_n for various choices of ρ. (Note that $\sqrt{n-2}\, r_n/\sqrt{1-r_n^2}$ only has a t-distribution when $\rho = 0$.)

16.9.2. The Multiple Correlation Coefficient

We now return to the general correlation model is which we observe $(Y_1, \mathbf{T}_1), \ldots, (Y_n, \mathbf{T}_n)$ independent $(s+1)$ dimensional random vectors such that

$$\begin{pmatrix} Y_i \\ \mathbf{T}_i \end{pmatrix} \sim N_{s+1}(\boldsymbol{\mu}, \boldsymbol{\Sigma}), \qquad \mu = \begin{pmatrix} \mu_1 \\ \boldsymbol{\mu}_2 \end{pmatrix} = \begin{pmatrix} \delta + \boldsymbol{\nu}'\gamma \\ \boldsymbol{\nu} \end{pmatrix},$$

$$\boldsymbol{\Sigma} = \begin{pmatrix} \Sigma_{11} & \boldsymbol{\Sigma}_{12} \\ \boldsymbol{\Sigma}_{21} & \boldsymbol{\Sigma}_{22} \end{pmatrix} = \begin{pmatrix} \sigma^2 + \gamma'\boldsymbol{\Xi}\gamma & \gamma'\boldsymbol{\Xi} \\ \boldsymbol{\Xi}\gamma & \gamma'\boldsymbol{\Xi}\gamma \end{pmatrix}$$

If $s > 1$, then there is no correlation coefficient between Y_i and $\mathbf{T}_i$. The *multiple correlation coefficient* $\bar{R}$ between Y_i and $\mathbf{T}_i$ is defined to be the correlation coefficient between Y_i and $EY_i \mid \mathbf{T}_i$. The *sample multiple correlation coefficient* R between Y_i and $\mathbf{T}_i$ is defined to be the MLE of the multiple correlation coefficient between Y_i and $\mathbf{T}_i$.

We first find expressions for these coefficients.

LEMMA 16.16.

$$\bar{R} = \sqrt{\frac{\boldsymbol{\Sigma}_{12}\boldsymbol{\Sigma}_{22}^{-1}\boldsymbol{\Sigma}_{21}}{\Sigma_{11}}} = \sqrt{\frac{\gamma'\boldsymbol{\Xi}\gamma}{\sigma^2 + \gamma'\boldsymbol{\Xi}\gamma}},$$

$$R = \sqrt{\frac{\hat{\boldsymbol{\Sigma}}_{12}\hat{\boldsymbol{\Sigma}}_{22}^{-1}\hat{\boldsymbol{\Sigma}}_{21}}{\hat{\Sigma}_{11}}} = \sqrt{\frac{\hat{\gamma}'\hat{\boldsymbol{\Xi}}\hat{\gamma}}{\frac{n-s-1}{n-1}\hat{\sigma}^2 + \hat{\gamma}'\hat{\boldsymbol{\Xi}}\hat{\gamma}}}$$

PROOF. We note that $U_i = EY_i | \mathbf{T}_i = \delta + \mathbf{T}_i\gamma$. Therefore,

$$\begin{pmatrix} Y_i \\ U_i \end{pmatrix} = \begin{pmatrix} 1 & \mathbf{0} \\ 0 & \gamma' \end{pmatrix} \begin{pmatrix} Y_i \\ \mathbf{T}_i \end{pmatrix} + \begin{pmatrix} 0 \\ \delta \end{pmatrix},$$

$$\operatorname{cov}\begin{pmatrix} Y_i \\ U_i \end{pmatrix} = \begin{pmatrix} 1 & \mathbf{0} \\ \mathbf{0} & \gamma' \end{pmatrix} \begin{pmatrix} \sigma^2 + \gamma'\Xi\gamma & \gamma'\Xi \\ \Xi\gamma & \Xi \end{pmatrix} \begin{pmatrix} 1 & \mathbf{0} \\ \mathbf{0} & \gamma \end{pmatrix}$$

$$= \begin{pmatrix} \sigma^2 + \gamma'\Xi\gamma & \gamma'\Xi\gamma \\ \gamma'\Xi\gamma & \gamma'\Xi\gamma \end{pmatrix} = \begin{pmatrix} \Sigma_{11} & \Sigma_{12}\Sigma_{22}^{-1}\Sigma_{21} \\ \Sigma_{12}\Sigma_{22}^{-1}\Sigma_{21} & \Sigma_{12}\Sigma_{22}^{-1}\Sigma_{21} \end{pmatrix}$$

Therefore, the correlation coefficient between Y_i and U_i is

$$\sqrt{\frac{\gamma'\Xi\gamma}{\sigma^2 + \gamma'\Xi\gamma}} = \sqrt{\frac{\Sigma_{12}\Sigma_{22}^{-1}\Sigma_{21}}{\Sigma_{11}}}$$

The formulas for the sample multiple correlation coefficient follow from substituting the MLEs for the parameters in the formulas for the multiple correlation coefficient. □

COROLLARY 1. $\bar{R} \geqslant 0$, $R \geqslant 0$. $\bar{R} = 0$ if and only if Y_i and $\mathbf{T}_i$ are independent.

COROLLARY 2. $R^2 = \|\tilde{\mathbf{T}}\hat{\gamma}\|^2 / \|\tilde{\mathbf{Y}}\|^2$.

Note that if $s = 1$ then the multiple correlation coefficient is just the absolute value of the correlation coefficient and the sample multiple correlation coefficient is the absolute value of the sample correlation coefficient. In light of Corollary 1, we think of the multiple correlation coefficient as measuring the strength of the dependence between Y_i and $\mathbf{T}_i$. If $\bar{R}$ is 0, then Y_i and $\mathbf{T}_i$ are independent. If $\bar{R}$ is near one, then Y_i and $\mathbf{T}_i$ are highly dependent. (Note that $\bar{R}$ is the correlation coefficient between U_i and Y_i and therefore $\bar{R} \leqslant 1$.)

We now consider the problem of testing that $\bar{R} = 0$ against $\bar{R} \geqslant 0$, or equivalently testing that Y_i and $\mathbf{T}_i$ are independent. From Lemma 16.16, we see that this problem is the same as testing that $\gamma = \mathbf{0}$ against $-\infty < \gamma < \infty$. A maximal invariant, parameter maximal invarient and UMP invariant size α test for this problem are F_1, θ_1, and ϕ_1 given in (16.12) and (16.14). We now give the relationship between (F_1, θ_1) and $(R, \bar{R})$.

LEMMA 16.17.

$$R = \sqrt{\frac{s}{n - s - 1 + sF_1}}, \qquad \bar{R} = \sqrt{\frac{\theta_1}{1 + \theta_1}},$$

$$F_1 = \frac{n - s - 1}{s} \frac{R^2}{1 - R^2}.$$

The distribution of R depends only on $\bar{R}$ [and (n,s)]. The test ϕ_1 can be rewritten as

$$\phi_1(R) = \begin{cases} 1 & \text{if } \dfrac{n-s-1}{s}\,\dfrac{R^2}{1-R^2} > F^{\alpha}_{s,n-s-1} \\ 0 & \text{if } \dfrac{n-s-1}{s}\,\dfrac{R^2}{1-R^2} \leqslant F^{\alpha}_{s,n-s-1} \end{cases}$$

PROOF. See Exercise B9. □

The distribution of R for general $\bar{R}$ is derived in Exercise C1. Percentage points of this distribution are given in Pearson and Hartley (1972, Table 52). (Note that $[(n-s-1)/s]R^2/(1-R^2)$ has an F distribution if and only if $\bar{R}=0$.) R is a maximal invariant under G_1, G_2, G_3 defined in Section 16.2. Therefore, R is invariant under the transformation

$$g(Y_i, T_i) = (bY_i + a_1, \mathbf{T}_i\mathbf{B} + \mathbf{a}_2)$$

where $b>0$, $\mathbf{B}$ is invertible. Therefore, R is not sensitive to the units of Y_i or $\mathbf{T}_i$.

We now discuss an alternative interpretation for $\bar{R}$.

LEMMA 16.18. Let $V_i = \mathbf{T}_i\mathbf{a} + b$, where $\mathbf{a} \in R^s$, $\mathbf{a} \neq 0$, $b \in R^1$. Then the correlation coefficient between Y_i and V_i is less than or equal to $\bar{R}$.

PROOF. See Exercise B10. □

We note also that $U_i = EY_i \mid \mathbf{T}_i = \mathbf{T}_i\gamma + \delta$. Therefore $\bar{R}$ is the maximum correlation coefficient between Y_i and affine functions of $\mathbf{T}_i$.

We now give another interpretation for R^2. Using Corollary 2 to Lemma 16.16., we see that $R^2 = \|\tilde{\mathbf{T}}\hat{\gamma}\|^2/\|\tilde{\mathbf{Y}}\|^2$. We think of $\|\tilde{\mathbf{Y}}\|^2$ as the total variation in the experiment, and $\|\tilde{\mathbf{Y}} - (\delta\mathbf{1} + \mathbf{T}\hat{\gamma})\|^2 = \|\tilde{\mathbf{Y}} - \tilde{\mathbf{T}}\hat{\gamma}\|^2$ as the variation that is not explained by including $\mathbf{T}$ in the analysis. Therefore, $\|\tilde{\mathbf{T}}\hat{\gamma}\|^2 = \|\tilde{\mathbf{Y}}\|^2 - \|\tilde{\mathbf{Y}} - \tilde{\mathbf{T}}\hat{\gamma}\|^2$ is the amount of variation that has been explained by including $\mathbf{T}$. Hence, R^2 is the proportion of the total variation in the Y's that is explained by using $\mathbf{T}$. R^2 is called the *coefficient of determination* for the problem and is often used, even in ordinary linear models as a measure of how well the model fits. Note that R^2 always increases as more variables are added (see Exercise B11), and so may be quite large in a particular problem only because a model has many variables. For this reason this coefficent is often adjusted in the following manner. We note that the Y_i are independently, identically, and normally distributed random variables. We can therefore think of $S^2 = \|\tilde{\mathbf{Y}}\|^2/(n-1)$ as an estimator of the variance if we do not include the $\mathbf{T}_i$ in the model. We can also think of $\hat{\sigma}^2 = \|\tilde{\mathbf{Y}} - \tilde{\mathbf{T}}\gamma\|^2/(n-s-1)$ as an estimator if we do include the $\mathbf{T}_i$ in the model. We define the *adjusted coefficient of* *determination (adjusted R^2)* to be

$$R^{*2} = \frac{S^2 - \hat{\sigma}^2}{S^2}.$$

Therefore, R^{*2} is a measure of the proportion of the marginal variance of the Y_i that is explained by including the $\mathbf{T}_i$'s in the model and is therefore another measure of how well the model fits. R^{*2} is preferable to R^2 because it does not necessarily increase as more variables are added to the model. In Exercise B12, you are asked to show that

$$R^{*2} = 1 - \frac{n-1}{n-s-1}(1 - R^2). \tag{16.26}$$

16.9.3. Partial Correlation Coefficients

Partition $\mathbf{T}_i$, γ, $\hat{\gamma}$, $\boldsymbol{\Xi}$ and $\hat{\boldsymbol{\Xi}}$ as in Section 16.3. That is, let

$$\mathbf{T}_i = \begin{pmatrix} \mathbf{T}_{i1} \\ \mathbf{T}_{i2} \end{pmatrix}, \quad \gamma = \begin{pmatrix} \gamma_1 \\ \gamma_2 \end{pmatrix}, \quad \hat{\gamma} = \begin{pmatrix} \hat{\gamma}_1 \\ \hat{\gamma}_2 \end{pmatrix}$$

$$\boldsymbol{\Xi} = \begin{pmatrix} \boldsymbol{\Xi}_{11} & \boldsymbol{\Xi}_{12} \\ \boldsymbol{\Xi}_{21} & \boldsymbol{\Xi}_{22} \end{pmatrix}, \quad \hat{\boldsymbol{\Xi}} = \begin{pmatrix} \hat{\boldsymbol{\Xi}}_{11} & \hat{\boldsymbol{\Xi}}_{12} \\ \hat{\boldsymbol{\Xi}}_{21} & \hat{\boldsymbol{\Xi}}_{22} \end{pmatrix}$$

where $\mathbf{T}_{i1}$, γ_1, and $\hat{\gamma}_1$ are $q \times 1$, $\boldsymbol{\Xi}_{11}$ and $\hat{\boldsymbol{\Xi}}_{11}$ are $q \times q$.

Before defining partial correlation coefficients and partial multiple correlation coefficients, we give the following lemma.

LEMMA 16.19. Let $\boldsymbol{\Xi}_{11.2} = \boldsymbol{\Xi}_{11} - \boldsymbol{\Xi}_{12}\boldsymbol{\Xi}_{22}^{-1}\boldsymbol{\Xi}_{21}$ be the conditional covariance matrix of $\mathbf{T}_{i1}$ given $\mathbf{T}_{i2}$. Then the conditional covariance matrix of $(Y_i, \mathbf{T}_{i1})$ given $\mathbf{T}_{i2}$ is

$$\begin{pmatrix} \sigma^2 + \gamma_1'\boldsymbol{\Xi}_{11.2}\gamma_1 & \gamma_1'\boldsymbol{\Xi}_{11.2} \\ \boldsymbol{\Xi}_{11.2}\gamma_1 & \boldsymbol{\Xi}_{11.2} \end{pmatrix}.$$

PROOF. By Lemma 16.1, we see that the covariance matrix of Y_i, $\mathbf{T}_{i1}$ and $\mathbf{T}_{i2}$ is

$$\begin{bmatrix} \sigma + \gamma'\boldsymbol{\Xi}\gamma & \gamma_1'\boldsymbol{\Xi}_{11} + \gamma_2'\boldsymbol{\Xi}_{21} & \gamma_1'\boldsymbol{\Xi}_{12}\gamma_2'\boldsymbol{\Xi}_{22} \\ \boldsymbol{\Xi}_{11}\gamma_1 + \boldsymbol{\Xi}_{12}\gamma_2 & \boldsymbol{\Xi}_{11} & \boldsymbol{\Xi}_{12} \\ \boldsymbol{\Xi}_{21}\gamma_1 + \boldsymbol{\Xi}_{22}\gamma_2 & \boldsymbol{\Xi}_{21} & \boldsymbol{\Xi}_{22} \end{bmatrix}.$$

Therefore, by Theorem 3.7c, we see that the conditional covariance matrix of Y_i and $\mathbf{T}_{i1}$ given $\mathbf{T}_{i2}$ is

$$\begin{pmatrix} \sigma^2 + \gamma'\Xi\gamma & \gamma_1'\Xi_{11} + \gamma_2'\Xi_{21} \\ \Xi_{11}\gamma_1 + \Xi_{12}\gamma_2 & \Xi_{11} \end{pmatrix}$$

$$- \begin{pmatrix} \gamma_1'\Xi_{12} + \gamma_2'\Xi_{22} \\ \Xi_{12} \end{pmatrix} \Xi_{22}^{-1}(\Xi_{21}\gamma_1 + \Xi_{22}\gamma_2, \Xi_{21})$$

which equals the matrix above. □

Note that the conditional covariance matrix can be computed by replacing the matrix Ξ with $\Xi_{11.2}$ and γ by γ_1 in the joint covariance matrix.

We consider first the case in which $q = 1$. In this case, we define the *partial correlation coefficient* ρ^* between Y_i and T_{i1} given $\mathbf{T}_{i2}$ to be the correlation coefficient computed from the conditional covariance matrix of (Y_i, T_{i1}) given $\mathbf{T}_{i2}$. The *sample partial correlation coefficient* r^* between Y_i and T_{i1} given $\mathbf{T}_{i2}$ is defined to be the MLE of ρ^*. We think of the absolute value of the partial correlation coefficient as measuring the strength of the residual dependence between Y_i and T_{i1} after accounting for their common dependence on $\mathbf{T}_{i2}$. The sign of the correlation coefficient tells whether that residual dependence is positive (high Y_i with high T_{i1}) or negative (high Y_i with low T_i). It can happen that Y_i and T_{i1} have a highly positive correlation coefficient (indicating a strong positive dependence), but have a partial correlation coefficient that is 0. This would indicate that the dependence between Y_i and T_{i1} is completely explained by their common dependence on $\mathbf{T}_{i2}$. It can even happen that the correlation coefficient is highly positive but the partial correlation coefficient is negative.

Using Lemma 16.19 and the usual methods for MLE's, we see that

$$\rho^* = \frac{\gamma_1\sqrt{\Xi_{11.2}}}{\sqrt{\sigma^2 + \gamma_1\Xi_{11.2}^2}},$$

$$r^* = \frac{\hat{\gamma}_1\sqrt{\hat{\Xi}_{11.2}}}{\sqrt{\frac{n-s-1}{n-1}\hat{\sigma}^2 + \hat{\gamma}_1^2\hat{\Xi}_{11.2}}} \qquad (\hat{\Xi}_{11.2} = \hat{\Xi}_{11} - \hat{\Xi}_{12}\hat{\Xi}_{22}^{-1}\hat{\Xi}_{21}) \tag{16.27}$$

Now, consider the problem of testing that $\rho = \text{o}$ against $\rho^* \geqslant 0$, or equivalently that conditionally on the $\mathbf{T}_{i2}$, that Y_i and T_{i1} are independent. From (16.27), we see that this problem is the same as testing that $\gamma_1 = 0$ against $\gamma_1 \geqslant 0$. A maximal invariant, parameter maximal invariant and UMP invariant size α test for this problem are t_2, ξ_2, and ϕ^* given in (16.21) and (16.22)

LEMMA 16.20.

$$r^* = \frac{t_2}{\sqrt{n-s-1+t_2^2}}, \qquad \rho^* = \frac{\xi_2}{\sqrt{1+\xi_2}}, \qquad t_2 = \frac{\sqrt{n-s-1}\,r^*}{\sqrt{1-r^{*2}}}$$

The distribution of r^* depends only on ρ^* [and (n, s)]. The test ϕ_2^* can be rewritten as

$$\phi_2(r^*) = \begin{cases} 1 & \text{if } \dfrac{\sqrt{n-s-1}\,r^*}{\sqrt{1-r^{*2}}} > t_{n-s-1}^{\alpha} \\ 0 & \text{if } \dfrac{\sqrt{n-s-1}\,r^*}{\sqrt{1-r^{*2}}} \leqslant t_{n-s-1}^{\alpha} \end{cases}$$

PROOF. See Exercise B14. □

We now state one final result about the partial correlation coefficient. Let $r_n(\rho)$ be the distribution of the sample correlation coefficient computed from a sample of size n from a bivariate normal distribution with correlation coefficient r. Then

$$r^* \sim r_{n-s-1}(\rho^*) \tag{16.28}$$

(see Exercise C1 of Chapter 18). Therefore we can use Fisher's Z transformation on the partial correlation coefficient. The tables of David (1938) are also useful for the partial correlation coefficient.

We now return to the general case in which $q > 1$. If $q > 1$, we cannot define the partial correlation coefficient between Y_i and $\mathbf{T}_{i1}$ given $\mathbf{T}_{i2}$. Instead we define the *partial multiple correlation coefficient* $\bar{R}^*$ between Y_i and $\mathbf{T}_{i1}$ given $\mathbf{T}_{i2}$ to be the multiple correlation coefficient computed from the conditional covariance matrix of $(Y_i, \mathbf{T}_{i1})$ given $\mathbf{T}_{i2}$. The *sample partial multiple correlation coefficient* R^* between Y_i and $\mathbf{T}_{i1}$ given $\mathbf{T}_{i2}$ is defined to be the MLE of $\bar{R}^*$. Using Lemma 16.19, we see that

$$\bar{R}^* = \sqrt{\frac{\gamma_1'\boldsymbol{\Xi}_{11.2}\gamma_1}{\sigma^2 + \gamma_1'\boldsymbol{\Xi}_{11.2}\gamma_1}}, \qquad R^* = \sqrt{\frac{\hat{\gamma}_1'\hat{\boldsymbol{\Xi}}_{11.2}\hat{\gamma}_1}{\frac{n-s-1}{n-1}\hat{\sigma}^2 + \hat{\gamma}_1'\hat{\boldsymbol{\Xi}}_{11.2}\hat{\gamma}_1}}$$

Now, consider testing that $\bar{R}^* = 0$ against $\bar{R}^* \geqslant 0$. This is the same as testing that $\gamma_1 = 0$ against $-\infty < \gamma_1 < \infty$. (This problem was considered in Section 16.3.) By arguments similar to those above, we could show that R^* and $\bar{R}^*$ are a maximal invariant and a parameter maximal invariant for this problem, that the distribution of R^* depends only on $\bar{R}^*$ [and (n, s, q)] and that the UMP invariant test can be rewritten in terms of R^*. Finally, let

$R_{n,s}(\bar{R})$ be the distribution of the sample multiple correlation coefficient computed from a sample of size n from an $(s+1)$-dimensional normal distribution with multiple correlation coefficient $\bar{R}$. Then

$$R^* \sim R_{n-s+q,q}(\bar{R}^*) \qquad (16.29)$$

(see Exercise C1 of Chapter 18.)

EXERCISES.

Type A

1. Consider the data of Exercise A1 of Chapter 5. Let $\mathbf{T}_i = (X_i, X_i^2)$. Assume that it is reasonable to assume that the $\mathbf{T}_i'$ have a normal distribution.

(a) Find the sample multiple correlation coefficient between the Y_i and the $\mathbf{T}_i$. Using $\alpha = 0.05$ test the hypothesis that the multiple correlation coefficient is 0.

(b) Find the sample partial correlation coefficient between Y_i and X_i given X_i^2. Using $\alpha = 0.05$ test the hypothesis that partial correlation coefficient is 0 against the alternative that it is nonnegative.

Type B

1. Derive Lemma 16.1.

2. Show that the MLEs given for δ and σ^2 in Theorem 16.3c are correct.

3. Derive Lemma 16.8.

4. In Section 16.3, show that $\bar{G}_5$ consists of transformations of the form

$$\bar{g}_5(\delta, \gamma_1, \gamma_2, \sigma^2, \boldsymbol{\nu}, \boldsymbol{\Xi}) = \left(\delta, \mathbf{E}^{-1}\gamma_1, \gamma_2, \sigma^2, \begin{pmatrix} \mathbf{E}'\mathbf{0} \\ \mathbf{0I} \end{pmatrix} \boldsymbol{\nu}, \begin{pmatrix} \mathbf{E}'\mathbf{0} \\ \mathbf{0I} \end{pmatrix} \boldsymbol{\Xi} \begin{pmatrix} \mathbf{E0} \\ \mathbf{0I} \end{pmatrix}\right).$$

5. In the context of Lemma 16.10, show that

(a) $g_3(\hat{\gamma}_1, \hat{\sigma}^2, \hat{\boldsymbol{\Xi}}) = (\hat{\gamma}_1, \hat{\sigma}^2, \left(\begin{smallmatrix} \mathbf{I} & \mathbf{C}' \\ \mathbf{0} & \mathbf{I} \end{smallmatrix}\right) \hat{\boldsymbol{\Xi}} \left(\begin{smallmatrix} \mathbf{I} & \mathbf{0} \\ \mathbf{C} & \mathbf{I} \end{smallmatrix}\right))$.

(b) $g_4(\hat{\gamma}_1, \hat{\sigma}^2, \mathbf{U}, \hat{\boldsymbol{\Xi}}_{22}) = (\hat{\gamma}_1, \hat{\sigma}^2, \mathbf{U}, \mathbf{D}'\hat{\boldsymbol{\Xi}}_{22}\mathbf{D})$.

(c) $g_5(\hat{\gamma}_1, \hat{\sigma}^2, \mathbf{U}) = (\mathbf{E}^{-1}\gamma_1, \hat{\sigma}^2, \mathbf{E}'\mathbf{U}\mathbf{E})$.

(d) F is a maximal invariant under $G_1, \ldots, G_6$.

6. Verify (16.23).

7. Show that the UMP invariant size α test for testing $\sigma^2 = c$ against $\sigma^2 \geqslant c$ rejects if $\hat{\sigma}^2$ is too large.

8. Verify the formulas for R given in Lemma 16.16.

9. Derive Lemma 16.17.

10. Prove Lemma 16.18.

11. (a) Let $S \subset V$ be subspaces. Show that $\|\mathbf{P}_S\mathbf{Y}\|^2 \leqslant \|\mathbf{P}_V\mathbf{Y}\|^2$.
(b) Show that R^2 increases as more variables are added to the model. (Hint: Let S be the subspace spanned by columns of $\tilde{\mathbf{T}}$. Then $\|\tilde{\mathbf{T}}\hat{\gamma}\|^2 = \|\mathbf{P}_S\tilde{\mathbf{Y}}\|^2$. Why?)

12. Verify (16.26).

13. Verify the formula in (16.27) for the sample partial correlation coefficient.

14. Derive Lemma 16.20.

Type C

1. The non-null distribution of F_1 and the multiple correlation coefficient. We follow the notation of Section 16.2.

(a) Let K and S be random variables such that K given S has a Poisson distribution with mean $\theta_1 S/2$ and $S \sim X^2_{n-1}(0)$. Show that K has density

$$\frac{\Gamma\left(\frac{n-1}{2}+k\right)}{\Gamma\left(\frac{n-1}{2}\right)k!}\left(\frac{\theta_1}{\theta_1+1}\right)^k\left(\frac{1}{\theta_1+1}\right)^{(n-1)/2}.$$

(Note that if n − 1 is even this is a negative binomial density.)

(b) Show that $\mathbf{T}_i\gamma$ are independent, $\mathbf{T}_i\gamma \sim N_1(\boldsymbol{\nu}'\gamma, \gamma'\Xi\gamma)$ and hence that

$$\mathbf{T}\gamma \sim N_n((\mathbf{v}'\gamma)\mathbf{1}, \gamma'\Xi\gamma\mathbf{I}), \qquad \mathbf{1}' = (1, \ldots, 1).$$

(c) Let $\mathbf{S}(\mathbf{T}) = (n-1)\gamma'\hat{\Xi}\gamma/\gamma'\Xi\gamma$. Show that

$$\mathbf{S}(\mathbf{T}) \sim X^2_{n-1}(0).$$

(Hint: Let U be the one-dimensional subspace spanned by $\mathbf{1}$. Then $\mathbf{S}(\mathbf{T}) = \|\mathbf{P}_{U^\perp}\mathbf{T}\|^2/\gamma\Xi\gamma$. Why?)

(d) Show that the density of F_1 can be represented in the following fashion:

$$\frac{s}{s+2K}F_1 \mid K \sim F_{s+2K,n-s-1}(0),$$

where K has the density given in part a. [Hint: Use (16.13]

(e) Find the density function of F_1 in terms of θ_1.

(f) Find the density of the sampe multiple correlation coefficient R in terms of the multiple correlation coefficient $\bar{R}$. (Hint: Use Lemma 16.17.)
Note that the distribution of F_1 can be again considered a central F on a random number of degrees of freedom, but that now the degrees of freedom have a negative binomial distribution instead of a Poisson distribution.

2. Fisher's Z transformation. Let $(Y_1, T_1), \ldots, (Y_n, T_n)$ be independent,

$$\begin{pmatrix} Y_i \\ T_i \end{pmatrix} \sim N_2\left[\begin{pmatrix} 0 \\ 0 \end{pmatrix}, \begin{pmatrix} 1 & \rho \\ \rho & 1 \end{pmatrix}\right].$$

(a) Let $\mathbf{U}_i' = (Y_i^2, T_i^2, Y_i T_i)$. Show that

$$\mathbf{m}(\rho) = E\mathbf{U}_i = \begin{bmatrix} 1 \\ 1 \\ \rho \end{bmatrix}, \qquad \mathbf{K}(\rho) = \operatorname{cov}(\mathbf{U}_i) = \begin{bmatrix} 2 & 2\rho^2 & 2\rho \\ 2\rho^2 & 2 & 2\rho \\ 2\rho & 2\rho & 1+\rho^2 \end{bmatrix}.$$

(b) Show that $\sqrt{n}\,(\overline{\mathbf{U}} - m(\rho)). \xrightarrow{d} N_3(\mathbf{0}, \mathbf{K}(\rho))$. (See Exercise C2 of Chapter 10).

(c) Show that $\sqrt{n}\,(\overline{Y}^2, \overline{T}^2, \overline{Y}\overline{T}) \xrightarrow{p} 0$. [Note that $\overline{Y} \xrightarrow{p} 0$, $\sqrt{n}\, Y \xrightarrow{d} N_1(0, 1)$.]

(d) Let

$$S_1 = \frac{\sum_{i=1}^n Y_i^2}{n} - \overline{Y}^2, \qquad S_2 = \frac{\sum_{i=1}^n T_i^2}{n} - \overline{T}^2, \qquad S_{12} = \frac{\Sigma Y_i T_i}{n} - \overline{Y}\overline{T}.$$

Let $S' = (S_1, S_2, S_{12})$. Show that $\sqrt{n}\,(\mathbf{S} - \mathbf{m}(\rho)) \xrightarrow{d} N_3(\mathbf{0}, \mathbf{K}(\rho))$.

(e) Let $r = S_{12}/\sqrt{S_1 S_2}$. Show that $\sqrt{n}\,(r - \rho) \xrightarrow{d} N_1(0, (1-\rho^2)^2)$. (See Exercise C3 of Chapter 10.)

(f) Let $h(x) = \frac{1}{2}\log[(1+x)/(1-x)]$. Show that $\sqrt{n}\,(h(r) - h(\rho)) \xrightarrow{d} N_1(0,1)$.

CHAPTER 17

The Distribution Theory for Multivariate Analysis

With this chapter, we begin the discussion of multivariate analysis. The models considered in Chapters 1–15 can be transformed to models in which all the random variables are independent. We consider these models univariate models. Models that cannot be transformed in this fashion we call multivariate models. We often generate multivariate models when we replace the random variables in univariate models with random vectors. The correlation model discussed in Chapter 16 is really a multivariate model by this definition, and is often treated in books on multivariate analysis. We have treated it with the univariate models to emphasize the similarity with the ordinary linear model and because, conditionally on the matrix $\mathbf{X}$, the observations Y_i are independent.

In this chapter, we develop the basic distribution theory used in multivariate models. After some elementary discussion of random matrices in Section 17.1, we study the matrix normal distribution in Section 17.2. This distribution plays the same role for problems in multivariate analysis that the vector normal distribution discussed in Chapter 3 does for the univariate models. In Sections 17.3 and 17.4, we discuss the Wishart distribution which is analogous to the χ^2-distribution used in univariate models. In Section 17.5, we study the distribution of a multivariate generalization of the t-statistic. In Section 17.6, we find the density function for the Wishart distribution.

17.1. RANDOM MATRICES.

For the remainder of this book, we write

$$\mathbf{A} = (A_{ij})$$

to mean that A_{ij} is the ijth element of A. A *random matrix* $\mathbf{X} = (X_{ij})$ is a matrix such that X_{ij} are random variables. The *expected value of* $\mathbf{X}$ is defined to be the matrix whose ijth element is the expected value of X_{ij}. The covariance structure of $\mathbf{X}$ is somewhat more difficult. For all i, j, i', j', we need to know

$$\operatorname{cov}(X_{ij}, X_{i'j'})$$

so that we would need a four-dimensional array to represent all the covariances. For this reason, we merely state the covariances and do not attempt to put them into a matrix. The *moment generating function* of $\mathbf{X}$ is defined to be the joint moment generating function of the X_{ij}. To put the moment generating function in a reasonable form we use the trace of a matrix. (See Section 1 of the appendix.) The following properties of the trace are used often in the remainder of this book. Let $\mathbf{A} = (a_{ij})$, $\mathbf{B} = (b_{ij})$ be $m \times n$ matrices. Then

$$\operatorname{tr}\mathbf{AB}' = \sum_i \sum_j a_{ij}b_{ij} = \operatorname{tr}\mathbf{B}'\mathbf{A} = \operatorname{tr}\mathbf{A}'\mathbf{B} = \operatorname{tr}\mathbf{BA}',$$

$$\operatorname{tr}\mathbf{A}'\mathbf{A} = \sum_i \sum_j a_{ij}^2 = \operatorname{tr}\mathbf{AA}' \tag{17.1}$$

(We think of $\operatorname{tr}\mathbf{AB}'$ as the inner product between $\mathbf{A}$ and $\mathbf{B}$ and $\operatorname{tr}\mathbf{A}'\mathbf{A}$ as the length squared of $\mathbf{A}$.) Therefore,

$$\operatorname{tr}(\mathbf{ABC}) = \operatorname{tr}(\mathbf{CAB}) = \operatorname{tr}(\mathbf{BCA}). \tag{17.2}$$

(It is not true, however, that $\operatorname{tr}(\mathbf{ABC}) = \operatorname{tr}(\mathbf{CBA})$.) Also, from the definition of the trace,

$$\operatorname{tr}\mathbf{A}' = \operatorname{tr}\mathbf{A} \tag{17.3}$$

Using the trace notation, we can define the moment generating function of a random matrix $\mathbf{X}$ (i.e., the joint moment generating function of the X_{ij}) by

$$M_{\mathbf{X}}(\mathbf{t}) = E\left[\exp\left(\sum_i \sum_j X_{ij}t_{ij}\right)\right] = E\left\{\exp\left[\operatorname{tr}(\mathbf{X}'\mathbf{t})\right]\right\}, \tag{17.4}$$

where $\mathbf{t} = (t_{ij})$ is a matrix that has the same dimensions as $\mathbf{X}$.

We now derive some elementary properties of these moment generating functions. Let $\mathbf{X}$ be an $n \times p$ matrix, and let a be a scalar. Then $\mathbf{W} = a\mathbf{X}$ has moment generating function

$$M_{\mathbf{W}}(\mathbf{t}) = E\left(\exp\left\{\operatorname{tr}\left[\mathbf{X}'(a\mathbf{t})\right]\right\}\right) = M_{\mathbf{X}}(a\mathbf{t}). \tag{17.5}$$

Let $\mathbf{A}$ be an $m \times n$ matrix, $\mathbf{B}$ be a $p \times r$ matrix and $\mathbf{C}$ be an $m \times r$ matrix. Let $\mathbf{Y} = \mathbf{AXB} + \mathbf{C}$. In Exercise B1 you are asked to show that

$$M_{\mathbf{Y}}(\mathbf{t}) = \exp\left[\operatorname{tr}(\mathbf{C}'\mathbf{t})\right]M_{\mathbf{X}}(\mathbf{A}'\mathbf{t}\mathbf{B}'). \tag{17.6}$$

Let $\mathbf{X} = (\mathbf{X}_1, \mathbf{X}_2)$ where $\mathbf{X}_1$ is $n \times p_1$. Similarly, let $\mathbf{t} = (\mathbf{t}_1, \mathbf{t}_2)$ where $\mathbf{t}_1$ is $n \times p_1$. Then the marginal moment generating function of $\mathbf{X}_1$ is given by

$$M_{\mathbf{X}_1}(\mathbf{t}_1) = M_{\mathbf{X}}(\mathbf{t}_1, \mathbf{0}). \tag{17.7}$$

That is, the marginal moment generating function of $\mathbf{X}_1$ is found by putting $\mathbf{0}$ for $\mathbf{t}_2$ in the moment generating function of $\mathbf{X}$. Also, $\mathbf{X}_1$ and $\mathbf{X}_2$ are independent if and only if

$$M_{\mathbf{X}}(\mathbf{t}_1, \mathbf{t}_2) = M_{\mathbf{X}}(\mathbf{t}_1, \mathbf{0}) M_{\mathbf{X}}(\mathbf{0}, \mathbf{t}_2). \tag{17.8}$$

(See Lemma 3.4d.) Finally,

$$M_{\mathbf{X}'}(\mathbf{t}) = E \exp \operatorname{tr} \mathbf{X}\mathbf{t} = E \exp \operatorname{tr} \mathbf{X}'\mathbf{t}' = M_{\mathbf{X}}(\mathbf{t}'). \tag{17.9}$$

17.2. THE MATRIX NORMAL DISTRIBUTION.

We now define the matrix normal distribution. As in Chapter 3, we work mostly with the moment generating function rather than the density function. As in that chapter, there are two reasons for this choice. The moment generating function is often easier to use than the density function, and there are singular normal distributions that have moment generating functions but no density functions. The derivations in this section are very similar to those in Section 3.4 for the vector normal distribution.

As in Chapter 3, we start with a special case. Let $\mathbf{Z} = (Z_{ij})$ be an $n \times p$ random matrix such that the Z_{ij} are independent and

$$Z_{ij} \sim N_1(0, 1).$$

The moment generating function and density function of $\mathbf{Z}$ are therefore given by

$$M_{\mathbf{Z}}(\mathbf{t}) = \prod_i \prod_j \exp\left(\tfrac{1}{2} t_{ij}^2\right) = \exp\left(\tfrac{1}{2} \operatorname{tr}(\mathbf{t}'\mathbf{t})\right) \tag{17.10}$$

and

$$f(\mathbf{z}) = \prod_i \prod_j \frac{1}{(2\pi)^{1/2}} \exp - \tfrac{1}{2} z_{ij}^2 = \frac{1}{(2\pi)^{np/2}} \exp\left[-\tfrac{1}{2} \operatorname{tr}(\mathbf{z}'\mathbf{z})\right]. \tag{17.11}$$

Now, let $\mathbf{A}$ be $m \times n$, $\mathbf{B}$ be $p \times r$ and $\boldsymbol{\mu}$ be $m \times r$. Define

$$\mathbf{Y} = \mathbf{A}\mathbf{Z}\mathbf{B} + \boldsymbol{\mu}.$$

Then by (17.6),

$$M_{\mathbf{Y}}(\mathbf{t}) = \exp(\operatorname{tr} \boldsymbol{\mu}'\mathbf{t} + \tfrac{1}{2} \operatorname{tr}(\mathbf{t}'\mathbf{A}\mathbf{A}'\mathbf{t}\mathbf{B}'\mathbf{B})). \tag{17.12}$$

Let $\boldsymbol{\Xi} = \mathbf{A}\mathbf{A}'$, $\boldsymbol{\Sigma} = \mathbf{B}'\mathbf{B}$. Then

$$M_{\mathbf{Y}}(\mathbf{t}) = \exp\left[\operatorname{tr} \boldsymbol{\mu}'\mathbf{t} + \tfrac{1}{2} \operatorname{tr}(\mathbf{t}'\boldsymbol{\Xi}\mathbf{t}\boldsymbol{\Sigma})\right]. \tag{17.13}$$

Note that $\boldsymbol{\Xi} \geqslant 0$ and $\boldsymbol{\Sigma} \geqslant 0$. Therefore, let $\boldsymbol{\mu}$ be an $m \times r$ matrix, $\boldsymbol{\Xi}$ and $\boldsymbol{\Sigma}$ be $m \times m$ and $r \times r$ nonnegative definite matrices. We say that $\mathbf{Y}$ has a *matrix normal distribution* with parameters $\boldsymbol{\mu}$, $\boldsymbol{\Xi}$ and $\boldsymbol{\Sigma}$ if $\mathbf{Y}$ is an $m \times r$ random matrix having moment generating function given by (17.13). We write

$$\mathbf{Y} \sim N_{m,r}(\boldsymbol{\mu}, \boldsymbol{\Xi}, \boldsymbol{\Sigma}).$$

Note that this distribution is not completely identified. If a is a positive scalar, $\mathbf{Y}$ also has the distribution $Y \sim N_{m,r}(\boldsymbol{\mu}, (1/a)\boldsymbol{\Xi}, a\boldsymbol{\Sigma})$. In the applications studied in this book, the lack of identification is not important because the matrix $\boldsymbol{\Xi}$ is a known matrix (often the identity) and $\boldsymbol{\Sigma}$ is an unknown parameter. In this setting (known $\boldsymbol{\Xi}$) the distribution is completely identified. Note also that for any nonnegative definite matrices $\boldsymbol{\Xi}$ and $\boldsymbol{\Sigma}$ there exist $\mathbf{A}$ and $\mathbf{B}$ such that $\boldsymbol{\Xi} = \mathbf{A}\mathbf{A}'$ and $\boldsymbol{\Sigma} = \mathbf{B}'\mathbf{B}$. Therefore, for any $\boldsymbol{\mu}$, $\boldsymbol{\Xi}$ and $\boldsymbol{\Sigma}$ of the appropriate dimensions such that $\boldsymbol{\Xi}$ and $\boldsymbol{\Sigma}$ are nonnegative definite, there exists a distribution having moment generating function given by (17.13).

We first find the means, variances and covariances of the Y_{ij}.

THEOREM 17.1. Let $\mathbf{Y} \sim N_{m,r}(\boldsymbol{\mu}, \boldsymbol{\Xi}, \boldsymbol{\Sigma})$. Let $\mathbf{Y} = (Y_{ij})$, $\boldsymbol{\mu} = (\mu_{ij})$, $\boldsymbol{\Xi} = (\Xi_{ij})$, $\boldsymbol{\Sigma} = (\Sigma_{ij})$. Then $EY_{ij} = \mu_{ij}$, $\operatorname{var}(Y_{ij}) = \Xi_{ii}\Sigma_{jj}$, and $\operatorname{cov}(Y_{ij}, Y_{i'j'}) = \Xi_{ii'}\Sigma_{jj'}$.

PROOF. Let $\mathbf{Z}$ be defined as before. Then $EZ_{ij} = 0$, $\operatorname{var}(Z_{ij}) = 1$, $\operatorname{cov}(Z_{ij}, Z_{i'j'}) = 0$ unless $i = i'$ and $j = j'$. Now let $\mathbf{A}$ and $\mathbf{B}$ be such that $\boldsymbol{\Xi} = \mathbf{A}\mathbf{A}'$, $\boldsymbol{\Sigma} = \mathbf{B}'\mathbf{B}$, and let $\mathbf{Y} = \mathbf{AZB} + \boldsymbol{\mu}$. Then, by the development above $\mathbf{Y} \sim N_{m,r}(\boldsymbol{\mu}, \boldsymbol{\Xi}, \boldsymbol{\Sigma})$.

$$Y_{ij} = \sum_k \sum_s a_{ik} z_{ks} b_{sj} + \mu_{ij}.$$

Therefore, $EY_{ij} = \mu_{ij}$, and

$$\begin{aligned}\operatorname{cov}(Y_{ij}, Y_{i'j'}) &= \sum_k \sum_s \sum_{k'} \sum_{s'} a_{ik} b_{sj} a_{i'k'} b_{s'j'} \operatorname{cov}(Z_{ks} Z_{k's'}) \\ &= \sum_k a_{ik} a_{i'k} \sum_s b_{sj} b_{sj'} = \Xi_{ii'} \Sigma_{jj'}.\end{aligned}$$

The formula for the variance follows from the fact that $\operatorname{var}((Y_{ij}) = \operatorname{cov}(Y_{ij}, Y_{ij})$. □

It helps to think of $\boldsymbol{\Xi}$ as representing the covariance between the rows of $\mathbf{Y}$ and $\boldsymbol{\Sigma}$ as representing the covariance between the columns. By Theorem 17.1, then, the covariance between two elements Y_{ij} and $Y_{i'j'}$, is just the covariance between the rows i and i' multiplied by the covariance between the columns j and j'.

We now state some elementary properties of the matrix normal distribution. These results parallel the results derived for the vector normal distribution.

THEOREM 17.2. Let $\mathbf{X} \sim N_{m,r}(\boldsymbol{\mu}, \boldsymbol{\Xi}, \boldsymbol{\Sigma})$.

a. If $r = 1$, $\Sigma = \sigma^2$ (a scalar), then $\mathbf{X} \sim N_m(\boldsymbol{\mu}, \sigma^2 \boldsymbol{\Xi})$. (This is the vector normal distribution discussed in Chapter 3.)

b. If a is a scalar, then $a\mathbf{X} \sim N_{m,r}(a\boldsymbol{\mu}, a^2\boldsymbol{\Xi}, \boldsymbol{\Sigma})$.

c.

$$\mathbf{X}' \sim N_{r,m}(\boldsymbol{\mu}', \boldsymbol{\Sigma}, \boldsymbol{\Xi}).$$

d. If $\mathbf{C}$ is $n \times m$, $\mathbf{D}$ is $r \times p$ and $\mathbf{E}$ is $n \times p$, then

$$\mathbf{CXD} + \mathbf{E} \sim N_{n,p}(\mathbf{C}\boldsymbol{\mu}\mathbf{D} + \mathbf{E}, \mathbf{C}\boldsymbol{\Xi}\mathbf{C}', \mathbf{D}'\boldsymbol{\Sigma}\mathbf{D}).$$

e. Let

$$\mathbf{X} = (\mathbf{X}_1, \mathbf{X}_2), \qquad \boldsymbol{\mu} = (\boldsymbol{\mu}_1, \boldsymbol{\mu}_2) \qquad \boldsymbol{\Sigma} = \begin{pmatrix} \boldsymbol{\Sigma}_{11} & \boldsymbol{\Sigma}_{12} \\ \boldsymbol{\Sigma}_{21} & \boldsymbol{\Sigma}_{22} \end{pmatrix},$$

where $\mathbf{X}_1$ and $\boldsymbol{\mu}_1$ are $m \times r_1$ and $\boldsymbol{\Sigma}_{11}$ is $r_1 \times r_1$. Then

$$\mathbf{X}_1 \sim N_{m,r_1}(\boldsymbol{\mu}_1, \boldsymbol{\Xi}, \boldsymbol{\Sigma}_{11}), \mathbf{X}_2 \sim N_{m,r-r_1}(\boldsymbol{\mu}_2, \boldsymbol{\Xi}, \boldsymbol{\Sigma}_{22}).$$

f. if $\boldsymbol{\Xi} \neq 0$, then $\mathbf{X}_1$ and $\mathbf{X}_2$ are independent if and only if $\boldsymbol{\Sigma}_{12} = \mathbf{0}$.

g. If $\boldsymbol{\Sigma}_{22} > 0$, then

$$\mathbf{X}_1 | \mathbf{X}_2 \sim N_{m,r_1}\left(\boldsymbol{\mu}_1 + (\mathbf{X}_2 - \boldsymbol{\mu}_2)\boldsymbol{\Sigma}_{22}^{-1}\boldsymbol{\Sigma}_{21}, \boldsymbol{\Xi}, \boldsymbol{\Sigma}_{11} - \boldsymbol{\Sigma}_{12}\boldsymbol{\Sigma}_{22}^{-1}\boldsymbol{\Sigma}_{21}\right).$$

PROOF. See Exercise B2. □

By using parts c, e, f, and g of Theorem 17.2, we can also establish results when $\mathbf{X}$ is partitioned as $\mathbf{X} = \binom{\mathbf{X}_1}{\mathbf{X}_2}$.

For most of the multivariate models discussed in the remainder of the book, we study a collection of individuals on each of which we measure p quantities, for example, height, weight, and head circumference. Let $\mathbf{X}_i$ be a p-dimensional row vector with the measurements for the ith individual. We assume that the $\mathbf{X}_i$ are independent and that they are normally distributed with mean vector $\boldsymbol{\mu}_i$, and common covariance matrix $\boldsymbol{\Sigma}$. We now show that this model is a particular case of the matrix normal model.

THEOREM 17.3. Let $\mathbf{X} \sim N_{n,p}(\boldsymbol{\mu}, \mathbf{I}, \boldsymbol{\Sigma})$. Let

$$\mathbf{X} = \begin{bmatrix} \mathbf{X}_1 \\ \vdots \\ \mathbf{X}_n \end{bmatrix}, \qquad \boldsymbol{\mu} = \begin{bmatrix} \boldsymbol{\mu}_1 \\ \vdots \\ \boldsymbol{\mu}_n \end{bmatrix}.$$

Then the $\mathbf{X}_i$ are independent and $\mathbf{X}_i' \sim N_p(\boldsymbol{\mu}_i', \boldsymbol{\Sigma})$.

PROOF. Suppose that the $\mathbf{X}_i$ are independent and $\mathbf{X}_i' \sim N_p(\boldsymbol{\mu}_i', \boldsymbol{\Sigma})$. We find the moment generating function of $\mathbf{X}$ and show that it has the form

(17.13) with **I** for Ξ. Let

$$\mathbf{t} = \begin{bmatrix} \mathbf{t}_1 \\ \vdots \\ \mathbf{t}_n \end{bmatrix}.$$

where the $\mathbf{t}_i$ are $1 \times p$. Then

$$M_{\mathbf{X}}(\mathbf{t}) = \prod_i \exp\left[\, \boldsymbol{\mu}_i \mathbf{t}_i + \tfrac{1}{2} \mathbf{t}_i' \boldsymbol{\Sigma} \mathbf{t}_i \right] = \exp\left(\sum_i \boldsymbol{\mu}_i \mathbf{t}_i + \tfrac{1}{2} \sum_i \mathbf{t}_i' \boldsymbol{\Sigma} \mathbf{t}_i \right)$$

$$= \exp\left[\operatorname{tr} \boldsymbol{\mu}' \mathbf{t} + \tfrac{1}{2} \operatorname{tr} \mathbf{t}' \boldsymbol{\Sigma} \mathbf{t} \right]. \quad \square$$

The final result derived in this section is the density function for the matrix normal distribution. If Ξ or $\boldsymbol{\Sigma}$ is not positive definite, then the distribution is singular and has no density function (see Exercise B3). We therefore assume that $\Xi > 0$ and $\boldsymbol{\Sigma} > 0$.

THEOREM 17.4. If $\mathbf{X} \sim N_{m,r}(\boldsymbol{\mu}, \Xi, \boldsymbol{\Sigma})$, $\Xi > 0$ and $\boldsymbol{\Sigma} > 0$, then **X** has joint density

$$f_{\mathbf{X}}(\mathbf{x}) = \frac{1}{(2\pi)^{mr/2} |\Xi|^{r/2} |\boldsymbol{\Sigma}|^{m/2}} \exp - \tfrac{1}{2} \operatorname{tr} \Xi^{-1} (\mathbf{x} - \boldsymbol{\mu}) \boldsymbol{\Sigma}^{-1} (\mathbf{x} - \boldsymbol{\mu})'.$$

PROOF. We start with **Z** defined above having joint density

$$f_{\mathbf{Z}}(\mathbf{z}) = \frac{1}{(2\pi)^{mr/2}} \exp - \tfrac{1}{2} \operatorname{tr} \mathbf{z}\mathbf{z}'.$$

Let **A** and **B** be invertible matrices such that $\Xi = \mathbf{A}\mathbf{A}'$, $\boldsymbol{\Sigma} = \mathbf{B}'\mathbf{B}$. This is possible because Ξ and $\boldsymbol{\Sigma}$ are positive definite. Let $\mathbf{X} = \mathbf{A}\mathbf{Z}\mathbf{B} + \boldsymbol{\mu} \sim N_{m,r}(\boldsymbol{\mu}, \Xi, \boldsymbol{\Sigma})$. Then **X** is an invertible function of **Z** and $\mathbf{Z} = \mathbf{A}^{-1}(\mathbf{X} - \boldsymbol{\mu})\mathbf{B}^{-1}$. The absolute value of the Jacobian of this transformation is given by

$$|\mathbf{J}| = \left| |\mathbf{A}^{-1}|^r |\mathbf{B}^{-1}|^m \right| = \frac{1}{|\Xi|^{r/2} |\boldsymbol{\Sigma}|^{m/2}}.$$

(See Theorem 16 of the appendix.) Therefore, **X** has density

$$f_{\mathbf{X}}(\mathbf{x}) = \frac{1}{|\Xi|^{r/2} |\boldsymbol{\Sigma}|^{m/2}} f_Z\left(\mathbf{A}^{-1} (\mathbf{x} - \boldsymbol{\mu}) \mathbf{B}^{-1} \right)$$

$$= \frac{1}{(2\pi)^{mr/2} |\Xi|^{r/2} |\boldsymbol{\Sigma}|^{m/2}}$$

$$\exp\left\{ - \tfrac{1}{2} \operatorname{tr}\left[\mathbf{A}^{-1} (\mathbf{x} - \boldsymbol{\mu}) \mathbf{B}^{-1} (\mathbf{B}^{-1})' (\mathbf{x} - \boldsymbol{\mu})' (\mathbf{A}^{-1})' \right] \right\}$$

$$= \frac{1}{(2\pi)^{mr/2} |\Xi|^{r/2} |\boldsymbol{\Sigma}|^{m/2}} \exp\left\{ - \tfrac{1}{2} \operatorname{tr}\left[\Xi^{-1} (\mathbf{x} - \boldsymbol{\mu}) \boldsymbol{\Sigma}^{-1} (\mathbf{x} - \boldsymbol{\mu})' \right] \right\}. \quad \square$$

17.3. THE WISHART DISTRIBUTION.

We now define the Wishart distribution and derive some of its elementary properties. The reader should think of the Wishart distribution as a generalized χ^2-distribution. Let

$$\mathbf{X} = \begin{bmatrix} \mathbf{X}_1 \\ \vdots \\ \mathbf{X}_n \end{bmatrix} \sim N_{n,p}(\boldsymbol{\mu}, \mathbf{I}, \boldsymbol{\Sigma}) \tag{17.14}$$

(i.e., the $\mathbf{X}_i$ are independent, $\mathbf{X}_i' \sim N_p(\boldsymbol{\mu}_i', \boldsymbol{\Sigma})$). Let

$$\mathbf{W} = \mathbf{X}'\mathbf{X} = \sum_{i=1}^{n} \mathbf{X}_i'\mathbf{X}_i. \tag{17.15}$$

Then $\mathbf{W}$ is said to have a *Wishart distribution*. Note that $\mathbf{W}$ is a $p \times p$ matrix and $\mathbf{W} \geqslant 0$.

LEMMA 17.5. The distribution of $\mathbf{W}$ depends on $\boldsymbol{\mu}$ only through $\boldsymbol{\mu}'\boldsymbol{\mu}$.

PROOF. We use an invariance argument to show this result. Let $\boldsymbol{\Gamma}$ be an $n \times n$ orthogonal matrix. Then

$$\mathbf{Y} = \boldsymbol{\Gamma}\mathbf{X} \sim N_{n,\,p}(\boldsymbol{\Gamma}\boldsymbol{\mu}, \mathbf{I}, \boldsymbol{\Sigma})$$

by Theorem 17.2d. In addition

$$\mathbf{W} = \mathbf{X}'\mathbf{X} = \mathbf{Y}'\mathbf{Y}$$

and the distribution of $\mathbf{W}$ is obviously independent of whether $\mathbf{W}$ is computed from $\mathbf{X}$ or from $\mathbf{Y}$. Therefore, if $F(\mathbf{W}; \boldsymbol{\mu})$ is the distribution function of $\mathbf{W}$ for a particular $\boldsymbol{\mu}$, then

$$F(\mathbf{W}; \boldsymbol{\Gamma}\boldsymbol{\mu}) = F(\mathbf{W}, \boldsymbol{\mu}).$$

Hence F is invariant under the group $\overline{G}$ of orthogonal transformations

$$\bar{g}(\boldsymbol{\mu}) = \boldsymbol{\Gamma}\boldsymbol{\mu}.$$

A maximal invariant under this group is $\boldsymbol{\delta}(\boldsymbol{\mu}) = \boldsymbol{\mu}'\boldsymbol{\mu}$, as the following argument shows.

(1) $\boldsymbol{\delta}(\boldsymbol{\Gamma}\boldsymbol{\mu}) = (\boldsymbol{\Gamma}\boldsymbol{\mu})'\boldsymbol{\Gamma}\boldsymbol{\mu} = \boldsymbol{\mu}'\boldsymbol{\mu} = \delta(\boldsymbol{\mu})$, so $\boldsymbol{\delta}$ is invariant.
(2) If $\boldsymbol{\delta}(\boldsymbol{\mu}) = \boldsymbol{\delta}(\boldsymbol{\nu})$, then $\boldsymbol{\mu}'\boldsymbol{\mu} = \boldsymbol{\nu}'\boldsymbol{\nu}$, By Theorem 11 of the appendix, there exists $\boldsymbol{\Gamma}$, orthogonal such that $\boldsymbol{\mu} = \boldsymbol{\Gamma}\boldsymbol{\nu}$. Therefore, $\boldsymbol{\delta}$ is maximal.

The distribution is invariant under $\overline{G}$ and hence by Lemma 1.11 must be a function of the maximal invariant under $\overline{G}$, which is $\boldsymbol{\delta} = \boldsymbol{\mu}'\boldsymbol{\mu}$. □

With Lemma 17.5 in mind, we say that the distribution of $\mathbf{W}$ defined by (17.14) and (17.15) is a *p-dimensional Wishart distribution with n degrees of freedom, on the covariance matrix* $\boldsymbol{\Sigma}$, *and with noncentrality matrix* $\boldsymbol{\delta} = \boldsymbol{\mu}'\boldsymbol{\mu}$.

We write

$$\mathbf{W} \sim W_p(n, \boldsymbol{\Sigma}, \boldsymbol{\delta}).$$

Note that both $\boldsymbol{\Sigma} \geqslant 0$ and $\boldsymbol{\delta} \geqslant 0$. If $\boldsymbol{\delta} = \mathbf{0}$, we say that W has a *central Wishart distribution*, and write

$$\mathbf{W} \sim W_p(n, \boldsymbol{\Sigma}).$$

If $\boldsymbol{\delta} \neq 0$, we say $\mathbf{W}$ has a *noncentral Wishart distribution.*

The following theorem summarizes some elementary properties of the Wishart distribution that are derived directly from the definition.

THEOREM 17.6. Let $\mathbf{W} \sim W_p(n, \boldsymbol{\Sigma}, \boldsymbol{\delta})$.

a. $\mathbf{EW} = n\boldsymbol{\Sigma} + \boldsymbol{\delta}$.

b. If $p = 1$, $\boldsymbol{\Sigma} = \sigma^2 > 0$, then $W \sim \sigma^2\chi_n^2(\delta/\sigma^2)$

c. If $a \geqslant 0$ is a scalar, then $a\mathbf{W} \sim W_p(n, a\boldsymbol{\Sigma}, a\boldsymbol{\delta})$.

d. If $\mathbf{A}$ is a $k \times p$ matrix, then $\mathbf{AWA'} \sim W_k(n, \mathbf{A}\boldsymbol{\Sigma}\mathbf{A'}, \mathbf{A}\boldsymbol{\delta}\mathbf{A'})$.

e. Let

$$\mathbf{W} = \begin{pmatrix} \mathbf{W}_{11} & \mathbf{W}_{12} \\ \mathbf{W}_{21} & \mathbf{W}_{22} \end{pmatrix}, \qquad \boldsymbol{\Sigma} = \begin{pmatrix} \boldsymbol{\Sigma}_{11} & \boldsymbol{\Sigma}_{12} \\ \boldsymbol{\Sigma}_{21} & \boldsymbol{\Sigma}_{22} \end{pmatrix}, \qquad \boldsymbol{\delta} = \begin{pmatrix} \boldsymbol{\delta}_{11} & \boldsymbol{\delta}_{12} \\ \boldsymbol{\delta}_{21} & \boldsymbol{\delta}_{22} \end{pmatrix}$$

where $\mathbf{W}_{11}, \boldsymbol{\Sigma}_{11}$, and $\boldsymbol{\delta}_{11}$ are $k \times k$. Then $\mathbf{W}_{11} \sim W_k(n, \boldsymbol{\Sigma}_{11}, \boldsymbol{\delta}_{11})$, $\mathbf{W}_{22} \sim W_{p-k}(n, \boldsymbol{\Sigma}_{22}, \boldsymbol{\delta}_{22})$.

PROOF. a. Let $\mathbf{W} = \sum_1^n \mathbf{X}_i'\mathbf{X}_i$, where the $\mathbf{X}_i$ are independent, $\mathbf{X}_i \sim N_p(\boldsymbol{\mu}_i, \boldsymbol{\Sigma})$. Then $E\mathbf{X}_i'\mathbf{X}_i = \boldsymbol{\Sigma} + \boldsymbol{\mu}_i'\boldsymbol{\mu}_i$. Therefore,

$$E\mathbf{W} = \sum_{i=1}^{n} E\mathbf{X}_i'\mathbf{X}_i = n\boldsymbol{\Sigma} + \sum_{i=1}^{n} \boldsymbol{\mu}_i'\boldsymbol{\mu}_i = n\boldsymbol{\Sigma} + \boldsymbol{\delta}.$$

b. and c. See Exercise B4.

d. Let $\mathbf{W} = \mathbf{X'X}$ where $\mathbf{X} \sim N_{n,\,p}(\boldsymbol{\mu}, \mathbf{I}, \boldsymbol{\Sigma})$. By Theorem 17.2d $\mathbf{XA'} \sim N_{n,\,k}(\boldsymbol{\mu}\mathbf{A'}, \mathbf{I}, \mathbf{A}\boldsymbol{\Sigma}\mathbf{A'})$. Therefore, by the definition of the Wishart distribution

$$\mathbf{AWA'} = (\mathbf{XA'})'\mathbf{XA'} \sim W_k\left[n, \mathbf{A}\boldsymbol{\Sigma}\mathbf{A}, (\boldsymbol{\mu}\mathbf{A'})'\boldsymbol{\mu}\mathbf{A'}\right], \qquad (\boldsymbol{\mu}\mathbf{A'})'\boldsymbol{\mu}\mathbf{A'} = \mathbf{A}\boldsymbol{\delta}\mathbf{A'}.$$

e. See Exercise B4. □

The next theorem establishes some other elementary results. These results are analogous to those derived in Chapter 3 for the χ^2 distributions. The proofs are identical for the multivariate case considered here.

THEOREM 17.7. Let $\mathbf{X} \sim N_{n,\,p}(\boldsymbol{\mu}, \mathbf{I}, \boldsymbol{\Sigma})$.

a. Let V be a k dimensional subspace of R^n. Then

$$\mathbf{X'P}_V\mathbf{X} \sim W_p(k, \boldsymbol{\Sigma}, \boldsymbol{\mu}'\mathbf{P}_V\boldsymbol{\mu})$$

[and hence if $\mathbf{A}$ is an $n \times n$ indempotent matrix of rank k, then

$$\mathbf{X'AX} \sim W_p(k, \boldsymbol{\Sigma}, \boldsymbol{\mu}'\mathbf{A}\boldsymbol{\mu})].$$

b. 1. $\mathbf{AX}$ and $\mathbf{BX}$ are independent if and only if $\mathbf{AB}' = \mathbf{0}$.
2. If $\mathbf{B} \geqslant 0$ and $\mathbf{AB} = \mathbf{0}$, then $\mathbf{AX}$ and $\mathbf{X}'\mathbf{BX}$ are independent.
3. If $\mathbf{A} \geqslant 0$, $\mathbf{B} \geqslant 0$ and $\mathbf{AB} = \mathbf{0}$, then $\mathbf{X}'\mathbf{AX}$ and $\mathbf{X}'\mathbf{BX}$ are independent.

PROOF. a. Let $\mathbf{U}$ be an orthornormal basis matrix for V. By Theorem 17.2d,

$$\mathbf{U}'\mathbf{X} \sim N_{k,\,p}(\mathbf{U}'\boldsymbol{\mu}, \mathbf{I}, \boldsymbol{\Sigma}).$$

Therefore, by the definition of the Wishart distribution

$$(\mathbf{U}'\mathbf{X})'(\mathbf{U}'\mathbf{X}) = \mathbf{X}'\mathbf{P}_V\mathbf{X} \sim W_p(k, \boldsymbol{\Sigma}, \boldsymbol{\mu}'\mathbf{U}'\mathbf{U}\boldsymbol{\mu}), \qquad \boldsymbol{\mu}'\mathbf{U}'\mathbf{U}\boldsymbol{\mu} = \boldsymbol{\mu}'\mathbf{P}_V\boldsymbol{\mu}.$$

b. See Exercise B5. □

There is one more very important property of the central Wishart distribution. By the definition of the Wishart distribution, $\mathbf{W} \geqslant 0$. We want to know when $\mathbf{W} > 0$, and hence when $\mathbf{W}$ is invertible. We first prove the following theorem. The second part of the theorem was used in Chapter 16.

THEOREM 17.8. Let $\mathbf{X} \sim N_{n,\,p}(\boldsymbol{\mu}, \boldsymbol{\Xi}, \boldsymbol{\Sigma})$, $\boldsymbol{\Xi} > 0$, $\boldsymbol{\Sigma} > 0$.
a. If $n \geqslant p$, then $P(\mathbf{X}$ has rank $p) = 1$.
b. Let $\mathbf{a} \in R^n$, $\mathbf{a} \neq \mathbf{0}$. If $n > p$, then $P[(\mathbf{a}, \mathbf{X})$ has rank $p + 1] = 1$.

PROOF. Let $\mathbf{X} = (\mathbf{X}_1, \ldots, \mathbf{X}_p)$. Let $S_k(\mathbf{X}_1, \ldots, \mathbf{X}_k)$ be the subspace of R^n spanned by $\mathbf{X}_1, \ldots, \mathbf{X}_k$. Since the conditional distribution of $\mathbf{X}_{k+1} \mid (\mathbf{X}_1, \ldots, \mathbf{X}_k)$ is a nonsingular (vector) normal distribution,

$$P\left[\mathbf{X}_{k+1} \in S_k(\mathbf{X}_1, \ldots, \mathbf{X}_k) \mid (\mathbf{X}_1, \ldots, \mathbf{X}_k)\right] = 0 \qquad \text{if } k < n.$$

(Note that S_k is a subspace of dimension at most k.) Therefore,

$$P\left[\mathbf{X}_{k+1} \in S_k(\mathbf{X}_1, \ldots, \mathbf{X}_k)\right] = E\left\{P\left[\mathbf{X}_{k+1} \in S_k(\mathbf{X}_1, \ldots, \mathbf{X}_k) \mid (\mathbf{X}_1, \ldots, \mathbf{X}_k)\right]\right\} = 0.$$

Finally,

$$P(\mathbf{X}_1, \ldots, \mathbf{X}_p \text{ are linearly dependent}) \leqslant \sum_{k=0}^{p-1} P\left[\mathbf{X}_{k+1} \in S_k(\mathbf{X}_1, \ldots, \mathbf{X}_k)\right] = 0.$$

Therefore, if $n \geqslant p$, $\boldsymbol{\Xi} > 0$, $\boldsymbol{\Sigma} > 0$, then $P(\mathbf{X}$ has rank $p) = 1$. The second part follows similarly and is left for Exercise B6. □

COROLLARY. Let $\mathbf{W} \sim W_p(n, \boldsymbol{\Sigma}, \boldsymbol{\delta})$ If $n \geqslant p$, and $\boldsymbol{\Sigma} > 0$, then

$$P(\mathbf{W} > 0) = 1.$$

PROOF. The rank of $W = \mathbf{X}'\mathbf{X}$ is the same as the rank of $\mathbf{X}$ (see Lemma 9 of the appendix). By Theorem 17.8, if $\boldsymbol{\Sigma} > 0$, $n \geqslant p$, then the rank of $\mathbf{W}$ is p

with probability 1. Since $\mathbf{W}$ is $p \times p$, it is invertible, and hence, with probability 1, $\mathbf{W} > 0$. □

Since $\mathbf{W}$ is invertible with probability 1 if $\boldsymbol{\Sigma} > 0$ and $n \geqslant p$, we call this case the *nonsingular central Wishart distribution*, which is the Wishart distribution that will be studied for most of this book. In addition, we ignore the null set on which $\mathbf{W}$ is not invertible and assume $\mathbf{W} > 0$. In Section 17.6, we derive the density function of the nonsingular central Wishart distribution. It is a rather interesting density, being over the space of positive definite matrices. In many books and references, if $n < p$ or $\boldsymbol{\Sigma}$ is not positive definite then the central Wishart distribution is called a pseudo-Wishart distribution. In this book we call such a distribution a *singular Wishart distribution.*

17.4. AN IMPORTANT LEMMA.

We have now finished the derivation of the elementary facts about the Wishart distribution. In the remaining sections of this chapter as well as in later chapters we need some deeper properties of the Wishart distribution. These properties will all be derived using the following lemma.

Let $\mathbf{W} \sim W_p(n, \boldsymbol{\Sigma})$ with $\boldsymbol{\Sigma} > 0$, $n \geqslant p$. Partition $\mathbf{W}$ and $\boldsymbol{\Sigma}$ as

$$\mathbf{W} = \begin{pmatrix} \mathbf{W}_{11} & \mathbf{W}_{12} \\ \mathbf{W}_{21} & \mathbf{W}_{22} \end{pmatrix}, \qquad \boldsymbol{\Sigma} = \begin{pmatrix} \boldsymbol{\Sigma}_{11} & \boldsymbol{\Sigma}_{12} \\ \boldsymbol{\Sigma}_{21} & \boldsymbol{\Sigma}_{22} \end{pmatrix} \tag{17.16}$$

where $\mathbf{W}_{11}$ and $\boldsymbol{\Sigma}_{11}$ are $q \times q$.

LEMMA 17.9. Let $\mathbf{T} = \mathbf{W}_{22}$, $\mathbf{U} = \mathbf{W}_{22}^{-1}\mathbf{W}_{21}$, $\mathbf{V} = \mathbf{W}_{11} - \mathbf{W}_{12}\mathbf{W}_{22}^{-1}\mathbf{W}_{21}$. Then

$$\mathbf{T} \sim W_{p-q}(n, \boldsymbol{\Sigma}_{22}), \qquad \mathbf{U} \mid \mathbf{T} \sim N_{p-q,q}\left(\boldsymbol{\Sigma}_{22}^{-1}\boldsymbol{\Sigma}_{21}, \mathbf{T}^{-1}, \boldsymbol{\Sigma}_{11} - \boldsymbol{\Sigma}_{12}\boldsymbol{\Sigma}_{22}^{-1}\boldsymbol{\Sigma}_{21}\right),$$
$$\mathbf{V} \mid (\mathbf{U}, \mathbf{T}) \sim W_q\left(n - p + q, \boldsymbol{\Sigma}_{11} - \boldsymbol{\Sigma}_{12}\boldsymbol{\Sigma}_{22}^{-1}\boldsymbol{\Sigma}_{21}\right).$$

PROOF. The marginal distribution of $\mathbf{T}$ follows from Theorem 17.6e. Now, let $(\mathbf{Y}\mathbf{X}) \sim N_{n,p}(\mathbf{0}, \mathbf{I}, \boldsymbol{\Sigma})$ where $\mathbf{Y}$ is $n \times q$, and let

$$\mathbf{W} = (\mathbf{Y}\mathbf{X})'(\mathbf{Y}\mathbf{X}).$$

Then $\mathbf{W} \sim W_p(n, \boldsymbol{\Sigma})$ and $\mathbf{W}_{11} = \mathbf{Y}'\mathbf{Y}$, $\mathbf{W}_{12} = \mathbf{Y}'\mathbf{X}$, $\mathbf{W}_{21} = \mathbf{X}'\mathbf{Y}$, $\mathbf{W}_{22} = \mathbf{X}'\mathbf{X}$. Therefore

$$\mathbf{T} = \mathbf{X}'\mathbf{X}, \mathbf{U} = (\mathbf{X}'\mathbf{X})^{-1}\mathbf{X}'\mathbf{Y}, \mathbf{V} = \mathbf{Y}'\left(\mathbf{I} - \mathbf{X}(\mathbf{X}'\mathbf{X})^{-1}\mathbf{X}'\right)\mathbf{Y}. \tag{17.17}$$

We now find the conditional distribution of $(\mathbf{U}, \mathbf{V})$ given $\mathbf{X}$. By Theorem 17.2g we see that

$$\mathbf{Y} \mid \mathbf{X} \sim N_{n,q}\left(\mathbf{X}\boldsymbol{\Sigma}_{22}^{-1}\boldsymbol{\Sigma}_{21}, \mathbf{I}, \boldsymbol{\Sigma}_{11} - \boldsymbol{\Sigma}_{12}\boldsymbol{\Sigma}_{22}^{-1}\boldsymbol{\Sigma}_{21}\right). \tag{17.18}$$

From Theorem 17.2d we see that

$$\mathbf{U}\,|\,\mathbf{X} \sim N_{p-q,q}\left[\boldsymbol{\Sigma}_{22}^{-1}\boldsymbol{\Sigma}_{21}, (\mathbf{X}'\mathbf{X})^{-1}, \boldsymbol{\Sigma}_{11} - \boldsymbol{\Sigma}_{12}\boldsymbol{\Sigma}_{22}^{-1}\boldsymbol{\Sigma}_{21}\right].$$

Since the conditional distribution of **U** given **X** depends on **X** only through $\mathbf{T} = \mathbf{X}'\mathbf{X}$, we see that

$$\mathbf{U}\,|\,\mathbf{T} \sim N_{p-q,q}\left(\boldsymbol{\Sigma}_{22}^{-1}\boldsymbol{\Sigma}_{21}, \mathbf{T}^{-1}, \boldsymbol{\Sigma}_{11} - \boldsymbol{\Sigma}_{12}\boldsymbol{\Sigma}_{22}^{-1}\boldsymbol{\Sigma}_{21}\right).$$

Now $\mathbf{I} - \mathbf{X}(\mathbf{X}'\mathbf{X})^{-1}\mathbf{X}'$ is an idempotent matrix of rank $n - (p - q)$ and

$$(\mathbf{X}\boldsymbol{\Sigma}_{22}^{-1}\boldsymbol{\Sigma}_{21})'\left[\mathbf{I} - \mathbf{X}(\mathbf{X}'\mathbf{X})^{-1}\mathbf{X}'\right](\mathbf{X}\boldsymbol{\Sigma}_{22}^{-1}\boldsymbol{\Sigma}_{21}) = 0.$$

Therefore, by Theorem 17.7a,

$$\mathbf{V}\,|\,\mathbf{X} \sim W_q\left(n - p + q, \boldsymbol{\Sigma}_{11} - \boldsymbol{\Sigma}_{12}\boldsymbol{\Sigma}_{22}^{-1}\boldsymbol{\Sigma}_{21}\right). \tag{17.19}$$

Finally,

$$\left[(\mathbf{X}'\mathbf{X})^{-1}\mathbf{X}'\right]\left[\mathbf{I} - \mathbf{X}(\mathbf{X}'\mathbf{X})^{-1}\mathbf{X}'\right] = \mathbf{0}.$$

Therefore, by Theorem 17.7b **U** and **V** are independent, conditionally on **X**. This fact together with (17.19) implies that **V** is independent of $(\mathbf{U}, \mathbf{T})$ and therefore

$$\mathbf{V}\,|\,(\mathbf{U}, \mathbf{T}) \sim W_q\left(n - p + q, \boldsymbol{\Sigma}_{11} - \boldsymbol{\Sigma}_{12}\boldsymbol{\Sigma}_{22}^{-1}\boldsymbol{\Sigma}_{21}\right). \quad \square$$

In understanding the derivation of this lemma, it is helpful to look at the case $q = 1$, in which case the conditional distribution of $(\mathbf{U}, V)$ given **T** is just a restatement of known results about the coordinatized ordinary linear model, as we now show. Let $(\mathbf{YX})$ be as defined in the proof of the lemma (with $q = 1$) and let

$$\boldsymbol{\beta} = \boldsymbol{\Sigma}_{22}^{-1}\boldsymbol{\Sigma}_{21}, \qquad \sigma^2 = \Sigma_{11} - \boldsymbol{\Sigma}_{12}\boldsymbol{\Sigma}_{22}^{-1}\boldsymbol{\Sigma}_{21}.$$

By (17.18),

$$\mathbf{Y}\,|\,\mathbf{X} \sim N_n(\mathbf{X}\boldsymbol{\beta}, \sigma^2\mathbf{I}).$$

Therefore the conditional distribution of **Y** given **X** is just that of a coordinatized linear model. Note that **X** is an $n \times (p - 1)$ matrix of rank $p - 1$. Also

$$\mathbf{U} = (\mathbf{X}'\mathbf{X})^{-1}\mathbf{X}'\mathbf{Y} = \hat{\boldsymbol{\beta}}, \qquad V = \mathbf{Y}'\left[\mathbf{I} - \mathbf{X}(\mathbf{X}'\mathbf{X})^{-1}\mathbf{X}'\right]\mathbf{Y} = \left[n - (p - 1)\right]\hat{\sigma}^2.$$

By known results for this coordinatized linear model, **U** and V are independent conditionally on **X** and

$$\mathbf{U}\,|\,\mathbf{X} \sim N_{p-1}\left[\boldsymbol{\beta}, \sigma^2(\mathbf{X}'\mathbf{X})^{-1}\right], \qquad \mathbf{V}\,|\,\mathbf{X} \sim \sigma^2\chi^2_{n-p-1}(0).$$

These results together with Theorem 17.6e are equivalent to the theorem for the case $q = 1$.

17.5. THE DISTRIBUTION OF HOTELLING'S T^2.

Let $\mathbf{X}$ and $\mathbf{W}$ be independent,

$$\mathbf{X} \sim N_p(\boldsymbol{\mu}, \boldsymbol{\Sigma}), \qquad \mathbf{W} \sim W_p(n, \Sigma)$$

where $n \geqslant p$, $\boldsymbol{\Sigma} > 0$. Then $P(\mathbf{W}$ is invertible$) = 1$. Define

$$F = \frac{n-p+1}{p} \mathbf{X}'\mathbf{W}^{-1}\mathbf{X}. \tag{17.21}$$

$T^2 = [np/(n-p+1)]\, F$ is called Hotelling's T^2. The next result in this chapter is to find the distribution of F and hence of T^2. If $p = 1$, then F has the more familiar expression

$$F = T^2 = \left(\frac{X}{\sqrt{W/n}} \right)^2,$$

that is, if $p = 1$, then F is the square of a random variable having a noncentral t-distribution, and $F \sim F_{1,n}(\mu^2/\Sigma)$. In this section we show that the F has an F distribution even when $p > 1$. We first need the following lemma.

LEMMA 17.10. Let $\mathbf{W} \sim W_p(n, \Sigma)$, $n \geqslant p$, $\boldsymbol{\Sigma} > 0$. Let $\mathbf{a} \in R^p$, $\mathbf{a} \neq \mathbf{0}$. Then

$$\frac{\mathbf{a}'\boldsymbol{\Sigma}^{-1}\mathbf{a}}{\mathbf{a}'\mathbf{W}^{-1}\mathbf{a}} \sim \chi^2_{n-p+1}(0).$$

PROOF. Let $\mathbf{a}_0$ be the p-dimensional vector given by

$$\mathbf{a}_0' = (1, 0, 0, \ldots, 0).$$

and let

$$\mathbf{W} = \begin{pmatrix} W_{11} & \mathbf{W}_{12} \\ \mathbf{W}_{21} & \mathbf{W}_{22} \end{pmatrix}, \boldsymbol{\Sigma} = \begin{pmatrix} \Sigma_{11} & \boldsymbol{\Sigma}_{12} \\ \boldsymbol{\Sigma}_{21} & \boldsymbol{\Sigma}_{22} \end{pmatrix}$$

where W_{11} and Σ_{11} are 1×1. Then

$$\mathbf{a}_0'\mathbf{W}^{-1}\mathbf{a}_0 = \left(W_{11} - \mathbf{W}_{12}\mathbf{W}_{22}^{-1}\mathbf{W}_{21} \right)^{-1}, \qquad \mathbf{a}_0'\Sigma^{-1}\mathbf{a}_0 = \left(\Sigma_{11} - \boldsymbol{\Sigma}_{12}\boldsymbol{\Sigma}_{22}^{-1}\boldsymbol{\Sigma}_{21} \right)^{-1}$$

(see Lemma 2 of the appendix). Therefore, by Lemma 17.9 (note $q = 1$) and Theorem 17.6b

$$\frac{\mathbf{a}_0\boldsymbol{\Sigma}^{-1}\mathbf{a}_0}{\mathbf{a}_0\mathbf{W}^{-1}\mathbf{a}_0} = \frac{W_{11} - \mathbf{W}_{12}\mathbf{W}_{22}^{-1}\mathbf{W}_{21}}{\Sigma_{11} - \boldsymbol{\Sigma}_{12}\boldsymbol{\Sigma}_{22}^{-1}\boldsymbol{\Sigma}_{21}} \sim \chi^2_{n-p+1}(0). \tag{17.22}$$

Hence the lemma is true for $\mathbf{a} = \mathbf{a}_0$. Now let $\mathbf{a} \in R^p$, $\mathbf{a} \neq \mathbf{0}$. Let $\mathbf{A}$ be an invertible matrix whose first row is $\mathbf{a}$. Then $\mathbf{a} = \mathbf{A}\mathbf{a}_0$. By Theorem 17.6d,

$$\mathbf{A}^{-1}\mathbf{W}(\mathbf{A}^{-1})' \sim W_p\left(n, \mathbf{A}^{-1}\boldsymbol{\Sigma}(\mathbf{A}^{-1})'\right),$$

and $\mathbf{A}^{-1}\boldsymbol{\Sigma}(\mathbf{A}^{-1})' > 0$. Therefore, by (17.22)

$$\frac{\mathbf{a}'\boldsymbol{\Sigma}^{-1}\mathbf{a}}{\mathbf{a}'\mathbf{W}^{-1}\mathbf{a}} = \frac{\mathbf{a}_0'\left(\mathbf{A}^{-1}\boldsymbol{\Sigma}(\mathbf{A}^{-1})'\right)^{-1}\mathbf{a}_0}{\mathbf{a}_0'\left(\mathbf{A}^{-1}\mathbf{W}(\mathbf{A}^{-1})'\right)^{-1}\mathbf{a}_0} \sim \chi^2_{n-p+1}(0). \quad \square$$

We are now ready to derive the distribution of F.

THEOREM 17.11. Let $\mathbf{X}$ and $\mathbf{W}$ be independent, $\mathbf{X} \sim N_p(\boldsymbol{\mu}, \boldsymbol{\Sigma})$, $\mathbf{W} \sim W_p(n, \boldsymbol{\Sigma})$, with $n \geqslant p, \boldsymbol{\Sigma} > 0$. Then

$$F = \frac{n-p+1}{p}\mathbf{X}'\mathbf{W}^{-1}\mathbf{X} \sim F_{p,n-p+1}(\boldsymbol{\mu}'\boldsymbol{\Sigma}^{-1}\boldsymbol{\mu}).$$

PROOF. Let $U = \mathbf{X}'\boldsymbol{\Sigma}^{-1}\mathbf{X}$, $V = \mathbf{X}'\boldsymbol{\Sigma}^{-1}\mathbf{X}/\mathbf{X}'\mathbf{W}^{-1}\mathbf{X}$. By Lemma 17.10, $V \mid \mathbf{X} \sim \chi^2_{n-p+1}(0)$. Therefore V is independent of $\mathbf{X}$ and hence also of U. By Theorem 3.10, $U \sim \chi^2_p(\boldsymbol{\mu}'\boldsymbol{\Sigma}^{-1}\boldsymbol{\mu})$. Therefore,

$$F = \frac{n-p+1}{p}\frac{U}{V} \sim F_{p,n-p+1}(\boldsymbol{\mu}'\boldsymbol{\Sigma}^{-1}\boldsymbol{\mu}). \quad \square$$

17.6. THE WISHART DENSITY FUNCTION.

In this section we find the density function for the nonsingular central Wishart distribution. To keep the notation simpler, we look first at the case $\boldsymbol{\Sigma} = \mathbf{I}$.

THEOREM 17.12. Let $\mathbf{W} \sim W_p(n, \mathbf{I})$, $n \geqslant p$. Then the density function of $\mathbf{W}$ is

$$f_{\mathbf{W}}(\mathbf{w}) = \frac{|\mathbf{w}|^{(n-p-1)/2}\exp\left[(-1/2)\operatorname{tr}\mathbf{w}\right]}{2^{np/2}\pi^{p(p-1)/4}\prod_{i=1}^{p}\Gamma\left(\frac{1}{2}(n+1-i)\right)}$$

for $\mathbf{w} > 0$, and $f_{\mathbf{W}}(\mathbf{w}) = 0$ elsewhere.

PROOF. We prove this theorem by induction on p the dimension of $\mathbf{W}$. When $p = 1$, the theorem is true. (See Exercise B7.) We now assume that the theorem is true for $p = k - 1$ and show that it is true for $p = k$. By Theorem 17.8, the density is 0 if $\mathbf{w}$ is not positive definite. We therefore assume that $\mathbf{w}$ is positive definite. Partition $\mathbf{W}$ and $\mathbf{w}$ as in (17.16) with $q = 1$, and define $\mathbf{T}$, $\mathbf{U}$ and V as in Lemma 17.9. We can find the joint density of $\mathbf{T}$, $\mathbf{U}$, and $\mathbf{V}$. (We can find the density of $\mathbf{T}$ by the induction hypothesis.) Note that $\mathbf{W}$ is an invertible function of $(\mathbf{T}, \mathbf{U}, \mathbf{V})$. The Jacobian matrix of the transformation from $\mathbf{W}$ to $(\mathbf{T}, \mathbf{U}, \mathbf{V})$ has the form

$$\begin{array}{cccc} & \mathbf{T} & \mathbf{U} & V \\ \mathbf{W}_{22} & \mathbf{I} & & \\ \mathbf{W}_{21} & \mathbf{0} & \mathbf{W}_{22}^{-1} & \\ \mathbf{W}_{11} & \mathbf{0} & \mathbf{0} & 1 \end{array}$$

where the omitted parts are not necessary to compute the Jacobian (the determinant of the Jacobian matrix). The Jacobian of this transformation is $|\mathbf{W}_{22}|^{-1}$. Therefore,

$$f_{\mathbf{W}}(\mathbf{w}) = |\mathbf{w}_{22}|^{-1} f_{\mathbf{T}}(\mathbf{w}_{22}) f_{\mathbf{U}\,|\,\mathbf{T}}(\mathbf{w}_{22}^{-1}\mathbf{w}_{21}\,|\,\mathbf{w}_{22})$$
$$\times f_{\mathbf{V}\,|\,(\mathbf{U},\mathbf{T})}\left[\mathbf{w}_{11} - \mathbf{w}_{12}\mathbf{w}_{22}^{-1}\mathbf{w}_{21}\,|\,(\mathbf{w}_{22}^{-1}\mathbf{w}_{21}, \mathbf{w}_{22})\right]$$

where $f_{\mathbf{T}}$, $f_{\mathbf{U}\,|\,\mathbf{T}}$ and $f_{\mathbf{V}\,|\,(\mathbf{U},\mathbf{T})}$ can be determined from Lemma 17.9 (with $q = 1$ and $\boldsymbol{\Sigma} = \mathbf{I}$). Therefore

$$\begin{aligned} f_{\mathbf{W}}(\mathbf{w}) &= \frac{1}{|\mathbf{w}_{22}|} \frac{|\mathbf{w}_{22}|^{(n-k)/2} \exp -\frac{1}{2}\operatorname{tr}\mathbf{w}_{22}}{2^{n(k-1)/2}\pi^{(k-1)(k-2)/4}\prod_{i=1}^{k-1}\Gamma\left(\frac{n+1-r}{2}\right)} \\ &\quad\times \frac{|\mathbf{w}_{22}|^{(1/2)}}{(2\pi)^{(k-1)/2}} \exp -\tfrac{1}{2}(\mathbf{w}_{22}^{-1}\mathbf{w}_{21})'\mathbf{w}_{22}(\mathbf{w}_{22}^{-1}\mathbf{w}_{21}) \\ &\quad\times \frac{(\mathbf{w}_{11} - \mathbf{w}_{12}\mathbf{w}_{22}^{-1}\mathbf{w}_{21})^{(n-k-1)/2}}{2^{(n-k+1)/2}\Gamma\left(\frac{n-k+1}{2}\right)} \exp -\tfrac{1}{2}(\mathbf{w}_{11} - \mathbf{w}_{12}\mathbf{w}_{22}^{-1}\mathbf{w}_{21}) \qquad (17.23) \\ &= \frac{\left[|\mathbf{w}_{22}|(\mathbf{w}_{11} - \mathbf{w}_{12}\mathbf{w}_{22}^{-1}\mathbf{w}_{21})\right]^{(n-k-1)/2}}{2^{nk/2}\pi^{k(k-1)/4}\prod_{i=1}^{k}\Gamma\left(\frac{n+1-i}{2}\right)} \exp\left[-\tfrac{1}{2}\operatorname{tr}\mathbf{w}_{22} - \tfrac{1}{2}\mathbf{w}_{11}\right]. \end{aligned}$$

However, $|\mathbf{w}_{22}|(\mathbf{w}_{11} - \mathbf{w}_{12}\mathbf{w}_{22}^{-1}\mathbf{w}_{21}) = |\mathbf{w}|$, and $\mathbf{w}_{11} + \operatorname{tr}\mathbf{w}_{22} = \operatorname{tr}\mathbf{w}$. Substituting these two facts into (17.23) shows that the density $f_{\mathbf{W}}$ is correct for $p = k$ if it is correct for $p = k - 1$, and hence it is correct for all p. □

The final result in this section is to find the density of $\mathbf{W}$ for general $\boldsymbol{\Sigma} > 0$.

COROLLARY. Let $\mathbf{W} \sim W_p(n, \boldsymbol{\Sigma})$, $n \geqslant p$, $\boldsymbol{\Sigma} > 0$. Then the density of $\mathbf{W}$ is given by

$$f_{\mathbf{W}}(\mathbf{w}) = \frac{|\mathbf{w}|^{(n-p-1)/2}\exp\left[-\frac{1}{2}\operatorname{tr}\boldsymbol{\Sigma}^{-1}\mathbf{w}\right]}{2^{np/2}\pi^{p(p-1)/4}|\boldsymbol{\Sigma}|^{n/2}\prod_{i=1}^{p}\Gamma\left[\frac{1}{2}(n+1-i)\right]}$$

for $\mathbf{w} > 0$, and $f_{\mathbf{W}}(\mathbf{w}) = 0$ elsewhere.

PROOF. By Theorem 8 of the appendix, there exists an invertible upper triangular matrix $\mathbf{C}$ such that $\mathbf{\Sigma} = \mathbf{CC}'$. Let $\mathbf{S} \sim W_p(n, \mathbf{I})$. Let

$$\mathbf{W} = \mathbf{CSC}' \sim W_p(n, \mathbf{\Sigma}).$$

The transformation from $\mathbf{S}$ to $\mathbf{W}$ is invertible, with $\mathbf{S} = \mathbf{C}^{-1}\mathbf{W}(\mathbf{C}')^{-1}$. The Jacobian of this transformation is given by (see Theorem 17 of the appendix),

$$J = |\mathbf{C}^{-1}|^{p+1} = \frac{1}{|\mathbf{\Sigma}|^{(p+1)/2}}.$$

Therefore,

$$f_{\mathbf{W}}(\mathbf{w}) = \frac{1}{|\mathbf{\Sigma}|^{(p+1)/2}} f_S\left(\mathbf{C}^{-1}\mathbf{w}(\mathbf{C}')^{-1}\right).$$

The remaining details are left for Exercise B8. □

17.7. OTHER RESULTS.

We now give some other results about the matrix normal and Wishart distributions. We first give two properties of the matrix normal distribution.

THEOREM 17.13. a. Let $\mathbf{X} \sim N_{m,r}(\boldsymbol{\mu}, \mathbf{\Xi}, \mathbf{\Sigma})$ and let $\mathbf{T}$ be an m × r matrix. Then $\operatorname{tr} \mathbf{T}'X \sim N_1(\operatorname{tr} \mathbf{T}'\boldsymbol{\mu}, \operatorname{tr} \mathbf{T}'\mathbf{\Xi}\mathbf{T}\mathbf{\Sigma})$.

b. Let $\mathbf{X} \sim N_{m,r}(\boldsymbol{\mu}, \mathbf{\Xi}, \mathbf{\Sigma})$, $\mathbf{\Xi} > 0$, $\mathbf{\Sigma} > 0$. Then

$$\operatorname{tr}(\mathbf{X} - \boldsymbol{\mu})'\mathbf{\Xi}^{-1}(\mathbf{X} - \boldsymbol{\mu})\mathbf{\Sigma}^{-1} \sim \chi^2_{mr},$$

and $\operatorname{tr} \mathbf{X}'\mathbf{\Xi}^{-1} \sim \chi^2_{mr}(\operatorname{tr} \boldsymbol{\mu}'\mathbf{\Xi}^{-1}\boldsymbol{\mu}\mathbf{\Sigma})$.

PROOF. See Exercises B9 and B10. □

We now give a result about the noncentral Wishart distribution.

THEOREM 17.14. Let $\mathbf{W} \sim W_p(n, \mathbf{\Sigma}, \boldsymbol{\delta})$. Then $\operatorname{tr} \mathbf{\Sigma}^{-1}\mathbf{W} \sim \chi^2_{np}(\operatorname{tr} \mathbf{\Sigma}^{-1}\boldsymbol{\delta})$.

PROOF. See exercise B11. □

Finally, we give several additional properties of the central Wishart distribution.

THEOREM 17.15. Let $\mathbf{W} \sim W_p(n, \mathbf{\Sigma})$, $\Sigma > 0$.

a. $\mathbf{W}$ has moment generating function $M_{\mathbf{W}}(\mathbf{T}) = |I - 2\mathbf{\Sigma}\mathbf{T}|^{-n/2}$ for all symmetric $\mathbf{T}$ such that $\mathbf{\Sigma}^{-1} - 2\mathbf{T} > 0$.

b. If $n \geqslant p$, then $|\mathbf{W}|/|\mathbf{\Sigma}| = |\mathbf{\Sigma}^{-1}\mathbf{W}|$ has the same distribution as $\prod_{i=1}^{p} U_i$ where the U_i are independent and $U_i \sim \chi^2_{n-i+1}$.

c. If $n \geqslant p$, then $E|\mathbf{W}| = (n-1)\ldots(n-p+1)|\mathbf{\Sigma}|$. If $n > p+1$, then $E|\mathbf{W}^{-1}| = E|\mathbf{W}|^{-1} = |\mathbf{\Sigma}^{-1}|/((n-2)(n-3)\ldots(n-p-1))$.

d. If $n > p+1$, then $E\mathbf{W}^{-1} = (1/(n-p-1))\mathbf{\Sigma}^{-1}$.

e. $E\,\mathrm{tr}\,\mathbf{W} = n\,\mathrm{tr}\,\boldsymbol{\Sigma}$. If $n > p+1$, then $E\,\mathrm{tr}\,\mathbf{W}^{-1} = \mathrm{tr}\,\boldsymbol{\Sigma}^{-1}/(n-p-1)$.

f. If $\mathbf{W}_1$ and $\mathbf{W}_2$ are independent, $\mathbf{W}_i \sim W_p(n_i, \boldsymbol{\Sigma})$, then $\mathbf{W}_1 + \mathbf{W}_2 \sim W_p(n_1+n_2, \boldsymbol{\Sigma})$.

g. If $n \geqslant p$ and $\mathbf{A}$ is a $q \times p$ matrix of rank q, then $(\mathbf{A}\mathbf{W}^{-1}\mathbf{A})^{-1} \sim W_q(n-p+q, (\mathbf{A}\boldsymbol{\Sigma}^{-1}\mathbf{A})^{-1})$.

PROOF. See Exercises B12–B17. □

17.8. FURTHER COMMENTS.

The distribution of Hotelling's T^2 in the case $\boldsymbol{\mu} = \mathbf{0}$ was first derived by Hotelling (1931). The distribution for general $\boldsymbol{\mu}$ was derived simultaneously by Bose and Roy (1938) and Hsu (1938). The Wishart density function was first derived by Wishart (1928). (It had been derived in the case $p = 2$ by Fisher (1915).) Various forms of the noncentral Wishart density function have been derived under various conditions. See James (1955) and Anderson (1946). Lemma 17.9, which is the basis for the derivation of the Wishart density and the distribution of Hotelling's T^2 was derived by Barlett (1933) for the case $q = 1$, which is the only case used in this chapter. (The case of $q > 1$ is used in Chapters 19, 20, and 21.) The derivations given in this chapter of the Wishart density function and the distribution of Hotelling's T^2 follow Stein (1969).

EXERCISES.

Type B

1. Derive (17.6), (17.7), and (17.8).

2. Prove Theorem 17.2.

3. Show that if either $\boldsymbol{\Xi}$ or $\boldsymbol{\Sigma}$ is not positive definite then there is no continuous density function for the matrix normal distribution. [Hint: $\mathrm{Var}(\mathbf{a}'\mathbf{X}\mathbf{b}) = \mathbf{a}'\boldsymbol{\Xi}\mathbf{a}\mathbf{b}'\boldsymbol{\Sigma}\mathbf{b}$.]

4. Prove Theorem 17.6, parts b, c, and e.

5. Derive Theorem 17.7b.

6. Derive Theorem 17.8b.

7. Show that the formula given in Theorem 17.12 is correct when $p = 1$. [If $p = 1$, then $W \sim \chi_n^2(0)$.]

8. Finish the derivation of the corollary to Theorem 17.12.

9. Prove Theorem 17.13a. [Hint: Let $U = \operatorname{tr}\mathbf{T}'\mathbf{X}$. Then $M_V(t) = M_{\mathbf{X}}(t\mathbf{T})$. Why?]

10. (a) Let $\mathbf{Z} = (Z_{ij})$, $\boldsymbol{\mu} = (\mu_{ij})$ where the Z_{ij} are independent, $Z_{ij} \sim N_1(\mu_{ij}, 1)$. Then $\mathbf{Z} \sim N_{mr}(\boldsymbol{\mu}, \mathbf{I}, \mathbf{I})$. Show that $\operatorname{tr}\mathbf{Z}'\mathbf{Z} \sim \chi^2_{mr}(\operatorname{tr}\boldsymbol{\mu}'\boldsymbol{\mu})$.

(b) Prove Theorem 17.13b.

11. Prove Theorem 17.14. [Hint: Use the definition of the Wishart distribution together with Theorem 17.13b.)

12. Prove Theorem 17.15a. [Hint: Find $E\exp(\mathbf{XTX}')$ where

$$\mathbf{X} \sim N_{n-p}(\mathbf{0}, \mathbf{I}, \boldsymbol{\Sigma})\Big].$$

13. (a) Prove Theorem 17.15b. [Hint: Use induction, Lemma 17.9 (with $q = 1$) and Lemma 2 of the appendix.]

(b) Prove Theorem 17.15c.

14. Let $\mathbf{W} \sim W_p(n, \mathbf{I})$, $n > p + 1$. Let $\mathbf{W}^{-1} = (W^{ij})$.

(a) Show that $(W^{ii})^{-1} \sim \chi^2_{n-p+1}$ [Hint: $W^{ii} = \mathbf{a}_i'\mathbf{W}^{-1}\mathbf{a}_i$ where $\mathbf{a}_i$ is the p-dimensional vector with a 1 in the ith place and 0's elsewhere.)

(b) Show that $E\mathbf{W}^{-1}$ exists. [Hint: $EW^{ii} = 1/(n-p-1)$ by part a. Since $E\mathbf{W}^{-1} > 0$, $(E\mathbf{W}^{ij})^2 \leqslant E\mathbf{W}^{ii}E\mathbf{W}^{jj}$.]

(c) Let $\mathbf{A} = E\mathbf{W}^{-1}$. Show that $\boldsymbol{\Gamma}\mathbf{A}\boldsymbol{\Gamma}' = \mathbf{A}$ for all orthogonal $\boldsymbol{\Gamma}$. [Hint: $\boldsymbol{\Gamma}\mathbf{W}\boldsymbol{\Gamma}'$ has the same distribution as $\mathbf{W}$. Why?)

(d) Show that $E\mathbf{W}^{-1} = (1/(n-p-1))\mathbf{I}$. [Hint: By part c and Corollary 4 to Theorem 1 of the appendix, $\mathbf{A} = c\mathbf{I}$ for some c. Use part a to find c.)

(e) Prove Theorem 17.15d.

15. (a) If $\mathbf{U}$ is a $p \times p$ matrix, show that $E(\operatorname{tr}\mathbf{U}) = \operatorname{tr}(E\mathbf{U})$.

(b) Prove Theorem 17.15e.

16. Prove Theorem 17.15f.

17. Prove Theorem 17.15g. [Hint: Follow the proof of Lemma 17.10.)

Type C

1. For any $q \times p$ matrix $\mathbf{A} = (\mathbf{A}_1, \ldots, \mathbf{A}_p)$ and any $q \times q$ and $p \times p$ matrices $\boldsymbol{\Xi} = (\xi_{ij})$ and $\boldsymbol{\Sigma}$, let $\tilde{\mathbf{A}}$ and $\boldsymbol{\Xi} \otimes \boldsymbol{\Sigma}$ be the qp-dimensional vector and

$qp \times qp$ matrix given by

$$\tilde{\mathbf{A}} = \begin{bmatrix} \mathbf{A}_1 \\ \vdots \\ \mathbf{A}_p \end{bmatrix}, \qquad \boldsymbol{\Xi} \otimes \boldsymbol{\Sigma} = \begin{bmatrix} \xi_{11}\boldsymbol{\Sigma} & \cdots & \xi_{1q}\boldsymbol{\Sigma} \\ \vdots & & \vdots \\ \xi_{q1}\boldsymbol{\Sigma} & \cdots & \xi_{qq}\boldsymbol{\Sigma} \end{bmatrix}$$

($\boldsymbol{\Xi} \otimes \boldsymbol{\Sigma}$ is called the Kronecker product of $\boldsymbol{\Xi}$ and $\boldsymbol{\Sigma}$. Note that $\boldsymbol{\Xi} \otimes \boldsymbol{\Sigma} \neq \boldsymbol{\Sigma} \otimes \boldsymbol{\Xi}$.)

(a) Let $\mathbf{X}$ and $\mathbf{t}$ be $q \times p$ matrices and let $\boldsymbol{\Xi}$ and $\boldsymbol{\Sigma}$ be $q \times q$ and $p \times p$ symmetric matrices. Show that $\operatorname{tr} \mathbf{t}'\mathbf{X} = \tilde{\mathbf{t}}'\tilde{X}$, $\operatorname{tr} \mathbf{t}'\boldsymbol{\Xi}\mathbf{t}\boldsymbol{\Sigma} = \tilde{\mathbf{t}}'(\boldsymbol{\Xi} \otimes \boldsymbol{\Sigma})\tilde{\mathbf{t}}$.

(b) Let $\mathbf{X} \sim N_{q,\,p}(\boldsymbol{\mu}, \boldsymbol{\Xi}, \boldsymbol{\Sigma})$. Show that $\tilde{\mathbf{X}} \sim N_{qp}(\tilde{\boldsymbol{\mu}}, \boldsymbol{\Xi} \otimes \boldsymbol{\Sigma})$. (Therefore, a matrix normal distribution can be rearranged into a vector normal distribution with a particular covariance matrix.)

(c) Let $\mathbf{U}$ be a $p \times q$ random matrix. Show that $\mathbf{U} \sim N_{q,\,p}(\boldsymbol{\mu}, \boldsymbol{\Xi}, \boldsymbol{\Sigma})$ if and only if $\operatorname{tr} \mathbf{t}'\mathbf{U} \sim N_1(\operatorname{tr} \mathbf{t}'\boldsymbol{\mu}, \operatorname{tr} \mathbf{t}'\boldsymbol{\Xi}\mathbf{t}\boldsymbol{\Sigma})$ for all $q \times p$ matrices $\mathbf{t} \neq 0$. [Hint: $\mathbf{U} \sim N_{q,\,p}(\boldsymbol{\mu}, \boldsymbol{\Xi}, \boldsymbol{\Sigma})$ if and only if $\tilde{\mathbf{U}} \sim N_{qp}(\tilde{\boldsymbol{\mu}}, \boldsymbol{\Xi} \otimes \boldsymbol{\Sigma})$. See Exercise C3 of Chapter 3.]

(d) Now, let $\mathbf{U}_1, \mathbf{U}_2, \ldots,$ be a sequence of $p \times q$ random matrices. Show that $\mathbf{U}_n \xrightarrow{d} N_{q,\,p}(\boldsymbol{\mu}, \boldsymbol{\Xi}, \boldsymbol{\Sigma})$ if and only if $\operatorname{tr} \mathbf{t}'\mathbf{U}_n \xrightarrow{d} N_1(\operatorname{tr} \mathbf{A}'\boldsymbol{\mu}, \operatorname{tr} \mathbf{t}'\boldsymbol{\Xi}\mathbf{t}\boldsymbol{\Sigma})$ for all $q \times p$ matrices $\mathbf{t} \neq 0$. [Hint: Use Theorem 10.2.) .

CHAPTER 18

The Multivariate One- and Two-Sample Models—Inference about the Mean Vector

The multivariate one sample model is one in which we observe $\mathbf{Y}_1, \ldots, \mathbf{Y}_n$ independently identically distributed r-dimensional random vectors such that

$$\mathbf{Y}_i \sim N_r(\boldsymbol{\mu}, \boldsymbol{\Sigma}), \qquad -\infty < \boldsymbol{\mu} < \infty, \qquad \boldsymbol{\Sigma} > 0.$$

Note that we are assuming the distribution of $\mathbf{Y}_i$ is a nonsingular normal distribution (such that $\boldsymbol{\Sigma} > 0$). This model would occur if we had r measurements (e.g., height, weight, income) on each of n individuals. In this chapter we are primarily concerned with inference about the mean vector $\boldsymbol{\mu}$. In Section 18.1, we find a complete sufficient statistic for this model and derive its distribution. In Section 18.2, we find minimum variance unbiased estimators and MLEs of $\boldsymbol{\mu}$ and $\boldsymbol{\Sigma}$ and the best invariant estimator of $\boldsymbol{\mu}$, and in Section 18.3, we find a James-Stein estimator of $\boldsymbol{\mu}$. We consider testing hypotheses about $\boldsymbol{\mu}$ in Section 18.4. Simultaneous confidence intervals for linear functions of $\boldsymbol{\mu}$ are given in 18.5. The asymptotic validity of the tests and simultaneous confidence intervals is established in Section 18.6. An application of these results to a generalized repeated measures model is given in Section 18.7. In Section 18.8, we consider the model in which we have two independent samples with the same covariance matrix.

The tests derived in this chapter are due to Hotelling (1931). Their UMP invariance is due to Hunt and Stein (1946). The simultaneous confidence intervals are due to Roy and Bose (1953).

18.1. A COMPLETE SUFFICIENT STATISTIC AND ITS DISTRIBUTION.

Let

$$\bar{\mathbf{Y}} = \frac{1}{n}\sum_{i=1}^{n} \mathbf{Y}_i,$$

$$\mathbf{S} = \frac{1}{n-1}\sum_{i=1}^{n} (\mathbf{Y}_i - \bar{\mathbf{Y}})(\mathbf{Y}_i - \bar{\mathbf{Y}})' = \frac{1}{n-1}\left(\sum_{i=1}^{n} \mathbf{Y}_i\mathbf{Y}_i' - n\bar{\mathbf{Y}}\bar{\mathbf{Y}}'\right). \quad (18.1)$$

$\bar{\mathbf{Y}}$ is called the *sample mean* and $\mathbf{S}$ is called the *sample covariance matrix*.

THEOREM 18.1. $(\bar{\mathbf{Y}}, \mathbf{S})$ is a complete sufficient statistic for the one sample model.

PROOF. We first show that $\mathbf{U} = \sum_{i=1}^{n}\mathbf{Y}_i$ and $\mathbf{V} = \sum_{i=1}^{n}\mathbf{Y}_i\mathbf{Y}_i'$ is complete and sufficient. The joint density of the $\mathbf{Y}_i$ is given by

$$f(\mathbf{Y}_1, \ldots, \mathbf{Y}_r; \boldsymbol{\mu}, \boldsymbol{\Sigma}) = \frac{1}{(2\pi)^{nr/2}|\boldsymbol{\Sigma}|^{n/2}} \exp\left(-\tfrac{1}{2}\sum_{i=1}^{n}(\mathbf{Y}_i - \boldsymbol{\mu})'\boldsymbol{\Sigma}^{-1}(\mathbf{Y}_i - \boldsymbol{\mu})\right)$$

$$= \frac{1}{(2\pi)^{nr/2}|\boldsymbol{\Sigma}|^{n/2}} \exp -\tfrac{n}{2}\,\boldsymbol{\mu}'\boldsymbol{\Sigma}^{-1}\boldsymbol{\mu}$$

$$\times \exp\left(\boldsymbol{\mu}'\boldsymbol{\Sigma}^{-1}\sum_{i=1}^{n}\mathbf{Y}_i - \tfrac{1}{2}\sum_{i=1}^{n}\mathbf{Y}_i'\boldsymbol{\Sigma}^{-1}\mathbf{Y}_i\right).$$

However,

$$\sum_{i=1}^{n}\mathbf{Y}_i'\boldsymbol{\Sigma}^{-1}\mathbf{Y}_i = \sum_{i=1}^{n}\operatorname{tr}\mathbf{Y}_i'\boldsymbol{\Sigma}^{-1}\mathbf{Y}_i = \sum_{i=1}^{n}\operatorname{tr}\boldsymbol{\Sigma}^{-1}\mathbf{Y}_i\mathbf{Y}_i' = \operatorname{tr}\boldsymbol{\Sigma}^{-1}\sum_{i=1}^{n}\mathbf{Y}_i\mathbf{Y}_i' = \operatorname{tr}\boldsymbol{\Sigma}^{-1}\mathbf{V}.$$

Therefore, the joint density of the $\mathbf{Y}_i$'s has the form

$$K(\boldsymbol{\mu}, \boldsymbol{\Sigma})\exp\left(\boldsymbol{\mu}'\boldsymbol{\Sigma}^{-1}\mathbf{U} - \tfrac{1}{2}\operatorname{tr}\boldsymbol{\Sigma}^{-1}\mathbf{V}\right).$$

We can see now that $\mathbf{U}$ and $\mathbf{V}$ are sufficient by the factorization criterion. To show the completeness, we must put the density in exponential form. Let $\Delta = (\Delta_{ij}) = \boldsymbol{\Sigma}^{-1}$, and let $\mathbf{V} = (\mathbf{V}_{ij})$. Define

$$\tilde{\boldsymbol{\Delta}}_1' = (\Delta_{11}, \ldots, \Delta_{rr}),$$

$$\tilde{\boldsymbol{\Delta}}_2' = (\Delta_{12}, \Delta_{13}, \ldots, \Delta_{1r}, \Delta_{23}, \ldots, \Delta_{2r}, \ldots, \Delta_{r-1,r})$$

$$\tilde{\mathbf{V}}_1' = (V_{11}, \ldots, V_{rr}),$$

$$\tilde{\mathbf{V}}_2' = (V_{12}, V_{13}, \ldots, V_{1r}, V_{23}, \ldots, V_{2r}, \ldots, V_{r-1,r})$$

That is, $\tilde{\boldsymbol{\Delta}}_1$ and $\tilde{\mathbf{V}}_1$ are r-dimensional vectors whose elements are the diagonal elements of $\boldsymbol{\Delta}$ and $\mathbf{V}$, whereas $\tilde{\boldsymbol{\Delta}}_2$ and $\tilde{\mathbf{V}}_2$ are $r(r-1)/2$ dimensional vectors consisting of the above diagonal elements of $\boldsymbol{\Delta}$ and $\mathbf{V}$. Then

$$\operatorname{tr}\boldsymbol{\Sigma}^{-1}\mathbf{V} = \operatorname{tr}\boldsymbol{\Delta}\mathbf{V} = \sum_{i=1}^{r}\sum_{j=1}^{r}\Delta_{ij}V_{ij} = \tilde{\boldsymbol{\Delta}}_1'\tilde{\mathbf{V}}_1 + 2\tilde{\boldsymbol{\Delta}}_2'\tilde{\mathbf{V}}_2$$

(using the symmetry of $\boldsymbol{\Delta}$ and $\mathbf{V}$). Now let

$$\mathbf{Q}(\boldsymbol{\mu},\boldsymbol{\Sigma}) = \begin{pmatrix} -\frac{1}{2}\tilde{\boldsymbol{\Delta}}_1 \\ -\tilde{\boldsymbol{\Delta}}_2 \\ \boldsymbol{\Sigma}^{-1}\boldsymbol{\mu}' \end{pmatrix}, \qquad \mathbf{T}(\mathbf{Y}_1,\ldots,\mathbf{Y}_n) = \begin{pmatrix} \tilde{\mathbf{V}}_1 \\ \tilde{\mathbf{V}}_2 \\ \mathbf{U} \end{pmatrix}$$

Then we see that

$$f(\mathbf{Y}_1,\ldots,\mathbf{Y}_n;\boldsymbol{\mu},\boldsymbol{\Sigma}) = K(\boldsymbol{\mu},\boldsymbol{\Sigma})\exp\mathbf{Q}'(\boldsymbol{\mu},\boldsymbol{\Sigma})\mathbf{T}(\mathbf{Y}_1,\ldots,\mathbf{Y}_n)$$

which is in exponential form. The only detail we need to check is that the image of $\mathbf{Q}(\boldsymbol{\mu},\boldsymbol{\Sigma})$ has an interior point. It can be shown that as $\boldsymbol{\Delta}$ ranges over the space of positive definite matrices, then $(\tilde{\boldsymbol{\Delta}}_1,\tilde{\boldsymbol{\Delta}}_2)$ ranges over a set that has interior points. For any $\boldsymbol{\Delta}$, as $\boldsymbol{\mu}$ ranges over R^p, $\boldsymbol{\Sigma}^{-1}\boldsymbol{\mu}$ ranges over R^p, which has interior points. Therefore $\mathbf{T}$ is a complete sufficient statistic for the one sample model, and hence $(\mathbf{U},\mathbf{V})$, which is an invertible function of $\mathbf{T}$ is also complete and sufficient. Finally, $(\overline{\mathbf{Y}},\mathbf{S})$ is an invertible function of $(\mathbf{U},\mathbf{V})$ and is therefore also complete and sufficient. □

We now find the joint distribution of $\overline{\mathbf{Y}}$ and $\mathbf{S}$.

THEOREM 18.2. $\overline{\mathbf{Y}}$ and $\mathbf{S}$ are independent, $\overline{\mathbf{Y}}\sim N_r(\boldsymbol{\mu},\ (1/n)\boldsymbol{\Sigma})$, $\mathbf{S}\sim W_r(n-1,\ [1/(n-1)]\boldsymbol{\Sigma})$.

PROOF. We first write the $\mathbf{Y}_i$ in a matrix, to use the results derived in the last chapter. To use these results, we write the $\mathbf{Y}_i$ as row vectors. Let

$$\mathbf{Y} = \begin{pmatrix} \mathbf{Y}_1' \\ \vdots \\ \mathbf{Y}_n' \end{pmatrix}, \qquad \boldsymbol{\theta} = \begin{pmatrix} \boldsymbol{\mu}' \\ \vdots \\ \boldsymbol{\mu}' \end{pmatrix}.$$

By Theorem 17.3, we see that $\mathbf{Y}\sim N_{n,r}(\boldsymbol{\theta},\mathbf{I},\boldsymbol{\Sigma})$. Now let

$$\mathbf{a} = \left[(1/n),\ldots,(1/n)\right].$$

Then

$$\overline{\mathbf{Y}}' = \mathbf{a}\mathbf{Y}, \qquad (n-1)\mathbf{S} = \mathbf{Y}'\mathbf{Y} - \mathbf{n}(\mathbf{a}\mathbf{Y})'\mathbf{a}\mathbf{Y} = \mathbf{Y}'(\mathbf{I}-\mathbf{n}\mathbf{a}'\mathbf{a})\mathbf{Y}.$$

By Theorem 17.2d, we see that

$$\overline{\mathbf{Y}}' = \mathbf{a}\mathbf{Y}\sim N_{1,r}(\mathbf{a}\boldsymbol{\theta},\mathbf{a}\mathbf{a}',\boldsymbol{\Sigma}).$$

However, $\mathbf{a}\boldsymbol{\theta} = \boldsymbol{\mu}'$ and $\mathbf{a}\mathbf{a}' = 1/n$. Therefore, by Theorem 17.2a and c, we see that

$$\bar{\mathbf{Y}} \sim N_r\left(\boldsymbol{\mu}, \frac{1}{n}\boldsymbol{\Sigma}\right).$$

Now, $\mathbf{I} - n\mathbf{a}'\mathbf{a}$ is an idempotent matrix of rank $n-1$ and

$$\boldsymbol{\theta}'(\mathbf{I} - n\mathbf{a}'\mathbf{a})\boldsymbol{\theta} = \mathbf{0}$$

(See Exercise B1.) Therefore, by Theorem 17.7a, we see that

$$(n-1)\mathbf{S} \sim W_r(n-1, \boldsymbol{\Sigma}),$$

and by Theorem 17.6c, that

$$\mathbf{S} \sim W_r\left(n-1, \frac{1}{n-1}\boldsymbol{\Sigma}\right).$$

We now show that $\bar{\mathbf{Y}}$ and $\mathbf{S}$ are independent. Since $\mathbf{I} - n\mathbf{a}'\mathbf{a}$ is idempotent, $\mathbf{I} - n\mathbf{a}'\mathbf{a} \geqslant 0$. Also $\mathbf{a}(\mathbf{I} - n\mathbf{a}'\mathbf{a}) = 0$. Therefore, by Theorem 17.7b part 2, we see that $\bar{\mathbf{Y}}' = \mathbf{a}\mathbf{Y}$ and $(n-1)\mathbf{S} = \mathbf{Y}'(\mathbf{I} - n\mathbf{a}'\mathbf{a})\mathbf{Y}$ are independent. Therefore $\bar{\mathbf{Y}}$ and $\mathbf{S}$ are independent. □

18.2. MINIMUM VARIANCE UNBIASED ESTIMATORS, MAXIMUM LIKELIHOOD ESTIMATORS AND BEST INVARIANT ESTIMATORS.

We now derive the classical estimators for $\boldsymbol{\mu}$ and $\boldsymbol{\Sigma}$. The James-Stein estimator for $\boldsymbol{\mu}$ is derived in the next section.

THEOREM 18.3. $\bar{\mathbf{Y}}$ and $\mathbf{S}$ are minimum variance unbiased estimators for $\boldsymbol{\mu}$ and $\boldsymbol{\Sigma}$.

PROOF. By Theorem 18.2, $E\bar{\mathbf{Y}} = \boldsymbol{\mu}, E\mathbf{S} = (n-1)(1/(n-1))\boldsymbol{\Sigma} = \boldsymbol{\Sigma}$. □

To find the maximum likelihood estimators for most multivariate models, we need the following lemma, which is proved in the appendix.

LEMMA 18.4. Let $\mathbf{U} > 0$ and $\mathbf{A} > 0$ be $r \times r$ matrices. Define

$$f(\mathbf{U}) = \frac{1}{|\mathbf{U}|^{n/2}} \exp - \tfrac{1}{2} \operatorname{tr} \mathbf{U}^{-1}\mathbf{A}.$$

Then $f(\mathbf{U})$ is maximized for $\mathbf{U} = (1/n)\mathbf{A}$.

PROOF. See Theorem 15 of the Appendix. □

We now use Lemma 18.4 to find the maximum likelihood estimators.

THEOREM 18.5. If $n > r$ then the maximum likelihood estimators of $\boldsymbol{\mu}$ and $\boldsymbol{\Sigma}$ are $\bar{\mathbf{Y}}$ and $[(n-1)/n]\mathbf{S}$.

PROOF. We first note the following identity

$$\sum_i (\mathbf{Y}_i - \boldsymbol{\mu})(\mathbf{Y}_i - \boldsymbol{\mu})' = (n-1)\mathbf{S} + n(\overline{\mathbf{Y}} - \boldsymbol{\mu})(\overline{\mathbf{Y}} - \boldsymbol{\mu})'$$

(see Exercise B2). Therefore,

$$\begin{aligned} f(\mathbf{Y}_1, \ldots, \mathbf{Y}_n; \boldsymbol{\mu}, \boldsymbol{\Sigma}) &= \frac{1}{(2\pi)^{nr/2}|\boldsymbol{\Sigma}|^{n/2}} \exp -\tfrac{1}{2}\operatorname{tr}\boldsymbol{\Sigma}^{-1}\sum_{i=1}^{n}(\mathbf{Y}_i - \boldsymbol{\mu})'(\mathbf{Y}_i - \boldsymbol{\mu}) \\ &= \frac{1}{(2\pi)^{nr/2}|\boldsymbol{\Sigma}|^{n/2}} \exp\Big\{-\tfrac{1}{2}\operatorname{tr}\boldsymbol{\Sigma}^{-1}\big[(n-1)\mathbf{S}\big] \\ &\qquad -\tfrac{n}{2}\operatorname{tr}\boldsymbol{\Sigma}^{-1}(\overline{\mathbf{Y}} - \boldsymbol{\mu})(\overline{\mathbf{Y}} - \boldsymbol{\mu})'\Big\} \\ &= \frac{1}{(2\pi)^{nr/2}|\boldsymbol{\Sigma}|^{n/2}} \exp\Big\{-\tfrac{1}{2}\operatorname{tr}\boldsymbol{\Sigma}^{-1}\big[(n-1)\mathbf{S}\big] \\ &\qquad -\tfrac{n}{2}(\overline{\mathbf{Y}} - \boldsymbol{\mu})'\boldsymbol{\Sigma}^{-1}(\overline{\mathbf{Y}} - \boldsymbol{\mu})\Big\}. \end{aligned}$$

Since $\boldsymbol{\Sigma} > 0$, $(\overline{\mathbf{Y}} - \boldsymbol{\mu})'\boldsymbol{\Sigma}^{-1}(\overline{\mathbf{Y}} - \boldsymbol{\mu})$ is minimized (for all $\boldsymbol{\Sigma}$) when $\boldsymbol{\mu} = \overline{\mathbf{Y}}$. Therefore, for all $\boldsymbol{\Sigma}$, $f(\mathbf{Y}_1, \ldots, \mathbf{Y}_n); \boldsymbol{\mu}, \boldsymbol{\Sigma})$ is maximized when $\boldsymbol{\mu} = \overline{\mathbf{Y}}$, and hence the maximum likelihood estimator of $\boldsymbol{\mu}$ is $\overline{\mathbf{Y}}$. To find the maximum likelihood estimator of $\boldsymbol{\Sigma}$ we must maximize

$$f(\mathbf{Y}_1, \ldots, \mathbf{Y}_n; \overline{\mathbf{Y}}, \boldsymbol{\Sigma}) = \frac{1}{(2\pi)^{nr/2}|\boldsymbol{\Sigma}|^{n/2}} \exp\Big\{-\tfrac{1}{2}\operatorname{tr}\boldsymbol{\Sigma}^{-1}\big[(n-1)\mathbf{S}\big]\Big\}.$$

If $n > r$, then $[(n-1)/n]\mathbf{S} > 0$. Therefore, by Lemma 18.4, this is maximized for $\boldsymbol{\Sigma} = [(n-1)/n)\mathbf{S}$. □

If $n \leqslant r$, then $[(n-1)/n]\mathbf{S}$ is not positive definite and hence cannot be the MLE for $\boldsymbol{\Sigma}$. In fact there is no MLE unless $n > r$.

As a final result in this section, we show that $\overline{\mathbf{Y}}$ is the best invariant estimator of $\boldsymbol{\mu}$ with respect to the Mahalanobis distance loss function

$$L(\mathbf{d}; (\boldsymbol{\mu}, \boldsymbol{\Sigma})) = n(\mathbf{d} - \boldsymbol{\mu})'\boldsymbol{\Sigma}^{-1}(\mathbf{d} - \boldsymbol{\mu})$$

[Note that $\overline{\mathbf{Y}} \sim N_r(\boldsymbol{\mu}, (1/n)\boldsymbol{\Sigma})$.] This estimation problem is invariant under the group G of transformations

$$g(\overline{\mathbf{Y}}, \mathbf{S}) = (\mathbf{A}\overline{\mathbf{Y}} + \mathbf{b}, \mathbf{ASA}'), \quad \bar{g}(\boldsymbol{\mu}, \boldsymbol{\Sigma}) = (\mathbf{A}\boldsymbol{\mu} + \mathbf{b}, \mathbf{A}\boldsymbol{\Sigma}\mathbf{A}'), \quad \tilde{g}(\mathbf{d}) = \mathbf{Ad} + \mathbf{b}$$

where $\mathbf{A}$ is an arbitrary invertible $r \times r$ matrix and $\mathbf{b}$ is an arbitrary $r \times 1$

vector. Note that

$$L(\tilde{g}(\mathbf{d});\ \bar{g}(\boldsymbol{\mu},\boldsymbol{\Sigma})) = n((\mathbf{Ad}+\mathbf{b})-(\mathbf{A}\boldsymbol{\mu}+\mathbf{b}))'(\mathbf{A}\boldsymbol{\Sigma}\mathbf{A}')^{-1}(\mathbf{Ad}+\mathbf{b}-(\mathbf{A}\boldsymbol{\mu}+\mathbf{b}))$$

$$= n(\mathbf{d}-\boldsymbol{\mu})'\mathbf{A}'(\mathbf{A}')^{-1}\boldsymbol{\Sigma}^{-1}\mathbf{A}^{-1}\mathbf{A}(\mathbf{d}-\boldsymbol{\mu})$$

$$= n(\mathbf{d}-\boldsymbol{\mu})'\boldsymbol{\Sigma}^{-1}(\mathbf{d}-\boldsymbol{\mu}) = L(\mathbf{d};(\boldsymbol{\mu},\boldsymbol{\Sigma}))$$

THEOREM 18.6. $\overline{\mathbf{Y}}$ is the best invariant estimator of $\boldsymbol{\mu}$ for the loss function L under the group G.

PROOF. An estimator $\mathbf{T}(\overline{\mathbf{Y}},\mathbf{S})$ is invariant if and only if

$$\mathbf{T}(\mathbf{A}\overline{\mathbf{Y}}+\mathbf{b},\mathbf{ASA}') = \mathbf{AT}(\overline{\mathbf{Y}},\mathbf{S})+\mathbf{b}$$

for all invertible $\mathbf{A}$ and all $\mathbf{b}$. Note that $\overline{\mathbf{Y}}$ is an invariant estimator. Now, let $\mathbf{T}$ be any other invariant estimator. Let $\mathbf{A}=\mathbf{S}^{-1/2}$, $\mathbf{B}=-\mathbf{S}^{-1/2}\overline{\mathbf{Y}}$. Since $\mathbf{T}$ is invariant, we see that

$$\mathbf{T}(\mathbf{0},\mathbf{I}) = \mathbf{S}^{-1/2}\mathbf{T}(\overline{\mathbf{Y}},\mathbf{S}) - \mathbf{S}^{-1/2}\overline{\mathbf{Y}}$$

or equivalently,

$$\mathbf{T}(\overline{\mathbf{Y}},\mathbf{S}) = \mathbf{S}^{1/2}\mathbf{T}(\mathbf{0},\mathbf{I}) + \overline{\mathbf{Y}}.$$

[Note that $\mathbf{T}$ is completely determined by $\mathbf{T}(\mathbf{0},\mathbf{I})$.] Now the risk function of $\mathbf{T}$ is

$$R(\mathbf{T};(\boldsymbol{\mu},\boldsymbol{\Sigma})) = nE\big(\mathbf{S}^{1/2}\mathbf{T}(\mathbf{0},\mathbf{I})+\overline{\mathbf{Y}}-\boldsymbol{\mu}\big)'\boldsymbol{\Sigma}^{-1}\big(\mathbf{S}^{1/2}\mathbf{T}(\mathbf{0},\mathbf{I})+\overline{\mathbf{Y}}-\boldsymbol{\mu}\big)$$

$$= nE\mathbf{T}'(\mathbf{0},\mathbf{I})\mathbf{S}^{1/2}\boldsymbol{\Sigma}^{-1}\mathbf{S}^{1/2}\mathbf{T}(\mathbf{0},\mathbf{I}) + nE\big(\overline{\mathbf{Y}}-\boldsymbol{\mu}\big)'\boldsymbol{\Sigma}^{-1}\big(\overline{\mathbf{Y}}-\boldsymbol{\mu}\big)$$

$$+ 2nE\mathbf{T}'(\mathbf{0},\mathbf{I})\mathbf{S}^{1/2}\boldsymbol{\Sigma}^{-1}\big(\overline{\mathbf{Y}}-\boldsymbol{\mu}\big)$$

However,

$$E\mathbf{T}'(\mathbf{0},\mathbf{I})\mathbf{S}^{1/2}\boldsymbol{\Sigma}^{-1}\big(\overline{\mathbf{Y}}-\boldsymbol{\mu}\big) = \mathbf{T}'(\mathbf{0},\mathbf{I})(E\mathbf{S}^{1/2})\boldsymbol{\Sigma}^{-1}E\big(\overline{\mathbf{Y}}-\boldsymbol{\mu}\big) = 0$$

(since $\overline{\mathbf{Y}}$ and $\mathbf{S}$ are independent). Also

$$nE\big(\overline{\mathbf{Y}}-\boldsymbol{\mu}\big)'\boldsymbol{\Sigma}^{-1}\big(\overline{\mathbf{Y}}-\boldsymbol{\mu}\big) = R\big(\overline{\mathbf{Y}};(\boldsymbol{\mu},\boldsymbol{\Sigma})\big).$$

Therefore,

$$R(\mathbf{T};(\boldsymbol{\mu},\boldsymbol{\Sigma})) = nE\mathbf{T}'(\mathbf{0},\mathbf{I})\mathbf{S}^{1/2}\boldsymbol{\Sigma}^{-1}\mathbf{S}^{1/2}\mathbf{T}(\mathbf{0},\mathbf{I}) + R\big(\overline{\mathbf{Y}};(\boldsymbol{\mu},\boldsymbol{\Sigma})\big).$$

Now, $\boldsymbol{\Sigma}^{-1}>0$ (see Theorem 5 of the appendix) $\mathbf{S}^{1/2}$ is invertible and $\mathbf{T}(\mathbf{0},\mathbf{I})\neq\mathbf{0}$ (since $\mathbf{T}\neq\overline{\mathbf{Y}}$). Therefore, $\mathbf{S}^{1/2}\mathbf{T}(\mathbf{0},\mathbf{I})\neq\mathbf{0}$ and

$$E\mathbf{T}'(\mathbf{0},\mathbf{I})\mathbf{S}^{1/2}\boldsymbol{\Sigma}^{-1}\mathbf{S}^{1/2}\mathbf{T}(\mathbf{0},\mathbf{I}) > 0.$$

Hence $\overline{\mathbf{Y}}$ has lower risk than any other invariant estimator and is the best invariant estimator. □

18.3. JAMES-STEIN ESTIMATORS FOR $\boldsymbol{\mu}$.

In this section, we find estimators for $\boldsymbol{\mu}$ that are better than $\overline{\mathbf{Y}}$ when $n > r > 2$. As an indication why it should be possible to improve on $\overline{\mathbf{Y}}$, we note that

$$E\|\overline{\mathbf{Y}}\|^2 = \|\boldsymbol{\mu}\|^2 + \frac{1}{n}\operatorname{tr}\boldsymbol{\Sigma}.$$

(See Exercise B3.) Therefore, if we use $\overline{\mathbf{Y}}$ to estimate $\boldsymbol{\mu}$, we are substantially overestimating $\|\boldsymbol{\mu}\|^2$. It seems that we should be able to improve on $\overline{\mathbf{Y}}$ by shrinking it.

The loss function that we use is the Mahalanobis distance between $\boldsymbol{\mu}$ and an estimate $\mathbf{d}$ of $\boldsymbol{\mu}$,

$$L(\mathbf{d},(\boldsymbol{\mu},\boldsymbol{\Sigma})) = n(\mathbf{d}-\boldsymbol{\mu})'\boldsymbol{\Sigma}^{-1}(\mathbf{d}-\boldsymbol{\mu}).$$

For $r = 1$, the estimator $\overline{\mathbf{Y}}$ has been known to be admissible (see Ferguson, 1967, p. 141). Stein (1955) showed that $\overline{\mathbf{Y}}$ is also admissible for $r = 2$, but that $\overline{\mathbf{Y}}$ is inadmissible for $r > 2$. James and Stein (1960) derived an estimator that is better than $\overline{\mathbf{Y}}$. We follow that derivation in this section. Throughout the section we assume that $r > 2$. We also assume that $n > r$ so that $\mathbf{S}$ is invertible. (See the corollary to Theorem 17.8.) We note that any estimator that is better than $\overline{\mathbf{Y}}$ cannot be invariant; nor can it be unbiased.

First, the risk function of $\overline{\mathbf{Y}}$ is derived.

LEMMA 18.7. $R(\overline{\mathbf{Y}};(\boldsymbol{\mu},\boldsymbol{\Sigma})) = r$

PROOF. $\overline{\mathbf{Y}} \sim N_r(\boldsymbol{\mu},(1/n)\boldsymbol{\Sigma})$. Therefore, by Lemma 3.8, $n(\overline{\mathbf{Y}}-\boldsymbol{\mu})'\boldsymbol{\Sigma}^{-1}(\overline{\mathbf{Y}}-\boldsymbol{\mu}) \sim \chi_r^2$. Hence, $nE(\overline{\mathbf{Y}}-\boldsymbol{\mu})'\boldsymbol{\Sigma}^{-1}(\overline{\mathbf{Y}}-\boldsymbol{\mu}) = r$. □

We now consider estimators $d_c(\overline{\mathbf{Y}},\mathbf{S})$ of the form

$$d_c(\overline{\mathbf{Y}},\mathbf{S}) = \left(1 - \frac{c}{\overline{\mathbf{Y}}'\mathbf{S}^{-1}\overline{\mathbf{Y}}}\right)\overline{\mathbf{Y}}.$$

We find the risk function for d_c and find the c that minimizes it.

LEMMA 18.8. Let K have a Poisson distribution with mean $n\boldsymbol{\mu}'\boldsymbol{\Sigma}^{-1}\boldsymbol{\mu}/2$. Then

$$R(d_c;(\boldsymbol{\mu},\boldsymbol{\Sigma}))$$
$$= r - \left[\frac{2nc(n-r)(r-2)}{n-1} - \frac{c^2n^2(n-r)(n-r+2)}{(n-1)^2}\right] E\frac{1}{r-2+2K}.$$

PROOF. We first note that Lemma 17.10 and Theorem 18.2 imply that

$$(n-1)\frac{\bar{\mathbf{Y}}'\boldsymbol{\Sigma}^{-1}\bar{\mathbf{Y}}}{\bar{\mathbf{Y}}'\mathbf{S}^{-1}\bar{\mathbf{Y}}} \sim \chi^2_{n-r}(0)$$

independently of $\bar{\mathbf{Y}}$. (See the proof of Theorem 17.11.) Therefore

$$R(d_c;(\boldsymbol{\mu},\boldsymbol{\Sigma})) = nE\left(\bar{\mathbf{Y}}-\boldsymbol{\mu}-\frac{c\bar{\mathbf{Y}}}{\bar{\mathbf{Y}}'\mathbf{S}^{-1}\bar{\mathbf{Y}}}\right)\boldsymbol{\Sigma}^{-1}\left(\bar{\mathbf{Y}}-\boldsymbol{\mu}-\frac{\mathbf{c}\bar{\mathbf{Y}}}{\bar{\mathbf{Y}}'\mathbf{S}^{-1}\bar{\mathbf{Y}}}\right)$$

$$= nE(\bar{\mathbf{Y}}-\boldsymbol{\mu})'\boldsymbol{\Sigma}^{-1}(\bar{\mathbf{Y}}-\boldsymbol{\mu}) - 2cnE\frac{\bar{\mathbf{Y}}'\boldsymbol{\Sigma}^{-1}(\bar{\mathbf{Y}}-\boldsymbol{\mu})}{\bar{\mathbf{Y}}'\mathbf{S}^{-1}\bar{\mathbf{Y}}}$$

$$+ c^2nE\frac{\bar{\mathbf{Y}}'\boldsymbol{\Sigma}^{-1}\bar{\mathbf{Y}}}{(\bar{\mathbf{Y}}'\mathbf{S}^{-1}\bar{\mathbf{Y}})^2}$$

$$= nE(\bar{\mathbf{Y}}-\boldsymbol{\mu})'\boldsymbol{\Sigma}^{-1}(\bar{\mathbf{Y}}-\boldsymbol{\mu})$$

$$-\frac{2cn}{n-1}E(n-1)\frac{\bar{\mathbf{Y}}'\boldsymbol{\Sigma}^{-1}\bar{\mathbf{Y}}}{\bar{\mathbf{Y}}'\mathbf{S}^{-1}\bar{\mathbf{Y}}}E\frac{\bar{\mathbf{Y}}'\boldsymbol{\Sigma}^{-1}(\bar{\mathbf{Y}}-\boldsymbol{\mu})}{\bar{\mathbf{Y}}'\boldsymbol{\Sigma}^{-1}\bar{\mathbf{Y}}}$$

$$+\frac{c^2n^2}{(n-1)^2}E\left(\frac{(n-1)\bar{\mathbf{Y}}'\boldsymbol{\Sigma}^{-1}\bar{\mathbf{Y}}}{\bar{\mathbf{Y}}'\mathbf{S}^{-1}\bar{\mathbf{Y}}}\right)^2 E\frac{1}{n\bar{\mathbf{Y}}'\boldsymbol{\Sigma}^{-1}\bar{\mathbf{Y}}},$$

using the independence of $(n-1)\bar{\mathbf{Y}}'\boldsymbol{\Sigma}^{-1}\bar{\mathbf{Y}}/\bar{\mathbf{Y}}'\mathbf{S}^{-1}\bar{\mathbf{Y}}$ and $\bar{\mathbf{Y}}$. By Lemma 18.7,

$$nE(\bar{\mathbf{Y}}-\boldsymbol{\mu})'\boldsymbol{\Sigma}^{-1}(\bar{\mathbf{Y}}-\boldsymbol{\mu}) = r.$$

Since $(n-1)\bar{\mathbf{Y}}'\boldsymbol{\Sigma}^{-1}\bar{\mathbf{Y}}/\bar{\mathbf{Y}}'\mathbf{S}^{-1}\bar{\mathbf{Y}} \sim \chi^2_{n-r}(0)$, we see that

$$E(n-1)\frac{\bar{\mathbf{Y}}'\boldsymbol{\Sigma}^{-1}\bar{\mathbf{Y}}}{\bar{\mathbf{Y}}'\mathbf{S}^{-1}\bar{\mathbf{Y}}} = n-r, \qquad E\left(\frac{(n-1)\bar{\mathbf{Y}}'\boldsymbol{\Sigma}^{-1}\bar{\mathbf{Y}}}{\bar{\mathbf{Y}}'\mathbf{S}^{-1}\bar{\mathbf{Y}}}\right)^2 = (n-r)(n-r+2).$$

Now, let $\mathbf{U} = \sqrt{n}\,\boldsymbol{\Sigma}^{-1/2}\bar{\mathbf{Y}}$, $\boldsymbol{\theta} = \sqrt{n}\,\boldsymbol{\Sigma}^{-1/2}\boldsymbol{\mu}$. Then $\mathbf{U} \sim N_r(\boldsymbol{\theta}, \mathbf{I})$, and

$$\|\boldsymbol{\theta}\|^2 = n\boldsymbol{\mu}'\boldsymbol{\Sigma}^{-1}\boldsymbol{\mu}, \qquad \frac{\bar{\mathbf{Y}}'\boldsymbol{\Sigma}^{-1}(\bar{\mathbf{Y}}-\boldsymbol{\mu})}{\bar{\mathbf{Y}}'\boldsymbol{\Sigma}^{-1}\bar{\mathbf{Y}}} = \frac{\mathbf{U}'(\mathbf{U}-\boldsymbol{\theta})}{\mathbf{U}'\mathbf{U}}, \qquad \frac{1}{n\bar{\mathbf{Y}}\boldsymbol{\Sigma}^{-1}\bar{\mathbf{Y}}} = \frac{1}{\mathbf{U}'\mathbf{U}}.$$

Therefore, by Lemma 11.3, we see that

$$E\frac{\bar{\mathbf{Y}}'\boldsymbol{\Sigma}^{-1}(\bar{\mathbf{Y}}-\boldsymbol{\mu})}{\bar{\mathbf{Y}}'\boldsymbol{\Sigma}^{-1}\bar{\mathbf{Y}}} = E\frac{r-2}{r-2+2K}, \qquad \frac{1}{n\bar{\mathbf{Y}}'\boldsymbol{\Sigma}^{-1}\bar{\mathbf{Y}}} = E\frac{1}{r-2+2K}.$$

Substitution of these results into the risk function gives

$$R(d_c;(\mu,\Sigma)) = r - \frac{2cn(n-r)(r-2)}{(n-1)} E\frac{1}{r-2+2K}$$

$$+ c^2 \frac{n^2(n-r)(n-r+2)}{(n-1)^2} E\frac{1}{r-2+2K}. \quad \square$$

We note that $E(r-2+2K)^{-1} > 0$ and hence $R(d_c;(\boldsymbol{\mu},\boldsymbol{\Sigma}))$ is minimized for all $\boldsymbol{\mu}$ and $\boldsymbol{\Sigma}$ by letting

$$c = \frac{(r-2)(n-1)}{(n-r+2)n}.$$

Therefore, define

$$\hat{\boldsymbol{\mu}} = \left[1 - \frac{(r-2)(n-1)}{(n-r+2)n\overline{\mathbf{Y}}'\mathbf{S}^{-1}\overline{\mathbf{Y}}}\right]\overline{\mathbf{Y}}.$$

$\hat{\boldsymbol{\mu}}$ is called the James-Stein estimator of $\boldsymbol{\mu}$. The following result follows directly from Lemma 18.8.

THEOREM 18.9. Let K have a Poisson distribution with mean $\delta = n\boldsymbol{\mu}'\boldsymbol{\Sigma}^{-1}\boldsymbol{\mu}$. Then

$$R(\hat{\boldsymbol{\mu}};(\boldsymbol{\mu},\boldsymbol{\Sigma})) = r - \frac{(r-2)^2(n-r)}{(n-r+2)} E\frac{1}{r-2+2K} < r.$$

If $\boldsymbol{\mu} = \mathbf{0}$, then $R(\hat{\boldsymbol{\mu}};(\mathbf{0},\boldsymbol{\Sigma})) = 2n/(n-r+2)$.

We therefore note that the risk of $\hat{\boldsymbol{\mu}}$ is about 2 if $\boldsymbol{\mu} = \mathbf{0}$ and is less than r for all $\boldsymbol{\mu}$ and $\boldsymbol{\Sigma}$. Therefore $\hat{\boldsymbol{\mu}}$ is better than $\overline{\mathbf{Y}}$ and is much better if $\boldsymbol{\mu}$ is near $\mathbf{0}$. Hence $\overline{\mathbf{Y}}$ is inadmissible. For large n, the risk function is approximately equal to

$$k_r\left(\frac{n\boldsymbol{\mu}'\boldsymbol{\Sigma}^{-1}\boldsymbol{\mu}}{2}\right) = r - (r-2)^2 E\frac{1}{r-2+2K}$$

This function is graphed in Figure 11.1.

We do not, of course, have to shrink toward the origin. Let $\mathbf{v} \in R^n$ and

$$\hat{\boldsymbol{\mu}}_{\mathbf{v}} = \left[1 - \frac{(r-2)(n-1)}{(n-r+2)n(\overline{\mathbf{Y}}-\mathbf{v})'\mathbf{S}^{-1}(\overline{\mathbf{Y}}-\mathbf{v})}\right](\overline{\mathbf{Y}}-\mathbf{v}) + \mathbf{v}.$$

Then it is straightforward to show that

$$R(\hat{\boldsymbol{\mu}}_{\mathbf{v}},(\boldsymbol{\mu},\boldsymbol{\Sigma})) = R(\hat{\boldsymbol{\mu}};(\boldsymbol{\mu}-\mathbf{v},\boldsymbol{\Sigma})).$$

Therefore $\hat{\boldsymbol{\mu}}_{\mathbf{v}}$ does very well if $\boldsymbol{\mu}$ is near $\mathbf{v}$. This method of estimation has the property that we can pick a point $\mathbf{v} \in R^r$ that we expect $\boldsymbol{\mu}$ to be near. We then compute $\hat{\boldsymbol{\mu}}_{\mathbf{v}}$. If we are correct, and $\boldsymbol{\mu}$ is near $\mathbf{v}$, then our estimator is much better than $\overline{\mathbf{Y}}$ (as we would expect). However, if we are wrong, our estimator is still moderately better than $\overline{\mathbf{Y}}$. Using this method we can guess a value and lose nothing if we are wrong.

Although the estimator $\hat{\boldsymbol{\mu}}_{\mathbf{v}}$ is not invariant, it does have the following invariance property:

$$\hat{\boldsymbol{\mu}}_{\mathbf{Av}+\mathbf{b}}(\mathbf{A}\overline{\mathbf{Y}} + \mathbf{b}, \mathbf{ASA}') = \mathbf{A}\hat{\boldsymbol{\mu}}_{\mathbf{v}}(\overline{\mathbf{Y}}, \mathbf{S}) + \mathbf{b}$$

That is, if we transform the point we are shrinking toward when we transform $\overline{\mathbf{Y}}$ and $\mathbf{S}$, the estimator transforms correctly.

As in Chapter 11, we can improve the estimator $\hat{\boldsymbol{\mu}}_{\mathbf{v}}$ by guaranteeing that it is in fact a shrinking estimator, that is, guaranteeing that it does not shrink $\overline{\mathbf{Y}}$ past $\mathbf{v}$. Let

$$\hat{\hat{\boldsymbol{\mu}}}_{\mathbf{v}} = \begin{cases} \hat{\boldsymbol{\mu}}_{\mathbf{v}} & \text{if } \left[1 - \dfrac{(r-2)(n-1)}{(n-r+2)n(\overline{\mathbf{Y}} - \mathbf{v})'\mathbf{S}^{-1}(\overline{\mathbf{Y}} - \mathbf{v})}\right] \geqslant 0 \\ \mathbf{v} & \text{if } \left[1 - \dfrac{(r-2)(n-1)}{(n-r+2)n(\overline{\mathbf{Y}} - \mathbf{v})'\mathbf{S}^{-1}(\overline{\mathbf{Y}} - \mathbf{v})}\right] < 0. \end{cases}$$

Then $\hat{\hat{\boldsymbol{\mu}}}_v$ is better than $\hat{\boldsymbol{\mu}}_v$ (see Stein, 1966).

18.4. TESTING HYPOTHESES ABOUT $\boldsymbol{\mu}$.

We now consider testing several hypotheses about $\boldsymbol{\mu}$. The groups that we use to reduce the problem will be stated first for $\mathbf{Y}_i$ and then for the sufficient statistic $(\overline{\mathbf{Y}}, \mathbf{S})$. Throughout the remainder of the chapter, we assume that $n > r$, so that $\mathbf{S}$ is invertible.

The first testing problem that we consider is testing that $\boldsymbol{\mu} = \mathbf{0}$ against $\boldsymbol{\mu}$ unrestricted. This problem is invariant under the group G of transformations g of the form

$$g(\mathbf{Y}_i) = \mathbf{AY}_i$$

where $\mathbf{A}$ is an arbitrary $r \times r$ invertible matrix. In terms of $(\overline{\mathbf{Y}}, \mathbf{S})$ the transformation g has the form

$$g(\overline{\mathbf{Y}}, \mathbf{S}) = (\mathbf{A}\overline{\mathbf{Y}}, \mathbf{ASA}').$$

The induced transformation on the parameter space is

$$\bar{g}(\boldsymbol{\mu}, \boldsymbol{\Sigma}) = (\mathbf{A}\boldsymbol{\mu}, \mathbf{A}\boldsymbol{\Sigma}\mathbf{A}').$$

We first find a maximal invariant under G.

Let

$$F_1 = \frac{(n-r)n}{r(n-1)} \overline{\mathbf{Y}}'\mathbf{S}^{-1}\overline{\mathbf{Y}}, \qquad \delta_1 = n\boldsymbol{\mu}'\boldsymbol{\Sigma}^{-1}\boldsymbol{\mu}. \tag{18.1}$$

LEMMA 18.10. F_1 is a maximal invariant and δ_1 is a parameter maximal invariant under G.

PROOF. Let

$$T(\overline{\mathbf{Y}}, \mathbf{S}) = \frac{(n-r)n}{r(n-1)} \overline{\mathbf{Y}}'\mathbf{S}^{-1}\overline{\mathbf{Y}}.$$

Then

$$\begin{aligned} T(\mathbf{A}\overline{\mathbf{Y}}, \mathbf{ASA}') &= \frac{(n-r)n}{r(n-1)} \overline{\mathbf{Y}}'\mathbf{A}'(\mathbf{A}')^{-1}\mathbf{S}^{-1}\mathbf{A}^{-1}\mathbf{A}\overline{\mathbf{Y}} \\ &= \frac{(n-r)n}{r(n-1)} \overline{\mathbf{Y}}'\mathbf{S}^{-1}\overline{\mathbf{Y}} = T(\overline{\mathbf{Y}}, \mathbf{S}), \end{aligned}$$

so that T is invariant. Now suppose that $T(\overline{\mathbf{Y}}^*, \mathbf{S}^*) = T(\overline{\mathbf{Y}}, \mathbf{S})$. Then $\overline{\mathbf{Y}}^{*\prime}\mathbf{S}^{*-1/2}\overline{\mathbf{Y}}^* = \overline{\mathbf{Y}}'\mathbf{S}^{-1}\overline{\mathbf{Y}}$. By Theorem 12 of the appendix there exists $\mathbf{A}$ invertible such that $\overline{\mathbf{Y}}^* = \mathbf{A}\overline{\mathbf{Y}}, \mathbf{S}^* = \mathbf{ASA}'$ and hence T is maximal. The proof for δ_1 follows similarly. □

Now, $\overline{\mathbf{Y}}$ and $\mathbf{S}$ are independent and

$$\sqrt{n}\,\overline{\mathbf{Y}} \sim N_r(\sqrt{n}\,\boldsymbol{\mu}, \boldsymbol{\Sigma}), \qquad (n-1)\mathbf{S} \sim W_r(n-1, \boldsymbol{\Sigma}).$$

Therefore, by Theorem 17.11,

$$F_1 = \frac{n-r}{r}(\sqrt{n}\,\overline{\mathbf{Y}})'\left[(n-1)\mathbf{S}\right]^{-1}(\sqrt{n}\,\overline{\mathbf{Y}}) \sim F_{r,n-r}(\delta_1).$$

Hence, after reducing by invariance, we have the problem in which we observe $F_1 \sim F_{r,n-r}(\delta_1)$ and are testing that $\delta_1 = 0$ against $\delta_1 \geqslant 0$. The UMP size α test for this reduced problem is therefore

$$\phi_1(F_1) = \begin{cases} 1 & \text{if } F_1 > F^{\alpha}_{r,\,n-r} \\ 0 & \text{if } F_1 \leqslant F^{\alpha}_{r,\,n-r} \end{cases}. \tag{18.2}$$

We have now proved the following theorem.

THEOREM 18.11. The UMP invariant size α test of the hypothesis that $\boldsymbol{\mu} = \mathbf{0}$ against the alternative that $\boldsymbol{\mu}$ is unrestricted is given by ϕ_1 defined by (18.1) and (18.2).

The test ϕ_1 is called the one sample Hotelling's T^2-test and is also the likelihood ratio test for this model (see Exercise B4).

Now let $\boldsymbol{\mu} = \binom{\boldsymbol{\mu}_1}{\boldsymbol{\mu}_2}$ where $\boldsymbol{\mu}_1$ is $s \times 1$. We now consider testing that $\boldsymbol{\mu}_1 = \mathbf{0}$ against $\boldsymbol{\mu}$ unrestricted. Let

$$\mathbf{Y}_i = \begin{pmatrix} \mathbf{Y}_{i1} \\ \mathbf{Y}_{i2} \end{pmatrix}, \qquad \overline{\mathbf{Y}} = \begin{bmatrix} \overline{\mathbf{Y}}_1 \\ \overline{\mathbf{Y}}_2 \end{bmatrix}, \qquad \mathbf{S} = \begin{pmatrix} \mathbf{S}_{11} & \mathbf{S}_{12} \\ \mathbf{S}_{21} & \mathbf{S}_{22} \end{pmatrix}, \qquad \boldsymbol{\Sigma} = \begin{pmatrix} \boldsymbol{\Sigma}_{11} & \boldsymbol{\Sigma}_{12} \\ \boldsymbol{\Sigma}_{21} & \boldsymbol{\Sigma}_{22} \end{pmatrix}.$$

where $\mathbf{Y}_{i1}$ and $\overline{\mathbf{Y}}_1$ are $s \times 1$ and $\mathbf{S}_{11}$ and $\boldsymbol{\Sigma}_{11}$ are $s \times s$. The problem is invariant under the following four groups.

(1) The first group G_1 consists of transformations of the form $g_1(\mathbf{Y}_{i1}, \mathbf{Y}_{i2}) = (\mathbf{Y}_{i1}, \mathbf{Y}_{i2} + \mathbf{a})$ where $\mathbf{a}$ is an arbitrary $(r - s) \times 1$ vector. In terms of $(\overline{\mathbf{Y}}, \mathbf{S})$, this transformation becomes $g_1(\overline{\mathbf{Y}}_1, \overline{\mathbf{Y}}_2, \mathbf{S}_{11}, \mathbf{S}_{12}, \mathbf{S}_{22}) = (\overline{\mathbf{Y}}_1, \overline{\mathbf{Y}}_2 + \mathbf{a}, \mathbf{S}_{11}, \mathbf{S}_{12}, \mathbf{S}_{22})$. The induced transformation on the parameter space is analogous.

(2) G_2 consists of transformations g_2 of the form $g_2(\mathbf{Y}_{i1}, \mathbf{Y}_{i2}) = (\mathbf{Y}_{i1}, \mathbf{Y}_{i2} + \mathbf{B}\mathbf{Y}_{i1})$ where $\mathbf{B}$ is an arbitrary $(r - s) \times s$ matrix. In terms of $(\overline{\mathbf{Y}}, \mathbf{S})$,

$$g_2(\overline{\mathbf{Y}}_1, \overline{\mathbf{Y}}_2, \mathbf{S}_{11}, \mathbf{S}_{12}, \mathbf{S}_{22}) = \big(\overline{\mathbf{Y}}_1, \overline{\mathbf{Y}}_2 + \mathbf{B}\overline{\mathbf{Y}}_1, \mathbf{S}_{11}, \mathbf{S}_{12} + \mathbf{S}_{11}\mathbf{B}', \mathbf{S}_{22} + \mathbf{S}_{12}'\mathbf{B}' + \mathbf{B}\mathbf{S}_{12} + \mathbf{B}\mathbf{S}_{11}\mathbf{B}'\big).$$

The induced transformation on the parameter space is analogous to this one.

(3) The group G_3 consists of transformations of the form $g_3(\mathbf{Y}_{i1}, \mathbf{Y}_{i2}) = (\mathbf{Y}_{i1}, \mathbf{C}\mathbf{Y}_{i2})$ where $\mathbf{C}$ is invertible. Equivalently, in terms of $(\overline{\mathbf{Y}}, \mathbf{S})$, $g_3(\overline{\mathbf{Y}}_1, \overline{\mathbf{Y}}_2, \mathbf{S}_{11}, \mathbf{S}_{12}, \mathbf{S}_{22}) = (\overline{\mathbf{Y}}_1, \mathbf{C}\overline{\mathbf{Y}}_2, \mathbf{S}_{11}, \mathbf{S}_{12}\mathbf{C}', \mathbf{C}\mathbf{S}_{22}\mathbf{C}')$. The parameter transformation is analogous.

(4) The final group G_4 consists of transformations of the form $g_4(\mathbf{Y}_{i1}, \mathbf{Y}_{i2}) = (\mathbf{D}\mathbf{Y}_{i1}, \mathbf{Y}_{i2})$ where $\mathbf{D}$ is invertible, Equivalently, $g_4(\overline{\mathbf{Y}}_1, \overline{\mathbf{Y}}_2, \mathbf{S}_{11}, \mathbf{S}_{12}, \mathbf{S}_{22}) = (\mathbf{D}\overline{\mathbf{Y}}_1, \overline{\mathbf{Y}}_2, \mathbf{D}\mathbf{S}_{11}\mathbf{D}', \mathbf{D}\mathbf{S}_{12}, \mathbf{S}_{22})$, with the parameter group acting similarly.

The reader should verify that this problem is invariant under the four groups and should also understand why it is not invariant under such groups as $g_5(\mathbf{Y}_{i1}, \mathbf{Y}_{i2}) = (\mathbf{Y}_{i1} + \mathbf{E}\mathbf{Y}_{i2}, \mathbf{Y}_{i2})$. Invariance under G_1, G_2, G_3, and G_4 is equivalent to invariance under the large group G of transformations of the form

$$g(\mathbf{Y}_{i1}, \mathbf{Y}_{i2}) = \begin{pmatrix} \mathbf{D} & \mathbf{0} \\ \mathbf{B} & \mathbf{C} \end{pmatrix} \begin{pmatrix} \mathbf{Y}_{1i} \\ \mathbf{Y}_{2i} \end{pmatrix} + \begin{pmatrix} \mathbf{0} \\ \mathbf{a} \end{pmatrix}.$$

It is easier, however, to reduce the problem under each of the smaller groups separately than under the larger group. The groups considered here are the same as groups considered in Chapter 16 for the correlation model. For that reason, the maximal invariant under each group will be stated with the derivations left for the reader. Define

$$F_2 = \frac{(n-s)n}{s(n-1)} \overline{\mathbf{Y}}_1' \mathbf{S}_{11}^{-1} \overline{\mathbf{Y}}_1, \qquad \delta_2 = n\boldsymbol{\mu}_1' \boldsymbol{\Sigma}_{11}^{-1} \boldsymbol{\mu}_1. \tag{18.3}$$

LEMMA 18.12. F_2 is a maximal invariant and δ_2 is a parameter maximal invariant.

PROOF. We reduce first by G_1, which consists of transformations of the form $g_1(\overline{\mathbf{Y}}_1, \overline{\mathbf{Y}}_2, \mathbf{S}_{11}, \mathbf{S}_{12}, \mathbf{S}_{22}) = (\overline{\mathbf{Y}}_1, \overline{\mathbf{Y}}_2 + \mathbf{a}, \mathbf{S}_{11}, \mathbf{S}_{12}, \mathbf{S}_{22})$. A maximal invariant under this group is

$$T_1(\overline{\mathbf{Y}}_1, \overline{\mathbf{Y}}_2, \mathbf{S}_{11}, \mathbf{S}_{12}, \mathbf{S}_{22}) = (\overline{\mathbf{Y}}_1, \mathbf{S}_{11}, \mathbf{S}_{12}, \mathbf{S}_{22}).$$

We now reduce by G_2, which is (in terms of T_1) given by

$$g_2(\overline{\mathbf{Y}}_1, \mathbf{S}_{11}, \mathbf{S}_{12}, \mathbf{S}_{22}) = (\overline{\mathbf{Y}}_1, \mathbf{S}_{11}, \mathbf{S}_{12} + \mathbf{S}_{12}\mathbf{B}', \mathbf{S}_{22} + \mathbf{S}_{12}'\mathbf{B}' + \mathbf{B}\mathbf{S}_{12} + \mathbf{B}\mathbf{S}_{11}\mathbf{B}').$$

A maximal invariant under G_2 is

$$T_2(\overline{\mathbf{Y}}_1, \mathbf{S}_{11}, \mathbf{S}_{12}, \mathbf{S}_{22}) = (\overline{\mathbf{Y}}_1, \mathbf{S}_{11}, \mathbf{S}_{22} - S_1\mathbf{S}_{11}^{-1}\mathbf{S}_{12}).$$

Now G_3 becomes transformations of the form $g_3(\overline{\mathbf{Y}}_1, \mathbf{S}_{11}, \mathbf{S}_{22} - \mathbf{S}_{21}\mathbf{S}_{11}^{-1}\mathbf{S}_{12}) = [\overline{\mathbf{Y}}_1, \mathbf{S}_{11}, \mathbf{C}(\mathbf{S}_{22} - \mathbf{S}_{21}\mathbf{S}_{11}^{-1}\mathbf{S}_{12})\mathbf{C}']$. A maximal invariant under this group is

$$T_3(\overline{\mathbf{Y}}_1, \mathbf{S}_{11}, \mathbf{S}_{22} - \mathbf{S}_{21}\mathbf{S}_{11}^{-1}\mathbf{S}_{12}) = (\overline{\mathbf{Y}}_1, \mathbf{S}_{11}).$$

The final group G_4 becomes transformations of the form $g_4(\overline{\mathbf{Y}}_1, \mathbf{S}_{11}) = (\mathbf{D}\overline{\mathbf{Y}}_1, \mathbf{D}\mathbf{S}_{11}\mathbf{D}')$. A maximal invariant under this group is

$$F_2 = \frac{(n-s)n}{s(n-1)} \overline{\mathbf{Y}}_1' \mathbf{S}_{11}^{-1} \overline{\mathbf{Y}}_1.$$

The argument for the parameter maximal invariant follows similarly. □

Now $\overline{\mathbf{Y}}_1$ and $\mathbf{S}_{11}$ are independent (since $\overline{\mathbf{Y}}$ and $\mathbf{S}$ are). Also, by Theorems 3.7a and 17.6e, we see that $\overline{\mathbf{Y}}_1 \sim N_s(\boldsymbol{\mu}_1, (1/n)\boldsymbol{\Sigma}_{11})$ and $(n-1)\mathbf{S}_{11} \sim W_s(n-1, \boldsymbol{\Sigma}_{11})$. Therefore, by Theorem 17.11, we see that $F_2 \sim F_{s, n-s}(\delta_2)$. We are testing that $\delta_2 = 0$ against $\delta_2 \geqslant 0$. Therefore, the UMP size α test for the reduced problem is given by

$$\phi_2(F_2) = \begin{cases} 1 & \text{if } F_2 > F^{\alpha}_{s,\, n-s} \\ 0 & \text{if } F_2 \leqslant F^{\alpha}_{s,\, n-s} \end{cases}. \tag{18.4}$$

We have therefore proved the following theorem.

THEOREM 18.13. Let $\phi_2(F_2)$ be defined by (18.3) and (18.4). Then ϕ_2 is the UMP invariant size α test for testing that $\boldsymbol{\mu}_1 = \mathbf{0}$ against $\boldsymbol{\mu}$ unrestricted in the one sample model.

The final problem that we consider is testing that $\mathbf{A}\boldsymbol{\mu} = \mathbf{b}$ where $\mathbf{A}$ is $s \times r$ of rank s. In Exercise B5 you are asked to transform this problem to the form considered above, and establish the following corollary. Let

$$F_2^* = \frac{(n-s)n}{s(n-1)}(\mathbf{A}\overline{\mathbf{Y}} - \mathbf{b})'(\mathbf{A}\mathbf{S}\mathbf{A}')^{-1}(\mathbf{A}\overline{\mathbf{Y}} - \mathbf{b}),$$

$$\phi_2^*(F_2^*) = \begin{cases} 1 & \text{if } F_2^* > F_{s,\,n-s}^{\alpha} \\ 0 & \text{if } F_2^* \leqslant F_{s,\,n-s}^{\alpha} \end{cases}. \tag{18.5}$$

COROLLARY. The UMP invariant size α test for testing that $\mathbf{A}\boldsymbol{\mu} = \mathbf{b}$ against $\boldsymbol{\mu}$ unrestricted in the one sample model is given by (18.5).

18.5. SIMULTANEOUS CONFIDENCE INTERVALS.

Let $\mathbf{A}$ be an $s \times r$ matrix of rank s. A *contrast* associated with testing that $\mathbf{A}\boldsymbol{\mu} = \mathbf{0}$ is a function $\mathbf{t}'\mathbf{A}\boldsymbol{\mu}$, $\mathbf{t} \in R^s$. We now find simultaneous confidence intervals for the set of all contrasts associated with testing that $\mathbf{A}\boldsymbol{\mu} = \mathbf{0}$. We first need the following lemma.

LEMMA 18.14. Let $\mathbf{T} > 0$ be a $k \times k$ matrix and let $\mathbf{v} \in R^k$. Then

$$\mathbf{v}'\mathbf{T}^{-1}\mathbf{v} = \sup_{\substack{\mathbf{u} \in R^k \\ \mathbf{u} \neq \mathbf{0}}} \frac{(\mathbf{u}'\mathbf{v})^2}{\mathbf{u}'\mathbf{T}\mathbf{u}}.$$

PROOF. By Lemma 8.1 (with $V = R^k$), we see that if $\mathbf{x} \in R^k$, then

$$\mathbf{x}'\mathbf{x} = \sup_{\substack{\mathbf{y} \in R^k \\ \mathbf{y} \neq \mathbf{0}}} \frac{(\mathbf{x}'\mathbf{y})^2}{\mathbf{y}'\mathbf{y}}.$$

Now, let $\mathbf{x} = \mathbf{T}^{-1/2}\mathbf{v}, \mathbf{y} = \mathbf{T}^{1/2}\mathbf{u}$. Then $\mathbf{y} \in R^k$, $\mathbf{y} \neq \mathbf{0}$ if and only if $\mathbf{u} \in R^k$, $\mathbf{u} \neq 0$. Therefore

$$\mathbf{v}'\mathbf{T}^{-1}\mathbf{v} = \mathbf{x}'\mathbf{x} = \sup_{\substack{\mathbf{y} \in R^k \\ \mathbf{y} \neq \mathbf{0}}} \frac{(\mathbf{x}'\mathbf{y})^2}{(\mathbf{y}'\mathbf{y})^2} = \sup_{\substack{\mathbf{u} \in R^k \\ \mathbf{u} \neq \mathbf{0}}} \frac{(\mathbf{u}'\mathbf{v})^2}{\mathbf{u}'\mathbf{T}\mathbf{u}}. \quad \square$$

We now give the main result of this section.

THEOREM 18.15. A set of $(1-\alpha)$ simultaneous confidence intervals for the set of all contrasts associated with testing that $\mathbf{A}\boldsymbol{\mu} = \mathbf{0}$ is given by

$$\mathbf{t}'\mathbf{A}\boldsymbol{\mu} \in \mathbf{t}'\mathbf{A}\overline{\mathbf{Y}} \pm \left[\frac{s(n-1)}{(n-s)n} F^{\alpha}_{s,\,n-s} \mathbf{t}'\mathbf{A}\mathbf{S}\mathbf{A}'\mathbf{t} \right]^{1/2}.$$

The hypothesis $\mathbf{A}\boldsymbol{\mu} = \mathbf{0}$ is rejected with the test ϕ_2^* given in (18.5) if and only if the above confidence interval for $\mathbf{t}'\mathbf{A}\boldsymbol{\mu}$ does not contain 0 for some $\mathbf{t} \in R^s$.

PROOF. See Exercise B6. □

These confidence intervals can be used after the F-test ϕ_2^* is rejected to see which linear functions $\mathbf{t}'\mathbf{A}\boldsymbol{\mu}$ are causing the hypothesis to be rejected. If we put $\mathbf{A} = \mathbf{I}$ in the above theorem, we have the following corollary.

COROLLARY. A set of simultaneous $(1-\alpha)$ confidence intervals for the set of all functions $\mathbf{t}'\boldsymbol{\mu}$ is given by

$$\mathbf{t}'\boldsymbol{\mu} \in \mathbf{t}'\overline{\mathbf{Y}} \pm \left[\frac{r(n-1)}{(n-r)n} F^{\alpha}_{r,\,n-r} \mathbf{t}'\mathbf{S}\mathbf{t} \right]^{1/2}.$$

The hypothesis $\boldsymbol{\mu} = \mathbf{0}$ is rejected with the test ϕ_1 defined in (18.2) if and only if at least one of the above confidence intervals does not contain 0.

18.6. ASYMPTOTIC VALIDITY OF PROCEDURES.

We now establish that the tests and simultaneous confidence intervals are asymptotically valid procedures as $n \to \infty$ even if the $\mathbf{Y}_i$ are not normally distributed. The model we consider in this section is one in which we observe $\mathbf{Y}_1, \mathbf{Y}_2, \ldots,$ independently and identically distributed r-dimensional random vectors with $E\mathbf{Y}_i = \boldsymbol{\mu}$, $\operatorname{cov}(\mathbf{Y}_i) = \boldsymbol{\Sigma}$. Let

$$\overline{\mathbf{Y}}_n = \frac{1}{n} \sum_{i=1}^{n} \mathbf{Y}_i, \qquad \mathbf{S}_n = \frac{1}{n-1} \sum_{i=1}^{n} (\mathbf{Y}_i - \overline{\mathbf{Y}}_n)(\mathbf{Y}_i - \overline{\mathbf{Y}}_n)'$$

be the sample mean vector and sample covariance matrix computed from the first $n\mathbf{Y}_i$'s. Let $\mathbf{A}$ be an $s \times r$ matrix of rank s and let

$$F_n = \frac{(n-s)n}{s(n-1)} \left[\mathbf{A}(\overline{\mathbf{Y}}_n - \boldsymbol{\mu}) \right]' (\mathbf{A}\mathbf{S}_n\mathbf{A}')^{-1} \mathbf{A}(\overline{\mathbf{Y}}_n - \boldsymbol{\mu}). \tag{18.6}$$

In this section we show that the asymptotic distribution of F_n does not depend on the distribution of $\mathbf{Y}_i$. If $\mathbf{A}\boldsymbol{\mu} = \mathbf{b}$, then F_n is just F_2^* defined in equation (18.5). Therefore, the null distribution of F_2^* is asymptotically independent of the normal distribution used in deriving it, and hence the size of ϕ_2^* is asymptotically independent of the normal assumption. Similarly, the sizes of ϕ_1 and ϕ_2 defined in Section 18.4 are asymptotically independent of the normal assumption. Finally, the $(1-\alpha)$ confidence intervals defined in the last section depend only on the distribution of F_n and are therefore asymptotically size α confidence intervals even if the $\mathbf{Y}_i$ are not normally distributed.

Before deriving the asymptotic distribution of F_n, we state some definitions and theorems from probability theory in addition to those already given in Chapter 10. Let $\mathbf{T}_1, \mathbf{T}_2, \ldots,$ be a sequence of $k \times s$ random matrices and let $\mathbf{A}$ be a $k \times s$ constant matrix. We say $\mathbf{T}_n \xrightarrow{p} \mathbf{A}$ if all the components of $\mathbf{T}_n$ converge in probability to the components of $\mathbf{A}$. Now, let $F_n(\mathbf{t})$ be the distribution function of T_n and $F(\mathbf{t})$ be a distribution function for $k \times s$ random matrices. We say that $\mathbf{T}_n \xrightarrow{d} F$ if

$$\lim_{m\to\infty} F_n(\mathbf{t}) = F(\mathbf{t})$$

for all $\mathbf{t}$ such that F is continuous at $\mathbf{t}$. The following theorem is a generalization of Theorems 10.1 and 10.2.

THEOREM 18.16. a. Weak law of large numbers. Let $\mathbf{T}_1, \mathbf{T}_2, \ldots,$ be a sequence of independently identically distributed $t \times q$ matrices with $E\mathbf{T}_i = \mathbf{A}$. Then

$$\overline{\mathbf{T}}_n = \frac{1}{n}\sum_{i=1}^{n} \mathbf{T}_i \xrightarrow{p} \mathbf{A}.$$

b. Central limit theorem. Let $\mathbf{U}_1, \mathbf{U}_2, \ldots,$ be a sequence of independently identically distributed q-dimensional random vectors with $E\mathbf{U}_i = \mathbf{b}$, $\operatorname{cov}(\mathbf{U}_i) = \mathbf{A}$. Let

$$\overline{\mathbf{U}}_n = \frac{1}{n}\sum_{i=1}^{n} \mathbf{U}_i.$$

Then $\sqrt{n}\,(\overline{\mathbf{U}}_n - \mathbf{b}) \xrightarrow{d} N_q(\mathbf{0}, \mathbf{A})$.

c. Let $\mathbf{T}_n$ and $\mathbf{U}_n$ be $t \times q$ random matrices such that $\mathbf{T}_n \xrightarrow{p} \mathbf{A}, \mathbf{U}_n \xrightarrow{p} \mathbf{B}$. Then $\mathbf{T}_n + \mathbf{U}_n \xrightarrow{p} \mathbf{A} + \mathbf{B}$.

d. Let $\mathbf{U}_1, \mathbf{U}_2, \ldots,$ be a sequence of random vectors or matrices such that $\mathbf{U}_n \xrightarrow{d} F$ and let $\mathbf{T}_1, \mathbf{T}_2, \ldots,$ be a sequence of random matrices such that $\mathbf{T}_n \xrightarrow{p} \mathbf{B}$. Let $g(\mathbf{U}, \mathbf{T})$ be a continuous function of $\mathbf{U}$ and $\mathbf{T}$ and let $F_{g,\mathbf{B}}$ be the distribution function of $g(\mathbf{U}, \mathbf{B})$ when $\mathbf{U}$ has distribution function F. Then

$$g(\mathbf{U}_n, \mathbf{T}_n) \xrightarrow{d} F_{g,\mathbf{B}}.$$

PROOF. Part a of this theorem follows directly from the weak law of large numbers for the components of $\mathbf{T}_n$. Part b follows from the central limit theorem for univariate random variables and Theorem 10.3e (see Exercise C2 of Chapter 10). The third part follows directly from Theorem 10.2a applied to the components of $\mathbf{T}_n$. The fourth part can be derived from Therorem 10.2d, since convergence in probability is the same as convergence in distribution to a degenerate distribution. □

We now use this theorem to derive the asymptotic distribution of F_n.

THEOREM 18.17. Let F_n be defined by (18.6). Then

$$F_n \xrightarrow{d} \chi_s^2/s$$

PROOF. By Theorem 18.16a and b,

$$\overline{\mathbf{Y}}_n \xrightarrow{p} \boldsymbol{\mu}, \qquad \sqrt{n}\left(\overline{\mathbf{Y}}_n - \boldsymbol{\mu}\right) \xrightarrow{d} N_r(\mathbf{0}, \boldsymbol{\Sigma}).$$

By Theorem 18.16a,

$$\frac{1}{n}\sum_{i=1}^{n} \mathbf{Y}_i\mathbf{Y}_i' \xrightarrow{p} \boldsymbol{\Sigma} + \boldsymbol{\mu}\boldsymbol{\mu}'.$$

Therefore,

$$\frac{n-1}{n-s}\mathbf{S}_n = \frac{n}{n-s}\left(\frac{1}{n}\sum_{i=1}^{n}\mathbf{Y}_i\mathbf{Y}_i' - \overline{\mathbf{Y}}_n\overline{\mathbf{Y}}_n'\right) \xrightarrow{p} \boldsymbol{\Sigma} + \boldsymbol{\mu}\boldsymbol{\mu}' - \boldsymbol{\mu}\boldsymbol{\mu}' = \boldsymbol{\Sigma}.$$

Now, if $\mathbf{U} \sim N_r(0, \Sigma)$, then $h(\mathbf{U}, \boldsymbol{\Sigma}) = (\mathbf{AU})'(\mathbf{A}\boldsymbol{\Sigma}\mathbf{A}')^{-1}\mathbf{AU} \sim \chi_s^2$. Therefore, by Theorem 18.16d

$$F_n = \frac{h\left(\sqrt{n}\,(\mathbf{Y}_n - \boldsymbol{\mu}), \dfrac{n-1}{n-s}\mathbf{S}\right)}{s}$$

$$= \frac{n\left(\mathbf{A}(\overline{\mathbf{Y}}_n - \boldsymbol{\mu})\right)'\left[\left(\dfrac{n-1}{n-s}\right)\mathbf{S}\right]^{-1}\mathbf{A}(\overline{\mathbf{Y}}_n - \boldsymbol{\mu})}{s} \xrightarrow{d} \frac{\chi_s^2}{s}. \quad \square$$

18.7. GENERALIZED REPEATED MEASURES MODELS.

We now show how the results in this chapter can be used to analyze repeated measures designs in which it is not possible to assume that the covariance matrix has the form assumed in Chapter 14. The repeated measures designs for which the techniques of this chapter are applicable are those for which each individual receives the same treatments, and hence has the same mean vector. A model in which different individuals receive different treatments is considered in Section 19.8.

We apply results from this chapter to Example 1 of Chapter 14 in which we observe $\mathbf{Y}_k = (Y_{1k}, \ldots, Y_{rk})', k = 1, \ldots, n$ with

$$Y_{ik} = \theta + \alpha_i + e_{ik}, \qquad \sum_i \alpha_i = 0, \qquad \mathbf{e}_k = (e_{1k}, \ldots, e_{rk})' \sim N_r(\mathbf{0}, \boldsymbol{\Sigma}),$$

and the $\mathbf{e}_k$ are independent. [Note that we have replaced the assumption that $\mathbf{e}_k \sim N_r(\mathbf{0}, \sigma^2\mathbf{A}(\rho))$ with the weaker assumption that $\mathbf{e}_k \sim N_r(\mathbf{0}, \boldsymbol{\Sigma})$.] We want to test that $\alpha_i = 0$. Let $\boldsymbol{\mu} = (\theta + \alpha_1, \ldots, \theta + \alpha_r)' = \mathbf{E}\mathbf{Y}_k$. Then $\mathbf{Y}_k \sim N_r(\boldsymbol{\mu}, \boldsymbol{\Sigma})$ and are independent. Let

$$\mathbf{A} = (\mathbf{I} - \mathbf{1}) \tag{18.7}$$

where $\mathbf{I}$ is an $(r-1) \times (r-1)$ identity matrix and $\mathbf{1}$ is an $(r-1) \times 1$ vector of 1's. Then

$$\mathbf{A}\boldsymbol{\mu} = (\alpha_1 - \alpha_r, \ldots, \alpha_{r-1} - \alpha_r).$$

Therefore, the $\alpha_i = 0$ if and only if $\mathbf{A}\boldsymbol{\mu} = \mathbf{0}$. Hence, the UMP invariant size α test that the $\alpha_i = 0$ for the generalized repeated measures model is given by (18.5) with $s = r - 1$ and $\mathbf{A}$ defined by (18.7). Note that this test is different from the test that was derived for the repeated measures model in Chapter 14. Note also that we need to assume that $n > r$ to do the test in this section, whereas it is only necessary to assume that $n > 1$ to do the test suggested in Chapter 14.

We could apply these techniques in a similar way to test the three hypotheses suggested in Example 2 of Chapter 14. In fact we could analyze any repeated measures design in which every individual gets the same treatments as long as the mean vector $\boldsymbol{\mu}$ is unrestricted under the alternative hypothesis.

18.8. THE TWO SAMPLE MODEL.

We now consider the two sample model in which we observe $\mathbf{Y}_{ij}$ independent random vectors, $j = 1, \ldots, n_{ij}$; $i = 1, 2$, such that

$$\mathbf{Y}_{ij} \sim N_r(\boldsymbol{\mu}_i, \boldsymbol{\Sigma}), \qquad \boldsymbol{\Sigma} > 0.$$

Note that the covariance matrix is the same for both samples. Since the derivation of results in this section is very similar to the derivation of results for the one sample model, that derivation is left for homework.

Let

$$\overline{\mathbf{Y}}_i = \frac{1}{n_i} \sum_{j=1}^{n_i} \mathbf{Y}_{ij},$$

$$\mathbf{S} = \frac{1}{n_1 + n_2 - 2} \left(\sum_{j=1}^{n_1} (\mathbf{Y}_{1j} - \overline{\mathbf{Y}}_1)(\mathbf{Y}_{1j} - \overline{\mathbf{Y}}_1)' + \sum_{j=1}^{n_2} (\mathbf{Y}_{2j} - \overline{\mathbf{Y}}_2)(\mathbf{Y}_{2j} - \overline{\mathbf{Y}}_2)' \right)$$

We call $\mathbf{S}$ the *pooled sample covariance matrix*.

THEOREM 18.18. a. $(\overline{\mathbf{Y}}_1, \overline{\mathbf{Y}}_2, \mathbf{S})$ is a complete sufficient statistic for the two-sample model.

b. $\overline{\mathbf{Y}}_1, \overline{\mathbf{Y}}_2$, and $\mathbf{S}$ are independent and

$$\overline{\mathbf{Y}}_1 \sim N_r\left(\boldsymbol{\mu}_1, \frac{1}{n_1}\boldsymbol{\Sigma}\right), \qquad \overline{\mathbf{Y}}_2 \sim N_r\left(\boldsymbol{\mu}_2, \frac{1}{n_2}\boldsymbol{\Sigma}\right),$$

$$\mathbf{S} \sim W_r\left(n_1 + n_2 - 2, \frac{1}{n_1 + n_2 - 2}\boldsymbol{\Sigma}\right).$$

c. $\overline{\mathbf{Y}}_1, \overline{\mathbf{Y}}_2$, and $\mathbf{S}$ are minimum variance unbiased estimators of $\boldsymbol{\mu}_1$, $\boldsymbol{\mu}_2$ and $\boldsymbol{\Sigma}$.

d. $\overline{\mathbf{Y}}_1, \overline{\mathbf{Y}}_2$ and $[(n_1 + n_2 - 2)/(n_1 + n_2)]\mathbf{S}$ are maximum likelihood estimators of $\boldsymbol{\mu}_1$, $\boldsymbol{\mu}_2$, and $\boldsymbol{\Sigma}$.

PROOF. See Exercise B7. $\square$

Now consider testing the hypothesis that $\boldsymbol{\mu}_1 = \boldsymbol{\mu}_2$ against $\boldsymbol{\mu}_1$ and $\boldsymbol{\mu}_2$ unrestricted. This problem is invariant under the group G of transformations g of the form

$$g(\overline{\mathbf{Y}}_1, \overline{\mathbf{Y}}_2, \mathbf{S}) = (\mathbf{A}\overline{\mathbf{Y}}_1 + \mathbf{b}, \mathbf{A}\overline{\mathbf{Y}}_2 + \mathbf{b}, \mathbf{A}\mathbf{S}\mathbf{A}')$$

where $\mathbf{A}$ is an invertible $r \times r$ matrix and $\mathbf{b} \in R^p$. We assume that $n_1 + n_2 - 2 \geqslant r$, so that $\mathbf{S}$ is invertible.

THEOREM 18.19. a. A maximal invariant under G is F_3 and a parameter maximal invariant is δ_3, where

$$F_3 = \frac{(n_1 + n_2 - 1 - r)n_1 n_2}{r(n_1 + n_2)(n_1 + n_2 - 2)}(\overline{\mathbf{Y}}_1 - \overline{\mathbf{Y}}_2)'\mathbf{S}^{-1}(\overline{\mathbf{Y}}_1 - \overline{\mathbf{Y}}_2),$$

$$\delta_3 = \frac{n_1 n_2}{n_1 + n_2}(\boldsymbol{\mu}_1 - \boldsymbol{\mu}_2)'\boldsymbol{\Sigma}^{-1}(\boldsymbol{\mu}_1 - \boldsymbol{\mu}_2)$$

b. $F_3 \sim F_{r, n_1 + n_2 - 1 - r}(\delta_3)$.

c. The UMP invariant size α test of the hypothesis that $\mu_1 = \mu_2$ in the two sample model is

$$\phi_3(F_3) = \begin{cases} 1 & \text{if } F_3 > F^{\alpha}_{r, n_1 + n_2 - 1 - r} \\ 0 & \text{if } F_3 \leqslant F^{\alpha}_{r, n_1 + n_2 - 1 - r} \end{cases}.$$

PROOF. See Exercise B8. $\square$

The test ϕ_3 is often called the two sample Hotelling's T^2-test.

18.9. OTHER COMMENTS.

We first make some comments about estimation. We have shown that the sample mean is an inadmissible estimator of $\boldsymbol{\mu}$ when $r \geqslant 3$. In a similar way, we could show for the two sample model that the sample means are inadmissible. The James-Stein estimator and the modified James-Stein estimator are also inadmissible, but they are better than $\bar{\mathbf{Y}}$. The estimator $\bar{\mathbf{Y}}$ is a minimax estimator for $\boldsymbol{\mu}$. Since $\bar{\mathbf{Y}}$ has constant risk, another estimator is minimax if and only if it is as good as $\bar{\mathbf{Y}}$. Therefore searching for estimators that improve on $\bar{\mathbf{Y}}$ is equivalent to searching for minimax estimators. Efron and Morris (1976) gives a family of minimax estimators for $\boldsymbol{\mu}$. Strawderman (1971) gives a family of minimax Bayes estimators for known $\boldsymbol{\Sigma}$. These estimators are all admissible (because they are Bayes) and they are all better than $\bar{\mathbf{Y}}$. Finally, Brown (1966) shows that for any continuous family of distribution functions the best invariant estimator of three or more location parameters is inadmissible.

We now discuss the testing problems. The tests ϕ_1, ϕ_2, and ϕ_3 are admissible (see Stein, 1956). Since they are UMP invariant they are unbiased. Salaevski (1968) showed tbat the test ϕ_1 is most stringent for testing that $\boldsymbol{\mu} = \mathbf{0}$. This result can be extended to show that ϕ_3 is also most stringent for testing that $\boldsymbol{\mu}_1 = \boldsymbol{\mu}_2$.

EXERCISES.

Type A

1. Consider a one-sample model in which we observe two-dimensional vectors

$$\begin{pmatrix}1\\2\end{pmatrix}, \begin{pmatrix}1\\3\end{pmatrix}, \begin{pmatrix}2\\3\end{pmatrix}, \begin{pmatrix}2\\1\end{pmatrix}, \begin{pmatrix}-1\\1\end{pmatrix}.$$

(a) Do a size 0.05 test that $\boldsymbol{\mu} = \mathbf{0}$.

(b) Do a size 0.05 test that $\boldsymbol{\mu} = (1, 1)'$.

(c) Find the general form for the simultaneous confidence intervals associated with testing that $\boldsymbol{\mu} = \mathbf{0}$. Find the particular versions for $\mu_1 - 2\mu_2$ and $\mu_1 + \mu_2$.

2. Consider a one-sample model in which we observe the three-dimensional vectors

$$\begin{bmatrix}1\\2\\3\end{bmatrix}, \begin{bmatrix}3\\2\\1\end{bmatrix}, \begin{bmatrix}1\\2\\2\end{bmatrix}, \begin{bmatrix}1\\3\\2\end{bmatrix}, \begin{bmatrix}-1\\1\\2\end{bmatrix}.$$

(a) Test the hypothesis that $\mu_1 + 3\mu_2 = 0$ and $2\mu_1 + \mu_3 = 0$. Use a size 0.05 test.

(b) Find the form of the simultaneous confidence intervals associated with testing this hypothesis. Find the particular interval for $3\mu_1 + 3\mu_2 + \mu_3$.

3. Consider the one-way generalized repeated measures model discussed in Section18.7. Use the data of Exercise A2 to test the hypothesis that the α_i are all 0. Use a 0.05 test.

4. Consider a two-sample model in which we observe the two-dimensional random vectors

$$Y_{1i} = \begin{pmatrix}1\\2\end{pmatrix}, \begin{pmatrix}1\\3\end{pmatrix}, \begin{pmatrix}1\\3\end{pmatrix}, \begin{pmatrix}-3\\4\end{pmatrix}, \qquad Y_{2i} = \begin{pmatrix}5\\6\end{pmatrix}, \begin{pmatrix}6\\7\end{pmatrix}, \begin{pmatrix}7\\8\end{pmatrix}.$$

Use a 0.05 test to test the equality of the means for the two samples.

Type B

1. Let $\boldsymbol{\mu}$ and $\boldsymbol{\theta}$ be defined as equal to the proof of Theorem 18.2.

(a) Show that $\mathbf{I} - n\mathbf{a}'\mathbf{a}$ is idempotent and has rank $n-1$. [Note that rank of $\mathbf{I} - n\mathbf{a}\mathbf{a}'$ equals $\operatorname{tr}(\mathbf{I} - n\mathbf{a}\mathbf{a}')$. Why?]

(b) Show that $\boldsymbol{\theta}'(\mathbf{I} - n\mathbf{a}\mathbf{a}')\boldsymbol{\theta} = 0$.

(c) Show that $\mathbf{a}'(\mathbf{I} - n\mathbf{a}\mathbf{a}') = 0$.

2. Show that $\sum_i(\mathbf{Y}_i - \boldsymbol{\mu})(\mathbf{Y}_i - \boldsymbol{\mu})' = (a-1)\mathbf{S} + a(\overline{\mathbf{Y}} - \boldsymbol{\mu})(\overline{\mathbf{Y}} - \boldsymbol{\mu})'$.

3. Show that $E\|\overline{\mathbf{Y}}\|^2 = \|\boldsymbol{\mu}\|^2 + (1/n)\operatorname{tr}\boldsymbol{\Sigma}$.

4. (a) Let $\mathbf{a} \in R^r$, $\mathbf{U} > 0$ be $r \times r$. Show that $|\mathbf{a}\mathbf{a}' + \mathbf{U}| = |\mathbf{U}|(1 + \mathbf{a}'S^{-1}\mathbf{a})$. [Hint: expand $\left|\begin{smallmatrix}1 & \mathbf{a}'\\ \mathbf{a} & \mathbf{U}\end{smallmatrix}\right|$ two different ways. See Lemma 2 of the appendix.]

(b) Show that the test ϕ_1 is the likelihood ratio test for testing that $\boldsymbol{\mu} = \mathbf{0}$.

5. Prove the corollary to Theorem 18.13.

6. Prove Theorem 18.15.

7. Prove Theorem 18.18.

8. Prove Theorem 18.19.

Type C

1. Let

$$\mathbf{Z}_i = \begin{pmatrix} Y_1 \\ \mathbf{T}_{11} \\ \mathbf{T}_{12} \end{pmatrix}$$

be independent, $\mathbf{Z}_i \sim N_{s+1}(\boldsymbol{\mu}, \boldsymbol{\Sigma})$ where Y_i is 1×1, and $\mathbf{T}_{i1}$ is $q \times 1$. Let $\mathbf{S}$ be the sample covariance matrix computed from the $\mathbf{Z}_i$. Partition $\mathbf{S}$ and $\boldsymbol{\Sigma}$

as

$$\mathbf{S} = \begin{pmatrix} \mathbf{S}_{11} & \mathbf{S}_{12} \\ \mathbf{S}_{21} & \mathbf{S}_{22} \end{pmatrix}, \qquad \mathbf{\Sigma} = \begin{pmatrix} \Sigma_{11} & \Sigma_{12} \\ \mathbf{\Sigma}_{12} & \mathbf{\Sigma}_{22} \end{pmatrix}$$

where S_{11} and Σ_{11} are $(q+1) \times (q+1)$.

(a) Show that $\mathbf{S}_{11} - \mathbf{S}_{12}\mathbf{S}_{22}^{-1}\mathbf{S}_{21} \sim W_{q+1}(n-s+q-1, \mathbf{\Sigma}_{11} - \mathbf{\Sigma}_{12}\mathbf{\Sigma}_{22}^{-1}\mathbf{\Sigma}_{21})$. (See Lemma 17.9.)

(b) Verify (16.29). [Note that $R_{n,s}(\bar{R})$ is the distribution of the sample multiple correlation coefficient computed from a matrix $\mathbf{W} \sim W_s(n-1, \mathbf{\Sigma})$ and $\bar{R}$ is the multiple correlation coefficient computed from $\mathbf{\Sigma}$. Now R^* is the sample multiple correlation coefficient computed from $\mathbf{S}_{11} - \mathbf{S}_{12}\mathbf{S}_{22}^{-1}\mathbf{S}_{21}$ and $\bar{R}^*$ is the multiple correlation coefficient computed from $\mathbf{\Sigma}_{11} - \mathbf{\Sigma}_{12}\mathbf{\Sigma}_{22}^{-1}\mathbf{\Sigma}_{21}$.]

(c) Verify (16.28).

CHAPTER 19

The Multivariate Linear Model

Let $\boldsymbol{\mu}$ be an $n \times r$ matrix, and let V be a p-dimensional subspace of R^n. We say that $\boldsymbol{\mu} \in V$ if the *columns* of $\boldsymbol{\mu}$ are in V. The multivariate linear model is the model in which we observe

$$\mathbf{Y} \sim N_{n,r}(\boldsymbol{\mu}, \mathbf{I}, \boldsymbol{\Sigma}), \qquad \boldsymbol{\mu} \in V, \qquad \boldsymbol{\Sigma} > 0. \tag{19.1}$$

Let

$$\mathbf{Y} = \begin{pmatrix} \mathbf{Y}_1 \\ \vdots \\ \mathbf{Y}_n \end{pmatrix}, \qquad \boldsymbol{\mu} = \begin{pmatrix} \boldsymbol{\mu}_1 \\ \vdots \\ \boldsymbol{\mu}_n \end{pmatrix}.$$

Then the $\mathbf{Y}_i$ are independent and

$$\mathbf{Y}_i' \sim N_r(\boldsymbol{\mu}_i', \boldsymbol{\Sigma}).$$

In the above formulation of the linear model, therefore, the independent replication is represented by the rows of the $\mathbf{Y}$ matrix. Many books and papers dealing with the multivariate linear model transpose $\mathbf{Y}$, and use the columns to represent the replication. We choose to represent the replications by rows to emphasize the similarity of the results for this model with those for the univariate linear model discussed in Chapters 4–12 (where the independent replications are also represented by the rows). In fact, if $r = 1$, then Σ is 1×1, and the multivariate linear model reduces to the univariate linear model (see Thereom 17.2a).

Now, suppose that V is the one-dimensional subspace of all vectors whose elements are the same. Then

$$\boldsymbol{\mu} = \begin{pmatrix} \boldsymbol{\theta} \\ \vdots \\ \boldsymbol{\theta} \end{pmatrix}$$

for some $1 \times r$ vector $\boldsymbol{\theta}$. Therefore, in this case, the Y_i are independent

$$\mathbf{Y}_i' \sim N_r(\boldsymbol{\theta}', \boldsymbol{\Sigma})$$

and the model reduces to the multivariate one sample model discussed in Chapter 18. Similarly, if $n = n_1 + n_2$, and V is the two-dimensional subspace of vectors whose first n_1 elements are the same and whose last n_2 elements are the same, then this model reduces to the two-sample model discussed in 18.9. The multivariate linear model therefore includes as special cases the univariate linear model and the multivariate one and two sample models.

Let $\mathbf{X}$ be a basis matrix for V. Then $\boldsymbol{\mu} \in V$ if and only if $\boldsymbol{\mu} = \mathbf{X}\boldsymbol{\beta}$ for some $p \times r$ matrix $\boldsymbol{\beta}$, and $\boldsymbol{\beta} = (\mathbf{X}'\mathbf{X})^{-1}\mathbf{X}'\boldsymbol{\mu}$ is unique. Therefore, an equivalent version of the multivariate linear model is the model in which we observe

$$\mathbf{Y} \sim N_{n,r}(\mathbf{X}\boldsymbol{\beta}, \mathbf{I}, \boldsymbol{\Sigma}), \qquad -\infty < \boldsymbol{\beta} < \infty, \qquad \boldsymbol{\Sigma} > 0 \tag{19.2}$$

where $\mathbf{X}$ is a known $n \times p$ matrix of rank p. We say that the version of the linear model defined by equation (19.1) is the *coordinate-free* version of the model and the version defined by (19.2) is the *coordinatized* version, since we have picked a basis matrix for V and hence have coordinatized it. The coordinatized version is often called the multivariate regression model.

In Section 19.1, we find a complete sufficient statistic for the multivariate linear model and find its joint distribution. We also find minimum variance unbiased estimators, MLEs, and the best invariant estimator of $\boldsymbol{\mu}$. In Section 19.2, we consider testing that $\boldsymbol{\mu} \in W$, a k-dimensional subspace of V. This problem is often called the problem of testing the multivariate linear hypothesis. In Section 19.3, we find simultaneous confidence intervals for contrasts associated with testing the multivariate linear hypothesis. In Section 19.4, we look at the problem of testing the generalized linear hypothesis that $\boldsymbol{\mu}\mathbf{A} \in W$, where $\mathbf{A}$ is a known matrix. This testing problem occurs in multivariate regression models and generalized repeated measures models. In Sections 19.5–19.7, we look at examples of the multivariate linear model: the multivariate regression model, some multivariate analysis of variance models, and a generalized repeated measures model. In Section 19.8, we discuss the asymptotic validity of tests and simultaneous confidence intervals for this model. In Section 19.9, we look at James-Stein estimators for this model. Finally, in Sections 19.10 and 19.11, we look at two generalizations of the multivariate linear model.

19.1. SUFFICIENCY AND ESTIMATION.

We now find a complete sufficient statistic for the multivariate linear model.

Define

$$\hat{\boldsymbol{\mu}} = \mathbf{P}_V \mathbf{Y},$$

$$\hat{\boldsymbol{\Sigma}} = \frac{1}{n-p}\mathbf{Y}'\mathbf{P}_{V^\perp}\mathbf{Y} = \frac{1}{n-p}(\mathbf{Y}-\hat{\boldsymbol{\mu}})'(\mathbf{Y}-\hat{\boldsymbol{\mu}}) = \frac{1}{n-p}(\mathbf{Y}'\mathbf{Y}-\hat{\boldsymbol{\mu}}'\hat{\boldsymbol{\mu}}). \tag{19.3}$$

(Note that we are using $\mathbf{P}_V$ and $\mathbf{P}_{V^\perp}$ as matrices in the above expression.)

THEOREM 19.1. a. $(\hat{\boldsymbol{\mu}}, \hat{\boldsymbol{\Sigma}})$ is a complete sufficient statistic for the multivariate linear model.

b. $\hat{\boldsymbol{\mu}}$ and $\hat{\boldsymbol{\Sigma}}$ are independent,

$$\hat{\boldsymbol{\mu}} \sim N_{n,r}(\boldsymbol{\mu}, \mathbf{P}_V, \boldsymbol{\Sigma}), \qquad \hat{\boldsymbol{\Sigma}} \sim W_r\left(n-p, \frac{1}{n-p}\boldsymbol{\Sigma}\right).$$

PROOF. a. Let $\mathbf{X}$ be an orthonormal basis matrix for V. We note that $\boldsymbol{\mu} \in V$, so that $\boldsymbol{\mu}' = (\mathbf{P}_V\boldsymbol{\mu})' = \boldsymbol{\mu}'\mathbf{X}\mathbf{X}'$. Therefore, the joint density of $\mathbf{Y}$ is

$$f(\mathbf{Y};(\boldsymbol{\mu},\boldsymbol{\Sigma})) = \frac{1}{(2\pi)^{rn/2}|\boldsymbol{\Sigma}|^{n/2}} \exp -\tfrac{1}{2}\operatorname{tr}\boldsymbol{\Sigma}^{-1}(\mathbf{Y}-\boldsymbol{\mu})'(\mathbf{Y}-\boldsymbol{\mu})$$

$$= h(\boldsymbol{\mu},\boldsymbol{\Sigma}) \exp\left(-\tfrac{1}{2}\operatorname{tr}\boldsymbol{\Sigma}^{-1}\mathbf{Y}'\mathbf{Y} + \operatorname{tr}\boldsymbol{\Sigma}^{-1}\boldsymbol{\mu}'\mathbf{X}\mathbf{X}'\mathbf{Y}\right).$$

Let

$$\boldsymbol{\Delta} = \boldsymbol{\Sigma}^{-1} = \begin{pmatrix} \Delta_{11} & \cdots & \Delta_{1r} \\ \vdots & & \vdots \\ \Delta_{r1} & \cdots & \Delta_{rr} \end{pmatrix}, \qquad \mathbf{T} = \mathbf{Y}'\mathbf{Y} = \begin{pmatrix} T_{11} & \cdots & T_{1r} \\ \vdots & & \vdots \\ T_{r1} & \cdots & T_{rr} \end{pmatrix}$$

$$\boldsymbol{\theta} = \mathbf{X}'\boldsymbol{\mu}\boldsymbol{\Sigma}^{-1} = \begin{pmatrix} \boldsymbol{\theta}_1 \\ \vdots \\ \boldsymbol{\theta}_p \end{pmatrix}, \qquad \mathbf{U} = \mathbf{X}'\mathbf{Y} = \begin{pmatrix} \mathbf{U}_1 \\ \vdots \\ \mathbf{U}_p \end{pmatrix}.$$

(Note that $\boldsymbol{\theta}_i$ and $\mathbf{U}_i$ are $1 \times r$.) As in Chapter 18, let

$$\tilde{\boldsymbol{\Delta}}_1' = (\Delta_{11}, \ldots, \Delta_{rr}), \qquad \tilde{\boldsymbol{\Delta}}_2' = (\Delta_{12}, \Delta_{13}, \ldots, \Delta_{1r}, \Delta_{23}, \ldots, \Delta_{r-1,r})$$

$$\tilde{\mathbf{T}}_1' = (T_{11}, \ldots, T_{rr}), \qquad \tilde{\mathbf{T}}_2' = (T_{12}, T_{13}, \ldots, T_{1r}, T_{23}, \ldots, T_{r-1,r})$$

$$\tilde{\boldsymbol{\theta}}' = (\boldsymbol{\theta}_1, \ldots, \boldsymbol{\theta}_p), \qquad \mathbf{U}' = (\mathbf{U}_1, \ldots, \mathbf{U}_p).$$

(That is, $\tilde{\boldsymbol{\theta}}$ and $\tilde{\mathbf{U}}$ are $pr \times 1$ vectors.) Then

$$\operatorname{tr}\boldsymbol{\Sigma}^{-1}\mathbf{Y}'\mathbf{Y} = \tilde{\boldsymbol{\Delta}}_1'\tilde{\mathbf{T}}_1 + 2\tilde{\boldsymbol{\Delta}}_2'\tilde{\mathbf{T}}_2, \qquad \operatorname{tr}\boldsymbol{\Sigma}^{-1}\boldsymbol{\mu}'\mathbf{X}\mathbf{X}'\mathbf{Y} = \tilde{\boldsymbol{\theta}}'\tilde{\mathbf{U}}.$$

Therefore, let

$$\mathbf{Q}(\boldsymbol{\mu},\boldsymbol{\Sigma}) = \begin{bmatrix} -\frac{1}{2}\tilde{\boldsymbol{\Delta}}_1 \\ -\tilde{\boldsymbol{\Delta}}_2 \\ \tilde{\boldsymbol{\theta}} \end{bmatrix}, \qquad \mathbf{S}(\mathbf{Y}) = \begin{bmatrix} \tilde{\mathbf{T}}_1 \\ \tilde{\mathbf{T}}_2 \\ \tilde{\mathbf{U}} \end{bmatrix}.$$

Then

$$f(\mathbf{Y};(\boldsymbol{\mu},\boldsymbol{\Sigma})) = h(\boldsymbol{\mu},\boldsymbol{\Sigma})\exp \mathbf{Q}'(\boldsymbol{\mu},\boldsymbol{\Sigma})\mathbf{S}(\mathbf{Y})$$

which is in the exponential form. As mentioned in the last chapter, the range of $(\tilde{\boldsymbol{\Delta}}_1,\tilde{\boldsymbol{\Delta}}_2)$ contains an open set. Also, $\mathbf{X}'\boldsymbol{\mu}$ ranges over the whole space of $p \times r$ matrices, and hence for all $(\tilde{\boldsymbol{\Delta}}_1,\tilde{\boldsymbol{\Delta}}_2)$, $\tilde{\boldsymbol{\theta}}$ ranges over all of R^{rp}. Therefore, the image of $\mathbf{Q}(\boldsymbol{\mu},\boldsymbol{\Sigma})$ contains an open rectangle and $\mathbf{S}(\mathbf{Y})$ is a complete sufficient statistic. Finally $(\hat{\boldsymbol{\mu}},\hat{\boldsymbol{\Sigma}})$ is an invertible function of $\mathbf{S}(\mathbf{Y})$ and is also complete and sufficient.

b. See Exercise B1. □

THEOREM 19.2. a. $\hat{\boldsymbol{\mu}}$ and $\hat{\boldsymbol{\Sigma}}$ are the minimum variance unbiased estimators of $\boldsymbol{\mu}$ and $\boldsymbol{\Sigma}$.

b. $\hat{\boldsymbol{\mu}}$ and $[(n-p)/n]\hat{\boldsymbol{\Sigma}}$ are the MLEs of $\boldsymbol{\mu}$ and $\boldsymbol{\Sigma}$.

c. $\hat{\boldsymbol{\mu}}$ is the best invariant estimator of $\boldsymbol{\mu}$ for loss functions of the form

$$\left[L(\mathbf{d};(\boldsymbol{\mu},\boldsymbol{\Sigma}))\right] = \operatorname{tr}\boldsymbol{\Sigma}^{-1}(\mathbf{d}-\boldsymbol{\mu})'\mathbf{H}(\mathbf{d}-\boldsymbol{\mu})$$

where $\mathbf{H} > 0$ is a known $n \times n$ matrix. (If $\mathbf{H} = \mathbf{I}$, then this loss function is the Mahalanobis distance loss function for this problem. See Exercise B15.)

PROOF. a. See Exercise B2.

b. The density function is given by

$$f(\mathbf{Y};(\boldsymbol{\mu},\boldsymbol{\Sigma})) = \frac{1}{(2\pi)^{nr/2}|\boldsymbol{\Sigma}|^{n/2}}\exp -\tfrac{1}{2}\operatorname{tr}\boldsymbol{\Sigma}^{-1}(\mathbf{Y}-\boldsymbol{\mu})'(\mathbf{Y}-\boldsymbol{\mu}).$$

For all $\boldsymbol{\Sigma}$, this function is maximized when

$$\operatorname{tr}\boldsymbol{\Sigma}^{-1}(\mathbf{Y}-\boldsymbol{\mu})'(\mathbf{Y}-\boldsymbol{\mu}) = \operatorname{tr}\boldsymbol{\Sigma}^{-1}(\mathbf{Y}-\hat{\boldsymbol{\mu}})'(\mathbf{Y}-\hat{\boldsymbol{\mu}}) + \operatorname{tr}\boldsymbol{\Sigma}^{-1}(\hat{\boldsymbol{\mu}}-\boldsymbol{\mu})'(\hat{\boldsymbol{\mu}}-\boldsymbol{\mu})$$

is minimized. Since

$$\operatorname{tr}\boldsymbol{\Sigma}^{-1}(\hat{\boldsymbol{\mu}}-\boldsymbol{\mu})'(\hat{\boldsymbol{\mu}}-\boldsymbol{\mu}) = \operatorname{tr}(\hat{\boldsymbol{\mu}}-\boldsymbol{\mu})\boldsymbol{\Sigma}^{-1}(\hat{\boldsymbol{\mu}}-\boldsymbol{\mu})' \geqslant 0$$

the density function is maximized for all $\boldsymbol{\Sigma}$ by $\boldsymbol{\mu} = \hat{\boldsymbol{\mu}}$. By Lemma 18.4, $f(\mathbf{Y};(\hat{\boldsymbol{\mu}},\boldsymbol{\Sigma}))$ is maximized by letting

$$\boldsymbol{\Sigma} = \frac{1}{n}(\mathbf{Y}-\hat{\boldsymbol{\mu}})'(\mathbf{Y}-\hat{\boldsymbol{\mu}}) = \frac{n-p}{n}\hat{\boldsymbol{\Sigma}}.$$

c. Let $\mathbf{d} \in V$, $\mathbf{d}$ $n \times r$. This problem is invariant under the group G of transformations given by

$$g(\hat{\boldsymbol{\mu}},\hat{\boldsymbol{\Sigma}}) = (\hat{\boldsymbol{\mu}}\mathbf{A}' + \mathbf{v}, \mathbf{A}\hat{\boldsymbol{\Sigma}}\mathbf{A}'),\ \bar{g}(\boldsymbol{\mu},\boldsymbol{\Sigma}) = (\boldsymbol{\mu}\mathbf{A}' + \mathbf{v}, \mathbf{A}\boldsymbol{\Sigma}\mathbf{A}'),\ \tilde{g}(\mathbf{d}) = \mathbf{d}\mathbf{A}' + \mathbf{v}$$

where $\mathbf{v}$ is $n \times r$, $\mathbf{v} \in V$, and $\mathbf{A}$ is $r \times r$, $\mathbf{A}$ is invertible. (Note that if $\boldsymbol{\mu} \in V$, $\mathbf{v} \in V$, then $\boldsymbol{\mu}\mathbf{A}' + \mathbf{v} \in V$. Why?) Note also that

$$\begin{aligned} L(\tilde{g}(\mathbf{d}), \bar{g}(\boldsymbol{\mu}, \boldsymbol{\Sigma})) &= \operatorname{tr}(\mathbf{A}\boldsymbol{\Sigma}\mathbf{A}')^{-1}(\mathbf{d}\mathbf{A}' + \mathbf{v} - (\boldsymbol{\mu}\mathbf{A}' + \mathbf{v}))'\mathbf{H}(\mathbf{d}\mathbf{A}' + \mathbf{v} - (\boldsymbol{\mu}\mathbf{A} + \mathbf{v})) \\ &= \operatorname{tr}(\mathbf{A}')^{-1}\boldsymbol{\Sigma}^{-1}\mathbf{A}^{-1}\mathbf{A}(\mathbf{d} - \boldsymbol{\mu})'\mathbf{H}(\mathbf{d} - \boldsymbol{\mu})\mathbf{A}' \\ &= \operatorname{tr}\boldsymbol{\Sigma}^{-1}(\mathbf{d} - \boldsymbol{\mu})'\mathbf{H}(\mathbf{d} - \boldsymbol{\mu}) = L(\mathbf{d}; (\boldsymbol{\mu}, \boldsymbol{\Sigma})). \end{aligned}$$

An estimator $\mathbf{T}$ is invariant if and only if

$$\mathbf{T}(\hat{\boldsymbol{\mu}}\mathbf{A}' + \mathbf{v}, \mathbf{A}\hat{\boldsymbol{\Sigma}}\mathbf{A}') = \mathbf{T}(\hat{\boldsymbol{\mu}}, \hat{\boldsymbol{\Sigma}})\mathbf{A}' + \mathbf{v}.$$

Note that $\hat{\boldsymbol{\mu}}$ is invariant. Now, let $\mathbf{T}$ be any other invariant estimator, and let $\mathbf{A} = \hat{\boldsymbol{\Sigma}}^{-1/2}, \mathbf{v} = -\hat{\boldsymbol{\mu}}\hat{\boldsymbol{\Sigma}}^{-1/2}$. Then

$$\mathbf{T}(\hat{\boldsymbol{\mu}}, \hat{\boldsymbol{\Sigma}}) = \mathbf{T}(\mathbf{0}, \mathbf{I})\hat{\boldsymbol{\Sigma}}^{1/2} + \hat{\boldsymbol{\mu}}.$$

[Note that $\mathbf{T}(\mathbf{0}, \mathbf{I}) \neq \mathbf{0}$ since $\mathbf{T} \neq \hat{\boldsymbol{\mu}}$.] Therefore, $\mathbf{T}$ has risk function

$$\begin{aligned} R(\mathbf{T}; (\boldsymbol{\mu}, \boldsymbol{\Sigma})) &= E \operatorname{tr}\boldsymbol{\Sigma}^{-1}\big(\mathbf{T}(\mathbf{0}, \mathbf{I})\hat{\boldsymbol{\Sigma}}^{1/2} + \hat{\boldsymbol{\mu}} - \boldsymbol{\mu}\big)'\mathbf{H}\big(\mathbf{T}(\mathbf{0}, \mathbf{I})\hat{\boldsymbol{\Sigma}}^{1/2} + \hat{\boldsymbol{\mu}} - \boldsymbol{\mu}\big) \\ &= E \operatorname{tr}\boldsymbol{\Sigma}^{-1}\hat{\boldsymbol{\Sigma}}^{1/2}\mathbf{T}'(\mathbf{0}, \mathbf{I})\mathbf{H}\mathbf{T}(\mathbf{0}, \mathbf{I})\hat{\boldsymbol{\Sigma}}^{1/2} \\ &\quad + E \operatorname{tr}\boldsymbol{\Sigma}^{-1}(\hat{\boldsymbol{\mu}} - \boldsymbol{\mu})\mathbf{H}(\hat{\boldsymbol{\mu}} - \boldsymbol{\mu}) \\ &\quad + 2E \operatorname{tr}\boldsymbol{\Sigma}^{-1}\hat{\boldsymbol{\Sigma}}^{1/2}\mathbf{T}'(\mathbf{0}, \mathbf{I})\mathbf{H}(\hat{\boldsymbol{\mu}} - \boldsymbol{\mu}). \end{aligned}$$

Now, $E \operatorname{tr}\mathbf{R} = \operatorname{tr} E\mathbf{R}$. (See Exercise B2.) Also $E(\hat{\boldsymbol{\mu}} - \boldsymbol{\mu}) = 0$ and $\hat{\boldsymbol{\mu}}$ and $\hat{\boldsymbol{\Sigma}}$ are independent. Therefore, the third term of the above expression is 0. The second term is just the risk for $\hat{\boldsymbol{\mu}}$. Finally,

$$\mathbf{U} = \boldsymbol{\Sigma}^{-1/2}\hat{\boldsymbol{\Sigma}}^{1/2}\mathbf{T}'(\mathbf{0}, \mathbf{I})\mathbf{H}\mathbf{T}(\mathbf{0}, \mathbf{I})\hat{\boldsymbol{\Sigma}}^{1/2}\boldsymbol{\Sigma}^{-1/2} \geqslant 0$$

and $\mathbf{U} \neq 0$. Therefore, the first term, which equals $E \operatorname{tr}\mathbf{U}$ is positive. Hence $R(\mathbf{T}; (\boldsymbol{\mu}, \boldsymbol{\Sigma})) > R(\hat{\boldsymbol{\mu}}; (\boldsymbol{\mu}, \boldsymbol{\Sigma}))$ and $\hat{\boldsymbol{\mu}}$ is the best invariant estimator of $\boldsymbol{\mu}$. □

19.2. TESTING THE MULTIVARIATE LINEAR HYPOTHESIS.

We now consider the problem of testing the null hypothesis that $\boldsymbol{\mu} \in W$ against the alternative that $\boldsymbol{\mu} \in V$, where W is a k-dimensional subspace of V. We assume that $n - p \geqslant r$, so that $\hat{\boldsymbol{\Sigma}}$ is positive definite. Since the derivations in this section are somewhat more complicated than in other sections, we now summarize the results to be derived.

Let

$$\mathbf{S}_2 = \mathbf{Y}'\mathbf{P}_{V|W}\mathbf{Y}, \qquad \mathbf{S}_3 = \mathbf{Y}'\mathbf{P}_{V^\perp}\mathbf{Y} = (n-p)\hat{\boldsymbol{\Sigma}}, \qquad \boldsymbol{\delta} = \boldsymbol{\mu}'\mathbf{P}_{V|W}\boldsymbol{\mu}.$$

(Note that $\mathbf{S}_2$, $\mathbf{S}_3$, and $\boldsymbol{\delta}$ are all $r \times r$ matrices.) Let $b = \min(p-k, r)$. In Section 19.2.3, we show that a maximal invariant for this testing problem is the set of ordered nonzero roots $(t_1 \geqslant \ldots \geqslant t_b)$ of

$$|\mathbf{S}_2 - t\mathbf{S}_3| = 0$$

and a parameter maximal invariant is the set of largest ordered roots $(\theta_1 \geqslant \ldots \geqslant \theta_b)$ of

$$|\boldsymbol{\delta} - \theta\boldsymbol{\Sigma}| = 0.$$

We are testing the null hypothesis that the θ_i are all 0 against the alternative hypothesis that they are all nonnegative.

In Section 19.2.4, we consider the case in which $b = 1$, that is, the case in which there is only one nonzero root t_1. When $b = 1$, either $r = 1$ or $p - k = 1$. We look first at the case in which $r = 1$. In that case

$$t_1 = \frac{\|\mathbf{P}_{V|W}\mathbf{Y}\|^2}{\|\mathbf{P}_{V^\perp}\mathbf{Y}\|^2}, \qquad \theta_1 = \frac{\|\mathbf{P}_{V|W}\boldsymbol{\mu}\|^2}{\Sigma}, \qquad \frac{(n-p)t_1}{p-k} \sim F_{p-k,\, n-p}(\theta_1)$$

and the UMP invariant size α test is

$$\phi(t_1) = \begin{cases} 1 & \text{if } t_1 > \dfrac{p-k}{n-p} F^{\alpha}_{p-k,\, n-p} \\ 0 & \text{if } t_1 \leqslant \dfrac{p-k}{n-p} F^{\alpha}_{p-k, n-p} \end{cases}$$

(Note that if $r = 1$ then the problem is one of testing the univariate linear hypothesis as discussed in Chapter 7.)

Now, consider the case in which $p - k = 1$. Let $\mathbf{X}_2$ be an orthonormal basis matrix for $V|W$ (i.e., let $\mathbf{X}_2$ be a unit vector in $V|W$). Then

$$t_1 = \mathbf{X}_2'\mathbf{Y}\mathbf{S}_3^{-1}\mathbf{Y}'\mathbf{X}_2, \qquad \theta_1 = \mathbf{X}_2'\boldsymbol{\mu}\boldsymbol{\Sigma}^{-1}\boldsymbol{\mu}'\mathbf{X}_2,$$

$$\frac{n-p-r+1}{r} t_1 \sim F_{r, n-p-r+1}(\theta_1)$$

and the UMP invariant size α test is given by

$$\phi(t_1) = \begin{cases} 1 & \text{if } t_1 > \dfrac{r}{n-p-r+1} F^{\alpha}_{r, n-p-r+1} \\ 0 & \text{if } t_1 \leqslant \dfrac{r}{n-p-r+1} F^{\alpha}_{r, n-p-r+1} \end{cases}$$

(Note that the one and two sample Hotelling's T^2-tests of the last chapter are special cases of this test.)

We now consider the case in which $b > 1$. In this situation, there is no UMP invariant size α test. Four tests which have been considered for this situation are now given. Let

$$\lambda_1 = \prod_{i=1}^{b}(1+t_i) = \frac{|\mathbf{S}_2+\mathbf{S}_3|}{|\mathbf{S}_3|}, \qquad \lambda_2 = \sum_{i=1}^{b} t_i = \operatorname{tr}\mathbf{S}_2\mathbf{S}_3^{-1} \qquad \lambda_3 = t_1,$$

$$\lambda_4 = \sum_{i=1}^{b} t_i/(1+t_i) = \operatorname{tr}\mathbf{S}_2(\mathbf{S}_2+\mathbf{S}_3)^{-1}$$

and for $i = 1,2,3,4$, let

$$\phi_i(\lambda_1) = \begin{cases} 1 & \text{if } \lambda_i > c_i^\alpha \\ 0 & \text{if } \lambda_i \leqslant c_i^\alpha \end{cases}$$

where c_i^α is the upper α point of the null distribution of λ_i. All four of these tests are invariant, unbiased (see Das Gupta, et al., 1964) and admissible (see Schwartz, 1967). At present there seems to be no theoretical reason to prefer one test over the others. (The test ϕ_3 has the advantage of having associated simultaneous confidence intervals, which are derived in Section 19.3). The test ϕ_1 is the likelihood ratio test and is called Wilks' test; ϕ_2 is derived by the substitution method and is called the Lawley-Hotelling test; ϕ_3 is derived by the union-intersection method and is called Roy's largest root test; and ϕ_4 is called Pillai's test. (For definitions of these methods, see Section 19.2.5.)

To use the tests defined above, it is necessary to know or approximate the null distribution of the test statistics. In Section 19.2.6, it is shown that

$$(n-p)\log\lambda_1 \xrightarrow{d} \chi^2_{r(p-k)}(\operatorname{tr}\boldsymbol{\delta}\boldsymbol{\Sigma}^{-1}), \qquad (n-p)\lambda_2 \xrightarrow{d} \chi^2_{r(p-k)}(\operatorname{tr}\boldsymbol{\delta}\boldsymbol{\Sigma}^{-1})$$

$$(n-p)\lambda_4 \xrightarrow{d} \chi^2_{r(p-k)}(\operatorname{tr}\boldsymbol{\delta}\boldsymbol{\Sigma}^{-1}).$$

and therefore, for large n,

$$\log c_i^\alpha \approx \frac{\chi^{2\,\alpha}_{r(p-k)}}{n-p}, \qquad c_2^\alpha \approx \frac{\chi^{2\,\alpha}_{r(p-k)}}{n-p} \qquad c_4^\alpha \approx \frac{\chi^{2\,\alpha}_{r(p-k)}}{n-p}.$$

(Note that the above result also implies that the asymptotic powers of the tests ϕ_1, ϕ_2 and ϕ_4 are the same. The asymptotic power of ϕ_3 is different from these.) No exact distributions are derived for λ_1, λ_2, and λ_4.

We now discuss a more accurate approximation to the distribution of λ_1. Bartlett (1938) suggests that the approximation can be improved for small samples by letting $m\log\lambda_1$ be approximately $\chi^2_{r(p-k)}$, where

$$m = n - p - \frac{r-(p-k)+1}{2} = n - \frac{r+p+k+1}{2}$$

and therefore that

$$\log c_1^\alpha \approx \frac{\chi^{2\,\alpha}_{r(p-k)}}{m}.$$

Box (1949) gives a still more accurate approximation and indicates that Barlett's approximation is accurate to three decimal places as long as

$$r^2 + (p-k)^2 \leqslant \frac{m}{3}.$$

Table 47 of Pearson and Hartley (1976) gives conversion factors to convert to exact percentiles the approximate 0.05 and 0.01 values found with Bartlett's approximation. (In using this table, it is helpful to note that the distribution of λ_1 is the same when $r = a$, $p - k = b$, and $n - p = c$, as when $r = b$, $p - k = a$, and $n - p = c - a + b$. See Exercise C1.)

We now look at $\lambda_3 = t_1$. Let $s_{a,c,d}$ be the distribution of the largest root s of

$$|\mathbf{W}_1 - s\mathbf{W}_2| = 0.$$

where $\mathbf{W}_1$ and $\mathbf{W}_2$ are independent, $\mathbf{W}_1 \sim W_a(c, \mathbf{I})$, $\mathbf{W}_2 \sim W_a(d, \mathbf{I})$. Let $s^{\alpha}_{a,c,d}$ be the upper α point for the distribution $s_{a,c,d}$. Tables of $s^{.05}_{a,c,d}$ and $s^{.01}_{a,c,d}$ are given in Pearson and Hartley (1972), Tables 48 and 49. (See also Morrison, 1976 pp. 379–403. In using these tables it is helpful to note that $s^{\alpha}_{a,c,d} = s^{\alpha}_{c,a,d-c+a}$. See Exercise C1.) In Section 19.2, it is shown that, under the null hypothesis,

$$\frac{t_1}{1+t_1} \sim s_{r,p-k,n-p}, \qquad c_3^{\alpha} = \frac{s^{\alpha}_{r,p-k,n-p}}{1 - s^{\alpha}_{r,p-k,n-p}}.$$

Finally, we discuss the asymptotic null distribution of t_1. Let $u_{a,c}$ be the distribution of the largest root u of

$$|\mathbf{W} - u\mathbf{I}| = 0$$

where $\mathbf{W} \sim W_a(c, \mathbf{I})$, and let $u^{\alpha}_{a,c}$ be the upper α point for this distribution. (Tables of $u^{\alpha}_{a,c}$ are given in Pearson and Hartley, 1972, Table 52.) In Section 19.2.6, it is shown that under the null hypothesis

$$(n-p)t_1 \xrightarrow{d} u_{r,p-k}$$

and therefore

$$c_3^{\alpha} \approx \frac{u^{\alpha}_{r,p-k}}{n-p}.$$

19.2.1. Some More Matrix Algebra

Before transforming this testing problem to canonical form, we state some matrix algebra results that are discussed in the appendix. The first result is the principal axis theorem.

LEMMA 19.3. Let $\mathbf{U} \geqslant 0$ be $m \times m$. Then

$$p(t) = |\mathbf{U} - t\mathbf{I}|$$

is an mth degree polynomial. The roots of $p(t)=0$ are all nonnegative real numbers. If $\mathbf{U}>0$, then the roots are all positive. Let $t_1 \geqslant \cdots \geqslant t_m$ be the ordered roots of $p(t)$. Then there exists $\boldsymbol{\Gamma}$ orthogonal such that

$$\boldsymbol{\Gamma}\mathbf{U}\boldsymbol{\Gamma}' = \begin{pmatrix} t_1 & & 0 \\ & \ddots & \\ 0 & & t_m \end{pmatrix}, \qquad t_1 = \sup_{\substack{\mathbf{v}\in R^m \\ \mathbf{v}\neq 0}} \frac{\mathbf{v}'\mathbf{U}\mathbf{v}}{\mathbf{v}'\mathbf{v}}.$$

PROOF. See Section 6 of the appendix. □

We called the roots $t_1, \ldots, t_m$ the *eigenvalues* of $\mathbf{U}$. In the appendix, we generalize this result to get the following lemma.

LEMMA 19.4. Let $\mathbf{U} \geqslant 0$ and $\mathbf{S} > 0$ be $m \times m$. Then

$$q(t) = |\mathbf{U} - t\mathbf{S}|$$

is an mth degree polynomial. The roots of $q(t)=0$ are all nonnegative real numbers. If $\mathbf{U}>0$, then the roots are all positive. Let $t_1 \geqslant \ldots \geqslant t_m$ be the ordered roots of $q(t)=0$. Then there exists $\mathbf{A}$ invertible such that

$$\mathbf{ASA}' = \mathbf{I}, \qquad \mathbf{AUA}' = \begin{pmatrix} t_1 & & 0 \\ & \ddots & \\ 0 & & t_m \end{pmatrix}, \qquad t_1 = \sup_{\substack{\mathbf{v}\in R^m \\ \mathbf{v}\neq \mathbf{0}}} \frac{\mathbf{v}'\mathbf{U}\mathbf{v}}{\mathbf{v}'\mathbf{S}\mathbf{v}}$$

PROOF. See Theorem 7 of the appendix. □

We need one more result before we consider the canonical form for testing the multivariate linear hypothesis.

LEMMA 19.5. Let $\mathbf{U}$ and $\mathbf{V}$ be $s \times m$ and let $\mathbf{S}>0$ and $\mathbf{T}>0$ be $m \times m$.

a. If $\mathbf{U}'\mathbf{U} = \mathbf{V}'\mathbf{V}$, then there exists an orthogonal matrix $\boldsymbol{\Gamma}$ such that $\mathbf{V} = \boldsymbol{\Gamma}\mathbf{U}$.

b. If $\mathbf{U}'\mathbf{S}^{-1}\mathbf{U} = \mathbf{V}'\mathbf{T}^{-1}\mathbf{V}$, then there exists an invertible matrix $\mathbf{A}$ such that $\mathbf{V} = \mathbf{AU}$, $\mathbf{T} = \mathbf{ASA}'$.

PROOF. See Theorems 11 and 12 of the appendix. □

19.2.2. The Canonical Form

We now transform the testing problem to canonical form. This derivation is very similar to that in Section 7.4a for the univariate linear model. Let $\mathbf{X}_1$,

$\mathbf{X}_2$, and $\mathbf{X}_3$ be orthonormal basis matrices for W, $V|W$ and $V^{\perp}$. Let

$$\mathbf{X} = (\mathbf{X}_1, \mathbf{X}_2, \mathbf{X}_3), \qquad \mathbf{Z} = \begin{pmatrix} \mathbf{Z}_1 \\ \mathbf{Z}_2 \\ \mathbf{Z}_3 \end{pmatrix} = \mathbf{X}'\mathbf{Y}, \qquad \boldsymbol{\nu} = \begin{pmatrix} \boldsymbol{\nu}_1 \\ \boldsymbol{\nu}_2 \\ \boldsymbol{\nu}_3 \end{pmatrix} = \mathbf{X}'\boldsymbol{\mu}, \qquad (19.4)$$

where $\mathbf{Z}_1$ and $\boldsymbol{\nu}_1$ are $k \times r$, $\mathbf{Z}_2$ and $\boldsymbol{\nu}_2$ are $(p-k)\times r$, and $\mathbf{Z}_3$ and $\boldsymbol{\nu}_3$ are $(n-p)\times r$. Since $\mathbf{X}'\mathbf{X} = \mathbf{I}$, we see that

$$\mathbf{Z} \sim N_{n,r}(\boldsymbol{\nu}, \mathbf{I}, \boldsymbol{\Sigma})$$

by Theorem 17.2d. Since $\mathbf{X}$ is an orthogonal matrix, and hence invertible, the model in which we observe $\mathbf{Z}$ is just a transformation of the model in which we observe $\mathbf{Y}$. The rows of $\mathbf{Z}$ are independent, and therefore, $\mathbf{Z}_1$, $\mathbf{Z}_2$, and $\mathbf{Z}_3$ are independent. Furthermore,

$$\mathbf{Z}_1 \sim N_{k,r}(\boldsymbol{\nu}_1, \mathbf{I}, \boldsymbol{\Sigma}), \qquad \mathbf{Z}_2 \sim N_{p-k,r}(\boldsymbol{\nu}_2, \mathbf{I}, \boldsymbol{\Sigma}), \qquad \mathbf{Z}_3 \sim N_{n-p,r}(\boldsymbol{\nu}_3, \mathbf{I}, \boldsymbol{\Sigma}). \qquad (19.5)$$

By Lemma 7.6 we see that $\boldsymbol{\mu} \in V$ (if and only if the columns of $\boldsymbol{\mu}$ are in V), if and only if the columns of $\boldsymbol{\nu}_3$ are $\mathbf{0}$ if and only if $\boldsymbol{\nu}_3 = \mathbf{0}$. Similarly, $\boldsymbol{\mu} \in W$ if and only if $\boldsymbol{\nu}_2 = \mathbf{0}$ and $\boldsymbol{\nu}_3 = \mathbf{0}$. Note that $\boldsymbol{\nu}_1$ is unrestricted under both hypotheses and that $\boldsymbol{\nu}_2$ is unrestricted under the alternative hypothesis. We have therefore transformed the problem of testing that $\boldsymbol{\mu} \in W$ to the problem in which we observe $\mathbf{Z}_1$, $\mathbf{Z}_2$, and $\mathbf{Z}_3$ independent, with $\mathbf{Z}_i$ having the distribution given in (19.5), and we want to test the null hypothesis that $\boldsymbol{\nu}_2 = 0$ and $\boldsymbol{\nu}_3 = 0$ against the alternative that $\boldsymbol{\nu}_3 = 0$. This transformed problem is called the *canonical form* for testing the multivariate linear hypothesis.

19.2.3. A Maximal Invariant

We now find a maximal invariant for the canonical form of the problem. The canonical form is invariant under the following three groups.

(1) G_1 consists of transformations of the form $g_1(\mathbf{Z}_1, \mathbf{Z}_2, \mathbf{Z}_3) = (\mathbf{Z}_1 + a, \mathbf{Z}_2, \mathbf{Z}_3)$ where $\mathbf{a}$ is an arbitrary $k \times r$ matrix. The induced transformation $\bar{g}_1$ in the parameter space is given by $\bar{g}_1(\boldsymbol{\nu}_1, \boldsymbol{\nu}_2, \boldsymbol{\nu}_3, \boldsymbol{\Sigma}) = (\boldsymbol{\nu}_1 + \mathbf{a}, \boldsymbol{\nu}_2, \boldsymbol{\nu}_3, \boldsymbol{\Sigma})$.

(2) G_2 consists of transformations of the form $g_2(\mathbf{Z}_1, \mathbf{Z}_2, \mathbf{Z}_3) = (\mathbf{Z}_1, \boldsymbol{\Gamma}\mathbf{Z}_2, \mathbf{Z}_3)$ where $\boldsymbol{\Gamma}$ is an arbitrary orthogonal $(p-k)\times(p-k)$ matrix. The induced transformation is $\bar{g}_2(\boldsymbol{\nu}_1, \boldsymbol{\nu}_2, \boldsymbol{\nu}_3, \boldsymbol{\Sigma}) = (\boldsymbol{\nu}_1, \boldsymbol{\Gamma}\boldsymbol{\nu}_2, \boldsymbol{\nu}_3, \boldsymbol{\Sigma})$.

(3) G_3 consists of transformations of the form $g_3(\mathbf{Z}_1, \mathbf{Z}_2, \mathbf{Z}_3) = (\mathbf{Z}_1\mathbf{A}, \mathbf{Z}_2\mathbf{A},$

$\mathbf{Z}_3\mathbf{A}$) where $\mathbf{A}$ is an arbitrary $r \times r$ invertible matrix. The induced transformation is $\bar{g}_3(\boldsymbol{\nu}_1, \boldsymbol{\nu}_2, \boldsymbol{\nu}_3, \boldsymbol{\Sigma}) = (\boldsymbol{\nu}_1\mathbf{A}, \boldsymbol{\nu}_2\mathbf{A}, \boldsymbol{\nu}_3\mathbf{A}, \mathbf{A}'\boldsymbol{\Sigma}\mathbf{A})$.

Note that these three groups are the natural generalizations of the groups used to reduce the canonical form of the univariate linear model. We now find a maximal invariant under these three groups.

THEOREM 19.6. a. Suppose that $p - k \geq r$. Let $t_1 \geq \ldots \geq t_r$ be the ordered roots of the equation

$$|\mathbf{Z}_2{}'\mathbf{Z}_2 - t\mathbf{Z}_3'\mathbf{Z}_3| = 0 \tag{19.6}$$

and $\theta_1 \geq \cdots \geq \theta_r$ be the ordered roots of the equation

$$|\boldsymbol{\nu}_2'\boldsymbol{\nu}_2 - \theta\boldsymbol{\Sigma}| = 0. \tag{19.7}$$

Then $(t_1, \ldots, t_r)$ is a maximal invariant and $(\theta_1, \ldots, \theta_r)$ is a parameter maximal invariant for the canonical form of the multivariate general linear hypothesis.

b. Suppose that $p - k \leq r$. Let $t_1 \geq \ldots \geq t_{p-k}$ be the ordered roots of the equation

$$|\mathbf{Z}_2(\mathbf{Z}_3'\mathbf{Z}_3)^{-1}\mathbf{Z}_2' - t\mathbf{I}| = 0 \tag{19.8}$$

and $\theta_1 \geq \ldots \geq \theta_{p-k}$ be the ordered roots of the equation

$$|\boldsymbol{\nu}_2\boldsymbol{\Sigma}^{-1}\boldsymbol{\nu}_2' - \theta\mathbf{I}| = 0. \tag{19.9}$$

Then $(t_1, \ldots, t_{p-k})$ is a maximal invariant and $(\theta_1, \ldots, \theta_{p-k})$ is a parameter maximal invariant for the canonical form of the general linear hypothesis.

PROOF. a. We first reduce by sufficiency, then by G_1, followed by G_2, and finally by G_3. Since $\boldsymbol{\nu}_3 = 0$ under both hypotheses, a sufficient statistic for the problem is $(\mathbf{Z}_1, \mathbf{Z}_2, \mathbf{Z}_3'\mathbf{Z}_3)$. The first group becomes $g_1(\mathbf{Z}_1, \mathbf{Z}_2, \mathbf{Z}_3'\mathbf{Z}_3) = (\mathbf{Z}_1 + \mathbf{a}, \mathbf{Z}_2, \mathbf{Z}_3'\mathbf{Z}_3)$ and a maximal invariant is $Q_1 = (\mathbf{Z}_2, \mathbf{Z}_3'\mathbf{Z}_3)$. The second group G_2 becomes $g_2(\mathbf{Z}_2, \mathbf{Z}_3'\mathbf{Z}_3) = (\boldsymbol{\Gamma}\mathbf{Z}_2, \mathbf{Z}_3'\mathbf{Z}_3)$ and a maximal invariant is $Q_2 = (\mathbf{Z}_2'\mathbf{Z}_2, \mathbf{Z}_3{}'\mathbf{Z}_3)$ using Lemma 19.5a. The third group, G_3, becomes $g_3(\mathbf{Z}_2'\mathbf{Z}_2, \mathbf{Z}_3'\mathbf{Z}_3) = (\mathbf{A}'\mathbf{Z}_2'\mathbf{Z}_2\mathbf{A}, \mathbf{A}'\mathbf{Z}_3'\mathbf{Z}_3\mathbf{A})$. Let $Q_3(\mathbf{Z}_2'\mathbf{Z}_2, \mathbf{Z}_3'\mathbf{Z}_3)$ be the set of ordered roots of (19.6). We now show that Q_3 is a maximal invariant under G_3. $Q_3(\mathbf{A}'\mathbf{Z}_2'\mathbf{Z}_2\mathbf{A}, \mathbf{A}'\mathbf{Z}_3'\mathbf{Z}_3\mathbf{A})$ is the set of roots of

$$0 = |\mathbf{A}'\mathbf{Z}_2'\mathbf{Z}_2\mathbf{A} - t\mathbf{A}'\mathbf{Z}_3'\mathbf{Z}_3\mathbf{A}| = |\mathbf{A}|^2|\mathbf{Z}_2'\mathbf{Z}_2 - t\mathbf{Z}_3'\mathbf{Z}_3|$$

which is the same as the set of roots $|\mathbf{Z}_2'\mathbf{Z}_2 - t\mathbf{Z}_3'\mathbf{Z}_3| = 0$, and hence Q_3 is invariant. Suppose that $Q_3(\mathbf{Z}_2'\mathbf{Z}_2, \mathbf{Z}_3'\mathbf{Z}_3) = Q_3(\mathbf{Z}_2^{*\prime}\mathbf{Z}_2^*, \mathbf{Z}_3^{*\prime}\mathbf{Z}_3^*)$. By Lemma

19.4, there exist $\mathbf{A}$ and $\mathbf{A}^*$ invertible such that

$$\mathbf{A}'\mathbf{Z}_3'\mathbf{Z}_3\mathbf{A} = \mathbf{I} = \mathbf{A}^{*\prime}\mathbf{Z}_3^{*\prime}\mathbf{Z}_3^*\mathbf{A}^*, \qquad \mathbf{A}'\mathbf{Z}_2'\mathbf{Z}_2\mathbf{A} = \begin{pmatrix} t_1 & & 0 \\ & \ddots & \\ 0 & & t_r \end{pmatrix} = \mathbf{A}^{*\prime}\mathbf{Z}_2^{*\prime}\mathbf{Z}_2^*\mathbf{A}^*.$$

Therefore, $\mathbf{Z}_2^{*\prime}\mathbf{Z}_2^* = (\mathbf{A}^{*\prime})^{-1}\mathbf{A}'(\mathbf{Z}_2'\mathbf{Z}_2)\mathbf{A}(\mathbf{A}^*)^{-1}$ and $\mathbf{Z}_3^{*\prime}\mathbf{Z}_3^* = (\mathbf{A}^{*\prime})^{-1}\mathbf{A}'(\mathbf{Z}_3'\mathbf{Z}_3)$ $\mathbf{A}(\mathbf{A}^*)^{-1}$ and hence Q_3 is maximal. We have now finished the derivation of the maximal invariant for this case. The parameter maximal invariant is derived similarly.

b. We now reduce in a different order, reducing by G_3 before reducing by G_2. By the argument given above, a maximal invariant under G_1 is $(\mathbf{Z}_2, \mathbf{Z}_3'\mathbf{Z}_3)$. Now G_3 becomes transformations of the form $g_3(\mathbf{Z}_2, \mathbf{Z}_3'\mathbf{Z}_3) = (\mathbf{Z}_2\mathbf{A}, \mathbf{A}'\mathbf{Z}_3'\mathbf{Z}_3\mathbf{A})$. Since $n - p \geqslant r, \mathbf{Z}_3'\mathbf{Z}_3 > 0$. We see by Lemma 19.5, that a maximal invariant under G_3 is $\mathbf{Z}_2(\mathbf{Z}_3'\mathbf{Z}_3)^{-1}\mathbf{Z}_2'$. Now G_2 becomes transformations of the form $g_2(\mathbf{Z}_2(\mathbf{Z}_3'\mathbf{Z}_3)^{-1}\mathbf{Z}_2') = \mathbf{\Gamma}\mathbf{Z}_2(\mathbf{Z}_3'\mathbf{Z}_3)^{-1}\mathbf{Z}_2'\mathbf{\Gamma}'$. Using Lemma 19.3, we see that a maximal invariant under this group is the set of roots of (19.8). The derivation for the parameter maximal invariant follows similarly. □

We now show that if $p - k \geqslant r$, then the roots of (19.6) are all positive and the nonzero roots of (19.8) are the same as the roots of (19.6). Therefore, the roots of (19.8) are just those of (19.6) with some 0's added, and the roots of (19.8) are hence an invertible function of those of (19.6). Similarly, if $p - k \leqslant r$, then the roots of (19.6) are just the roots of (19.8) with some 0's added.

LEMMA 19.7. The nonzero roots of (19.6) and (19.8) are the same. If $p - k \geqslant r$, then the roots of (19.6) are all positive (with probability 1), and if $p - k \leqslant r$, then the roots of (19.8) are all positive (with probability 1).

PROOF. In Exercise B3, you are asked to show that if $t \neq 0$, then

$$|\mathbf{Z}_2'\mathbf{Z}_2 - t\mathbf{Z}_3'\mathbf{Z}_3| = (-t)^{r-p+k}|\mathbf{Z}_3'\mathbf{Z}_3||\mathbf{Z}_2(\mathbf{Z}_3'\mathbf{Z}_3)^{-1}\mathbf{Z}_2' - t\mathbf{I}| \qquad (19.10)$$

and therefore the nonzero roots of (19.6) and (19.8) are the same. Since $(n - p) \geqslant r$, $\mathbf{Z}_3$ has rank r (See Theorem 17.8). Therefore, $\mathbf{Z}_3'\mathbf{Z}_3 > 0$. If $p - k \geqslant r$, then $\mathbf{Z}_2'\mathbf{Z}_2 > 0$ by the same argument, and by Lemma 19.4, the roots of (19.6) are all positive. If $p - k \leqslant r$, then $\mathbf{Z}_2$ has rank $p - k$ and therefore, $\mathbf{Z}_2(\mathbf{Z}_3'\mathbf{Z}_3)^{-1}\mathbf{Z}_2' > 0$ (see Exercise B4). By Lemma 19.3 the roots of (19.8) are all positive in this case. □

COROLLARY. Let b be the minimum of $p - k$ and r, and let $t_1 \geqslant \dots \geqslant t_b$ be the nonzero roots of (19.6) or equivalently of (19.8). Then

$(t_1, \ldots, t_b)$ is a maximal invariant for the canonical form of the general linear hypothesis.

In a similar way, we could characterize the parameter maximal invariant as the b largest roots $\theta_1 \geqslant \ldots \geqslant \theta_b$ of either (19.7) or (19.9). (We could not characterize them as the nonzero roots, since the roots are all 0 under the null hypothesis.) We are testing that the θ_i are 0 against the alternative that $\theta_1 \geqslant 0$. Since the θ_i are 0 under the null hypothesis, the null distribution of $t_1, \ldots, t_b$ does not depend on any unknown parameters, and hence the distribution of any invariant test is completely specified under the null hypothesis.

We now give formulas for the t_i and θ_i in terms of the original observations and the original parameters $(\boldsymbol{\mu}, \boldsymbol{\Sigma})$. Let $\mathbf{S}_2 = \mathbf{Z}_2'\mathbf{Z}_2$, $\mathbf{S}_3 = \mathbf{Z}_3'\mathbf{Z}_3, \boldsymbol{\delta} = \mathbf{v}_2'\mathbf{v}_2$. Then, by arguments similar to those in Section 7.6.1,

$$\mathbf{S}_2 = \mathbf{Y}'\mathbf{P}_{V|W}\mathbf{Y}, \mathbf{S}_3 = \mathbf{Y}'\mathbf{P}_{V^\perp}\mathbf{Y}, \boldsymbol{\delta} = \boldsymbol{\mu}'\mathbf{P}_{V|W}\boldsymbol{\mu}. \tag{19.11}$$

We see that $t_1, \ldots, t_b$ are the nonzero roots of

$$|\mathbf{S}_2 - t\mathbf{S}_3| = 0$$

and $\theta_1, \ldots, \theta_b$ are the largest roots of

$$|\boldsymbol{\delta} - \theta\boldsymbol{\Sigma}| = 0.$$

19.2.4. *UMP Invariant Tests when b = 1*

We now consider the two cases in which $b = 1$, (i.e., when $r = 1$ or $p - k = 1$).

If $r = 1$, then $\mathbf{Z}_2'\mathbf{Z}_2$ and $\mathbf{Z}_3'\mathbf{Z}_3$ are both 1×1, as is Σ. Hence

$$t_1 = \frac{\mathbf{Z}_2'\mathbf{Z}_2}{\mathbf{Z}_3'\mathbf{Z}_3}, \qquad \theta_1 = \frac{\boldsymbol{\nu}_2'\boldsymbol{\nu}_2}{\Sigma}.$$

Also $\mathbf{Z}_2 \sim N_{p-k}(\boldsymbol{\nu}_2, \Sigma\mathbf{I})$, and hence $\mathbf{Z}_2'\mathbf{Z}_2/\Sigma \sim \chi^2_{p-k}(\theta_1)$. Similarly $\mathbf{Z}_3'\mathbf{Z}_3/\Sigma \sim \chi^2_{n-p}(0)$. Since $\mathbf{Z}_2$ and $\mathbf{Z}_3$ are independent, we see that

$$\frac{n-p}{p-k} t_1 \sim F_{p-k,n-p}(\theta_1).$$

We are testing that $\theta_1 = 0$ against $\theta_1 \geqslant 0$. The UMP invariant size α test is therefore the test

$$\phi(t_1) = \begin{cases} 1 & \text{if } t_1 > \dfrac{p-k}{n-p} F^{\alpha}_{p-k,n-p} \\ 0 & \text{if } t_1 \leqslant \dfrac{p-k}{n-p} F^{\sigma}_{p-k,n-p} \end{cases}.$$

Note that if $r = 1$, the model is a univariate linear model as discussed in Chapters 4–12, and the test given above is the same as the F-test derived in Chapter 7.

If $p - k = 1$, then $\mathbf{Z}_2(\mathbf{Z}_3'\mathbf{Z}_3)^{-1}\mathbf{Z}_2'$ is 1×1 as in $\boldsymbol{\nu}_2\boldsymbol{\Sigma}^{-1}\boldsymbol{\nu}_2'$. Therefore

$$t_1 = \mathbf{Z}_2(\mathbf{Z}_3'\mathbf{Z}_3)^{-1}\mathbf{Z}_2', \qquad \theta_1 = \boldsymbol{\nu}_2\boldsymbol{\Sigma}^{-1}\boldsymbol{\nu}_2'.$$

Note that $\mathbf{Z}_3'\mathbf{Z}_3 \sim W_r(n-p, \boldsymbol{\Sigma})$, $\mathbf{Z}_2' \sim N_r(\boldsymbol{\nu}_2', \boldsymbol{\Sigma})$ and $\mathbf{Z}_2$ is independent of $\mathbf{Z}_3$. Therefore, by Theorem 17.11, we see that

$$\frac{n-p-r+1}{r} t_1 \sim F_{r, n-p-r+1}(\theta_1)$$

and we are again testing that $\theta_1 = 0$ against $\theta_1 \geqslant 0$. Therefore the UMP invariant size α test is

$$\phi(t_1) = \begin{cases} 1 & \text{if } t_1 > \dfrac{r}{n-p-r+1} F^{\alpha}_{r, n-p-r+1} \\ 0 & \text{if } t_1 \leqslant \dfrac{r}{n-p-r+1} F^{\alpha}_{r, n-p-r+1} \end{cases}.$$

The one and two sample Hotelling's T^2-tests derived in the last chapter are special cases of tests of this type.

19.2.5. *Some Invariant Tests when $b > 1$*

We now look at the case in which

$$b = \min(p-k, r) > 1.$$

Let $Q_{r,\, p-k, n-p}(\theta_1, \ldots, \theta_b)$ be the joint distribution of $(t_1, \ldots, t_b)$. After reducing by sufficiency and invariance, the problem of testing the multivariate linear hypothesis becomes the problem in which we observe $(t_1, \ldots, t_b)$ having joint distribution $Q_{r,\, p-k, n-p}(\theta_1, \ldots, \theta_b)$ and we are testing that $\theta_1 = \cdots = \theta_b = 0$ against $\theta_1 \geqslant 0, \ldots, \theta_b \geqslant 0$. The density function of the Q distribution has been determined (see Constantine, 1963, and James, 1964). However, this density function does not seem helpful in finding sensible tests for this problem. Under the null hypothesis, the θ_i are all 0, so that the null distribution of any invariant function could be completely determined. (That is, it does not depend on any unknown parameters.) One interesting property of the Q-distribution that may simplify discussion of this problem is that

$$Q_{r,\, p-k, n-p}(\theta_1, \ldots, \theta_b) = Q_{p-k, r, n-k-r}(\theta_1, \ldots, \theta_b) \qquad (19.12)$$

(see Exercise C1) so that it is only necessary to consider this model when $r \leqslant p-k$ or $p-k \leqslant r$, but not both.

We now look at four tests that have been considered for this model. Let

$$\lambda_i = \prod_{i=1}^{b}(1 + t_i), \qquad \lambda_2 = \sum_{i=1}^{b} t_i, \qquad \lambda_3 = t_1, \qquad \lambda_4 = \sum_{i=1}^{b} \frac{t_i}{1 + t_i},$$

and for $i = 1, 2, 3, 4$, let

$$\phi_i(\lambda_i) = \begin{cases} 1 & \text{if } \lambda_i > c_i^\alpha \\ 0 & \text{if } \lambda_i \leqslant c_i^\alpha \end{cases}$$

where c_i^α is the upper α point of the null distribution of λ_i. Note that if $b = 1$, then $\phi_1 = \phi_2 = \phi_3 = \phi_4$. Otherwise they are all different. Since they are all functions of the complete sufficient statistic, they have different power functions (see Exercise B5). They are all known to be admissible (see Schwartz, 1967), so that no invariant test can be as good as all four tests. Therefore, there is no UMP-invariant size α test. Furthermore, they are all known to be unbiased (see Das Gupta et al., 1964). Therefore, there can be no test that is UMP among unbiased tests, or among tests that are both invariant and unbiased. Certainly, there are many more tests that are admissible, invariant and unbiased. At this point, there does not seem to be any way we can decide which test to use in most practical situations, at least from a theoretical perspective.

We now find new expressions for the statistics λ_1, λ_2, and λ_4 in terms of the original matrix $\mathbf{Y}$, rather than in terms of $\mathbf{Z}$.

LEMMA 19.8. Let $\mathbf{S}_2 = \mathbf{Y}'\mathbf{P}_{V|W}\mathbf{Y}$, $\mathbf{S}_3 = \mathbf{Y}'\mathbf{P}_{V^\perp}\mathbf{Y}$. Then

$$\lambda_1 = \frac{|\mathbf{S}_2 + \mathbf{S}_3|}{|S_3|}, \qquad \lambda_2 = \operatorname{tr} \mathbf{S}_2\mathbf{S}_3^{-1}, \qquad \lambda_4 = \operatorname{tr} \mathbf{S}_2(\mathbf{S}_2 + \mathbf{S}_3)^{-1}.$$

PROOF. We use the fact that if $\mathbf{A}$ is an $m \times m$ symmetric matrix with eigenvalues $s_1, \ldots, s_m$, then $|\mathbf{A}| = \prod_{i=1}^m s_i$ and $\operatorname{tr} \mathbf{A} = \sum_{i=1}^m s_i$. (See Corollary 2 to Theorem 1 of the appendix.) Let $s_1, \ldots, s_{p-k}$ be the eigenvalues of $\mathbf{U} = \mathbf{Z}_2(\mathbf{Z}_3'\mathbf{Z}_3)^{-1}\mathbf{Z}_2'$ [i.e., the roots of (19.8)]. If $p - k \leqslant r$, then $b = p - k$ and $s_1 = t_1, \ldots, s_{p-k} = t_{p-k}$. If $p - k \geqslant r$, then $b = r$, and $s_1 = t_1, \ldots, s_r = t_r$, $s_{r+1} = 0, \ldots, s_{p-k} = 0$, by Lemma 19.5. We now look at λ_1. It is easily verified that the eigenvalues of $\mathbf{U} + \mathbf{I}$ are $s_1 + 1, \ldots, s_{p-k} + 1$. Therefore,

$$\lambda_1 = \prod_{i=1}^{b}(1 + t_i) = \prod_{i=1}^{p-k}(1 + s_i) = |\mathbf{I} + \mathbf{U}| = |\mathbf{Z}_2(\mathbf{Z}_3'\mathbf{Z}_3)^{-1}\mathbf{Z}_2' + \mathbf{I}|$$

$$= \frac{|\mathbf{Z}_2'\mathbf{Z}_2 + \mathbf{Z}_3'\mathbf{Z}_3|}{|\mathbf{Z}_3'\mathbf{Z}_3|} = \frac{|\mathbf{S}_2 + \mathbf{S}_3|}{|\mathbf{S}_3|}$$

[using (19.10) with $t = -1$]. We now consider λ_2. We see that

$$\lambda_2 = \sum_{i=1}^{b} t_i = \sum_{i=1}^{p-k} s_i = \operatorname{tr}\mathbf{U} = \operatorname{tr}\mathbf{Z}_2(\mathbf{Z}_3'\mathbf{Z}_3)^{-1}\mathbf{Z}_2' = \operatorname{tr}\mathbf{Z}_2'\mathbf{Z}_2(\mathbf{Z}_3'\mathbf{Z}_3) = \operatorname{tr}\mathbf{S}_2\mathbf{S}_3^{-1}.$$

Finally, consider λ_4. We note that $s_1/(1+s_1), \ldots, s_{p-k}/(1+s_{p-k})$ are the eigenvalues of $\mathbf{Z}_2(\mathbf{Z}_2'\mathbf{Z}_2 + \mathbf{Z}_3'\mathbf{Z}_3)^{-1}\mathbf{Z}_2'$ (see Exercise B6). Therefore,

$$\lambda_4 = \sum_{i=1}^{b} \frac{t_i}{1+t_i} = \sum_{i=1}^{p-k} \frac{s_i}{1+s_i} = \operatorname{tr}\mathbf{Z}_2(\mathbf{Z}_2'\mathbf{Z}_2 + \mathbf{Z}_3'\mathbf{Z}_3)^{-1}\mathbf{Z}_2'$$
$$= \operatorname{tr}\mathbf{S}_2(\mathbf{S}_2 + \mathbf{S}_3)^{-1}. \quad \square$$

We now mention three methods that are sometimes used to find sensible tests in situations in which no optimal tests have been found.

(1) *The Likelihood Ratio Method.* In this method we find the likelihood ratio test.

(2) *The Substitution Method.* In this method, we first assume that some of the parameters are known and find the optimal test under this assumption. We then substitute estimators for these parameters and find a test that does not depend on any unknown parameters.

(3) *The Union-Intersection Method.* Suppose that we are testing that $\theta \in A$ aginst $\theta \in B$, where $A \subset B$, and that $A = \bigcap C_\delta$, $\delta \in \Delta$. Suppose further that for each $\delta \in \Delta$ we can find an optimal test ϕ_δ of the hypothesis $\theta \in C_\delta$ against $\theta \in B$. Let ϕ be the test that accepts $\theta \in A$ if d only if ϕ_δ accepts $\theta \in C_\delta$ for all δ. Then the acceptance region of ϕ is just the intersection of the acceptance regions for the ϕ_δ, and the critical region for ϕ is the union of the critical regions for the ϕ_δ (hence the name union-intersection method). In other words, ϕ accepts the hypothesis that $\phi \in A$ if and only ϕ_δ accepts the hypothesis that $\theta \in C_\delta$ for all δ. ϕ is called the union-intersection test. The union-intersection method is due to Roy (1953). It has the advantage that we can often find simultaneous confidence intervals associated with the union-intersection test as we shall see in Sections 19.3 and 21.2.

For the testing problems considered earlier in the book these three methods would, in general, lead to the same tests, namely the UMP invariant test that has been derived for those problems. For the problem considered in this section, there is no UMP invariant test, and the three methods lead to three different tests.

(1) *Wilks Test.* Wilks (1932) suggested the use of the test ϕ_1, which is therefore called the Wilks test. He showed that $\lambda_1^{-n/2}$ is the likelihood

ratio test statistic (see Exercise B7), and hence ϕ_1 is the likelihood ratio test.

(2) *The Lawley-Hotelling Test*. The test ϕ_2 was suggested by Lawley (1938) and Hotelling (1951) and is called the Lawley-Hotelling test. If $\mathbf{\Sigma}$ is known, then the UMP invariant size α test that $\boldsymbol{\nu}_2 = \mathbf{0}$ is to reject if $\operatorname{tr} \mathbf{S}_2 \mathbf{\Sigma}^{-1}$ is too large, (see Exercise B8). Hence, the substitution method would imply that we reject if $\operatorname{tr} \mathbf{S}_2 \hat{\mathbf{\Sigma}}^{-1} = (n-p)\lambda_2$ is too large. The substitution method, therefore leads to the test ϕ_2.

(3) *Roy's Largest Root Test*. The test ϕ_3 is usually called Roy's largest root test and was suggested in Roy (1953). It is derived by the union-intersection method, as the following argument shows. We first note that $\boldsymbol{\nu}_2 = \mathbf{0}$ if and only if $\mathbf{a}'\boldsymbol{\nu}_2 = \mathbf{0}$ for all $\mathbf{a} \neq \mathbf{0}$ [$\mathbf{a}$ is $(p-k) \times 1$]. Consider testing that $\mathbf{a}'\boldsymbol{\nu}_2 = \mathbf{0}$. This is a case of testing the multivariate linear hypothesis in which $b = 1$. Therefore there is a UMP invariant test for testing that $\mathbf{a}'\boldsymbol{\nu}_2 = \mathbf{0}$. In fact, the UMP invariant size α test rejects when

$$\frac{\mathbf{a}'\mathbf{Z}_2(\mathbf{Z}_3'\mathbf{Z}_3)^{-1}\mathbf{Z}_2'\mathbf{a}}{\mathbf{a}'\mathbf{a}} > \frac{r}{n-p-r+1} F^{\alpha}_{r,\, n-p-r+1} = C$$

Therefore, the union-intersection method would lead to the test which rejects if

$$\sup_{\mathbf{a} \neq \mathbf{0}} \frac{\mathbf{a}'\mathbf{Z}_2(\mathbf{Z}_3'\mathbf{Z}_3)^{-1}\mathbf{Z}_2'\mathbf{a}}{\mathbf{a}'\mathbf{a}} > C.$$

However

$$\sup_{\mathbf{a} \neq \mathbf{0}} \frac{\mathbf{a}'\mathbf{Z}_2(\mathbf{Z}_3'\mathbf{Z}_3)^{-1}\mathbf{Z}_2'\mathbf{a}}{\mathbf{a}'\mathbf{a}} = t_1.$$

(See Lemma 19.3.) Therefore, the union-intersection test would reject when $t_1 > C$, and hence is the test ϕ_3.

(4) Pillai's Test. The statistic λ_4 was suggested by Pillai (1955) and is called Pillai's trace.

We have now discussed four tests for testing the multivariate linear hypothesis when $b > 1$. All four tests are admissible, unbiased and invariant. It seems likely that there are many more such admissible, invariant, unbiased tests. At this time, there seems to be no reason to prefer any test over the others. (One advantage of ϕ_3 is that there are simultaneous confidence intervals associated with it, as we see in Section 19.3.) In the next section, we show that the test statistics $(n-p)\log\lambda_1$, $(n-p)\lambda_2$ and $(n-p)\lambda_4$ all converge in probability to $\operatorname{tr} \mathbf{S}_2\mathbf{\Sigma}^{-1}$, the maximal invariant for testing the multivariate linear hypothesis when $\mathbf{\Sigma}$ is known. Therefore, for large n, the tests ϕ_1, ϕ_2, and ϕ_4 should be about the same. The test statistic $(n-p)\lambda_3$ converges in probability to the largest root r of $|\mathbf{S}_2 - r\boldsymbol{\Sigma}| = 0$, and hence ϕ_3 would be different, even asymptotically.

Several studies have been made comparing the powers of these four tests and others in particular situations (see Mikhail, 1965, Pillai and Jayachandran, 1967, Ito, 1962, Smith et al., 1962, and Gabriel, 1969b). About the only clear conclusion from these studies is that ϕ_3 is more powerful than the others when the largest root θ_1 is much larger than the other roots.

19.2.6. The Distributions of the Statistics

It is not possible to use the tests mentioned in the last two sections without knowing the null distribution or approximate null distributions of the statistics λ_i. Since the distribution of $t_1, \ldots, t_b$ depends only on $\theta_1, \ldots, \theta_b$ and the θ_i are all 0 under the null hypothesis, the null distribution of any invariant test does not depend on any unknown parameters, and could be completely specified. In addition, by (19.12) we need only specify the null distribution for $r \leqslant p-k$. The null distribution of $(t_1, \ldots, t_b)$ was derived simultaneously by Fisher (1939), Hsu (1939), Girshik (1939), and Roy (1939). (See also Anderson, 1958, pp. 307–318, and Giri, 1977, pp. 211–213.) We do not use that distribution in this book.

As mentioned in the beginning of this section, Table 47 of Pearson and Hartley (1972) gives conversion factors for determining some exact percentage points for the distribution of λ_1. The only other statistic for which exact percentage points are available is the statistic λ_3, which we now discuss. Let $s_{r,\, p-k,\, n-p}$ be the distribution of the largest root of

$$r(s) = |\mathbf{W}_1 - s(\mathbf{W}_1 + \mathbf{W}_2)| = 0$$

where $\mathbf{W}_1$ and $\mathbf{W}_2$ are independent, $\mathbf{W}_1 \sim W_r(p-k, \mathbf{I})$, $\mathbf{W}_2 \sim W_r(n-p, \mathbf{I})$. Let $s^{\alpha}_{r,\, p-k,\, n-p}$ be the upper α point for this distribution. Tables of $s^{\alpha}_{r,\, p-k,\, n-p}$ are given in Pearson and Hartley (1972) Tables 48 and 49. (See also Morrison, 1976, pp. 379–403.) Under the null hypothesis

$$\frac{t_1}{1+t_1} \sim s_{r,\, p-k,\, n-p}, \qquad c_3^{\alpha} = \frac{s^{\alpha}_{r,\, p-k,\, n-p}}{1 - s^{\alpha}_{r,\, p-k,\, n-p}},$$

(see Exercise B6). Therefore, we can use the s-distribution to design size 0.01 and size 0.05 versions of ϕ_3.

We now discuss the asymptotic distributions of λ_1, λ_2, and λ_4. We first need the following lemma.

LEMMA 19.9. Let $\mathbf{W}_m \sim W_r(m, \boldsymbol{\Sigma})$. Then $(1/m)\mathbf{W}_m \xrightarrow{p} \boldsymbol{\Sigma}$ as $m \to \infty$.

PROOF. Let $W_m = \mathbf{X}'_m\mathbf{X}_m = \sum_{i=1}^m \mathbf{X}_{mi}{}'\mathbf{X}_{mi}$ where

$$\mathbf{X}_m = \begin{pmatrix} \mathbf{X}_{m1} \\ \vdots \\ \mathbf{X}_{mm} \end{pmatrix} \sim N_{m,r}(\mathbf{0}, \mathbf{I}, \boldsymbol{\Sigma}).$$

Then the $\mathbf{X}_{mi}$ are independent and $\mathbf{X}'_{mi} \sim N_r(\mathbf{0}, \boldsymbol{\Sigma})$. Hence the $\mathbf{X}'_{mi}\mathbf{X}_{mi}$ are independently identically distributed and $\mathbf{E}\mathbf{X}'_{mi}\mathbf{X}_{mi} = \boldsymbol{\Sigma}$. Therefore, by the weak law of large numbers,

$$\frac{1}{m}\mathbf{W}_m \xrightarrow{P} \boldsymbol{\Sigma}. \quad \square$$

We now study the asymptotic distribution of the test statistics as $n \to \infty$, but p, k, and r remain fixed. We also assume that $\boldsymbol{\delta} = \boldsymbol{\mu}'\mathbf{P}_{V|W}\boldsymbol{\mu}$ stays fixed. Let $\mathbf{S}_2(n)$ and $\mathbf{S}_3(n)$ be the matrices $\mathbf{S}_2$ and $\mathbf{S}_3$ defined earlier and let $t_1(n), \ldots, t_b(n)$ be the nonzero roots of

$$|\mathbf{S}_2(n) - t\mathbf{S}_3(n)| = 0.$$

Finally, let $s_1(n), \ldots, s_b(n)$ be the nonzero roots of

$$|\mathbf{S}_2(n) - s\boldsymbol{\Sigma}| = 0.$$

We now show that $(n-p)t_i(n)$ converges to $s_i(n)$.

THEOREM 19.10. As $n \to \infty$, $(n-p)t_i(n) - s_i(n) \xrightarrow{P} 0$.

PROOF. We note that $\mathbf{S}_3(n) \sim W_r(n-p, \boldsymbol{\Sigma})$, and therefore

$$\frac{1}{n-p}\mathbf{S}_3(n) \xrightarrow{P} \boldsymbol{\Sigma}.$$

Also $\mathbf{S}_2(n) \sim W_r(p-k, \boldsymbol{\Sigma}, \boldsymbol{\delta})$ and hence

$$\mathbf{S}_2(n) \xrightarrow{d} W_r(p-k, \boldsymbol{\Sigma}, \boldsymbol{\delta}).$$

For any $r \times r$ matrices $\mathbf{U}$, $\mathbf{V}$ and $\mathbf{W}$, let q_i and r_i be the ith largest roots of

$$|\mathbf{U} - q\mathbf{V}| = 0 \qquad |\mathbf{U} - r\mathbf{W}| = 0$$

and let $h_i(\mathbf{U}, \mathbf{V}, \mathbf{W}) = q_i - r_i$. Then

$$(n-p)t_i(n) - s_i(n) = h_i\big(\mathbf{S}_2(n), (n-p)^{-1}\mathbf{S}_3(n), \boldsymbol{\Sigma}\big).$$

Since h_i is a continuous function, the limiting distribution of $(n-p)t_i(n) - s_i(n)$ is the distribution of $h_i(\mathbf{U}, \boldsymbol{\Sigma}, \boldsymbol{\Sigma})$, where $\mathbf{U} \sim W_r(p-k, \boldsymbol{\Sigma}, \delta)$. However, $h_i(\mathbf{U}, \boldsymbol{\Sigma}, \boldsymbol{\Sigma}) \equiv 0$. Therefore $(n-p)t_i(n) - s_i(n)$ converges in distribution to the distribution that is degenerate at 0 and hence converges in probability to 0. $\square$

We now show that $(n-p)\log\lambda_1$, $(n-p)\lambda_2$, and $(n-p)\lambda_4$ all converge to $\operatorname{tr}\mathbf{S}_2(n)\boldsymbol{\Sigma}^{-1}$, the maximal invariant for testing the linear hypothesis when $\boldsymbol{\Sigma}$ is known.

THEOREM 19.11. As $n\to\infty$,

$$(n-p)\log\lambda_1 - \operatorname{tr}\mathbf{S}_2(n)\boldsymbol{\Sigma}^{-1}\xrightarrow{P}0,\ (n-p)\lambda_2-\operatorname{tr}\mathbf{S}_2(n)\boldsymbol{\Sigma}^{-1}\xrightarrow{P}0,$$

and $(n-p)\lambda_4-\operatorname{tr}\mathbf{S}_2(n)\boldsymbol{\Sigma}^{-1}\xrightarrow{P}0$.

PROOF. By arguments similar to those in the last section, $\operatorname{tr}\mathbf{S}_2(n)\boldsymbol{\Sigma}^{-1} = \sum_{i=1}^{b}s_i(n)$. Now,

$$(n-p)\lambda_2-\operatorname{tr}\mathbf{S}_2(n)\boldsymbol{\Sigma}^{-1}=\textstyle\sum_{i=1}^{b}\left[(n-p)t_i(n)-s_i(n)\right]\xrightarrow{P}0.$$

Similarly,

$$(n-p)\lambda_4-\operatorname{tr}\mathbf{S}_2(n)\boldsymbol{\Sigma}^{-1}=\sum_{i=1}^{b}\frac{\left[(n-p)t_i(n)-s_i(n)\right]}{1+t_i(n)}-\frac{s_i(n)t_i(n)}{1+t_i(n)}.$$

Now, $(n-p)t_i(n)$ converges in distribution to the distribution of $s_i(n)$. [This distribution does not depend on n, since the distribution of $\mathbf{S}_2(n)$ does not.] Therefore, $t_i(n)\xrightarrow{P}0$. Hence $1+t_i(n)\xrightarrow{P}1$ and $s_i(n)t_i(n)/[1+t_i(n)]\xrightarrow{P}0$. Therefore $(n-p)\lambda_4-\operatorname{tr}\mathbf{S}_2\boldsymbol{\Sigma}^{-1}\xrightarrow{P}0$. It can be shown (see Exercise C3) that $(n-p)\log[1+t_i(n)]-s_i(n)\xrightarrow{P}0$. Therefore,

$$(n-p)\log\lambda_1-\operatorname{tr}\mathbf{S}_2(n)\boldsymbol{\Sigma}^{-1}=\sum_{i=1}^{b}\left\{(n-p)\log\left[1+t_i(n)\right]\right\}-s_i(n)\xrightarrow{P}0. \quad \square$$

COROLLARY.

$$(n-p)\log\lambda_1\xrightarrow{d}\chi^2_{r(p-k)}(\operatorname{tr}\boldsymbol{\delta\Sigma}^{-1}),\qquad (n-p)\lambda_2\xrightarrow{d}\chi^2_{r(p-k)}(\operatorname{tr}\boldsymbol{\delta\Sigma}^{-1}),$$

$$(n-p)\lambda_4\xrightarrow{d}\chi^2_{r(p-k)}(\operatorname{tr}\boldsymbol{\delta\Sigma}^{-1}).$$

PROOF. $(n-p)\log\lambda_1=\operatorname{tr}[\mathbf{S}_2(n)\boldsymbol{\Sigma}^{-1}]+\{(n-p)\log\lambda_1-\operatorname{tr}[\mathbf{S}_2(n)\boldsymbol{\Sigma}^{-1}]\}$. In addition $\operatorname{tr}[\mathbf{S}_2(n)\Sigma^{-1}]\sim\chi^2_{r(p-k)}(\operatorname{tr}\boldsymbol{\delta\Sigma}^{-1})$ (see Exercise B8) and the result follows from Theorem 10.2b. The proofs for λ_2 and λ_4 follow similarly. $\square$

Therefore, we see that if n is large then

$$c_1^\alpha\approx\exp\frac{\chi^{2\alpha}_{r(p-k)}}{n-p},\qquad c_2^\alpha\approx\frac{\chi^{2\alpha}_{r(p-k)}}{n-p},\qquad c_4^\alpha\approx\frac{\chi^{2\alpha}_{r(p-k)}}{n-p}.$$

We now consider the asymptotic distribution of λ_3. Let $u_{r,p-k}$ be the distribution of the largest root of

$$q(u)=|\mathbf{W}-u\mathbf{I}|=0$$

where $\mathbf{W} \sim W_r(p-k, \mathbf{I})$. Under the null hypothesis $s_1(n) \sim u_{r,p-k}$ and hence under the null hypothesis

$$(n-p)\lambda_3 \xrightarrow{d} u_{r,p-k}$$

Let $u_{r,p-k}^{\alpha}$ be the upper α point of the distribution of $u_{r,p-k}$. Therefore for large n,

$$c_3^{\alpha} \approx \frac{u_{r,p-k}^{\alpha}}{n-p}.$$

Tables of $u_{r,p-k}^{\alpha}$ are given in Pearson and Hartley (1972, Table 52).

The above corollary also implies that the three tests ϕ_1, ϕ_2, and ϕ_4 have the same asymptotic power function, which is the same as the power function of the UMP invariant test when $\boldsymbol{\Sigma}$ is assumed known. The test ϕ_3 has a different asymptotic power function. The power funciton of ϕ_3 converges to the power function of the test that rejects if the largest root of $|\mathbf{S}_2^{(n)} - s\boldsymbol{\Sigma}| = 0$ is too large. This test is not an invariant test for the multivariate linear hypothesis when $\boldsymbol{\Sigma}$ is known.

The asymptotic derivations in this section are somewhat different from those presented earlier in the book in that we are assuming throughout the derivation that the assumptions of the model are met. In Section 19.8, we discuss asymptotic results when we relax the normal assumption.

19.3. SIMULTANEOUS CONFIDENCE INTERVALS FOR CONTRASTS.

A *contrast* associated with testing the hypothesis that $\boldsymbol{\mu} \in W$ against $\boldsymbol{\mu} \in V$ is a linear function of the form $\mathbf{d}'\boldsymbol{\mu}\mathbf{q}$, $\mathbf{d} \in V|W$, $\mathbf{q} \in R^r$. (Note that this definition reduces to the definition used in Chapter 8 for the univariate linear model.) We now find a set of simultaneous confidence intervals for the set of all contrasts associated with testing that $\boldsymbol{\mu} \in W$. The intervals are due to Roy and Bose (1953). We also show that the largest root test ϕ_3 rejects the hypothesis $\boldsymbol{\mu} \in W$ if and only if at least one of those confidence intervals does not contain 0. Therefore, after rejecting the hypothesis that $\boldsymbol{\mu} \in W$, with the test ϕ_3, these confidence intervals can be used to determine which contrasts are causing the hypothesis to be rejected.

Let s be the largest root of

$$q(s) = |(\hat{\boldsymbol{\mu}} - \boldsymbol{\mu})'\mathbf{P}_{V|W}(\hat{\boldsymbol{\mu}} - \boldsymbol{\mu}) - s(n-p)\hat{\boldsymbol{\Sigma}}| = 0.$$

Now, $(\hat{\boldsymbol{\mu}} - \boldsymbol{\mu})'\mathbf{P}_{V|W}(\hat{\boldsymbol{\mu}} - \boldsymbol{\mu}) = (\mathbf{Y} - \boldsymbol{\mu})'\mathbf{P}_{V|W}(\mathbf{Y} - \boldsymbol{\mu}) \sim W_r(p-k, \boldsymbol{\Sigma})$, $(n-p)\hat{\boldsymbol{\Sigma}} \sim W_r(n-p, \boldsymbol{\Sigma})$, and $(\hat{\boldsymbol{\mu}} - \boldsymbol{\mu})\mathbf{P}_{V|W}(\hat{\boldsymbol{\mu}} - \boldsymbol{\mu})$ and $(n-p)\hat{\boldsymbol{\Sigma}}$ are independent.

Therefore s has the same distribution as the null distribution of λ_3. Hence

$$P(s \leqslant c_3^\alpha) = 1 - \alpha.$$

where c_3^α is defined in the last section. We now find an alternative expression for s.

LEMMA 19.12.

$$s = \sup_{\substack{\mathbf{d} \in V|W,\, \mathbf{q} \in R^r \\ \mathbf{d} \neq \mathbf{0},\, \mathbf{q} \neq \mathbf{0}}} \frac{\left[\mathbf{d}'(\hat{\boldsymbol{\mu}} - \boldsymbol{\mu})\mathbf{q}\right]^2}{(n-p)\|\mathbf{d}\|^2 \mathbf{q}'\hat{\boldsymbol{\Sigma}}\mathbf{q}}$$

PROOF. By Lemma 19.4, we see that

$$s = \sup_{\substack{\mathbf{q} \in R^r \\ \mathbf{q} \neq \mathbf{0}}} \frac{\mathbf{q}'(\hat{\boldsymbol{\mu}} - \boldsymbol{\mu})\mathbf{P}_{V|W}(\hat{\boldsymbol{\mu}} - \boldsymbol{\mu})\mathbf{q}}{(n-p)\mathbf{q}'\hat{\boldsymbol{\Sigma}}\mathbf{q}}.$$

Now,

$$\mathbf{q}'(\hat{\boldsymbol{\mu}} - \boldsymbol{\mu})\mathbf{P}_{V|W}(\hat{\boldsymbol{\mu}} - \boldsymbol{\mu})\mathbf{q} = \|\mathbf{P}_{V|W}(\hat{\boldsymbol{\mu}} - \boldsymbol{\mu})\mathbf{q}\|^2.$$

By Lemma 8.1,

$$\|\mathbf{P}_{V|W}(\hat{\boldsymbol{\mu}} - \boldsymbol{\mu})\mathbf{q}\|^2 = \sup_{\substack{\mathbf{d} \in V|W \\ \mathbf{d} \neq \mathbf{0}}} \frac{\left[\mathbf{d}'(\hat{\boldsymbol{\mu}} - \boldsymbol{\mu})\mathbf{q}\right]^2}{\|\mathbf{d}\|^2}.$$

The result follows directly. □

We are now ready for the simultaneous confidence intervals.

THEOREM 19.13. The set of intervals

$$\mathbf{d}'\boldsymbol{\mu}\mathbf{q} \in \mathbf{d}'\hat{\boldsymbol{\mu}}\mathbf{q} \pm \left[(n-p)c_3^\alpha \|\mathbf{d}\|^2 \mathbf{q}'\hat{\boldsymbol{\Sigma}}\mathbf{q}\right]^{1/2}$$

is a set of simultaneous $(1-\alpha)$ confidence intervals for the set of all contrasts associated with testing that $\boldsymbol{\mu} \in W$ against $\boldsymbol{\mu} \in V$. The test ϕ_3 rejects the null hypothesis $\boldsymbol{\mu} \in W$ if and only if at least one of these confidence intervals does not contain 0.

PROOF. The intervals associated with $\mathbf{d} = \mathbf{0}$ and $\mathbf{q} = \mathbf{0}$ are satisfied trivially. Therefore,

$$P\left(\mathbf{d}'\boldsymbol{\mu}\mathbf{q} \in \mathbf{d}'\hat{\boldsymbol{\mu}}\mathbf{q} \pm \left[(n-p)c_3^\alpha \|\mathbf{d}\|^2 \mathbf{q}'\hat{\boldsymbol{\Sigma}}\mathbf{q}\right]^{1/2} \quad \text{for all } \mathbf{d} \in V|W,\, \mathbf{q} \in R^r\right)$$

$$= P\left[\frac{(\mathbf{d}'(\hat{\boldsymbol{\mu}} - \boldsymbol{\mu})\mathbf{q})^2}{(n-p)\|\mathbf{d}\|^2 \mathbf{q}'\hat{\boldsymbol{\Sigma}}\mathbf{q}} \leqslant c_3^\alpha \quad \text{for all } \mathbf{d} \in V|W, \mathbf{q} \in R^r, \mathbf{d} \neq \mathbf{0}, \mathbf{q} \neq \mathbf{0}\right]$$

$$= P(s \leqslant c_3^\alpha) = 1 - \alpha,$$

and therefore the set of intervals is a set of simultaneous $(1-\alpha)$ confidence intervals. Note that the hypothesis $\boldsymbol{\mu} \in W$ is accepted with ϕ_3 if and only if $t_1 \leqslant c_3^{\alpha}$. However, by an argument similar to that of Lemma 19.12, we see that

$$t_1 = \sup_{\substack{\mathbf{d} \in V \mid W, \mathbf{q} \in R^r \\ \mathbf{d} \neq 0, \mathbf{q} \neq 0}} \frac{(\mathbf{d}'\hat{\boldsymbol{\mu}}\mathbf{q})^2}{(n-p)\|\mathbf{d}\|^2 \mathbf{q}'\hat{\boldsymbol{\Sigma}}\mathbf{q}}$$

and hence the hypothesis is accepted if and only if

$$(\mathbf{d}'\hat{\boldsymbol{\mu}}\mathbf{q})^2 \leqslant (n-p)c_3^{\alpha}\|\mathbf{d}\|^2 \mathbf{q}'\hat{\boldsymbol{\Sigma}}\mathbf{q}$$

that is, if and only if

$$0 \in \mathbf{d}'\hat{\boldsymbol{\mu}}\mathbf{q} \pm \left((n-p)c_3^{\alpha}\|\mathbf{d}\|^2 \mathbf{q}'\hat{\boldsymbol{\Sigma}}\mathbf{q}\right)^{1/2}. \quad \square$$

The simultaneous confidence intervals derived in this section are consistent with the test ϕ_3 in the sense that ϕ_3 rejects if and only if at least one confidence interval does not contain 0. No simultaneous confidence intervals have been derived that are consistent with the other tests defined in the last section.

19.4. TESTING THE GENERALIZED MULTIVARIATE LINEAR HYPOTHESIS.

We now consider a generalization of the testing problem considered in Section 19.2. It is useful in multivariate regression problems and generalized repeated measures problems. Let $\mathbf{A}$ be an $r \times s$ matrix of rank s. The problem of testing the generalized multivariate linear hypothesis is the problem of testing that $\boldsymbol{\mu}\mathbf{A} \in W$ and $\boldsymbol{\mu} \in V$ against the alternative that $\boldsymbol{\mu} \in V$. If $\mathbf{A} = \mathbf{I}$, then this is the problem considered in the previous section. (If $p = 1$, then $\mathbf{A}$ is a scalar, and $\boldsymbol{\mu}\mathbf{A} \in W$ if and only if $\boldsymbol{\mu} \in W$, so that it is not necessary to consider this problem in the univariate case.)

In Exercise C4 we define several groups which leave this problem invariant and the following theorem is established.

THEOREM 19.14. Let $\mathbf{S}_1$, $\mathbf{S}_2$ and $\boldsymbol{\delta}$ be defined by (19.11) and let $b^* = \min(s, p-k)$. A maximal invariant for the problem of testing the generalized linear hypothesis is the set of nonzero roots, $t_1^* \geqslant \cdots \geqslant t_{b^*}^*$ of

$$|\mathbf{A}'\mathbf{S}_2\mathbf{A} - t^*\mathbf{A}'\mathbf{S}_3\mathbf{A}| = 0$$

and a parameter maximal invariant is the set of b^* largest roots, $\theta_1^* \geqslant \cdots$

$\geqslant \theta^*_{b^*}$ of

$$|\mathbf{A}'\boldsymbol{\delta}\mathbf{A} - \theta^*\mathbf{A}'\boldsymbol{\Sigma}\mathbf{A}| = 0.$$

PROOF. See Exercise C4. □

We now note that

$$(t_1^*, \ldots, t_{b^*}^*) \sim Q_{s,\, p-k,\, n-p}(\theta_1^*, \ldots, \theta_{b^*}^*).$$

Therefore, after reducing by invariance we are in the same situation as in testing the multivariate linear model (with r replaced by s). If $b^* = 1$, then there is a UMP invariant size α test which is an F-test. Otherwise, there is no UMP invariant test. However, the following four tests are available:

$$\lambda_1^* = \prod_{i=1}^{b^*} (1 + t_i^*) = \frac{|\mathbf{A}'\mathbf{S}_2\mathbf{A} + \mathbf{A}'\mathbf{S}_3\mathbf{A}|}{|\mathbf{A}\mathbf{S}_3\mathbf{A}|}; \qquad \lambda_2^* = \sum_{i=1}^{b^*} t_i^* = \operatorname{tr} \mathbf{A}'\mathbf{S}_2\mathbf{A}(\mathbf{A}'\mathbf{S}_3\mathbf{A})^{-1}$$

$$\lambda_3^* = t_1^*, \qquad \lambda_4^* = \sum_{i=1}^{b^*} \frac{t_i^*}{1 + t_i^*} = \operatorname{tr} \mathbf{A}'\mathbf{S}_2\mathbf{A}(\mathbf{A}'\mathbf{S}_2\mathbf{A} + \mathbf{A}'\mathbf{S}_3\mathbf{A})^{-1}$$

$$\phi_i^*(\lambda_i^*) = \begin{cases} 1 & \text{if } \lambda_i^* > c_i^{*\alpha} \\ 0 & \text{if } \lambda_i^* < c^{*\alpha}_i \end{cases}$$

where $c_i^{*\alpha}$ is the upper α point of distribution of λ_i^*. We note that $c^{*\alpha}_i$ can be computed from c_i^α by replacing r by s.

In Exercise B9 simultaneous confidence intervals associated with contrasts are derived. These are compatible with the test ϕ_3^*.

19.5. MULTIVARIATE REGRESSION.

In the multivariate regression model, we have a basis matrix $\mathbf{X}$ for the space V. Let $\boldsymbol{\beta}$ and $\hat{\boldsymbol{\beta}}$ be the unique solutions to the equations

$$\boldsymbol{\mu} = \mathbf{X}\boldsymbol{\beta}, \qquad \hat{\boldsymbol{\mu}} = \mathbf{X}\hat{\boldsymbol{\beta}}.$$

In the regression model we are primarily interested in inference about the $p \times r$ matrix $\boldsymbol{\beta}$. We note first that

$$\boldsymbol{\beta} = (\mathbf{X}'\mathbf{X})^{-1}\mathbf{X}'\boldsymbol{\mu}, \qquad \hat{\boldsymbol{\beta}} = (\mathbf{X}'\mathbf{X})^{-1}\mathbf{X}'\hat{\boldsymbol{\mu}} = (\mathbf{X}'\mathbf{X})^{-1}\mathbf{X}'\mathbf{Y}.$$

From Theorem 19.1 we see that $(\hat{\boldsymbol{\beta}}, \hat{\boldsymbol{\Sigma}})$ is a complete sufficient statitic (since it is an invertible function of $(\hat{\boldsymbol{\mu}}, \hat{\boldsymbol{\Sigma}})$) and that $\hat{\boldsymbol{\beta}}$ and $\hat{\boldsymbol{\Sigma}}$ are independent. We note that $\hat{\boldsymbol{\beta}} \sim N_{p,r}(\boldsymbol{\beta}, (\mathbf{X}'\mathbf{X})^{-1}, \boldsymbol{\Sigma})$, and hence $\hat{\boldsymbol{\beta}}$ is the minimum variance unbiased estimator of $\boldsymbol{\beta}$. Finally, $\hat{\boldsymbol{\beta}}$ is the same function of $\hat{\boldsymbol{\mu}}$ that $\boldsymbol{\beta}$ is of $\boldsymbol{\mu}$, and hence by Theorem 19.2, $\hat{\boldsymbol{\beta}}$ is the MLE of $\boldsymbol{\beta}$.

We now consider the problem of testing that

$$\mathbf{C}\boldsymbol{\beta}\mathbf{A}=0$$

against $\boldsymbol{\beta}$ unspecified, where $\mathbf{C}$ is a $(p-k)\times p$ matrix of rank $p-k$ and $\mathbf{A}$ is an $r\times s$ matrix of rank s. This is equivalent to testing that

$$\mathbf{C}(\mathbf{X}'\mathbf{X})^{-1}\mathbf{X}'\boldsymbol{\mu}\mathbf{A}=0.$$

Therefore, let Q be the subspace spanned by the columns $\mathbf{X}(\mathbf{X}'\mathbf{X})^{-1}\mathbf{C}'$. Then $Q\subset V$. Let $W=V|\,Q$. Then $\mathbf{C}\boldsymbol{\beta}\mathbf{A}=\mathbf{0}$ if and only if $\boldsymbol{\mu}\mathbf{A}\in W$. By Lemma 7.3 we see that $\mathbf{X}(\mathbf{X}'\mathbf{X})^{-1}\mathbf{C}'$ is a basis matrix for $V|\,W$, and hence W has dimension $p-(p-k)=k$. This problem is therefore in the framework of testing the generalized multivariate linear hypothesis. Let b^* be the minimum of $p-k$ and s, and let

$$S_2^*=\mathbf{A}'\hat{\boldsymbol{\beta}}'\mathbf{C}'\left(\mathbf{C}(\mathbf{X}'\mathbf{X})^{-1}\mathbf{C}'\right)^{-1}\mathbf{C}\hat{\boldsymbol{\beta}}\mathbf{A}=\mathbf{A}'\hat{\boldsymbol{\mu}}'\mathbf{P}_{V|W}\hat{\boldsymbol{\mu}}\mathbf{A}$$

$$S_3^*=(n-p)\mathbf{A}'\hat{\boldsymbol{\Sigma}}\mathbf{A}.$$

By Theorem 19.14, a maximal invariant for this problem is the set $t_1^* \geqslant \cdots \geqslant t_{b^*}^*$ of nonzero roots of $|\mathbf{S}_2^*-t^*\mathbf{S}_3^*|=0$. If $p-k=1$, then there is only one nonzero root and the UMP invariant size α test is one that rejects if t_1^* is greater than $(s/(n-p-s+1)F_{s,n-p-s+1}$. Similarly, if $s=1$, then there is only one root, and the UMP invariant size α test is one which rejects if t_1^* is greater than $[(p-k)/(n-p)]F^{\alpha}_{p-k,\,n-p}$. If $b^*>1$, then there is no UMP invariant test, but the tests $\phi_1^*,\phi_2^*,\phi_3^*,\phi_4^*$, defined in Section 19.5, are available.

19.6. MULTIVARIATE ANALYSIS OF VARIANCE.

We now consider multivariate extensions of the models discussed in Section 7.3. All the problems we consider can be put into the form of testing a multivariate linear hypothesis as discussed in Section 19.2. For each model, we therefore find the matrices $\mathbf{S}_2$ and $\mathbf{S}_3$. The maximal invariant is then the set of nonzero roots of

$$|\mathbf{S}_2-t\mathbf{S}_3|=0.$$

We first look at the one way analysis of variance model in which we observe $\mathbf{Y}_{ij}$ independent, with

$$\mathbf{Y}_{ij}\sim N_r(\boldsymbol{\theta}+\boldsymbol{\alpha}_i,\boldsymbol{\Sigma}),\qquad i=1,\ldots,p,\qquad j=1,\ldots,n_i,\qquad \sum_i n_i\boldsymbol{\alpha}_i=0.$$

(Note that $\boldsymbol{\theta}$ and $\boldsymbol{\alpha}_i$ are now r-dimensional vectors.) We want to test that

$\boldsymbol{\alpha}_i = \mathbf{0}$. Let $N = \sum_i n_i$, let $\mathbf{Y}$ be the $N \times r$ matrix

$$\mathbf{Y} = \begin{pmatrix} \mathbf{Y}'_{11} \\ \vdots \\ \mathbf{Y}'_{1n_1} \\ \vdots \\ \mathbf{Y}'_{p,n_p} \end{pmatrix}$$

and let $\boldsymbol{\mu} = E\mathbf{Y}$. Then $\mathbf{Y} \sim N_{N,r}(\boldsymbol{\mu}, \mathbf{I}, \boldsymbol{\Sigma})$. Now, let V be the p-dimensional subspace of R^N in which the first n_1 elements are the same, the next n_2 elements are the same, and so on, and let W be the one-dimensional subspace in which all the elements are the same. We are testing that $\boldsymbol{\mu} \in W$ against $\boldsymbol{\mu} \in V$. So this is in the form of multivariate hypothesis considered in Section 19.2. Note that V and W are the same for the multivariate model as for the univariate model, and hence $\mathbf{P}_{V|W}$ and $\mathbf{P}_{V^\perp}$ are the same for the multivariate model as for the univariate one. Therefore, it is easily seen that

$$\mathbf{S}_2 = \mathbf{Y}'\mathbf{P}_{V|W}\mathbf{Y} = \sum_i n_i(\overline{\mathbf{Y}}_{i.} - \overline{\mathbf{Y}}_{..})(\overline{\mathbf{Y}}_{i.} - \overline{\mathbf{Y}}_{..})',$$

$$\mathbf{S}_3 = \mathbf{Y}'\mathbf{P}_{V^\perp}\mathbf{Y} = \sum_i \sum_j (\mathbf{Y}_{ij} - \overline{\mathbf{Y}}_{i.})(\mathbf{Y}_{ij} - \overline{\mathbf{Y}}_{i.})'.$$

(Note that $\mathbf{S}_2$ and $\mathbf{S}_3$ are $r \times r$.) In this model, $k = 1$, so that b is the minimum of $p - 1$ and r. If $p - 1$ is 1, the UMP invariant test reduces to the F-test derived in Chapter 18 for the two-sample problem.

From the above example, it should be clear that to find formulas for a multivariate analysis of variance problem, we can let $\mathbf{S}_2$ and $\mathbf{S}_3$ be the formulas for the univariate numerator and denominator sums of squares with u^2 replaced by uu'.

We now consider simultaneous confidence intervals for contrasts for this model. Since $V|W$ is the same for the multivariate one-way analysis of variance model as for the univariate one-way analysis of variance model, a contrast for the multivariate model can be found from a contrast for the univariate model by multiplying by $\mathbf{q}$ on the right. Hence the contrasts for this model are functions of the from $\sum_i b_i \boldsymbol{\alpha}'_i \mathbf{q}$, where $\sum_i b_i = 0$ and $\mathbf{q} \in R^r$. Similarly the simultaneous confidence intervals for contrasts for the multivariate model can be derived for those in the univariate model by multiplying both the contrast and the estimated contrast on the left by $\mathbf{q}$ and replacing

$$\hat{\sigma}^2(p - k)F^{\alpha}_{N-p,\, p-1} \quad \text{with } (N - p)\mathbf{q}'\hat{\boldsymbol{\Sigma}}\mathbf{q}c_3^{\alpha} = \mathbf{q}'\mathbf{S}_3\mathbf{q}c_3^{\alpha}.$$

Therefore, the simultaneous confidence intervals for the set of all contrasts for this model is given by

$$\sum_i b_i \boldsymbol{\alpha}_i' \mathbf{q} \in \sum_i b_i \overline{\mathbf{Y}}_{i.} \mathbf{q} \pm \left(c_3^{\alpha} \mathbf{q}' \mathbf{S}_3 \mathbf{q} \sum_i \frac{b_i^2}{n_i} \right)^{1/2}.$$

As a final example, we consider the two-way analysis of variance model, in which we observe $\mathbf{Y}_{ijk}$ independent, with

$$\mathbf{Y}_{ijk} \sim N_r(\boldsymbol{\theta} + \alpha_i + \boldsymbol{\beta}_j + \gamma_{ij}, \boldsymbol{\Sigma}), \qquad i = 1, \ldots, d; \qquad j = 1, \ldots, c;$$

$$k = 1, \ldots, m. \qquad \sum_i \boldsymbol{\alpha}_i = \mathbf{0}, \qquad \sum_j \boldsymbol{\beta}_j = \mathbf{0}, \quad \sum_i \boldsymbol{\gamma}_{ij} = \mathbf{0}, \qquad \sum_j \boldsymbol{\gamma}_{ij} = \mathbf{0}.$$

Suppose that we want to test that the $\boldsymbol{\alpha}_i = 0$. Then, we see that

$$\mathbf{S}_2 = c \sum_i (\overline{\mathbf{Y}}_{i..} - \overline{\mathbf{Y}}_{...})(\overline{\mathbf{Y}}_{i..} - \overline{\mathbf{Y}}_{...})', \qquad \mathbf{S}_3 = \sum_i \sum_j \sum_k (\mathbf{Y}_{ijk} - \overline{\mathbf{Y}}_{ij.})(\mathbf{Y}_{ijk} - \overline{\mathbf{Y}}_{ij.})'.$$

In this model, $p = dc$, $p - k = d - 1$, so that b is the minimum of $a - 1$ and r.

Finally consider testing that the $\gamma_{ij} = 0$. Then $\mathbf{S}_3$ is the same as above,

$$\mathbf{S}_2 = m \sum_i \sum_j (\overline{\mathbf{Y}}_{ij.} - \overline{\mathbf{Y}}_{i..} - \overline{\mathbf{Y}}_{.j.} + \overline{\mathbf{Y}}_{...})(\overline{\mathbf{Y}}_{ij.} - \overline{\mathbf{Y}}_{i..} - \overline{\mathbf{Y}}_{.j.} + \overline{\mathbf{Y}}_{...})'$$

and b is the minimum of $(d - 1)(c - 1)$ and r.

19.7. THE GENERALIZED REPEATED MEASURES MODEL.

The procedures derived in this chapter can often be used to test hypotheses in the generalized repeated measures model that occurs in the repeated measures model when we replace the assumption that $\boldsymbol{\Sigma} = \sigma^2 \mathbf{A}(\rho) > 0$ with the weaker assumption that $\boldsymbol{\Sigma} > 0$. In this section we illustrate this sort of application with Example 3 of Chapter 14. In that example, we observe $\mathbf{Y}_{jk} = (Y_{1jk}, \ldots, \mathbf{Y}_{djk})'$, where

$$\mathbf{Y}_{ijk} = \theta + \alpha_i + \beta_j + \gamma_{ij} + e_{ijk}; \qquad i = 1, \ldots, d; \qquad j = 1, \ldots, c;$$

$$k = 1, \ldots, m; \qquad \sum_i \alpha_i = 0, \qquad \sum_j \beta_j = 0,$$

$$\sum_i \gamma_{ij} = 0, \qquad \sum_j \gamma_{ij} = 0,$$

$$\mathbf{e}_{jk} = (e_{1jk}, \ldots, e_{djk})' \sim N_d(\mathbf{0}, \boldsymbol{\Sigma}),$$

and the $\mathbf{e}_{jk}$ are independent. (Note that α_i, β_j, and γ_{ij} are all scalars for this model.) Now, let

$$\mathbf{Y} = \begin{pmatrix} \mathbf{Y}'_{11} \\ \vdots \\ \mathbf{Y}'_{1m} \\ \mathbf{Y}'_{21} \\ \vdots \\ \mathbf{Y}'_{cm} \end{pmatrix}, \qquad \boldsymbol{\mu} = E\mathbf{Y}.$$

Then $\mathbf{Y} \sim N_{cm,\,d}(\boldsymbol{\mu}, \mathbf{I}, \boldsymbol{\Sigma})$. Let V be the c-dimensional subspace of R^{cm} consisting of vectors whose first m components are the same, whose next m components are the same, etc. Then the only restriction on $\boldsymbol{\mu}$ is that $\boldsymbol{\mu} \in V$, so that this model is a multivariate linear model.

We first consider testing the hypothesis that $\boldsymbol{\beta}_j = 0$. As before, let $\mathbf{1}_a$ be the a-dimensional vector of 1's. Then

$$\boldsymbol{\mu}\mathbf{1}'_d = d \begin{pmatrix} \theta + \beta_1 \\ \vdots \\ \theta + \beta_c \end{pmatrix}.$$

Now, let W_1 be the one-dimensional subspace of V consisting of vectors all of whose elements are the same. Then $\beta_j = 0$ if and only if $\boldsymbol{\mu}\mathbf{1}'_d \in W_2$. Therefore, this hypothesis is in the form of the generalized multivariate linear hypothesis. Also $s = 1$ for this model so that there is a UMP invariant size α test for this model. Let

$$\mathbf{S}_2 = \mathbf{Y}'P_{V \mid W_1}\mathbf{Y} = \sum_i^d m\left(\overline{\mathbf{Y}}_{i.} - \overline{\mathbf{Y}}_{..}\right)\left(\overline{\mathbf{Y}}_{i.} - \overline{\mathbf{Y}}_{..}\right)',$$

$$\mathbf{S}_3 = \mathbf{Y}'P_{V^\perp}\mathbf{Y} = \sum_i \sum_j \left(\mathbf{Y}_{ij} - \overline{\mathbf{Y}}_{i.}\right)\left(\mathbf{Y}_{ij} - \overline{\mathbf{Y}}_{i.}\right)'.$$

(Note that W_1 and V are the same subspaces as considered in the multivariate one-way analysis of variance model.) By Theorem 19.14, we see that a maximal invariant for this model is

$$t = \frac{\mathbf{1}'_d\mathbf{S}_2\mathbf{1}_d}{\mathbf{1}'_d\mathbf{S}_3\mathbf{1}_d}.$$

Since $\dim V \mid W_1 = c - 1$, $\dim V^\perp = cm - c = c(m-1)$, the UMP invariant size α test is given by

$$\phi(t) = \begin{cases} 1 & \text{if } t > \dfrac{c-1}{c(m-1)} F^{\alpha}_{c-1,\, c(m-1)} \\ 0 & \text{if } t \leqslant \dfrac{c-1}{c(m-1)} F^{\alpha}_{c-1,\, c(m-1)} \end{cases}$$

Now, consider testing that $\alpha_i = 0$. Let $\mathbf{X}$ be a basis matrix for the space orthogonal to $\mathbf{1}_d$. Then

$$\boldsymbol{\mu}\mathbf{X} = \begin{bmatrix} \alpha_1 + \gamma_{11}, \ldots, \alpha_d + \gamma_{d1} \\ \alpha_1 + \gamma_{11}, \ldots, \alpha_d + \gamma_{d1} \\ \alpha_1 + \gamma_{12}, \ldots, \alpha_d + \gamma_{d2} \\ \vdots \\ \alpha_1 + \gamma_{1c}, \ldots, \alpha_d + \gamma_{dc} \end{bmatrix} \mathbf{X}.$$

Let W_2 be the $(c-1)$-dimensional subspace consisting of vectors whose components sum to 0. Then $\alpha_i = 0$ if and only if $\boldsymbol{\mu}\mathbf{X} \in W_2$, so that this hypothesis is also in the form of the generalized multivariate linear hypothesis. In this case, $p - k = 1$, so that there is again a UMP invariant size α F-test for this hypothesis. A maximal invariant for this problem is the one nonzero root of

$$|\mathbf{X}'\mathbf{Y}'\mathbf{P}_{V \mid W_2}\mathbf{Y}\mathbf{X} - t\mathbf{X}'\mathbf{S}_3\mathbf{X}| = 0.$$

In Exercise B13, you are asked to show that this root is given by

$$t = \frac{\mathbf{1}'_{cm}\mathbf{Y}\mathbf{X}(\mathbf{X}'\mathbf{S}_3\mathbf{X})^{-1}\mathbf{X}'\mathbf{Y}'\mathbf{1}_{cm}}{cm} \tag{19.13}$$

and the UMP invariant size α test for this model is given by

$$\phi(t) = \begin{cases} 1 & \text{if } t > \dfrac{(d-1)}{c(m-1)-d} F^{\alpha}_{d-1, c(m-1)-d} \\ 0 & \text{if } t \leqslant \dfrac{(d-1)}{c(m-1)-d} F^{\alpha}_{d-1, c(m-1)-d} \end{cases}$$

Finally, consider testing that $\gamma_{ij} = 0$. Let W_1 be defined as in testing that $\beta_j = 0$. Then $\gamma_{ij} = 0$ if and only if $\boldsymbol{\mu}\mathbf{X} \in W_1$. Therefore, this hypothesis is also a generalized multivariate linear hypothesis. For this problem b^* is the minimum of $c - 1$ and $d - 1$, so that there is no UMP invariant size α test. A maximal invariant is the set of nonzero roots of

$$|\mathbf{X}'\mathbf{S}_2\mathbf{X} - t^*\mathbf{X}'\mathbf{S}_3\mathbf{X}| = 0$$

where $\mathbf{S}_2$ and $\mathbf{S}_3$ are defined above. The tests ϕ_1^*, ϕ_2^*, ϕ_3^*, and ϕ_4^* are all available for this hypothesis.

It might appear that the maximal invariants for the last two testing problems could depend on the choice of the basis matrix **X**. In Exercise B14, you are asked to show that they do not. In particular, it is not necessary to choose an orthonormal basis matrix.

We now look at the problem of testing that $\alpha_i = 0$ in more detail. We consider the tests that have been derived for this problem for three different models: the generalized repeated measures model, the repeated measures model and the ordinary linear model. For all three models there is a UMP invaariant size α test that is an F-test. If the ordinary linear model is correct, then the noncentrality parameters for the three F-statistics are the same, and if the repeated measures model is correct, then the noncentrality parameters for the F-statistics for the repeated measures model and the generalized repeated measures model are the same. We also note that the numerator degrees of freedom is the same for all three models $(d - 1)$. However, the denominator degrees of freedom for the ordinary linear model is $dc(m - 1)$, the denominator degrees of freedom for the repeated measures model is $c(d - 1)(m - 1)$, and the denominator degrees of freedom for the generalized repeated measures model is $c(m - 1) - d$. We note that the drop in degrees of freedom from the ordinary linear model to the repeated measures model is much smaller than the drop from the repeated measures model to the generalized repeated measures model. All these facts indicate that if the ordinary linear model is true and we analyze the data using the repeated measures model we lose a moderate amount of power. However, if the repeated measures model is true, and we analyze the data using the generalized repeated measures model we lose considerably more power.

Now consider testing that the $\beta_j = 0$. For this problem, we see that the drop in denominator degrees of freedom from the ordinary linear model to the repeated measures model is much greater than it is for the example considered above, $[dc(m - 1)$ to $c(m - 1)]$, but there is no drop at all from the repeated measures model to the generalized repeated measures model. This example indicates that for testing type B hypotheses, the major drop in power occurs between the repeated measures model and the generalized repeated measures model, while for testing type A hypotheses, the only drop occurs between the ordinary linear model and the repeated measures model.

The test that $\beta_j = 0$ for the generalized repeated measures model treated in this section is the same as the test for the repeated measures derived in Chapter 14 (see Exercise B12). In fact, it can be shown that for any type A hypothesis the test for the generalized repeated measures model is the same as the test for repeated measures model. (Once we have reduced to the Z_k, the averages of the observations on particular individuals, the covariance structure on the observations for a given individual is irrelevant, as long as it is the same for each individual).

19.8. THE ASYMPTOTIC VALIDITY OF THE PROCEDURES.

We now consider the asymptotic validity of the procedures defined in Sections 19.2–19.4 for testing the multivariate linear hypothesis and generalized multivariate linear hypothesis and for finding simultaneous confidence intervals for contrasts associated with those testing problems. We show that these procedures are asymptotically insensitive to the normal assumption used in deriving them in exactly the same situation that procedures for the univariate linear model are asymptotically unaffected by nonnormality. That is, we show that these procedures are asymptotically valid as long as Huber's condition is satisfied. (See Section 10.3 for a discussion of Huber's condition.) The derivations in this section are very similar to those in Chapter 10 for the univariate model.

We first state the following elementary generalization of the Cramer-Wold theorem (Theorem 10.2e.).

THEOREM 19.15. Let $\mathbf{U}_1, \mathbf{U}_2, \ldots,$ be a sequence of $m \times q$ random matrices. Then $\mathbf{U}_n \xrightarrow{d} N_{m,q}(\boldsymbol{\theta}, \boldsymbol{\Xi}, \boldsymbol{\Sigma})$ if and only if

$$\operatorname{tr} \mathbf{T}'\mathbf{U}_n \xrightarrow{d} N_1(\operatorname{tr} \mathbf{T}'\boldsymbol{\theta}, \operatorname{tr} \boldsymbol{\Xi}\mathbf{T}\boldsymbol{\Sigma}\mathbf{T}')$$

for all $m \times q$ matrices $\mathbf{T}$.

PROOF. By Theorem 17.13a, if $\mathbf{X} \sim N_{m,q}(\boldsymbol{\theta}, \boldsymbol{\Xi}, \boldsymbol{\Sigma})$ then

$$\operatorname{tr} \mathbf{T}'\mathbf{X} \sim N_1(\operatorname{tr} \mathbf{T}'\boldsymbol{\theta}, \operatorname{tr} \mathbf{T}'\boldsymbol{\Xi}\mathbf{T}\boldsymbol{\Sigma}).$$

Therefore, if $\mathbf{U}_n \xrightarrow{d} N_{m,q}(\boldsymbol{\theta}, \boldsymbol{\Xi}, \boldsymbol{\Sigma})$, then $\operatorname{tr} \mathbf{T}'\mathbf{U}_n \xrightarrow{d} N_1(\operatorname{tr} \mathbf{T}'\boldsymbol{\theta}, \operatorname{tr} \mathbf{T}'\boldsymbol{\Xi}\mathbf{T}\boldsymbol{\Sigma})$. Conversely, suppose that $Y_n = \operatorname{tr} \mathbf{T}'\mathbf{U}_n \xrightarrow{d} N_1(\operatorname{tr} \mathbf{T}'\boldsymbol{\theta}, \operatorname{tr} \mathbf{T}'\boldsymbol{\Xi}\mathbf{T}\boldsymbol{\Sigma})$ for all $\mathbf{T}$. Suppose that $\mathbf{U}_n$ has moment generating function $M_n(\mathbf{T})$. (If $\mathbf{U}_n$ has no moment generating function, then the characteristic can be substituted for it.) Let Y_n have moment generating function $M_n^*(t)$. Then

$$M_n(\mathbf{T}) = \mathbf{E} \exp(\operatorname{tr} \mathbf{T}'\mathbf{U}_n) = E \exp Y_n = M_n^*(1) \to \exp(\operatorname{tr} \mathbf{T}'\boldsymbol{\theta} + \operatorname{tr} \mathbf{T}'\boldsymbol{\Xi}\mathbf{T}\boldsymbol{\Sigma}/2).$$

Therefore, the moment generating function of $\mathbf{U}_n$ converges to that for $N_{m,q}(\boldsymbol{\theta}, \boldsymbol{\Xi}, \boldsymbol{\Sigma})$ and hence $\mathbf{U}_n \xrightarrow{d} N_{m,q}(\boldsymbol{\theta}, \boldsymbol{\Xi}, \boldsymbol{\Sigma})$. □

In this section we consider a sequence of models (over n) in which we observe the $n \times r$ matrix $\mathbf{Y}_n$ such that

$$\mathbf{Y}_n = \boldsymbol{\mu}_n + \mathbf{e}_n, \ \boldsymbol{\mu}_n \in V_n, \mathbf{e}_n' = (\mathbf{f}_1, \ldots, \mathbf{f}_n)$$

Where $\boldsymbol{\mu}_n$ is an unknown constant matrix, V_n is a known p-dimensional subspace of R^n and $\mathbf{f}_1, \mathbf{f}_2, \ldots,$ is a sequence of independently, identically distributed r-dimensional random row vectors such that

$$E\mathbf{f}_i' = \mathbf{0}, \qquad \operatorname{cov}(\mathbf{f}_i') = \boldsymbol{\Sigma}.$$

(If we add the assumption that the $\mathbf{f}_i$ are normally distributed, then this becomes a sequence of multivariate linear models. We do not add that assumption in this section.) Note that we are assuming that r and p are fixed but n goes to ∞. The results in this section are therefore appropriate when the number of observations n is much larger than the dimension of the subspace p and the dimension of the observation vectors r.

As in Chapter 10, for any matrix $\mathbf{A}$ with components a_{ij}, define

$$m(\mathbf{A}) = \max_{i,j} |a_{ij}|.$$

As in the univariate model, we say that *Huber's condition* is satisfied for this sequence of models if $m(\mathbf{P}_{V_n}) \to 0$ as $n \to \infty$.

The first result derived for this model is a generalization of Theorem 10.3.

THEOREM 19.16. Let $\mathbf{A}_1, \mathbf{A}_2, \ldots,$ be a sequence of constant matrices such that $\mathbf{A}_n$ is $n \times r$, $\operatorname{tr} \mathbf{A}_n \boldsymbol{\Sigma} \mathbf{A}_n{}' = c$ and $m(\mathbf{A}_n) \to 0$. Then

$$\operatorname{tr} \mathbf{A}_n' \mathbf{e}_n \xrightarrow{d} N_1(0, c).$$

PROOF. We follow the proof of Theorem 10.3, using the Lindeberg-Feller and bounded convergence theorems (Theorem 10.9 a and b). Let $m_n = m(\mathbf{A}_n \boldsymbol{\Sigma} \mathbf{A}_n')$. Then $m_n \leqslant r^2 m(\boldsymbol{\Sigma})[m(\mathbf{A}_n)]^2 \to 0$ [see (10.5)]. Let $\mathbf{A}_n' = (\mathbf{a}_{n1}, \ldots, \mathbf{a}_{nn})$. (Note that $\mathbf{a}_{nj}$ is $r \times 1$.) Finally, let $W_{ni} = \mathbf{a}_{ni}{}'\mathbf{f}_i$. Since the $\mathbf{f}_i$ are independent the W_{ni} are independent for fixed n. Also

$$\operatorname{tr} \mathbf{A}_n' \mathbf{e}_n = \sum_{i=1}^{n} \mathbf{a}_{ni}{}'\mathbf{f}_i = \sum_{i=1}^{n} W_{ni}, \qquad EW_{ni} = 0, \qquad \operatorname{var}(W_{ni}) = \mathbf{a}_{ni}{}'\boldsymbol{\Sigma}\mathbf{a}_{ni},$$

$$\sum_{i=1}^{n} \operatorname{var}(W_{ni}) = \sum_{i=1}^{n} \mathbf{a}_{ni}{}'\boldsymbol{\Sigma}\mathbf{a}_{ni} = \operatorname{tr} \mathbf{A}_n \boldsymbol{\Sigma} \mathbf{A}_n' = c,$$

$$\max_{i \leqslant n} \operatorname{var}(W_{ni}) = \max_{i \leqslant n} \mathbf{a}_{ni}{}'\boldsymbol{\Sigma}\mathbf{a}_{ni} = m_n \to 0.$$

By the Lindeberg-Feller theorem, we will be finished when we show that for all $\epsilon > 0$

$$C_n = E \sum_{i=1}^{n} W_{ni}^2 I_\epsilon(W_{ni}) \to 0, \qquad I_\epsilon(W_{ni}) = \begin{cases} 1 & \text{if } |W_{ni}| > \epsilon \\ 0 & \text{if } |W_{ni}| \leqslant \epsilon \end{cases}.$$

Let $U = \mathbf{f}_1' \boldsymbol{\Sigma}^{-1} \mathbf{f}_1$. In Exercise B20, it is shown that

$$0 \leqslant C_n \leqslant cEUI_{\epsilon^2/m_n}(U).$$

Also, $UI_{\epsilon^2/m_n}(U) \leqslant U$ and $EU = E\|\boldsymbol{\Sigma}^{-1/2}\mathbf{f}_1\|^2 = p$. Therefore, by the bounded convergence theorem

$$\lim_{n\to\infty} EUI_{\epsilon^2/m_n}(U) = E \lim_{n\to\infty} UI_{\epsilon^2/m_n}(U) = 0.$$

Hence C_n is trapped between 0 and a sequence which is converging to 0 and must also converge to 0. □

We now generalize Theorem 10.4.

THEOREM 19.17. Let $W_1, W_2, \ldots,$ be a sequence of k-dimensional subspaces such that $W_n \subset R^n$ and $m(\mathbf{P}_{W_n}) \to 0$. Then

$$\mathbf{e}_n'\mathbf{P}_{W_n}\mathbf{e}_n \xrightarrow{d} W_r(k, \boldsymbol{\Sigma}).$$

PROOF. Let $\mathbf{X}_n$ be an orthonormal basis matrix for W_n. We first show that $\mathbf{X}_n'\mathbf{e}_n \xrightarrow{d} N_{k,r}(\mathbf{0}, \mathbf{I}, \boldsymbol{\Sigma})$. By Theorem 19.15, this is equivalent to showing that $\operatorname{tr}\mathbf{T}'\mathbf{X}_n'\mathbf{e}_n \xrightarrow{d} N_1(0, \operatorname{tr}\mathbf{T}\boldsymbol{\Sigma}\mathbf{T}')$ for all $k \times r$ matrices $\mathbf{T}$. Let $\mathbf{A}_n = \mathbf{X}_n\mathbf{T}$. Then

$$\operatorname{tr}\mathbf{T}'\mathbf{X}_n'\mathbf{e}_n = \operatorname{tr}\mathbf{A}_n'\mathbf{e}_n, \operatorname{tr}\mathbf{A}_n\boldsymbol{\Sigma}\mathbf{A}_n' = \operatorname{tr}\mathbf{T}\boldsymbol{\Sigma}\mathbf{T}'\mathbf{X}_n'\mathbf{X}_n = \operatorname{tr}\mathbf{T}\boldsymbol{\Sigma}\mathbf{T}',$$

since $\mathbf{X}_n$ is an orthonormal basis matrix for W_n. Also,

$$m(\mathbf{A}_n) \leqslant km(\mathbf{T})m(\mathbf{X}_n) \leqslant km(\mathbf{T})\left[m(\mathbf{P}_{W_n})\right]^{1/2} \to 0.$$

By Theorem 19.16, therefore, $\operatorname{tr}\mathbf{T}'\mathbf{X}_n'\mathbf{e}_n \xrightarrow{d} N_1(0, \operatorname{tr}\mathbf{T}\boldsymbol{\Sigma}\mathbf{T}')$ and by Theorem 19.15, $\mathbf{X}_n'\mathbf{e}_n \xrightarrow{d} N_{k,r}(\mathbf{0}, \mathbf{I}, \boldsymbol{\Sigma})$. Finally, by Theorem 10.2d, we see that

$$\mathbf{e}_n'\mathbf{P}_{W_n}\mathbf{e}_n = (\mathbf{X}_n'\mathbf{e}_n)'(\mathbf{X}_n'\mathbf{e}_n) \xrightarrow{d} W_r(k, \boldsymbol{\Sigma}). \quad \square$$

Now, as in previous sections, let

$$\hat{\boldsymbol{\Sigma}}_n = \frac{1}{n-p}\mathbf{Y}_n'\mathbf{P}_{V_n^\perp}\mathbf{Y}_n$$

THEOREM 19.18. If $m(\mathbf{P}_{V_n}) \to 0$, then $\hat{\boldsymbol{\Sigma}}_n \xrightarrow{p} \boldsymbol{\Sigma}$.

PROOF. We first note that since $\boldsymbol{\mu}_n \in V_n$,

$$\hat{\boldsymbol{\Sigma}}_n = \frac{1}{n-p}\mathbf{e}_n'\mathbf{P}_{V_n^\perp}\mathbf{e}_n = \frac{n}{n-p}\left(\frac{1}{n}\mathbf{e}_n'\mathbf{e}_n - \frac{1}{n}\mathbf{e}_n'\mathbf{P}_{V_n}\mathbf{e}_n\right).$$

By the weak law of large numbers (Theorem 18.18a.)

$$\frac{1}{n}\mathbf{e}_n'\mathbf{e}_n = \frac{1}{n}\sum_{i=1}^{n}\mathbf{f}_i\mathbf{f}_i' \xrightarrow{p} \boldsymbol{\Sigma}.$$

By Theorem 19.17, $\mathbf{e}_n'\mathbf{P}_{V_n}\mathbf{e}_n \xrightarrow{d} W_r(p, \mathbf{\Sigma})$ and therefore

$$\frac{1}{n}\mathbf{e}_n'\mathbf{P}_{V_n}\mathbf{e}_n \xrightarrow{p} \mathbf{0}.$$

Finally, $(n-p)/n \to 1$. Therefore, $\hat{\mathbf{\Sigma}}_n \to \mathbf{\Sigma}$. □

Now, let $R_{a,c}$ be the distribution of the nonzero roots $r_1 \geqslant \cdots \geqslant r_b$ of $|\mathbf{R} - r\mathbf{I}| = 0$ where $\mathbf{R} \sim W_a(c, \mathbf{I})$ and $b = \min(a, c)$.

THEOREM 19.19. Let $W_1, W_2, \ldots,$ be a sequence of k-dimensional subspaces $W_n \subset V_n$. Let $\mathbf{T}_n = \mathbf{e}_n'\mathbf{P}_{V_n|W_n}\mathbf{e}_n$ and let $u_1 \geqslant, \ldots, \geqslant u_b$ be the nonzero roots of

$$|\mathbf{T}_n - u\hat{\mathbf{\Sigma}}_n| = 0$$

[where $b = \min(r, p-k)$]. If $m(\mathbf{P}_{V_n}) \to 0$, then

$$(u_1, \ldots, u_b) \xrightarrow{d} R_{r,\, p-k}.$$

PROOF. We note first that $m(\mathbf{P}_{V_n|W_n}) = m(\mathbf{P}_{V_n} - \mathbf{P}_{W_n}) \leqslant m(\mathbf{P}_{V_n})$ and therefore $m(\mathbf{P}_{V_n|W_n}) \to 0$. By Theorem 19.17, $\mathbf{T}_n \xrightarrow{d} W_r(p-k, \mathbf{\Sigma})$ and by Theorem 19.18, $\hat{\mathbf{\Sigma}}_n \xrightarrow{p} \mathbf{\Sigma}$. Therefore,

$$\mathbf{\Sigma}^{-1/2}\mathbf{T}_n\mathbf{\Sigma}^{-1/2} \xrightarrow{d} W(p-k, \mathbf{I}), \mathbf{\Sigma}^{-1/2}\hat{\mathbf{\Sigma}}_n\mathbf{\Sigma}^{-1/2} \xrightarrow{p} \mathbf{I}.$$

Finally, $(u_1, \ldots, u_b)$ are the nonzero roots of

$$|\mathbf{\Sigma}^{-1/2}\mathbf{T}_n\mathbf{\Sigma}^{-1/2} - u\mathbf{\Sigma}^{-1/2}\hat{\mathbf{\Sigma}}_n\mathbf{\Sigma}^{-1/2}| = |\mathbf{\Sigma}^{-1/2}||\mathbf{T}_n - \mathbf{u}\hat{\mathbf{\Sigma}}_n||\mathbf{\Sigma}^{-1/2}| = 0.$$

Since $(u_1, \ldots, u_b)$ is a continuous function of

$$(\mathbf{\Sigma}^{-1/2}\mathbf{T}_n\mathbf{\Sigma}^{-1/2}, \mathbf{\Sigma}^{-1/2}\hat{\mathbf{\Sigma}}_n\mathbf{\Sigma}^{-1/2})$$

the result follows from Theorem 18.16d. □

COROLLARY 1. If $m(\mathbf{P}_{V_n}) \to 0$, then the size of any invariant test of a multivariate linear hypothesis is asymptotically unaffected by nonnormality.

PROOF. Let $\mathbf{S}_2(n) = \mathbf{Y}_n'\mathbf{P}_{V_n|W_n}\mathbf{Y}_n$, $\mathbf{S}_3(n) = \mathbf{Y}_n'\mathbf{P}_{V_n^\perp}\mathbf{Y}_n = (n-p)\hat{\mathbf{\Sigma}}_n$. Any invariant procedure for testing that multivariate linear hypothesis that $\boldsymbol{\mu}_n \in W_n$ depends on the nonzero roots $t_1 \geqslant \cdots \geqslant t_b$ of

$$0 = |\mathbf{S}_2(n) - t\mathbf{S}_3(n)|$$

However, $\mathbf{S}_3(n) = (n-p)\hat{\mathbf{\Sigma}}_n$ and, under the null hypothesis, $\mathbf{S}_2(n) = \mathbf{T}_n$. Therefore, under the null hypothesis, $(n-p)(t_1, \ldots, t_b) = (u_1, \ldots, u_b)$.

Since the asymptotic distribution of $(u_1, \ldots, u_b)$ is unaffected by nonnormality, neither is the asymptotic null distribution of $(t_1, \ldots, t_b)$. Therefore, the asymptotic null distribution of any invariant test is unaffected by nonnormality. □

COROLLARY 2. If $m(\mathbf{P}_{V_n}) \to 0$, then the confidence coefficient for the simultaneous confidence intervals for contrasts associated with testing a multivariate linear hypothesis is also asymptotically unaffected by nonnormality.

PROOF. The confidence coefficient for such confidence intervals depends only on the distribution of the largest root u_1 of

$$|\mathbf{T}_n - u(n-p)\hat{\boldsymbol{\Sigma}}_n| = 0$$

and by Theorem 19.19, this distribution is asymptotically unaffected by nonnormality. □

In Exercise B21, these results are extended to show that the size of any invariant test of a generalized multivariate linear hypothesis is also unaffected by nonnormality, nor is the confidence coefficient for the simultaneous confidence intervals for contrasts associated with a generalized multivariate linear hypothesis. Furthermore, by an argument similar to that in Exercise C1 of Chapter 10, we could show that the power functions of these tests are asymptotically unaffected by nonnormality. These results imply that if n is much larger than r and p and the diagonal elements of $\mathbf{P}_{V_n}$ are all small, then the procedures derived in Sections 19.2–19.4 are not too sensitive to the normal assumption used in deriving them.

19.9. JAMES-STEIN ESTIMATION.

We now give a discussion of James-Stein estimation for the multivariate linear model. Although we cannot prove that any estimator improves on the estimator $\hat{\boldsymbol{\mu}}$, we do show that the estimator can be substantiallly improved if we know $\boldsymbol{\Sigma}$. This would indicate that substantial improvement should also be possible for the case of unknown $\boldsymbol{\Sigma}$.

We first discuss the loss function which will be used in this section, which is

$$L(\mathbf{d}; (\boldsymbol{\mu}, \boldsymbol{\Sigma})) = \operatorname{tr}(\mathbf{d} - \boldsymbol{\mu})\boldsymbol{\Sigma}^{-1}(\mathbf{d} - \boldsymbol{\mu})' = \operatorname{tr}(\mathbf{d} - \boldsymbol{\mu})'(\mathbf{d} - \boldsymbol{\mu})\boldsymbol{\Sigma}^{-1},$$

where $\mathbf{d}$ is an estimate of $\boldsymbol{\mu}$, (i.e., $\mathbf{d} \in V$ and $\mathbf{d}$ is $n \times r$). To motivate this loss function, we first look at two special cases that we have already studied, the univariate linear model and the one-sample model. In the univariate model, $r = 1$, and

$$L(\mathbf{d}; (\boldsymbol{\mu}, \Sigma)) = \frac{(\mathbf{d} - \boldsymbol{\mu})'(\mathbf{d} - \boldsymbol{\mu})}{\Sigma}$$

which is the loss function used in Chapter 11 for this model. In the one sample model, V is the one-dimensional subsapce of R^n consisting of vectors all of whose components are the same. Therefore

$$\boldsymbol{\mu} = \begin{bmatrix} \boldsymbol{\theta} \\ \vdots \\ \boldsymbol{\theta} \end{bmatrix}, \qquad \mathbf{d} = \begin{bmatrix} \mathbf{D} \\ \vdots \\ \mathbf{D} \end{bmatrix}$$

for some $\boldsymbol{\theta}$ and $\mathbf{D}$. Hence

$$L(\mathbf{d}; (\boldsymbol{\theta}, \boldsymbol{\Sigma})) = n(\mathbf{D} - \boldsymbol{\theta})\boldsymbol{\Sigma}^{-1}(\mathbf{D} - \boldsymbol{\theta})'$$

which is the loss function used in Chapter 18. Therefore, L reduces to the loss function discussed earlier in the two cases that have already been studied. In Exercise B15, you show that this loss function is the Mahalanobis distance for this situation.

We now find an estimator for $\boldsymbol{\mu}$ that is better than $\hat{\boldsymbol{\mu}}$ for the case of known $\boldsymbol{\Sigma}$. We look at the estimator

$$\hat{\hat{\boldsymbol{\mu}}}_c = \left(1 - \frac{c}{\operatorname{tr} \hat{\boldsymbol{\mu}}\boldsymbol{\Sigma}^{-1}\hat{\boldsymbol{\mu}}'}\right)\hat{\boldsymbol{\mu}}.$$

We assume that $rp > 2$, and show that the optimal choice for c is $rp - 2$. Although it is hard to imagine a situation in which we would know $\boldsymbol{\Sigma}$, this result indicates that there should be an improved estimator for the case of unknown $\boldsymbol{\Sigma}$.

THEOREM 19.20. a. $R(\hat{\boldsymbol{\mu}}; (\boldsymbol{\mu}, \boldsymbol{\Sigma})) = rp$

b. $R(\hat{\hat{\boldsymbol{\mu}}}_c; (\boldsymbol{\mu}, \boldsymbol{\Sigma}))$ is minimized for $c_0 = rp - 2$,

c. $R(\hat{\hat{\boldsymbol{\mu}}}_{c_0}; (\boldsymbol{\mu}, \boldsymbol{\Sigma})) = rp - (rp - 2)^2 E(rp - 2 + 2K)^{-1}$, where K has a Poisson distribution with mean $\delta = \frac{1}{2}\operatorname{tr} \boldsymbol{\mu}\boldsymbol{\Sigma}^{-1}\boldsymbol{\mu}'$.

PROOF. a. See Exercise B16.

b. and c. We now find $R(\hat{\hat{\boldsymbol{\mu}}}_c; (\boldsymbol{\mu}, \boldsymbol{\Sigma}))$.

$$L(\hat{\hat{\boldsymbol{\mu}}}_c; (\boldsymbol{\mu}, \boldsymbol{\Sigma})) = \operatorname{tr}(\hat{\boldsymbol{\mu}} - \boldsymbol{\mu})\boldsymbol{\Sigma}^{-1}(\hat{\boldsymbol{\mu}} - \boldsymbol{\mu}) - 2c\frac{\operatorname{tr} \hat{\boldsymbol{\mu}}\boldsymbol{\Sigma}^{-1}(\hat{\boldsymbol{\mu}} - \boldsymbol{\mu})'}{\operatorname{tr} \hat{\boldsymbol{\mu}}\boldsymbol{\Sigma}^{-1}\hat{\boldsymbol{\mu}}'} + \frac{c^2}{\operatorname{tr} \hat{\boldsymbol{\mu}}\boldsymbol{\Sigma}^{-1}\hat{\boldsymbol{\mu}}'}.$$

By part a, $E \operatorname{tr}(\hat{\boldsymbol{\mu}} - \boldsymbol{\mu})\boldsymbol{\Sigma}^{-1}(\hat{\boldsymbol{\mu}} - \boldsymbol{\mu})' = rp$. To compute the other parts of the risk, let $\mathbf{X}$ be an orthonormal basis matrix for V. Define

$$\mathbf{U} = \begin{pmatrix} \mathbf{U}_1 \\ \vdots \\ \mathbf{U}_p \end{pmatrix} = \mathbf{X}'\mathbf{Y}\boldsymbol{\Sigma}^{-1/2}, \qquad \boldsymbol{\nu} = \begin{pmatrix} \boldsymbol{\nu}_1 \\ \vdots \\ \boldsymbol{\nu}_p \end{pmatrix} = \mathbf{X}'\boldsymbol{\mu}\boldsymbol{\Sigma}^{-1/2}.$$

Then $\mathbf{U} \sim N_{p,r}(\boldsymbol{\nu}, \mathbf{I}, \mathbf{I})$. Let

$$\tilde{\mathbf{U}} = \begin{pmatrix} \mathbf{U}_1' \\ \vdots \\ \mathbf{U}_p' \end{pmatrix}, \qquad \tilde{\boldsymbol{\nu}} = \begin{pmatrix} \boldsymbol{\nu}_1' \\ \vdots \\ \boldsymbol{\nu}_p' \end{pmatrix}.$$

Then

$$\tilde{\mathbf{U}} \sim N_{rp}(\tilde{\boldsymbol{\nu}}, \mathbf{I}).$$

We note that

$$\begin{aligned} \operatorname{tr} \hat{\boldsymbol{\mu}}'\boldsymbol{\Sigma}^{-1}\hat{\boldsymbol{\mu}} &= \operatorname{tr} \mathbf{X}\mathbf{X}'\mathbf{Y}\boldsymbol{\Sigma}^{-1}\mathbf{Y}'\mathbf{X}\mathbf{X}' = \operatorname{tr} \mathbf{X}'\mathbf{Y}\boldsymbol{\Sigma}^{-1/2}(\mathbf{X}\mathbf{Y}\boldsymbol{\Sigma}^{-1/2})' \\ &= \operatorname{tr} \mathbf{U}\mathbf{U}' = \sum_i \mathbf{U}_i\mathbf{U}_i' = \tilde{\mathbf{U}}'\tilde{\mathbf{U}} \end{aligned}$$

(since $\mathbf{X}'\mathbf{X} = \mathbf{I}$). Similarly,

$$\operatorname{tr} \hat{\boldsymbol{\mu}}\boldsymbol{\Sigma}^{-1}(\hat{\boldsymbol{\mu}} - \boldsymbol{\mu}) = \tilde{\mathbf{U}}'(\tilde{\mathbf{U}} - \tilde{\boldsymbol{\nu}}).$$

Therefore, by Lemma 10.2, we see that

$$\frac{E \operatorname{tr} \hat{\boldsymbol{\mu}}\boldsymbol{\Sigma}^{-1}(\hat{\boldsymbol{\mu}} - \boldsymbol{\mu})'}{\operatorname{tr} \hat{\boldsymbol{\mu}}\boldsymbol{\Sigma}^{-1}\hat{\boldsymbol{\mu}}} = E\frac{\tilde{\mathbf{U}}'(\tilde{\mathbf{U}} - \tilde{\boldsymbol{\nu}})}{\tilde{\mathbf{U}}'\tilde{\mathbf{U}}} = E\frac{rp-2}{rp-2+2K},$$

$$E\frac{1}{\operatorname{tr} \hat{\boldsymbol{\mu}}\boldsymbol{\Sigma}^{-1}\hat{\boldsymbol{\mu}}} = E\frac{1}{\tilde{\mathbf{U}}'\tilde{\mathbf{U}}} = E\frac{1}{rp-2+2K}$$

where K has a Poisson distribution with mean

$$\begin{aligned} \tfrac{1}{2}\tilde{\boldsymbol{\nu}}\tilde{\boldsymbol{\nu}} &= \tfrac{1}{2}\operatorname{tr} \boldsymbol{\nu}\boldsymbol{\nu}' = \tfrac{1}{2}\operatorname{tr} \mathbf{X}'\boldsymbol{\mu}\boldsymbol{\Sigma}^{-1}\boldsymbol{\mu}'\mathbf{X} = \tfrac{1}{2}\operatorname{tr} \mathbf{X}\mathbf{X}'\boldsymbol{\mu}\boldsymbol{\Sigma}^{-1}\boldsymbol{\mu}' \\ &= \tfrac{1}{2}\operatorname{tr} \mathbf{P}_V\boldsymbol{\mu}\boldsymbol{\Sigma}^{-1}\boldsymbol{\mu}' = \tfrac{1}{2}\operatorname{tr} \boldsymbol{\mu}\boldsymbol{\Sigma}^{-1}\boldsymbol{\mu}' = \delta, \end{aligned}$$

(since $\boldsymbol{\mu} \in V$). Therefore

$$R(\hat{\hat{\boldsymbol{\mu}}}_c; (\boldsymbol{\mu}, \boldsymbol{\Sigma})) = rp - \left(2c(rp-2) - c^2\right)E\frac{1}{p-2+2K}.$$

This is minimized for $c_0 = rp - 2$, and

$$R(\hat{\hat{\boldsymbol{\mu}}}_{c_0}; (\boldsymbol{\mu}, \boldsymbol{\Sigma})) = rp - (rp-2)^2 E\frac{1}{p-2+2K}. \quad \square$$

From this theorem, we see that if $rp > 2$, we can improve on the estimator $\hat{\boldsymbol{\mu}}$, at least when $\boldsymbol{\Sigma}$ is known. We note that $R(\hat{\boldsymbol{\mu}}_c; (\mathbf{0}, \boldsymbol{\Sigma})) = 2$, whereas $R(\hat{\boldsymbol{\mu}}; (\mathbf{0}, \boldsymbol{\Sigma})) = rp$ so that if $\hat{\boldsymbol{\mu}}$ is near 0 and rp is much greater then 2, we can make a substantial improvement on $\hat{\boldsymbol{\mu}}$ when $\boldsymbol{\Sigma}$ is known. When $\boldsymbol{\Sigma}$ is not known, the result suggests looking at estimators of the form

$$\left(1 - \frac{c}{\operatorname{tr} \hat{\boldsymbol{\mu}}'\hat{\boldsymbol{\Sigma}}^{-1}\hat{\boldsymbol{\mu}}}\right)\hat{\boldsymbol{\mu}}$$

and finding the optimal choice for c. In the univariate case, (when $r = 1$) we see from Chapter 11 that the optimal choice is $c = (n - p)(p - 2)/(n - p + 2)$ and in the one-sample model (when $p = 1$) we see that the optimal choice is $c = (n - r)(r - 2)/(n - r + 2)$. At this time I cannot find the risk function for estimators of the above form when $p > 1$ and $r > 1$ and therefore cannot find the optimal choice for c unless $p = 1$ or $r = 1$.

19.10. THE GROWTH CURVES MODEL.

We now consider a generalization of the multivariate linear model suggested in Potthoff and Roy (1964). This generalization is useful in analyzing growth of plants and is called the growth curves model. In this model we observe the $n \times r$ random matrix $\mathbf{Y}$ such that

$$\mathbf{Y} \sim N_{n,r}(\mathbf{X}\boldsymbol{\gamma}\mathbf{Z}, \mathbf{I}, \boldsymbol{\Sigma})$$

where $\mathbf{X}$ and $\mathbf{Z}$ are known $n \times p$ and $q \times r$ matrices of rank p and q, $p < n$, $q < r$. $\boldsymbol{\gamma}$ is an unknown $p \times q$ parameter and $\boldsymbol{\Sigma} > 0$ is an unknown $r \times r$ parameter. This model is much more difficult to analyze than the multivariate linear model (when $\mathbf{Z} = \mathbf{I}$). There is no complete sufficient statistic for this model and so minimum variance unbaised estimators are unlikely. Formulas for the MLEs have not been determined. In testing hypotheses, the likelihood ratio tests have not been determined (since the MLEs have not) and invariance is not much help. Therefore, we take a somewhat different approach to this model from the approach taken for earlier models. We first find optimal procedures for the growth curves model with known $\boldsymbol{\Sigma}$. We then find a sensible estimator $\hat{\boldsymbol{\Sigma}}$ for $\boldsymbol{\Sigma}$ and substitute $\hat{\boldsymbol{\Sigma}}$ into the procedures derived under the assumption of known $\boldsymbol{\Sigma}$.

Before considering the growth curves model with known $\boldsymbol{\Sigma}$, we discuss some more matrix algebra. Let $\mathbf{A} = (\mathbf{A}_1 \ldots \mathbf{A}_t)$ be an $s \times t$ matrix and let $\mathbf{B} = (b_{ij})$ be a $q \times r$ matrix. Define

$$\tilde{\mathbf{A}} = \begin{pmatrix} \mathbf{A}_1 \\ \vdots \\ \mathbf{A}_t \end{pmatrix}, \qquad \mathbf{A} \otimes \mathbf{B} = \begin{pmatrix} \mathbf{A}b_{11} & \cdots & \mathbf{A}b_{1r} \\ \vdots & & \vdots \\ \mathbf{A}b_{q1} & & \mathbf{A}b_{qr} \end{pmatrix}$$

(Note that $\tilde{\mathbf{A}}$ is an st-dimensional vector and $\mathbf{A}\otimes\mathbf{B}$ is an $sq\times tr$ matrix. $\mathbf{A}\otimes\mathbf{B}$ is called the Kronecker product of $\mathbf{A}$ and $\mathbf{B}$. $\mathbf{A}\otimes\mathbf{B}\neq\mathbf{B}\otimes\mathbf{A}$.) The definitions above can be used to derive the following elementary facts:

$$\tilde{\mathbf{A}}'\tilde{\mathbf{A}}=\operatorname{tr}\mathbf{A}'\mathbf{A}, \widetilde{\mathbf{ABC}}=(\mathbf{A}\otimes\mathbf{C}')\tilde{\mathbf{B}}, (\mathbf{A}\otimes\mathbf{B})(\mathbf{C}\otimes\mathbf{D})=\mathbf{AC}\otimes\mathbf{BD}$$

$$(\mathbf{A}\otimes\mathbf{B})'=\mathbf{A}'\otimes\mathbf{B}', (\mathbf{A}\otimes\mathbf{B})^{-1}=\mathbf{A}^{-1}\otimes\mathbf{B}^{-1}. \tag{19.14}$$

(provided inverses and multiplications are well-defined). In addition, if $\mathbf{U}\sim N_{n,r}(\mathbf{C},\mathbf{D},\mathbf{E})$, then

$$\tilde{\mathbf{U}}\sim N_{nr}(\tilde{\mathbf{C}},\mathbf{D}\otimes\mathbf{E}) \tag{19.15}$$

(see Exercise C1 of Chapter 17). That is, $\tilde{\mathbf{U}}$ has an nr dimensional vector normal distribution with covariance matrix $\mathbf{D}\otimes\mathbf{E}$.

We now apply these results to the growth curves model with known $\boldsymbol{\Sigma}$. By (19.15), we see that

$$\tilde{\mathbf{Y}}\sim N_{nr}(\widetilde{\mathbf{X}\gamma\mathbf{Z}},\mathbf{I}\otimes\boldsymbol{\Sigma}).$$

However, by (19.14)

$$\widetilde{\mathbf{X}\gamma\mathbf{Z}}=(\mathbf{X}\otimes\mathbf{Z}')\tilde{\gamma}.$$

Let $\mathbf{T}=\mathbf{X}\otimes\mathbf{Z}', \mathbf{A}=\mathbf{I}\otimes\boldsymbol{\Sigma}$. Then

$$\tilde{\mathbf{Y}}\sim N_{nr}(\mathbf{T}\gamma,\mathbf{A}).$$

and this model is a generalized linear model (with known σ^2). This model is treated in Exercise B10 of Chapter 13. A complete sufficient statistic for this model is

$$\hat{\tilde{\gamma}}=(\mathbf{T}'\mathbf{A}^{-1}\mathbf{T})^{-1}\mathbf{T}'\mathbf{A}^{-1}\tilde{\mathbf{Y}}. \tag{19.16}$$

Now, let

$$\hat{\gamma}=(\mathbf{X}'\mathbf{X})^{-1}\mathbf{X}'\mathbf{Y}\boldsymbol{\Sigma}^{-1}\mathbf{Z}'(\mathbf{Z}\boldsymbol{\Sigma}^{-1}\mathbf{Z}')^{-1}.$$

Then

$$\hat{\tilde{\gamma}}=\tilde{\hat{\gamma}} \tag{19.17}$$

(see Exercise B18). Hence $\hat{\gamma}$ is an invertible function of $\hat{\tilde{\gamma}}$, and is a complete sufficient statistic for the growth curves model. By Theorem 17.2,

$$\hat{\gamma}\sim N_{p,q}\left(\gamma,(\mathbf{X}'\mathbf{X})^{-1},(\mathbf{Z}\boldsymbol{\Sigma}^{-1}\mathbf{Z}')^{-1}\right).$$

Therefore, $E\hat{\gamma}=\gamma$ and $\hat{\gamma}$ is the minimum variance unbiased estimator of γ. Also $\hat{\tilde{\gamma}}$ is the MLE of $\tilde{\gamma}$, and $\hat{\gamma}$ is the same function of $\hat{\tilde{\gamma}}$ as γ is of $\tilde{\gamma}$. Therefore, $\hat{\gamma}$ is the MLE of γ in the growth curves model with known $\boldsymbol{\Sigma}$. Now, consider testing that $\mathbf{A}\gamma\mathbf{B}=\mathbf{0}$ where $\mathbf{A}$ and $\mathbf{B}$ are known $s\times p$ and $q\times t$ matrices of rank s and t. By (19.15) $\widetilde{\mathbf{A}\gamma\mathbf{B}}=(\mathbf{A}\otimes\mathbf{B}')\tilde{\gamma}$. Now, let

$\mathbf{C} = \mathbf{A} \otimes \mathbf{B}'$. Testing that $\mathbf{A}\gamma\mathbf{B} = 0$ in the growth curves model with known $\boldsymbol{\Sigma}$ is the same as testing that $\mathbf{C}\tilde{\gamma} = \mathbf{0}$ in the generalized linear model with known variance. A maximal invariant and parameter maximal invariant for this problem are

$$U = \hat{\tilde{\gamma}}'\mathbf{C}'\left(\mathbf{C}(\mathbf{T}'\mathbf{A}^{-1}\mathbf{T})^{-1}\mathbf{C}'\right)^{-1}\mathbf{C}\hat{\tilde{\gamma}}, \delta = \tilde{\gamma}'\mathbf{C}'\left(\mathbf{C}(\mathbf{T}'\mathbf{A}^{-1}\mathbf{T})^{-1}\mathbf{C}'\right)^{-1}\mathbf{C}\tilde{\gamma}$$

and $U \sim \chi^2_{st}(\delta)$. The UMP invariant test for this problem is given by

$$\phi(U) = \begin{cases} 1 & \text{if } U > \chi^{2\,\alpha}_{st} \\ 0 & \text{if } U \leqslant \chi^{2\,\alpha}_{st} \end{cases}. \tag{19.18}$$

In Exercise B18, it is shown that

$$U = \operatorname{tr} \mathbf{A}\hat{\gamma}\mathbf{B}\left(\mathbf{B}'(\mathbf{Z}\boldsymbol{\Sigma}^{-1}\mathbf{Z}')^{-1}\mathbf{B}\right)^{-1}\mathbf{B}'\hat{\gamma}'\mathbf{A}'\left(\mathbf{A}(\mathbf{X}'\mathbf{X})^{-1}\mathbf{A}'\right)^{-1}$$

$$\delta = \operatorname{tr} \mathbf{A}\gamma\mathbf{B}\left(\mathbf{B}'(\mathbf{Z}\boldsymbol{\Sigma}^{-1}\mathbf{Z}')^{-1}\mathbf{B}\right)^{-1}\mathbf{B}'\gamma'\mathbf{A}'\left(\mathbf{A}(\mathbf{X}'\mathbf{X})^{-1}\mathbf{A}'\right)^{-1}. \tag{19.19}$$

We have now proved the following result.

LEMMA 19.21. Let $\hat{\gamma}$, U, δ, and ϕ be defined by (19.16), (19.18), and (19.19). Then $\hat{\gamma}$ is a complete sufficient statistic for the growth curves model with known $\boldsymbol{\Sigma}$ and is the minimum variance unbiased estimator and MLE of γ for that model. U and δ are the maximal invariant and parameter maximal invariant for testing that $\mathbf{A}\gamma\mathbf{B} = \mathbf{0}$ and ϕ is the UMP invariant size α test for that problem.

We now return to the more practical problem in which $\boldsymbol{\Sigma}$ is not known. As mentioned earlier, our approach to that model is to find a "sensible" estimator of $\boldsymbol{\Sigma}$ and substitute that estimator in for $\boldsymbol{\Sigma}$ in the above procedures. Let $\boldsymbol{\beta} = \gamma\mathbf{Z}$. Then $\mathbf{Y} \sim N_{n,r}(\mathbf{X}\boldsymbol{\beta}, \mathbf{I}, \boldsymbol{\Sigma})$. (That is, $\mathbf{Y}$ has the distribution of a coordinatized linear model, although the parameter space is more restricted than for that model since $\boldsymbol{\beta}'$ is restricted to the space spanned by the rows of $\mathbf{Z}$.) Let

$$\hat{\boldsymbol{\beta}} = (\mathbf{X}'\mathbf{X})^{-1}\mathbf{X}'\mathbf{Y}, \qquad \hat{\boldsymbol{\Sigma}} = \frac{1}{n-p}(\mathbf{Y} - \mathbf{X}\hat{\boldsymbol{\beta}})'(\mathbf{Y} - \mathbf{X}\hat{\boldsymbol{\beta}}).$$

Then $(\hat{\boldsymbol{\beta}}, \hat{\boldsymbol{\Sigma}})$ is a sufficient statistic for the growth curves model, since it is sufficient for the linear model in which $\boldsymbol{\beta}$ is unrestricted. It is not however complete for the growth curves model (see Exercise B19). In fact there is no complete sufficient statistic for this model. From results for the multivariate linear model, we see that $\hat{\boldsymbol{\beta}}$ and $\hat{\boldsymbol{\Sigma}}$ are independent and

$$\hat{\boldsymbol{\beta}} \sim N_{p,r}\left(\boldsymbol{\beta}, (\mathbf{X}'\mathbf{X})^{-1}, \boldsymbol{\Sigma}\right), \qquad \hat{\boldsymbol{\Sigma}} \sim W_p\left(n-p, \frac{1}{n-p}\boldsymbol{\Sigma}\right).$$

Therefore, $\hat{\boldsymbol{\Sigma}}$ is an unbiased estimator of $\boldsymbol{\Sigma}$ and $\tilde{\boldsymbol{\Sigma}} \xrightarrow{p} \boldsymbol{\Sigma}$ as $n \to \infty$ (see Lemma 19.9). If n is large, then $\hat{\boldsymbol{\Sigma}}$ would be a "sensible" estimator of $\boldsymbol{\Sigma}$. We now consider estimating $\boldsymbol{\gamma}$. If we substitute $\hat{\boldsymbol{\Sigma}}$ in for $\boldsymbol{\Sigma}$ in $\hat{\gamma}$ we get the estimator

$$\hat{\hat{\gamma}} = (\mathbf{X}'\mathbf{X})^{-1}\mathbf{X}'\mathbf{Y}\hat{\boldsymbol{\Sigma}}^{-1}\mathbf{Z}'(\mathbf{Z}\hat{\boldsymbol{\Sigma}}^{-1}\mathbf{Z}')^{-1} = \hat{\boldsymbol{\beta}}\hat{\boldsymbol{\Sigma}}^{-1}\mathbf{Z}'(\mathbf{Z}\hat{\boldsymbol{\Sigma}}^{-1}\mathbf{Z}')^{-1}.$$

LEMMA 19.22. $\hat{\hat{\gamma}}$ is an unbiased estimator of $\boldsymbol{\gamma}$.

PROOF. Since $\hat{\boldsymbol{\beta}}$ and $\hat{\boldsymbol{\Sigma}}$ are independent,

$$E\hat{\hat{\gamma}} = E\hat{\boldsymbol{\beta}}\mathbf{E}\hat{\boldsymbol{\Sigma}}^{-1}\mathbf{Z}'(\mathbf{Z}\hat{\boldsymbol{\Sigma}}^{-1}\mathbf{Z}')^{-1} = \boldsymbol{\gamma}\mathbf{Z}\mathbf{E}\hat{\boldsymbol{\Sigma}}^{-1}\mathbf{Z}'(\mathbf{Z}\hat{\boldsymbol{\Sigma}}^{-1}\mathbf{Z})^{-1}$$
$$= \boldsymbol{\gamma}\mathbf{E}(\mathbf{Z}\hat{\boldsymbol{\Sigma}}^{-1}\mathbf{Z}')(\mathbf{Z}\hat{\boldsymbol{\Sigma}}^{-1}\mathbf{Z}')^{-1} = \boldsymbol{\gamma}. \quad \square$$

We note that $\hat{\hat{\gamma}}$ is not a minimum variance unbiased estimator of $\boldsymbol{\gamma}$. Since there is no complete sufficient statistic for this model, minimum variance unbiased estimators seem unlikely. However, if n is large, then $\hat{\boldsymbol{\Sigma}}$ would be near $\boldsymbol{\Sigma}$, and this estimator would be near the minimum variance unbiased estimator if we knew $\boldsymbol{\Sigma}$.

Now consider testing that $\mathbf{A}\boldsymbol{\gamma}\mathbf{B} = \mathbf{0}$. Let $\hat{\mathrm{U}}$ be the test statistic computed when we substitute $\hat{\boldsymbol{\Sigma}}$ for $\boldsymbol{\Sigma}$ in U. Then

$$\hat{U} - U \xrightarrow{p} 0, \qquad \hat{U} \xrightarrow{d} \chi^2_{st}(\delta)$$

as $n \to \infty$ (see Exercise C5). Therefore, an approximate size α test for large n is to reject if $\hat{U} > \chi^{2\,\alpha}_{st}$. For large n the power function of this test is nearly the same as the UMP invariant size α test for the model with known $\boldsymbol{\Sigma}$.

19.11. ANOTHER GENERALIZATION OF THE LINEAR MODEL.

One aspect of the multivariate linear model that is often bothersome is that all the columns of the mean matrix $\boldsymbol{\mu}$ are assumed to lie in the same subspace. We now present a generalization of the linear model due to Zellner (1962) in which the columns of $\boldsymbol{\mu}$ are allowed to lie in different subspaces. This generalization is much harder to analyze than the linear model. There is no apparent complete sufficient statistic for this model and hence minimum variance unbiased estimators seem unlikely. The MLEs have not been determined for this model. We limit discussion to estimation for this generalization of the linear model. We follow an approach similar to that of the last section in which we find the optimal estimator of $\boldsymbol{\mu}$ for known $\boldsymbol{\Sigma}$. We then find a sensible estimator of $\boldsymbol{\Sigma}$ and substitute that in for $\boldsymbol{\Sigma}$ in the optimal estimator of $\boldsymbol{\mu}$.

The model we now consider is one in which we observe

$$\mathbf{Y} \sim N_{n,r}(\boldsymbol{\mu}, \mathbf{I}, \boldsymbol{\Sigma}), \qquad \boldsymbol{\mu} = (\boldsymbol{\mu}_1, \ldots, \boldsymbol{\mu}_r), \qquad \boldsymbol{\mu}_i \in V_i, \qquad \boldsymbol{\Sigma} > 0$$

where V_i is a known p_i-dimensional subsapce of R^n, $i = 1, \ldots, r$; and $\boldsymbol{\mu}$ and $\boldsymbol{\Sigma}$ are unknown parameters. Let $\tilde{\mathbf{Y}}$, $\tilde{\boldsymbol{\mu}}$, and $\mathbf{I} \otimes \boldsymbol{\Sigma}$ be defined in (19.14). Let $V = V_1 \times V_2 \times \cdots \times V_r \subset R^{nr}$, that is,

$$V = \left[\begin{pmatrix} \mathbf{v}_1 \\ \vdots \\ \mathbf{v}_r \end{pmatrix}, \mathbf{v}_i \in V_i \right]$$

Then

$$\tilde{\mathbf{Y}} \sim N_{nr}(\tilde{\boldsymbol{\mu}}, \mathbf{I} \otimes \boldsymbol{\Sigma}), \qquad \tilde{\boldsymbol{\mu}} \in V, \qquad \boldsymbol{\Sigma} > 0$$

[see (19.15)]. If $\boldsymbol{\Sigma}$ is known, then this model is a univariate generalized linear model with known σ^2. Let $\mathbf{X}_i$ be a basis matrix for V_i, $i = 1, \ldots, r$; and let

$$\mathbf{X} = \begin{pmatrix} \mathbf{X}_1 & 0 & \cdots & 0 \\ 0 & \mathbf{X}_2 & \cdots & 0 \\ \vdots & \vdots & & \vdots \\ 0 & 0 & \cdots & \mathbf{X}_r \end{pmatrix}.$$

Then $\mathbf{X}$ is a basis for matrix V. A complete sufficient statistic for this generalized linear model is

$$\hat{\tilde{\boldsymbol{\mu}}} = \begin{pmatrix} \hat{\boldsymbol{\mu}}_1 \\ \vdots \\ \hat{\boldsymbol{\mu}}_r \end{pmatrix} = \mathbf{X}\left(\mathbf{X}'(\mathbf{I} \otimes \boldsymbol{\Sigma})^{-1}\mathbf{X}\right)^{-1}\mathbf{X}'(\mathbf{I} \otimes \boldsymbol{\Sigma})^{-1}\mathbf{Y}$$

$$\sim N_{nr}\left(\tilde{\boldsymbol{\mu}}, \mathbf{X}\left(\mathbf{X}'(\mathbf{I} \otimes \boldsymbol{\Sigma})^{-1}\mathbf{X}\right)^{-1}\mathbf{X}'\right)$$

and $\hat{\tilde{\boldsymbol{\mu}}}$ is the minimum variance unbiased estimator and MLE of $\tilde{\boldsymbol{\mu}}$. Therefore, $\hat{\boldsymbol{\mu}}_i$ is the minimum variance unbiased estimator and MLE of $\boldsymbol{\mu}_i$, $i = 1, \ldots, r$ for known $\boldsymbol{\Sigma}$.

Now let $\boldsymbol{\beta}_i$ and $\hat{\boldsymbol{\beta}}_i$ be the unique solutions to $\boldsymbol{\mu}_i = \mathbf{X}_i\boldsymbol{\beta}_i$, $\hat{\boldsymbol{\mu}}_i = \mathbf{X}_i\hat{\boldsymbol{\beta}}_i$. Then $\hat{\boldsymbol{\beta}}_i$ is the minimum variance unbiased estimator and MLE of $\boldsymbol{\beta}_i$. Now,

$$\hat{\boldsymbol{\beta}} = \begin{pmatrix} \hat{\boldsymbol{\beta}}_1 \\ \vdots \\ \hat{\boldsymbol{\beta}}_r \end{pmatrix} = \left[\mathbf{X}'(\mathbf{I} \otimes \boldsymbol{\Sigma})^{-1}\mathbf{X}\right]^{-1}\mathbf{X}'(\mathbf{I} \otimes \boldsymbol{\Sigma})^{-1}\mathbf{Y}.$$

We have now found the optimal estimators of $\boldsymbol{\mu}$ and $\boldsymbol{\beta}$ for this model as long as $\boldsymbol{\Sigma}$ is known.

We now return to the more realistic model in which $\boldsymbol{\Sigma}$ is not known. Let V^* be the smallest subspace containing $V_1, \ldots, V_r$. That is

$$V^* = \{\Sigma_i \mathbf{v}_i : \mathbf{v}_i \in V_i\}$$

(V^* is often called the sum of the V_i. See Exercise C1 of Chapter 2.) Let V^* have dimension $q \leqslant \Sigma p_i$. Then $\mathbf{Y} \sim N_{n,r}(\boldsymbol{\mu}, \mathbf{I}, \boldsymbol{\Sigma})$ and $\boldsymbol{\mu} \in V^*$ (since the columns of $\boldsymbol{\mu}$ are V^*). Let

$$\hat{\boldsymbol{\Sigma}} = \frac{1}{n-q} \mathbf{Y}'\mathbf{P}_{V^\perp}\mathbf{Y}.$$

Then $\hat{\boldsymbol{\Sigma}} \sim W_r(n-q, [1/(n-q)]\boldsymbol{\Sigma})$ and therefore $E\hat{\boldsymbol{\Sigma}} = \boldsymbol{\Sigma}$ and $\hat{\boldsymbol{\Sigma}} \xrightarrow{p} \boldsymbol{\Sigma}$ as $n \to \infty$ (see Lemma 19.9). Therefore, if $n - q$ is large then this estimator would be a sensible estimator of $\boldsymbol{\Sigma}$. (Note that we need that $n - q \geqslant r$ to guarantee the invertibility of $\hat{\boldsymbol{\Sigma}}$.) Let $\mathbf{X}^*$ be a matrix whose columns are a maximal set of linearly independent columns of $(\mathbf{X}_1, \ldots, \mathbf{X}_r)$. Then $\mathbf{X}^*$ is a basis matrix for V^* and

$$\hat{\boldsymbol{\Sigma}} = \frac{1}{n-q} \mathbf{Y}'\left(\mathbf{I} - \mathbf{X}^*(\mathbf{X}^{*\prime}\mathbf{X}^*)^{-1}\mathbf{X}^{*\prime}\right)\mathbf{Y}.$$

Let $\hat{\hat{\boldsymbol{\mu}}}_i$ (or $\hat{\hat{\boldsymbol{\beta}}}_i$) be the estimator of $\boldsymbol{\mu}_i$ (or $\boldsymbol{\beta}_i$) that is created when we substitute $\hat{\boldsymbol{\Sigma}}$ for $\boldsymbol{\Sigma}$ in $\hat{\boldsymbol{\mu}}_i$ (or $\hat{\boldsymbol{\beta}}_i$). $\hat{\hat{\boldsymbol{\mu}}}_i$ (or $\hat{\hat{\boldsymbol{\beta}}}_i$) is not an unbiased estimator of $\boldsymbol{\mu}_i$ (or $\boldsymbol{\beta}_i$). However, if $n - q$ is large, it should be very near the optimal estimator for known $\boldsymbol{\Sigma}$ and should therefore be a sensible estimator for $\boldsymbol{\mu}_i$ (or $\boldsymbol{\beta}_i$).

Another estimator that has been suggested for this model is now given. Let

$$\mathbf{Y} = (\mathbf{Y}_1, \ldots, \mathbf{Y}_r), \qquad \boldsymbol{\Sigma} = \begin{bmatrix} \Sigma_{11} & \cdots & \Sigma_{1r} \\ \vdots & & \vdots \\ \Sigma_{r1} & \cdots & \Sigma_{rr} \end{bmatrix}.$$

Then $\mathbf{Y}_i \sim N_n(\boldsymbol{\mu}_i, \Sigma_{ii}\mathbf{I})$. Therefore, alternative estimators for $\boldsymbol{\mu}_i$ and $\boldsymbol{\beta}_i$ are

$$\hat{\boldsymbol{\mu}}_i^* = \mathbf{P}_{V_i}\mathbf{Y}_i, \qquad \hat{\boldsymbol{\beta}}_i^* = (\mathbf{X}_i'\mathbf{X}_i)^{-1}\mathbf{X}_i'\mathbf{Y}_i.$$

These estimators treat each column of $\mathbf{Y}$ as a separate univariate linear model. Zellner (1962) gives some examples which indicate that these estimators $\hat{\boldsymbol{\mu}}_i^*$ and $\hat{\boldsymbol{\beta}}_i^*$ are inferior to $\hat{\hat{\boldsymbol{\mu}}}_i$ and $\hat{\hat{\boldsymbol{\beta}}}_i$ derived above.

EXERCISES.

Type A

1. Consider a one-way multivariate regression model in which

$$Y_{ij} = \beta_0 + \beta_{1j}X_i + \beta_{2j}X_i^2 + e_{ij}, \qquad i = 1, \ldots, 9; \qquad j = 1, 2$$

where $\mathbf{e}_i = (\mathbf{e}_{i1}, e_{i2})$ are independent, $\mathbf{e}_i' \sim N_2(\mathbf{0}, \boldsymbol{\Sigma})$. Suppose we observe X_i given by $-4, -3, -2, -1, 0, 1, 2, 3, 4$, we observe Y_{i1} given by $2, 8, 9, 5, 7, 6, 5, 6, 7$, and we observe Y_{i2} given by $7, 8, 9, 6, 7, 5, 4, 7, 3$.

(a) Find the minimum variance unbiased estimators of $\boldsymbol{\beta}$ and $\boldsymbol{\Sigma}$.

(b) Use a size 0.05 test to test that $\beta_{02} = \beta_{12} = 0$. Note that

$$\begin{pmatrix} \beta_{02} \\ \beta_{12} \end{pmatrix} = \begin{pmatrix} 1 & 0 & 0 \\ 0 & 1 & 0 \end{pmatrix} \boldsymbol{\beta} \begin{pmatrix} 0 \\ 1 \end{pmatrix}$$

2. Consider the problem of testing the multivariate linear hypothesis in which $r = 2$, $k = 3$, $p = 12$, $n = 115$. Suppose that

$$\mathbf{S}_2 = Y'P_{V|W}Y = \begin{pmatrix} 4 & 2 \\ 2 & 1 \end{pmatrix}, \qquad \mathbf{S}_3 = Y'P_{V^\perp}Y = \begin{pmatrix} 10 & 10 \\ 10 & 20 \end{pmatrix}.$$

(a) Use a size 0.05 Wilks test to test this hypothesis.

(b) Use a size 0.05 Lawley-Hotelling test to test this hypothesis.

(c) Use a size 0.05 Pillai test to test this hypothesis.

(d) Use a 0.05 Roy test to test this hypothesis. [Note that $s^{0.05}_{2,9,103} = 0.265$ and therefore $c_3^{0.05} = 0.265/(1 - 0.265)$.]

3. Consider a one-way multivariate analysis of variance problem in which $r = 2$, $p = 3$, $n_i = 31$. Suppose that

$$\overline{\mathbf{Y}}_1 = \begin{pmatrix} 5 \\ 2 \end{pmatrix}, \quad \overline{\mathbf{Y}}_2 = \begin{pmatrix} 6 \\ 2 \end{pmatrix}, \quad \overline{\mathbf{Y}}_3 = \begin{pmatrix} 10 \\ -1 \end{pmatrix}, \quad \hat{\boldsymbol{\Sigma}} = \frac{1}{90}\ \mathbf{S}_3 = \begin{pmatrix} 10 & -6 \\ -6 & 4 \end{pmatrix}.$$

(a) Use a size 0.05 Wilks test to test that the means are equal.

(b) Use a size 0.05 Lawley-Hotelling test.

(c) Use a size 0.05 Pillai test.

Type B

1. Prove Theorem 19.1b.

2. Prove Theorem 19.2a. and show that $E\,\mathrm{tr}\,\mathbf{R} = \mathrm{tr}\,E\mathbf{R}$ for any $r \times r$ matrix $\mathbf{R}$.

3. Prove (19.10). (Hint: Expand

$$\begin{vmatrix} \mathbf{I} & \mathbf{Z}_2 \\ \mathbf{Z}_2' & t\mathbf{Z}_3'\mathbf{Z}_3 \end{vmatrix}$$

two different ways.)

4. Let $\mathbf{T} > 0$ be $r \times r$ and let $\mathbf{U}$ be $s \times r$ of rank s. Show that $\mathbf{U}'\mathbf{T}\mathbf{U} > 0$.

5. Let $\phi_1(\hat{\boldsymbol{\mu}}, \hat{\boldsymbol{\Sigma}})$ and $\phi_2(\hat{\boldsymbol{\mu}}, \hat{\boldsymbol{\Sigma}})$ be two tests that have the same power functions. Show that $P[\phi_1(\hat{\mu}, \hat{\Sigma}) = \phi_2(\hat{\mu}, \hat{\Sigma})] = 1$. [Hint: use the completeness of $(\hat{\mu}, \hat{\Sigma})$.]

6. (a) Show that $t_1/(1+t_1), \ldots, t_b/(1+t_b)$ are the nonzero roots of both the following equations:

$$|\mathbf{Z}_2(\mathbf{Z}_2'\mathbf{Z}_2 + \mathbf{Z}_3'\mathbf{Z}_3)^{-1}\mathbf{Z}_2' - u\mathbf{I}| = 0$$

$$|\mathbf{Z}_2'\mathbf{Z}_2 - u(\mathbf{Z}_2'\mathbf{Z}_2 + \mathbf{Z}_3'\mathbf{Z}_3)| = 0.$$

(b) Show that $\operatorname{tr}[\mathbf{S}_2(\mathbf{S}_2 + \mathbf{S}_3)^{-1}] = \Sigma t_i/(1+t_i)$.

(c) Show that under the null hypothesis $t_1/(1+t_1) \sim s_{r,\, p-k,\, n-p}$, (see Section 19.2.6. Note that the null distribution does not depend on any unknown parameters and hence does not depend on $\mathbf{\Sigma}$.)

(d) Show that $c_3^\alpha = s^\alpha_{r,\, p-k,\, n-p}/(1 - s^\alpha_{r,\, p-k,\, n-p})$.

7. Show that $\lambda_1^{-n/2}$ is the likelihood ratio test statistic for testing the multivariate linear hypothesis.

8. Testing the multivariate linear model for known $\mathbf{\Sigma}$. Let $\mathbf{Z}_1$, $\mathbf{Z}_2$ and $\mathbf{Z}_3$ be as given in Section 19.2. In this problem we consider testing that $\boldsymbol{\nu}_2 = \mathbf{0}$, $\boldsymbol{\nu}_3 = \mathbf{0}$ against $\boldsymbol{\nu}_3 = \mathbf{0}$ when $\mathbf{\Sigma}$ is known. Let $\mathbf{Z}_i^* = (\mathbf{Z}_{i1}^*, \ldots, \mathbf{Z}^*_{ir}) = \mathbf{Z}_i\mathbf{\Sigma}^{-1/2}$. $\boldsymbol{\nu}_i^* = (\boldsymbol{\nu}_{i1}^*, \ldots, \boldsymbol{\nu}_{ir}^*) = \boldsymbol{\nu}_i\mathbf{\Sigma}^{-1/2}$. Finally, let

$$\tilde{\mathbf{Z}}_i = \begin{pmatrix} \mathbf{Z}_{i1}^* \\ \vdots \\ \mathbf{Z}_{ir}^* \end{pmatrix}, \qquad \tilde{\boldsymbol{\nu}}_i = \begin{pmatrix} \boldsymbol{\nu}_{i1}^* \\ \vdots \\ \boldsymbol{\nu}_{ir}^* \end{pmatrix}.$$

(a) Show that $\tilde{\mathbf{Z}}_1$, $\tilde{\mathbf{Z}}_2$, and $\tilde{\mathbf{Z}}_3$ are independent, that $\tilde{\mathbf{Z}}_1 \sim N_{rk}(\tilde{\boldsymbol{\nu}}_1, \mathbf{I})$,
$\tilde{\mathbf{Z}}_2 \sim N_{r(p-k)}(\tilde{\boldsymbol{\nu}}_2, \mathbf{I}), \tilde{\mathbf{Z}}_3 \sim N_{r(n-p)}(\tilde{\boldsymbol{\nu}}_3, \mathbf{I})$.

(b) Show that a sufficient statistic is $(\tilde{\mathbf{Z}}_1, \tilde{\mathbf{Z}}_2)$.

(c) Show that this testing problem is invariant under the following two groups of transformations: (1) $g_1(\tilde{\mathbf{Z}}_1, \tilde{\mathbf{Z}}_2) = (\tilde{\mathbf{Z}}_1 + \mathbf{a}, \tilde{\mathbf{Z}}_2), \mathbf{a} \in R^{rk}$; (2) $g_2(\tilde{\mathbf{Z}}_1, \tilde{\mathbf{Z}}_2) = (\tilde{\mathbf{Z}}_1, \mathbf{\Gamma}\tilde{\mathbf{Z}}_2)$ where $\mathbf{\Gamma}$ is an arbitrary $r(p-k) \times r(p-k)$ dimensional orthogonal matrix.

(d) Show that a maximal invariant under this group is $U = \tilde{\mathbf{Z}}_2'\tilde{\mathbf{Z}}_2$ and $U \sim \chi^2_{r(p-k)}(\tilde{\boldsymbol{\nu}}_2'\tilde{\boldsymbol{\nu}}_2)$.

(e) Show that the UMP invariant size α test rejects if $\mathbf{U} > \chi^{2\,\alpha}_{r(p-k)}$.

(f) Show that $U = \operatorname{tr}(\mathbf{S}_2\mathbf{\Sigma}^{-1}), \tilde{\boldsymbol{\nu}}_2'\tilde{\boldsymbol{\nu}}_2 = \operatorname{tr}(\boldsymbol{\nu}_2\mathbf{\Sigma}^{-1}\boldsymbol{\nu}_2') = \operatorname{tr}(\boldsymbol{\delta}\mathbf{\Sigma}^{-1})$.

9. A *contrast* associated with testing the generalized multivariate linear hypothesis is a function $\mathbf{d}'\boldsymbol{\mu}\mathbf{A}\mathbf{q}$ where $\mathbf{d} \in V/W$ and $\mathbf{q} \in R^s$. Find a set of simultaneous $(1-\alpha)$ confidence intervals for the set of all contrasts associated with testing the generalized multivariate linear hypothesis and show that the test ϕ_3^* rejects if and only if at least one of the confidence intervals does not contain 0.

10. Find a set of simultaneous confidence intervals for the set of all contrasts associated with testing that $\boldsymbol{\alpha}_i = \mathbf{0}$ in the balanced two-way multivariate analysis of variance model.

11. Consider the multivariate nested analysis of variance model in which we observe $\mathbf{Y}_{ijk}$ that are independent r-dimensional random vectors such that $\mathbf{Y}_{ijk} = \boldsymbol{\theta} + \boldsymbol{\alpha}_i + \boldsymbol{\gamma}_{ij} + \mathbf{e}_{ijk}$, $\sum_i \boldsymbol{\alpha}_i = 0$, $\sum_j \boldsymbol{\gamma}_{ij} = 0$ and $\mathbf{e}_{ijk} \sim N_r(\mathbf{0}, \boldsymbol{\Sigma})$, for $i = 1, \ldots, a$; $j = 1, \ldots, c$; $k = 1, \ldots, d$. Find the maximal invariant for testing that $\boldsymbol{\alpha}_i = 0$ and the maximal invariant for testing that $\boldsymbol{\gamma}_{ij} = 0$.

12. Show that the test defined in Section 19.7 for testing that $\beta_j = 0$ in the generalized repeated measures model is the same as the test defined in Chapter 14 for testing that $\beta_j = 0$ in the repeated measures model.

13. Show that the maximal invariant for testing that $\alpha_i = 0$ in the generalized repeated measures model of Section 19.7 has the form given in (19.13).

14. Show that the tests derived in Section 19.7 for testing that $\alpha_i = 0$ and $\gamma_{ij} = 0$ in the generalized repeated measures model do not depend on which basis matrix $\mathbf{X}$ is chosen. (Hint: If $\mathbf{X}$ and $\mathbf{X}^*$ are basis matrices for the same space, then $\mathbf{X}^* = \mathbf{XB}$ for some invertible matrix $\mathbf{B}$. Why?)

15. Let $\mathbf{Y} \sim N_{n,r}(\boldsymbol{\mu}, \mathbf{I}, \boldsymbol{\Sigma})$ and let $\mathbf{d}$ be an $n \times r$ matrix. Define $\mathbf{Y}^*$, $\boldsymbol{\mu}^*$, $\mathbf{d}^*$ and $\boldsymbol{\Sigma}^*$ by

$$\mathbf{Y} = \begin{pmatrix} \mathbf{Y}_1 \\ \vdots \\ \mathbf{Y}_n \end{pmatrix}, \quad \boldsymbol{\mu} = \begin{pmatrix} \mu_1 \\ \vdots \\ \mu_n \end{pmatrix}, \quad \mathbf{d} = \begin{pmatrix} \mathbf{d}_1 \\ \vdots \\ \mathbf{d}_n \end{pmatrix}, \quad \mathbf{Y}^* = \begin{pmatrix} \mathbf{Y}_1' \\ \vdots \\ \mathbf{Y}_n' \end{pmatrix},$$

$$\boldsymbol{\mu}^* \begin{pmatrix} \mu_1' \\ \vdots \\ \mu_n' \end{pmatrix}, \quad \mathbf{d}^* = \begin{pmatrix} \mathbf{d}_1' \\ \vdots \\ \mathbf{d}_2' \end{pmatrix}, \quad \boldsymbol{\Sigma}^* = \begin{pmatrix} \boldsymbol{\Sigma} & \mathbf{0} & \cdots & \mathbf{0} \\ \mathbf{0} & \boldsymbol{\Sigma} & \cdots & \mathbf{0} \\ \vdots & \vdots & \vdots & \vdots \\ \mathbf{0} & \mathbf{0} & \cdots & \boldsymbol{\Sigma} \end{pmatrix}$$

[where $\boldsymbol{\Sigma}^*$ is $(nr) \times (nr)$]. We have therefore strung $\mathbf{Y}$, $\boldsymbol{\mu}$ and $\mathbf{d}$ out as nr-dimensional vectors.

(a) Show that $\mathbf{Y}^* \sim N_{nr}(\boldsymbol{\mu}^*, \boldsymbol{\Sigma}^*)$.

(b) Show that $(\mathbf{d}^* - \boldsymbol{\mu}^*)(\boldsymbol{\Sigma}^*)^{-1}(\mathbf{d}^* - \boldsymbol{\mu}^*) = \operatorname{tr}(\mathbf{d} - \boldsymbol{\mu})\boldsymbol{\Sigma}^{-1}(\mathbf{d} - \boldsymbol{\mu})$, and hence L is the Mahalanobis distance for estimating $\boldsymbol{\mu}$.

16. Prove Theorem 19.20a.

17. Verify (19.14).

18. Verify (19.17) and (19.19).

19. Show that $(\hat{\boldsymbol{\beta}}, \hat{\boldsymbol{\Sigma}})$ is not a complete sufficient statistic for the growth curves model. (Hint: There exists a vector $\mathbf{v}$ such that $\boldsymbol{\beta}\mathbf{v} = 0$. Why? What is $E\hat{\boldsymbol{\beta}}\mathbf{v}$?)

20. We follow the notation of Theorem 19.17.

(a) Show that $W_{ni}^2 \leqslant (\mathbf{a}_{ni}'\boldsymbol{\Sigma}\mathbf{a}_{ni})(\mathbf{f}_i\boldsymbol{\Sigma}^{-1}\mathbf{f}_i)$. {Hint: $W_{ni}^2 = [(\boldsymbol{\Sigma}^{1/2}\mathbf{a}_{ni})'(\boldsymbol{\Sigma}^{-1/2}\mathbf{f}_i)]^2$. Use Cauchy-Schwartz.}

(b) Show that $C_n \leqslant \sum_{i=1}^n (\mathbf{a}_{ni}'\boldsymbol{\Sigma}\mathbf{a}_{ni})E(\mathbf{f}_i\boldsymbol{\Sigma}^{-1}\mathbf{f}_i)I_{\epsilon^2/m_n}(\mathbf{f}_i'\boldsymbol{\Sigma}^{-1}\mathbf{f}_i)$
$= cEUI_{\epsilon^2 m_n}(U)$.

21. Let $\mathbf{T}_n$ and $\hat{\boldsymbol{\Sigma}}_n$ be defined as in Theorem 19.19 and let $\mathbf{A}$ be a constant $r \times s$ matrix of rank s. Suppose that Huber's condition is satisfied.

(a) Show that $\mathbf{A}'\mathbf{T}_n\mathbf{A} \xrightarrow{d} W_s(p-k, \mathbf{A}'\boldsymbol{\Sigma}\mathbf{A})$ and $\mathbf{A}'\hat{\boldsymbol{\Sigma}}_n\mathbf{A} \xrightarrow{p} \mathbf{A}'\boldsymbol{\Sigma}\mathbf{A}$.

(b) Show that the size of any invariant test of a generalized multivariate linear hypothesis is asymptotically unaffected by nonnormality.

(c) Show that the confidence coefficient for the simultaneous confidence intervals for contrasts associated with testing the generlized multivariate linear hypothesis are asymptotically unaffected by nonnormality.

Type C

1. Equality of $Q_{r,s,m}$ and $Q_{s,r,m-r+s}$. Let $\mathbf{U}$ and $\mathbf{W}$ be independent, $\mathbf{U} \sim N_{s,r}(\boldsymbol{\nu}, \mathbf{I}, \mathbf{I})$, $\mathbf{W} \sim W_r(m, \mathbf{I})$. Let $b = \min(r,s)$. Let $t_1 \geqslant \cdots \geqslant t_b$ be the nonzero roots of $|\mathbf{U}'\mathbf{U} - t\mathbf{W}| = 0$ or equivalently of $|\mathbf{U}\mathbf{W}^{-1}\mathbf{U}' - t\mathbf{I}| = 0$. Also, let $\theta_1 \geqslant \cdots \geqslant \theta_b$ be the largest roots of $|\boldsymbol{\nu}_2'\boldsymbol{\nu}_2 - \theta\mathbf{I}| = 0$ or equivalently $|\boldsymbol{\nu}_2\boldsymbol{\nu}_2' - \theta\mathbf{I}| = 0$.

(a) Show that $(t_1, \ldots, t_b) \sim Q_{r,s,m}(\theta_1, \ldots, \theta_b)$.

(b) Let $s \leqslant r$. Then $\mathbf{U}\mathbf{U}' > 0$. Let $\mathbf{S} = (\mathbf{U}\mathbf{U}')^{1/2}(\mathbf{U}\mathbf{W}^{-1}\mathbf{U}')^{-1}(\mathbf{U}\mathbf{U}')^{1/2}$. Show that $\mathbf{S}\,|\,\mathbf{U} \sim W_s(m-r+s, \mathbf{I})$. (Hint: see Theorem 17.15g.)

(c) Let $q_1, \ldots, q_b$ be the nonzero roots of $|\mathbf{U}\mathbf{U}' - q\mathbf{S}| = 0$ or equivalently $|\mathbf{U}'\mathbf{S}^{-1}\mathbf{U} - qI| = 0$. Show that

$$(q_1, \ldots, q_b) \sim Q_{s,r,m-r+s}(\theta_1, \ldots, \theta_b).$$

(d) Show that $t_1 = q_1, \ldots, t_b = q_b$. Therefore, if $s \leqslant r$, then

$$Q_{r,s,m} = Q_{s,r,m-r+s}.$$

(e) Show that if $s \geqslant r$, then $Q_{s,r,m-r+s} = Q_{r,s,m}$.

2. The moments of $1/\lambda_1$. Let $\mathbf{U}, \mathbf{V}$ and $\mathbf{W}$ be independent, $\mathbf{U} \sim W_r(\mathbf{a}, \mathbf{I})$,

$\mathbf{V} \sim W_r(c, \mathbf{I})$, $\mathbf{W} \sim W_r(c + 2h, \mathbf{I})$, where $c \geqslant r$. Let

$$d(n) = 2^{nr} \prod_{i=1}^{r} \Gamma\left(\tfrac{1}{2}n + 1 - i\right), \qquad M = \frac{|\mathbf{V}|}{|\mathbf{U} + \mathbf{V}|}.$$

(a) Use the Wishart density function to show that

$$EM^h \,|\, \mathbf{U} = \frac{d(c + 2h)}{d(c)} E \frac{1}{|\mathbf{U} + \mathbf{W}|^h} |\mathbf{U}.$$

(b) Use the Wishart density function to show that

$$E \frac{1}{|\mathbf{U} + \mathbf{W}|^h} = \frac{d(a + c)}{d(a + c + 2h)}.$$

[Hint: $\mathbf{U} + \mathbf{W} \sim W_r((a + c + 2h), \mathbf{I})$. Why?]

(c) Find EM^h.

(d) Let $r_i \sim Be(\frac{1}{2}(c - i + 1), \frac{1}{2}a)$, r_i independent. Show that

$$EM^h = E\left(\prod_{i=1}^{p} r_i\right)^h.$$

(Since $0 \leqslant M \leqslant 1$, this proves that M has the same distribution as Πr_i.)

(e) Find $E\lambda_1^{-h}$. Show under the null hypothesis that λ_1^{-1} has the same distribution as the product of beta random variables. What are the parameters?

3. (a) Show that $\log(1 + x) = x - (1 + h)^{-2}x^2/2$ for some h such that $0 \leqslant h \leqslant x$. [Hint: Do a Taylor series expansion of $\log(1 + x)$ about $x = 0$.]

(b) Let U_n and V_n be sequences of random variables such that $U_n \xrightarrow{d} F$, $V_n \xrightarrow{p} 0$. Show that

$$(n - p)\log\left(1 + \frac{U_n + V_n}{n - p}\right) - U_n = V_n - \frac{1}{(1 + h_n)^2} \frac{(U_n + V_n)^2}{n - p},$$

for some h_n such that $0 \leqslant h_n \leqslant (U_n + V_n)/(n - p)$.

(c) Show that

$$\frac{U_n + V_n}{n - p} \xrightarrow{p} 0, \quad \frac{(U_n + V_n)^2}{n - p} \xrightarrow{p} 0.$$

(d) Show that $h_n \xrightarrow{p} 0$.

(e) Show that

$$(n-p)\log\left(1+\frac{U_n+V_n}{n-p}\right)-U_n \xrightarrow{P} 0.$$

(f) Let $s_i(n)$ and $t_i(n)$ be as defined in Section 19.2.6. Show that

$$(n-p)\log\left[1+t_i(n)\right]-s_i(n)\xrightarrow{P}0.$$

[Hint: let $U_n = s_i(n)$, $V_n = (n-p)t_i(n) - s_i(n)$.]

4. Consider testing the generalized linear hypothesis as described in Section 19.4. Let **B** be a matrix such that (**A B**) is invertible and let $\mathbf{X}_1$, $\mathbf{X}_2$, and $\mathbf{X}_3$ be orthonormal basis matrices for W, $V|W$ and V. Let

$$\mathbf{U}=\begin{bmatrix}\mathbf{U}_{11} & \mathbf{U}_{12}\\ \mathbf{U}_{21} & \mathbf{U}_{22}\\ \mathbf{U}_{31} & \mathbf{U}_{32}\end{bmatrix}=\begin{bmatrix}\mathbf{X}_1'\\ \mathbf{X}_2'\\ \mathbf{X}_3'\end{bmatrix}\mathbf{Y}(\mathbf{A}\ \mathbf{B}),\ \boldsymbol{\delta}=\begin{bmatrix}\boldsymbol{\delta}_{11} & \boldsymbol{\delta}_{12}\\ \boldsymbol{\delta}_{21} & \boldsymbol{\delta}_{22}\\ \boldsymbol{\delta}_{31} & \boldsymbol{\delta}_{32}\end{bmatrix}=\begin{bmatrix}\mathbf{X}_1'\\ \mathbf{X}_2'\\ \mathbf{X}_3'\end{bmatrix}\boldsymbol{\mu}(\mathbf{A}\ \mathbf{B}),$$

$$\boldsymbol{\Xi}=\begin{pmatrix}\mathbf{A}'\\ \mathbf{B}'\end{pmatrix}\boldsymbol{\Sigma}(\mathbf{A}\ \mathbf{B}),\ \mathbf{T}=\begin{pmatrix}\mathbf{T}_{11} & \mathbf{T}_{12}\\ \mathbf{T}_{21} & \mathbf{T}_{22}\end{pmatrix}=\begin{pmatrix}\mathbf{U}_{31}'\mathbf{U}_{31} & \mathbf{U}_{31}'\mathbf{U}_{32}\\ \mathbf{U}_{32}'\mathbf{U}_{31} & \mathbf{U}_{32}'\mathbf{U}_{32}\end{pmatrix}.$$

(a) Show that $\mathbf{U}\sim N_{n,r}(\boldsymbol{\delta}, I, \boldsymbol{\Xi})$, that $\boldsymbol{\mu}\in V$ if and only if $\boldsymbol{\delta}_{31}=\mathbf{0}$, $\boldsymbol{\delta}_{32}=\mathbf{0}$, and that $\boldsymbol{\mu}\mathbf{A}\in W$ and $\boldsymbol{\mu}\in V$ if and only if $\boldsymbol{\delta}_{21}=\mathbf{0}$, $\boldsymbol{\delta}_{31}=\mathbf{0}$, $\boldsymbol{\delta}_{32}=\mathbf{0}$. (This is the canonical form for the generalized multivariate linear hypothesis.)

(b) Show that this problem is invariant under the following five groups of transformations:
(i) $g_1(\mathbf{U}_{11},\mathbf{U}_{12},\mathbf{U}_{21},\mathbf{U}_{22},\mathbf{U}_{31},\mathbf{U}_{32})=(\mathbf{U}_{11}+\mathbf{D}_{11},\mathbf{U}_{12}+\mathbf{D}_{12},\mathbf{U}_{21},\mathbf{U}_{22}+\mathbf{D}_{22},\mathbf{U}_{31},\mathbf{U}_{32})$ where $\mathbf{D}_{11}$, $\mathbf{D}_{12}$ and $\mathbf{D}_{22}$ are arbitrary matrices.
(ii) $g_2(\mathbf{U}_{11},\mathbf{U}_{12},\mathbf{U}_{21},\mathbf{U}_{22},\mathbf{U}_{31},\mathbf{U}_{32})=(\mathbf{U}_{11},\mathbf{U}_{12}+\mathbf{U}_{11}\mathbf{E},\mathbf{U}_{21},\mathbf{U}_{22}+\mathbf{U}_{21}\mathbf{E},\mathbf{U}_{31},\mathbf{U}_{32}+\mathbf{U}_{31}\mathbf{E})$ where **E** is an arbitrary matrix.
(iii) $g_3(\mathbf{U}_{11},\mathbf{U}_{12},\mathbf{U}_{21},\mathbf{U}_{22},\mathbf{U}_{31},\mathbf{U}_{32})=(\mathbf{U}_{11},\mathbf{U}_{12}\mathbf{F},\mathbf{U}_{21},\mathbf{U}_{22}\mathbf{F},\mathbf{U}_{31},\mathbf{U}_{32}\mathbf{F})$ where **F** is an invertible matrix.
(iv) $g_4(\mathbf{U}_{11},\mathbf{U}_{12},\mathbf{U}_{21},\mathbf{U}_{22},\mathbf{U}_{31},\mathbf{U}_{32})=(\mathbf{U}_{11},\mathbf{U}_{12},\boldsymbol{\Gamma}\mathbf{U}_{21},\boldsymbol{\Gamma}\mathbf{U}_{22},\mathbf{U}_{31},\mathbf{U}_{32})$ where $\boldsymbol{\Gamma}$ is an orthogonal matrix.
(v) $g_5(\mathbf{U}_{11},\mathbf{U}_{12},\mathbf{U}_{21},\mathbf{U}_{22},\mathbf{U}_{31},\mathbf{U}_{32})=(\mathbf{U}_{11}\mathbf{G},\mathbf{U}_{12},\mathbf{U}_{21}\mathbf{G},\mathbf{U}_{22},\mathbf{U}_{31}\mathbf{G},\mathbf{U}_{32})$ where **G** is an invertible matrix.
What are the induced transformations on the parameter space?

(c) Show that $(\mathbf{U}_{11},\mathbf{U}_{12},\mathbf{U}_{21},\mathbf{U}_{22},\mathbf{T}_{11},\mathbf{T}_{12},\mathbf{T}_{22})$ is a sufficient statistic for this model.

(d) Show that a maximal invariant under the five above groups is the set of roots of $|\mathbf{U}_{21}'\mathbf{U}_{21}-t^*\mathbf{T}_{11}|=0$. [Reducing by the groups in the above order, the maximal invariants at each stage are $(\mathbf{U}_{21},\mathbf{T}_{11},\mathbf{T}_{12},\mathbf{T}_{22})$,

$(\mathbf{U}_{21}, \mathbf{T}_{11}, \mathbf{T}_{22} - \mathbf{T}'_{12}\mathbf{T}_{11}^{-1}\mathbf{T}_{12})$, $(\mathbf{U}_{21}, \mathbf{T}_{11})$, $(\mathbf{U}'_{21}\mathbf{U}_{21}, \mathbf{T}_{11})$, and the roots of the above equation.]

(e) Show that $\mathbf{U}'_{21}\mathbf{U}_{21} = \mathbf{A}'\mathbf{Y}'\mathbf{P}_{V|W}\mathbf{Y}\mathbf{A}$ and $\mathbf{T}_{11} = \mathbf{A}'\mathbf{Y}'\mathbf{P}_{V^\perp}\mathbf{Y}\mathbf{A}$.

5. The asymptotic distribution of U for the growth curves model. We follow the notation of Section 19.10. We consider the model as $n \to \infty$ but p, q, s, and t remain fixed. Let

$$\mathbf{S}_n = \hat{\boldsymbol{\beta}}'\mathbf{A}'\left(\mathbf{A}(\mathbf{X}'\mathbf{X})^{-1}\mathbf{A}'\right)^{-1}\mathbf{A}\hat{\boldsymbol{\beta}}, \qquad \mathbf{W} = \mathbf{Z}'\boldsymbol{\gamma}'\mathbf{A}'\left(\mathbf{A}(\mathbf{X}'\mathbf{X})^{-1}\mathbf{A}'\right)^{-1}\mathbf{A}\boldsymbol{\gamma}\mathbf{Z}$$

$$\mathbf{T}_n = \hat{\boldsymbol{\Sigma}}^{-1}\mathbf{Z}'(\mathbf{Z}\hat{\boldsymbol{\Sigma}}^{-1}\mathbf{Z}')^{-1}\mathbf{B}\left(\mathbf{B}'(\mathbf{Z}\hat{\boldsymbol{\Sigma}}^{-1}\mathbf{Z})^{-1}\mathbf{B}\right)^{-1}\mathbf{B}(\mathbf{Z}\hat{\boldsymbol{\Sigma}}^{-1}\mathbf{Z}')^{-1}\mathbf{Z}\hat{\boldsymbol{\Sigma}}^{-1}$$

$$\boldsymbol{\Xi} = \boldsymbol{\Sigma}^{-1}\mathbf{Z}(\mathbf{Z}\boldsymbol{\Sigma}^{-1}\mathbf{Z}')^{-1}\mathbf{B}\left(\mathbf{B}'(\mathbf{Z}\boldsymbol{\Sigma}^{-}\mathbf{Z})^{-1}\mathbf{B}\right)^{-1}\mathbf{B}(\mathbf{Z}\boldsymbol{\Sigma}^{-1}\mathbf{Z}')^{-1}\mathbf{Z}\boldsymbol{\Sigma}^{-1}.$$

We also assume that $\mathbf{W}$ does not depend on n.

(a) Show that $U = \operatorname{tr} \mathbf{S}_n\boldsymbol{\Xi}$ and $\hat{U}_n = \operatorname{tr} \mathbf{S}_n\mathbf{T}_n$.

(b) Show that $\mathbf{S}_n \sim W_r(s, \boldsymbol{\Sigma}, \boldsymbol{\gamma})$ and hence $\mathbf{S}_n \xrightarrow{d} \mathrm{W}_r(s, \boldsymbol{\Sigma}, \mathbf{W})$.

(c) Show that $\mathbf{T}_n \xrightarrow{p} \boldsymbol{\Xi}$.

(d) Show that $\hat{U}_n - U_n \xrightarrow{p} 0$.

(e) Show that $U_n \xrightarrow{d} \chi^2_{st}(\delta)$.

CHAPTER 20

Discriminant Analysis

In this chapter, we consider the problem in which we observe an r-dimensional random vector $\mathbf{X}$ on an individual, and on the basis of this random vector, we must decide to which of two populations the individual belongs. As an example, consider the problem of deciding on the basis of medical test results whether a person has a particular disease, or not. Other examples would include the problem of deciding from what tribe a skeleton came, or whether an applicant should be given a job, admitted to graduate school, and so on.

We assume that $\mathbf{X} \sim N_r(\boldsymbol{\mu}, \boldsymbol{\Sigma})$, and $\boldsymbol{\mu} = \boldsymbol{\mu}_i$ if $\mathbf{X}$ came from the ith population. We are therefore testing that $\boldsymbol{\mu} = \boldsymbol{\mu}_1$ against $\boldsymbol{\mu} = \boldsymbol{\mu}_2$. This problem is rather different from the other testing problems considered in this book in that we view the hypotheses symmetrically. There is no clear null hypothesis for this problem. In Section 20.1, we give some discussion of this different perspective. Then in Section 20.2, we consider the simple model in which $\boldsymbol{\mu}_1$, $\boldsymbol{\mu}_2$, and $\boldsymbol{\Sigma}$ are known, and find the optimal procedures. In Section 20.3, we look at the more reasonable model in which we do not know $\boldsymbol{\mu}_1$, $\boldsymbol{\mu}_2$, or $\boldsymbol{\Sigma}$, but have two samples, $\mathbf{Y}_{1j}$ and $\mathbf{Y}_{2j}$ such that $\mathbf{Y}_{ij} \sim N_r(\boldsymbol{\mu}_i, \boldsymbol{\Sigma})$. We want to test that $\boldsymbol{\mu} = \boldsymbol{\mu}_1$ against $\boldsymbol{\mu} = \boldsymbol{\mu}_2$. Methods for estimating the probabilities of misclassification are discussed in Section 20.4. It is often true in discriminant analysis problems that we have some prior knowledge of the prior probability that an individual will be in the ith population. A Bayesian analysis which utilizes these probabilities is given in Section 20.5. In Section 20.6, we consider the problem of testing that some of the variables of $\mathbf{X}$ are not contributing to the discrimination. In Section 20.7 we discuss discrimination among more than two populations.

20.1. SYMMETRIC HYPOTHESES.

In many of the problems considered in this chapter, we consider the hypotheses to be symmetric in that there is no reason to call one hypothesis the null hypothesis. In problems considered earlier in the book, we set the probability of type I error to be α and then found procedures which had small probability of type II error. This approach is clearly asymmetric in the hypotheses and does not seem sensible for these symmetric problems. In this section, we discuss symmetric hypotheses and indicate that some of the concepts used in previous chapters for asymmetric problems are also sensible for symmetric ones.

We first state a fairly general version of a symmetric problem. In this problem we observe the random vector $\mathbf{X}$ having density $f(\mathbf{x};\boldsymbol{\theta})$ and are testing that $\boldsymbol{\theta} \in \omega$ against $\boldsymbol{\theta} \in \xi$, where ω and ξ have empty intersection. In the notation used in earlier chapters, the null set is contained in the alternative set. So, let $\Omega = \omega \cup \xi$. Then we are testing that $\boldsymbol{\theta} \in \omega$ against $\boldsymbol{\theta} \in \Omega$. When we view the hypotheses symmetrically, then, we mean that we view ω and $\xi = \Omega - \omega$ symmetrically.

We now look at the problem of testing a simple null hypothesis against a simple alternative, i.e., testing that $\boldsymbol{\theta} = \boldsymbol{\theta}_0$ against $\boldsymbol{\theta} = \boldsymbol{\theta}_1$ (or in the notation of earlier chapters, testing that $\boldsymbol{\theta} = \boldsymbol{\theta}_0$ against $\boldsymbol{\theta} = \boldsymbol{\theta}_0$ or $\boldsymbol{\theta} = \boldsymbol{\theta}_1$). For simplicity, we look at the case in which the densities $f(\mathbf{x};\boldsymbol{\theta})$ are continuous and the support of $f(\mathbf{x};\boldsymbol{\theta})$ does not depend on $\boldsymbol{\theta}$. In this case, by the Neymann-Pearson theorem, we see that the UMP tests are given by

$$\phi_k(\mathbf{X}) = \begin{cases} 1 & \text{if } \dfrac{f(\mathbf{X};\boldsymbol{\theta}_1)}{f(\mathbf{X};\boldsymbol{\theta}_0)} > k \\ 0 & \text{if } \dfrac{f(\mathbf{X};\boldsymbol{\theta}_1)}{f(\mathbf{X};\boldsymbol{\theta}_0)} \leqslant k \end{cases} \tag{20.1}$$

for $0 \leqslant k \leqslant \infty$ ($\phi_\infty = 0$ and is the most powerful size 0 test). We now make some general definitions. For any rule, ϕ, let $\alpha_\phi = E_{\theta_0}\phi$ be the probability of type I error, and $\beta_\phi = E_{\theta_1}(1-\phi)$ be the probability of type II error. We say that ϕ is *as good as* ϕ^* if $\alpha_\phi \leqslant \alpha_{\phi^*}$ and $\beta_\phi \leqslant \beta_{\phi^*}$. If at least one of the inequalities is strict, we say that ϕ is *better then* ϕ^*. Let C be a class of rules. Then C is *essentially complete* if for any rule ϕ there is a rule in C that is as good as ϕ. A rule ϕ is *admissible* if there is no rule that is better than ϕ. The following theorem is proved in Ferguson (1967, pp. 198–204).

THEOREM 20.1. The rule ϕ_k is admissible for all k. Let $C = \{\phi_k;\ 0 \leqslant k \leqslant \infty\}$. Then C is an essentially complete class for the problem of testing that $\theta = \theta_0$ against $\theta = \theta_1$. This theorem implies that there is no reason to use any rule that is not a Neyman-Pearson rule.

We now return to the more general situation in which we are testing the composite hypotheses that $\boldsymbol{\theta} \in \omega$ against $\boldsymbol{\theta} \in \xi$. We now discuss invariance. We note that property of invariance is symmetric in ω and ξ. (Since $\bar{g}$ is invertible, $\bar{g}(\Omega) = \Omega$, and $\bar{g}(\omega) = \omega$, we see that $\bar{g}(\xi) = \bar{g}(\Omega - \omega) = \bar{g}(\Omega) - \bar{g}(\omega) = \Omega - \omega = \xi$.) Therefore, it seems reasonable to ask that tests be invariant for symmetric problems, and hence that they be functions only of the maximal invariant.

As a final topic, we look at the likelihood ratio test. The version of the likelihood ratio test discussed in Chapter 1 is not symmetric. A more reasonable version of the likelihood ratio tests for symmetric problems is

$$\lambda^* = \frac{\sup_{\boldsymbol{\theta} \in \omega} f(\mathbf{X}; \boldsymbol{\theta})}{\sup_{\boldsymbol{\theta} \in \xi} f(\mathbf{X}; \boldsymbol{\theta})}, \qquad \phi(\lambda^*) = \begin{cases} 1 & \text{if } \lambda^* > c \\ 0 & \text{if } \lambda^* \leqslant c \end{cases}.$$

(Let λ be the likelihood ratio test statistic as defined in Chapter 1. Then λ is the maximum of 1 and λ^* so that the likelihood ratio test defined in this chapter is the same as the likelihood ratio test defined in Chapter 1 as long as $c < 1$.)

20.2. THE CASE OF KNOWN PARAMETERS.

We now consider the problem in which we observe $\mathbf{X} \sim N_r(\boldsymbol{\mu}, \boldsymbol{\Sigma})$ and are testing that $\boldsymbol{\mu} = \boldsymbol{\mu}_1$ against $\boldsymbol{\mu} = \boldsymbol{\mu}_2$, where $\boldsymbol{\mu}_1$, $\boldsymbol{\mu}_2$, and $\boldsymbol{\Sigma}$ are all known. (In other words, we are testing that the individual comes from population 1 against the hypothesis that he comes from population 2). This is a problem of testing a simple null hypothesis aginst a simple alternative. Let

$$W(\mathbf{X}) = \log \frac{f(\mathbf{X};\, \boldsymbol{\mu}_2, \boldsymbol{\Sigma})}{f(\mathbf{X};\, \boldsymbol{\mu}_1, \boldsymbol{\Sigma})} = (\boldsymbol{\mu}_2 - \boldsymbol{\mu}_1)'\boldsymbol{\Sigma}^{-1}(\mathbf{X} - \tfrac{1}{2}(\boldsymbol{\mu}_1 + \boldsymbol{\mu}_2))',$$

$$\phi_k(W) = \begin{cases} 1 & \text{if } W > \log k \\ 0 & \text{if } W \leqslant \log k \end{cases}. \tag{20.2}$$

[By this rule, we classify the individual into population 1 if $W(\mathbf{X}) \leqslant \log k$ and in population 2 if $W(\mathbf{X}) > \log k$.] The function $W(\mathbf{X})$ is called the *linear discriminant function*. By Theorem 20.1, the set of all ϕ_k is an essentially complete class for this problem, so that there is no reason to use a rule that is not of this form. Rules of this form were first suggested by Fisher (1936).

To select k, it is helpful to have formulas for $\alpha_k = E_{\theta_0}\phi_k$ the probability of type I error and $\beta_k = E_{\theta_1}(1 - \phi_k)$ the probability of type II error. We use the following lemma which follows directly from Theorem 3.6c.

LEMMA 20.2. $W \sim N_1((\boldsymbol{\mu}_2 - \boldsymbol{\mu}_1)'\boldsymbol{\Sigma}^{-1}(\boldsymbol{\mu} - \frac{1}{2}(\boldsymbol{\mu}_1 + \boldsymbol{\mu}_2)),\ (\boldsymbol{\mu}_1 - \boldsymbol{\mu}_2)'\boldsymbol{\Sigma}^{-1}(\boldsymbol{\mu}_1 - \boldsymbol{\mu}_2))$.

PROOF. See Exercise B1. □

Now let $\delta = (\boldsymbol{\mu}_1 - \boldsymbol{\mu}_2)'\boldsymbol{\Sigma}^{-1}(\boldsymbol{\mu}_1 - \boldsymbol{\mu}_2)$. (Note that δ is assumed known in this section.)

COROLLARY.

$$\alpha_k = 1 - N\left(\frac{\log k + \delta/2}{\sqrt{\delta}}\right), \qquad \beta_k = N\left(\frac{\log k - \delta/2}{\sqrt{\delta}}\right),$$

where $N(x)$ is the distribution function of a normal random variable with mean 0 and variance 1.

Note that $\alpha_k \to 0$ and $\beta_k \to 1$ as $k \to \infty$ and that $\alpha_k \to 1$ and $\beta_k \to 0$ as $k \to 0$. Also α_k and β_k are continuous functions of k. Therefore, we can achieve any value for the ratio α_k/β_k. A researcher in the setting of this section would choose a value for α_k/β_k, find α_k and β_k by trial and error using the corollary, and then apply ϕ_k defined by (20.2). (Another approach to determining k is given in Section 20.5.)

20.3. THE CASE OF UNKNOWN PARAMETERS.

We now consider the discrimination problem in the case in which the parameters $\boldsymbol{\mu}_1$, $\boldsymbol{\mu}_2$, and $\boldsymbol{\Sigma}$ are unknown but we have samples from population 1 and population 2, and so can estimate the parameters. In this problem, therefore, we observe $\mathbf{X}, \mathbf{Y}_{11}, \ldots, \mathbf{Y}_{1n}, \mathbf{Y}_{21}, \ldots, \mathbf{Y}_{2m}$ all independent such that

$$\mathbf{X} \sim N_r(\boldsymbol{\mu}, \boldsymbol{\Sigma}), \qquad \mathbf{Y}_{1j} \sim N_r(\boldsymbol{\mu}_1, \boldsymbol{\Sigma}), \qquad \mathbf{Y}_{2j} \sim N_r(\boldsymbol{\mu}_2, \boldsymbol{\Sigma})$$

and we are testing that $\boldsymbol{\mu} = \boldsymbol{\mu}_1$ against $\boldsymbol{\mu} = \boldsymbol{\mu}_2$. Let

$$\overline{\mathbf{Y}}_1 = \frac{1}{n}\sum_{i=1}^{n} \mathbf{Y}_{1i}, \qquad \overline{\mathbf{Y}}_2 = \frac{1}{m}\sum_{j=1}^{m} \mathbf{Y}_{2j},$$

$$\mathbf{S} = \frac{1}{n+m-2}\left(\sum_{i=1}^{n}(\mathbf{Y}_{1i} - \overline{\mathbf{Y}}_1)(\mathbf{Y}_{1i} - \overline{\mathbf{Y}}_1)' + \sum_{j=1}^{m}(\mathbf{Y}_{2j} - \overline{\mathbf{Y}}_2)(\mathbf{Y}_{2j} - \overline{\mathbf{Y}}_2)'\right).$$

We assume that $n + m - 2 \geqslant r$ so that $\mathbf{S}$ is invertible. It is easily verified that $(\mathbf{X}, \overline{\mathbf{Y}}_1, \overline{\mathbf{Y}}_2, \mathbf{S})$ is a sufficient statistic for this model and that $\mathbf{X}, \overline{\mathbf{Y}}_1, \overline{\mathbf{Y}}_2$, and $\mathbf{S}$ are unbiased estimators of $\boldsymbol{\mu}$, $\boldsymbol{\mu}_1$, $\boldsymbol{\mu}_2$, and $\boldsymbol{\Sigma}$.

We now reduce the problem by invariance. The problem is invariant under the following two groups.

(1) $g_1(\mathbf{X}, \overline{\mathbf{Y}}_1, \overline{\mathbf{Y}}_2, \mathbf{S}) = (\mathbf{X} + \mathbf{a}, \overline{\mathbf{Y}}_1 + \mathbf{a}, \overline{\mathbf{Y}}_2 + \mathbf{a}, \mathbf{S})$ where $\mathbf{a} \in R^r$;
(2) $g_2(\mathbf{X}, \overline{\mathbf{Y}}_1, \overline{\mathbf{Y}}_2, \mathbf{S}) = (\mathbf{AX}, \mathbf{A}\overline{\mathbf{Y}}_1, \mathbf{A}\overline{\mathbf{Y}}_2, \mathbf{ASA}')$ where $\mathbf{A}$ is an invertible $r \times r$ matrix.

Let

$$\mathbf{T} = \begin{pmatrix} T_{11} & T_{12} \\ T_{21} & T_{22} \end{pmatrix} = (\mathbf{X} - \overline{\mathbf{Y}}_1, \mathbf{X} - \overline{\mathbf{Y}}_2)'\mathbf{S}^{-1}(\mathbf{X} - \overline{\mathbf{Y}}_1, \mathbf{X} - \overline{\mathbf{Y}}_2),$$

$$\boldsymbol{\theta} = \begin{pmatrix} \theta_{11} & \theta_{12} \\ \theta_{21} & \theta_{22} \end{pmatrix} = (\boldsymbol{\mu} - \boldsymbol{\mu}_1, \boldsymbol{\mu} - \boldsymbol{\mu}_2)'\boldsymbol{\Sigma}^{-1}(\boldsymbol{\mu} - \boldsymbol{\mu}_1, \boldsymbol{\mu} - \boldsymbol{\mu}_2).$$

THEOREM 20.3. $\mathbf{T}$ is a maximal invariant for the discrimination problem and $\boldsymbol{\theta}$ is a parameter maximal invariant.

PROOF. See Exercise B2. □

Note that $\theta_{12} = 0$ under both the null and alternative hypotheses. We are therefore testing that $\theta_{11} = 0$, $\theta_{12} = 0$ against $\theta_{22} = 0$, $\theta_{12} = 0$. There is no UMP invariant test for this problem. Since $\theta_{12} = 0$ under both hypotheses, it seems likely that sensible procedures would depend only on T_{11} and T_{22}. In this section we consider two procedures, neither of which depends on T_{12}.

The most commonly used procedure is based on the substitution principle discussed in Section 19.2.4. In the last section, we showed that if $\boldsymbol{\mu}_1$, $\boldsymbol{\mu}_2$, and $\boldsymbol{\Sigma}$ were known then optimal procedures had the form

$$W = (\boldsymbol{\mu}_2 - \boldsymbol{\mu}_1)'\boldsymbol{\Sigma}^{-1}(\mathbf{X} - \tfrac{1}{2}(\boldsymbol{\mu}_1 + \boldsymbol{\mu}_2)), \qquad \phi_k(W) = \begin{cases} 1 & \text{if } W > \log k \\ 0 & \text{if } W \leqslant \log k \end{cases}.$$

By the substitution principle, we should substitute $\overline{\mathbf{Y}}_1, \overline{\mathbf{Y}}_2$, and $\mathbf{S}$ for $\boldsymbol{\mu}_1$, $\boldsymbol{\mu}_2$, and $\boldsymbol{\Sigma}$ in this procedure. This leads to

$$V = (\overline{\mathbf{Y}}_2 - \overline{\mathbf{Y}}_1)\mathbf{S}^{-1}\left(\mathbf{X} - \tfrac{1}{2}(\overline{\mathbf{Y}}_1 + \overline{\mathbf{Y}}_2)\right) = \tfrac{1}{2}(T_{11} - T_{22}),$$

$$\phi_k^*(V) = \begin{cases} 1 & \text{if } V > \log k \\ 0 & \text{if } V \leqslant \log k \end{cases}. \tag{20.3}$$

V is called the *sample linear discriminant function.* Since $\overline{\mathbf{Y}}_1 \xrightarrow{p} \boldsymbol{\mu}_1$, $\overline{\mathbf{Y}}_2 \xrightarrow{p} \boldsymbol{\mu}_2$ and $\mathbf{S} \xrightarrow{p} \boldsymbol{\Sigma}$ as $n \to \infty$ and $m \to \infty$, this procedure is nearly as good as the optimal procedure derived in the last section when n and m are quite large. This test was suggested by Wald (1944) and its properties studied by Wald (1944) and Anderson (1951). It is often called the Wald-Anderson statistic or the Anderson statistic.

Another procedure that is used is the likelihood ratio test, which is given by

$$U = \frac{n + m - 2 + \frac{m}{m+1}(\mathbf{X} - \overline{\mathbf{Y}}_2)'\mathbf{S}^{-1}(\mathbf{X} - \overline{\mathbf{Y}}_2)}{n + m - 2 + \frac{n}{n+1}(\mathbf{X} - \overline{\mathbf{Y}}_1)'\mathbf{S}^{-1}(\mathbf{X} - \overline{\mathbf{Y}}_1)}$$

$$= \frac{n + m - 2 + \frac{m}{m+1} T_{22}}{n + m - 2 + \frac{n}{n+1} T_{11}}$$

$$\phi_k^{**}(U) = \begin{cases} 1 & \text{if } U > k \\ 0 & \text{if } U \leqslant k \end{cases} \tag{20.4}$$

(see Exercise B3). Kiefer and Schwartz (1965) show that this procedure is an admissible procedure.

To use either of these procedures, it is necessary to choose the cut-off point k. To do this sensibly, we need to have some idea of the probabilities of error for particular choices of k. We discuss methods of estimating these probabilities in the next section.

20.4. PROBABILITIES OF MISCLASSIFICATION.

In Section 20.2, we find exact expressions for the misclassification probabilities for the optimal rules in the case when the parameters are known. Unfortunately, in the case of unknown parameters, there are not useful exact expressions for these probabilitiies. We now consider several methods for approximating these probabilities. Although methods 2 and 3 are applicable to general rules, we restrict attention to the rule based on the sample linear discriminant function defined in (20.3). We want to estimate the two probabilities α and β, where α is the probability of classifying someone who belongs to population 1 as belonging to population 2 (i.e., the probability of type I error), and β is the probability of classifying someone from population 2 as belonging to population 1 (i.e., the probability of type II error).

METHOD 1. In Section 20.2, we found that the misclassification probabilities for the optimal rule based on W were

$$\alpha_W = 1 - N\left(\frac{\log k + \frac{1}{2}\delta}{\sqrt{\delta}}\right), \qquad \beta_W = N\left(\frac{\log k - \frac{1}{2}\delta}{\sqrt{\delta}}\right)$$

where $\delta = (\boldsymbol{\mu}_1 - \boldsymbol{\mu}_2)\boldsymbol{\Sigma}^{-1}(\boldsymbol{\mu}_1 - \boldsymbol{\mu}_2)$ and N is the standard normal distribution function. We note that sample discriminant function V is just an estimator of W, and hence the probabilities of error should be about the same. We now estimate δ from the data. Let $\hat{\delta} = (\overline{\mathbf{Y}}_1 - \overline{\mathbf{Y}}_2)'\mathbf{S}^{-1}(\overline{\mathbf{Y}}_1 - \overline{\mathbf{Y}}_2)$. $\hat{\delta}$ would seem to be a sensible estimator of δ, and therefore, we could estimate α and β by

$$\hat{\alpha}_1 = 1 - N\left(\frac{\log k + \frac{1}{2}\hat{\delta}}{\sqrt{\hat{\delta}}}\right), \qquad \hat{\beta}_1 = N\left(\frac{\log k - \frac{1}{2}\hat{\delta}}{\sqrt{\hat{\delta}}}\right)$$

In Exercise B4, you are asked to show that

$$E\hat{\delta} = \frac{n+m-2}{n+m-r-3}\left[\delta + r\left(\frac{1}{n} + \frac{1}{m}\right)\right] > \delta. \tag{20.5}$$

Therefore, $\hat{\delta}$ would tend to overestimate δ, and α and β would be underestimated. A better estimator of δ would be

$$\hat{\hat{\delta}} = \frac{n+m-r-3}{n+m-2}\hat{\delta} - r\left(\frac{1}{n} + \frac{1}{m}\right).$$

Then α and β would be estimated by

$$\hat{\hat{\alpha}}_1 = 1 - N\left[\frac{\log k + \frac{1}{2}\hat{\hat{\delta}}}{\sqrt{\hat{\hat{\delta}}}}\right], \qquad \hat{\hat{\beta}}_1 = N\left[\frac{\log k - \frac{1}{2}\hat{\hat{\delta}}}{\sqrt{\hat{\hat{\delta}}}}\right].$$

METHOD 2 (Resubstitution). In this method, we just apply the classification rule, ϕ, to each of the observations whose classifications we already know. We define $\hat{\alpha}_2$ to be the proportion of the $\mathbf{Y}_{1j}$ that the rule would incorrectly classify and let $\hat{\beta}_2$ be the proportion of the $\mathbf{Y}_{2j}$ that the rule would classify incorrectly. These estimators are obviously rather crude, since the classification rule is assessed with the same data used to define it.

METHOD 3 (Modified resubstitution). In this method, we eliminate one observation from sample 1, and recompute the discriminant rule. Then we see if this rule would classify that observation correctly. Let $\overline{\mathbf{Y}}_{1j}^*$ and $\mathbf{S}_j^*$ be the sample mean and pooled sample covariance matrix computed without $\mathbf{Y}_{1j}$, and let

$$V_{1j} = \left(\overline{\mathbf{Y}}_2 - \overline{\mathbf{Y}}_{1j}^*\right)\mathbf{S}_j^{*-1}\left[\mathbf{Y}_{1j} - \left(\overline{\mathbf{Y}}_{1j}^* + \overline{\mathbf{Y}}_2\right)\right].$$

If $V_{1j} \leqslant \log k$, we would have classified $\mathbf{Y}_{1j}$ correctly, and if $V_{1j} > k$, we would have classified it incorectly. We then define $\hat{\alpha}_3$ to be the proportion of the $\mathbf{Y}_{1j}$ that would have been classified incorrectly. Then we compute V_{2j} in a similar fashion. If $V_{2j} > \log k$, then we would have classified $\mathbf{Y}_{2j}$ correctly, and otherwise incorrectly. We define $\hat{\beta}_3$ to be the proportion of the $\mathbf{Y}_{2j}$ that would have been classified incorrectly.

METHOD 4 (Lachenbruch and Mickey, 1968). Let V_{ij} be defined as in method 3. In this method, we look at $V_{11}, \ldots, V_{1n}$ as a sample from the distribution of $V(\mathbf{X})$ if the observation $\mathbf{X}$ came from population 1. Let $\overline{V}_1$ and s_1^2 be the sample mean and sample variance of the V_{1j}. Then, if $\mathbf{X}$ comes from population 1, $[V(\mathbf{X}) - \overline{V}_1]/s_1$ should be approximately normally distributed with mean 0 and variance 1. Therefore, we approximate α by

$$\hat{\alpha}_4 = 1 - N\left(\frac{\log k - \overline{V}_1}{s_1}\right).$$

We could approximate β in a similar fashion.

METHOD 5 (Asymptotic method). Okamoto (1963) has given an asymptotic expansion of the distribution of $V(\mathbf{X})$ under the null and alternative hypotheses. The distribution depends on the unknown parameters. Estimators for these unknown parameters are then substituted for the parameters, and the probabilities of approximate probabilities of misclassification are computed.

Lachenbruch and Mickey (1968) have done a Monte Carlo study to compare the effectiveness of these five methods and some variations on them. The conclusion of that study is that method 5 is the best for large samples, with methods 3 and 4 very close to it. Methods 1 and 2 seem to be quite poor. Method 3 has the advantage of not using the assumed normality, so that $\hat{\alpha}_3$ and $\hat{\beta}_3$ are fairly good estimators even if the data from the two populations do not come from a normal distribution.

20.5. A BAYESIAN PERSPECTIVE.

When doing discriminant analysis, we often have prior knowledge of the probability that an individual comes from a particular population. In this section, we discuss a Bayesian approach to discrimination that utilizes these probabilities to help choose the cut-off point in the test based on the linear discriminant function and sample linear discriminant function. We assume that the reader is familiar with the Bayes test for a problem with a simple null and a simple alternative (see, for example, Mood et al., 1974, pp. 417–418).

Let p be the prior probability that an individual comes from population 1. We also assume two numbers $c(1\,|\,2)$ and $c(2\,|\,1)$, where $c(1\,|\,2)$ represents the loss if we classify an individual from population 2 into population 1 (i.e., the loss for a type II error), and $c(2\,|\,1)$ represents the loss if we classify an individual from population 1 into population 2 (i.e., the loss for a type I error).

We first look at the model in which we assume that $\boldsymbol{\mu}_1$, $\boldsymbol{\mu}_2$, and $\boldsymbol{\Sigma}$ are known. The Bayes rule is defined to be the rule that minimizes the expected loss, that is, the rule that minimizes

$$pc(2\,|\,1)E_{\mu_1}\phi + (1-p)c(1\,|\,2)E_{\mu_2}(1-\phi).$$

When $\boldsymbol{\mu}_1$, $\boldsymbol{\mu}_2$, and $\boldsymbol{\Sigma}$ are known, the testing problem is just the problem of testing a sample null against a simple alternative, and the Bayes rule is given by

$$\lambda = \frac{f(\mathbf{X};\,\boldsymbol{\mu}_2,\boldsymbol{\Sigma})}{f(\mathbf{X};\,\boldsymbol{\mu}_1,\boldsymbol{\Sigma})}, \qquad k = \frac{pc(2\,|\,1)}{(1-p)c(1\,|\,2)}, \qquad \phi(\lambda) = \begin{cases} 1 & \text{if } \lambda > k \\ 0 & \text{if } \lambda \leqslant k \end{cases} \tag{20.6}$$

Substituting the densities into (20.6) leads to the rule $\phi_k(W)$ defined in (20.2), with k defined by (20.6). Therefore, knowledge of p, $c(1\,|\,2)$ and $c(2\,|\,1)$ allows us to choose k in an optimal fashion.

Now suppose that $\boldsymbol{\mu}_1$, $\boldsymbol{\mu}_2$, and $\boldsymbol{\Sigma}$ are unknown. The substitution principle leads to the rule $\phi_k(V)$ defined by (20.3) with k defined in (20.6). It should be mentioned that this rule is not a Bayes rule for this problem since we have not chosen a prior for $\boldsymbol{\mu}_1$, $\boldsymbol{\mu}_2$, and $\boldsymbol{\Sigma}$.

Often when we do not know the prior probability p, the observations $\mathbf{Y}_{ij}$ can be considered as a sample from a larger population. In this case n is a random variable and $n/(n+m)$ is an unbiased estimator of p. Therefore we could substitute $n/(n+m)$ for p in the above procedure.

20.6. TESTING A HYPOTHESIS ABOUT THE DISCRIMINANT COEFFICIENTS.

Let

$$\mathbf{X} = \begin{pmatrix} \mathbf{X}_1 \\ \mathbf{X}_2 \end{pmatrix}, \qquad \boldsymbol{\nu} = \boldsymbol{\Sigma}^{-1}(\boldsymbol{\mu}_1 - \boldsymbol{\mu}_2) = \begin{pmatrix} \boldsymbol{\nu}_1 \\ \boldsymbol{\nu}_2 \end{pmatrix}$$

where $\mathbf{X}_1$ and $\boldsymbol{\nu}_1$ are $q \times 1$. Then the linear discrimiant function is given by

$$W = \boldsymbol{\nu}_1'\mathbf{X}_1 + \boldsymbol{\nu}_2'\mathbf{X}_2 - \boldsymbol{\nu}'\frac{(\boldsymbol{\mu}_1 + \boldsymbol{\mu}_2)}{2}.$$

If $\boldsymbol{\nu}_1 = \mathbf{0}$, this function does not depend on $\mathbf{X}_1$, and hence, under this condition, $\mathbf{X}_1$ is not contributing to the discrimination and could be eliminated. In this section, we consider testing that $\boldsymbol{\nu}_1 = 0$ against $\boldsymbol{\nu}_1$ unrestricted. (Note that this hypothesis is not a symmetric hypothesis as discussed in Section 20.1. Therefore, we treat it as we have treated other testing problems in this book.)

Let $\overline{\mathbf{Y}}_1, \overline{\mathbf{Y}}_2$, and $\mathbf{S}$ be as defined previously. We now reduce this problem by invariance. We first note this problem is invariant under the group G_1 of transformations of the form

$$g_1(\overline{\mathbf{Y}}_1, \overline{\mathbf{Y}}_2, \mathbf{S}) = (\overline{\mathbf{Y}}_1 + \mathbf{a}, \overline{\mathbf{Y}}_2 + \mathbf{a}, \mathbf{S}).$$

Let

$$\mathbf{U} = \left(\frac{1}{n} + \frac{1}{m}\right)(\overline{\mathbf{Y}}_2 - \overline{\mathbf{Y}}_1), \qquad \boldsymbol{\theta} = \left(\frac{1}{n} + \frac{1}{m}\right)(\boldsymbol{\mu}_1 - \boldsymbol{\mu}_2).$$

Then $(\mathbf{U}, \mathbf{S})$ is a maximal invariant for this group and $(\boldsymbol{\theta}, \boldsymbol{\Sigma})$ is a parameter maximal invariant. In addition $\mathbf{U}$ and $\mathbf{S}$ are independent,

$$\mathbf{U} \sim N_r(\boldsymbol{\theta}, \boldsymbol{\Sigma}), \qquad (n+m-2)\mathbf{S} \sim W_r((n+m-2), \boldsymbol{\Sigma}),$$

$$\left(\frac{1}{n} + \frac{1}{m}\right)\boldsymbol{\nu} = \boldsymbol{\Sigma}^{-1}\boldsymbol{\theta}.$$

Before listing three more groups that leave this problem invariant, we note if $\bar{g}(\boldsymbol{\theta}, \boldsymbol{\Sigma}) = (\mathbf{A}\boldsymbol{\theta}, \mathbf{A}\boldsymbol{\Sigma}\mathbf{A}')$, then $\bar{g}(\boldsymbol{\nu}) = (\mathbf{A}')^{-1}\boldsymbol{\nu}$. Using this fact, we can see that the problem of testing that $\boldsymbol{\nu}_1 = \mathbf{0}$ is invariant under the three additional groups G_2, G_3, and G_4 consisting of transformations of the following form. Let

$$\mathbf{U} = \begin{pmatrix} \mathbf{U}_1 \\ \mathbf{U}_2 \end{pmatrix}, \qquad \mathbf{S} = \begin{pmatrix} \mathbf{S}_{11} & \mathbf{S}_{12} \\ \mathbf{S}_{21} & \mathbf{S}_{22} \end{pmatrix}, \qquad \boldsymbol{\theta} = \begin{pmatrix} \boldsymbol{\theta}_1 \\ \boldsymbol{\theta}_2 \end{pmatrix}, \qquad \boldsymbol{\Sigma} = \begin{pmatrix} \boldsymbol{\Sigma}_{11} & \boldsymbol{\Sigma}_{12} \\ \boldsymbol{\Sigma}_{21} & \boldsymbol{\Sigma}_{22} \end{pmatrix}$$

where $\mathbf{U}_1$ and $\boldsymbol{\theta}_1$ are $q \times 1$ and $\mathbf{S}_{11}$ and $\boldsymbol{\Sigma}_{11}$ are $q \times q$.

(1) G_2 consists of transformations of the form

$$g_2(\mathbf{U}_1, \mathbf{U}_2, \mathbf{S}_{11}, \mathbf{S}_{12}, \mathbf{S}_{22}) = (\mathbf{U}_1 + \mathbf{C}\mathbf{U}_2, \mathbf{U}_2, \mathbf{S}_{11} + \mathbf{C}\mathbf{S}_{21} + \mathbf{S}_{12}\mathbf{C}' + \mathbf{C}\mathbf{S}_{22}\mathbf{C}', \mathbf{S}_{12} + \mathbf{C}\mathbf{S}_{22}, \mathbf{S}_{22}),$$

where $\mathbf{C}$ is an arbitrary $q \times (r-q)$ matrix. Note that

$$\begin{pmatrix} \mathbf{I} & \mathbf{0} \\ \mathbf{C}' & \mathbf{I} \end{pmatrix}^{-1} = \begin{pmatrix} \mathbf{I} & \mathbf{0} \\ -\mathbf{C}' & \mathbf{I} \end{pmatrix}.$$

(2) G_3 consists of transformations of the form

$$g_3(\mathbf{U}_1, \mathbf{U}_2, \mathbf{S}_{11}, \mathbf{S}_{12}, \mathbf{S}_{22}) = (\mathbf{U}_1, \mathbf{D}\mathbf{U}_2, \mathbf{S}_{11}, \mathbf{S}_{12}\mathbf{D}', \mathbf{D}\mathbf{S}_{22}\mathbf{D}')$$

where $\mathbf{D}$ is an arbitrary $(r-q) \times (r-q)$ invertible matrix.

(3) G_4 consists of transformtions of the form

$$g_4(\mathbf{U}_1, \mathbf{U}_2, \mathbf{S}_{11}, \mathbf{S}_{12}, \mathbf{S}_{22}) = (\mathbf{E}\mathbf{U}_1, \mathbf{U}_2, \mathbf{E}\mathbf{S}_{11}\mathbf{E}', \mathbf{E}\mathbf{S}_{12}, \mathbf{S}_{22})$$

where $\mathbf{E}$ is an arbitrary $q \times q$ invertible matrix. We are now ready to find a maximal invariant for this problem. Let

$$W_1 = \frac{\mathbf{U}_2'\mathbf{S}_{22}^{-1}\mathbf{U}_2}{n+m-2},$$

$$W_2 = \frac{(\mathbf{U}_1 - \mathbf{S}_{12}\mathbf{S}_{22}^{-1}\mathbf{U}_2)'(\mathbf{S}_{11} - \mathbf{S}_{12}\mathbf{S}_{22}^{-1}\mathbf{S}_{21})^{-1}(\mathbf{U}_1 - \mathbf{S}_{12}\mathbf{S}_{22}^{-1}\mathbf{U}_2)}{(n+m-2)(1+W_1)}$$

$$\delta_1 = \theta_2'\boldsymbol{\Sigma}_{22}^{-1}\boldsymbol{\theta}_2,$$

$$\delta_2 = (\boldsymbol{\theta}_1 - \boldsymbol{\Sigma}_{12}\boldsymbol{\Sigma}_{22}^{-1}\boldsymbol{\theta}_2)'(\boldsymbol{\Sigma}_{11} - \boldsymbol{\Sigma}_{12}\boldsymbol{\Sigma}_{22}^{-1}\boldsymbol{\Sigma}_{21})^{-1}(\boldsymbol{\theta}_1 - \boldsymbol{\Sigma}_{12}\boldsymbol{\Sigma}_{22}^{-1}\boldsymbol{\theta}_2).$$

LEMMA 20.4. A maximal invariant under the groups G_1, G_2, G_3, and G_4 is (W_1, W_2). A parameter maximal invariant is (δ_1, δ_2).

PROOF. A maximal invariant under G_2 is

$$(\mathbf{U}_1 - \mathbf{S}_{12}\mathbf{S}_{22}^{-1}\mathbf{U}_2, \mathbf{U}_2, \mathbf{S}_{11} - \mathbf{S}_{12}\mathbf{S}_{22}^{-1}\mathbf{S}_{21}, \mathbf{S}_{22})$$

(see Exercise B5). Now G_3 becomes transformations of the following form

$$g_3(\mathbf{U}_1 - \mathbf{S}_{12}\mathbf{S}_{22}^{-1}\mathbf{U}_2, \mathbf{U}_2, \mathbf{S}_{11} - \mathbf{S}_{12}\mathbf{S}_{22}^{-1}\mathbf{S}_{21}, \mathbf{S}_{22})$$
$$= (\mathbf{U}_1 - \mathbf{S}_{12}\mathbf{S}_{22}^{-1}\mathbf{U}_2, \mathbf{D}\mathbf{U}_2, \mathbf{S}_{11} - \mathbf{S}_{12}\mathbf{S}_{22}^{-1}\mathbf{S}_{21}, \mathbf{D}\mathbf{S}_{22}\mathbf{D}')$$

and a maximal invariant is $(W_1, \mathbf{U}_1 - \mathbf{S}_{12}\mathbf{S}_{22}^{-1}\mathbf{U}_2, \mathbf{S}_{11} - \mathbf{S}_{12}\mathbf{S}_{22}^{-1}\mathbf{S}_{21})$. Now, G_4 becomes transformations of the form

$$g_4(W_1, \mathbf{U}_1 - \mathbf{S}_{12}\mathbf{S}_{22}^{-1}\mathbf{U}_2, \mathbf{S}_{11} - \mathbf{S}_{12}\mathbf{S}_{22}^{-1}\mathbf{S}_{21})$$
$$= \left(W_1, \mathbf{E}(\mathbf{U}_1 - \mathbf{S}_{12}\mathbf{S}_{22}^{-1}\mathbf{U}_2), \mathbf{E}(\mathbf{S}_{11} - \mathbf{S}_{12}\mathbf{S}_{22}^{-1}\mathbf{S}_{21})\mathbf{E}'\right)$$

and a maximal invariant is

$$\left[W_1, (\mathbf{U}_1 - \mathbf{S}_{12}\mathbf{S}_{22}^{-1}\mathbf{U}_2)'(\mathbf{S}_{11} + \mathbf{S}_{12}\mathbf{S}_{22}^{-1}\mathbf{S}_{21})^{-1}(\mathbf{U}_1 - \mathbf{S}_{12}\mathbf{S}_{22}^{-1}\mathbf{U}_2)\right].$$

(W_1, W_2) is an invertible function of this maximal invariant and is therefore also a maximal invariant for this problem. The proof for the parameter maximal invariant follows similarly. □

We now find the joint distribution of W_1 and W_2.

LEMMA 20.5. The joint distribution of W_1 and W_2 is given by

$$\frac{n+m-r+q-1}{r-q} W_1 \sim F_{r-q,\, n+m-r+q-1}(\delta_1),$$

$$\frac{(n+m-r-1)W_2}{q} \,|\, W_1 \sim F_{q,\, n+m-r-1}\left(\frac{\delta_2}{1+W_1}\right).$$

PROOF. Let

$$\mathbf{T} = \begin{pmatrix} \mathbf{T}_{11} & \mathbf{T}_{12} \\ \mathbf{T}_{21} & \mathbf{T}_{22} \end{pmatrix} = (n + m - 2)\mathbf{S}, \qquad k = n + m - 2.$$

Then

$$\mathbf{W}_1 = \mathbf{U}_2'\mathbf{T}_{22}^{-1}\mathbf{U}_2,$$

$$W_2 = \frac{(\mathbf{U}_1 - \mathbf{T}_{12}\mathbf{T}_{22}^{-1}\mathbf{U}_2)'(\mathbf{T}_{11} - \mathbf{T}_{12}\mathbf{T}_{22}^{-1}\mathbf{T}_{21})^{-1}(\mathbf{U}_1 - \mathbf{T}_{12}\mathbf{T}_{22}^{-1}\mathbf{U}_2)}{1 + W_1}$$

Now, $\mathbf{U}$ and $\mathbf{T}$ are independent, $\mathbf{U} \sim N_r(\boldsymbol{\theta}, \boldsymbol{\Sigma}), \mathbf{T} \sim W_r(k, \boldsymbol{\Sigma})$. By Theorem 3.7, we see that

$$\mathbf{U}_2 \sim N_{r-q}(\boldsymbol{\theta}_2, \boldsymbol{\Sigma}_{22}),$$

$$\mathbf{U}_1 \mid \mathbf{U}_2 \sim N_q(\boldsymbol{\theta}_1 - \boldsymbol{\Sigma}_{12}\boldsymbol{\Sigma}_{22}^{-1}\boldsymbol{\theta}_2 + \boldsymbol{\Sigma}_{12}\boldsymbol{\Sigma}_{22}^{-1}\mathbf{U}_2, \boldsymbol{\Sigma}_{11} - \boldsymbol{\Sigma}_{12}\boldsymbol{\Sigma}_{22}^{-1}\boldsymbol{\Sigma}_{21}). \quad (20.7)$$

By Lemma 17.9, we see that

$$\mathbf{T}_{22} \sim W_{r-q}(k, \boldsymbol{\Sigma}_{22}),$$

$$\mathbf{T}_{12}\mathbf{T}_{22}^{-1} \mid \mathbf{T}_{22} \sim N_{q,r-q}(\boldsymbol{\Sigma}_{12}\boldsymbol{\Sigma}_{22}^{-1}, \boldsymbol{\Sigma}_{11} - \boldsymbol{\Sigma}_{12}\boldsymbol{\Sigma}_{22}^{-1}\boldsymbol{\Sigma}_{21}, \mathbf{T}_{22}^{-1}), \quad (20.8)$$

$$\mathbf{T}_{11} - \mathbf{T}_{12}\mathbf{T}_{22}^{-1}\mathbf{T}_{21} \mid (\mathbf{T}_{12}, \mathbf{T}_{22}) \sim W_q(k - r + q, \boldsymbol{\Sigma}_{11} - \boldsymbol{\Sigma}_{12}\boldsymbol{\Sigma}_{22}^{-1}\boldsymbol{\Sigma}_{21}). \quad (20.9)$$

By Theorem 17.11, we see that

$$\frac{k - r + q + 1}{r - q} W_1 \sim F_{r-q, k-r+q+1}(\delta_1).$$

We now find the conditional distribution of W_2 given $(\mathbf{U}_2, \mathbf{T}_{22})$. From (20.8), we see that $\mathbf{T}_{12}\mathbf{T}_{22}^{-1}\mathbf{U}_2 \mid (\mathbf{U}_2, \mathbf{T}_{22}) \sim \mathrm{N}_{q,1}(\boldsymbol{\Sigma}_{12}\boldsymbol{\Sigma}_{22}^{-1}\mathbf{U}_2, \boldsymbol{\Sigma}_{11} - \boldsymbol{\Sigma}_{12}\boldsymbol{\Sigma}_{22}^{-1}\boldsymbol{\Sigma}_{21}, W_1)$. Also, $\mathbf{U}_1$ and $\mathbf{T}_{12}\mathbf{T}_{22}^{-1}\mathbf{U}_2$ are independent conditionally on $(\mathbf{U}_2, \mathbf{T}_{22})$. Therefore,

$$(\mathbf{U}_1 - \mathbf{T}_{12}\mathbf{T}_{22}^{-1}\mathbf{U}_2) \mid (\mathbf{U}_2, \mathbf{T}_{22})$$
$$\sim N_q(\boldsymbol{\theta}_1 - \boldsymbol{\Sigma}_{12}\boldsymbol{\Sigma}_{22}^{-1}\boldsymbol{\theta}_2, (1 + W_1)(\boldsymbol{\Sigma}_{11} - \boldsymbol{\Sigma}_{12}\boldsymbol{\Sigma}_{22}^{-1}\boldsymbol{\Sigma}_{21})).$$

Finally, by (20.9), we see that conditionally on $(\mathbf{U}_2, \mathbf{T}_{22})$, $\mathbf{T}_{11} - \mathbf{T}_{12}\mathbf{T}_{22}^{-1}\mathbf{T}_{21} \sim W_q(k - r + q, \boldsymbol{\Sigma}_{11} - \boldsymbol{\Sigma}_{12}\boldsymbol{\Sigma}_{22}^{-1}\boldsymbol{\Sigma}_{21})$ independently of $\mathbf{U}_1 - \mathbf{T}_{12}\mathbf{T}_{22}^{-1}\mathbf{U}_2$. Therefore, by Theorem 17.11, we see that

$$\frac{k - r + 1}{q} W_2 \mid (\mathbf{U}_2, \mathbf{T}_{22}) \sim F_{q, k-r+q}\left(\frac{\delta_2}{1 + W_1}\right).$$

Since the conditional distribution of W_2 given $(\mathbf{U}_2, \mathbf{T}_{22})$ depends on $\mathbf{U}_2$ and $\mathbf{T}_{22}$ only through W_1, we see that

$$\frac{k - r + 1}{q} W_2 \mid W_1 \sim F_{q, k-r+1}\left(\frac{\delta_2}{1 + W_1}\right). \qquad \square$$

We now need to interpret the hypotheses in terms of δ_1 and δ_2. We first note that

$$\delta_2 = \boldsymbol{\nu}_1'(\boldsymbol{\Sigma}_{11} - \boldsymbol{\Sigma}_{12}\boldsymbol{\Sigma}_{22}^{-1}\boldsymbol{\Sigma}_{21})\boldsymbol{\nu}_1\left(\frac{1}{n} + \frac{1}{m}\right)^2 \tag{20.10}$$

(see Exercise B6). Therefore $\boldsymbol{\nu}_1 = \mathbf{0}$ if and only if $\delta_2 = 0$. Since $\delta_1 = \boldsymbol{\mu}_2'\boldsymbol{\Sigma}_{22}^{-1}\boldsymbol{\mu}_2$, δ_1 is unrestricted under both null and alternative hypotheses. Therefore, after reducing by invariance, we have the testing problem in which we observe W_1 and W_2 having the joint distribution described in Lemma 20.5, and we are testing that $\delta_2 = 0$, $\delta_1 \geqslant 0$ against $\delta_2 \geqslant 0$, $\delta_1 \geqslant 0$. There is no UMP invariant size α test for this problem. However, one sensible test is to reject if W_2 is too large. Therefore, let

$$\phi(W_2) = \begin{cases} 1 & \text{if } W_2 > \dfrac{q}{n+m-r-1} F^{\alpha}_{q,\, n+m-r-1} \\ 0 & \text{if } W_2 < \dfrac{q}{n+m-r-1} F^{\alpha}_{q,\, n+m-r-1}. \end{cases}$$

We note that this test is a size α test since $(n+m-r-1)W_2/q \sim F_{q,\, n+m-r-1}(0)$ under the null hypothesis. The alternative distribution is *not* a noncentral F-distribution. However, we can use the unbiasedness of the F-test to show that ϕ is unbiased as the following calculation shows:

$$E_{\delta_1, \delta_2}\phi(W_2) = E_{\delta_1}E_{\delta_2}\phi(W_2) \,|\, W_1 \geqslant E_{\delta_1}\alpha = \alpha.$$

In fact the test ϕ is UMP among all invariant unbiased size α tests. (See Giri, 1977, pp. 263–270 for details.)

20.7. DISCRIMINATION AMONG SEVERAL POPULATIONS.

In this section we consider the problem of discriminating among k populations. We assume that if an individual comes from the ith population, the r-dimensional random vector of his measurements would be normally distributed with mean vector $\boldsymbol{\mu}_i$, and covariance matrix $\boldsymbol{\Sigma}$ (the same for all populations). We have a vector of measurements $\mathbf{X}$ on an individual and we need to decide into which population to classify him. Throughout this section, we use a Bayesian approach similar to that used in Section 20.5.

We first look at the case in which $\boldsymbol{\mu}_1, \ldots, \boldsymbol{\mu}_k$ and $\boldsymbol{\Sigma}$ are known. Let p_i be the prior probability that an individual is from population i, and let $c(i \,|\, j)$ be the cost of classifying an individual into population i when he comes from population j. We assume that $c(i \,|\, i) = 0$.

A procedure S is a partition $S = (S_1, \ldots, S_k)$ (i.e., $S_i \cap S_j = \phi$ and $\cup S_i = R^r$). By this rule S, if $\mathbf{X} \in S_i$, we classify the person into the ith population.

Now let $f_i(\mathbf{x})$ be the density of $\mathbf{X}$ if the individual comes from population i. Define

$$R_S(i) = \sum_{j \neq i} c(j \mid i) \int_{S_j} f_i(\mathbf{x})\, d\mathbf{x}.$$

That is $R_S(i)$ is the expected loss if we use procedure S and the observation came from population i. The Bayes risk of S is defined to be

$$r_S = \sum_i p_i R_S(i),$$

that is, the risk averaged over the prior distribution. The *Bayes rule* is the rule that minimizes the Bayes risk. We now find the Bayes rule for this problem. Let

$$h_j(\mathbf{x}) = \sum_{i \neq j} p_i c(j \mid i) f_i(\mathbf{x}).$$

THEOREM 20.6. a. Let $S = (S_1, \ldots, S_k)$ be such that if $\mathbf{x} \in S_j$, then $h_j(\mathbf{x}) = \min_i h_i(\mathbf{x})$. Then S is a Bayes rule.

b. If $U = (U_1, \ldots, U_k)$ is a Bayes rule, then $P(\mathbf{X} \in U_j,\ h_j \neq \min_i h_i(\mathbf{X})) = 0$, for $j = 1, \ldots, k$.

PROOF. Let $U = (U_1, \ldots, U_k)$ be a rule. Then

$$r_U = \sum_i p_i R_U(i) = \sum_i \sum_{j \neq i} p_i c(j \mid i) \int_{U_j} f_i(\mathbf{x})\, \mathbf{dx}$$
$$= \sum_j \int_{U_j} \sum_{i \neq j} p_i c(j \mid i) f_i(\mathbf{x})\, \mathbf{dx} = \sum_j \int_{U_j} h_j(\mathbf{x})\, \mathbf{dx}.$$

Now let S be a rule such that if $\mathbf{x} \in S_j$, then $h_j(\mathbf{x}) = \min_i h_i(\mathbf{x})$. Then

$$r_S = \sum_j \int_{S_j} \min_i h_i(\mathbf{x})\, d\mathbf{x} = \int_{R^r} \min_i h_i(\mathbf{x})\, d\mathbf{x} = \sum_j \int_{U_j} \min_i h_i(\mathbf{x})\, d\mathbf{x}.$$

Therefore

$$r_U = \sum_j \int_{U_j} h_j(\mathbf{x})\, \mathbf{dx} \geqslant \sum_j \int_{U_j} \min_i h_i(\mathbf{x})\, d\mathbf{x} = r_S$$

and hence S is a Bayes rule. Now U will be a Bayes rule only if it has the same Bayes risk as S, that is, only if

$$P\Big(\mathbf{X} \in U_j, h_j(\mathbf{X}) \neq \min_i h_i(\mathbf{X})\Big) = 0. \quad \square$$

This theorem implies that the Bayes rules have the following form (up to sets of measure 0). We assign $\mathbf{X}$ to population j if j minimizes $h_j(\mathbf{X})$. If there is a tie for the minimum, it does not matter to which of the tied populations $\mathbf{X}$ is assigned.

We now apply this theorem to the problem in this section. We see that

$$h_j(\mathbf{X}) = \frac{1}{(2\pi)^{r/2}|\boldsymbol{\Sigma}|} \sum_{i \neq j} p_i c(j \mid i) \exp - \tfrac{1}{2}(\mathbf{X} - \boldsymbol{\mu}_i)'\boldsymbol{\Sigma}^{-1}(\mathbf{X} - \boldsymbol{\mu}_i).$$

Finding j that minimizes $h_j(X)$ is equivalent to finding j that minimizes

$$h_j^*(\mathbf{X}) = \sum_{i \neq j} p_i c(j \mid i) \exp - \tfrac{1}{2}(\mathbf{X} - \boldsymbol{\mu}_i)'\boldsymbol{\Sigma}^{-1}(\mathbf{X} - \boldsymbol{\mu}_i).$$

Therefore, the Bayes rule finds j that minimzes $h_j^*(\mathbf{X})$ and assigns the individual to that class. This rule can be somewhat simplified in the case in which $c(j \mid i) = d_i$ for $j \neq i$ (i.e., in which the cost of misclassification depends on the population to which the individual belongs, but not the population to which he is misclassified). In this case

$$h_j^*(\mathbf{X}) = \left[\sum_i p_i d_i \exp - \tfrac{1}{2}(\mathbf{X} - \boldsymbol{\mu}_i)'\boldsymbol{\Sigma}^{-1}(\mathbf{X} - \boldsymbol{\mu}_i) \right]$$
$$- p_j d_j \exp - \tfrac{1}{2}(\mathbf{X} - \boldsymbol{\mu}_j)'\boldsymbol{\Sigma}^{-1}(\mathbf{X} - \boldsymbol{\mu}_j).$$

Since the first term of this expression does not depend on j, we see that minimizing $h_j(X)$ is equivalent to maximizing

$$p_j d_j \exp - \tfrac{1}{2}(\mathbf{X} - \boldsymbol{\mu}_j)'\boldsymbol{\Sigma}^{-1}(\mathbf{X} - \boldsymbol{\mu}_j)$$

or equivalently minimizing

$$k_j(\mathbf{X}) = -2 \log p_j d_j + (\mathbf{X} - \boldsymbol{\mu}_j)'\boldsymbol{\Sigma}^{-1}(\mathbf{X} - \boldsymbol{\mu}_j).$$

If $p_j d_j = c$, this rule has a particularly simple interpretation. In that case the rule classifies the individual to that class whose mean has the shortest Mahalanobis distance from $\mathbf{X}$.

Now suppose that $\boldsymbol{\mu}_1, \ldots, \boldsymbol{\mu}_k$ and $\boldsymbol{\Sigma}$ are unknown, but that we have independent samples $\mathbf{Y}_{ij}$ such that $\mathbf{Y}_{ij} \sim N_r(\boldsymbol{\mu}_i, \boldsymbol{\Sigma})$, $j = 1, \ldots, n_i$. Let $N = \Sigma n_i$.

$$\overline{Y}_i = \frac{1}{n_i} \sum_j \mathbf{Y}_{ij}, \qquad \mathbf{S} = \frac{1}{N-k} \sum_i \sum_j (\mathbf{Y}_{ij} - \overline{\mathbf{Y}}_i)(\mathbf{Y}_{ij} - \overline{\mathbf{Y}}_i)'$$

be the ith sample mean and the pooled sample covariance matrix. The substitution principle implies that we should substitute $\overline{\mathbf{Y}}_i$ for $\boldsymbol{\mu}_i$ and $\mathbf{S}$ for $\boldsymbol{\Sigma}$ in $h_j^*(\mathbf{x})$, obtaining

$$\hat{h}_j^*(\mathbf{X}) = \sum_{i \neq j} p_i c(j \mid i) \exp - \tfrac{1}{2}(\mathbf{X} - \overline{\mathbf{Y}}_i)'\mathbf{S}^{-1}(\mathbf{X} - \overline{\mathbf{Y}}_i).$$

We now observe $\mathbf{X}$, find the j that minimizes $\hat{h}_j^*(\mathbf{X})$ and classify the individual into that population. If $c(j \mid i) = d_i$, this rule can be put in a simpler form.

We now discuss the model in which we have no prior probabilities p_i. Consider first the case in which the parameters $\boldsymbol{\mu}_1, \ldots, \boldsymbol{\mu}_k$ and $\boldsymbol{\Sigma}$ are known. Let C be the class of all rules of the form given in Theorem 20.6 for some prior distribution. It can be shown that C is an essentially complete class, so that it is not necessary to look outside this class. Furthermore, if the $p_i > 0$, then the Bayes rule is admissible. See Ferguson (1967) for derivations of these facts in a more general setting. However, these facts do not help us find a sensible procedure for a particular problem. One approach that is sometimes useful is the minimax approach. We say that S is a *minimax procedure* if

$$\max_i R_S(i) \leqslant \max_i R_U(i)$$

for all procedures U. That is, the minimax procedure minimizes the maximum risk. Such a procedure would be a conservative one, but in the absence of other information would seem sensible. Unfortunately, it is often easy to establish the existence of a minimax procedure, but it is usually difficult to find one. One fact that is often helpful is that a Bayes procedure S such that $R_S(i) = R_S(j)$ for all i and j is minimax (see Exercise C1).

Now consider the case in which the parameters are not known, nor are the priors. By the substitution principle, we would act as though the sample means were the population means and the pooled sample covariance matrix were the common covariance matrix, and proceed as suggested in the last paragraph.

20.8. FURTHER COMMENTS.

Our discussion of discrimination has been limited to discrimination between normal populations with the same covariance matrix. (The appropriate procedure for the case of known parameters but unequal covariance matrices is derived in Exercise B7. This procedure leads to the use of a quadratic discriminant function rather than a linear one.) However, it is often necessary to discriminate between nonnormal populations or populations with unequal covariance matrices and much work has been done on procedures for these situations. See Das Gupta (1973) for an extensive bibliography of such procedures.

EXERCISES.

Type A

1. Consider a discrimination problem with unknown parameters. Use the data of Exercise A4 of Chapter 18 to estimate the parameters.

(a) Find the sample linear discrimination function.

(b) If $k = 1$, how do we classify $\mathbf{X} = (7, 8)'$ using this function?

(c) Find the rule associated with probability of the first class being $1/3$ and $c(1|2) = 2$, $c(2|1) = 3$. How do we classify $\mathbf{X} = (7, 8)'$?

(d) Find the LRT associated with $k = 1$. How do we classify $\mathbf{X} = (7, 8)'$ with this rule?

2. Continue with the data from the last problem and the rule in part b.

(a) Use method 1 to estimate the probabilities of misclassification.

(b) Use method 2 to estimate the probabilities of misclassification.

(c) Use method 3 to estimate the probabilities of misclassification.

(d) Use method 4 to estimate the probabilities of misclassification.

3. Using the data from the above problems, test the hypothesis that the first coordinate is not contributing to the discrimination.

4. Now consider the problem of discriminating between 3 classes. Use the data of Exercise A2 of Chapter 19. Choose $c(j|1) = 2$, $c(j|2) = 1$, $c(j|3) = 2$ for all j. Let $p_1 = 1/2$, $p_2 = 1/4$ and $p_3 = 1/4$. How do we classify $\mathbf{X} = (7, 8)'$?

Type B

1. Prove Lemma 20.2 and its corollary.

2. Prove Theorem 20.3.

3. Show that the likelihood ratio tests for discrimination between two groups in the case of unknown parameters is given in (20.4).

4. Verify (20.5). (Note that $\hat{\delta}$ is a constant times a noncentral F variable. Use Lemma 1.8.)

5. (a) Show that the testing problem considered in Section 20.6 is invariant under the group G_2 defined there.

(b) Show that the maximal invariant under G_1 and G_2 is

$$(\mathbf{U}_1 - \mathbf{S}_{12}\mathbf{S}_{22}^{-1}\mathbf{U}_2, \mathbf{U}_2, \mathbf{S}_{11} - \mathbf{S}_{12}\mathbf{S}_{22}^{-1}\mathbf{S}_{21}, \mathbf{S}_{22}).$$

6. (a) Verify (20.10).

(b) Show that

$$W_2 = \frac{\mathbf{U}'\mathbf{S}^{-1}\mathbf{U} - \mathbf{U}_1\mathbf{S}_{11}^{-1}\mathbf{U}_1}{(n+m-2)(1+W_1)}.$$

7. Let $\mathbf{X} \sim N_r(\boldsymbol{\mu}, \boldsymbol{\Sigma})$ and consider testing that $\boldsymbol{\mu} = \boldsymbol{\mu}_2$, $\boldsymbol{\Sigma} = \boldsymbol{\Sigma}_1$ against

$\boldsymbol{\mu} = \boldsymbol{\mu}_2$, $\boldsymbol{\Sigma} = \boldsymbol{\Sigma}_2$. Let

$$T = \log \frac{|\boldsymbol{\Sigma}_2|}{|\boldsymbol{\Sigma}_1|} - \frac{(\boldsymbol{\mu}_1'\boldsymbol{\Sigma}_1\boldsymbol{\mu}_1 - \boldsymbol{\mu}_2'\boldsymbol{\Sigma}_2\boldsymbol{\mu}_2)}{2}$$

$$+ \mathbf{X}'(\boldsymbol{\Sigma}_1^{-1}\boldsymbol{\mu}_1 - \boldsymbol{\Sigma}_2^{-1}\boldsymbol{\mu}_2) - \frac{\mathbf{X}'(\boldsymbol{\Sigma}_1^{-1} - \boldsymbol{\Sigma}_2^{-1})\mathbf{X}}{2}$$

Show that the Neymann-Pearson test would have the form

$$\phi(T) = \begin{cases} 1 & \text{if } T > \log k \\ 0 & \text{if } T \leqslant \log k \end{cases}.$$

Suggest a procedure for testing this hypothesis when the parameters are not known but there are two samples; one from $N_r(\boldsymbol{\mu}_1, \boldsymbol{\Sigma}_1)$ and one from $N_r(\boldsymbol{\mu}_2, \boldsymbol{\Sigma}_2)$.

Type C

1. Suppose that a rule with constant risk is not minimax. Show that it cannot be Bayes for any prior distribution. Therefore, a constant risk Bayes rule must be minimax.

CHAPTER 21

Testing Hypotheses about the Covariance Matrix

In this chapter we consider various testing problems concerning covariance matrices. In the first section, we consider testing the independence of two random vectors. In Section 21.2, we consider testing that the covariance matrix $\mathbf{\Sigma}$ is equal to a specified matrix $\mathbf{\Sigma}_0$. We also find simultaneous confidence intervals for the set of all $\mathbf{d}'\mathbf{\Sigma}\mathbf{d}$. In Section 21.3 we look at testing that two covariance matrices are equal. In Section 21.4 we consider testing that the covariance matrix $\mathbf{\Sigma}$ has the form $\mathbf{\Sigma} = \sigma^2\mathbf{I}$. If $\mathbf{\Sigma} = \sigma^2\mathbf{I}$, then the data can be rearranged into a univariate linear model. The procedures for the univariate linear model are much more powerful than those for the multivariate linear model. In this section, we are therefore testing the null hypothesis that the univariate linear model is appropriate. Similarly, in Section 21.5, we consider testing that the repeated measures model is valid. If it is, then we can use the more powerful procedures designed for that model in place of the procedures for the multivariate linear model.

For convenience in notation, in the first three sections, we look only at one-sample and two-sample models. In the last two sections, we consider the tests in the context of the multivariate linear model, since that is where they would arise.

In connection with the last two sections it should be emphasized that hypotheses of interest are null hypotheses. Accepting the null hypothesis that $\mathbf{\Sigma} = \sigma^2\mathbf{I}$ is not really evidence that the univariate linear model is valid so much as a lack of evidence that it is not. On the other hand, rejecting the hypothesis that $\mathbf{\Sigma} = \sigma^2\mathbf{I}$ would be evidence that the univariate linear model is not valid.

Another caution about the procedures in this chapter is that with the exception of those in the first section, they are very sensitive to the normal assumption used in deriving them. (The first problem can be considered conditionally as a problem in which we are testing a multivariate linear hypothesis and therefore the procedures are less sensitive to the normal assumptions.) Most of these procedures should be used only when the experimenter is fairly certain that the observations have a normal distribution.

In this chapter we often use the following fact about the likelihood ratio test statistic λ. If we are testing that k independent functions of the parameter are 0, then $-2 \log \lambda$ is approximately distributed as a χ_k^2 if there are a large number of observations.

21.1. TESTING FOR INDEPENDENCE.

Let

$$\begin{pmatrix} \mathbf{X}_1 \\ \mathbf{Y}_1 \end{pmatrix}, \ldots, \begin{pmatrix} \mathbf{X}_n \\ \mathbf{Y}_n \end{pmatrix}$$

be independently, identically distributed random vectors in which $\mathbf{Y}_i$ is $p \times 1$ and $\mathbf{X}_i$ is $q \times 1$, $p \leqslant q$ and $n > p + q$. We assume that

$$\begin{pmatrix} \mathbf{Y}_i \\ \mathbf{X}_i \end{pmatrix} \sim N_{p+q}\left(\begin{pmatrix} \boldsymbol{\mu}_1 \\ \boldsymbol{\mu}_2 \end{pmatrix}, \begin{pmatrix} \boldsymbol{\Sigma}_{11} & \boldsymbol{\Sigma}_{12} \\ \boldsymbol{\Sigma}_{21} & \boldsymbol{\Sigma}_{22} \end{pmatrix} \right), \quad \begin{pmatrix} \boldsymbol{\Sigma}_{11} & \boldsymbol{\Sigma}_{12} \\ \boldsymbol{\Sigma}_{21} & \boldsymbol{\Sigma}_{22} \end{pmatrix} > 0$$

where $\boldsymbol{\mu}_1$ is $p \times 1$ and $\boldsymbol{\Sigma}_{11}$ is $p \times p$. We now consider the problem of testing that the $\mathbf{Y}_i$ and $\mathbf{X}_i$ are independent. By Theorem 3.7b, this is equivalent to testing that $\boldsymbol{\Sigma}_{12} = 0$ against $\boldsymbol{\Sigma}$ unspecified. If $p = 1$, this is the problem considered in Section 16.8.2 involving the multiple correlation coefficient.

This problem is invariant under the following three groups.

(1) G_1 consists of transformations of the form

$$g_1(\mathbf{Y}_i, \mathbf{X}_i) = (\mathbf{Y}_i + \mathbf{a}, \mathbf{X}_i + \mathbf{b}),$$

where $\mathbf{a} \in R^p$, $\mathbf{b} \in R^q$.

(2) G_2 consists of transformations of the form $g_2(\mathbf{Y}_i, \mathbf{X}_i) = (\mathbf{Y}_i, \mathbf{A}\mathbf{X}_i)$ where $\mathbf{A}$ is a $q \times q$ invertible matrix.

(3) G_3 consists of transformations of the form $g_3(\mathbf{Y}_i, \mathbf{X}_i) = (\mathbf{B}\mathbf{Y}_i, \mathbf{X}_i)$ where $\mathbf{B}$ is a $p \times p$ invertible matrix.

Let

$$\mathbf{S} = \frac{1}{n-1} \sum_{i=1}^{n} \begin{pmatrix} \mathbf{Y}_i - \mathbf{Y} \\ \mathbf{X}_i - \mathbf{X} \end{pmatrix} \begin{pmatrix} \mathbf{Y}_i - \mathbf{Y} \\ \mathbf{X}_i - \mathbf{X} \end{pmatrix}' = \begin{pmatrix} \mathbf{S}_{11} & \mathbf{S}_{12} \\ \mathbf{S}_{21} & \mathbf{S}_{22} \end{pmatrix},$$

(where $\mathbf{S}_{11}$ is $p \times p$) be the joint sample covariance matrix for the $\binom{\mathbf{Y}_i}{\mathbf{X}_i}$.

Now, let $t_1 \geqslant \cdots \geqslant t_p$ be the roots of

$$|\mathbf{S}_{12}\mathbf{S}_{22}^{-1}\mathbf{S}_{21} - t\mathbf{S}_{11}| = 0 \tag{21.1}$$

and $\tau_1 \geqslant \cdots \geqslant \tau_p$ be the roots of

$$|\mathbf{\Sigma}_{12}\mathbf{\Sigma}_{22}^{-1}\mathbf{\Sigma}_{21} - \tau\mathbf{\Sigma}_{11}| = 0. \tag{21.2}$$

THEOREM 21.1. The maximal invariant under G_1, G_2, and G_3 for testing that $\mathbf{\Sigma}_{12} = \mathbf{0}$ is the set of roots of (21.1). The parameter maximal invariant is the set of roots of (21.2).

PROOF. This model is just a one sample model. Therefore a sufficient statistic for this model is $(\overline{\mathbf{Y}}, \overline{\mathbf{X}}, \mathbf{S}_{11}, \mathbf{S}_{21}, \mathbf{S}_{22})$. In terms of the sufficient statistic G_1 becomes transformations of the form

$$g_1(\overline{\mathbf{Y}}, \overline{\mathbf{X}}, \mathbf{S}_{11}, \mathbf{S}_{21}, \mathbf{S}_{22}) = (\overline{\mathbf{Y}} + \mathbf{a}, \overline{\mathbf{X}} + \mathbf{b}, \mathbf{S}_{11}, \mathbf{S}_{21}, \mathbf{S}_{22})$$

and a maximal invariant is $(\mathbf{S}_{11}, \mathbf{S}_{21}, \mathbf{S}_{22})$. In terms of this maximal invariant, G_2 becomes transformations of the form

$$g_2(\mathbf{S}_{11}, \mathbf{S}_{21}, \mathbf{S}_{22}) = (\mathbf{S}_{11}, \mathbf{A}\mathbf{S}_{21}, \mathbf{A}\mathbf{S}_{22}\mathbf{A}')$$

and a maximal invariant is $(\mathbf{S}_{11}, \mathbf{S}_{12}\mathbf{S}_{22}^{-1}\mathbf{S}_{21})$. In terms of this maximal invariant, G_3 becomes transformations of the form

$$g_3(\mathbf{S}_{11}, \mathbf{S}_{12}\mathbf{S}_{22}^{-1}\mathbf{S}_{21}) = (\mathbf{B}\mathbf{S}_{11}\mathbf{B}', \mathbf{B}\mathbf{S}_{12}\mathbf{S}_{22}^{-1}\mathbf{S}_{21}\mathbf{B}')$$

and a maximal invariant is the set of roots of (21.1). The proof for the parameter maximal invariant follows similarly. □

There is no UMP invariant test for this problem. The likelihood ratio test is given by the following

$$\lambda = \frac{|\mathbf{S}|^{n/2}}{|\mathbf{S}_{11}|^{n/2}|\mathbf{S}_{22}|^{n/2}} = |\mathbf{I} - \mathbf{S}_{11}^{-1/2}\mathbf{S}_{12}\mathbf{S}_{22}^{-1}\mathbf{S}_{21}\mathbf{S}_{11}^{-1/2}|^{n/2} = \prod_{i=1}^{p} (1 - t_i)^{n/2} \tag{21.3}$$

$$\phi(\lambda) = \begin{cases} 1 & \text{if } \lambda < c \\ 0 & \text{if } \lambda \geqslant c \end{cases}.$$

(Note that the t_i are the eigenvalues of $\mathbf{S}_{11}^{-1/2}\mathbf{S}_{12}\mathbf{S}_{22}^{-1}\mathbf{S}_{21}\mathbf{S}_{11}^{-1/2}$.) This test is unbiased. (See Anderson and Das Gupta, 1964.) Since $\mathbf{\Sigma}_{12}$ has pq independent parameters, $-2\log\lambda$ has an approximate χ^2_{pq} distribution under the null hypothesis. We give a more refined approximation below.

We now look at the union-intersection test for this problem. Let $\mathbf{a} \in R^p$, $\mathbf{b} \in R^q$, $\mathbf{a} \neq 0$, $\mathbf{b} \neq 0$, and let $U_i^{\alpha} = \mathbf{a}'\mathbf{Y}_i$, $V_i^{\mathbf{b}} = \mathbf{b}'\mathbf{X}_i$. Then $\mathbf{\Sigma}_{12} = \mathbf{0}$ if and only if $U_i^{\mathbf{a}}$ and $V_i^{\mathbf{b}}$ are independent for all $\mathbf{a}$ and $\mathbf{b}$. Now, (U_i, V_i) are independently identically distributed and

$$\begin{pmatrix} U_i \\ V_i \end{pmatrix} \sim N_2\left(\begin{pmatrix} \mathbf{a}' & \boldsymbol{\mu}_1 \\ \mathbf{b}' & \boldsymbol{\mu}_2 \end{pmatrix}, \begin{pmatrix} \mathbf{a}'\boldsymbol{\Sigma}_{11}\mathbf{a} & \mathbf{a}'\boldsymbol{\Sigma}_{12}\mathbf{b} \\ \mathbf{b}'\boldsymbol{\Sigma}_{21}\mathbf{a} & \mathbf{b}'\boldsymbol{\Sigma}_{22}\mathbf{b} \end{pmatrix}\right).$$

Therefore, the problem of testing that $U_i^{\mathbf{a}}$ and $V_i^{\mathbf{b}}$ are independent reduces to the problem of testing that a simple correlation coefficient is 0. Let $r_{\mathbf{a},\mathbf{b}}$ be the sample correlation coefficient between $U_i^{\mathbf{a}}$ and $V_i^{\mathbf{b}}$. Then the UMP invariant test that $U_i^{\mathbf{a}}$ and $V_i^{\mathbf{b}}$ are independent is to reject if $r_{\mathbf{a},\mathbf{b}}^2$ is too large. Hence, the union-intersection test for testing that $\boldsymbol{\Sigma}_{12} = \mathbf{0}$ would reject if

$$\sup_{\mathbf{a}\neq\mathbf{0},\,\mathbf{b}\neq\mathbf{0}} r_{\mathbf{a},\mathbf{b}}^2$$

is too large. Now,

$$r_{\mathbf{a},\mathbf{b}}^2 = \frac{(\mathbf{a}'\mathbf{S}_{12}\mathbf{b})^2}{\mathbf{a}'\mathbf{S}_{11}\mathbf{a}\mathbf{b}'\mathbf{S}_{22}\mathbf{b}}. \tag{21.4}$$

Since t_1 is the largest root of (21.1), we see (using Lemmas 18.14 and 19.4) that

$$t_1 = \sup_{\mathbf{a}\neq\mathbf{0}} \frac{(\mathbf{a}'\mathbf{S}_{12})\mathbf{S}_{22}^{-1}(\mathbf{S}_{21}\mathbf{a})}{\mathbf{a}'\mathbf{S}_{11}\mathbf{a}} = \sup_{\mathbf{a}\neq\mathbf{0},\,\mathbf{b}\neq\mathbf{0}} \frac{(\mathbf{a}'\mathbf{S}_{12}\mathbf{b})^2}{\mathbf{a}'\mathbf{S}_{11}\mathbf{a}\mathbf{b}'\mathbf{S}_{22}\mathbf{b}}.$$

Therefore, the union intersection test is the one given by

$$\phi^*(t_1) = \begin{cases} 1 & \text{if } t_1 > c^* \\ 0 & \text{if } t_1 \leqslant c \end{cases}.$$

This test is also unbiased (see Mikhail and Roy, 1961).

In order to use union-intersection test, we need to find the null distribution of t_1. Let

$$\mathbf{V} = \mathbf{S}_{11} - \mathbf{S}_{12}\mathbf{S}_{22}^{-1}\mathbf{S}_{21}, \qquad \mathbf{Q} = \mathbf{S}_{12}\mathbf{S}_{22}^{-2}\mathbf{S}_{21}.$$

LEMMA 21.2. If $\boldsymbol{\Sigma}_{12} = 0$, then $\mathbf{V}$ and $\mathbf{Q}$ are independent and

$$\mathbf{V} \sim W_p\left(n-q-1, \frac{1}{n-1}\boldsymbol{\Sigma}_{11}\right), \qquad \mathbf{Q} \sim W_p\left(q, \frac{1}{n-1}\boldsymbol{\Sigma}_{11}\right).$$

PROOF. We note that $\mathbf{S} \sim W_{p+q}(n-1, [1/(n-1)]\boldsymbol{\Sigma})$. By Lemma 18.9, $\mathbf{V}$ is independent of $\mathbf{S}_{22}^{-1}\mathbf{S}_{21}$ and $\mathbf{S}_{22}$, and is hence independent of $\mathbf{Q}$. The distribution of $\mathbf{V}$ follows directly from Lemma 17.9 also. Finally, by Lemma 17.9 again, if $\boldsymbol{\Sigma}_{12} = \mathbf{0}$, then

$$\mathbf{S}_{22}^{-1}\mathbf{S}_{21} \mid \mathbf{S}_{22} \sim N_{q,p}\left(\mathbf{0}, \mathbf{S}_{22}^{-1}, \frac{1}{n-1}\boldsymbol{\Sigma}_{11}\right)$$

and hence

$$\mathbf{S}_{22}^{-1/2}\mathbf{S}_{21} = \mathbf{S}_{22}^{1/2}(\mathbf{S}_{22}^{-1}\mathbf{S}_{21}) \sim N_{q,\,p}\left(\mathbf{0}, \mathbf{I}, \frac{1}{n-1}\boldsymbol{\Sigma}_{11}\right).$$

Therefore, by the definition of the Wishart distribution, we see that

$$Q = (\mathbf{S}_{22}^{-1/2}\mathbf{S}_{21})'(\mathbf{S}_{22}^{-1/2}\mathbf{S}_{21}) \sim W_p\left(q, \frac{1}{n-1}\boldsymbol{\Sigma}_{11}\right). \quad \square$$

Now, t_1 is the largest root of

$$0 = |\mathbf{S}_{12}\mathbf{S}_{22}^{-1}\mathbf{S}_{21} - t\mathbf{S}_{11}| = |\mathbf{Q} - t(\mathbf{V} + \mathbf{Q})|.$$

Therefore, by the definition of the s-distribution (see Section 19.2.6),

$$t_1 \sim s_{p,\,q,\,n-q-1}$$

and the critical point for the size α largest root test can be determined from the tables of the s-distribution mentioned in Section 19.2.6.

Now, λ defined in (21.3) can be written as

$$\lambda^{-2/n} = |\mathbf{S}_{11} - \mathbf{S}_{12}\mathbf{S}_{22}^{-1}\mathbf{S}_{21}|/|\mathbf{S}_{11}| = |\mathbf{V}|/|\mathbf{V} + \mathbf{Q}|.$$

Therefore, $\lambda^{-2/n}$ has the same null distribution as λ_1, defined in Section 19.2 for testing the multivariate linear hypothesis (with r replaced by p, $p - k$ replaced by q, and $n - p$ replaced by $n - q - 1$). Hence any results about the null distribution of λ_1 for that model are also applicable to the distribution of $\lambda^{-2/n}$ for this model. In particular, we see that a better approximation to the null distribution of λ is given by $-2\rho \log\lambda$ is approximately χ^2_{pq}, where

$$\rho = 1 - \frac{p + q + 3}{2n}.$$

This approximation is accurate to three decimal places as long as

$$p^2 + q^2 \leqslant np/3$$

The exact conversion factors of Table 47 of Pearson and Hartley (1972) are also available.

There is a relationship between the testing problem considered in this section and the problem of testing the multivariate linear hypothesis considered in Section 19.2. This relationship is essentially the same as that between the univariate linear model and the correlation model discussed in Chapter 16. We now discus this relationship. Let

$$\mathbf{Y} = \begin{pmatrix} \mathbf{Y}_1' \\ \vdots \\ \mathbf{Y}_n' \end{pmatrix}, \qquad \mathbf{X} = \begin{pmatrix} \mathbf{X}_1' \\ \vdots \\ \mathbf{X}_n' \end{pmatrix}$$

Then $(\mathbf{Y}\ \mathbf{X}) \sim N_{n,\ p+q}(\mathbf{1}\boldsymbol{\mu}', \mathbf{I}, \boldsymbol{\Sigma})$, where $\mathbf{1}' = (1, \ldots, 1)$. Now, let

$$\mathbf{T} = (\mathbf{1X}), \qquad \boldsymbol{\alpha} = \boldsymbol{\mu}_1' - \boldsymbol{\mu}_2'\boldsymbol{\Sigma}_{11}^{-1}\boldsymbol{\Sigma}_{21}, \qquad \boldsymbol{\gamma} = \boldsymbol{\Sigma}_{22}^{-1}\boldsymbol{\Sigma}_{21}, \qquad \boldsymbol{\beta} = \begin{pmatrix}\boldsymbol{\alpha}\\ \boldsymbol{\gamma}\end{pmatrix},$$

$$\boldsymbol{\Xi} = \boldsymbol{\Sigma}_{11} - \boldsymbol{\Sigma}_{12}\boldsymbol{\Sigma}_{22}^{-1}\boldsymbol{\Sigma}_{21}.$$

Using Theorem 17.2g, we see that

$$\mathbf{Y}|\mathbf{X} \sim N_{n,\ p}(\mathbf{T}\boldsymbol{\beta}, \mathbf{I}, \boldsymbol{\Xi}).$$

Therefore, conditionally on $\mathbf{X}$, this model is a multivariate linear model. In addition, testing that $\boldsymbol{\Sigma}_{12} = \mathbf{0}$ for the unconditional model is equivalent to testing that $\boldsymbol{\gamma} = \mathbf{0}$ for the conditional model. We now make some elementary comments about the relationship between tests for the unconditional model considered in this section and tests for the conditional model considered in Section 19.2. First, let $(r_1, \ldots, r_p)$ be the maximal invariant derived in Section 19.2 for the conditional linear model and let $(t_1, \ldots, t_p)$ be the maximal invariant derived in this section for the unconditional model. Then $(t_1, \ldots, t_p)$ is an invertible function of $(r_1, \ldots, r_p)$ $(t_i = r_i/(1 - r_i))$ and therefore is also a maximal invariant for that conditional model. Hence a test is invariant for the conditional model if and only if it is invariant for the unconditional one. Furthermore, the (conditional) null distribution of $(r_1, \ldots, r_p)$ for the conditional model does not depend on $\mathbf{X}$, and therefore the null distribution of any invariant test statistic is the same for the two models, as is the size of any invariant test. (Note that the alternative distribution and power function are different.) Finally, the likelihood ratio tests are the same for the two models as are the union-intersection tests.

As a final comment in this section, we note if $m(\mathbf{T}(\mathbf{T}'\mathbf{T})^{-1}\mathbf{T}') \to 0$, then the sizes of any invariant procedures for the multivariate linear model are asymptotically insensitive to nonnormality in the conditional distribution of $\mathbf{Y}\,|\,\mathbf{X}$ (see Section 19.8). In Arnold (1980a) it is shown that if the $\mathbf{X}_i$ are a sample from a distribution with a finite covariance matrix, the $m(\mathbf{T}(\mathbf{T}'\mathbf{T})^{-1}\mathbf{T}')$ goes to 0 with probability 1. Therefore, the sizes of invariant procedures for this problem are asymptotically insensitive to nonnormality in either the conditional distribution of $\mathbf{Y}\,|\,\mathbf{X}$ or in the marginal distribution of $\mathbf{X}$.

21.2. TESTING THAT $\boldsymbol{\Sigma} = \boldsymbol{\Sigma}_0$.

We now consider the problem of testing that $\boldsymbol{\Sigma} = \boldsymbol{\Sigma}_0$ (a specified positive definite matrix) in the one sample model. That is, we observe $\mathbf{X}_1, \ldots, \mathbf{X}_n$ independently, identically distributed random vectors with $\mathbf{X}_i \sim N_r(\boldsymbol{\mu}, \boldsymbol{\Sigma})$,

$\boldsymbol{\Sigma} > 0$ and we are testing that $\boldsymbol{\Sigma} = \boldsymbol{\Sigma}_0$ against $\boldsymbol{\Sigma}$ unspecified. As usual, we assume that $n > r$.

We first transform the problem to one that is easier to analyze. Let $\mathbf{Y}_i = \boldsymbol{\Sigma}_0^{-1/2}\mathbf{X}_i$, $\boldsymbol{\nu} = \boldsymbol{\Sigma}_0^{-1/2}\boldsymbol{\mu}$, and $\boldsymbol{\Xi} = \boldsymbol{\Sigma}_0^{-1/2}\boldsymbol{\Sigma}\boldsymbol{\Sigma}_0^{-1/2}$. Since $\boldsymbol{\Sigma}_0^{-1/2}$ is invertible, observing the $\mathbf{Y}_i$ is equivalent to observing the $\mathbf{X}_i$. Also $\mathbf{Y}_i \sim N_r(\boldsymbol{\nu}, \boldsymbol{\Xi})$ and $\boldsymbol{\Sigma} = \boldsymbol{\Sigma}_0$ if and only if $\boldsymbol{\Xi} = \mathbf{I}$. Therefore, in the transformed problem, we are testing that $\boldsymbol{\Xi} = \mathbf{I}$ against $\boldsymbol{\Xi}$ unspecified.

The transformed version of the problem is invariant under the following two groups.

(1) G_1 consists of transformations of the form $g_1(\mathbf{Y}_i) = \mathbf{Y}_i + a$, $\mathbf{a} \in R^r$.
(2) G_2 consists of transformations of the form $g_2(\mathbf{Y}_i) = \boldsymbol{\Gamma}\mathbf{Y}_i$ where $\boldsymbol{\Gamma}$ is an orthogonal $r \times r$ matrix.

Let $\mathbf{T}$ be the sample covariance matrix for the $\mathbf{Y}_i$'s, that is,

$$\mathbf{T} = \frac{1}{n-1}\Sigma(\mathbf{Y}_i - \overline{\mathbf{Y}})(\mathbf{Y}_i - \overline{\mathbf{Y}})'.$$

THEOREM 21.3. A maximal invariant under G_1 and G_2 for testing that $\boldsymbol{\Xi} = \mathbf{I}$ is the set of roots $t_1 \geqslant \cdots \geqslant t_r$ of

$$|\mathbf{T} - t\mathbf{I}| = 0. \tag{21.5}$$

A parameter maximal invariant of the set of roots $\tau_1 \geqslant \cdots \geqslant \tau_r$ of

$$|\boldsymbol{\Xi} - \tau\mathbf{I}| = 0. \tag{21.6}$$

PROOF. By Theorem 18.1, $(\overline{\mathbf{Y}}, \mathbf{T})$ is a sufficient statistic. The first group G_1 consists of transformations of the form $g_1(\overline{\mathbf{Y}}, \mathbf{T}) = (\overline{\mathbf{Y}} + \mathbf{a}, \mathbf{T})$, and a maximal invariant is $\mathbf{T}$. G_2 consists of transformations of the form $g_2(\mathbf{T}) = \boldsymbol{\Gamma}\mathbf{T}\boldsymbol{\Gamma}'$. A maximal invariant is the set of roots of (21.5). The proof for the parameter maximal invariant follows similarly. □

We now express $t_1, \ldots, t_r$ and $\tau_1, \ldots, \tau_r$ in terms of the original random vectors X_i and the original parameters $\boldsymbol{\mu}$ and $\boldsymbol{\Sigma}$. Let

$$\mathbf{S} = \frac{1}{n-1}\Sigma_i(\mathbf{X}_i - \overline{\mathbf{X}})(\mathbf{X}_i - \overline{\mathbf{X}})'$$

be the sample covariance matrix of the $\mathbf{X}_i$'s.

LEMMA 21.4. The roots $(t_1, \ldots, t_r)$ of (21.3) are the roots of

$$|\mathbf{S} - t\boldsymbol{\Sigma}_0| = 0 \tag{21.7}$$

and the roots $(\tau_1, \ldots, \tau_r)$ of (21.4) are the roots of

$$|\boldsymbol{\Sigma} - \tau\boldsymbol{\Sigma}_0| = 0. \tag{21.8}$$

PROOF. $\mathbf{T} = \Sigma_0^{-1/2}\mathbf{S}\Sigma_0^{-1/2}$. Therefore $|\mathbf{T} - t\mathbf{I}| = 0$ if and only if

$$|\mathbf{S} - t\Sigma_0| = |\Sigma_0^{-1/2}||\mathbf{T} - t\mathbf{I}||\Sigma_0^{-1/2}| = 0.$$

The proof for the parameter maximal invariant is similar. □

We have now found a maximal invariant for this problem. There is no UMP invariant test in this setting. The likelihood ratio test is

$$\lambda = \frac{e^{-nr/2}\left(\frac{n-1}{n}\right)^{n/2}|\mathbf{S}|^{n/2}}{|\Sigma_0|^{n/2}} \exp -\tfrac{1}{2}(n-1)\operatorname{tr}\Sigma_0^{-1}\mathbf{S}$$

$$= e^{-nr/2}\left(\frac{n-1}{n}\right)^{n/2}|\mathbf{T}|^{n/2}\exp -\tfrac{1}{2}(n-1)\operatorname{tr}\mathbf{T}$$

$$= e^{-nr/2}\left(\frac{n-1}{n}\right)^{n/2}(\Pi t_i)^{n/2}\exp -\tfrac{1}{2}(n-1)\Sigma t_i$$

$$\phi(\lambda) = \begin{cases} 1 & \text{if } \lambda < c \\ 0 & \text{if } \lambda \geqslant c \end{cases}. \tag{21.9}$$

This test is not unbiased (see Das Gupta, 1969). However, the modified likelihood ratio test given by

$$\lambda^* = \frac{e^{-(n-1)r/2}|\mathbf{S}|^{(n-1)/2}}{|\Sigma_0|^{(n-1)/2}} \exp -\tfrac{1}{2}\operatorname{tr}(n-1)\Sigma_0^{-1}\mathbf{S}$$

$$= e^{-(n-1)r/2}(\Pi d_i)^{(n-1)/2}\exp -\tfrac{1}{2}(n-1)\Sigma d_i$$

$$\phi^*(\lambda^*) = \begin{cases} 1 & \text{if } \lambda^* < c^* \\ 0 & \text{if } \lambda^* \geqslant c^* \end{cases}$$

is unbiased (see Suguira and Nagao, 1968). Note that this test can be computed from the likelihood ratio test by replacing n by $n - 1$ everywhere, including in the expression for S. The size of these tests can be approximated by noting that both $-2\log(\lambda)$ and $-2\log(\lambda^*)$ have the approximate $\chi^2_{p(p+1)/2}$ distributions if n is large under the null hypothesis (since Σ has $p(p+1)/2$ independent parameters). Korin (1968) has shown that a better approximation to the null distribution of λ^* is that $-2\rho\log\lambda^*$ has approximately a $\chi^2_{p(p+1)/2}$ distribution where

$$\rho = 1 - \frac{2p + 1 - 2/(p+1)}{6(n-1)}$$

and also derives a still more exact approximation. Some exact conversion factors for the null distribution of λ^* are given in Pearson and Hartley (1972), Table 53.

We now apply the union-intersection principle to this problem. We first note that $\mathbf{\Sigma} = \mathbf{\Sigma}_0$ if and only if the variance of $U_i^{\mathbf{a}} = \mathbf{a}'\mathbf{X}_i$ is $\mathbf{a}'\mathbf{\Sigma}_0\mathbf{a}$ for all $\mathbf{a} \neq \mathbf{0}$. The sample variance of the $U_i^{\mathbf{a}}$ is $\mathbf{a}'\mathbf{S}\mathbf{a}$, and

$$\frac{(n-1)\mathbf{a}'\mathbf{S}\mathbf{a}}{\mathbf{a}'\mathbf{\Sigma}\mathbf{a}} \sim \chi^2_{n-1}(0).$$

Therefore, a test that the variance of $U_i^{\mathbf{a}}$ is $\mathbf{a}'\mathbf{\Sigma}_0\mathbf{a}$ would be

$$\phi_{\mathbf{a}}\left(\frac{\mathbf{a}'\mathbf{S}\mathbf{a}}{\mathbf{a}'\mathbf{\Sigma}_0\mathbf{a}}\right) = \begin{cases} 1 & \text{if } \dfrac{\mathbf{a}'\mathbf{S}\mathbf{a}}{\mathbf{a}'\mathbf{\Sigma}_0\mathbf{a}} > c_2 \quad \text{or} \quad \dfrac{\mathbf{a}'\mathbf{S}\mathbf{a}}{\mathbf{a}'\mathbf{\Sigma}_0\mathbf{a}} < c_1 \\ 0 & \text{if } c_1 \leqslant \dfrac{\mathbf{a}'\mathbf{S}\mathbf{a}}{\mathbf{a}'\mathbf{\Sigma}_0\mathbf{a}} \leqslant c_2 \end{cases}.$$

The union-intersection principle therefore leads to the test

$$\phi^{**} = \begin{cases} 1 & \text{if } \sup\limits_{a \neq 0} \dfrac{\mathbf{a}'\mathbf{S}\mathbf{a}}{\mathbf{a}'\mathbf{\Sigma}_0\mathbf{a}} > c_1 \quad \text{or} \quad \inf\limits_{a \neq 0} \dfrac{\mathbf{a}'\mathbf{S}\mathbf{a}}{\mathbf{a}'\mathbf{\Sigma}_0\mathbf{a}} < c_2 \\ 0 & \text{if } c_2 \leqslant \inf\limits_{\mathbf{a} \neq \mathbf{0}} \dfrac{\mathbf{a}'\mathbf{S}\mathbf{a}}{\mathbf{a}'\mathbf{\Sigma}_0\mathbf{a}} \leqslant \sup\limits_{a \neq 0} \dfrac{\mathbf{a}'\mathbf{S}\mathbf{a}}{\mathbf{a}'\mathbf{\Sigma}_0\mathbf{a}} \leqslant c_1 \end{cases}.$$

By Lemma 19.4, we see that

$$t_1 = \sup_{\mathbf{a} \neq \mathbf{0}} \frac{\mathbf{a}'\mathbf{S}\mathbf{a}}{\mathbf{a}'\mathbf{\Sigma}_0\mathbf{a}}, \qquad t_r = \frac{1}{\sup\limits_{\mathbf{a} \neq 0} \dfrac{\mathbf{a}'\mathbf{\Sigma}_0\mathbf{a}}{\mathbf{a}'\mathbf{S}\mathbf{a}}} = \inf_{a \neq 0} \frac{\mathbf{a}'\mathbf{S}\mathbf{a}}{\mathbf{a}'\mathbf{\Sigma}_0\mathbf{a}}.$$

(Note that $1/t_r$ is the largest root of $|\mathbf{\Sigma}_0 - \mathbf{q}\mathbf{S}| = 0$.) Therefore, the union-intersection principle leads to the test

$$\phi^{**}(t_1, \ldots, t_r) = \begin{cases} 1 & \text{if } t_1 > c_1 \quad \text{or } t_r < c_2 \\ 0 & \text{if } c_2 \leqslant t_r \leqslant t_1 \leqslant c_1, \end{cases}$$

that is, the test that rejects if the largest root t_1 is too large or the smallest root t_r is too small.

To use the union-intersection test, it it necessary to known the null distribution of t_1 and t_r. Tables of this distribution are given in Pearson and Hartley (1972, Table 51).

We can also find simultaneous confidence intervals for the set of all $\mathbf{d}'\mathbf{\Sigma}\mathbf{d}$. Let $q_1, \ldots, q_r$ be the roots of

$$|\mathbf{S} - q\mathbf{\Sigma}| = 0.$$

Then

$$q_1 = \sup_{d \neq 0} \frac{\mathbf{d}'\mathbf{S}\mathbf{d}}{\mathbf{d}'\mathbf{\Sigma}\mathbf{d}}, \qquad q_r = \inf_{d \neq 0} \frac{\mathbf{d}'\mathbf{S}\mathbf{d}}{\mathbf{d}'\mathbf{\Sigma}\mathbf{d}}.$$

Further, the distribution of (q_1, q_r) is the same as the null distribution of (t_1, t_r) (since $q_1 = t_1$ and $q_r = t_r$ under the null hypothesis). Let c_1 and c_2 be chosen so that the union-intersection test ϕ^{**} has size α. Then

$$1 - \alpha = P(c_2 \leqslant q_r \leqslant q_1 \leqslant c_1) = P\left(c_2 \leqslant \frac{\mathbf{d}'\mathbf{S}\mathbf{d}}{\mathbf{d}'\boldsymbol{\Sigma}\mathbf{d}} \leqslant c_1 \quad \text{for all } \mathbf{d} \neq \mathbf{0}\right)$$

$$= P\left(\frac{\mathbf{d}'\mathbf{S}\mathbf{d}}{c_2} \geqslant \mathbf{d}'\boldsymbol{\Sigma}\mathbf{d} \geqslant \frac{\mathbf{d}'\mathbf{S}\mathbf{d}}{c_1} \quad \text{for all } \mathbf{d}\right).$$

Therefore, the intervals

$$\frac{\mathbf{d}'\mathbf{S}\mathbf{d}}{c_2} \geqslant \mathbf{d}'\boldsymbol{\Sigma}\mathbf{d} \geqslant \frac{\mathbf{d}'\mathbf{S}\mathbf{d}}{c_1}$$

is a set of simultaneous $(1 - \alpha)$ confidence intervals for the set of all $\mathbf{d}'\boldsymbol{\Sigma}\mathbf{d}$. In addition, the hypothesis $\boldsymbol{\Sigma} = \boldsymbol{\Sigma}_0$ is rejected with the test ϕ^{**} if and only if there exists at least one $\mathbf{d}$ such that $\mathbf{d}'\boldsymbol{\Sigma}_0\mathbf{d}$ is not in the interval given above for $\mathbf{d}'\boldsymbol{\Sigma}\mathbf{d}$. After rejecting with the test ϕ^{**}, we can therefore look for those $\mathbf{d}$ such that $\mathbf{d}'\boldsymbol{\Sigma}_0\mathbf{d}$ is not in the interval for $\mathbf{d}'\boldsymbol{\Sigma}\mathbf{d}$. We then consider such $\mathbf{d}'\boldsymbol{\Sigma}\mathbf{d}$ to be the functions that are causing the rejection of the hypothesis. Since $\operatorname{var}(\mathbf{d}'\mathbf{X}_i) = \mathbf{d}'\boldsymbol{\Sigma}\mathbf{d}$, we can think of such $\mathbf{d}'\mathbf{X}_i$ as the linear combinations of the variables that are causing the hypothesis to be rejected.

21.3. TESTING THE EQUALITY OF COVARIANCE MATRICES.

We now consider the problem of testing the equality of two covariance matrices. We observe $\mathbf{X}_1, \ldots, \mathbf{X}_n, \mathbf{Y}_1, \ldots, \mathbf{Y}_m$ independent with

$$\mathbf{X}_i \sim N_r(\boldsymbol{\mu}, \boldsymbol{\Sigma}), \qquad \mathbf{Y}_i \sim N_r(\boldsymbol{\nu}, \boldsymbol{\Xi}), \qquad \boldsymbol{\Sigma} > 0, \qquad \boldsymbol{\Xi} > 0.$$

We are testing that $\boldsymbol{\Sigma} = \boldsymbol{\Xi}$ against $\boldsymbol{\Sigma} > 0$ and $\boldsymbol{\Xi} > 0$. We assume that $n \geqslant r$, $m \geqslant r$. Let

$$\mathbf{S} = \frac{1}{n-1}\sum_{i=1}^{n}(\mathbf{X}_i - \overline{\mathbf{X}})(\mathbf{X}_i - \overline{\mathbf{X}})', \qquad \mathbf{T} = \frac{1}{m-1}\sum_{i=1}^{m}(\mathbf{Y}_i - \overline{\mathbf{Y}})(\mathbf{Y}_i - \overline{\mathbf{Y}})'$$

be the sample covariance matrices for the $\mathbf{X}$ and $\mathbf{Y}$ samples. Then

$$(\overline{\mathbf{X}}, \overline{\mathbf{Y}}, \mathbf{S}, \mathbf{T})$$

is a sufficient statistic for this problem. The problem is invariant under the following two groups of transformations:

(1) G_1 consists of the transformations $g_1(\overline{\mathbf{X}}, \overline{\mathbf{Y}}, \mathbf{S}\mathbf{T}) = (\overline{\mathbf{X}} + \mathbf{a}, \overline{\mathbf{Y}} + \mathbf{b}, \mathbf{S}, \mathbf{T})$ for $\mathbf{a}, \mathbf{b} \in R^r$.

(2) G_2 consists of transformations $g_2(\overline{\mathbf{X}}, \overline{\mathbf{Y}}, \mathbf{S}, \mathbf{T}) = (\mathbf{A}\overline{\mathbf{X}}, \mathbf{A}\overline{\mathbf{Y}}, \mathbf{A}\mathbf{S}\mathbf{A}', \mathbf{A}\mathbf{T}\mathbf{A}')$ where $\mathbf{A}$ is an $r \times r$ invertible matrix.

Let $t_1 \geqslant \cdots \geqslant t_r$ be the roots of

$$|\mathbf{S} - t\mathbf{T}| = 0 \tag{21.10}$$

and let $\tau_1 \cdots \tau_r$ be the roots of

$$|\boldsymbol{\Sigma} - \tau\boldsymbol{\Xi}| = 0. \tag{21.11}$$

THEOREM 21.5. $(t_1, \ldots, t_r)$ is a maximal invariant under G_1 and G_2 and $(\tau_1, \ldots, \tau_r)$ is a parameter maximal invariant for testing that $\boldsymbol{\Sigma} = \boldsymbol{\Xi}$.

PROOF. See Exercise B4. □

There is no UMP invariant test for this problem. The likelihood ratio test is given by

$$\lambda = \frac{\left|\frac{n-1}{n}\mathbf{S}\right|^{n/2}\left|\frac{m-1}{m}\mathbf{T}\right|^{m/2}}{\left|\frac{n-1}{n+m}\mathbf{S} + \frac{m-1}{m+n}\mathbf{T}\right|^{(n+m)/2}}$$

$$= \frac{\left(\frac{n-1}{n}\right)^{nr/2}\left(\frac{m-1}{m}\right)^{mr/2}|\mathbf{T}^{-1/2}\mathbf{S}\mathbf{T}^{-1/2}|^{m/2}}{\left|\frac{n-1}{n+m}\mathbf{T}^{-1/2}\mathbf{S}\mathbf{T}^{-1/2} + \frac{m-1}{n+m}\mathbf{I}\right|^{(n+m)/2}}$$

$$= \frac{\left(\frac{n-1}{n}\right)^{mr/2}\left(\frac{m-1}{m}\right)^{mr/2}\prod_{i=1}^{r} t_i^{m/2}}{\prod_{i=1}^{r}\left(\frac{(n-1)t_i + m - 1}{n+m}\right)^{(m+n)/2}}. \tag{21.12}$$

$$\phi(\lambda) = \begin{cases} 1 & \text{if } \lambda < c \\ 0 & \text{if } \lambda \geqslant c \end{cases},$$

but this test is again not unbiased unless $n = m$ (see Das Gupta, 1969). The modified likelihood ratio test

$$\lambda^* = \frac{|\mathbf{S}|^{(n-1)/2}|\mathbf{T}|^{(m-1)/2}}{\left|\frac{n-1}{n+m-2}\mathbf{S} + \frac{m-1}{n+m-2}\mathbf{T}\right|^{(n+m-2)/2}}$$

$$= \frac{\prod t_i^{(n-1)/2}}{\prod\left(\frac{(n-1)t_i + (m-1)}{n+m-2}\right)^{(m+n-2)/2}},$$

$$\phi^*(\lambda^*) = \begin{cases} 1 & \text{if } \lambda^* < c^* \\ 0 & \text{if } \lambda^* \geqslant c^* \end{cases}$$

is unbiased (see Suguira and Nagao (1968)) and can be determined by substituting $n-1$ for n and $m-1$ for m in the likelihood ratio test (including in S and T). (Note that if $n=m$, the $\lambda^* = \lambda^{(n-1)/n}$ and the tests ϕ and ϕ^* are the same.) The sizes of these tests can be approximated by using the fact that both $-2\log(\lambda)$ and $-2\log(\lambda^*)$ have approximate $\chi^2_{p(p+1)/2}$ distributions when n and m are large, under the null hypothesis. Box (1949) shows that a better approximation to the null distribution of λ^* is that $-2\rho\log\lambda^*$ has approximately a $\chi^2_{p(p+1)/2}$ distribution where

$$\rho = 1 - \left(\frac{2p^2+3p-1}{6(p+1)}\right)\left(\frac{1}{n-1} + \frac{1}{m-1} - \frac{1}{n+m-2}\right).$$

He also gives a more exact approximation. Some conversion factors for exact percentiles when $n=m$ are given in Pearson and Hartley (1972), Table 50.

We now look at the union-intersection approach to this problem. Let $U_i^{\mathbf{a}} = \mathbf{a}'\mathbf{X}_i$ and $V_i^{\mathbf{a}} = \mathbf{a}'\mathbf{Y}_i$. Then $\boldsymbol{\Xi} = \boldsymbol{\Sigma}$ if and only if $U_i^{\mathbf{a}}$ and $V_i^{\mathbf{a}}$ have the same variance for all $\mathbf{a} \neq \mathbf{0}$. So we now consider the problem of testing the equality of two variances. The sample variance for the $U_i^{\mathbf{a}}$ is $\mathbf{a}'\mathbf{S}\mathbf{a}$ and the sample variance for the $V_i^{\mathbf{a}}$ is $\mathbf{a}'\mathbf{T}\mathbf{a}$. Therefore, an F-test for this hypothesis has the form

$$\phi_{\mathbf{a}}\left(\frac{\mathbf{a}'\mathbf{S}\mathbf{a}}{\mathbf{a}'\mathbf{T}\mathbf{a}}\right) = \begin{cases} 1 & \text{if } \dfrac{\mathbf{a}'\mathbf{S}\mathbf{a}}{\mathbf{a}'\mathbf{T}\mathbf{a}} > c_1 \quad \text{or} \quad \dfrac{\mathbf{a}'\mathbf{S}\mathbf{a}}{\mathbf{a}'\mathbf{T}\mathbf{a}} < c_2 \\ 0 & \text{if } c_2 \leqslant \dfrac{\mathbf{a}'\mathbf{S}\mathbf{a}}{\mathbf{a}'\mathbf{T}\mathbf{a}} \leqslant c_1 \end{cases}.$$

Therefore, the union-intersection principle leads to the test

$$\phi^{**} = \begin{cases} 1 & \text{if } \sup\limits_{\mathbf{a}\neq\mathbf{0}} \dfrac{\mathbf{a}'\mathbf{S}\mathbf{a}}{\mathbf{a}'\mathbf{T}\mathbf{a}} > c_1 \quad \text{or } \inf\limits_{\mathbf{a}\neq\mathbf{0}} \dfrac{\mathbf{a}'\mathbf{S}\mathbf{a}}{\mathbf{a}'\mathbf{T}\mathbf{a}} < c_2 \\ 0 & \text{if } c_2 \leqslant \inf\limits_{\mathbf{a}\neq\mathbf{0}} \dfrac{\mathbf{a}'\mathbf{S}\mathbf{a}}{\mathbf{a}'\mathbf{T}\mathbf{a}} \leqslant \sup\limits_{\mathbf{a}\neq\mathbf{0}} \dfrac{\mathbf{a}'\mathbf{S}\mathbf{a}}{\mathbf{a}'\mathbf{T}\mathbf{a}} \leqslant c_1 \end{cases}.$$

As in the previous section,

$$\sup_{\mathbf{a}\neq\mathbf{0}} \frac{\mathbf{a}'\mathbf{S}\mathbf{a}}{\mathbf{a}'\mathbf{T}\mathbf{a}} = t_1, \qquad \inf_{\mathbf{a}\neq\mathbf{0}} \frac{\mathbf{a}'\mathbf{S}\mathbf{a}}{\mathbf{a}'\mathbf{T}\mathbf{a}} = t_r.$$

Therefore, the union-intersection test is given by

$$\phi^*(t_1, \ldots, t_r) = \begin{cases} 1 & \text{if } t_r < c_1 \quad \text{or } t_1 > c_2 \\ 0 & \text{if } c_1 \leqslant t_r \leqslant t_1 \leqslant c_2 \end{cases},$$

that is, the union-intersection test rejects if the largest root is too large or the smallest root is too small.

To use the union-intersection test for this problem, it is necessary to know the null distributions of t_1 and t_r. We first note that

$$(n-1)\mathbf{S} \sim W_r(n-1, \boldsymbol{\Sigma}), \qquad (m-1)\mathbf{T} \sim W_r(m-1, \boldsymbol{\Xi}).$$

Under the null hypothesis $\boldsymbol{\Sigma} = \boldsymbol{\Xi}$. Now, let

$$v_1(t_1) = \frac{\dfrac{n-1}{m-1} t_1}{1 + \dfrac{n-1}{m-1} t_1}, \qquad v_r(t_r) = \frac{\dfrac{m-1}{n-1} \dfrac{1}{t_r}}{1 + \dfrac{m-1}{n-1} \dfrac{1}{t_r}}.$$

Then $v_1(t_1)$ is the largest root of

$$|(n-1)\mathbf{S} - v((n-1)\mathbf{S} + (m-1)\mathbf{T})| = 0.$$

Therefore, under the null hypothesis

$$v_1(t_1) \sim s_{r, n-1, m-1}$$

(see Section 19.2). Since v_1 is an increasing function of t_1, the upper α point for the null distribution of t_1 can be determined from that for v_1. Now, $v_r(t_r)$ is the largest root of

$$|(m-1)\mathbf{T} - v((m-1)\mathbf{T} + (n-1)\mathbf{S})| = 0$$

and therefore

$$v_r(t_r) \sim s_{r, m-1, n-1}.$$

Since $v_r(t_r)$ is a decreasing function of t_r, the upper $(1-\alpha)$ point for the null distribution t_r can be found from the upper α point of the distribution of v_r. (Note that the upper $1-\alpha$ point of t_r is what we need for a size 2α test).

21.4. TESTING THE VALIDITY OF THE ORDINARY LINEAR MODEL.

The testing problem that we consider in this section is one in which we observe $\mathbf{Y} \sim N_{n,r}(\boldsymbol{\mu}, \mathbf{I}, \boldsymbol{\Sigma})$, $\boldsymbol{\mu} \in V$, and we are testing that $\boldsymbol{\Sigma} = \sigma^2\mathbf{I}$ against the alternative that $\boldsymbol{\Sigma} > 0$. Let

$$\mathbf{Y} = (\mathbf{Y}_1, \ldots, \mathbf{Y}_r), \qquad \tilde{\mathbf{Y}} = \begin{pmatrix} \mathbf{Y}_1 \\ \vdots \\ \mathbf{Y}_r \end{pmatrix}, \qquad \boldsymbol{\mu} = (\boldsymbol{\mu}_1, \ldots, \boldsymbol{\mu}_r), \qquad \tilde{\boldsymbol{\mu}} = \begin{pmatrix} \boldsymbol{\mu}_1 \\ \vdots \\ \boldsymbol{\mu}_r \end{pmatrix}.$$

If $\boldsymbol{\Sigma} = \sigma^2\mathbf{I}$, then $\tilde{\mathbf{Y}} \sim N_{nr}(\tilde{\boldsymbol{\mu}}, \sigma^2\mathbf{I})$. The parameter space is the set in which $\tilde{\boldsymbol{\mu}} \in V^n = (V \times V \cdots \times V)$ and $\sigma^2 > 0$ (see Exercise B6). Therefore, when $\boldsymbol{\Sigma} = \sigma^2\mathbf{I}$, we could use procedures derived for the ordinary linear model to test hypotheses about $\boldsymbol{\mu}$. These procedures for the ordinary linear model are much more powerful than those for the multivariate linear model. In this section we are considering testing the null hypothesis that the ordinary linear model is correct against the alternative that the multivariate linear model is correct. It should be emphasized that the hypothesis that the ordinary linear model is correct is a null hypothesis, so accepting this hypothesis should not be construed as proof that it is true. However, rejecting the hypothesis could be considered proof that the procedures for the ordinary linear model are not valid. The hypothesis could also be rejected because that data is not normally distributed and the procedures for testing hypotheses about means in the ordinary linear model are not too sensitive to the normal assumption used in deriving them (see Chapter 10). Therefore one should use these procedures with caution.

This problem is invariant under the following three groups:

(1) $g_1(\mathbf{Y}) = \mathbf{Y} + \mathbf{A}$, where $\mathbf{A} \in V$ is an $n \times r$ matrix.
(2) $g_2(\mathbf{Y}) = \mathbf{Y\Gamma}$ where $\mathbf{\Gamma}$ is an $r \times r$ orthogonal matrix.
(3) $g_3(\mathbf{Y}) = \mathbf{c}\mathbf{Y}$, where $\mathbf{c}$ is a scalar.

(As usual, the reader should determine the induced groups on the parameter space and determine that they do not change the null set for the problem.) A sufficient statistic for this model is $\hat{\boldsymbol{\mu}} = \mathbf{P}_V\mathbf{Y}, \hat{\mathbf{\Sigma}} = [1/(n-p)]\mathbf{Y}'\mathbf{P}_{V^\perp}\mathbf{Y}$. The first group becomes $g_1(\hat{\boldsymbol{\mu}}, \hat{\mathbf{\Sigma}}) = (\hat{\boldsymbol{\mu}} + \mathbf{A}, \hat{\mathbf{\Sigma}})$, and a maximal invariant is $\hat{\mathbf{\Sigma}}$. The second group becomes transformations of the form $g_2(\hat{\mathbf{\Sigma}}) = \mathbf{\Gamma}'\mathbf{\Sigma}\mathbf{\Gamma}$, and by Lemma 19.4 a maximal invariant is the set of roots $t_1 \geqslant \cdots \geqslant t_r$ of $|\hat{\mathbf{\Sigma}} - t\mathbf{I}| = 0$. Since $n - p \geqslant r$, we see that $\hat{\mathbf{\Sigma}} > 0$, and therefore by Lemma 19.4 $t_r > 0$. Now the third group operates by $g_3(t_1, \ldots, t_r) = (at_1, \ldots, at_r)$. A maximal invariant under this group is $t_1/t_r, \ldots, t_{r-1}/t_r$. We have now proved the following theorem.

THEOREM 21.6. Let $t_1 \geqslant \cdots \geqslant t_r$ be the roots of $|\hat{\mathbf{\Sigma}} - t\mathbf{I}| = 0$. Then a maximal invariant for the problem of testing that $\mathbf{\Sigma} = \sigma^2\mathbf{I}$ is the set $(t_1/t_r, \ldots, t_{r-1}/t_r)$.

There is no UMP invariant size α test for this problem. The likelihood ratio test is given by

$$\lambda_1 = \frac{|\hat{\mathbf{\Sigma}}|^{n/2}}{\left(\dfrac{\operatorname{tr}\hat{\mathbf{\Sigma}}}{r}\right)^{nr/2}} = \frac{\Pi(t_i/t_r)^{n/2}}{\left(\dfrac{1 + \Sigma(t_i/t_r)}{r}\right)^{nr/2}} \tag{21.13}$$

$$\phi_1(\lambda_1) = \begin{cases} 1 & \text{if } \lambda_1 < c \\ 0 & \text{if } \lambda_1 \geqslant c \end{cases}.$$

and is unbiased (see Suguira and Nagao, 1968). If n is large, the usual approximation for the null distribution of the likelihood ratio statistic implies that $-2\log\lambda$ is approximately $\chi^2_{[r(r+1)/2]-1}$. Anderson (1958) shows that a more refined approximation to the null distribution of λ_1 is that $-2\rho((n-p)/n)\log\lambda_1$ is approximately distributed as a $\chi^2_{r(r+1)/2-1}$, where

$$\rho = 1 - \frac{2r^2 + r + 2}{6r(n-p)}.$$

He also derives a more exact approximation (see pp. 263–264).

The union-intersection method leads to the test

$$\lambda_2 = \frac{t_1}{t_r}, \qquad \phi(\lambda_2) = \begin{cases} 1 & \text{if } \lambda_2 > d \\ 0 & \text{if } \lambda_2 \leqslant d \end{cases}.$$

21.5. TESTING THE VALIDITY OF THE REPEATED MEASURES MODEL.

We continue with the multivariate linear model in which we observe $\mathbf{Y} \sim N_{n,r}(\boldsymbol{\mu}, \mathbf{I}, \boldsymbol{\Sigma})$, $\boldsymbol{\mu} \in V$, $\boldsymbol{\Sigma} > 0$. In this section we consider testing that $\boldsymbol{\Sigma} = \sigma^2\mathbf{A}(\rho)$, where

$$\mathbf{A}(\rho) = \begin{pmatrix} 1 & \rho & \cdots & \rho \\ \rho & 1 & \cdots & \rho \\ \vdots & \vdots & \ddots & \vdots \\ \rho & \rho & \cdots & 1 \end{pmatrix}.$$

Let

$$\mathbf{Y} = \begin{pmatrix} \mathbf{Y}_1 \\ \vdots \\ \mathbf{Y}_n \end{pmatrix}, \qquad \boldsymbol{\mu} = \begin{pmatrix} \boldsymbol{\mu}_1 \\ \vdots \\ \boldsymbol{\mu}_n \end{pmatrix}.$$

Then the $\mathbf{Y}_i'$ are independent, $\mathbf{Y}_i' \sim N_r(\boldsymbol{\mu}_i', \boldsymbol{\Sigma})$. If $\boldsymbol{\Sigma} = \sigma^2\mathbf{A}(\rho)$, then this model is a repeated measures model as discussed in Chapter 14, and the procedures derived in that chapter can be used to test hypotheses about the $\boldsymbol{\mu}_i$. These procedures are more powerful than the procedures derived for the multivariate linear model. In this section, we are testing the null hypothesis that the repeated measures model is valid against the alternative hypothesis that the multivariate linear model is valid. It should again be mentioned that acceptance of this null hypothesis should not be construed as proof that the null hypothesis is valid. However, rejection of this hypothesis could be construed as proof that the repeated measures model is invalid. Rejection might also occur because that data is nonnormal and the procedures for testing hypotheses about the means for the repeated measures model are often not too sensitive to the normal assumption.

As in Chapter 14, let $\mathbf{C}$ be an $(r-1)\times r$ matrix such that

$$\boldsymbol{\Gamma} = \begin{pmatrix} r^{-1/2}\mathbf{1}' \\ \mathbf{C} \end{pmatrix}$$

is orthogonal, and let

$$\mathbf{Y}^* = \mathbf{Y}\boldsymbol{\Gamma}', \qquad \boldsymbol{\mu}^* = \boldsymbol{\mu}\boldsymbol{\Gamma}', \qquad \boldsymbol{\Sigma}^* = \boldsymbol{\Gamma}\boldsymbol{\Sigma}\boldsymbol{\Gamma}'. \tag{21.14}$$

Since $\boldsymbol{\Gamma}$ is invertible, observing $\mathbf{Y}$ is equivalent to observing $\mathbf{Y}^*$. In addition $\boldsymbol{\mu} \in V$ if and only if $\boldsymbol{\mu}\boldsymbol{\Gamma}' \in V$ (see Exercise B8). Therefore

$$\mathbf{Y}^* \sim N_{n,r}(\boldsymbol{\mu}^*, \mathbf{I}, \boldsymbol{\Sigma}^*), \qquad \boldsymbol{\mu}^* \in V, \qquad \boldsymbol{\Sigma}^* > 0.$$

Finally, by Lemma 14.3, we see that $\boldsymbol{\Sigma} = \sigma^2\mathbf{A}(\rho)$ if and only if

$$\boldsymbol{\Sigma}^* = \begin{pmatrix} \tau_1^2 & \mathbf{0} \\ \mathbf{0} & \tau_2^2\mathbf{I} \end{pmatrix} \tag{21.15}$$

for some $\tau_1^2 > 0$, $\tau_2^2 > 0$, where $\tau_1^2 = \sigma^2(1+(r-1)\rho)$, $\tau_2^2 = \sigma^2(1-\rho)$. Therefore, in the transformed problem we are testing that $\boldsymbol{\Sigma}^*$ has the form given in (21.15) against the alternative that $\boldsymbol{\Sigma}^* > 0$.

This problem is invariant under several groups, but there does not seem to be any nice form for the maximal invariant under these groups. Therefore, we limit discussion to the likelihood ratio test.

As usual, let

$$\hat{\boldsymbol{\mu}}^* = \mathbf{P}_V\mathbf{Y}^*, \qquad \hat{\boldsymbol{\Sigma}}^* = \frac{1}{n-p}\mathbf{Y}^{*\prime}\mathbf{P}_{V^\perp}\mathbf{Y}^*.$$

Under the alternative hypothesis, the model is a multivariate linear model, and therefore the maximum likelihood estimators of $\boldsymbol{\mu}^*$ and $\boldsymbol{\Sigma}^*$ are $\hat{\boldsymbol{\mu}}^*$ and $[(n-p/n]\hat{\boldsymbol{\Sigma}}^*$. Under the null hypothesis, the maximum likelihood estimator of $\boldsymbol{\mu}^*$ is $\hat{\boldsymbol{\mu}}^*$, since $\hat{\boldsymbol{\mu}}^*$ maximizes the likelihood for all $\boldsymbol{\Sigma}^*$. Now let

$$\hat{\boldsymbol{\Sigma}}^* = \begin{pmatrix} \hat{\Sigma}_{11}^* & \hat{\boldsymbol{\Sigma}}_{12}^* \\ \hat{\boldsymbol{\Sigma}}_{21}^* & \hat{\boldsymbol{\Sigma}}_{22}^* \end{pmatrix}$$

where $\hat{\Sigma}_{11}^*$ is 1×1. Under the null hypothesis, $\Sigma_{11}^* = \tau_1^2$, $\boldsymbol{\Sigma}_{12} = \mathbf{0}$, and $\boldsymbol{\Sigma}_{22} = \tau_2^2\mathbf{I}$. The maximum likelihood estimators of τ_1^2 and τ_2^2 under the null hypothesis are $[(n-p)/n]\hat{\Sigma}_{11}^*$ and $[(n-p)/n(r-1)]\operatorname{tr}\hat{\boldsymbol{\Sigma}}_{22}^*$ (see Exercise B9). Therefore the likelihood ratio test statistic is

$$\lambda = \frac{|\hat{\boldsymbol{\Sigma}}^*|^{n/2}}{(\hat{\Sigma}_{11}^*)^{n/2}\left(\dfrac{1}{r-1}\operatorname{tr}\hat{\boldsymbol{\Sigma}}_{22}^*\right)^{n(r-1)/2}}$$

and the likelihood ratio test is

$$\phi(\lambda) = \begin{cases} 1 & \text{if } \lambda < c \\ 0 & \text{if } \lambda \geqslant c \end{cases}.$$

If n is large, then $-2\log\lambda$ is approximately $\chi^2_{[r(r+1)/2]-1}$. Box (1950) gives the more refined approximation that $-2\rho((n-p)/n)\log\lambda_1$ is approximately $\chi^2_{r(r+1)/2-1}$, where

$$\rho = 1 - \frac{r(r+1)^2(2r-3)}{6(r-1)(r^2+r-4)(n-p)}$$

and he also gives a more refined approximation.

We now express λ in terms of the orginal random matrix $\mathbf{Y}$. Let

$$\hat{\mu} = \mathbf{P}_V\mathbf{Y}, \qquad \hat{\boldsymbol{\Sigma}} = \frac{1}{n-p}\mathbf{Y}'\mathbf{P}_{V^\perp}\mathbf{Y}.$$

Then $\hat{\boldsymbol{\Sigma}}^* = \boldsymbol{\Gamma}\hat{\boldsymbol{\Sigma}}\boldsymbol{\Gamma}'$. Therefore

$$|\hat{\boldsymbol{\Sigma}}^*| = |\boldsymbol{\Gamma}||\hat{\boldsymbol{\Sigma}}||\boldsymbol{\Gamma}'| = |\boldsymbol{\Gamma}'\boldsymbol{\Gamma}||\hat{\boldsymbol{\Sigma}}| = |\hat{\boldsymbol{\Sigma}}|, \qquad \operatorname{tr}\hat{\boldsymbol{\Sigma}}^* = \operatorname{tr}\hat{\boldsymbol{\Sigma}}\boldsymbol{\Gamma}'\boldsymbol{\Gamma} = \operatorname{tr}\hat{\boldsymbol{\Sigma}}.$$

Also, by the definition of $\boldsymbol{\Gamma}$, we see that

$$\hat{\Sigma}^*_{11} = \frac{1}{r}1'\hat{\Sigma}1 = \frac{1}{r}\sum_{i=1}^{r}\sum_{j=1}^{r}\hat{\sigma}_{ij}, \qquad \text{where } \hat{\boldsymbol{\Sigma}} = \begin{bmatrix} \hat{\sigma}_{11} & \cdots & \hat{\sigma}_{1r} \\ \vdots & & \\ \hat{\sigma}_{r1} & & \hat{\sigma}_{rr} \end{bmatrix}. \tag{21.16}$$

Finally,

$$\operatorname{tr}\hat{\boldsymbol{\Sigma}}^*_{22} = \operatorname{tr}\hat{\boldsymbol{\Sigma}}^* - \hat{\Sigma}^*_{11} = \operatorname{tr}\hat{\boldsymbol{\Sigma}} - \hat{\Sigma}^*_{11}.$$

Therefore

$$\lambda^{2/n} = \frac{(r-1)^{r-1}|\hat{\boldsymbol{\Sigma}}|}{\hat{\Sigma}^*_{11}(\operatorname{tr}\hat{\boldsymbol{\Sigma}} - \hat{\Sigma}^*_{11})^{r-1}}$$

where $\hat{\Sigma}^*_{11}$ is defined in (21.16).

It is not necessary for $\boldsymbol{\Sigma} = \boldsymbol{\sigma}^2\mathbf{A}(\rho)$ to guarantee the validity of procedures for testing hypotheses of type A or type B for the repeated measures model. Procedures for testing hypotheses of type A depend only on $\mathbf{Y}^*_1$, which has 1×1 covariance matrix Σ^*_{11}. Therefore, the procedures for testing hypotheses of type A are valid no matter what the structure of $\boldsymbol{\Sigma}$. Procedures for testing type B hypotheses depend only on $\mathbf{Y}^*_2$. They are therefore valid as long as $\boldsymbol{\Sigma}^*_{22} = \tau_2^2\mathbf{I}$. Therefore, the procedures for testing type A and B hypotheses are valid as long as $\boldsymbol{\Sigma}^*_{22} = \tau_2^2\mathbf{I}$. (They would not necessarily be optimal unless $\boldsymbol{\Sigma} = \sigma^2\mathbf{A}(\rho)$, but they would be much more powerful than those procedures derived with no assumption about $\boldsymbol{\Sigma}$.) We now consider testing that $\boldsymbol{\Sigma}^*_{22} = \tau_2^2\mathbf{I}$ against $\boldsymbol{\Sigma} > 0$. Let $\mathbf{Y}^* = (\mathbf{Y}^*_1, \mathbf{Y}^*_2)$ where $\mathbf{Y}^*_1$ is $n\times 1$.

This problem is invariant under the following five groups.

(1) $g_1(\mathbf{Y}^*_1, \mathbf{Y}^*_2) = (\mathbf{Y}^*_1, \mathbf{Y}^*_2) + \mathbf{A}$ where $\mathbf{A} \in V$ is an $n\times r$ matrix.
(2) $g_2(\mathbf{Y}^*_1, \mathbf{Y}^*_2) = (b\mathbf{Y}^*_1, \mathbf{Y}^*_2)$, where $b \neq 0$ is 1×1.
(3) $g_3(\mathbf{Y}^*_1, \mathbf{Y}^*_2) = (\mathbf{Y}^*_1 + \mathbf{Y}^*_2\mathbf{D}, \mathbf{Y}^*_2)$ where $\mathbf{D}$ is $(r-1)\times 1$.

(4) $g_4(\mathbf{Y}_1^*, \mathbf{Y}_2^*) = (\mathbf{Y}_1^*, \mathbf{Y}_2^*\mathbf{E})$ where $\mathbf{E}$ is an $(r-1) \times (r-1)$ orthogonal matix.

(5) $g_5(\mathbf{Y}_1^*, \mathbf{Y}_2^*) = (\mathbf{Y}_1^*, f\mathbf{Y}_2^*)$ where $f > 0$ is 1×1.

As usual, let $\hat{\boldsymbol{\Sigma}} = [1/(n-p)]\mathbf{Y}'\mathbf{P}_{V^\perp}\mathbf{Y}$, and let $\hat{\boldsymbol{\Sigma}}_{22}^* = \mathbf{C}\hat{\boldsymbol{\Sigma}}\mathbf{C}'$.

THEOREM 21.7. Let $t_1^* \geqslant \cdots \geqslant t_{r-1}^*$ be the roots of $|\hat{\boldsymbol{\Sigma}}_{22}^* - t^*\mathbf{I}| = 0$ maximal invariant for the problem of testing that $\boldsymbol{\Sigma}_{22}^* = \tau_2^2\mathbf{I}$ is the set $(t_1^*/t_{r-1}^*, \ldots, t_{r-2}^*/t_{r-1}^*)$.

PROOF. See Exercise B10. □

Two tests that have been suggested for this problem (analogous to those in the last section) are

$$\lambda_1^* = \frac{|\hat{\boldsymbol{\Sigma}}_{22}^*|^{n/2}}{\left(\dfrac{\operatorname{tr}\hat{\boldsymbol{\Sigma}}_{22}^*}{r-1}\right)^{n(r-1)/2}}, \qquad \phi_1^*(\lambda_1^*) = \begin{cases} 1 & \text{if } \lambda_1^* < c^* \\ 0 & \text{if } \lambda_1^* \geqslant c^* \end{cases},$$

$$\lambda_2^* = \frac{t_1^*}{t_{r-1}^*}, \qquad \phi_2^*(\lambda_2^*) = \begin{cases} 1 & \text{if } \lambda_2^* > d^* \\ 0 & \text{if } \lambda_2^* \leqslant d^* \end{cases}.$$

Note that c^* and d^* can be computed from c and d of the last section by replacing r by $r-1$.

EXERCISES.

Type A

1. Consider the model of Section 21.1, with $p = q = 2$, $n = 106$. Suppose that

$$S = \begin{pmatrix} 2 & 5 & 1 & 1 \\ 5 & 15 & 1 & 2 \\ 1 & 1 & 5 & 3 \\ 1 & 2 & 3 & 2 \end{pmatrix}$$

(a) Do a size 0.05 LRT that $\boldsymbol{\Sigma}_{12} = 0$.

(b) Do a size 0.05 union intersection test that $\boldsymbol{\Sigma}_{12} = 0$. (Note that $s_{2,2,103} = 0.85$.)

2. Consider the model of Section 2.12 with $p = 2$, $n = 100$. Suppose that

$$\mathbf{S} = \begin{pmatrix} 2 & 5 \\ 5 & 15 \end{pmatrix}, \qquad \boldsymbol{\Sigma}_0 = \begin{pmatrix} 3 & 4 \\ 4 & 12 \end{pmatrix}$$

(a) Use a size 0.05 LRT to test that $\boldsymbol{\Sigma} = \boldsymbol{\Sigma}_0$.

(b) Use a size 0.05 LRT modified to be unbiased to test that $\boldsymbol{\Sigma} = \boldsymbol{\Sigma}_0$.

(c) Compute the value of the test statistic for the union-intersection test.

3. Consider the model of Section 21.3 with $r = 2$, $n = 40$, $m = 60$. Suppose that

$$\mathbf{S} = \begin{pmatrix} 1 & 2 \\ 2 & 5 \end{pmatrix}, \qquad \mathbf{T} = \begin{pmatrix} 2 & 1 \\ 1 & 5 \end{pmatrix}$$

(a) Use a size 0.05 LRT to test the quality of the covariance matrices.

(b) Use a size 0.05 LRT modified to be unbiased to test the equality of covariance matrices.

(c) Compute the value of the union-intersection test statistic for this problem.

4. Consider the problem of Section 21.4 with $r = 2$, $n = 100$. Suppose that

$$\mathbf{S} = \begin{pmatrix} 2 & 1 \\ 1 & 3 \end{pmatrix}$$

(a) Use a size 0.05 LRT to test that $\boldsymbol{\Sigma} = \sigma^2\mathbf{I}$.

(b) Compute the value of the union-intersection test statistic for this problem.

Type B

1. Verify that the likelihood ratio test for testing that $\boldsymbol{\Sigma}_{12} = \mathbf{0}$ is given by (21.3).

2. Verify (21.4).

3. Show that the likelihood ratio test for testing that $\boldsymbol{\Sigma} = \boldsymbol{\Sigma}_0$ is given by (21.9).

4. Prove Theorem 21.5.

5. Verify that the likelihood ratio test for testing that $\boldsymbol{\Sigma} = \boldsymbol{\Xi}$ is given by (21.12).

6. In the notation of Section 21.4, show that $\boldsymbol{\mu} \in V$ if and only if $\tilde{\boldsymbol{\mu}} \in V^n$.

7. Verify that the likelihood ratio test for testing that $\boldsymbol{\Sigma} = \sigma^2\mathbf{I}$ is given by (21.13). Verify the formula for the degrees of freedom in the asymptotic distribution of $-2\log\lambda$.

8. Let $\boldsymbol{\mu}$ be an $n \times r$ matrix, V a p-dimensional subspace of R^n and $\mathbf{A}$ an invertible $r \times r$ matrix. Show that $\boldsymbol{\mu}\mathbf{A} \in V$ if and only if $\boldsymbol{\mu} \in V$.

9. Verify the formula for $\hat{\boldsymbol{\Sigma}}^*_{22}$ in the derivation of the likelihood ratio test in Section 21.5.

10. Prove Theorem 21.7.

CHAPTER 22

Simplifying the Structure of the Covariance Matrix

In this chapter, we discuss two methods of simplifying the structure of the covariance matrix: principal components analysis and canonical analysis. These methods allow us to transform the data to a form in which the covariance matrix is more easily interpreted. These methods are often useful in shrinking the dimension of the data so that most of relevant information in the data is contained in a data set of smaller dimension. Both these methods were suggested by Hotelling (1933 and 1936).

One important method for bringing out the structure of the covariance is not presented in this book. That method is factor analysis. While factor analysis is more important than either of the methods discussed in this chapter, a short treatment of it would raise more questions than it would answer and a careful treatment would take a separate book. In addition, the methods used in factor analysis are quite removed from the other methods discussed in this book. For these reasons we do not study factor analysis. For a nice treatment of factor analysis at about the same level as this book, see Lawley and Maxwell (1971).

22.1. PRINCIPAL COMPONENTS.

Let $\mathbf{X}$ be an r-dimensional random vector with covariance matrix $\mathbf{\Sigma} \geqslant 0$. In this section we discuss an orthogonal transformation on $\mathbf{X}$ that leads to a simpler covariance structure. We assume for now that $\mathbf{\Sigma}$ is known. We use the following theorem.

THEOREM 22.1. Let $\mathbf{T} \geqslant 0$ be an $r \times r$ matrix and let $d_1 \geqslant \cdots \geqslant d_r$ be the ordered roots of

$$|\mathbf{T} - d\mathbf{I}| = 0. \tag{22.1}$$

Then $d_i \geqslant 0$ and there exists an orthogonal matrix $\mathbf{C}$ such that

$$\mathbf{T} = \mathbf{CDC}', \qquad \mathbf{D} = \begin{bmatrix} d_1 & & 0 \\ & \ddots & \\ 0 & & d_r \end{bmatrix}.$$

PROOF. See Section 6 of the appendix. □

Now $\boldsymbol{\Sigma} \geqslant 0$. Therefore, let $\delta_1 \geqslant \cdots \geqslant \delta_r \geqslant 0$ be the ordered roots of

$$|\boldsymbol{\Sigma} - \delta\mathbf{I}| = 0$$

and let $\boldsymbol{\Gamma}$ be an orthogonal matrix such that

$$\boldsymbol{\Sigma} = \boldsymbol{\Gamma\Delta\Gamma}', \qquad \boldsymbol{\Delta} = \begin{bmatrix} \delta_1 & & 0 \\ & \ddots & \\ 0 & & \delta_r \end{bmatrix}.$$

Define

$$\mathbf{U} = \begin{bmatrix} U_1 \\ \vdots \\ U_r \end{bmatrix} = \boldsymbol{\Gamma}'\mathbf{X}.$$

Then $\mathbf{U}$ has covariance matrix $\boldsymbol{\Gamma}'\boldsymbol{\Sigma\Gamma} = \boldsymbol{\Delta}$. Therefore the U_i are uncorrelated and the variance of U_i is δ_i. $\mathbf{U}$ is called a *vector of principal components of* $\mathbf{X}$ and U_i is called an *ith principal component of* $\mathbf{X}$. Since $\boldsymbol{\Gamma}$ is orthogonal, observing $\mathbf{U}$ is equivalent to observing $\mathbf{X}$.

We first discuss the uniqueness of principal components. If U_i is an ith principal component of $\mathbf{X}$ then so is $-U_i$. If the δ_i are different then this is the only nonuniqueness in the definition of $\mathbf{U}$. Therefore if the δ_i are distinct, then the principal components are defined up to sign changes. We now look at the case in which there are repeated roots to (22.1). Suppose that $\delta_{s+1} = \cdots = \delta_{s+t}$. Let $\mathbf{B}$ be an orthogonal $t \times t$ matrix and let

$$\begin{bmatrix} U^*_{s+1} \\ \vdots \\ U^*_{s+t} \end{bmatrix} = \mathbf{B} \begin{bmatrix} U_{s+1} \\ \vdots \\ U_{s+t} \end{bmatrix}.$$

Then $(U_1, \ldots, U_s, U^*_{s+1}, \ldots, U^*_{s+t}, U_{s+t+1}, \ldots, U_r)'$ is also a vector of principal components of $\mathbf{X}$. In particular, if $\boldsymbol{\Sigma} = \mathbf{I}$, then $\mathbf{CX}$ is a vector of principal components of $\mathbf{X}$ for any orthogonal $\mathbf{C}$. In order to get uniqueness, at least up to sign changes, we therefore usually assume that the δ_i are distinct when we use principal components.

An alternative approach to principal components that is quite helpful in understanding them is given by the following theorem.

THEOREM 22.2. Let $\mathbf{U} = (U_1, \ldots, U_r)'$ be a vector of principal components of $\mathbf{X}$. Then $U_i = \mathbf{a}_i'\mathbf{X}$ for some $\mathbf{a}_i$ such that $\|\mathbf{a}_i\| = 1$. Let $\mathbf{a} \in R^r$ such that $\|\mathbf{a}\| = 1$. The $\operatorname{var} \mathbf{a}'\mathbf{X} \leqslant \operatorname{var} U_1$. If $\mathbf{a}'\mathbf{X}$ is uncorrelated with $U_1, \ldots, U_{i-1}$ then $\operatorname{var} \mathbf{a}'\mathbf{X} \leqslant \operatorname{var} U_i$.

PROOF. Let $\boldsymbol{\Gamma} = (\boldsymbol{\Gamma}_1, \ldots, \boldsymbol{\Gamma}_n)$. Then $U_i = \mathbf{a}_i'\mathbf{X}$ where $\mathbf{a}_i = \boldsymbol{\Gamma}_i$. Since $\boldsymbol{\Gamma}$ is orthogonal, $\|\mathbf{a}_i\| = \|\boldsymbol{\Gamma}_i\| = 1$. Now, let $\mathbf{a} \in R^r$, $\|\mathbf{a}\| = 1$. Define $\mathbf{b} = (b_1, \ldots, b_r)' = \boldsymbol{\Gamma}\mathbf{a}$. Then $\|\mathbf{b}\| = \|\mathbf{a}\| = 1$ and $\mathbf{a}'\mathbf{X} = \mathbf{b}'\mathbf{U} = \sum_{j=1}^r b_j U_j$. Since the U_j are independent, $\operatorname{var} U_j = \delta_j$, we see that

$$\operatorname{var} \mathbf{a}'\mathbf{X} = \operatorname{var} \mathbf{b}'\mathbf{U} = \sum_{j=1}^{r} \delta_j b_j^2 \leqslant \delta_1 \sum_{j=1}^{r} b_j^2 = \delta_1 \|\mathbf{b}\|^2 = \delta_1 = \operatorname{var} U_1.$$

The covariance between $\mathbf{a}'\mathbf{X} = \sum_{j=1}^r b_j U_j$ and U_k is b_k. Therefore $\mathbf{a}'\mathbf{X}$ is uncorrelated with $U_1, \ldots, U_{i-1}$ if and only if $b_1 = b_2 = \cdots = b_{i-1} = 0$. Hence if $\mathbf{a}'\mathbf{X}$ is uncorrelated with $U_1, \ldots, U_{i-1}$, then

$$\operatorname{var} \mathbf{a}'\mathbf{X} = \operatorname{var} \mathbf{b}'\mathbf{U} = \sum_{j=1}^{r} \delta_j b_j^2 \leqslant \delta_i \sum_{j=1}^{r} b_j^2 = \delta_i \|\mathbf{b}\|^2 = \delta_i = \operatorname{var} U_i. \quad \square$$

This theorem implies that we can find principal components sequentially, starting with U_1. We find a first principal component by finding a linear combination with maximum variance among all linear combinations $\mathbf{a}'\mathbf{X}$ such that $\|\mathbf{a}\| = 1$. We then find a second principal component by finding a linear combination of maximum variance among all linear combinations $\mathbf{a}'\mathbf{X}$ such that $\mathbf{a}'\mathbf{X}$ is uncorrelated with $\mathbf{U}_1$ and $\|\mathbf{a}\| = 1$. We can continue in this fashion until we have found r principal components.

In practice, of course, we rarely know $\boldsymbol{\Sigma}$, the covariance matrix of $\mathbf{X}$, and hence cannot find the principal components of $\mathbf{X}$. We now discuss estimation of these principal components. Let $\mathbf{Y}_1, \ldots, \mathbf{Y}_n$ be a sample from the distribution of $\mathbf{X}$, and let

$$\mathbf{S} = \frac{1}{n-1} \Sigma(\mathbf{Y}_i - \overline{\mathbf{Y}})(\mathbf{Y}_i - \overline{\mathbf{Y}})'$$

be the sample covariance matrix of the $\mathbf{Y}$'s. Then $\mathbf{S} \geqslant 0$. Therefore, let $d_1 \geqslant \cdots \geqslant d_r$ be the roots of

$$|\mathbf{S} - d\mathbf{I}| = 0 \tag{24.2}$$

and let $\mathbf{C}$ be an orthogonal matrix such that

$$\mathbf{S} = \mathbf{CDC}', \quad \mathbf{D} = \begin{bmatrix} d_1 & & 0 \\ & \ddots & \\ 0 & & d_r \end{bmatrix}.$$

Define

$$\hat{\mathbf{U}} = \begin{bmatrix} \hat{U}_1 \\ \vdots \\ \hat{U}_n \end{bmatrix} = \mathbf{C}'\mathbf{X}.$$

Then $\hat{\mathbf{U}}$ is called a *vector of sample principal components* and $\hat{U}_i$ is called an *ith sample principal component*. If $n > r$ and $\mathbf{\Sigma} > 0$, it can be shown that the d_i are distinct with probability 1, even if the δ_i are not, so that the sample principal components are defined up to sign changes in this case. If $n \leqslant r$ but $\mathbf{\Sigma} > 0$, then $\mathbf{S}$ has rank $n-1$ and $d_n = d_{n+1} = \cdots = d_r = 0$. In this case, however, $d_1, \ldots, d_n$ are distinct, and therefore $U_1, \ldots, U_{n-1}$ are defined up to a sign change. Also $\operatorname{var} U_n = \operatorname{var} U_{n+1} = \cdots = \operatorname{var} U_r = 0$. Finally if $\mathbf{\Sigma}$ is not positive definite, but has rank k, then $\mathbf{S}$ has rank s, the minimum of k and $n-1$. Therefore $U_1, \ldots, U_s$ are defined up to sign changes and the remaining components have variance 0.

We now assume that $\mathbf{X}$ is normally distributed, and hence the $\mathbf{Y}_i$ are a sample from a multivariate normal distribution. (Up to this point the only assumption made about the distribution of $\mathbf{X}$ is that it have a finite covariance matrix.) Then $[(n-1)/n]\mathbf{S}$ is the maximum likelihood estimator of $\mathbf{\Sigma}$. Since

$$\frac{n-1}{n}\mathbf{S} = \mathbf{C}\left(\frac{n-1}{n}\mathbf{D}\right)\mathbf{C}',$$

the sample principal components are the same whether computed from the unbiased estimator $\mathbf{S}$ or the maximum likelihood estimator $[(n-1)/n]\mathbf{S}$ and are the MLE's for the principal components.

It is often the case that for some k, $\delta_1, \ldots, \delta_k$ are all moderately large but that $\delta_{k+1}, \ldots, \delta_r$ are all quite small (or that $d_1, \ldots, d_k$ are all moderately large but that $d_{k+1}, \ldots, d_r$ are all quite small). In this case, nearly all the variation in $\mathbf{X}$ can be explained by the first k principal components $(U_1, \ldots, U_k)'$ (or by the first k sample principal components $(\hat{U}_1, \ldots, \hat{U}_k)'$). If $\|\mathbf{a}\| = 1$ and $\mathbf{W} = \mathbf{a}'\mathbf{X}$ is uncorrelated with $U_1, \ldots, U_k$, then $\operatorname{var}(W) \leqslant \delta_{k+1}$ (by Theorem 22.2) which we are assuming to be quite small. Therefore, we can replace the original r-dimensional vector $\mathbf{X}$ with the k-dimensional vector $(U_1, \ldots, U_k)'$ (or $(\hat{U}_1, \ldots, \hat{U}_k)'$) without sacrificing much of the variation in $\mathbf{X}$. If k is substantially smaller than r, there can be a large decrease in the dimension of the problem by this technique. For this reason, the principal components of $\mathbf{X}$ are often extracted sequentially beginning with the first one. In this situation, we need a measure that indicates how much of the variation is explained by the first k principal components. A measure that is often used is

$$t_k = \frac{\sum_{i=1}^{k}\delta_i}{\sum_{i=1}^{n}\delta_i} = \frac{\sum_{i=1}^{k}\delta_i}{\operatorname{tr}\boldsymbol{\Sigma}} \qquad \left(\text{or } \hat{t}_k = \frac{\sum_{i=1}^{k}d_i}{\operatorname{tr}\mathbf{S}}\right).$$

We consider t_k or $\hat{t}_k$ to be the proportion of total variation explained by the first k principal components of X. When t_k or $\hat{t}_k$ is large enough we stop extracting principal components. (Note that in computing t_k or $\hat{t}_k$ it is not necessary to know $\delta_{k+1}, \ldots, \delta_r$ or $d_{k+1}, \ldots, d_r$.)

Principal components analysis is often performed before some of the procedures described earlier in the book to shrink the size of the data sets. For example, consider a discrimination problem in which we have n observations in the first population and m observations from the second population, but $n + m - 2 < r$ (the dimension of the random vectors). Then the pooled sample covariance matrix is singular, and it is not possible to compute the sample linear discrimination function. One approach that can be used is to find the sample principal components of the observations from the pooled covariance matrix. Replace each r-dimensional vector of observations $\mathbf{X}_{ij}$ with the p-dimensional vector of its first p principal components $\mathbf{U}_{ij}$. If p is much less than $n + m - 2$, then the discrimination can be performed on the $\mathbf{U}_{ij}$ instead of the $\mathbf{X}_{ij}$. This type of procedure should be used with great caution. In the first place, the principal components analysis is performed with the same data as the discriminant analysis, which complicates the analysis of the results. (In particular, the $\mathbf{U}_{ij}$ are not independent, even if the $\mathbf{X}_{ij}$ were.) A more serious complication is that the principal components are chosen to have maximum variance, in some sense (see Theorem 22.2). They are not chosen to be good discriminators. It may be that some linear combination of the observations with low variance is the crucial combination for discrimination.

One unfortunate aspect of prinicpal components analysis is that the principal components are very dependent on the units used in defining $\mathbf{X}$. Let $\mathbf{X}^* = \mathbf{BX}$ where $\mathbf{B}$ is an invertible matrix. If $\mathbf{B}$ is a diagonal matrix with positive diagonal elements, then $\mathbf{B}$ would just represent a change of scale for the components of $\mathbf{X}$. The covariance matrix of $\mathbf{X}^*$ is

$$\mathbf{B\Sigma B}' = \mathbf{BC\Delta C'B}'.$$

Unless $\mathbf{B} = c\mathbf{I}$ (i.e., unless the units are all changed by the same multiple) there is no way to determine the principal components for $\mathbf{X}^*$ from those for $\mathbf{X}$. To determine the principal components for $\mathbf{X}^*$, we need to know the

δ_i also. Therefore, two researchers who observe the same data but used different units would find completely different principal components. Because of this problem the components of $\mathbf{X}$ are often standardized to have mean 0 and variance 1 before principal components are found. The covariance matrix of the standardized variables is, of course, just the correlation matrix of $\mathbf{X}$. Performing principal components on the standardized variables eliminates the problem of units, but the results are somewhat difficult to interpret. If $\mathbf{W} = (W_1, \ldots, W_r)'$ is a vector of standardized variables, the correlation between W_i and W_j is ρ_{ij} and $\|a\| = 1$, then

$$\operatorname{var}(\mathbf{a}'\mathbf{W}) = 1 + \sum_{i \neq j} \sum a_i a_j \rho_{ij}$$

and it is not clear exactly why we should want to maximize this quantity. Because of this problem with units, any principal components analysis on variables that are not measured in the same units should be used with caution.

For a more detailed discussion of principal components analysis, see Kshirsagar (1972, pp. 424–465).

22.2. CANONICAL ANALYSIS.

Let $\mathbf{X}$ and $\mathbf{Y}$ be p- and q-dimensional random vectors with $p \leqslant q$. In this section we find two invertible linear transformations $\mathbf{AX}$ and $\mathbf{BY}$ that help to clarify the correlation structure between $\mathbf{X}$ and $\mathbf{Y}$. We use the following theorem.

THEOREM 22.3. Let $\mathbf{R} > 0$ and $\mathbf{T} > 0$ be $p \times p$ and $q \times q$ matrices with $p \leqslant q$ and let $\mathbf{V}$ be $p \times q$. Let $d_1 \geqslant \cdots \geqslant d_p$ be p largest roots of

$$\begin{vmatrix} -d\mathbf{R} & \mathbf{V} \\ \mathbf{V}' & -d\mathbf{T} \end{vmatrix} = 0.$$

Then $d_i \geqslant 0$. There exist invertible matrices $\mathbf{A}$ and $\mathbf{B}$ with $\mathbf{A}$ $p \times p$ and $\mathbf{B}$ $q \times q$ such that

$$\mathbf{ARA}' = \mathbf{I}, \qquad \mathbf{BTB}' = \mathbf{I}, \qquad \mathbf{AVB}' = (\mathbf{D}\ \mathbf{0}), \qquad \mathbf{D} = \begin{bmatrix} d_1 & & 0 \\ & \ddots & \\ 0 & & d_p \end{bmatrix}.$$

PROOF. See Theorem 13 of the appendix. □

Now, let $\mathbf{\Sigma}$ be the joint covariance matrix of $\mathbf{X}$ and $\mathbf{Y}$,

$$\mathbf{\Sigma} = \begin{pmatrix} \mathbf{\Sigma}_{11} & \mathbf{\Sigma}_{12} \\ \mathbf{\Sigma}_{21} & \mathbf{\Sigma}_{22} \end{pmatrix}$$

where $\mathbf{\Sigma}_{11}$ is $p \times p$. Throughout this section, we assume that $\mathbf{\Sigma} > 0$. For now we also assume that $\mathbf{\Sigma}$ is known. Since $\mathbf{\Sigma} > 0$, we see that $\mathbf{\Sigma}_{11} > 0$ and $\mathbf{\Sigma}_{22} > 0$. Therefore, let $\delta_1 \geqslant \cdots \geqslant \delta_p$ be the largest roots of

$$\begin{vmatrix} -\delta\mathbf{\Sigma}_{11} & \mathbf{\Sigma}_{12} \\ \mathbf{\Sigma}_{21} & -\delta\mathbf{\Sigma}_{22} \end{vmatrix} = 0 \tag{22.3}$$

and let $\mathbf{A}$ and $\mathbf{B}$ be such that

$$\mathbf{A}\mathbf{\Sigma}_{11}\mathbf{A}' = \mathbf{I}, \qquad \mathbf{B}\mathbf{\Sigma}_{22}\mathbf{B}' = \mathbf{I}, \qquad \mathbf{A}\mathbf{\Sigma}_{12}\mathbf{B}' = (\mathbf{\Delta}\,\mathbf{0}), \qquad \mathbf{\Delta} = \begin{pmatrix} \delta_1 & & 0 \\ & \ddots & \\ 0 & & \delta_p \end{pmatrix}.$$

Define

$$\mathbf{U} = \begin{pmatrix} U_1 \\ \vdots \\ U_p \end{pmatrix} = \mathbf{AX}, \qquad \mathbf{V} = \begin{pmatrix} V_1 \\ \vdots \\ V_q \end{pmatrix} = \mathbf{BY}.$$

Then the joint covariance matrix of $\mathbf{U}$ and $\mathbf{V}$ is

$$\begin{pmatrix} \mathbf{A} \\ \mathbf{B} \end{pmatrix} \begin{pmatrix} \mathbf{\Sigma}_{11} & \mathbf{\Sigma}_{12} \\ \mathbf{\Sigma}_{21} & \mathbf{\Sigma}_{22} \end{pmatrix} (\mathbf{A}'\,\mathbf{B}') = \begin{pmatrix} \mathbf{I} & (\mathbf{\Delta}\,\mathbf{0}) \\ \binom{\mathbf{\Delta}}{\mathbf{0}} & \mathbf{I} \end{pmatrix}.$$

Therefore the U_i are uncorrelated and have variance 1. Similarly the V_j are uncorrelated and have variance 1. In addition, U_i and V_j are uncorrelated unless $i = j$. Finally, the correlation between U_i and V_i is δ_i. Hence the entire correlation structure between U and V is summarized by the pairs (U_i, V_i) and the correlations δ_i. The pair $(U_i . V_i)$ is called a *pair of ith canonical variables*. δ_i is called the *ith canonical correlation*. Note that the canonical correlations are all nonnegative. (The variables $V_{p+1}, \ldots, V_q$ are uncorrelated with all the U_i and the $V_1, \ldots, V_p$ and are ignored.)

Before discussing the uses of canonical correlations and variables, we discuss the uniqueness of them. The canonical correlations are unique since they are the p largest ordered roots of (22.3). The canonical variables are not uniquely defined. If (U_i, V_i) is a pair of ith canonical variables, then so is $(-U_i, -V_i)$. If the δ_i are all different, then this is the only ambiguity in the definition. If some of the δ_i are the same, then the ambiguity is more complicated.

We now provide an alternative approach to canonical correlations and canonical variables that is useful in understanding them.

THEOREM 22.4. Let $(U_1, V_1), \ldots, (U_p, V_p)$ be a set of canonical variables for $\mathbf{X}$ and $\mathbf{Y}$, and let δ_i be the ith canonical correlation. Let $\mathbf{a} \in R^p$, $\mathbf{b} \in R^q$. Then the correlation between $\mathbf{a}'\mathbf{X}$ and $\mathbf{b}'\mathbf{Y}$ is no greater than δ_1, the correlation between U_1 and V_1. If $\mathbf{a}'\mathbf{X}$ is uncorrelated with $U_1, \ldots, U_{i-1}$ and $\mathbf{b}'\mathbf{Y}$ is uncorrelated with $V_1, \ldots, V_{i-1}$, then the correlation between $\mathbf{a}'\mathbf{X}$ and $\mathbf{b}'\mathbf{Y}$ is no greater than δ_i, the correlation between U_i and V_i.

PROOF. See Exercise B1. □

Therefore, we can interpret a first pair of canonical variables (U_1, V_1) as a pair of variables that have the maximum correlation out of all variables $\mathbf{a}'\mathbf{X}$ and $\mathbf{b}'\mathbf{Y}$ that have variance 1. (The restriction to variance 1 is necessary for definiteness, since the correlation between U_1 and V_1 is the same as the correlation between cU_1 and dV_1 for any $c > 0$, $d > 0$.) The ith canonical variables are defined recursively by the following scheme. An ith pair of canonical variables (U_i, V_i) is a pair that has the maximum correlation out of all pairs $\mathbf{a}'\mathbf{X}$ and $\mathbf{b}'\mathbf{Y}$ that have variance 1 and such that $\mathbf{a}'\mathbf{X}$ is uncorrelated with $U_1, \ldots, U_{i-1}$ and $\mathbf{b}'\mathbf{Y}$ is uncorrelated with $V_1, \ldots, V_{i-1}$. The ith canonical correlation is defined as the correlation between a pair of ith canonical variables. In particular, this approach makes it clear that if $p = 1$, then the first (and only) canonical correlation is just the multiple correlation coefficient. (See Theorem 16.18.)

In practice, of course, we rarely know $\boldsymbol{\Sigma}$. Therefore, let $\mathbf{T}_1, \ldots, \mathbf{T}_n$ be a sample from the joint distribution of $\mathbf{X}$ and $\mathbf{Y}$ (i.e., $\mathbf{T}_i$ is $(p+q)$-dimensional), and let $\mathbf{S}$ be the sample covariance matrix of the $\mathbf{T}_i$'s,

$$\mathbf{S} = \begin{pmatrix} \mathbf{S}_{11} & \mathbf{S}_{12} \\ \mathbf{S}_{21} & \mathbf{S}_{22} \end{pmatrix} = \frac{1}{n-1} \sum_i (\mathbf{T}_i - \bar{\mathbf{T}})(\mathbf{T}_i - \bar{\mathbf{T}})'$$

where $\mathbf{S}_{11}$ is $p \times p$. We assume that $n > p + q$, so that $S > 0$ (with probability 1), and hence $\mathbf{S}_{11} > 0$ and $\mathbf{S}_{22} > 0$. Therefore, let $d_1 \geqslant \cdots \geqslant d_p$ be the largest roots of

$$\begin{vmatrix} -d\mathbf{S}_{11} & \mathbf{S}_{12} \\ \mathbf{S}_{21} & -d\mathbf{S}_{22} \end{vmatrix} = 0 \qquad (22.4)$$

and let $\mathbf{G}$ and $\mathbf{H}$ be invertible $p \times p$ and $q \times q$ matrices such that

$$\mathbf{G}\mathbf{S}_{11}\mathbf{G}' = \mathbf{I}, \qquad \mathbf{H}\mathbf{S}_{22}\mathbf{H}' = \mathbf{I}, \qquad \mathbf{G}\mathbf{S}_{12}\mathbf{H}' = (\mathbf{D}\,\mathbf{0}), \qquad \mathbf{D} = \begin{bmatrix} d_1 & & 0 \\ & \ddots & \\ 0 & & d_p \end{bmatrix}.$$

Define

$$\hat{\mathbf{U}} = \begin{pmatrix} \hat{U}_1 \\ \vdots \\ \hat{U}_p \end{pmatrix} = \mathbf{GX}, \qquad \hat{\mathbf{V}} = \begin{pmatrix} \hat{V}_1 \\ \vdots \\ \hat{V}_q \end{pmatrix} = \mathbf{HY}.$$

Then $(\hat{U}_i, \hat{V}_i)$ is called an *ith pair of sample canonical variables* and d_i is called the *ith sample canonical correlation*. It can be shown that the d_i are distinct with probability 1 (even if the δ_i are not), and therefore the sample canonical variables are defined up to a sign change in each pair. If $\mathbf{X}$ and $\mathbf{Y}$ are jointly normally distributed, then the maximum likelihood estimator of $\boldsymbol{\Sigma}$ is $[(n-1)/n]S$. In Exercise B2, you are asked to show that the canonical correlations are the same whether computed from the unbiased estimator $\mathbf{S}$ or the maximum likelihood estimator $[(n-1)/n]\mathbf{S}$, but that the sample canonical varaibles computed from $[(n-1)/n]\mathbf{S}$ are $[(n-1)/n]$ times as great as those computed from $\mathbf{S}$.

It is often the case that while $\delta_1, \ldots, \delta_k$ are all moderately large, $\delta_{k+1}, \ldots, \delta_p$ are all quite small (or $d_1, \ldots, d_k$ are all moderately large, but $d_{k+1}, \ldots, d_p$ are all quite small). In this case, nearly all the correlation structure between $\mathbf{U}$ and $\mathbf{V}$ (and hence between $\mathbf{X}$ and $\mathbf{Y}$) can be explained by the pairs $(U_1, V_1), \ldots, (U_k, V_k)$ (or $(\hat{U}_1, \hat{V}_1), \ldots, (\hat{U}_k, \hat{V}_k)$). We have reduced the dimension from $p+q$ to $2k$ without losing much information about the correlation structure between $\mathbf{X}$ and $\mathbf{Y}$. The following measures are often used to indicate how much of the correlation structure has been explained by the first k variables:

$$t_k = \frac{\sum_{i=1}^k \delta_i^2}{\sum_{i=1}^n \delta_i^2} = \frac{\sum_{i=1}^k \delta_i^2}{\operatorname{tr} \boldsymbol{\Sigma}_{11}^{-1}\boldsymbol{\Sigma}_{12}\boldsymbol{\Sigma}_{22}^{-1}\boldsymbol{\Sigma}_{21}}, \qquad \hat{t}_k = \frac{\sum_{i=1}^k d_i^2}{\sum_{i=1}^n d_i^2} = \frac{\sum_{i=1}^k d_i^2}{\operatorname{tr} \mathbf{S}_{11}^{-1}\mathbf{S}_{12}\mathbf{S}_{22}^{-1}\mathbf{S}_{21}}.$$

(Note that the δ_i^2 are the eigenvalues of $\boldsymbol{\Sigma}_{11}^{-1/2}\boldsymbol{\Sigma}_{12}\boldsymbol{\Sigma}_{22}^{-1}\boldsymbol{\Sigma}_{21}\boldsymbol{\Sigma}_{11}^{-1/2}$. See Exercise B3.)

In Exercise B4, you are asked to show that the canonical correlations are unchanged by a change of units in $\mathbf{X}$ and/or $\mathbf{Y}$, and the canonical variables are also the same after rescaling. One reason that we would expect that canonical correlations and canonical variables would not be sensitive to scale changes is given in Theorem 22.4. We see from that theorem that the canonical variables are variables that maximize a correlation, and corelations are unit-free. Also the canonical correlations are themselves correlations between variables and therefore are also unit-free.

As a final comment, we note $(d_1^2, \ldots, d_p^2)$ is the maximal invariant and $(\delta_1^2, \ldots, \delta_p^2)$ is the parameter maximal invariant for testing that $\boldsymbol{\Sigma}_{12} = 0$

(see Exercise B3). Since $(d_1, \ldots, d_p)$ is an invertible function of $(d_1^2, \ldots, d_p^2)$ it is also a maximal invariant.

For a more detailed discussion of canonical correlations, the reader is referred to Kshirsager (1972, pp. 247–288).

EXERCISES.

Type B

1. Prove Theorem 22.4.

2. Show that the canonical correlations are unchanged when computed from the MLE $[(n-1)/n]\mathbf{S}$ instead of from $\mathbf{S}$, but that the canonical variables are multiplied by $[(n-1)/n]$.

3. (a) Let δ_i be the ith canonical correlation between $\mathbf{X}$ and $\mathbf{Y}$. Show that the δ_i^2 are the roots of

$$q(t) = |\boldsymbol{\Sigma}_{11} - t\boldsymbol{\Sigma}_{12}\boldsymbol{\Sigma}_{22}^{-1}\boldsymbol{\Sigma}_{21}| = 0$$

[hence $(\delta_1^2, \ldots, \delta_p^2)$ is the parameter maximal invariant for testing that $\boldsymbol{\Sigma}_{12} = 0$.]

(b) Show that the δ_i^2 are the eigenvalues of $\boldsymbol{\Sigma}_{11}^{-1/2}\boldsymbol{\Sigma}_{12}\boldsymbol{\Sigma}_{22}^{-1}\boldsymbol{\Sigma}_{21}\boldsymbol{\Sigma}_{11}^{-1/2}$.

4. Show that the canonical correlations and canonical variables are not affected by unit changes in $\mathbf{X}$ and $\mathbf{Y}$. That is, if $\mathbf{X}^* = \mathbf{CX}$ and $\mathbf{Y}^* = \mathbf{DY}$, where $\mathbf{C}$ and $\mathbf{D}$ are diagonal matrices with positive elements and $\mathbf{a}_i'\mathbf{X}$ and $\mathbf{b}_i'\mathbf{Y}$ are a pair of ith canonical variables for X and Y then $(\mathbf{C}^{-1}\mathbf{a}_i)'\mathbf{X}^* = \mathbf{a}_i'\mathbf{X}$ and $(\mathbf{D}^{-1}\mathbf{b}_i)'\mathbf{Y}^* = \mathbf{b}_i'\mathbf{Y}$ are a pair of ith canonical variables for $\mathbf{X}^*$ and $\mathbf{Y}^*$.

APPENDIX: SOME MATRIX ALGEBRA

In this appendix we derive those matrix algebra results used in the text that are often not covered in a basic course in matrix algebra. In Section A.1, we state the basic definitions and elementary results that should be covered in a basic course in matrix algebra and also state the principal axis theorem, which is the basis for many of the later derivations. In Section A.2, we present results about partitioned matrices.

The basic theorems about positive definite and nonnegative definite matrices are derived in Section A.3, and other results from matrix algebra are derived in Section A.4. In Section A.5, we derive the result that is the basis for maximum likelihood estimators in multivariate models, and also find two Jacobians used for deriving densities in Chapter 17. Finally in Section A.6, we give a proof of the principal axis theorem, since that theorem is not often proved in basic matrix algebra classes.

A.1. DEFINITIONS AND BASIC PROPERTIES.

An *$n \times p$ matrix* $\mathbf{A}$ is a rectangular array of real numbers with n rows and p columns, that is,

$$\mathbf{A} = \begin{bmatrix} a_{11} & \cdots & a_{1p} \\ \vdots & & \vdots \\ a_{n1} & \cdots & a_{np} \end{bmatrix}. \tag{A.1}$$

We write $\mathbf{A} = (a_{ij})$ for (A.1). The real number a_{ij} is called the *i, j element* of $\mathbf{A}$. The numbers n and p are called the *dimensions* of $\mathbf{A}$. If $\mathbf{A}$ is an $n \times 1$ matrix, we say that $\mathbf{A}$ is an *n-dimensional vector*.

Let $\mathbf{A} = (a_{ij})$ and $\mathbf{B} = (b_{ij})$ be $n \times p$ matrices. The *sum* of $\mathbf{A}$ and $\mathbf{B}$, written $\mathbf{A} + \mathbf{B}$ is the matrix $\mathbf{C} = (c_{ij})$ where $c_{ij} = a_{ij} + b_{ij}$. Let s be a real number (called a scalar). The *scalar product* of s and $\mathbf{A}$ written $s\mathbf{A}$ is the matrix $\mathbf{D} = (d_{ij})$ where $d_{ij} = sa_{ij}$. Some properties of sums and scalar multiplications are the following:

$$\mathbf{A} + \mathbf{B} = \mathbf{B} + \mathbf{A}, \qquad (\mathbf{A} + \mathbf{B}) + \mathbf{C} = \mathbf{A} + (\mathbf{B} + \mathbf{C}),$$

$$(s + t)\mathbf{A} = s\mathbf{A} + t\mathbf{A}, \qquad s(\mathbf{A} + \mathbf{B}) = s\mathbf{A} + s\mathbf{B}, \qquad (st)\mathbf{A} = s(t\mathbf{A}). \quad (A.2)$$

If $\mathbf{A} = (a_{ij})$ is $n \times p$ and $\mathbf{B} = (b_{ij})$ is $p \times r$, then the *product* of $\mathbf{A}$ and $\mathbf{B}$, written $\mathbf{AB}$, is the $n \times r$ matrix $\mathbf{C} = (c_{ij})$ where $c_{ij} = \sum_{k=1}^{p} a_{ik} b_{kj}$. Note that this product is only defined if the number of columns of $\mathbf{A}$ is the same as the number of rows of $\mathbf{B}$. Note also that $\mathbf{AB} \neq \mathbf{BA}$ even if both are defined.

The important properties of the product are the following:

$$(\mathbf{AB})\mathbf{C} = \mathbf{A}(\mathbf{BC}), \qquad \mathbf{A}(\mathbf{B} + \mathbf{C}) = (\mathbf{AB}) + (\mathbf{AC}),$$
$$(\mathbf{A} + \mathbf{B})\mathbf{C} = (\mathbf{AC}) + (\mathbf{BC}) \quad (A.3)$$

Let $\mathbf{A} = (a_{ij})$ be an $n \times p$ matrix. Then the *transpose* of $\mathbf{A}$, written $\mathbf{A}'$, is the $p \times n$ matrix $\mathbf{C} = (c_{ij})$, where $c_{ij} = a_{ji}$. We note the following properties of transposes:

$$(\mathbf{A}')' = \mathbf{A}, \qquad (\mathbf{A} + \mathbf{B})' = \mathbf{A}' + \mathbf{B}', \qquad (\mathbf{AB})' = \mathbf{B}'\mathbf{A}' \quad (A.4)$$

Let $\mathbf{u}_i$ be p-dimensional vectors. We say that the $\mathbf{u}_i$ are *linearly independent* if $\sum_i a_i \mathbf{u}_i = \mathbf{0}$ implies that all the $a_i = 0$. Now let $\mathbf{A}$ be an $n \times p$ matrix. The *rank* of $\mathbf{A}$ is defined to be the largest number of linearly independent columns of $\mathbf{A}$. The rank of $\mathbf{A}$ is the same as the rank of $\mathbf{A}'$, and hence, the rank could also be defined as the largest number of linearly independent rows of $\mathbf{A}$. We note that the rank of $\mathbf{A}$ is therefore less than or equal to the minimum of the dimensions of $\mathbf{A}$. It is also true that the rank of $\mathbf{AB}$ is less than or equal to the minimum of the ranks of $\mathbf{A}$ and $\mathbf{B}$.

A *zero matrix*, written $\mathbf{0}$, is a matrix all of whose elements are zero. We note that for any matrix $\mathbf{A}$,

$$\mathbf{A} + \mathbf{0} = \mathbf{0} + \mathbf{A} = \mathbf{A}, \qquad \mathbf{0A} = \mathbf{0}, \qquad \mathbf{A} + (-1)\mathbf{A} = \mathbf{0}. \quad (A.5)$$

An *identity matrix* is a $p \times p$ matrix $\mathbf{I}$ of the following form

$$\mathbf{I} = \begin{pmatrix} 1 & 0 & \cdots & 0 \\ 0 & 1 & \cdots & 0 \\ & \vdots & \ddots & \vdots \\ 0 & 0 & \cdots & 1 \end{pmatrix}.$$

We note that

$$\mathbf{IA} = \mathbf{AI} = \mathbf{A}. \quad (A.6)$$

The $p \times p$ matrix $\mathbf{A}$ is *invertible* if there exists a $p \times p$ matrix $\mathbf{B}$ such that $\mathbf{AB} = \mathbf{BA} = \mathbf{I}$. If there is such a $\mathbf{B}$, it is unique and is called the *inverse* of $\mathbf{A}$. We write $\mathbf{A}^{-1}$ for the inverse of $\mathbf{A}$. The following are the basic properties of inverses:

$$(\mathbf{A}')^{-1} = (\mathbf{A}^{-1})', \qquad (\mathbf{AB})^{-1} = (\mathbf{B}^{-1})(\mathbf{A}^{-1}). \tag{A.7}$$

The $p \times p$ matrix $\mathbf{A}$ is invertible if and only if it has rank p. If $\mathbf{A}$ is invertible then the rank of $\mathbf{AB}$ is the same as the rank of $\mathbf{B}$.

Let $\mathbf{A} = (a_{ij})$ be a $p \times p$ matrix. Then the *determinant* of $\mathbf{A}$, written $|\mathbf{A}|$, is defined to be

$$\sum_{\pi} c(\pi) a_{1\pi(1)} \cdots a_{p\pi(p)}$$

where π runs over all $p!$ permutations of $(1, \ldots, p)$ and $c(\pi)$ is 1 if the number of inversions to get from $(1, \ldots, p)$ to $[\pi(1), \ldots, \pi(p)]$ is even and is -1 if the number of inversions is odd. {Note that there are many choices for the inversions to get from $(1, \ldots, p)$ to $[\pi(1), \ldots, \pi(p)]$, and hence many choices for the number of necessary inversions. However, for a particular permutation, the number of inversions is always even or it is always odd, so that $c(\pi)$ is well defined.} We use the following basic properties of determinants

$$|\mathbf{I}| = 1, \qquad |\mathbf{A}'| = |\mathbf{A}|, \qquad |\mathbf{A}^{-1}| = \frac{1}{|\mathbf{A}|}, \qquad |\mathbf{AB}| = |\mathbf{A}||\mathbf{B}|, \qquad |c\mathbf{A}| = c^p|\mathbf{A}| \tag{A.8}$$

(where c is a scalar and $\mathbf{A}$ is $p \times p$). We note also that $\mathbf{A}$ is invertible if and only if $|\mathbf{A}| \neq 0$.

If $\mathbf{A} = (a_{ij})$ is a $p \times p$ matrix, then the *trace of* $\mathbf{A}$, written $\operatorname{tr} \mathbf{A}$ is defined to be $\operatorname{tr} \mathbf{A} = \sum_i a_{ii}$. The properties of the trace are

$$\operatorname{tr} \mathbf{AB} = \operatorname{tr} \mathbf{BA}, \qquad \operatorname{tr} \mathbf{A}' = \operatorname{tr} \mathbf{A}. \tag{A.9}$$

(Note that $\operatorname{tr} \mathbf{AB} = \operatorname{tr} \mathbf{BA}$ even if $\mathbf{A}$ and $\mathbf{B}$ are not square.)

A $p \times p$ matrix $\mathbf{A}$ is *symmetric* if $\mathbf{A}' = \mathbf{A}$, and is *orthogonal* if $\mathbf{A}^{-1} = \mathbf{A}'$ (i.e., if $\mathbf{AA}' = \mathbf{A}'\mathbf{A} = \mathbf{I}$). The $p \times p$ matrix $\mathbf{A} = (a_{ij})$ is *diagonal* if $a_{ij} = 0$ when $i \neq j$. If $\mathbf{A}$ is orthogonal then $|\mathbf{A}| = \pm 1$. If $\mathbf{A}$ is diagonal then $|\mathbf{A}|$ is the product of the diagonal elements of $\mathbf{A}$.

We are now ready for the principal axis theorem, which is the basis for many of the other matrix algebra results used in this book. It is a much deeper result than those stated above. A proof is given in Section A.6 of this appendix. Let $\mathbf{S}$ be a $p \times p$ matrix. Define $q(t) = |\mathbf{S} - t\mathbf{I}|$. By the definition of the determinant, we note that $q(t)$ is a pth degree polynomial, and so has at most p real roots.

THEOREM A.1. Let $\mathbf{S}$ be a symmetric $p \times p$ matrix. Let

$$q(t) = |\mathbf{S} - t\mathbf{I}|.$$

Then there are p real roots to the equation $q(t) = 0$. Let $t_1 \geqslant \cdots \geqslant t_p$ be the roots of $q(t) = 0$. Then there exists an orthogonal matrix $\mathbf{\Gamma}$ such that

$$\mathbf{S} = \mathbf{\Gamma D \Gamma}', \qquad \mathbf{D} = \begin{bmatrix} t_1 & & 0 \\ & \ddots & \\ 0 & & t_p \end{bmatrix}, \qquad t_1 = \sup_{\substack{\mathbf{X} \in R^p \\ \mathbf{X} \neq \mathbf{0}}} \frac{\mathbf{X}'\mathbf{S}\mathbf{X}}{\|\mathbf{X}\|^2}$$

PROOF. See Section A.6. □

The numbers $t_1, \ldots, t_p$ are called the *eigenvalues* of $\mathbf{S}$. Note that the t_i are the roots of $q(t) = 0$ and are uniquely defined. Therefore, the diagonal matrix $\mathbf{D}$ is uniquely defined (at least if we make the convention that $d_{ii} \geqslant d_{jj}$ when $i \leqslant j$). However, the matrix $\mathbf{\Gamma}$ is not uniquely defined. We can multiply any row of $\mathbf{\Gamma}$ by -1 and get a new matrix that satisfies the conditions of the theorem. If some of the t_i are the same, then more complicated situations occur. In particular, if $\mathbf{S} = \mathbf{I}$, then any $p \times p$ orthogonal matrix $\mathbf{\Gamma}$ satisfies the conditions of the theorem.

We now give several corollaries to this theorem.

COROLLARY 1. The rank of $\mathbf{S}$ is equal to the number of nonzero eigenvalues of $\mathbf{S}$. $\mathbf{S}$ is invertible if and only if all of the eigenvalues of $\mathbf{S}$ are nonzero.

PROOF. Since $\mathbf{\Gamma}$ is orthogonal and hence invertible, the rank of $\mathbf{S}$ is equal to the rank of $\mathbf{D}$, which is the number of nonzero eigenvalues of $\mathbf{S}$. In addition, $\mathbf{S}$ is invertible if and only if $\mathbf{D}$ is invertible, if and only if the eigenvalues of $\mathbf{S}$ are all nonzero. □

COROLLARY 2. The determinant of $\mathbf{S}$ is equal to the product of the eigenvalues of $\mathbf{S}$ and the trace of $\mathbf{S}$ is the sum of the eigenvalues of $\mathbf{S}$.

PROOF. By (A.8), we see that

$$|\mathbf{S}| = |\mathbf{\Gamma}||\mathbf{D}||\mathbf{\Gamma}'| = |\mathbf{\Gamma\Gamma}'||\mathbf{D}| = |\mathbf{I}||\mathbf{D}| = |\mathbf{D}|.$$

Therefore, the determinant of $\mathbf{S}$ is equal to the determinant of $\mathbf{D}$ which is the product of the eigenvalues of $\mathbf{S}$. Similarly, by (A.9) we see that

$$\operatorname{tr}\mathbf{S} = \operatorname{tr}\mathbf{\Gamma D \Gamma}' = \operatorname{tr}\mathbf{\Gamma}'\mathbf{\Gamma D} = \operatorname{tr}\mathbf{D}. \quad \square$$

COROLLARY 3. The eigenvalues of $\mathbf{S}^{-1}$ are the inverses of the eigenvalues of $\mathbf{S}$.

PROOF. $|\mathbf{S}^{-1} - \lambda\mathbf{I}| = |-\lambda\mathbf{S}^{-1}||\mathbf{S} - (1/\lambda)\mathbf{I}|$. Since $\mathbf{S}$ is invertible, the eigenvalues of $\mathbf{S}$ are not 0 and $|\mathbf{S}| \neq 0$. Therefore we see that $|\mathbf{S}^{-1} - \lambda\mathbf{I}| = 0$ if and only if $|\mathbf{S} - (1/\lambda)\mathbf{I}| = 0$. □

COROLLARY 4. If $\mathbf{S}$ is a symmetric matrix and $\mathbf{\Gamma S \Gamma}' = \mathbf{S}$ for all orthogonal $\mathbf{\Gamma}$, then $\mathbf{S} = a\mathbf{I}$ for some scalar a.

PROOF. Let $\mathbf{S} = \mathbf{\Gamma D \Gamma}'$. Then $\mathbf{S} = \mathbf{\Gamma}'\mathbf{S\Gamma} = \mathbf{D}$ and therefore $\mathbf{S}$ is diagonal. In addition, $\mathbf{PDP}' = \mathbf{D}$ for all permutation matrices $\mathbf{P}$ (since these matrices are orthogonal). Therefore, the diagonal elements of $\mathbf{D}$ are all equal and $\mathbf{S} = \mathbf{D} = a\mathbf{I}$ for some a. □

A.2. PARTITIONED MATRICES.

We now study matrices that have been partioned into submatrices. In adding or multiplying partioned matrices, the important principle is the following. If all the additions or multiplications are well-defined, then we may act formally as though the submatrices were numbers. For example, suppose that $\mathbf{A}$ and $\mathbf{B}$ are $n \times p$ matrices and that

$$\mathbf{A} = \begin{pmatrix} \mathbf{A}_{11} & \mathbf{A}_{12} \\ \mathbf{A}_{21} & \mathbf{A}_{22} \end{pmatrix}, \qquad \mathbf{B} = \begin{pmatrix} \mathbf{B}_{11} & \mathbf{B}_{12} \\ \mathbf{B}_{21} & \mathbf{B}_{22} \end{pmatrix}$$

where $\mathbf{A}_{ij}$ and $\mathbf{B}_{ij}$ are $n_i \times p_j$ submatrices (and therefore, $n_1 + n_2 = n$ and $p_1 + p_2 = p$). Then

$$\mathbf{A} + \mathbf{B} = \begin{pmatrix} \mathbf{A}_{11} + \mathbf{B}_{11} & \mathbf{A}_{12} + \mathbf{B}_{12} \\ \mathbf{A}_{21} + \mathbf{B}_{21} & \mathbf{A}_{22} + \mathbf{B}_{22} \end{pmatrix} \tag{A.10}$$

Further, suppose that

$$\mathbf{C} = \begin{pmatrix} \mathbf{C}_{11} & \mathbf{C}_{12} \\ \mathbf{C}_{21} & \mathbf{C}_{22} \end{pmatrix}$$

is $p \times r$ and $\mathbf{C}_{ij}$ is $p_i \times r_j$ submatrix (and hence $r_1 + r_2 = r$). The product of $\mathbf{A}$ and $\mathbf{C}$ is defined and

$$\mathbf{AC} = \begin{pmatrix} \mathbf{A}_{11}\mathbf{C}_{11} + \mathbf{A}_{12}\mathbf{C}_{21} & \mathbf{A}_{11}\mathbf{C}_{12} + \mathbf{A}_{12}\mathbf{C}_{22} \\ \mathbf{A}_{21}\mathbf{C}_{11} + \mathbf{A}_{22}\mathbf{C}_{22} & \mathbf{A}_{21}\mathbf{C}_{12} + \mathbf{A}_{22}\mathbf{C}_{22} \end{pmatrix} \tag{A.11}$$

(A.10) and (A.11) can be derived directly from the definitions of addition and multiplication of matrices. More complicated partitions could be considered, but the basic rule is to treat the submatrices as if they were numbers. For example, if

$$\mathbf{A} = \begin{pmatrix} \mathbf{A}_{11} & \mathbf{A}_{12} & \mathbf{A}_{13} \\ \mathbf{A}_{21} & \mathbf{A}_{22} & \mathbf{A}_{23} \end{pmatrix}, \qquad \mathbf{C} = \begin{bmatrix} \mathbf{C}_{11} & \mathbf{C}_{12} & \mathbf{C}_{13} \\ \mathbf{C}_{21} & \mathbf{C}_{22} & \mathbf{C}_{23} \\ \mathbf{C}_{31} & \mathbf{C}_{32} & \mathbf{C}_{33} \end{bmatrix}$$

where $\mathbf{A}_{ij}$ is $n_i \times p_j$ and $\mathbf{C}_{ij}$ is $p_i \times r_j$, then

$$\mathbf{AC} = \begin{pmatrix} \mathbf{A}_{11}\mathbf{C}_{11} + \mathbf{A}_{12}\mathbf{C}_{21} + \mathbf{A}_{13}\mathbf{C}_{31} & \mathbf{A}_{11}\mathbf{C}_{12} + \mathbf{A}_{12}\mathbf{C}_{22} + \mathbf{A}_{13}\mathbf{C}_{32} & \mathbf{A}_{11}\mathbf{C}_{13} + \mathbf{A}_{12}\mathbf{C}_{23} + \mathbf{A}_{13}\mathbf{C}_{33} \\ \mathbf{A}_{21}\mathbf{C}_{11} + \mathbf{A}_{22}\mathbf{C}_{21} + \mathbf{A}_{23}\mathbf{C}_{31} & \mathbf{A}_{21}\mathbf{C}_{12} + \mathbf{A}_{22}\mathbf{C}_{22} + \mathbf{A}_{23}\mathbf{C}_{32} & \mathbf{A}_{21}\mathbf{C}_{13} + \mathbf{A}_{22}\mathbf{C}_{23} + \mathbf{A}_{23}\mathbf{C}_{33} \end{pmatrix}.$$

Using (A.11), we see that if $\mathbf{A}_{11}$ and $\mathbf{A}_{22}$ are invertible, then

$$\begin{pmatrix} \mathbf{A}_{11} & \mathbf{0} \\ \mathbf{0} & \mathbf{A}_{22} \end{pmatrix}^{-1} = \begin{pmatrix} \mathbf{A}_{11}^{-1} & \mathbf{0} \\ \mathbf{0} & \mathbf{A}_{22}^{-1} \end{pmatrix}, \qquad \begin{pmatrix} \mathbf{I} & \mathbf{C} \\ \mathbf{0} & \mathbf{I} \end{pmatrix}^{-1} = \begin{pmatrix} \mathbf{I} & -\mathbf{C} \\ \mathbf{0} & \mathbf{I} \end{pmatrix} \quad \text{(A.12)}$$

The final result in this section is a formula for the determinant of a symmetric partitioned matrix.

LEMMA A.2. Let

$$\mathbf{S} = \begin{pmatrix} \mathbf{S}_{11} & \mathbf{S}_{12} \\ \mathbf{S}_{21} & \mathbf{S}_{22} \end{pmatrix}$$

be a partitioned matrix with $\mathbf{S}_{ij}$ being $p_i \times p_j$.

a. If $\mathbf{S}_{22}$ is invertible, then

$$\mathbf{S} = |\mathbf{S}_{22}||\mathbf{S}_{11} - \mathbf{S}_{12}\mathbf{S}_{22}^{-1}\mathbf{S}_{21}|.$$

b. Let $\mathbf{S}$ be invertible and

$$\mathbf{V} = \begin{pmatrix} \mathbf{V}_{11} & \mathbf{V}_{12} \\ \mathbf{V}_{21} & \mathbf{V}_{22} \end{pmatrix} = \mathbf{S}^{-1}$$

where $\mathbf{V}_{ij}$ is $p_i \times p_j$. Then $\mathbf{V}_{11}^{-1} = \mathbf{S}_{11} - \mathbf{S}_{12}\mathbf{S}_{22}^{-1}\mathbf{S}_{21}$.

PROOF. a. Let

$$\mathbf{T} = \begin{pmatrix} \mathbf{I} & \mathbf{S}_{12}\mathbf{S}_{22}^{-1} \\ \mathbf{0} & \mathbf{I} \end{pmatrix}, \qquad \mathbf{U} = \begin{pmatrix} \mathbf{S}_{11} - \mathbf{S}_{12}\mathbf{S}_{22}^{-1}\mathbf{S}_{21} & \mathbf{0} \\ \mathbf{0} & \mathbf{S}_{22} \end{pmatrix},$$

$$\mathbf{T}^* = \begin{pmatrix} \mathbf{I} & \mathbf{0} \\ \mathbf{S}_{22}^{-1}\mathbf{S}_{21} & \mathbf{I} \end{pmatrix}.$$

Using (A.11) we see that $\mathbf{S} = \mathbf{TUT}^*$. Using the definition of the determinant we see that $|\mathbf{T}| = |\mathbf{T}^*| = 1$. Now,

$$\mathbf{U} = \begin{pmatrix} \mathbf{S}_{11} - \mathbf{S}_{12}\mathbf{S}_{22}^{-1}\mathbf{S}_{21} & \mathbf{0} \\ \mathbf{0} & \mathbf{I} \end{pmatrix} \begin{pmatrix} \mathbf{I} & \mathbf{0} \\ \mathbf{0} & \mathbf{S}_{22} \end{pmatrix}$$

and therefore [using (A.8) and the definition of the determinant]

$$|\mathbf{U}| = \begin{vmatrix} \mathbf{S}_{11} - \mathbf{S}_{12}\mathbf{S}_{22}^{-1}\mathbf{S}_{21} & \mathbf{0} \\ \mathbf{0} & \mathbf{I} \end{vmatrix} \begin{vmatrix} \mathbf{I} & \mathbf{0} \\ \mathbf{0} & \mathbf{S}_{22} \end{vmatrix} = |\mathbf{S}_{11} - \mathbf{S}_{12}\mathbf{S}_{22}^{-1}\mathbf{S}_{21}||\mathbf{S}_{22}|.$$

Therefore

$$|\mathbf{S}| = |\mathbf{T}||\mathbf{U}||\mathbf{T}^*| = |\mathbf{S}_{11} - \mathbf{S}_{12}\mathbf{S}_{22}^{-1}\mathbf{S}_{21}||\mathbf{S}_{22}|.$$

b. If S is invertible, then $\mathbf{S}_{22}$ is invertible. We see from part a that if $\mathbf{S}$ is invertible so is $\mathbf{S}_{11} - \mathbf{S}_{12}\mathbf{S}_{22}^{-1}\mathbf{S}_{21}$ (since a square matrix is invertible if and only if its determinant is not 0). Using (A.12), we see that

$$\mathbf{T}^{-1} = \begin{pmatrix} \mathbf{I} & -\mathbf{S}_{12}\mathbf{S}_{22}^{-1} \\ \mathbf{0} & \mathbf{I} \end{pmatrix}, \qquad \mathbf{U}^{-1} = \begin{bmatrix} (\mathbf{S}_{11} - \mathbf{S}_{12}\mathbf{S}_{22}^{-1}\mathbf{S}_{21})^{-1} & \mathbf{0} \\ \mathbf{0} & \mathbf{S}_{22}^{-1} \end{bmatrix},$$

$$(\mathbf{T}^*)^{-1} = \begin{pmatrix} \mathbf{I} & \mathbf{0} \\ -\mathbf{S}_{22}^{-1}\mathbf{S}_{21} & \mathbf{I} \end{pmatrix}.$$

Therefore, using (A.7)

$$\mathbf{V} = \mathbf{S}^{-1} = (\mathbf{T}^*)^{-1}\mathbf{U}^{-1}\mathbf{T}^{-1}$$

$$= \begin{pmatrix} \mathbf{I} & \mathbf{0} \\ -\mathbf{S}_{22}^{-1}\mathbf{S}_{21} & \mathbf{I} \end{pmatrix} \begin{bmatrix} (\mathbf{S}_{11} - \mathbf{S}_{12}\mathbf{S}_{22}^{-1}\mathbf{S}_{21})^{-1} & \mathbf{0} \\ \mathbf{0} & \mathbf{S}_{22}^{-1} \end{bmatrix} \begin{pmatrix} \mathbf{I} & -\mathbf{S}_{22}^{-1}\mathbf{S}_{21} \\ \mathbf{0} & \mathbf{I} \end{pmatrix}$$

and hence $\mathbf{V}_{11} = (\mathbf{S}_{11} - \mathbf{S}_{12}\mathbf{S}_{22}^{-1}\mathbf{S}_{21})^{-1}$. □

By a similar proof we could show that

$$|\mathbf{S}| = |\mathbf{S}_{11}||\mathbf{S}_{22} - \mathbf{S}_{21}\mathbf{S}_{11}^{-1}\mathbf{S}_{12}|, \qquad \mathbf{V}_{22}^{-1} = \mathbf{S}_{22} - \mathbf{S}_{21}\mathbf{S}_{11}^{-1}\mathbf{S}_{12} \qquad \text{(A.13)}$$

(provided $\mathbf{S}_{11}$ is invertible in the first equality and $\mathbf{S}$ is invertible in the second one). We also note by part a that if $\mathbf{S}_{22}$ and $\mathbf{S}_{11} - \mathbf{S}_{12}\mathbf{S}_{22}^{-1}\mathbf{S}_{21}$ are invertible then $\mathbf{S}$ is invertible. It is not enough that $\mathbf{S}_{11}$ and $\mathbf{S}_{22}$ be invertible.

A.3. POSITIVE DEFINITE AND NONNEGATIVE DEFINITE MATRICES.

Let $\mathbf{S}$ be a $p \times p$ symmetric matrix. We say that $\mathbf{S}$ is *nonnegative definite* and write $\mathbf{S} \geqslant 0$ if $\mathbf{a}'\mathbf{S}\mathbf{a} \geqslant 0$ for all p-dimensional vectors a. We say that $\mathbf{S}$ is *positive definite* and write $\mathbf{S} > 0$ if $\mathbf{a}'\mathbf{S}\mathbf{a} > 0$ for all nonzero p-dimensional vectors $\mathbf{a}$. We first state some results that follow directly from the definitions.

THEOREM A.3. Let $\mathbf{X}$ be an $n \times p$ matrix. Then $\mathbf{X}'\mathbf{X} \geqslant 0$. If $\mathbf{X}$ has rank p, then $\mathbf{X}'\mathbf{X} > 0$.

PROOF. Let $\mathbf{a}$ be a p-dimensional vector. Then

$$\mathbf{a}'(\mathbf{X}'\mathbf{X})\mathbf{a} = \|\mathbf{X}\mathbf{a}\|^2 \geqslant 0,$$

so that $\mathbf{X}'\mathbf{X} \geqslant 0$. Now suppose that $\mathbf{X}$ has rank p and $\mathbf{a}'(\mathbf{X}'\mathbf{X})\mathbf{a} = 0$. Then

$$0 = \mathbf{a}'(\mathbf{X}'\mathbf{X})\mathbf{a} = \|\mathbf{X}\mathbf{a}\|^2$$

and hence $\mathbf{Xa} = \mathbf{0}$. Now let $\mathbf{X} = (\mathbf{X}_1, \ldots, \mathbf{X}_p)$, $\mathbf{a}' = (a_1, \ldots, a_p)$. Then $\mathbf{0} = \mathbf{Xa} = \sum_i a_i \mathbf{X}_i$. However, $\mathbf{X}$ has rank p and has p columns. Therefore, the columns of $\mathbf{X}$ (i.e., the $\mathbf{X}_i$) are linearly independent. Hence $a_i = 0$ and $\mathbf{a} = \mathbf{0}$. Therefore if $\mathbf{X}$ has rank p, then $\mathbf{X}'\mathbf{X} > 0$. □

We now discuss the eigenvalues of nonnegative definite and positive definite matrices.

LEMMA A.4. Let $\mathbf{S}$ be a symmetric $p \times p$ matrix. Then $\mathbf{S} \geqslant 0$ if and only if the eigenvalues for $\mathbf{S}$ are nonnegative. $\mathbf{S} > 0$ if and only if the eigenvalues of $\mathbf{S}$ are positive.

PROOF. Let

$$\mathbf{S} = \mathbf{\Gamma D \Gamma}', \qquad \mathbf{D} = \begin{pmatrix} t_1 & & \mathbf{0} \\ & \ddots & \\ \mathbf{0} & & t_p \end{pmatrix}.$$

Let $\mathbf{a}$ be a p-dimensional vector, and let $\mathbf{b} = (b_1, \ldots, b_p)' = \mathbf{\Gamma}'\mathbf{a}$. Then

$$\mathbf{a}'\mathbf{S}\mathbf{a} = \mathbf{b}'\mathbf{D}\mathbf{b} = \sum_i b_i^2 t_i.$$

Therefore, $\mathbf{S} \geqslant 0$ if and only if $t_i \geqslant 0$ and $S > 0$ if and only if $t_i > 0$. □

We now use Lemma A.4 to derive the fundamental properties of these matrices.

THEOREM A.5. a. If $\mathbf{S} \geqslant 0$, then there exists a unique $\mathbf{V} \geqslant 0$ such that $\mathbf{S} = \mathbf{V}^2$. If $\mathbf{S} > 0$, then $\mathbf{V} > 0$.

b. If $\mathbf{S} \geqslant 0$ is $p \times p$ of rank r, then there exists $\mathbf{C}$ $p \times r$ of rank r such that $\mathbf{S} = \mathbf{CC}'$.

c. If $\mathbf{S} \geqslant 0$, then $\mathbf{S}$ is invertible if and only if $\mathbf{S} > 0$. If $\mathbf{S} > 0$, then $|\mathbf{S}| > 0$ and $\mathbf{S}^{-1} > 0$.

PROOF. a. Let

$$\mathbf{S} = \mathbf{\Gamma D \Gamma}', \qquad \mathbf{D} = \begin{pmatrix} t_1 & & \mathbf{0} \\ & \ddots & \\ \mathbf{0} & & t_p \end{pmatrix}.$$

By Lemma A.4, the $t_i \geqslant 0$. Therefore, let

$$\mathbf{D}^{1/2} = \begin{pmatrix} t_1^{1/2} & & \mathbf{0} \\ & \ddots & \\ \mathbf{0} & & t_p^{1/2} \end{pmatrix}$$

and let $\mathbf{V} = \mathbf{\Gamma D}^{1/2}\mathbf{\Gamma}'$. Then

$$\mathbf{V}^2 = \mathbf{\Gamma D}^{1/2}\mathbf{\Gamma}'\mathbf{\Gamma D}^{1/2}\mathbf{\Gamma}' = \mathbf{\Gamma D \Gamma}' = \mathbf{S}$$

Furthermore $\mathbf{V}$ is a symmetric matrix and by (A.8)

$$|\mathbf{V} - t\mathbf{I}| = |\mathbf{\Gamma}||\mathbf{D}^{1/2} - t\mathbf{I}||\mathbf{\Gamma}'| = |\mathbf{D}^{1/2} - t\mathbf{I}|.$$

Therefore, the eigenvalues of $\mathbf{V}$ are $t_1^{1/2}, \ldots, t_p^{1/2}$ and $t_i^{1/2} \geqslant 0$. Therefore $\mathbf{V} \geqslant 0$. If $\mathbf{S} > 0$, then $t_i > 0$, and the eigenvalues are positive. By Lemma A.4, therefore, if $\mathbf{S} > 0$, then $\mathbf{V} > 0$.

We now need to establish the uniqueness of $\mathbf{V}$. We first make two elementary comments. First, let $\mathbf{D}$ be a diagonal matrix with diagonal elements $d_i \geqslant 0$ and let $\mathbf{A} = (a_{ij})$. The i, jth elements of $\mathbf{DA}$ and $\mathbf{AD}$ are $d_i a_{ij}$ and $d_j a_{ij}$ and hence $\mathbf{DA} = \mathbf{AD}$ if and only if $a_{ij} = 0$ for all i and j such that $d_i \neq d_j$. This implies that $\mathbf{D}^2\mathbf{A} = \mathbf{AD}^2$ if and only if $\mathbf{DA} = \mathbf{AD}$. Secondly, let $\mathbf{B} = \mathbf{\Gamma D \Gamma}'$ where $\mathbf{\Gamma}$ is orthogonal and $\mathbf{D}$ is diagonal. Then $\mathbf{B}^2 = \mathbf{\Gamma D}^2\mathbf{\Gamma}'$, and therefore the eigenvalues of $\mathbf{B}^2$ are just the squares of the eigenvalues of $\mathbf{B}$. Now, suppose that $\mathbf{S} = \mathbf{V}^2 = \mathbf{U}^2$ where $\mathbf{U} \geqslant 0$ and $\mathbf{V} \geqslant 0$. We now show that $\mathbf{U} = \mathbf{V}$. By the second comment above, the eigenvalues of $\mathbf{U}$ and $\mathbf{V}$ are the same (since they are nonnegative). Therefore, there exist orthogonal matrices $\mathbf{\Gamma}_1$ and $\mathbf{\Gamma}_2$ and diagonal matrix $\mathbf{D}$ such that

$$\mathbf{U} = \mathbf{\Gamma}_1\mathbf{D\Gamma}_1', \qquad \mathbf{V} = \mathbf{\Gamma}_2\mathbf{D\Gamma}_2'$$

Also

$$\mathbf{\Gamma}_1\mathbf{D}^2\mathbf{\Gamma}_1' = \mathbf{U}^2 = \mathbf{V}^2 = \mathbf{\Gamma}_2\mathbf{D}^2\mathbf{\Gamma}_2'$$

Therefore

$$\mathbf{D}^2\mathbf{\Gamma}_1'\mathbf{\Gamma}_2 = \mathbf{\Gamma}_1'\mathbf{\Gamma}_2\mathbf{D}^2$$

By the first comment above,

$$\mathbf{D\Gamma}_1'\mathbf{\Gamma}_2 = \mathbf{\Gamma}_1'\mathbf{\Gamma}_2\mathbf{D}$$

and hence

$$\mathbf{U} = \mathbf{\Gamma}_1\mathbf{D\Gamma}_1' = \mathbf{\Gamma}_2\mathbf{D\Gamma}_2' = \mathbf{V}.$$

b. The rank of $\mathbf{S}$ is the number of nonzero eigenvalues of $\mathbf{S}$. Therefore,

$$\mathbf{D} = \begin{bmatrix} \begin{bmatrix} t_1 & & 0 \\ & \ddots & \\ 0 & & t_r \end{bmatrix} & \mathbf{0} \\ \mathbf{0} & \mathbf{0} \end{bmatrix} = \begin{pmatrix} \mathbf{T} & \mathbf{0} \\ \mathbf{0} & \mathbf{0} \end{pmatrix}$$

and $t_i > 0$. Now let

$$\mathbf{T}^{1/2} = \begin{bmatrix} t_1^{1/2} & & \mathbf{0} \\ & \ddots & \\ \mathbf{0} & & t_r^{1/2} \end{bmatrix}, \qquad \boldsymbol{\Gamma} = (\boldsymbol{\Gamma}_1 \boldsymbol{\Gamma}_2)$$

where $\boldsymbol{\Gamma}_1$ is $p \times r$. Define $\mathbf{C} = \boldsymbol{\Gamma}_1 \mathbf{T}^{1/2}$. Then $\mathbf{C}$ is $p \times r$ and

$$\mathbf{CC}' = \boldsymbol{\Gamma}_1 \mathbf{T}^{1/2}\mathbf{T}^{1/2}\boldsymbol{\Gamma}_1' = \boldsymbol{\Gamma}_1 \mathbf{T}\boldsymbol{\Gamma}_1' = \boldsymbol{\Gamma}\mathbf{D}\boldsymbol{\Gamma}' = \mathbf{S}.$$

Finally the rank of $\mathbf{C}$ is at most r, since $\mathbf{C}$ has r columns, and the rank of $\mathbf{C}$ is at least r since $\mathbf{S} = \mathbf{CC}'$ has rank r.

c. This follows directly from the corollaries to Theorem A.1. □

If $\mathbf{S} \geqslant 0$, we call the unique $\mathbf{V} \geqslant 0$ such that $\mathbf{S} = \mathbf{V}^2$ the *square root* of $\mathbf{S}$ and write $\mathbf{V} = \mathbf{S}^{1/2}$. If $\mathbf{S} > 0$ then $\mathbf{S}^{-1} > 0$. We let $\mathbf{S}^{-1/2}$ be the square root of $\mathbf{S}^{-1}$. Then

$$\mathbf{S}^{-1/2} = (\mathbf{S}^{1/2})^{-1}, \qquad |\mathbf{S}^{1/2}| = |\mathbf{S}|^{1/2}, \qquad |\mathbf{S}^{-1/2}| = |\mathbf{S}|^{-1/2} \tag{A.14}$$

(see Exercise B3 of Chapter 3).

We now look at partitioned matrices.

THEOREM A.6. Let

$$\mathbf{S} = \begin{pmatrix} \mathbf{S}_{11} & \mathbf{S}_{12} \\ \mathbf{S}_{21} & \mathbf{S}_{22} \end{pmatrix} > 0$$

where the $\mathbf{S}_{ij}$ are $p_i \times p_j$. Then $\mathbf{S}_{11} > 0$, $\mathbf{S}_{22} > 0$ and $\mathbf{S}_{11} - \mathbf{S}_{12}\mathbf{S}_{22}^{-1}\mathbf{S}_{21} > 0$.

PROOF. Let $\mathbf{a} \neq \mathbf{0}$ be a p-dimensional vector and let $\mathbf{b}$ be the $(p_1 + p_2)$-dimensional vector $(\mathbf{a}'\,\mathbf{0})'$. Since $\mathbf{S} > 0$, we see that

$$\mathbf{0} < \mathbf{b}'\mathbf{S}\mathbf{b} = \mathbf{a}'\mathbf{S}_{11}\mathbf{a}.$$

Therefore $\mathbf{S}_{11} > 0$. Similarly $\mathbf{S}_{22} > 0$. Now let $\mathbf{c}$ be the vector

$$\begin{pmatrix} \mathbf{a} \\ -\mathbf{S}_{22}^{-1}\mathbf{S}_{21}\mathbf{a} \end{pmatrix}.$$

Then

$$0 < \mathbf{c}'\mathbf{S}\mathbf{c} = \mathbf{a}'(\mathbf{S}_{11} - \mathbf{S}_{12}\mathbf{S}_{22}^{-1}\mathbf{S}_{21})\mathbf{a}$$

and therefore

$$\mathbf{S}_{11} - \mathbf{S}_{12}\mathbf{S}_{22}^{-1}\mathbf{S}_{21} > 0. \quad \square$$

The next result in this section is an extension of the principal axis theorem to the form used in Chapter 19.

THEOREM A.7. Let $\mathbf{S} > 0$ and $\mathbf{U} \geqslant 0$ be $p \times p$ matrices. Let

$$r(t) = |\mathbf{U} - t\mathbf{S}|.$$

Then $r(t)$ is a pth degree polynomial. The equation $r(t) = 0$ has p nonnegative roots, $t_1 \geqslant \cdots \geqslant t_p$. If $\mathbf{U} > 0$, the roots are positive. There exists $\mathbf{A}$, invertible, such that

$$\mathbf{ASA'} = \mathbf{I}, \qquad \mathbf{AUA'} = \mathbf{D} = \begin{bmatrix} t_1 & & 0 \\ & \ddots & \\ 0 & & t_p \end{bmatrix}, \qquad t_1 = \sup_{\substack{\mathbf{X} \in R^p \\ \mathbf{X} \neq \mathbf{0}}} \frac{\mathbf{X'UX}}{\mathbf{X'SX}}$$

PROOF. By (A.8), we see that

$$|\mathbf{U} - t\mathbf{S}| = |\mathbf{S}^{1/2}||\mathbf{S}^{-1/2}\mathbf{US}^{1/2} - t\mathbf{I}||\mathbf{S}^{1/2}| = |\mathbf{S}^{1/2}| = |\mathbf{S}^{1/2}|^2|\mathbf{S}^{1/2}\mathbf{US}^{-1/2} - t\mathbf{I}|$$

Also $S^{1/2} > 0$ and therefore $|\mathbf{S}^{1/2}| > 0$. Now $\mathbf{S}^{-1/2}\mathbf{US}^{-1/2}$ is easily seen to be nonnegative definite. Therefore by Theorem A.1 and Lemma A.4, we see that $r(t)$ is a pth degree polynomial with p nonnegative roots. If $\mathbf{U} > 0$ then $\mathbf{S}^{-1/2}\mathbf{US}^{-1/2} > 0$ and the roots are positive. By Theorem A.1, there exists $\mathbf{\Gamma}$ orthogonal such that $\mathbf{S}^{-1/2}\mathbf{US}^{-1/2} = \mathbf{\Gamma D \Gamma'}$. Hence

$$\mathbf{\Gamma S}^{-1/2}\mathbf{U}(\mathbf{\Gamma S}^{-1/2})' = \mathbf{D}.$$

Also

$$\mathbf{\Gamma S}^{-1/2}\mathbf{S}(\mathbf{\Gamma S}^{-1})' = \mathbf{\Gamma \Gamma} = \mathbf{I}$$

and the theorem is satisfied with $\mathbf{A} = \mathbf{\Gamma S}^{-1/2}$.

Finally

$$t_1 = \sup_{\substack{\mathbf{Z} \in R^p \\ \mathbf{Z} \neq 0}} \frac{\mathbf{Z'S}^{-1/2}\mathbf{US}^{-1/2}\mathbf{Z}}{\|\mathbf{Z}\|^2} = \sup_{\substack{\mathbf{X} \in R^p \\ \mathbf{X} \neq 0}} \frac{\mathbf{X'UX}}{(\mathbf{S}^{1/2}\mathbf{X})'(\mathbf{S}^{1/2}\mathbf{X})} = \sup_{\substack{\mathbf{X} \in R^p \\ \mathbf{X} \neq 0}} \frac{\mathbf{X'UX}}{\mathbf{X'SX}}$$

(with $\mathbf{Z} = \mathbf{S}^{1/2}\mathbf{X}$). □

The last result in this section is used in the derivation of the central Wishart density function for general covariance matrix. A $p \times p$ matrix $\mathbf{C} = (c_{ij})$ is called *upper triangular* if $c_{ij} = 0$ when $i > j$.

THEOREM A.8. Let $\mathbf{S} \geqslant 0$. Then there exists $\mathbf{C}$ upper triangular such that $\mathbf{S} = \mathbf{CC'}$.

PROOF. Let $\mathbf{S}$ be $p \times p$. By Theorem A.5, there exists $\mathbf{B}$ $p \times p$ such that $\mathbf{S} = \mathbf{BB'} = (\mathbf{B\Gamma})(\mathbf{B\Gamma})'$ for any orthogonal matrix $\mathbf{\Gamma}$. Therefore, the result will be shown if we show that for any $p \times p$ matrix $\mathbf{B}$ there exists $\mathbf{\Gamma}$ orthogonal such that $\mathbf{B\Gamma}$ is upper triangular. Let the rows of $\mathbf{B}$ be $\mathbf{B}_1, \ldots, \mathbf{B}_p$. Let $\mathbf{\Gamma}_1$ be a vector of length 1 such that $\mathbf{\Gamma}_1$ is orthogonal to the subspace spanned by $\mathbf{B}_2', \ldots, \mathbf{B}_p'$. (Such a vector is possible since that subspace has dimension at most $p - 1$.) Then

$$\mathbf{B\Gamma}_1 = \begin{pmatrix} c_{11} \\ 0 \\ \vdots \\ 0 \end{pmatrix}.$$

Now, let $\mathbf{\Gamma}_2$ be a vector of length 1 that is orthogonal to the subspace spanned by $\mathbf{\Gamma}_1$ and $\mathbf{B}'_3, \ldots, \mathbf{B}'_p$. Then $\mathbf{\Gamma}'_1\mathbf{\Gamma}_2 = 0$ and

$$\mathbf{B\Gamma}_2 = \begin{pmatrix} c_{11} \\ c_{22} \\ 0 \\ \vdots \\ 0 \end{pmatrix}.$$

We continue in this fashion. Let $\mathbf{\Gamma}_i$ be a vector of length 1 orthogonal to the subspace spanned by $\mathbf{\Gamma}_1, \ldots, \mathbf{\Gamma}_{i-1}$ and $\mathbf{B}'_{i+1}, \ldots, \mathbf{B}'_p$. Then $\mathbf{\Gamma}'_i\mathbf{\Gamma}_j = 0$ for $j < i$ and

$$\mathbf{B\Gamma}_i = \begin{pmatrix} c_{1i} \\ \vdots \\ c_{ii} \\ 0 \\ \vdots \\ 0 \end{pmatrix}.$$

Now, let $\mathbf{\Gamma} = (\mathbf{\Gamma}_1, \ldots, \mathbf{\Gamma}_p)$. Since the $\mathbf{\Gamma}_i$ have length 1 and $\mathbf{\Gamma}'_i\mathbf{\Gamma}_j = 0$, $\mathbf{\Gamma}$ is orthogonal. Finally,

$$\mathbf{B\Gamma} = \begin{pmatrix} c_{11} & c_{12} & c_{13} & \cdots & c_{1p} \\ 0 & c_{22} & c_{23} & \cdots & c_{2p} \\ 0 & 0 & c_{33} & \cdots & c_{3p} \\ \vdots & \vdots & & \ddots & \vdots \\ 0 & 0 & 0 & \cdots & c_{pp} \end{pmatrix}$$

and therefore $\mathbf{B\Gamma}$ is upper triangular. □

A.4. OTHER RESULTS ABOUT MATRICES.

In this section we derive two results that are very useful in proving the maximality of various invariants, as well as the result used in the definition of canonical correlations. We first establish two lemmas.

LEMMA A.9. Let $\mathbf{X}$ be a $p \times k$ matrix of rank r. Then $\mathbf{X}'\mathbf{X}$ has rank r and there exists an $r \times k$ matrix $\mathbf{U}$ of rank r such that $\mathbf{X}'\mathbf{X} = \mathbf{U}'\mathbf{U}$.

PROOF. We use several times in this proof the fact that the rank of a product is always less than or equal to the ranks of its factors. This implies that $\mathbf{X}'\mathbf{X}$ has rank at most r. Now let $\mathbf{A}$ be a $p \times r$ matrix whose columns are a maximal set of linearly independent columns of $\mathbf{X}$. Then every column of $\mathbf{X}$ can be written as a linear combination of the columns of $\mathbf{A}$, and hence $\mathbf{X} = \mathbf{AB}$ for some $r \times k$ matrix $\mathbf{B}$. Since $\mathbf{B}$ has only r rows and the rank of $\mathbf{X}$ is r, we see that $\mathbf{B}$ has rank r, as does $\mathbf{A}$. Therefore, by Theorem A.3, $\mathbf{A}'\mathbf{A} > 0$ and $\mathbf{BB}' > 0$, and hence both $\mathbf{A}'\mathbf{A}$ and $\mathbf{BB}'$ have rank r, and both are invertible, Now,

$$(\mathbf{BB}')^{-1}\mathbf{BX}'\mathbf{XB}'(\mathbf{BB}')^{-1} = \mathbf{A}'\mathbf{A}.$$

Since $\mathbf{A}'\mathbf{A}$ has rank r, $\mathbf{X}'\mathbf{X}$ must have rank at least r, and hence the rank of $\mathbf{X}'\mathbf{X}$ is r. By Theorem A.3, $\mathbf{X}'\mathbf{X} \geqslant 0$. Therefore, there exists an $r \times k$ matrix $\mathbf{U}$ of rank r such that $\mathbf{X}'\mathbf{X} = \mathbf{U}'\mathbf{U}$, by Theorem A.5. □

LEMMA A.10. Let $\mathbf{X}$ be a $p \times k$ matrix of rank r and let $\mathbf{U}$ be an $r \times k$ matrix of rank r such that $\mathbf{X}'\mathbf{X} = \mathbf{U}'\mathbf{U}$. Then there exists $\mathbf{\Gamma}$ orthogonal such that

$$\mathbf{\Gamma X} = \begin{pmatrix} \mathbf{U} \\ \mathbf{0} \end{pmatrix}.$$

PROOF. Let V be the subspace spanned by the columns of $\mathbf{X}$ and let $\mathbf{R}$ be an orthonormal basis matrix for $V^{\perp}$. Since $\mathbf{X}$ has rank r, V has dimension r and $V^{\perp}$ has dimension $p - r$. Therefore, $\mathbf{R}$ is $p \times (p - \mathrm{r})$. Also $\mathbf{UU}' > 0$ (see Theorem A.3) and is therefore invertible. Let

$$\mathbf{\Gamma} = \begin{pmatrix} (\mathbf{UU}')^{-1}\mathbf{UX}' \\ \mathbf{R}' \end{pmatrix}.$$

Then $\mathbf{\Gamma}$ is $p \times p$ and

$$\mathbf{\Gamma\Gamma}' = \begin{bmatrix} (\mathbf{UU}')^{-1}\mathbf{UU}'\mathbf{UU}'(\mathbf{UU}')^{-1} & (\mathbf{UU}')^{-1}\mathbf{UX}'\mathbf{R} \\ \mathbf{R}'\mathbf{XU}'(\mathbf{UU}')^{-1} & \mathbf{R}'\mathbf{R} \end{bmatrix}$$

$$= \begin{pmatrix} \mathbf{I} & \mathbf{0} \\ \mathbf{0} & \mathbf{I} \end{pmatrix} = \mathbf{I}$$

since $\mathbf{X}'\mathbf{X} = \mathbf{U}'\mathbf{U}$, $\mathbf{X}'\mathbf{R} = \mathbf{0}$ (the columns of $\mathbf{X}$ are in V, whereas those of $\mathbf{R}$ are in $V^{\perp}$), and $\mathbf{R}'\mathbf{R} = \mathbf{I}$ ($\mathbf{R}$ is an orthonormal basis matrix for $V^{\perp}$).

Therefore $\boldsymbol{\Gamma}$ is $p \times p$ matrix such that $\boldsymbol{\Gamma\Gamma}' = \mathbf{I}$, and is orthogonal. Also

$$\boldsymbol{\Gamma}\mathbf{X} = \begin{pmatrix} (\mathbf{UU}')^{-1}\mathbf{UU}'\mathbf{U} \\ \mathbf{R}'\mathbf{X} \end{pmatrix} = \begin{pmatrix} \mathbf{U} \\ \mathbf{0} \end{pmatrix}. \quad \square$$

We are now ready for the three main results of this section.

THEOREM A.11. Let $\mathbf{X}$ and $\mathbf{Y}$ be $p \times k$ matrices such that $\mathbf{X}'\mathbf{X} = \mathbf{Y}'\mathbf{Y}$. Then there exists a $p \times p$ orthogonal matrix $\boldsymbol{\Gamma}$ such that $\mathbf{Y} = \boldsymbol{\Gamma}\mathbf{X}$.

PROOF. Since $\mathbf{X}'\mathbf{X} = \mathbf{Y}'\mathbf{Y}$, $\mathbf{X}$ and $\mathbf{Y}$ have the same rank, say r. Let $\mathbf{U}$ be a $k \times r$ matrix such that $\mathbf{U}'\mathbf{U} = \mathbf{X}'\mathbf{X} = \mathbf{Y}'\mathbf{Y}$. By Lemma A.10, there exist $\boldsymbol{\Gamma}_1$ and $\boldsymbol{\Gamma}_2$ such that

$$\boldsymbol{\Gamma}_1\mathbf{X} = \begin{pmatrix} \mathbf{U} \\ \mathbf{0} \end{pmatrix} = \boldsymbol{\Gamma}_2\mathbf{Y}.$$

Therefore, $\mathbf{Y} = (\boldsymbol{\Gamma}_2'\boldsymbol{\Gamma}_1)\mathbf{X}$. Since $\boldsymbol{\Gamma}_2'\boldsymbol{\Gamma}_1$ is orthogonal, the theorem is proved with $\boldsymbol{\Gamma} = \boldsymbol{\Gamma}_2'\boldsymbol{\Gamma}_1$. $\square$

THEOREM A.12. Let $\mathbf{X}$ and $\mathbf{Y}$ be $p \times k$ matrices and let $\mathbf{S} > 0$ and $\mathbf{T} > 0$ be $p \times p$. If $\mathbf{X}'\mathbf{S}^{-1}\mathbf{X} = \mathbf{Y}'\mathbf{T}^{-1}\mathbf{Y}$, then there exists a $p \times p$ invertible matrix $\mathbf{A}$ such that $\mathbf{Y} = \mathbf{AX}$ and $\mathbf{T} = \mathbf{ASA}'$.

PROOF. We apply Theorem A.11 to the vectors $\mathbf{S}^{-1/2}\mathbf{X}$ and $\mathbf{T}^{-1/2}\mathbf{Y}$. We note first that

$$(\mathbf{S}^{-1/2}\mathbf{X})'(\mathbf{S}^{-1/2}\mathbf{X}) = \mathbf{X}'\mathbf{S}^{-1}\mathbf{X} = \mathbf{Y}'\mathbf{T}^{-1}\mathbf{Y} = (\mathbf{T}^{-1/2}\mathbf{Y})'(\mathbf{T}^{-1/2}\mathbf{Y}).$$

Therefore, there exists $\boldsymbol{\Gamma}$ orthogonal such that $\mathbf{T}^{-1/2}\mathbf{Y} = \boldsymbol{\Gamma}\mathbf{S}^{-1/2}\mathbf{X}$, or equivalently, $\mathbf{Y} = (\mathbf{T}^{1/2}\boldsymbol{\Gamma}\mathbf{S}^{-1/2})\mathbf{X}$. Now

$$(\mathbf{T}^{1/2}\boldsymbol{\Gamma}\mathbf{S}^{-1/2})\mathbf{S}(\mathbf{T}^{1/2}\boldsymbol{\Gamma}\mathbf{S}^{-1/2})' = \mathbf{T}^{1/2}\boldsymbol{\Gamma}\mathbf{I}\boldsymbol{\Gamma}'\mathbf{T}^{1/2} = \mathbf{T}.$$

Therefore, the theorem is proved with $\mathbf{A} = \mathbf{T}^{1/2}\boldsymbol{\Gamma}\mathbf{S}^{-1/2}$. $\square$

The last result in this section is the basic result used to define canonical correlations.

THEOREM A.13. Let $\mathbf{R} > 0$ and $\mathbf{S} > 0$ be $p \times p$ and $q \times q$ with $p \leqslant q$. Let

$$r(t) = \begin{vmatrix} -t\mathbf{R} & \mathbf{V} \\ \mathbf{V}' & -t\mathbf{S} \end{vmatrix}.$$

Then $r(t)$ is a polynomial of degree $p + q$. Let $t_1 \geqslant \cdots \geqslant t_p$ be the p largest roots of $r(t) = 0$. Then $t_i \geqslant 0$ and there exists invertible matrices $\mathbf{A}$ and $\mathbf{B}$ such that

$$\mathbf{ARA}' = \mathbf{I}, \qquad \mathbf{BSB}' = \mathbf{I}, \qquad \mathbf{AVB}' = (\mathbf{D}\,\mathbf{0}), \qquad \mathbf{D} = \begin{bmatrix} t_1 & & \mathbf{0} \\ & \ddots & \\ \mathbf{0} & & t_p \end{bmatrix}.$$

PROOF. Let $h(t) = |\mathbf{VS}^{-1}\mathbf{V}' - t\mathbf{R}|$. By Lemma A.2, we see that

$$r(t) = |-t\mathbf{S}||-t\mathbf{R} + \frac{1}{t}\mathbf{VS}^{-1}\mathbf{V}'| = (-t)^q t^{-p}|\mathbf{S}||\mathbf{VS}^{-1}\mathbf{V}' - t^2\mathbf{R}|$$
$$= (-1)^q|\mathbf{S}|t^{q-p}h(t^2). \tag{A.15}$$

By Theorem A.7, we see that $h(t)$ is a pth degree polynomial, and so $r(t)$ is a polynomial of degree $2p + q - p = p + q$. Also, by Theorem A.7, $h(t) = 0$ has p nonnegative roots, $s_1 \geqslant \cdots \geqslant s_p$. By (A.15) we see that $r(t) = 0$ has roots $\pm s_i^{1/2}$ and $q - p$ zeros. Therefore $t_i = s_i^{1/2} \geqslant 0$. By Theorem A.7, again, we see that there exists $\mathbf{A}$ invertible such that

$$\mathbf{ARA}' = \mathbf{I}, \qquad \mathbf{AVS}^{-1}\mathbf{V}'\mathbf{A}' = \mathbf{D}^2.$$

Using Theorem A.12 with $\mathbf{X} = \mathbf{V}'\mathbf{A}'$, $\mathbf{Y} = \binom{\mathbf{D}}{\mathbf{0}}$, and $\mathbf{T} = \mathbf{I}$, we see that there exists $\mathbf{B}$ invertible such that

$$\mathbf{BSB}' = \mathbf{I}, \qquad \mathbf{BV}'\mathbf{A}' = \begin{pmatrix}\mathbf{D}\\ \mathbf{0}\end{pmatrix}. \qquad \square$$

A.5. A MAXIMIZATION PROBLEM AND TWO JACOBIANS.

In this section, we solve a maximization problem that is basic to finding maximum likelihood estimators in most multivariate problems. We then find the Jacobians of two transformations that are used to find the matrix valued normal density and the general central Wishart density. Before deriving the maximization result, we derive the following lemma.

LEMMA A.14. Let $\mathbf{U} > 0$ and

$$h(\mathbf{U}) = \frac{1}{|\mathbf{U}|^{n/2}} \exp -\tfrac{1}{2}\operatorname{tr}\mathbf{U}^{-1}.$$

Then $h(\mathbf{U})$ is maximized when $\mathbf{U} = (1/\mathrm{n})\mathbf{I}$.

PROOF. Let $\mathbf{U}$ be $p \times p$ and let $\mathbf{U}^{-1}$ have eigenvalues $t_1 \geqslant \cdots t_p$. Then

$$h\ (\mathbf{U}\ ) = \left(\prod_i t_i\right)^{n/2} \exp -\tfrac{1}{2}\sum_i t_i = g\ (t_1, \ldots, t_p\ ).$$

We now find $t_1, \ldots, t_p$ that maximize $g(t_1, \ldots, t_p)$ over the set where $t_i \geqslant 0$, and then find the matrix associated with those eigenvalues. We note first that $g(t_1, \ldots, t_p) \to 0$ as $t_i \to \infty$, so the maximum occurs when

$$\frac{\delta}{\delta t_i} g(t_1, \ldots, t_p) = 0.$$

However,

$$\frac{\partial}{\partial t_i} g(t_1, \ldots, t_p) = \left(\frac{n}{2t_i} - \frac{1}{2}\right) g(t_1, \ldots, t_p).$$

Therefore, $(\partial/\partial t_i)g(t_1, \ldots, t_p) = 0$ if and only if $t_1 = \cdots = t_p = n$. Therefore $\mathbf{U}^{-1} = \boldsymbol{\Gamma} n \mathbf{I} \boldsymbol{\Gamma}' = n\mathbf{I}$. Hence $\mathbf{U} = (1/n)\mathbf{I}$. □

We are now ready for the result that is the basis for derivations of many maximum likelihood estimators in the multivariate case.

THEOREM A.15. Let $\mathbf{U} > 0$, $\mathbf{A} > 0$ be $p \times p$ matrices. Define

$$f(\mathbf{U}) = \frac{1}{|\mathbf{U}|^{n/2}} \exp - \tfrac{1}{2} \operatorname{tr} \mathbf{U}^{-1}\mathbf{A}.$$

Then $f(\mathbf{U})$ is maximized when $\mathbf{U} = (1/n)\mathbf{A}$.

PROOF. Note that

$$f(\mathbf{U}) = \frac{1}{|\mathbf{U}|^{n/2}} \exp - \tfrac{1}{2} \operatorname{tr} \mathbf{U}^{-1}\mathbf{A} = \frac{1}{|\mathbf{U}|^{n/2}} \exp - \tfrac{1}{2} \operatorname{tr} \mathbf{A}^{1/2}\mathbf{U}^{-1}\mathbf{A}^{1/2}$$

$$= \frac{1}{|\mathbf{A}^{1/2}|^{n} |\mathbf{A}^{-1/2}\mathbf{U}\mathbf{A}^{-1/2}|^{n/2}} \exp \tfrac{1}{2} (\mathbf{A}^{-1/2}\mathbf{U}\mathbf{A}^{-1/2})^{-1}$$

$$= \frac{1}{|\mathbf{A}^{1/2}|^{n}} h(\mathbf{A}^{-1/2}\mathbf{U}\mathbf{A}^{-1/2})$$

where h is defined in Lemma A.14. Therefore $f(\mathbf{U})$ is maximized when $\mathbf{A}^{-1/2}\mathbf{U}\mathbf{A}^{-1/2} = (1/n)\mathbf{I}$, or equivalently, when $\mathbf{U} = (1/n)\mathbf{A}$. □

We now find the Jacobians of two transformations. In the first we have a transformation from the space of $p \times m$ matrices to the space of $p \times m$ matrices. To find the Jacobian of such a transformation, we must list both the original and the transformed variables into vectors, reinterpret the transformation in terms of these vectors, and finally find the Jacobian of this reinterpreted transformation. That is what is meant by the Jacobian of the original transformation. In the second case, we have a transformation from the space of positive definite $p \times p$ matrices to itself. In that case, we list only the upper triangles of the original variables and the induced variables into vectors, since these variables completely determine the positive definite matrices. We then reinterpret the transformation in terms of these variables and find its Jacobian, which is the Jacobian of the original transformation from the space of positive definite matrices to itself.

THEOREM A.16. Let $\mathbf{X}$ be a $p \times m$ matrix and let $f(\mathbf{X}) = \mathbf{A}(\mathbf{X} - \mathbf{b})\mathbf{B}$ where $\mathbf{A}$ and $\mathbf{B}$ are invertible $p \times p$ and $m \times m$ matrices and $\mathbf{b}$ is a $p \times m$ matrix. Then f has Jacobian $J = |\mathbf{A}|^m |\mathbf{B}|^p$.

PROOF. The function $f(\mathbf{X})$ is a composition of three functions. In this derivation, we use the fact that the Jacobian of a composition of several functions is just the product of the Jacobians of the functions. Let $g(\mathbf{X})$

$= \mathbf{X} - \mathbf{b}$, $h(\mathbf{X}) = \mathbf{AX}$ and $k(\mathbf{X}) = \mathbf{XB}$. Then

$$f(\mathbf{X}) = k(h(g(\mathbf{X}))),$$

and therefore the Jacobian of f is just the product of the Jacobians of g, h, and k. The Jacobian of g is 1, since g just involves adding a constant. We now find the Jacobian of h. Let $\mathbf{Y} = h(\mathbf{X})$, and let $\mathbf{Y} = (\mathbf{Y}_1, \ldots, \mathbf{Y}_m), \mathbf{X} = (\mathbf{X}_1, \ldots, \mathbf{X}_m)$, and let

$$\mathbf{Y}^* = \begin{pmatrix} \mathbf{Y}_1 \\ \vdots \\ \mathbf{Y}_m \end{pmatrix}, \qquad \mathbf{X}^* = \begin{pmatrix} \mathbf{X}_1 \\ \vdots \\ \mathbf{X}_m \end{pmatrix}.$$

Then $\mathbf{Y}^*$ and $\mathbf{X}^*$ are just relistings of $\mathbf{X}$ and $\mathbf{Y}$ into pm-dimensional vectors. Also $\mathbf{Y} = h(\mathbf{X})$ if and only if

$$\mathbf{Y}^* = \begin{pmatrix} \mathbf{A} & & \mathbf{0} \\ & \ddots & \\ \mathbf{0} & & \mathbf{A} \end{pmatrix} \mathbf{X}^*$$

Therefore, the Jacobian of h is

$$\begin{vmatrix} \mathbf{A} & & \mathbf{0} \\ & \ddots & \\ \mathbf{0} & & \mathbf{A} \end{vmatrix} = |\mathbf{A}|^m$$

Similarly, the Jacobian of $k(X) = |\mathbf{B}|^p$. Therefore, the Jacobian of f is

$$1 \times |\mathbf{A}|^m \times |\mathbf{B}|^p. \quad \square$$

THEOREM A.17. Let $\mathbf{S} > 0$ be $p \times p$, and let $\mathbf{C}$ be a $p \times p$ invertible upper triangular matrix. Define $f(\mathbf{S}) = \mathbf{CSC}'$. Then f has Jacobian $|\mathbf{C}|^{p+1}$.

PROOF. Let $\mathbf{T} = (t_{ij}) = f(\mathbf{S}) = \mathbf{CSC}'$. Let $\mathbf{S} = (s_{ij})$, $\mathbf{C} = (c_{ij})$. Define

$$\mathbf{T}^* = \begin{pmatrix} t_{11} \\ \vdots \\ t_{1p} \\ t_{21} \\ \vdots \\ t_{pp} \end{pmatrix}, \qquad \mathbf{S}^* = \begin{pmatrix} s_{11} \\ \vdots \\ s_{1p} \\ s_{21} \\ \vdots \\ s_{pp} \end{pmatrix}.$$

Then $\mathbf{T}^*$ and $\mathbf{S}^*$ are just relistings of the upper triangles of $\mathbf{S}$ and $\mathbf{T}$ into vectors. Also

$$t_{ij} = \sum_{k \geqslant i} \sum_{m \geqslant j} c_{ik} c_{jm} s_{km}.$$

Suppose that $j > i$. Let $t_j^* = t_{kr}$ and $s_i^* = s_{mn}$. Since $i < j$, either $m < k$ or $n < r$. In either case, t_j^* does not depend on s_i^*, and hence

$$\frac{\partial t_j^*}{\partial s_i^*} = 0$$

Therefore the Jacobian matrix of $\mathbf{T}^*$ with respect to $\mathbf{S}^*$ is upper triangular. It is easily verified that the determinant of an upper triangular matrix is the product of its diagonal elements. Therefore, the Jacobian of this transformation is just the product of $\partial t_{ij}/\partial s_{ij}$ for all $j \geqslant i$. However,

$$\frac{\partial t_{ij}}{\partial s_{ij}} = \mathbf{c}_{ii}\mathbf{c}_{jj}.$$

Therefore the Jacobian is

$$\prod_{j \geqslant i} \mathbf{c}_{ii}\mathbf{c}_{jj} = \left(\prod \mathbf{c}_{ii}\right)^{p+1} = |\mathbf{C}|^{p+1}$$

using again the fact that the determinant of an upper triangular matrix is the product of its diagonal elements. □

A.6. A PROOF OF THE PRINCIPAL AXIS THEOREM.

In this section, we prove Theorem A.1 of this appendix. The derivation is more difficult than those in earlier sections, since it involves concepts from advanced calculus.

THEOREM A.18. Let $\mathbf{S}$ be a symmetric $p \times p$ matrix and let

$$q(t) = |\mathbf{S} - t\mathbf{I}|.$$

Then there are p real roots to the equation $q(t) = 0$. Let $t_1 \geqslant \cdots \geqslant t_p$ be these roots. Then there exists $\boldsymbol{\Gamma}$ orthogonal such that

$$\mathbf{S} = \boldsymbol{\Gamma}\mathbf{D}\boldsymbol{\Gamma}', \qquad \mathbf{D} = \begin{pmatrix} t_1 & & 0 \\ & \ddots & \\ 0 & & t_p \end{pmatrix}, \qquad t_1 = \sup_{\mathbf{X} \in R^p} \frac{\mathbf{X}'\mathbf{S}\mathbf{X}}{\|\mathbf{X}\|^2}$$

PROOF. Let $f(\mathbf{X}) = \mathbf{X}'\mathbf{S}\mathbf{X}$. Consider the problem of maximizing $f(\mathbf{X})$ subject to $\mathbf{X}'\mathbf{X} = 1$. Since $f(\mathbf{X})$ is a continuous function and the set in which $\mathbf{X}'\mathbf{X} = 1$ is a compact set, there is a maximum on this set, say $\boldsymbol{\Gamma}_1$. Let

$$g_\lambda(\mathbf{X}) = \mathbf{X}'\mathbf{S}\mathbf{X} - \lambda\mathbf{X}'\mathbf{X}.$$

By the method of LaGrange multipliers, there exists t_1 such that the gradient of $g_{t_1}(\mathbf{X})$ at $\mathbf{X} = \boldsymbol{\Gamma}_1$ is $\mathbf{0}$. Therefore

$$\mathbf{0} = \nabla g_{t_1}(\boldsymbol{\Gamma}_1) = 2\mathbf{S}\boldsymbol{\Gamma}_1 - 2t_1\boldsymbol{\Gamma}_1.$$

Therefore, there exists a scalar t_1 such that

$$\mathbf{S}\boldsymbol{\Gamma}_1 = t_1\boldsymbol{\Gamma}_1.$$

Now consider maximizing $f(\mathbf{X})$ subject to $\mathbf{X}'\mathbf{X} = 1$ and $\boldsymbol{\Gamma}_1'\mathbf{X} = 0$. This set is again a compact set and so the maximum, say $\boldsymbol{\Gamma}_2$, exists. Define

$$h_{\lambda,\delta}(\mathbf{X}) = \mathbf{X}'\mathbf{S}\mathbf{X} - \lambda\mathbf{X}'\mathbf{X} - \delta\boldsymbol{\Gamma}_2'\mathbf{X}.$$

By the method of LaGrange multipliers again, there exist δ and t_2 such that

$$\mathbf{0} = \nabla h_{t_2,\delta}(\boldsymbol{\Gamma}_2) = 2\mathbf{S}\boldsymbol{\Gamma}_2 - 2t_2\boldsymbol{\Gamma}_2 - \delta\boldsymbol{\Gamma}_1.$$

Note first that

$$\boldsymbol{\Gamma}_2'\mathbf{S}\boldsymbol{\Gamma}_1 = t_1\boldsymbol{\Gamma}_2'\boldsymbol{\Gamma}_1 = 0.$$

Therefore,

$$0 = 2\boldsymbol{\Gamma}_1'\mathbf{S}\boldsymbol{\Gamma}_2 - 2t_2\boldsymbol{\Gamma}_1'\boldsymbol{\Gamma}_2 - \delta\boldsymbol{\Gamma}_1'\boldsymbol{\Gamma}_1 = -\delta$$

and hence

$$\mathbf{S}\boldsymbol{\Gamma}_2 = t_2\boldsymbol{\Gamma}_2.$$

We continue in this fashion finding $\boldsymbol{\Gamma}_i$ that maximizes $f(\mathbf{X})$ subject to $\boldsymbol{\Gamma}_i'\boldsymbol{\Gamma}_i = 1$ and $\boldsymbol{\Gamma}_i'\boldsymbol{\Gamma}_j = 0$ for $j < i$. In this fashion, we find $\boldsymbol{\Gamma}_1, \ldots, \boldsymbol{\Gamma}_p$ and $t_1, \ldots, t_p$ such that

$$\boldsymbol{\Gamma}_i'\boldsymbol{\Gamma}_i = 1, \qquad \boldsymbol{\Gamma}_i'\boldsymbol{\Gamma}_j = 0, \qquad i \neq j, \qquad \mathbf{S}\boldsymbol{\Gamma}_i = t_i\boldsymbol{\Gamma}_i.$$

Define $\boldsymbol{\Gamma} = (\boldsymbol{\Gamma}_1, \ldots, \boldsymbol{\Gamma}_p)$. The equations above imply that $\boldsymbol{\Gamma}$ is orthogonal. Also

$$\boldsymbol{\Gamma}_i'\mathbf{S}\boldsymbol{\Gamma}_i = t_i\boldsymbol{\Gamma}_i'\boldsymbol{\Gamma}_i = t_i, \qquad \boldsymbol{\Gamma}_i'\mathbf{S}\boldsymbol{\Gamma}_j = t_j\boldsymbol{\Gamma}_i'\boldsymbol{\Gamma}_j = 0.$$

Therefore

$$\boldsymbol{\Gamma}'\mathbf{S}\boldsymbol{\Gamma} = \begin{pmatrix} t_1 & & \mathbf{0} \\ & \ddots & \\ \mathbf{0} & & t_p \end{pmatrix} = \mathbf{D}$$

and hence $\mathbf{S} = \boldsymbol{\Gamma}\mathbf{D}\boldsymbol{\Gamma}'$. Finally

$$q(t) = |\mathbf{S} - t\mathbf{I}| = |\boldsymbol{\Gamma}\mathbf{D}\boldsymbol{\Gamma}' - t\boldsymbol{\Gamma}\boldsymbol{\Gamma}'| = |\boldsymbol{\Gamma}||\mathbf{D} - t\mathbf{I}||\boldsymbol{\Gamma}'|$$

$$= |\mathbf{D} - t\mathbf{I}| = \prod_{j=1}^{p} (t_i - t)$$

and therefore, the t_i are the roots of $q(t)$. Now let $\mathbf{X} \in R^p$, $\mathbf{X} \neq 0$. Let $\mathbf{Z} = (1/\|\mathbf{X}\|)\mathbf{X}$. Then $\|\mathbf{Z}\| = 1$. By the derivation above

$$t_1 = \sup_{\|\mathbf{Z}\|=1} \mathbf{Z}'\mathbf{S}\mathbf{Z} = \sup_{\substack{\mathbf{X}\in R^p \\ \mathbf{X}\neq 0}} \frac{\mathbf{X}'\mathbf{S}\mathbf{X}}{\|\mathbf{X}\|^2}. \quad \square$$

Bibliography

Aitken, A. C. 1935. On least squares and linear combinations of observations, *Proc. Roy. Soc. Edinb.*, **55**, 42–48.

Anderson, T. W. 1946. The non-central Wishart distribution and certain properties of multivariate analysis. *Ann. Math. Stat.*, **17**, 409–431.

———. 1951, Classification by multivariate analysis, *Psychometrika*, **16**, 631–650.

———. 1956. *An Introduction to Multivariate Statistical Analysis*, Wiley, New York.

———, and S. Das Gupta 1964. Monotonicity of the power function of some tests of independence between two sets of variates, *Ann. Math. Stat.*, **35**, 206–207.

Andrews, D. F. 1974. A robust method of multivariate linear regression, *Technometrics*, **16**, 523–531.

Arnold, S. F. 1979a. Linear models with exchangeable errors, *J.A.S.A.*, **74**, 194–199.

———. 1979b. A coordinate-free approach to repeated measures designs, *Ann. of Stat*, **7**, 812–822.

———. 1980a. Asymptotic validity of *F*-tests for the ordinary linear model and the multiple correlation model, *J.A.S.A.* (To appear in Dec. 1980).

———. 1980b. The asymptotic validity of invariant procedures for the repeated measures model and multivariate linear model, *Penn. State Tech. Report*.

Arveson, J. and M. Layard 1975. Asymptotically robust tests in unbalanced variance components, *Ann. Stat.*, **3**, 1122–1134.

Baranchik, A. 1970. A family of minimax estimators of the mean of a multivariate normal distribution, *Ann. Math. Stat.*, **41**, 642–645.

———. 1973. Inadmissibility of maximum likelihood estimators in some multiple regression problems with three or more independent variables, *Ann. Stat.*, **1**, 312–321.7–1136.

Bartlett, M. S. 1930. On the theory of statistical regression, *Proc. Roy. Soc. Edinb.*, **53**, 260–283.

———. 1938. Further aspects of the theory of multiple regression, *Proc. Cam. Phil. Soc.*, **34**, 33–40.

Boardman, T. S. 1974. Confidence intervals for variance components—a comparative Monte Carlo study, *Biometrics*, **30**, 251–262.

Bose, R. C. and S. N. Roy 1938. The exact distribution of the studentized D^2 statistic, *Sankya*, **4**, 19–38.

Box, G. E. B. 1949. A general distribution theory for a class of likelihood criteria, *Biometrika*, **36**, 317–346.

———. 1950. Problems in the analysis of growth and wear curves, *Biometrics*, **6**, 362–389.

Brown, L. 1966. On the admissibility of estimators of one or more location parameters, *Ann. Math. Stat.*, **37**, 1087–1136.

———. 1973. Estimation with incompletely specified loss functions, *J.A.S.A*, **70**, 417–427.

Chung, K. L. 1974. *A Course in Probability Theory*, 2nd ed., Academic, New York.

Constantine, A. G. 1963. Some non-central distribution problems in multivariate analysis, *Ann. Math. Stat.*, **34**, 1270–1285.

David, F. N. 1938. *Tables of the Ordinates and Probability Intergrals of the Distribution of the Correlation Coefficient in Small Samples*, Biometrika, London.

Das Gupta, S. 1969. Properties of power functions of some tests concerning dispersion matrices, *Ann. Math. Stat.*, **40**, 697–702.

———. 1973. Classification procedures—a review, in *Discriminant Analysis and Applications*, T. Cacoullos, Ed., Academic, New York.

———, T. W. Anderson, and G. S. Maudholkar 1964. Monotonicity of the power function of some tests of the multivariate linear hypothesis, *Ann. Math. Stat.*, **35**, 200–205.

Dempster, A., M. Schatzoff, and N. Wermuth 1977. A simulation of alternatives to ordinary least squares, *J.A.S.A.*, **72**, 77–91.

Duncan, D. B. 1955. Multiple range and multiple *F* tests, *Biometrics*, **11**, 1–42.

Einot, I., and K. R. Gabriel 1975. A study of the power of several methods of multiple comparisons, *J.A.S.A.*, **70**, 574–583.

Eisenhart, C. 1947. The assumptions underlying the analysis of variance, *Biometrics*, **3**, 1–21.

Efron, B., and C. Morris 1973. Stein's estimation rule and its competitors, *J.A.S.A*, **68**, 117–130.

———. 1975. Data analysis using Stein's estimator and its generalizations, *J.A.S.A.*, **70**, 311–319.

———. 1976. Families of minimax estimators of the mean of a multivariate normal distribution, *Ann. Stat.*, **4**, 11–21.

Feller, W. 1957. *An Introduction to Probability Theory and its Applications*, 2nd ed., Wiley, New York.

Ferguson, T. 1967. *Mathematical Statistics, Decision Theoretic Approach*, Academic, New York.

Fisher, R. A. 1915. Frequency distribution of the values of the correlation coefficient in samples from an infinitely large sample, *Biometrika*, **10**, 507–521.

———. 1918. The correlation between relatives on the supposition of Mendelian inheritance, *Trans. Roy. Soc. Edinb.*, **52**, 399–433.

———. 1925. *Statistical Methods for Research Workers*, 1st ed., Oliver and Boyd, Edinburgh.

———. 1935. *The Design of Experiments*, Oliver and Boyd, Edinburgh.

———. 1936. Use of multiple measurements in taxonomic problems, *Ann. Eugenics*, **7**, 179–184.

———. 1939. The sampling distribution of some statistics obtained from non-linear experimentation, *Ann. Eugenics*, **9**, 238–249.

Gabriel, K. E. 1969a. Simultaneous test procedures—some theory of multiple comparisons, *Ann. Math. Stat.*, **40**, 224–250.

———. 1969b. Comparison of some methods of simultaneous inference in MANOVA, *Multivariate Analysis, Vol. 2*, Academic, New York.

Gauss, K. F. 1809. *Werke*, **4**, 1–93, Gottingen.

Ghosh, M. N., (1964) On the admissibility of some tests in MANOVA, *Ann. Math. Stat.*, **35**, 789–794.

Giri, N. 1977. *Multivariate Statistical Inference*, Academic, New York.

Girshick, M. A. 1939. On the sampling theory of roots of determinental equations, *Ann. Math. Stat.*, **10**, 203–224.

Graybill, F. A. 1976. *Theory and Applications of the Linear Model*, Duxbury, North Scituate, Mass.

Hall, W. J., R. A. Wijsman, and J. K. Ghosh 1965. The relationship between sufficiency and invariance with applications in sequential analysis, *Ann. Math. Stat.*, **36**, 575–614.

Halperin, M. 1951. Normal regression theory in the presence of intraclass correlation, *Ann. Math. Stat.*, **22**, 573–580.

Harville, D. A. 1977. Maximum likelihood approaches to variance component estimation and to related problems, *J.A.S.A.*, **72**, 320–337.

Hettmannsperger, T. P., and J. W. McKean 1977. A robust alternative based on ranks for least squares in analyzing linear models, *Technometrics*, **19**, 275–284.

Hoerl, A., and R. Kennard 1970a. Ridge regression: biased estimation for non-orthogonal problems, *Technometrics*, **12**, 53–63.

———. 1970b. Ridge regression: applications to non-orthogonal problems, *Technometrics*, **12**, 69–82.

Hogg, R. V., and A. T. Craig 1970. *Introduction to Mathematical Statistics*, 3rd ed., Macmillan, London.

Hotelling, H. 1931. The generalization of student's ratio, *Ann. Math. Stat.*, **2**, 139–142.

———. 1933. Analysis of a complex of statistical variables with principal components, *J. Educ. Psych.*, **24**, 417–441.

———. 1936. Relations between two sets of variates, *Biometrika*, **28**, 321–377.

———. 1951. A generalized *T*-test and measure of multivariate dispersion, *Proc. 2nd Berkeley Symp. Math. Stat. Prob.*, University of California Press, Berkeley, CA, 23–41.

Hsu, P. L. 1938. Notes on Hotelling's generalized T^2, *Ann. Math. Stat.*, **9**, 231–243.

———. 1939. On the distribution of the roots of certain determinental equations, *Ann. Eugenics*, **9**, 250–258.

Huber, P. J. 1973. Robust, regression, asymptotics, conjectures and Monte Carlo, *Ann. Stat.*, **1**, 799–821.

———. 1977. *Robust Statistical Procedures*, SIAM, Philadelphia.

Hunt, G., and C. Stein 1946. Most stringent tests of hypotheses. Unpublished.

Ito, K. 1962. A comparison of powers of two multivariate analysis of variance tests, *Biometrika*, **49**, 455–462.

James, A. 1955. The non-central Wishart distribution, *Proc. Roy. Soc. London A*, **229**, 364–366.

———. 1964. Distribution of matrix variates and latent roots derived from normal samples, *Ann Math. Stat.*, **35**, 475–501.

James, W., and C. Stein 1960. Estimation with quadratic loss, *Proc. 4th Berekely Symp. Math. Stat. Prob.*, **1**, 361–379.

Johnston, J. 1972. *Econometric Methods*, McGraw-Hill, New York.

Keuls, M. 1952. The uses of the studentized range in connection with an analysis of variance, *Euphytica*, **1**, 112–122.

Kiefer, J. 1957. Invariance, minimax sequential estimation and continuous time processes, *Ann. Math. Stat.*, **28**, 573–601.

———. and R. Schwartz 1965. Admissible Bayes character of T^2, R^2 and other full invariant tests for classical multivariate problems, *Ann. Math. Stat.*, **36**, 747–770.

Korin, G. P. 1968. On the distribution of a statistic used for testing a covariance matrix, *Biometrika*, **55**, 171–178.

Kruskal, W. 1961. The coordinate free approach to Gauss-Markov estimation and its application to missing and extra observations, *Proc. Fourth Berkeley Symp. Math. Stat. Prob.*, **1**, 435–461.

———. 1968. When are Gauss-Markov and least squares estimators identical? A co-ordinate-free approach, *Ann. Math. Stat.*, **39**, 70–75.

Kshirsagar, A. 1972. *Multivariate Analysis*, Marcel Decker, New York.

Lachenbruch, P. A., and M. R. Mickey 1968. Estimation of error rates in discriminant analysis, *Technometrics*, **10**, 1–11.

Lawley, D. N. 1938. A generalization of Fisher's Z-test, *Biometrika*, **30**, 180–187.

———, and A. E. Maxwell 1971. *Factor Analysis as a Statistical Method*, 2nd ed., American Elsevier, New York.

Legndre, A. M. 1806. *Nouvelles Méthodes pour le determination des orbites des comètes; avec Supplément Contenant Divers Perfectionnements de ces Méthodes et Leur Application aux deux Comètes de 1805*, Courcier, Paris.

Lehmann, E. L. 1959. *Testing Statistical Hypotheses*, Wiley, New York.

———. 1975. *Non-parametrics*, Holden-Day, San Francisco.

——— and C. Stein (1953) The admissibility of certain invariant statistical test involving a translation parameter, *Ann. Math. Stat.*, **24**, 473–479.

Mahalanobis, P. C. 1936. On the generalized distance in statistics, *Proc. Nat. Inst. Sci. India*, **12**, 49–55.

Markoff, A. A. 1900. *Wahrscheinlichkeitsrechnung*, Telner, Leipzig.

McElroy, F. W. 1967. A necessary and sufficient condition for ordinary least squares estimators to be best linear unbiased, *J.A.S.A.*, **67**, 1302–1304.

Mikhail, M. N. 1965. A comparison of tests of Wilks-Lawley hypothesis in multivariate analysis, *Biometrika*, **52**, 149.

Miller, R. 1966. *Simultaneous Statistical Inference*, McGraw-Hill, New York.

Mood, A. M., F. A. Graybill, and D. C. Boes 1974. *Introduction to the Theory of Statistics*, 3rd ed., McGraw-Hill, New York.

Morrison, D. 1976. *Multivariate Statistical Methods*, 2nd ed., McGraw-Hill, New York.

Neter, J., and W. Wasserman 1974. *Applied Linear Statistical Models*, Irwin, Homewood, Ill.

Newman, D. 1939. The distribution of the range in samples from the normal population, expressed in terms of an independent estimate of standard deviation, *Biometrika*, **31**, 20–30.

Okamoto, M. 1963. An asymptotic expansion for the distribution of the linear discriminant function, *Ann. Math. Stat.*, **34**, 1358–1359.

Pearson, E. S., and H. O. Hartley 1972. *Biometrika Tables for Statisticians*, Vol. 2, Cambridge University Press for Biometrika trustees, Cambridge, England.

Pillai, K. C. S. 1955. Some new test criteria in multivariate analysis, *Ann. Math. Stat.*, **26**, 117–121.

———, and Jayachandran, K. 1967. Power comparisons of tests of two multivariate hypotheses based on four criteria, *Biometrika*, **56**, 108–118.

Pitman, E. J. C. 1939. The estimation of location and scale parameters of a continuous population of any form, *Biometrika*, **30**, 391–421.

Pottoff, R. H., and S. N. Roy 1964. A generalized multivariate analysis of variance model useful especially for growth curve problems, *Biometrika*, **51**, 313–326.

Ramsey, P. 1978. Power differences between pairwise multiple compairsons, *J.A.S.A.*, **73**, 479–485.

Rao, C. R. 1965. *Linear Statistical Inference and Its Applications*, Wiley, New York.

Roy, S. M. 1939. *P*-statistics or some generalizations in the analysis of variance appropriate to multivariate problems, *Sankya*, **4**, 381–396.

———. 1953. On a heuristic method of test construction and its use in multivariate analysis, *Ann. Math. Stat.*, **24**, 220–238.

———, and R. C. Bose 1953. Simultaneous confidence interval estimation, *Ann. Math. Stat.*, **24**, 513–536.

———, and W. F. Mikhail 1961. On the montonic character of the power functions of two multivariate tests, *Ann. Math. Stat.*, **32**, 1145–1151.

Ryan, T. A. 1960. Significance tests for multiple comparisons of proportions, variances and other statistics, *Psychological Bulletin*, **57**, 318–328.

Sampson, A. R. 1974. A tale of two regressions, *J.A.S.A.*, **69**, 682–689.

Satterthwaite, E. E. 1946. An approximate distribution of estimates of variance components, *Biometrics*, **2**, 110–114.

Scheffe, H. 1953. A method of judging all contrasts in analysis of variance, *Biometrika*, **40**, 87–104.

———. 1959. *The Analysis of Variance*, Wiley, New York.

Schwartz, R. 1967. Admissible tests in multivariate analysis, *Ann. Math. Stat.*, **38**, 698–710.

Searle, S. R. 1971. *Linear Models*, Wiley, New York.

Seber, G. A. F. 1977. *Linear Regression Analysis*, Wiley, New York.

Smith, H., R. Gnanadesikan, and J. Hughes 1962. Multivariate analysis of variance, *Biometrika*, **32**, 70–80.

Snedecor, G., and W. Cochran 1967. *Statistical Methods*, 6th ed., Iowa State Univeristy Press, Ames, Iowa.

Salaevskii, O, V. 1968. Minimax character of Hotelling's T^2. *Sov. Math. Dokl.*, **9**, 733–735.

Stein, C. 1955. Inadmissibility of the usual estimator for the mean of a multivariate normal distribution, *Proc. 3rd Berekeley Symp. Math. Stat. Prob.*, **1**, 197–206.

———. 1956. The admissibility of Hotelling's T^2 test, *Ann. Math. Stat.*, **27**, 616–623.

———. 1960. Multiple Regression. *Contributions to Probability and Statistics*, Stanford University Press, Stanford, CA.

———. 1964. Inadmissibility of the usual estimator for the variance of a normal distribution with unknown mean, *Ann. Inst. Math.*, **16**, 155–160.

———. 1966. An approach to the recovery of inter-block information in balanced incomplete block designs, in *Festshrift for J. Neymann*, F. N. David, Ed., Wiley, New York, pp. 351–366.

———. 1969. *Multivariate Analysis I*, (Notes by M. L. Eaton.) Tech. Rep. #42, Dept. of Stat., Stanford University.

Strawderman, W. 1971. Proper Bayes minimax estimators of the multivariate normal mean, *Ann. Math. Stat.*, **42**, 385–388.

———. 1973. Proper Bayes minimax estimators of the multivariate normal mean vector for the case of common unknown variances, *Ann. Stat.*, **1**, 1189–1194.

Sugiura, N., and Nagao, H. 1968. Unbiasedness of some test criteria for the equality of one or two covariance matrices, *Ann. Math. Stat.*, **39**, 1689–1692.

Tukey, J. 1953. The problem of multiple comparisons. Unpublished paper, Princeton University.

Wald, A. 1944. On a statistical problem arising in the classification of an individual into one of two groups, *Ann. Math. Stat.*, **15**, 145–162.

Welsch, R. 1977. Stepwise multiple comparison procedures, *J.A.S.A.*, **72**, 566–575.

Wilks, S. S. 1932. Certain generalizations in the analysis of variance, *Biometrika*, **24**, 471–494.

Winer, B. J. 1971. *Statistical Principles in the Design of Experiments*, McGraw-Hill, New York.

Wishart, J. 1928. The generalized product moment distribution in samples from a normal multivariate population, *Biometrika*, **20**, 32–52.

Yates, F. 1934. The analysis of multiple classifications with unequal numbers in different classes, *J.A.S.A.*, **29**, 51–66.

Zellner, A. 1962. An efficient method of estimating seemingly unrelated regressions and test for aggregate bias, *J.A.S.A.*, **57**, 201–208.

Index

Applied Probability and Statistics (Continued)

FLEISS • Statistical Methods for Rates and Proportions, *Second Edition*
GALAMBOS • The Asymptotic Theory of Extreme Order Statistics
GIBBONS, OLKIN, and SOBEL • Selecting and Ordering Populations: A New Statistical Methodology
GNANADESIKAN • Methods for Statistical Data Analysis of Multivariate Observations
GOLDBERGER • Econometric Theory
GOLDSTEIN and DILLON • Discrete Discriminant Analysis
GROSS and CLARK • Survival Distributions: Reliability Applications in the Biomedical Sciences
GROSS and HARRIS • Fundamentals of Queueing Theory
GUPTA and PANCHAPAKESAN • Multiple Decision Procedures: Theory and Methodology of Selecting and Ranking Populations
GUTTMAN, WILKS, and HUNTER • Introductory Engineering Statistics, *Second Edition*
HAHN and SHAPIRO • Statistical Models in Engineering
HALD • Statistical Tables and Formulas
HALD • Statistical Theory with Engineering Applications
HARTIGAN • Clustering Algorithms
HILDEBRAND, LAING, and ROSENTHAL • Prediction Analysis of Cross Classifications
HOEL • Elementary Statistics, *Fourth Edition*
HOLLANDER and WOLFE • Nonparametric Statistical Methods
JAGERS • Branching Processes with Biological Applications
JESSEN • Statistical Survey Techniques
JOHNSON and KOTZ • Distributions in Statistics
Discrete Distributions
Continuous Univariate Distributions—1
Continuous Univariate Distributions—2
Continuous Multivariate Distributions
JOHNSON and KOTZ • Urn Models and Their Application: An Approach to Modern Discrete Probability Theory
JOHNSON and LEONE • Statistics and Experimental Design in Engineering and the Physical Sciences, Volumes I and II, *Second Edition*
JUDGE, GRIFFITHS, HILL and LEE • The Theory and Practice of Econometrics
KALBFLEISCH and PRENTICE • The Statistical Analysis of Failure Time Data
KEENEY and RAIFFA • Decisions with Multiple Objectives
LANCASTER • An Introduction to Medical Statistics
LEAMER • Specification Searches: Ad Hoc Inference with Nonexperimental Data
McNEIL • Interactive Data Analysis
MANN, SCHAFER and SINGPURWALLA • Methods for Statistical Analysis of Reliability and Life Data
MEYER • Data Analysis for Scientists and Engineers
MILLER, EFRON, BROWN, and MOSES • Biostatistics Casebook
OTNES and ENOCHSON • Applied Time Series Analysis: Volume I, Basic Techniques
OTNES and ENOCHSON • Digital Time Series Analysis
POLLOCK • The Algebra of Econometrics
PRENTER • Splines and Variational Methods
RAO and MITRA • Generalized Inverse of Matrices and Its Applications
RIPLEY • Spatial Statistics
SCHUSS • Theory and Applications of Stochastic Differential Equations
SEAL • Survival Probabilities: The Goal of Risk Theory
SEARLE • Linear Models
SPRINGER • The Algebra of Random Variables
UPTON • The Analysis of Cross-Tabulated Data
WEISBERG • Applied Linear Regression
WHITTLE • Optimization Under Constraints